BASIC AC CIRCUITS

Stanley R. Fulton, Ed.D.
Instructor, Mountain View College
Dallas County Community College District

John Clayton Rawlins, M.S.
Instructor, Eastfield College
Dallas County Community College District

With Contributions by
Gerald Luecke

Edited by
Robert E. Sawyer

This book is part of the BASIC ELECTRICITY SERIES
from the TEXAS INSTRUMENTS LEARNING CENTER consisting of:

Basic Electricity and DC Circuits
Basic AC Circuits

HOWARD W. SAMS & COMPANY
A Division of Macmillan, Inc.
4300 West 62nd Street
Indianapolis, Indiana 46268 USA

The BASIC AC CIRCUITS materials were developed by:

For the Book:

The Staff of the Texas Instruments Information Publishing Center

Managing Editor – Gerald Luecke
Technical Editor – Robert E. Sawyer
Word Processing – Vicki Seale
With Contributions by:
Ross Wise
Charles Battle

For the Course:

The Staff of the Texas Instruments Media Center:

Manager – Al M. Bond
Video Producers – Steve Floyd
– Dave Woody
Video Director – Greg Rardin
Video Recording – Harold Wallace
Talent – Tony Garrett
Video Art – Billy Von Kalow
– Sheila Reece
Camera – Blake Murray
– Charles Venable
With Contributions by:
Robbie Fletcher
Gerald Johnson

In cooperation with:

The Wisconsin Foundation for Vocational, Technical and Adult Education, Inc.
Grants Officer – Loren Brumm
Milwaukee Area Technical College
Associate Dean– Terry Adams
Technical Editor– Charles H. Evans

Texas Instruments Information Publishing Center — Orm F. Henning

Acknowledgements: Oscilloscope photographs and equipment courtesy of Tektronix®, Inc.

Artwork and Layout By: Plunk & Associates

First Edition, Ninth Printing - 1989

ISBN 0-672-27025-0
Library of Congress Catalog Number 80-54793

Printed in the United States of America

Contents

Contents

■ Contents

Preface

This textbook is designed for the person continuing a study of basic electricity. It provides the general information, theory, and problem-solving techniques required for an analysis and application of alternating current (ac) circuits—from the simplest to the most complex. Its aim is to help provide a solid foundation for the person interested in learning the concepts and principles of basic electricity and then, if the opportunity arises, in advancing in the studies of physics, electronics, computer science, and engineering. The book is written both as a textbook for classroom students and instructor and as a textbook for individualized learning. It has been written and organized in a way that will help the reader best acquire the knowledge necessary to analyze ac circuits, and to gain a basic understanding of some of the measurement techniques used in such analysis. Each lesson has specific objectives, detailing exactly what the reader should know or be able to do after completing the lesson.

Even though a knowledge of simple algebra in mathematics and a basic understanding of direct current (dc) circuits is assumed, new concepts and terms and the necessary mathematics are introduced as needed with illustrative examples. Detailed computations are carried out step by step so that algebraic methods, the use of exponents and scientific notation are easily understood.

The book begins with an introduction to alternating current, how an alternating voltage is generated and why the sine function describes the characteristics of such a voltage and the resulting circuit current.

Since the oscilloscope is such a basic measuring tool for observing alternating voltages and currents a lesson is devoted to describing it before continuing with resistive ac circuit analysis.

New concepts of capacitance and inductance and their action in ac circuits are discussed and then respective circuit analysis techniques are covered for series and parallel resistive-capacitive and resistive-inductive circuits. Contained within the inductance lesson is the important discussion of mutual inductance and transformers.

Since the reaction of inductance and capacitance in circuits that have time-varying voltages that produce time-varying currents is so important to ac circuit analysis, a lesson on time constants for resistive-capacitive and resistive-inductive circuits is included.

Building on prior lessons, more complex circuit analysis continues by including resistive, capacitive and inductive components all at the same time in series, parallel and combination series-parallel ac circuits. Right triangle analysis is replaced with phasor algebra with the introduction of imaginary numbers so that the techniques of analyzing the most complex ac circuit are presented.

A discussion of the special case in ac circuits where the ac frequency is varied to resonance completes the text.

The text is not written to support theories with complete rigid mathematical proofs but is written to explain the fundamental ideas so the reader may understand how to put electricity to use in practical ways. We hope we have succeeded in that goal.

S.F.
C.R.

Features of This Book

As stated in the preface this book is designed primarily for the person continuing a study of basic electricity—the entry level student. It assumes that the reader has a basic knowledge of the principles of direct current (dc) electricity and a basic mathematics background that includes algebra.

There are several features of this book specifically designed to increase its efficiency and help the reader grasp the principles of analyzing ac circuits.

1. At the beginning of each lesson, detailed objectives are listed. These objectives state what new things you should be able to do upon successful completion of the lesson. It is suggested that these objectives be read before beginning the lesson.

2. Although this book is designed to be a stand-alone text, it is also an essential part of a Texas Instruments Learning Center videotape course. This book, however, contains supplemental material not included in the videotape. For this reason, the book has been organized in a special way. All of the figures which are in the videotape are enclosed in a heavy-line border; figures which are book-only have no border. Equations appearing on videotape are in bold-face (darker) type; book-only equations are in standard typeface. The person using the book as a stand-alone text needs to know of these variations but should disregard them. They are not important for understanding the material.

3. At the end of each lesson there are three types of diagnostic material to help the reader understand better the basic concepts and apply the principles and techniques discussed in the lesson. One is a set of worked-through examples, with detailed step-by-step solutions. These worked-through examples apply the theory in each lesson to typical applications involving ac circuits. In this way you are led from the *knowledge* of the concepts you need to know, to the application of those concepts to ac circuit problems and applications.

   Another is a set of practice problems with answers provided at the end of the text material. These problems give you the opportunity to try your new knowledge and test your accuracy applying that knowledge.

   Finally, there is a quiz consisting of questions with answers provided at the end of the text material. The quiz can be taken in a relatively short period of time, and it will indicate areas in which you are proficient and areas in which you need to further review key concepts and principles.

LESSON 1

Introduction to Alternating Current

This lesson is an introductory lesson. Alternating current (ac) is defined and compared to direct current (dc), and the operation of an ac generator is discussed. Time, frequency and cyclic characteristics of the ac waveform are analyzed with examples provided for each concept.

■ Objectives

At the end of this lesson you should be able to:

1. Define an ac waveform and identify dc and ac waveforms from diagrams provided.

2. Describe how an ac generator produces an ac waveform.

3. Identify a cycle and the period of an ac waveform.

4. Given the time of one cycle, calculate the frequency of the waveform.

5. Given the frequency of a waveform, calculate the time of one cycle.

■ **Definition of Alternating Current**
■ **Generating An AC Waveform**

INTRODUCTION

The action of alternating currents in circuits is the subject of this book. The electromagnetic wave displayed on an oscilloscope in *Figure 1.1* is an electronic picture of alternating current and is one of the most useful and mysterious of all phenomena known to man. Waveforms such as this are radiating from radio, TV, telephone and other communication system antennas around the world each day. The alternating current in the antenna is a primary man-made source of electromagnetic waves. Words, music, TV pictures, other sounds are alternating currents amplified by various electronic circuits and applied to antennas to radiate through space and communicate information.

It is a textbook designed to provide the general information, theory, and problem-solving techniques required for an analysis of ac circuits from the simplest to the most complex. This first lesson provides an operational definition for ac with comparisons of ac and dc waveforms; theorizes and demonstrates the generation of an ac waveform; and introduces period and frequency relationships of ac waveforms.

AC VOLTAGE AND CURRENT

Definition of Alternating Current

AC is the abbreviation for alternating current. Alternating current is an electrical current which changes in both magnitude and direction. The term, magnitude, refers to the quantitative value of the current in a circuit — in other words, how much current is flowing. The term, direction, refers to the direction current flows in a circuit.

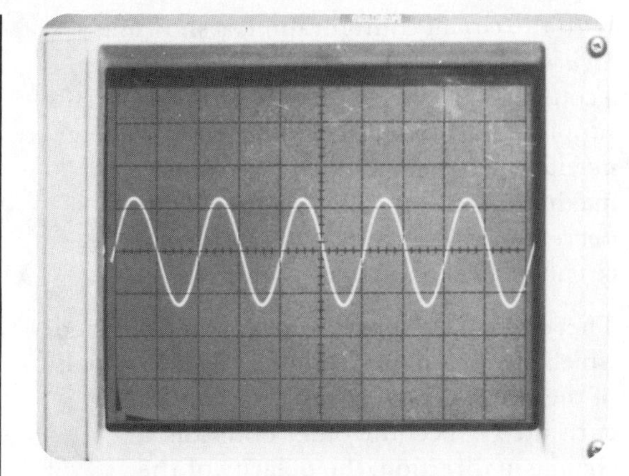

Figure 1.1 *An AC Waveform Displayed on an Oscilloscope*

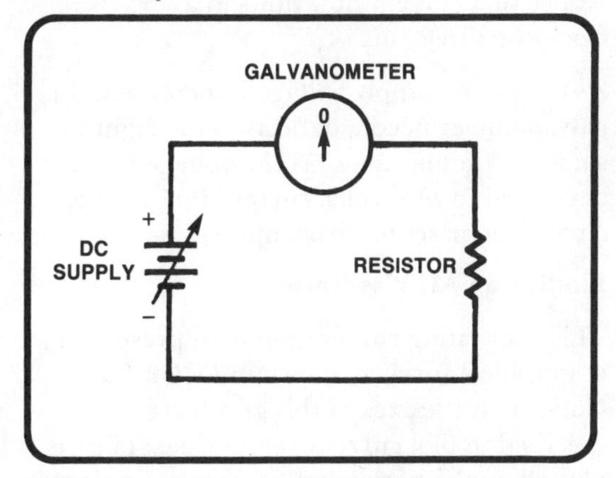

Figure 1.2 *A Simple DC Circuit With a Galvanometer*

Generating An AC Waveform

The simple dc circuit in *Figure 1.2* can be used to simulate alternating current. The circuit consists of a variable dc power supply, a resistor, and a galvanometer. The galvanometer is an ammeter with a center scale value of zero amperes. If current flows in the circuit in a counter-clockwise direction, the meter needle will deflect to the left. If current flows in a clockwise direction, the meter needle will deflect to the right.

■ Plotting An AC Waveform

With the circuit configuration as shown in *Figure 1.3,* electron current flow will be in a counter-clockwise direction. If the power supply voltage is increased, the galvanometer needle will deflect to the left to some maximum current value. As the voltage is decreased to zero volts, current flow in the circuit will decrease to zero amperes.

Therefore, a current flow has been predicted which changes in magnitude. This meets one of the two specified criteria for alternating current. To meet the other criterion, a change in direction, the polarity of the battery can be reversed as in *Figure 1.4*. Notice that current now flows in a clockwise direction.

As the power supply voltage is increased, the galvanometer needle deflects to the right to some maximum value. As the voltage is decreased to zero volts, current flow in the circuit decreases to zero amperes.

Plotting an AC Waveform

This alternating current can be represented in graphical form, as shown in *Figure 1.5*. Notice that the axes of this graph are specified to plot current versus time. Time is plotted on the horizontal, or X axis. Current is plotted on the vertical, or Y axis. The vertical axis is divided into a positive (+) current value above the X axis and a negative (−) current value below the X axis. This polarity designation is used simply to differentiate between direction of current flow.

For this application, current flow in a counter-clockwise direction will be designated as positive current, and current flow in the opposite, clockwise direction, will be designated a negative current. The polarity and direction selections are arbitrary.

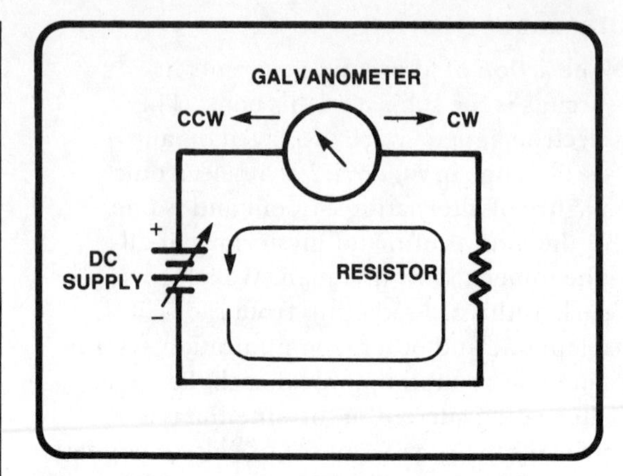

Figure 1.3 *Current Flow in a Counter-clockwise Direction*

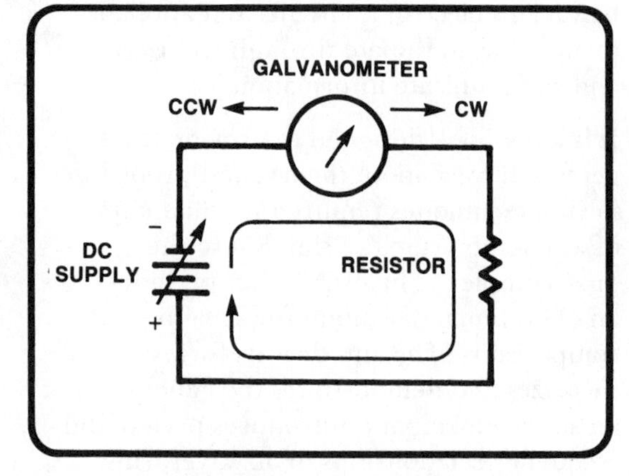

Figure 1.4 *Current Flow in Clockwise Direction*

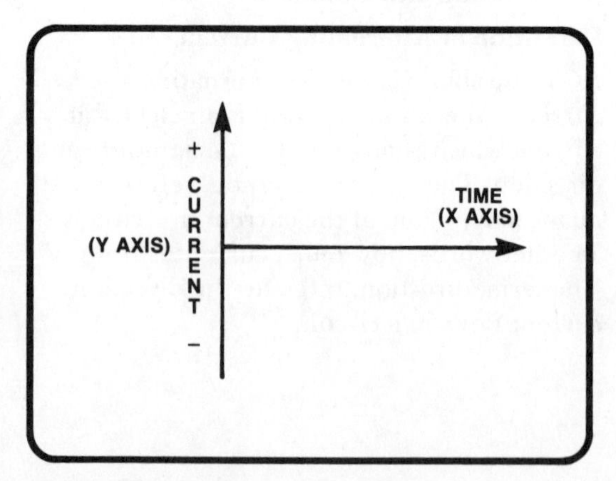

Figure 1.5 *Graph Used to Plot AC Current*

■ AC Voltages

With the circuit connected as in *Figure 1.3*, current, I, flows in a counter-clockwise direction, with current flow increasing and decreasing. Note that the direction of current stays the same. Only the value of the current is changed. Current in this direction is plotted in the top half of the graph to indicate positive current as shown in *Figure 1.6*.

With the battery reversed as shown in *Figure 1.4*, current now flows in a clockwise direction. The current always flows in the same direction. The magnitude of the current increases and decreases following the magnitude of the applied voltage, and the current is plotted in the negative (−) portion of the graph of *Figure 1.7*. This indicates negative current or, more precisely, a current that is flowing in a direction opposite to the direction originally chosen for positive current. Note the two distinguishing characteristics of this waveform. First, there is a *change in current value*—in this example, the change is continuous. Second, the *direction of current flow has changed.* This change in direction is indicated by the waveform crossing the X axis into the negative half of the graph. *If these two criteria are met, the waveform can be categorized as an ac waveform.*

The particular waveform shown in *Figure 1.7* is only one type of an ac waveform, a sine wave. Other ac waveforms which meet the specified criteria for an ac waveform will be introduced later.

AC Voltages

If there is current flow in a circuit, a difference in potential, or voltage, must be present. The voltage, E, that produces an alternating current must change in the same manner as the current as shown by the diagrams in *Figure 1.8*.

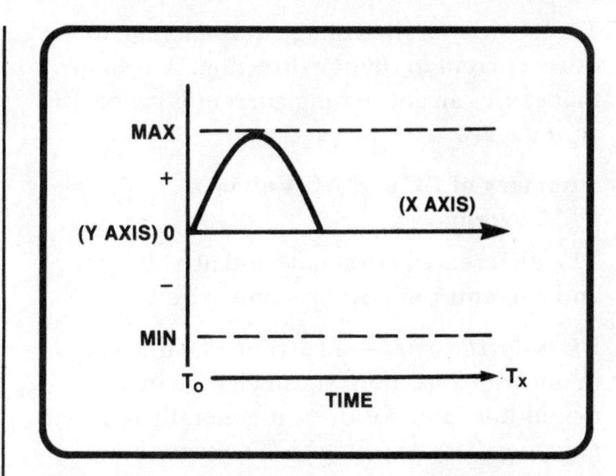

Figure 1.6 *Current Plotted in Positive Direction*

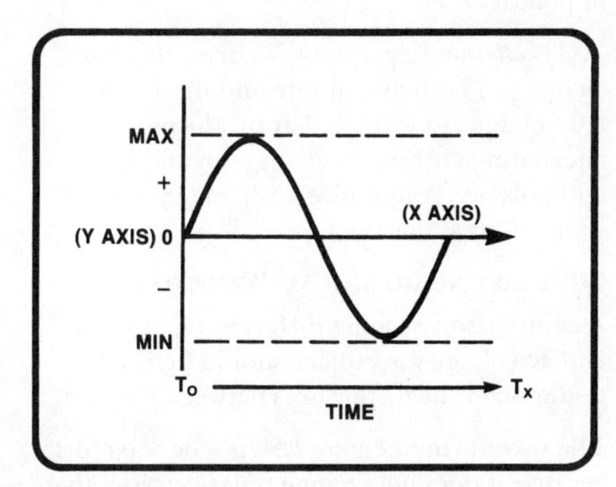

Figure 1.7 *Current Plotted in Positive and Negative Directions*

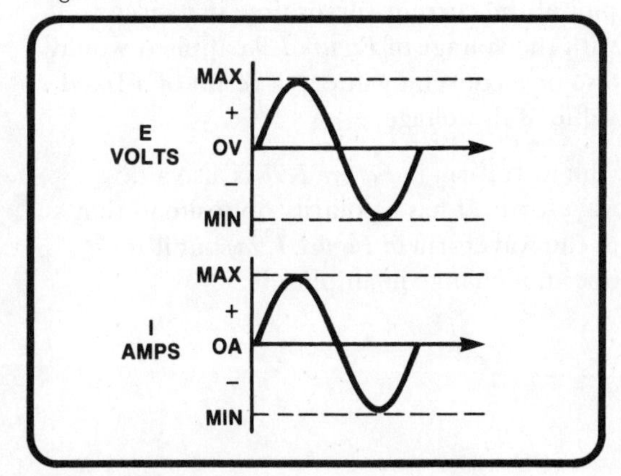

Figure 1.8 *Comparison of Alternating Voltage and Alternating Current*

■ Summary of DC and AC Voltages and Currents

The polarity of the voltage must change to cause current to change direction. A voltage that causes an alternating current is called an ac voltage.

Summary of DC and AC Voltages and Currents

The difference between dc and ac voltages and currents can now be summarized.

DC is *direct current*—a current which flows in only one direction. It can change in magnitude, and if it does, it generally is called *pulsating dc.* A dc voltage is a voltage that produces a direct current. It does not change in polarity.

AC is *alternating current*—a current which changes in both magnitude and direction. AC voltage is a voltage that produces an alternating current. It changes in amplitude and polarity. Amplitude is the magnitude or value of an ac voltage.

CONTRASTING DC AND AC WAVEFORMS

A comparison of some different types of dc and ac voltage waveforms should help you understand the differences between the two.

The waveform in *Figure 1.9a* is a dc waveform because it does not change polarity. Note that the amplitude remains at a constant level. A plot of the current versus time in a circuit with the voltage of *Figure 1.9a* applied would also be a constant value as a result of a fixed value of dc voltage.

The waveform in *Figure 1.9b* is also a dc waveform. It has a polarity opposite to that of the waveform in *Figure 1.9a,* but it too does not change in amplitude.

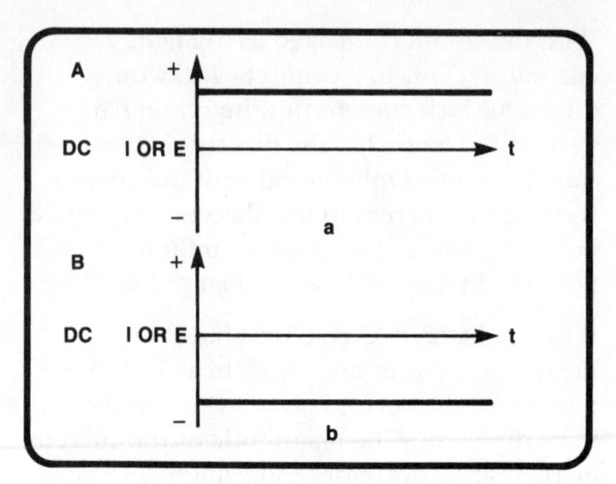

Figure 1.9 *a. A DC Waveform; b. A DC Waveform of Polarity Opposite to That of the One in a.*

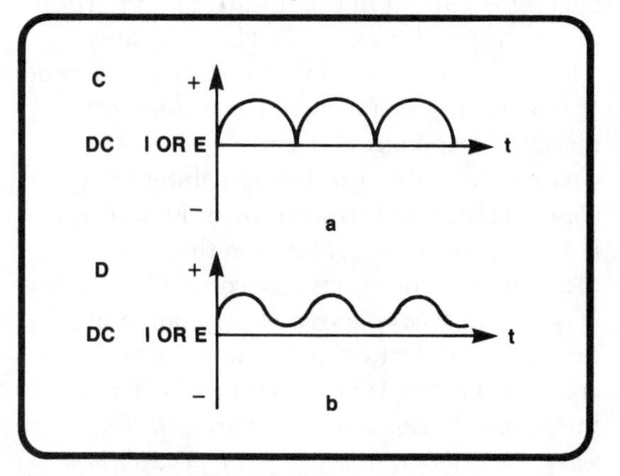

Figure 1.10 *a. A Pulsating DC Waveform; b. A DC Waveform Constantly Changing Amplitude*

The waveform in *Figure 1.10a* is a dc waveform, and it is a pulsating waveform. The entire waveform is in the positive portion of the graph, and never crosses the X axis. If the line graph had crossed the X axis into the opposite half of the graph, and if this voltage were applied to a circuit, then it would have caused the circuit current to change direction, and it would no longer be considered a dc voltage. This is the most important point in distinguishing between dc and ac waveforms.

The waveform in *Figure 1.10b* is a dc waveform that is constantly changing amplitude. This type of waveform is often referred to as an ac waveform riding on a steady state or constant dc voltage or current (indicated by the dotted line). The output from signal–amplifier circuits often looks like the one in *Figure 1.10b.* There is a steady-state dc voltage at the output displaced from zero when there is no signal being amplified. When a signal is amplified it rides on top of the dc voltage and swings the dc voltage above and below its steady-state value. In many circuits, resistive and capacitive coupling circuits are used to remove, or block, the dc component of the waveform. The resultant waveform then looks like *Figure 1.11a.*

Both waveforms in *Figure 1.11* are ac waveforms. Both are constantly changing in amplitude and direction. *Figure 1.11a* shows a typical sine waveform. *Figure 1.11b* shows an ac square waveform, commonly called a square wave. Note that the square wave maintains a constant amplitude for a period of time and then almost instantly changes to the same constant amplitude of opposite polarity for the same period of time. So the periods of time are equal and the constant amplitudes of opposite polarity are equal, and the changes are almost instantaneous or step-like.

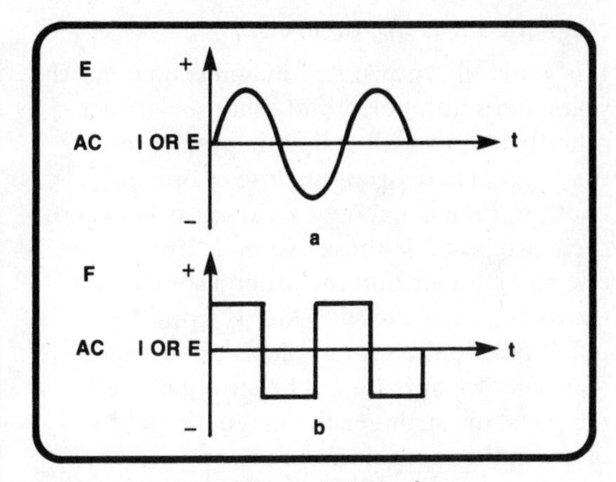

Figure 1.11 *Two Different AC Waveforms*

GENERATING AN AC WAVEFORM

Now that the differences in dc and ac have been determined and a definition of ac has been established, how alternating current is produced can be discussed.

Alternating current can be produced by periodically reversing the connections from a dc power source to the circuit. This, however, is impractical. For example, typical household alternating current resulting from a source voltage of 110 VAC, 60 hertz, reverses polarity 60 times every second. Reversing the connections to a dc power source 60 times per second is virtually impossible. A more practical method of generating alternating current is with an ac generator.

The AC Generator

An ac generator is a device which generates an ac voltage by rotating a loop of conductor material through a magnetic field. To understand the operation of an ac generator, some basic understanding of magnetic theory is necessary.

Magnetic Lines of Force

It is generally known that magnets have north poles and south poles, and that they attract other materials with magnetic properties. If two magnets are brought close to one another, then a magnetic field exists between their two poles. If these two poles are unlike, one a north pole and the other a south pole, the direction of the flux lines is from the north pole to the south pole, as shown in the diagram of *Figure 1.12*. The stronger the magnets, the stronger the magnetic field between the two poles.

Iron filings can be used to indicate the presence of the flux lines between north and south magnetic poles as shown in *Figure 1.13*. The magnets are placed on the table with north and south poles as shown, a sheet of plexiglas is placed over them and iron filings are sprinkled on top. The iron fillings, because they are magnetic material, align themselves with the flux lines of the magnetic field. These lines are important because they are used to explain how to generate ac.

Electromagnetic Induction

In 1831, Michael Faraday, a British physicist, discovered that if a moving magnetic field passes through a conductor, a voltage will be induced in the conductor and if the conductor is connected in a circuit a current will flow.

Conversely, if a moving conductor passes through a magnetic field, a voltage will also be induced in the conductor causing current to flow as shown in *Figure 1.14*, by generating an electromotive force by a process called electromagnetic induction. The direction of current flow in the conductor depends on the direction of the magnetic field and the direction of motion of the conductor through the field. This fact is summarized in what is known as the left-hand rule for generators.

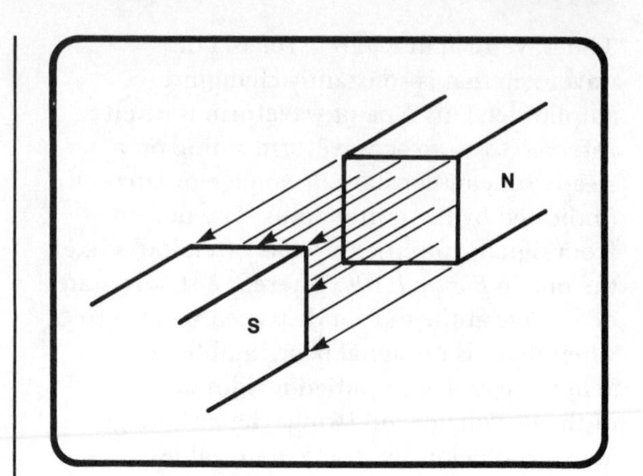

Figure 1.12 *Flux Lines Between Unlike Magnetic Poles*

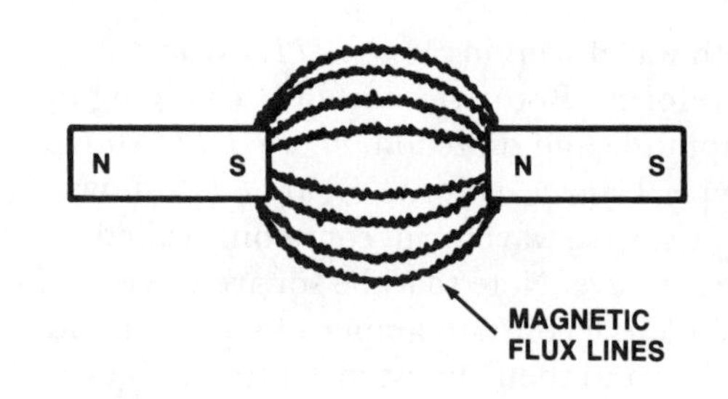

Figure 1.13 *Magnetic Flux Line Around Two Magnets*

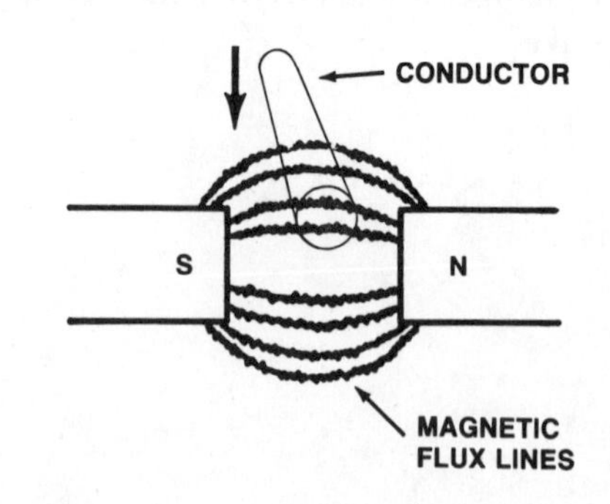

Figure 1.14 *Electromagnetic Induction*

■ Electromagnetic Induction

Figure 1.15 is a drawing illustrating this rule. The thumb points in the direction of motion of the conductor through the magnetic field. The forefinger points in the direction of the lines of magnetic field. And the index finger points in the direction of *electron current* flow. Electron current flow will be used throughout this book unless otherwise noted. Electron flow with its moving negative charges is opposite from the conventional current flow direction which was used by Ben Franklin when he assumed that positive charges were moving to constitute current flow.

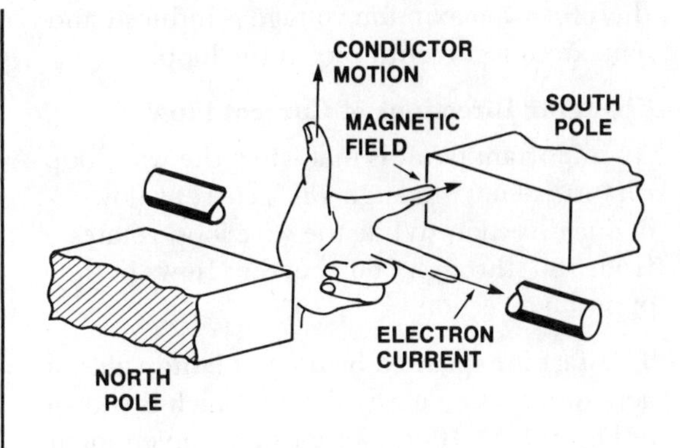

Figure 1.15 *Left-hand Rule for Generators*

A practical application of this concept is the ac single loop rotary generator shown in *Figure 1.16*. In this type of generator, a single loop of wire is rotated in a circular motion in a magnetic field. The direction of the magnetic field flux lines is as shown.

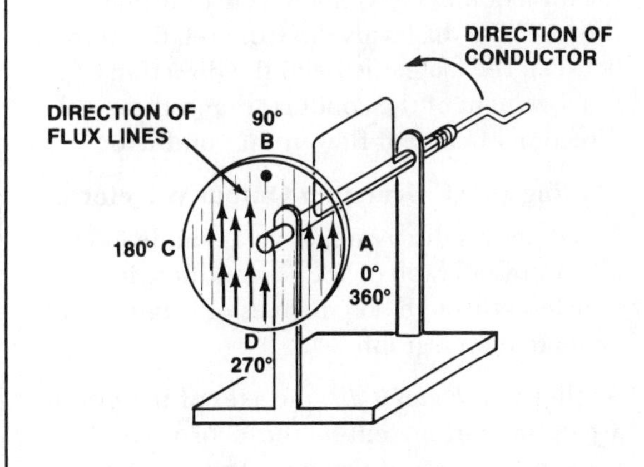

Figure 1.16 *AC Single-Loop Rotary Generator*

There are 360 degrees in any circle, and various points of rotation can be defined in terms of degrees. For this example, the loop is assumed closed and the current flow in the loop as a result of the induced voltage at four points of rotation of the loop, A, B, C and D as shown in *Figure 1.16,* will be analyzed.

The circular motion of the loop will be started at point A or 0 degrees; continue to point B at 90 degrees; go to point C, 180 degrees; pass through point D, 270 degrees; and return to point A at 360 degrees, or 0 degrees. The arrow in the drawing of *Figure 1.16* indicates the direction in which the conductor is moving through the stationary magnetic field.

As shown in *Figure 1.17*, at point A, 0°, and point C, 180°, the conductor is moving parallel to the lines of flux. It cuts no flux lines; therefore, no voltage is induced and no current will flow in the wire. However, at point B, 90° and point D, 270°, the conductor is moving perpendicular to the flux lines;

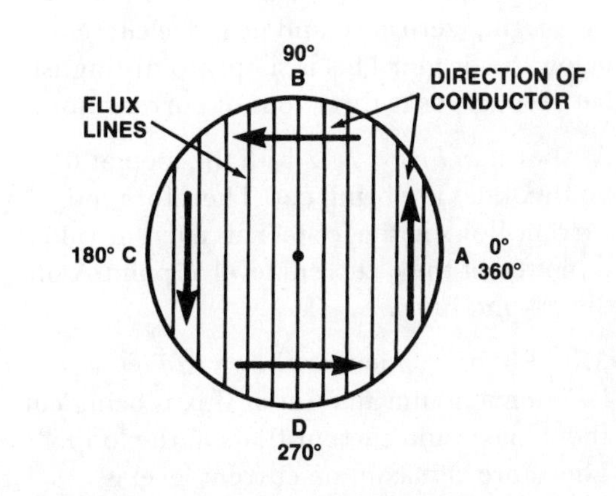

Figure 1.17 *Relationship of Flux Lines and Direction of Conductor*

therefore, a maximum voltage is induced and a maximum current flows in the loop.

Changing Directions of Current Flow

An important point is that while the wire loop rotates from 0° through 180°, current flows in one direction. While the wire loop rotates from 180° through 360°, current flows in the opposite direction.

This fact is explained by the left hand rule for generators using electron flow which is shown in *Figure 1.15*. If one were using conventional current then the right hand rule for generators as shown in *Figure 1.18* explains the direction of conventional current that flows. These diagrams show the relationship between the magnetic field, the direction of movement of the conductor, and the direction of current flow in the conductor.

Plotting an AC Generator Output Waveform

Based on this discussion of an ac single loop generator, a graph of the current flow in the wire loop through 360 degrees of rotation will be plotted on a graph.

As shown in *Figure 1.19*, the axes of the graph are defined in current on the Y, or vertical axis, and degrees of rotation at points A,B,C, and D on the X, or horizontal axis. Notice that the Y axis shows positive current above the origin, (zero axis) and negative current below the origin. This is simply to distingush between opposite directions of current flow.

As shown in *Figure 1.17* with the loop at 0°, no flux lines are being cut. Therefore, no current flows in the loop. This current value is plotted at the 0 current level at point A on the graph of *Figure 1.19*.

After 90° of rotation, as shown in *Figure 1.17*, a maximum amount of flux is being cut and a maximum current flows in the loop. Therefore, a maximum current level is plotted at point B on the graph.

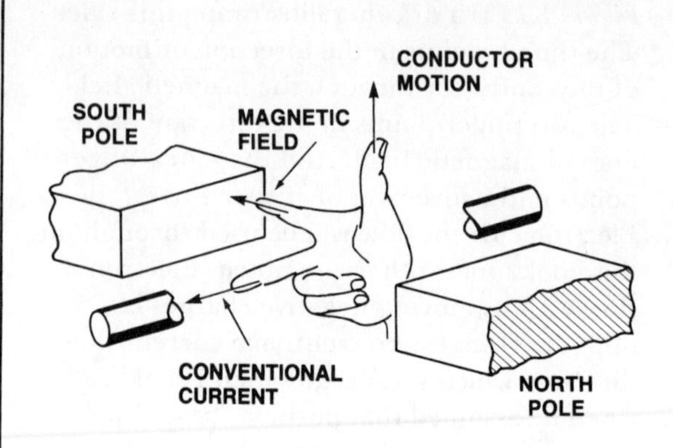

Figure 1.18 *Right-hand Rule for Generators*

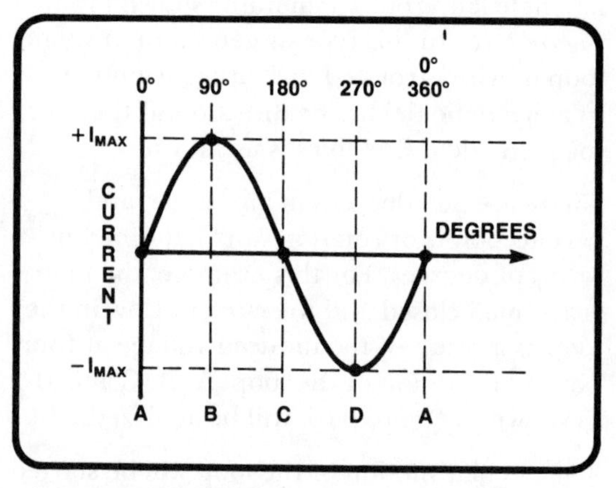

Figure 1.19 *Plot of One Complete Revolution of Loop*

At 180° of rotation, the loop is again moving parallel to the flux lines. Therefore, no current flows. The value is plotted at point C on the graph.

At 270°, a maximum current again flows, but now the current flows in the opposite direction to the direction at point B because the conductor is moving in the opposite direction. This maximum current is plotted at point D in the negative area of the graph to distinguish it from current flow in the opposite direction during the previous 180 degrees of rotation.

■ Cycle Alternations

At 360°, current again decreases to zero and is plotted at 0° at point A.

If the rotation should continue through another 360 degrees, the current flow would be identical to the previous rotation and identical points would be plotted on the graph.

Between the four points plotted for each cycle of loop rotation, current flow is not linear. It's a smooth, continuously changing waveform called a sine wave. When the current values at the four points A, B, C and D are connected as shown in *Figure 1.19,* a waveform called a *sinusoidal waveform* is the result.

It is the type of waveform most commonly found in ac, and it will be the primary type of waveform studied throughout this book. The characteristics of the waveform will be explained in detail in following lessons.

IDENTIFICATION OF WAVEFORM CYCLES

With an ac generator, when the loop makes one complete revolution, it generates a voltage that produces a current that progressively increases in value to a maximum, then decreases to zero, goes to a negative maximum, then returns to zero. As the loop begins another rotation within the magnetic field, the output is an exact duplicate of the previously generated waveform, provided the generator continues to rotate at a constant speed. Thus, the output repeats itself again and again every 360° of rotation as shown in *Figure 1.20.* Each 360° of rotation produces one cycle.

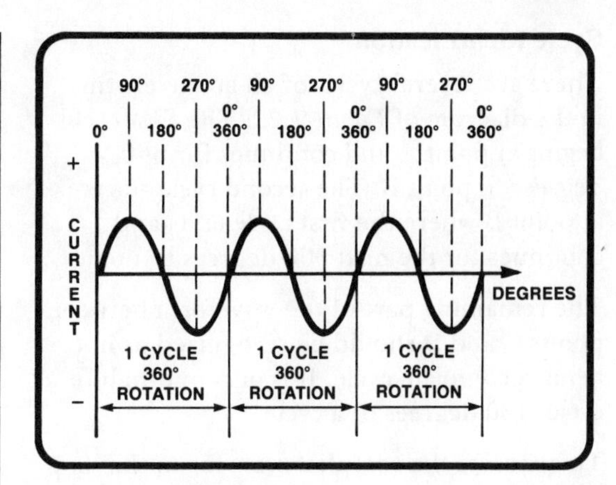

Figure 1.20 *Multiple Cycles*

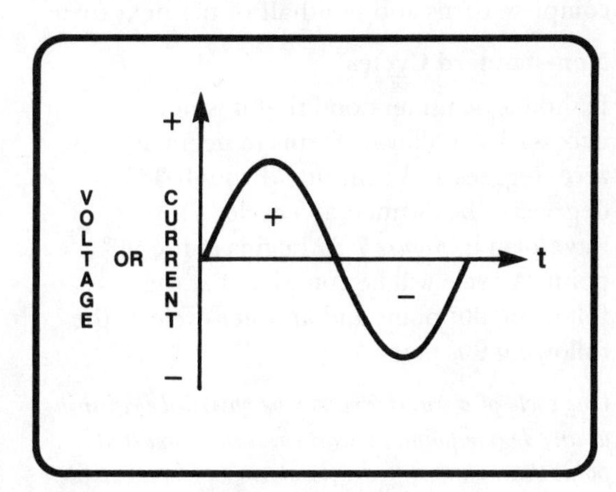

Figure 1.21 *Alternations of a Cycle*

Cycle Alternations

One cycle of a sinusoidal waveform can also be described in terms of *alternations.* Each cycle has two alternations: a positive alternation and a negative alternation. The positive alternation occurs while current is flowing in the same direction which is defined as the positive direction. The negative alternation occurs while current flows in the opposite direction.

Cycle alternations are identified in *Figure 1.21.*

■ **Cycle Identification**
■ **Non-standard Cycles**

Cycle Identification

There are several cycles of an ac waveform in the diagram of *Figure 1.22*. The first cycle begins at point A and continues for 360 degrees to point B. The second cycle starts at point B where the first cycle ends and continues for the next 360 degrees to point C.

The remaining part of the waveform between points C and D should be recognized as not being a complete cycle. It is only one-half of a cycle: 180 degrees of a cycle.

To generate the complete waveform, the loop of the generator would have to be rotated two complete turns and one-half of the next turn.

Non-standard Cycles

It should be understood that it is not necessary for all waveforms to begin at zero degrees and continue through 360 degrees to be defined as a cycle. The waveform in *Figure 1.23* begins at the 90° point. A cycle will be completed at the following 90° point and another cycle at the following 90° point.

One cycle of a waveform may be observed beginning at any degree point. It must continue from that point through a 360-degree change to be a complete cycle. For example, observe the waveform in *Figure 1.24* which begins at the 135° point. A cycle will be completed when it progresses through a 360-degree change and returns to 135°.

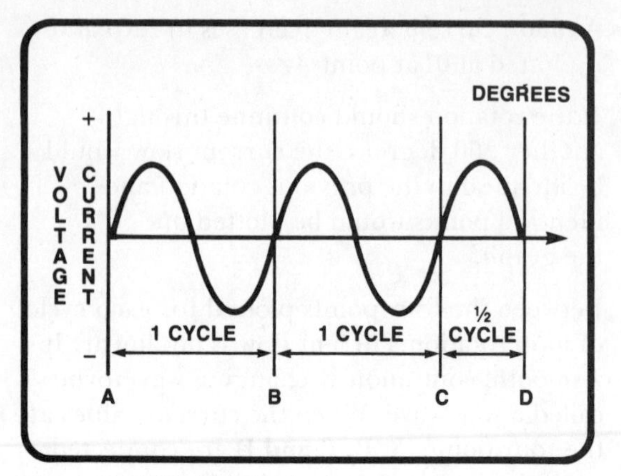

Figure 1.22 *Multiple Cycles*

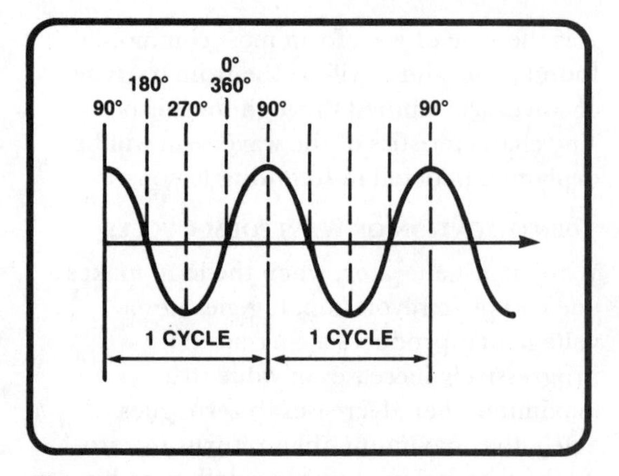

Figure 1.23 *Cycles Beginning at 90°-Point*

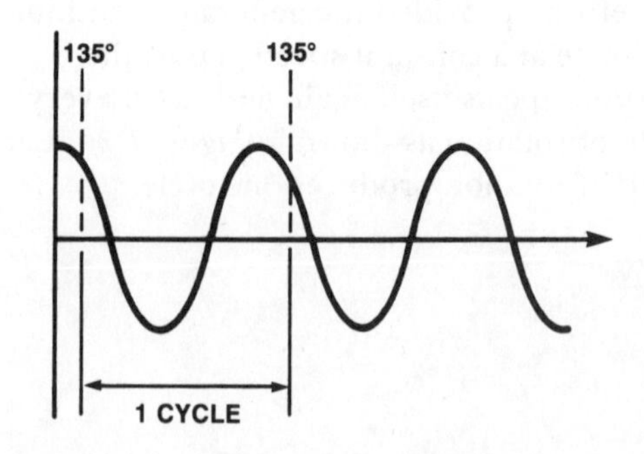

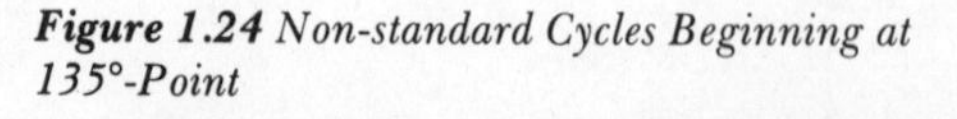

Figure 1.24 *Non-standard Cycles Beginning at 135°-Point*

■ Cycle Defined

If a waveform is observed beginning at a negative peak as in *Figure 1.25*, one cycle would be completed at the next negative peak.

Cycle Defined

Repetitious waveforms which are not sinusoidal have cyclic properties as shown in *Figure 1.26*. In each example the waveform has completed one cycle when it reaches a point where repetition of the waveform begins.

A cycle of a waveform can now be defined as a waveform that begins at any electrical degree point and progresses through a 360-degree change.

The square waves in *Figure 1.27* also have cyclic characteristics.

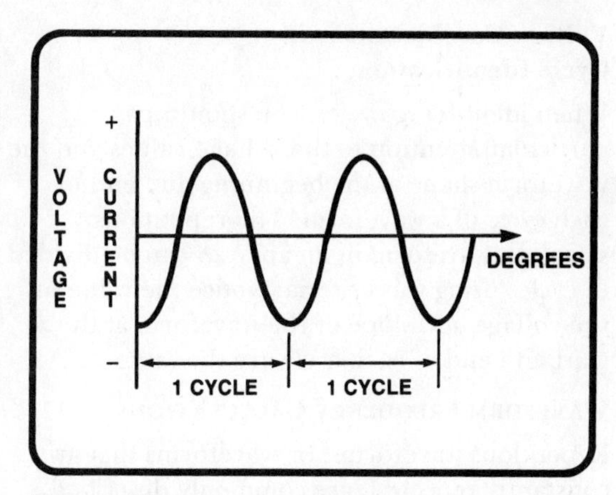

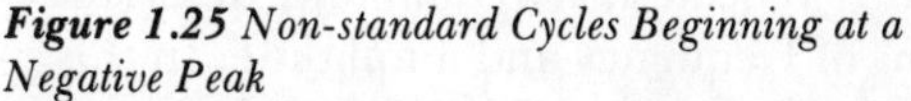
Figure 1.25 *Non-standard Cycles Beginning at a Negative Peak*

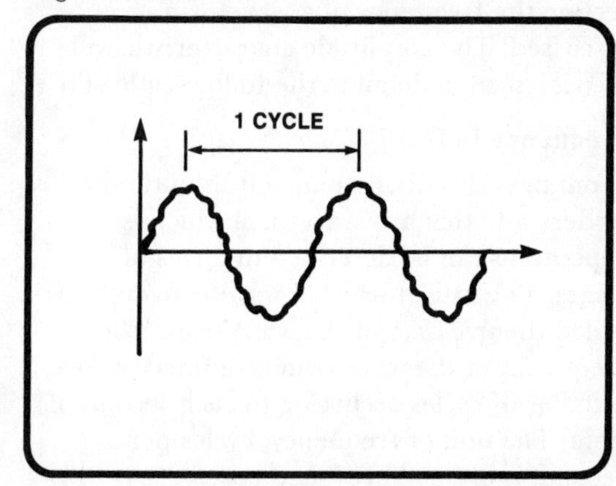

Figure 1.26 *Repetitious Non-sinusoidal Waveform*

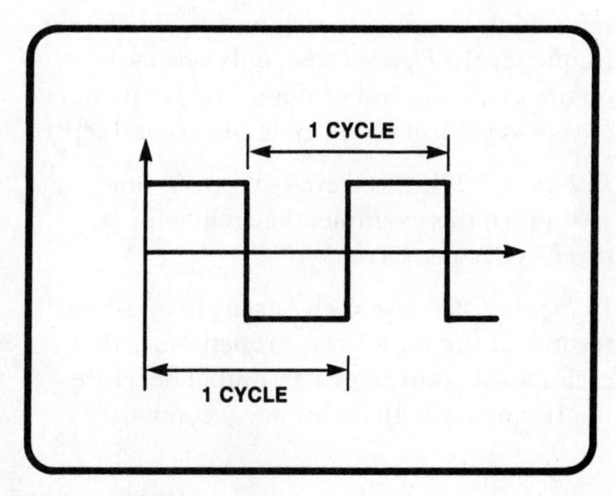

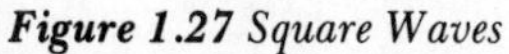
Figure 1.27 *Square Waves*

■ **Voltage Considerations in Cycle Identification**
■ **Frequency Defined**
■ **Unit of Frequency: Hertz**

Voltage Considerations in Cycle Identification

When identifying cycles, you should pay particular attention to the voltage values and the waveform shape at the beginning and end of each cycle of a waveform. The repetitious sawtooth waveform in *Figure 1.28* can be divided in cycles using this criteria. Notice the value of the voltage and slope of the waveform at the start and end of each cycle are the same.

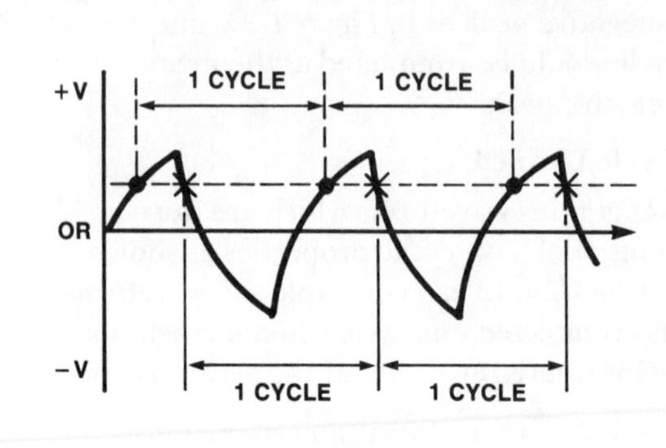

Figure 1.28 *Sawtooth Waveform*

WAVEFORM FREQUENCY CALCULATIONS

Repetitious waveforms, or waveforms that are constantly repeated, are commonly described in terms of frequency and amplitude. In this section the frequency of a waveform is discussed. The amplitude characteristic will be described in detail in the following lesson.

Frequency Defined

From previous discussion, you should now understand that any waveform which is repetitious can be described in terms of cycles. The rate at which a waveform cycles is called the *frequency* of the waveform. The frequency of the waveform is defined as the number of cycles occurring in each second of time. The unit of frequency, cycles per second, is often abbreviated cps.

For example, *Figure 1.29* shows three sinusoidal waveforms, each having a different frequency. In *Figure 1.29a*, only one cycle occurs in one second of time. The frequency of this waveform is one cycle per second (cps).

In *Figure 1.29b,* three cycles occur in one second. In this example, the frequency is three cycles per second.

In *Figure 1.29c*, five cycles occur in one-half second. If the waveform is repetitious, 10 cycles must occur in one second. Therefore, the frequency is 10 cycles per second.

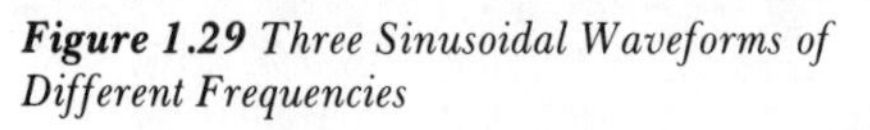

Figure 1.29 *Three Sinusoidal Waveforms of Different Frequencies*

Unit of Frequency: Hertz

In recent years, the unit of frequency, cycles per second, has been replaced by the term hertz, abbreviated Hz. The unit hertz was adopted to honor the German physicist, Heinrich Hertz, who made important discoveries in the area of electro-magnetic waves in the late nineteenth century.

One hertz is simply one cycle per second. Applying the newer terminology to the waveform of *Figure 1.29,* the waveform of *Figure 1.29a* has a frequency of one hertz, the waveform of *Figure 1.29b* has a frequency of three hertz, and the frequency of *Figure 1.29c* is 10 hertz.

Frequency Prefixes

Prefixes are often used to simplify the writing of high frequencies. The common prefixes used are:

k for kilo or thousand
M for mega or million, and
G for giga or billion

For example, a radio station broadcasting at 820,000 hertz (Hz) can have its frequency described as 820 kilohertz (kHz).

A 1,210,000 hertz signal could be written as 1.21 megahertz (MHz).

A radar system operating at 27,000,000,000 hertz may be specified as 27 gigahertz (GHz).

WAVEFORM PERIOD CALCULATIONS

Period Defined

The frequency of a waveform is determined by the lengh of time of one cycle. This time of one cycle of a waveform is defined as the period of the waveform.

The symbol T is used to represent time and period. Remember that the terms are synonomous. The period or time of a waveform is simply *the time required to complete one cycle of the waveform.* T has units of seconds.

Period Calculation Examples

Let's return to the examples in *Figure 1.29* and determine the period of each of the three waveforms.

The waveform in *Figure 1.29a* has a frequency of one hertz. One cycle occurs in one second of time. Therefore, the time of one cycle, the period, of this waveform is one second.

In *Figure 1.29b* the waveform has a frequency of three hertz. One cycle occurs during each one-third of a second time interval. Therefore, the period of one cycle is one-third second or 0.333 second. The waveform in *Figure 1.29c* has five cycles occurring in one-half of a second. The waveform repeats itself every one-tenth of a second. Each cycle occupies a one-tenth second time interval; therefore the period, T, of the waveform is one-tenth (0.1) second.

Period or Time Equation

There is a mathematical relationship between the period and frequency of a waveform. This mathematical relationship is expressed by this equation:

$$\textbf{period or time(s)} = \frac{\textbf{1}}{\textbf{frequency (Hz)}} \qquad (1\text{–}1)$$

The time of one cycle or period, T, of a waveform in seconds can be determined by dividing the number one by the frequency of the waveform in hertz. This equation is normally simplified by writing it in this form:

$$T = \frac{1}{f} \qquad (1\text{–}2)$$

You must remember that to obtain the period of a waveform in *seconds,* the frequency in *hertz* must be used in the equation. Attempts to calculate the period of a waveform by leaving the zeroes out of the frequency value will result in an incorrrect answer.

■ Period or Time Equation

Using equation *1–2* to find the period of a waveform when the frequency is known is a common technical electronic solution. For example, find the time of one cycle when the frequency is known to be 50 hertz.

$$\begin{aligned} \mathbf{T_{(S)}} &= \frac{\mathbf{1}}{\mathbf{f\ (Hz)}} \\ &= \frac{\mathbf{1}}{\mathbf{50}} \\ &= \mathbf{0.02} \\ &= \mathbf{20ms} \end{aligned}$$

The time of one cycle is 20 milliseconds. If the known frequency is a larger value, the solution can at first appear difficult, but the technique is identical to the previous example. For example, what is the time of one cycle when the frequency is 650,000 hertz? Using equation *1–2:*

$$\begin{aligned} \mathbf{T_{(S)}} &= \frac{\mathbf{1}}{\mathbf{f\ (Hz)}} \\ &= \frac{\mathbf{1}}{\mathbf{650{,}000Hz}} \\ &= \mathbf{0.00000154} \\ &= \mathbf{1.54\mu s} \end{aligned}$$

The time of one cycle is 1.54 microseconds.

CONVERTING FROM TIME TO FREQUENCY MEASUREMENTS

A typical method of determining the frequency of a waveform is to measure the time of one cycle on an oscilloscope and calculate the frequency from the period measurement.
The period equation can be manipulated in another form to allow the calculation of a frequency when the period is known.

The frequency of the waveform in hertz can be found by dividing the number one by the period of waveform in seconds.

$$\mathbf{f} = \frac{\mathbf{1}}{\mathbf{T}} \qquad (1\text{–}3)$$

Notice once again that the unit values of the variables are very important. To obtain a correct answer, you *must* keep the time value in its original form. Attempts to modify the time-value to simplify the mathematical manipulation will often result in an incorrect answer.

Using equation *1–3* to find the frequency of a waveform when the period of the waveform is known is a common technical electronic solution.

For example, find the frequency when the period of the waveform is known to be 0.05 second.

$$\begin{aligned} \mathbf{f} &= \frac{\mathbf{1}}{\mathbf{T}} \\ &= \frac{\mathbf{1}}{\mathbf{0.05s}} \\ &= \mathbf{20Hz} \end{aligned}$$

If the known time of one cycle is a smaller value, the solution at first can appear more difficult, but the technique is identical to the previous example. Here's another example. The period of a waveform is measured to be 0.000002 seconds or 2 microseconds (2×10^{-6} seconds).

Therefore,

$$f = \frac{1}{2 \times 10^{-6}}$$
$$f = 0.5 \times 10^{6}$$
$$f = 0.5 \text{ megahertz}$$

Another example: The period has been measured as 100 seconds. Therefore, the frequency is

$$f = \frac{1}{T}$$
$$f = \frac{1}{100}$$
$$f = 0.01\text{Hz}$$

■ Summary

SUMMARY

In this lesson, ac was defined and dc and ac waveforms were compared. The operation of an ac generator was discussed and the ac waveform was described in terms of time (period) and frequency. Examples were provided using the time and frequency equations.

Additional time and frequency conversion examples will be provided in the following lesson, and the amplitude characteristics of sinewaves will be discussed.

■ Worked-Out Examples

1. For each of the diagrams below, identify the waveform as an ac or dc waveform.

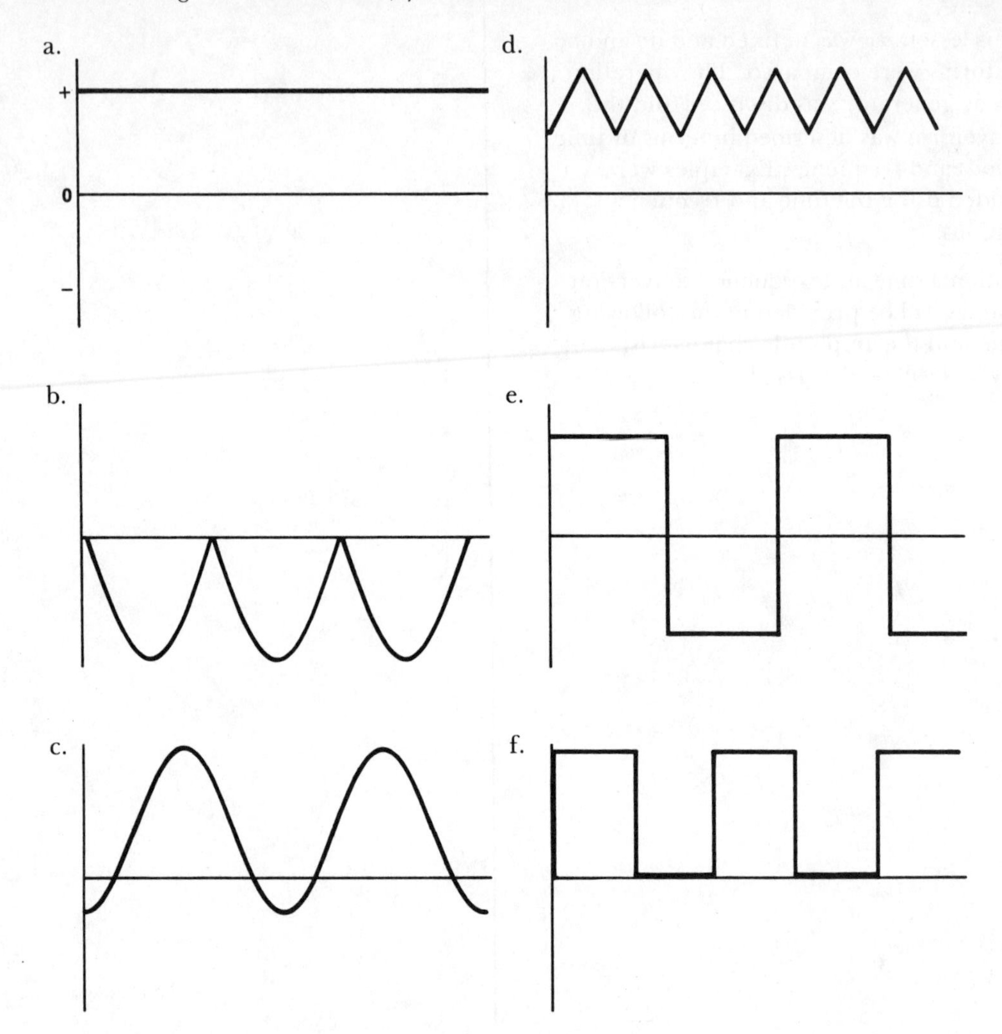

Solution: Recall the criteria for identifying an ac waveform; the magnitude and direction of current flow must change. Therefore, the polarity of the voltage causing the current flow must change and the waveform will be plotted in both the plus and minus portions of the graph.

a. dc
b. pulsating dc
c. ac
d. pulsating dc
e. ac square wave
f. dc square wave

■ **Worked-Out Examples**

2. Identify the number of cycles in each of the following diagrams:

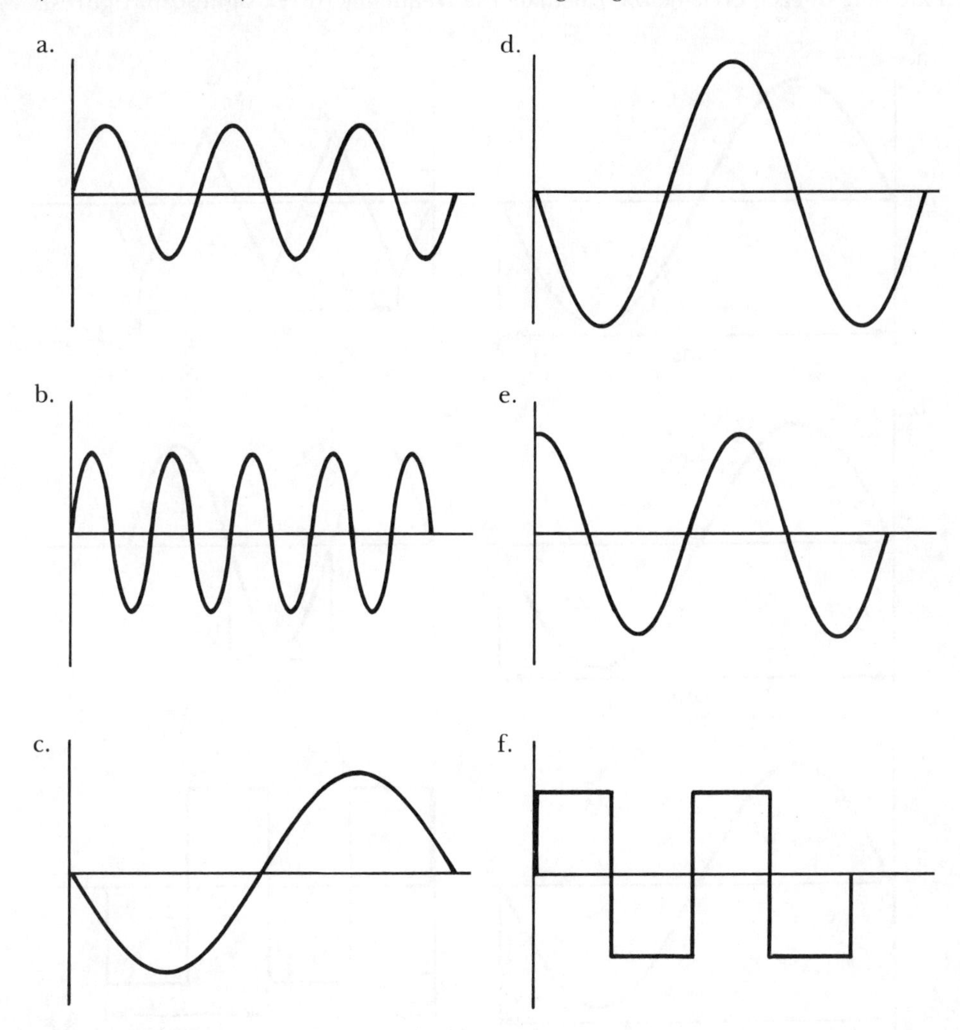

Solution: Remember the definition of a cycle; the waveform must make a complete 360 degree change. The starting point of the waveform is significant.

a. 3 cycles
b. 4½ cycles
c. 1 cycle
d. 1½ cycles
e. 1¾ cycles
f. 2 cycles

■ Worked-Out Examples

3. Given the time of each cycle below, calculate the frequency (three significant figures).

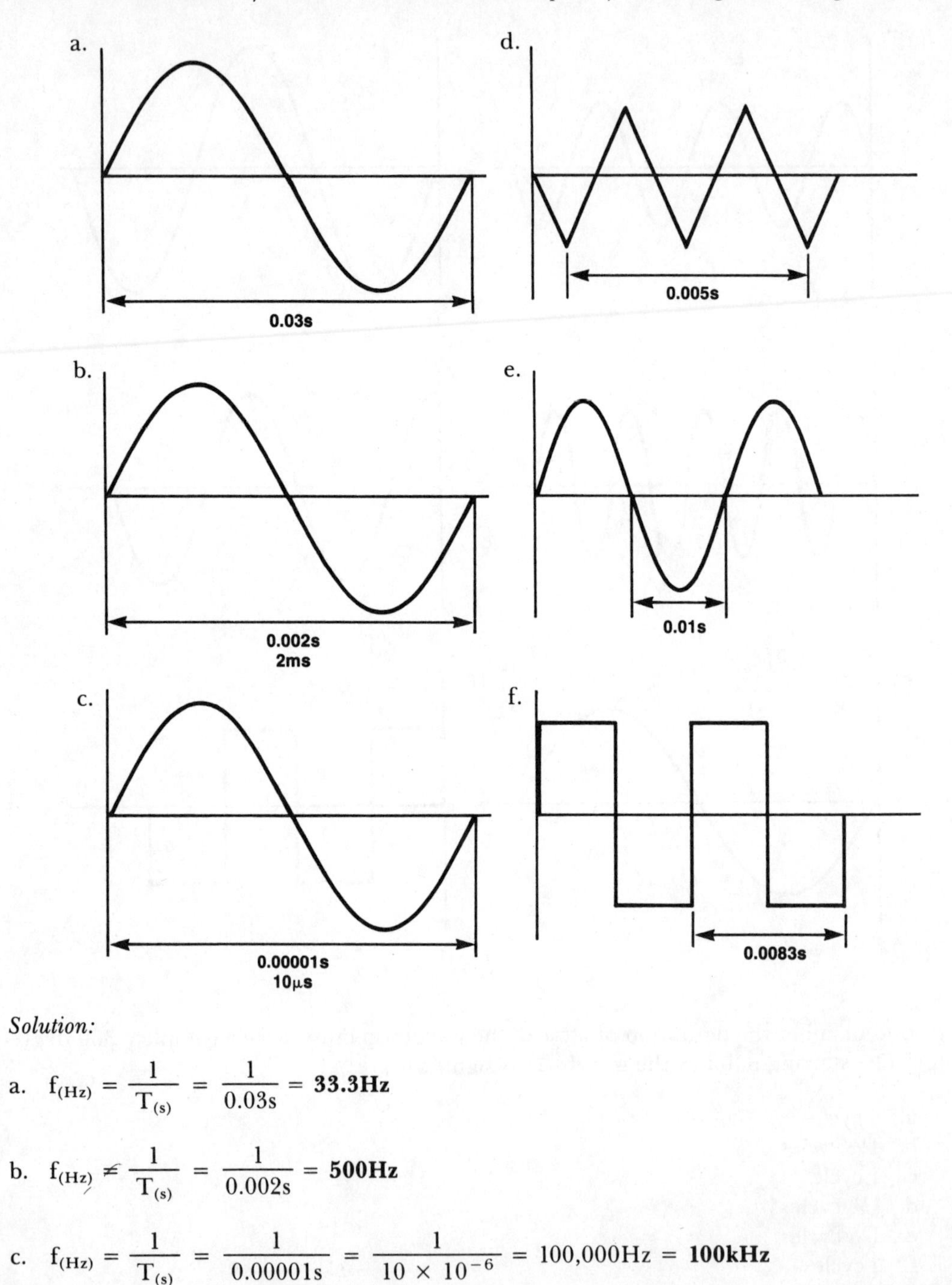

Solution:

a. $f_{(Hz)} = \frac{1}{T_{(s)}} = \frac{1}{0.03s} = \mathbf{33.3Hz}$

b. $f_{(Hz)} = \frac{1}{T_{(s)}} = \frac{1}{0.002s} = \mathbf{500Hz}$

c. $f_{(Hz)} = \frac{1}{T_{(s)}} = \frac{1}{0.00001s} = \frac{1}{10 \times 10^{-6}} = 100{,}000Hz = \mathbf{100kHz}$

■ Worked-Out Examples

d. 2 cycles = 0.005s

∴ 1 cycle = 0.0025s

$$f_{(Hz)} = \frac{1}{0.0025s} = \mathbf{400Hz}$$

e. ½ cycle = 0.01s

∴ 1 cycle = 0.02s

$$f_{(Hz)} = \frac{1}{0.02s} = \mathbf{50Hz}$$

f. $$f_{(Hz)} = \frac{1}{0.0083s} = \mathbf{121Hz}$$

4. Given the following frequencies, calculate the period of the waveform (three significant figures).

a. f = 105Hz
b. f = 60Hz
c. f = 8500Hz
d. f = 16.8kHz
e. f = 320kHz
f. f = 6.1MHz

Solution: Remember that period is the time of one cycle.

a. $$T_{(s)} = \frac{1}{f_{(Hz)}} = \frac{1}{105Hz} = \mathbf{0.00952s}$$

b. $$T_{(s)} = \frac{1}{f_{(Hz)}} = \frac{1}{60Hz} = 0.0167s = \mathbf{16.7ms}$$

c. $$T_{(s)} = \frac{1}{f_{(Hz)}} = \frac{1}{8500Hz} = 0.000118s = \mathbf{118\mu s}$$

d. $$T_{(s)} = \frac{1}{16.8kHz} = \frac{1}{16.8 \times 10^3} = 5.95 \times 10^{-5}s = \mathbf{59.5\mu s}$$

e. $$T_{(s)} = \frac{1}{320kHz} = \frac{1}{320 \times 10^3} = 0.00000313s = \mathbf{3.13\mu s}$$

f. $$T_{(s)} = \frac{1}{6.1MHz} = \frac{1}{6.1 \times 10^6} = 1.64 \times 10^{-7}s = \mathbf{0.164\mu s}$$

■ Worked-Out Examples

5. On the graph below, draw three cycles of an ac sinewave.

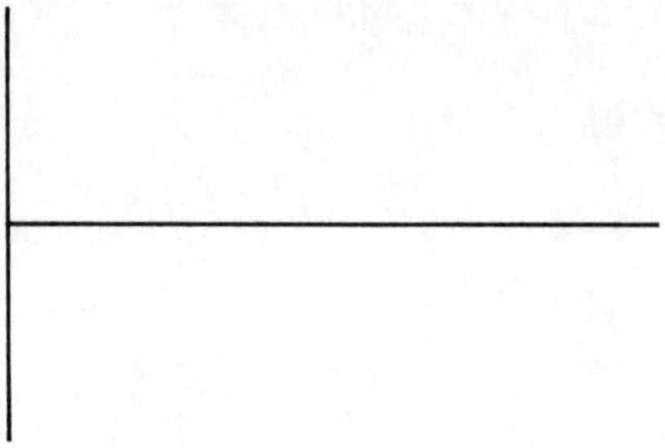

Solution:

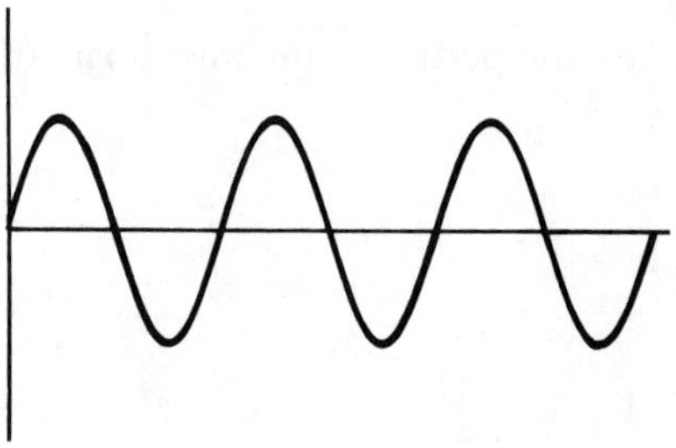

■ **Practice Problems**

1. In each of the six following diagrams, identify the waveform as an ac or dc waveform.

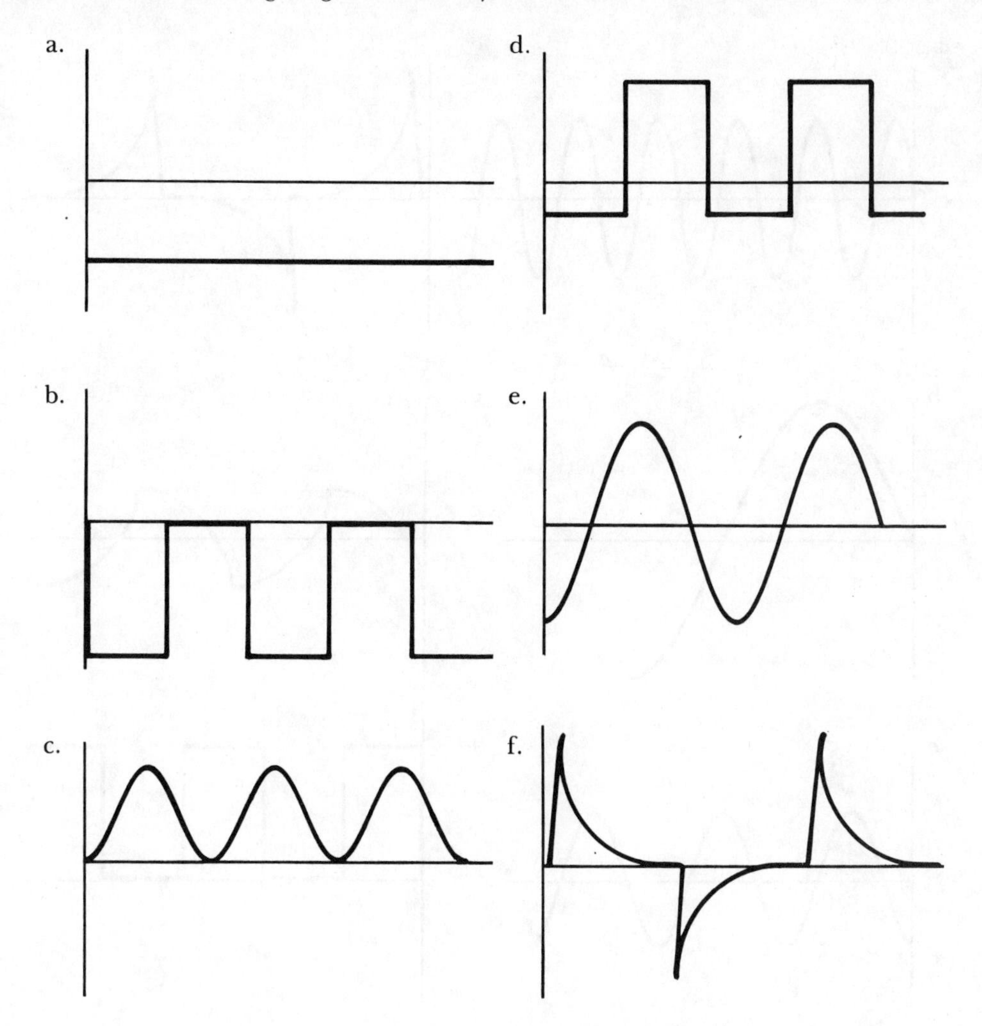

■ Practice Problems

2. Identify the number of cycles in each of the following diagrams:

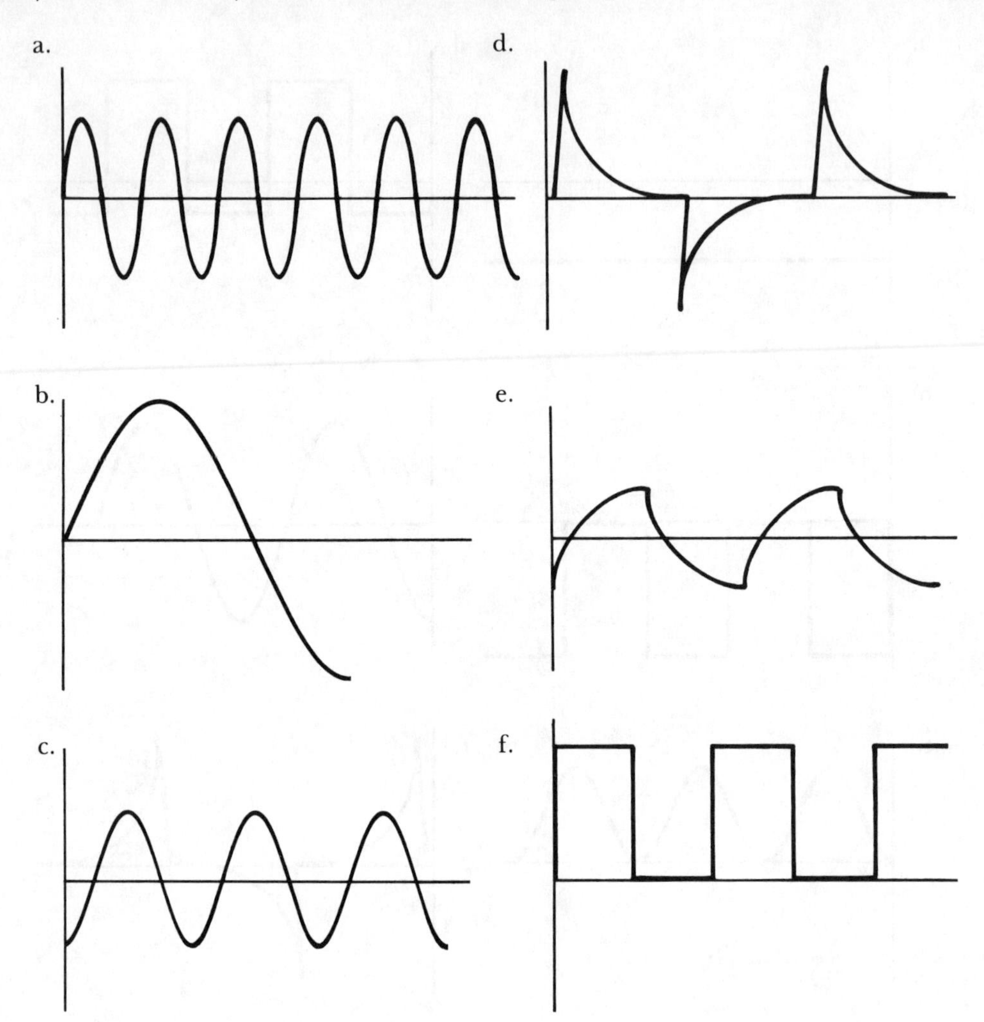

■ Practice Problems

3. Given the time of each waveform as shown, calculate the frequency (three significant figures).

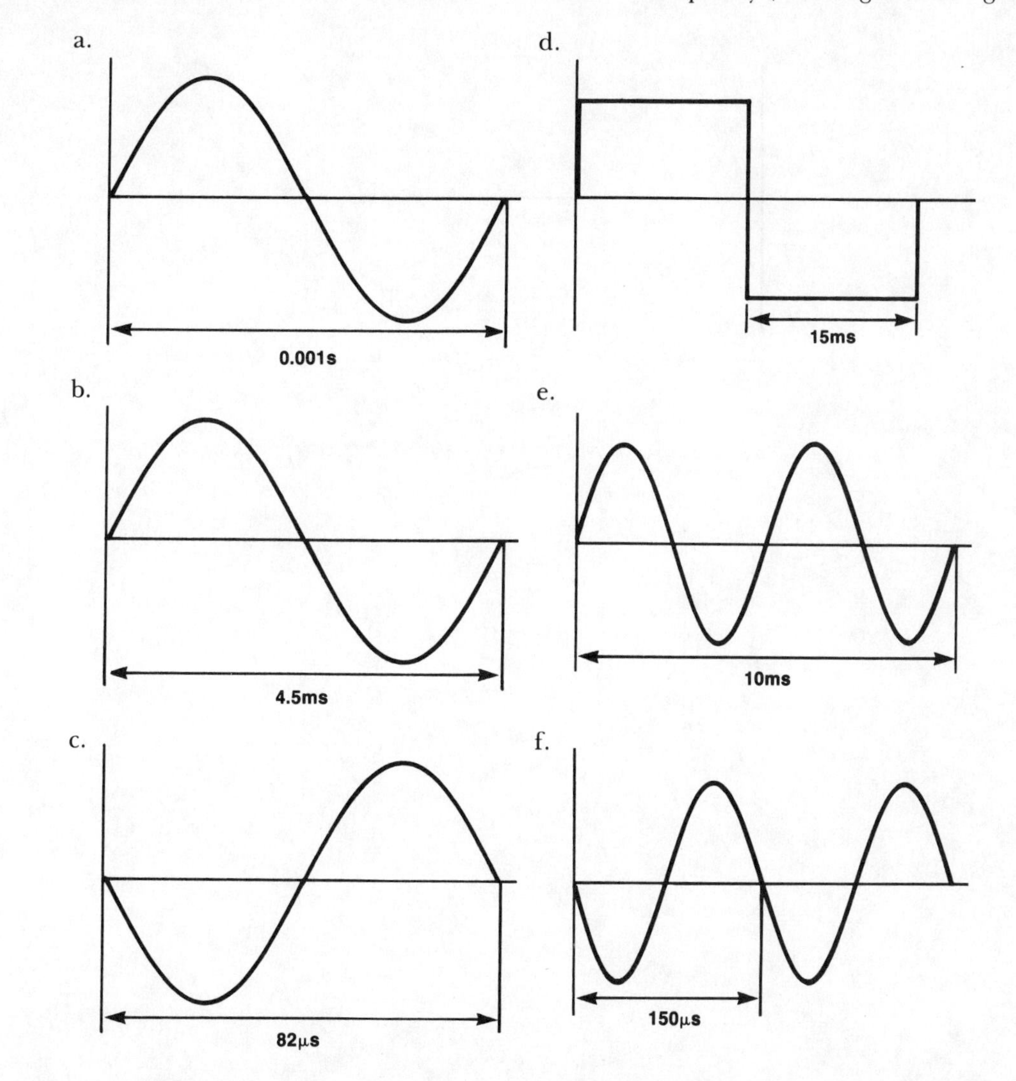

4. Given the following frequencies, calculate the time of one cycle (three significant figures).

a. f = 40Hz
b. f = 1.5kHz
c. f = 63kHz
d. f = 118kHz
e. f = 0.8MHz
f. f = 1.5MHz

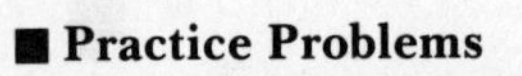

■ Practice Problems

5. On the graph, draw three cycles of positive dc square waveform.

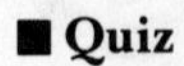

■ Quiz

1. In each of the following diagrams, identify the waveforms as an ac or dc waveform.

a. + 0 – dc

b. dc

c. ac

d. ac

e. dc

f. ac

2. Identify the number of cycles in each of the following diagrams.

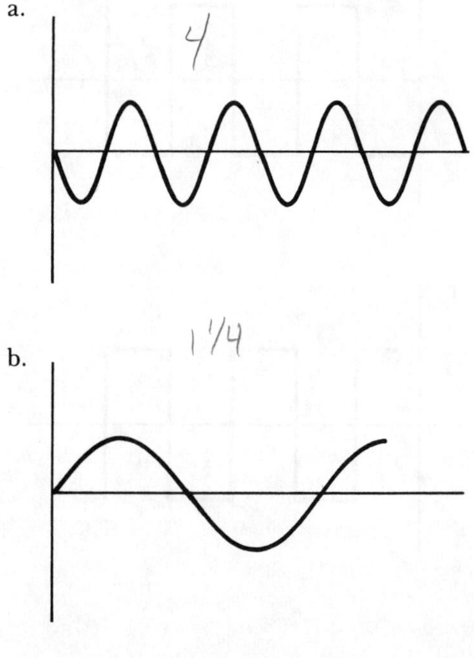

■ Quiz

c.

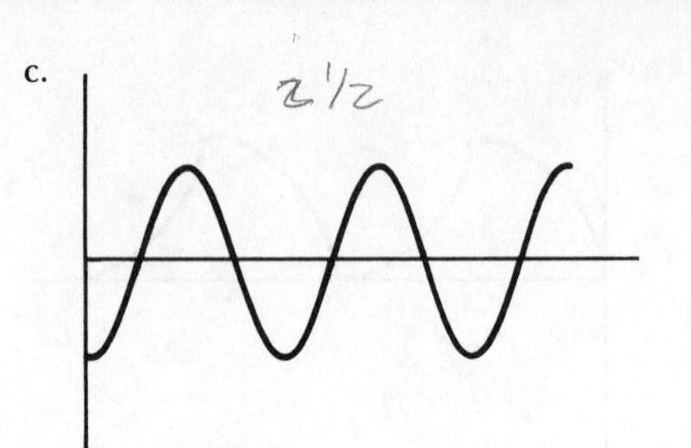

d.

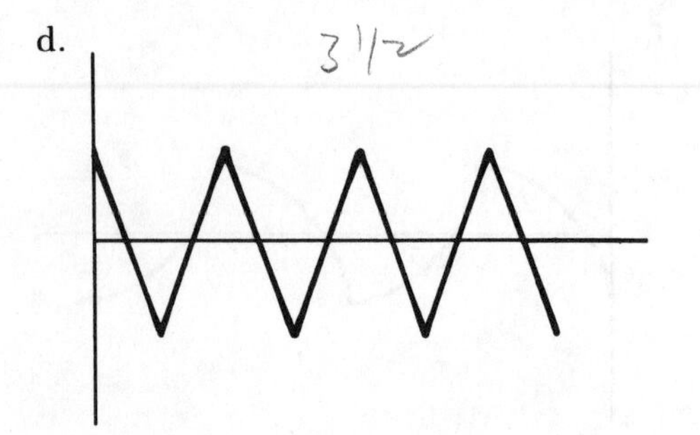

e.

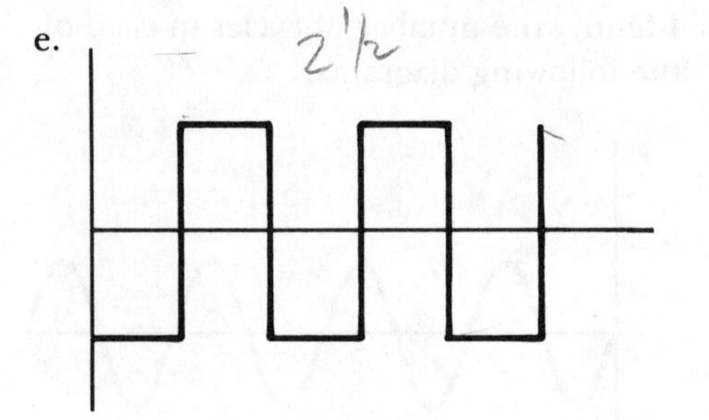

f.

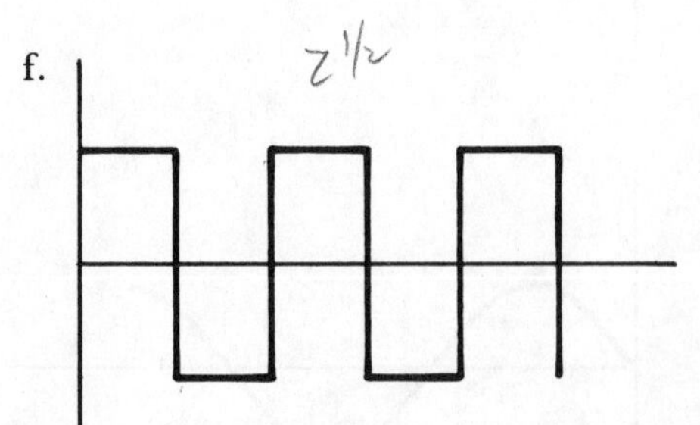

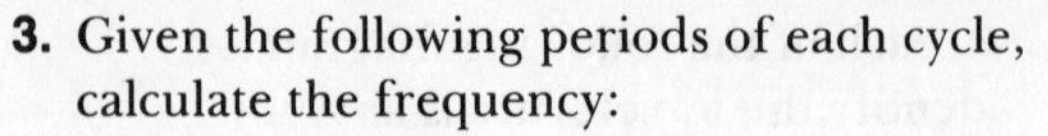

3. Given the following periods of each cycle, calculate the frequency:

a. T = 0.015s f = 66.7 Hz
b. T = 37.2ms 26.9 Hz
c. T = 37.2μs 26.9 kHz
d. T = 0.000008s 125 kHz
e. T = 48ms 20.8 Hz
f. T = 7.35μs 136 kHz

4. Given the following frequencies, calculate the period of the waveform:

a. f = 30Hz T = 33.3 ms
b. f = 300Hz 3.33 ms
c. f = 63.8kHz 16 μs
d. f = 20kHz 50.0 μs
e. f = 1700Hz 588 μs
f. f = 22MHz 45.5 ns

5. Draw two cycles of an ac square waveform on this graph.

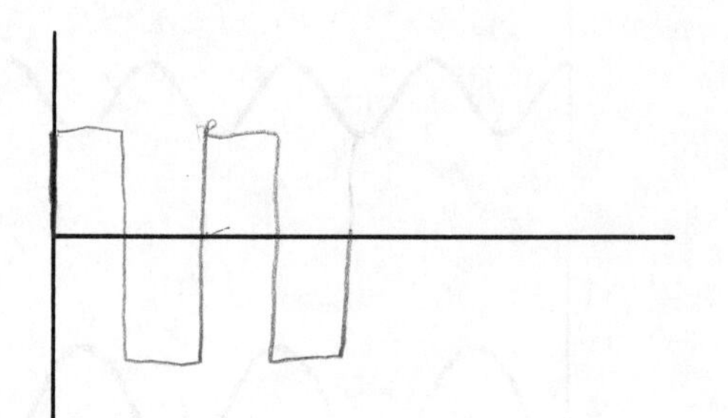

LESSON 2

AC and the Sine Wave

In Lesson 1, the sine wave was introduced by describing the development of an ac waveform with a single loop generator. The ac waveform was described in terms of its frequency or period, and another waveform parameter, the amplitude of the waveform, was mentioned briefly.

This lesson discusses in detail the amplitude descriptions of a sinusoidal waveform. Also, the time and frequency measurement of a waveform are reviewed, and an introduction to the trigonometric function will be presented.

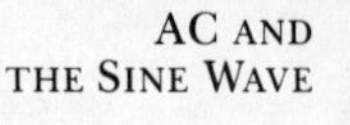

■ Objectives

This lesson discusses the amplitude descriptions of sinusoid a waveform and introduces the trigonometric functions which are essential to an analysis and understanding of ac circuits. At the end of this lesson, you should be able to:

1. Explain the three ways to express the amplitude of a sinusoidal waveform and the relationship between them.
2. Understand the importance of the 0.707 constant and how it is derived.
3. Convert peak, peak-to-peak, and rms voltage and current values from one value to another.
4. Explain the sine, cosine, and tangent trigonometric functions.
5. Explain the relationship between the sine trigonometric function and an ac waveform.
6. Calculate the value of the sine of any angle between 0° and 360°.

- **Frequency and Period Relationships**
- **Frequency and Period Calculations**

REVIEW OF WAVEFORM FREQUENCY AND PERIOD RELATIONSHIPS

Any sinusoidal waveform can be described completely by identifying either time or frequency parameters and one of three possible amplitude specifications.

The *frequency* of a waveform is the number of cycles of the waveform which occur in one second of time. Common unit of measurement is hertz (Hz).

The *period* of a waveform, which sometimes is called its time, is the time required to complete one cycle of a waveform. It is measured in units of seconds, such as seconds, tenths of seconds, milliseconds, or microseconds.

If a waveform is to be properly described in terms of its period or frequency, it must be a repetitious waveform. A repetitious waveform is one in which each following cycle is identical to the previous cycle.

Frequency and Period Relationships

The frequency of a waveform in cycles per second is mathematically described in terms of the period, T, of the waveform as:

$$\mathbf{f = \frac{1}{T}} \qquad (2\text{–}1)$$

where f is the frequency of the waveform in hertz, and T is the time of one cycle of the waveform in seconds.

Frequency and Period Calculations

If the time of one cycle, or period, of a waveform is known, the frequency of the waveform can be calculated using equation *2–1*. For example, *Figure 2.1* shows one cycle

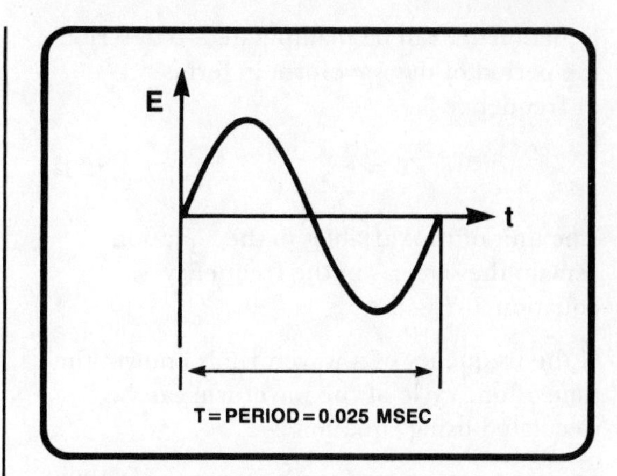

Figure 2.1 *One Cycle of Waveform with a Period of 0.025 Milliseconds*

of a waveform in which the cycle occurs in 0.025 milliseconds. Its frequency is calculated:

$$\begin{aligned}\mathbf{f} &= \mathbf{\frac{1}{T}}\\ &= \mathbf{\frac{1}{0.025 \times 10^{-3}}}\\ &= \mathbf{\frac{1}{25 \times 10^{-6}}}\\ &= \mathbf{0.04 \times 10^{6}}\\ &= \mathbf{40 \times 10^{3}}\\ &= \mathbf{40kHz}\end{aligned}$$

The frequency of the waveform is 40 kilohertz.

Or, for another example, a waveform has one cycle occurring every 10 milliseconds. Its frequency is calculated using equation *2–1*.

$$\begin{aligned}f &= \frac{1}{T}\\ &= \frac{1}{10 \times 10^{-3}}\\ &= 1 \times 10^{2}\\ &= 100\text{Hz}\end{aligned}$$

The frequency of the waveform is 100 hertz.

■ Additional Frequency Calculation Examples

Equation *2-1* can be manipulated to describe the period of the waveform in terms of frequency.

$$T = \frac{1}{f} \qquad (2\text{–}2)$$

The unit of the variables in the equation remain the same as in the frequency equation *2–1*.

If the frequency of a waveform is known, the time of one cycle of the waveform can be calculated using equation *2–2*.

For example, a waveform's frequency is 120 hertz. The time of one cycle of that waveform is calculated:

$$\begin{aligned} T &= \frac{1}{f} \\ &= \frac{1}{120} \\ &= 0.00833 \\ &= 8.33\text{ms} \end{aligned}$$

The time of one cycle, or the period, is 8.33 milliseconds.

Or, for another example, a waveform has a frequency of 80 kilohertz. The period of one cycle is calculated:

$$\begin{aligned} T &= \frac{1}{f} \\ &= \frac{1}{80 \times 10^3} \\ &= 0.0000125 \\ &= 12.5 \times 10^{-6} \\ &= 12.5\mu\text{s} \end{aligned}$$

The period of one cycle of the waveform is 12.5 microseconds.

Additional Frequency Calculation Examples

The following worked-out problems are additional examples of how to use equations *2–1* and *2–2* to calculate the frequency or period of a waveform.

For example, the frequency of a waveform is 8 hertz. Its period is calculated:

$$\begin{aligned} T &= \frac{1}{f} \\ &= \frac{1}{8} \\ &= 125\text{ms} \end{aligned}$$

The period of the waveform is 125 milliseconds.

As another example, assume that the period of a waveform is 250 milliseconds. Therefore,

$$\begin{aligned} f &= \frac{1}{T} \\ &= \frac{1}{250 \times 10^{-3}} \\ &= \frac{1}{0.250} \\ &= 4\text{Hz} \end{aligned}$$

The frequency of the waveform is calculated to be 4 hertz, one-half of the frequency of the waveform with a period of 125 milliseconds. Since the period is twice as large the frequency is half as much. Thus the accuracy of the previous calculations is confirmed.

These two examples should show you vividly the frequency and period relationships of ac waveforms, and also, that one equation can be used to verify your calculations performed using the other.

WAVEFORM AMPLITUDE SPECIFICATIONS

In addition to frequency and period values, a third major specification of a waveform is the *amplitude* or height of the wave. There are three possible ways to express the amplitude of a sinusoidal waveform: peak, peak-to-peak, and root-mean-square (rms).

Peak Amplitude Specifications

The peak amplitude of a sinusoidal waveform is the maximum positive or negative deviation of a waveform from its zero reference level. Recall from the discussion of the single-loop generator in Lesson 1 that this maximum voltage or current occurs as the loop of wire cut the magnetic flux at a 90-degree angle.

The sinusoidal waveform is a symmetrical waveform, so the *positive* peak value is the same as the *negative* peak value as shown in *Figure 2.2*. If the positive peak has a value of 10 volts, then the negative peak will also have a value of 10 volts. When measuring the peak value of a waveform, either positive or negative peaks can be used.

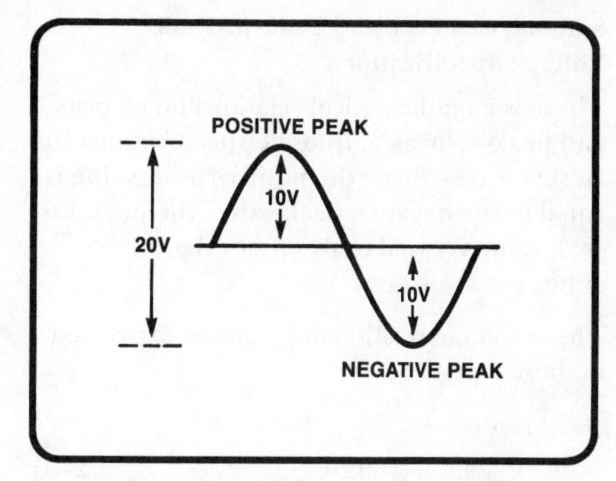

Figure 2.2 *A Sinusoidal Waveform is Symmetrical*

Peak-to-peak Amplitude Specifications

The peak-to-peak amplitude is simply a measurement of the amplitude of a waveform taken from its positive peak to its negative peak as shown in *Figure 2.3*.

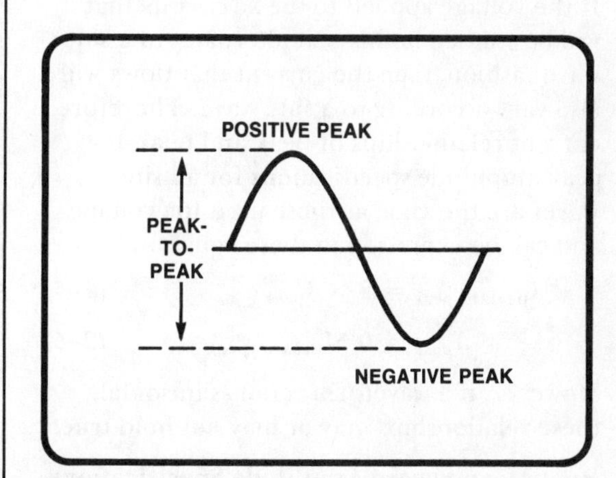

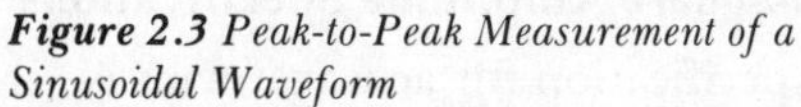

Figure 2.3 *Peak-to-Peak Measurement of a Sinusoidal Waveform*

For the non-sinusoidal waveform shown in *Figure 2.4*, the peak-to-peak value of the voltage can be determined by *adding* the magnitude of the *positive* and the *negative* peak. In this example, the peak-to-peak amplitude is 18 volts plus 2 volts for a total of 20 volts, peak-to-peak.

For sinusoidal waveforms, if the positive peak value is 10 volts in magnitude, then the negative peak value of the same waveform is also 10 volts. Measuring from peak-to-peak, there is a total of 20 volts. Therefore, the value of the sinusoidal waveform in *Figure 2.2* can be specified as either 10 volts peak or 20 volts peak-to-peak.

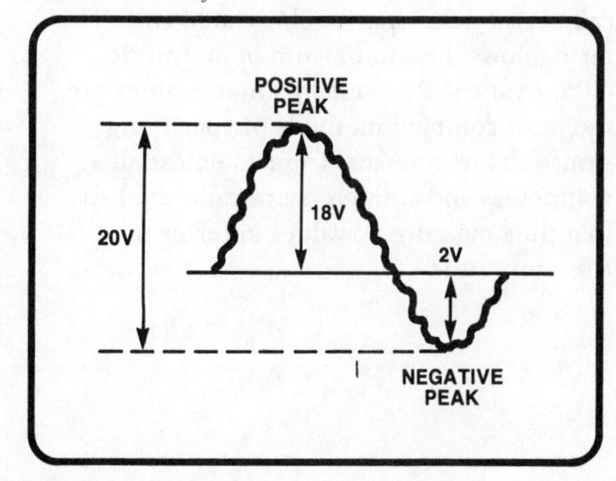

Figure 2.4 *Peak-to-Peak Measurement of a Non-Sinusoidal Waveform*

Summary of Peak and Peak-to-Peak Voltage Specifications

There is a mathematical relationship of peak and peak-to-peak amplitude specifications for all sine waves. Since the positive peak value is equal to the negative peak value, the peak-to-peak value is equal to two times the value of either peak voltage.

These voltage relationships can be expressed in these equations:

$$E_{peak\text{-}to\text{-}peak} = 2 \times E_{peak} \qquad (2\text{–}3)$$

$$E_{peak} = 0.5E_{peak\text{-}to\text{-}peak} \qquad (2\text{–}4)$$

If the voltage applied to the ac circuits that will be studied in this text alternates in a sine wave fashion, then the current that flows will also vary according to a sine wave. Therefore, current relationships of peak and peak-to-peak amplitude specifications for all sine waves are the same as those used for voltage and can be expressed in these equations:

$$I_{peak\text{-}to\text{-}peak} = 2 \times I_{peak} \qquad (2\text{–}5)$$

$$I_{peak} = 0.5I_{peak\text{-}to\text{-}peak} \qquad (2\text{–}6)$$

However, if a waveform is non-sinusoidal, these relationships may or may not hold true.

Root-Mean-Square Amplitude Specifications

The third specification for ac waveforms is called *root-mean-square* abbreviated rms. This term allows the comparison of ac and dc circuit values. Root-mean-square values are the most common methods of specifying sinusoidal waveforms. In fact, almost all ac voltmeters and ammeters are calibrated so that they measure ac values in terms of rms amplitude.

RMS Relation to DC Heating Effect

An rms value is also known as the *effective* value and is defined in terms of the equivalent heating effect of direct current. The rms value of a sinusoidal voltage *is equivalent to the value of a dc voltage which causes an equal amount of heat (power dissipation) due to the circuit current flowing through a resistance.* Since heating effect is independent of the direction of current flow, resistive power dissipation can be used as the basis for comparison of ac and dc values.

In other words, applying an ac voltage with a particular rms value to a resistive circuit will dissipate as much power in the resistors as a dc voltage would that has the same value.

The rms value of a sinusoidal voltage or current waveform is 70.7 percent or 0.707 of its peak amplitude value.

$$E_{rms} = 0.707E_{peak} \qquad (2\text{–}7)$$

$$I_{rms} = 0.707I_{peak} \qquad (2\text{–}8)$$

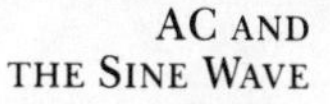

That is, as shown in *Figure 2.5,* a sinusoidal voltage with a peak amplitude of 1 volt has the same *effect* as a dc voltage of 0.707 volts as far as its ability to reproduce the same amount of heat in a resistance. Because the ac voltage of 1 volt peak or 0.707 volts rms is as *effective* as a dc voltage of 0.707 volts, the rms value of voltage is also referred to as the *effective* value.

Effective value and rms value are used interchangeably in electronics terminology. The rms value, however, is used more extensively and therefore it is the designation which will be used throughout this book.

Determining the 0.707 Constant

How is the 70.7 percent of peak-value constant derived? Essentially, the words root-mean-square tell how because they define the mathematical procedure used to determine the constant.

The word square comes from the square of instantaneous values of the ac waveform. Recall from dc circuit calculations that power dissipated in a circuit is equal to E, the voltage applied, times I, the current flowing in the circuit, as expressed in equation *2–9*.

$$P = EI \qquad (2\text{–}9)$$

By Ohm's law the applied voltage, E, is equal to the circuit current I times the circuit resistance. This is expressed in equation *2–10*.

$$E = IR \qquad (2\text{–}10)$$

Substituting equation *2–10* for E in equation *2–9* provides equation *2–11*, which allows the power dissipated in a circuit to be calculated if the circuit current and circuit resistance are given.

$$P = IRI = I^2R \qquad (2\text{–}11)$$

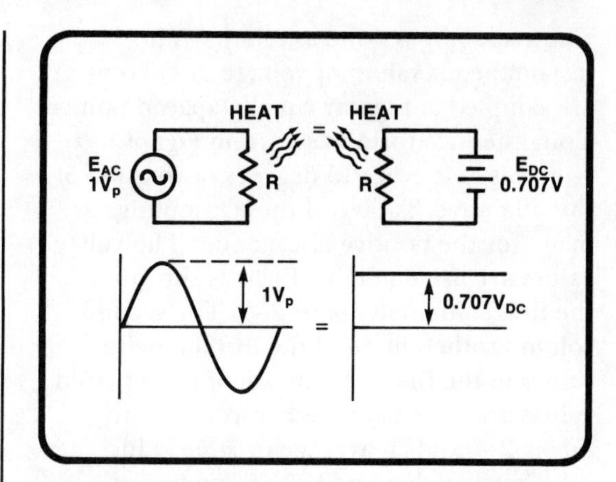

Figure 2.5 *Relationship of a Sinusoidal Voltage to a DC Voltage*

If the Ohm's law equation *2–10* is rearranged then equation *2–12* results

$$I = \frac{E}{R} \qquad (2\text{–}12)$$

Substituting equation *2–12* in equation *2–9* for I provides equation *2–13*, which allows the power dissipated in a circuit to be calculated if the applied voltage E and total circuit resistance are given.

$$P = E\left(\frac{E}{R}\right) \qquad (2\text{–}13)$$

$$P = \frac{E^2}{R}$$

Note that in both equation *2–11* and *2–13* power is determined by the *square* of either the voltage or current values.

Power calculations in ac circuits are somewhat different. Since the applied voltage and the resulting circuit current are both sine waves, they are constantly changing. Therefore, the power dissipated is constantly changing. For this reason a means of averaging the constantly changing values was derived.

Determining the 0.707 Constant

The averaging is done as follows: The instantaneous values of voltage or current are sampled at regular equally-spaced points along the waveform as shown in *Figure 2–6* — in this case every 15 degrees of rotation of the sine wave. Twelve of these samplings are made for the positive alternation. The voltage values are listed in table form as shown in the first column in *Figure 2–7*. The second column is the square of the instantaneous values in the first column. All of the squared values are summed together as shown in *Figure 2–7* and an average or *mean* value calculated by dividing by the number of samples taken, in this case, twelve.

Since the *mean* value is an average of squared values, the square root of the mean is the single value that is equivalent to a steady-state dc value. Thus, the result is the name root-mean-square since it is the square *root* of the *mean* of *squared* values. *Mean* is an average of the sum of the *squares* of instantaneous values of the voltage or current waveform. *Root* is the square root of the *mean*.

Power is dissipated on both the positive and negative alternation; it makes no difference which direction the current is flowing. However, squaring the instantaneous values eliminates any concern of direction or polarity.

Equation *2–14* is an expression for the mean value of the voltage waveform of *Figure 2.6*.

$$\text{mean} = \frac{\text{sum of } E_i^2 \text{ values}}{\text{total number of } E_i^2 \text{ values}} \qquad (2\text{–}14)$$

The mean value is the sum of the squared instantaneous voltage values divided by the number of squared instantaneous voltage values.

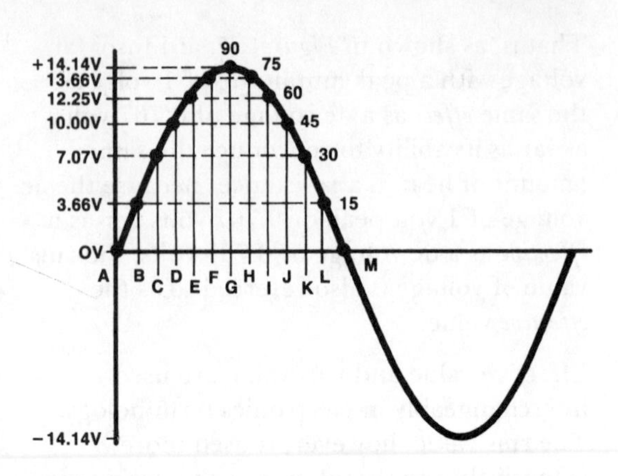

Figure 2.6 *Instantaneous Voltage Values of a Voltage Sine Wave*

RMS Calculation

Degrees Rotation	Point	Instantaneous Voltage Values (Volts)	Square of Instantaneous Voltage ($Volts^2$)
0	A	0.0	0.0
15	B	3.66	13.39
30	C	7.07	49.98
45	D	10.00	100.00
60	E	12.25	149.95
75	F	13.66	186.55
90	G	14.14	199.94
105	H	13.66	186.55
120	I	12.25	149.95
135	J	10.00	100.00
150	K	7.07	49.98
165	L	3.66	13.39
180	M	0.0	0.0
		TOTAL =	1199.66
		MEAN =	1199.66 / 12

Figure 2.7 *Instantaneous Voltage Values for the Voltage Sine Wave of Figure 2.6*

Therefore, for the E_i^2 values of *Figure 2.7*,

$$\text{mean} = \frac{1199.66}{12} = 99.97$$

1199.66 volts divided by 12 provides a mean value of 99.97 volts.

The square root of the mean, 99.97 volts, is approximately 10.0 volts:

$$E_{rms} = \sqrt{\text{mean}} \qquad (2\text{–}15)$$

$$= \sqrt{99.97}$$

$$\cong 10V$$

Thus, the rms voltage value of the waveform of *Figure 2.6* is approximately 10 volts. In the example, if a larger number of instantaneous points had been selected rather than 12, the rms value would have been exactly 10.0 volts.

The calculated rms voltage value can be proved to be 70.7 percent or 0.707 times the measured peak value of voltage of 14.14 volts.

$$\frac{E_{rms}}{E_{peak}} = \frac{10.0V}{14.14V} = 0.707 \qquad (2\text{–}16)$$

Converting the 0.707 ratio of rms voltage to peak voltage to a percentage gives equation *2–17*.

$$E_{rms} = (0.707)(100\%)E_{peak} \qquad (2\text{-}17)$$

$$\mathbf{E_{rms} = 70.7\%E_{peak}}$$

Relationship of an AC Waveform and DC Waveform

Graphically, the relationship of a dc waveform to an ac waveform is as shown in *Figure 2.8*. The rms value, or 0.707 of the peak value, is located about three-quarters of the way up the ac waveform. And as you can see, the peak value of the ac waveform is considerably higher than the level of the dc waveform. This should not be surprising since an ac voltage is at its peak only momentarily and then drops back down.

The rms value of the wave can be determined if the peak voltage or current is known by rearranging the rms ratio equation *2–16*. This can be rewritten as

$$\mathbf{E_{rms} = 0.707E_{peak}} \qquad (2\text{–}7)$$

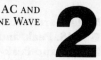

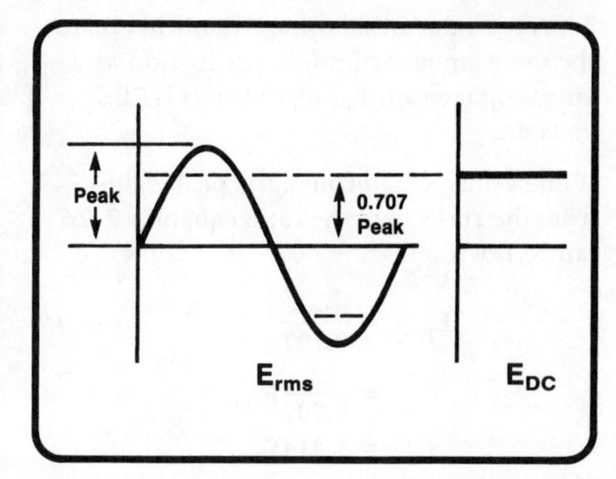

Figure 2.8 Relationship of a DC Waveform to an AC Waveform

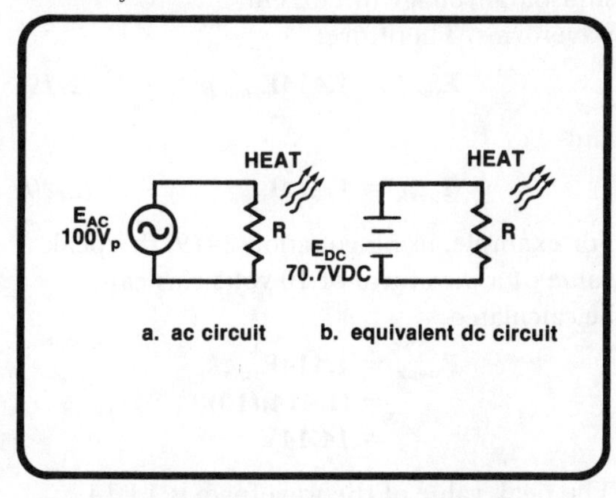

Figure 2.9 Relationship of Sinusoidal Voltage to DC Voltage

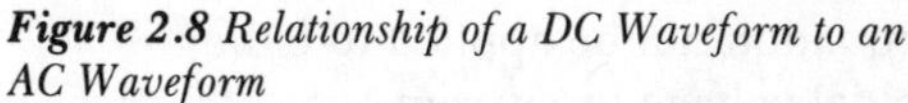

It's important to remember that rms or peak values can be used when referring to either voltage or current. For example, as shown in *Figure 2.9*, suppose a 100-volt peak sinusoidal voltage is applied across a resistor and one wants to know the value of dc voltage which could be applied across the resistor to create the same amount of power dissipation. The value of the ac voltage equals 0.707 of the peak value. Therefore, this can be calculated:

$$\mathbf{E_{rms} = 0.707 \times 100V}$$
$$\mathbf{E_{rms} = 70.7\ volts}$$

Thus the value of dc voltage that will create the same amount of power dissipation as an ac voltage with E_{pk} of 100 volts is 70.7 volts dc.

If one wants to determine the peak value from the rms value, the ratio equation *2–16* can be rewritten:

$$E_{peak} = \frac{E_{rms}}{0.707} = \frac{1}{0.707}E_{rms} = 1.414E_{rms} \quad (2\text{-}18)$$

This equation can be applied to either sinusoidal voltage or current waveforms. Therefore:

$$E_{peak} = 1.414E_{rms} \quad (2\text{-}19)$$

and

$$I_{peak} = 1.414I_{rms} \quad (2\text{-}20)$$

For example, using equation *2–19*, the peak value of a waveform of 10 volts rms can be calculated:

$$E_{peak} = 1.414E_{rms} = (1.414)(10) = 14.14V$$

The peak value of the waveform is 14.14 volts. Therefore, as shown in *Figure 2.10* 14.14 volts peak sinusoidal voltage is required to create the same amount of heat or power as that caused by 10 volts dc.

Standard Notation of Voltage RMS Values

Because ac voltage values are commonly specified as rms values, you will normally see voltages such as standard 60-hertz power voltages written as 115VAC. The VAC notation is a simplified way of specifying rms voltage values.

Any ac voltage listed as VAC can be assumed to be an rms value, and where the type of specification is not provided, you can also assume rms values.

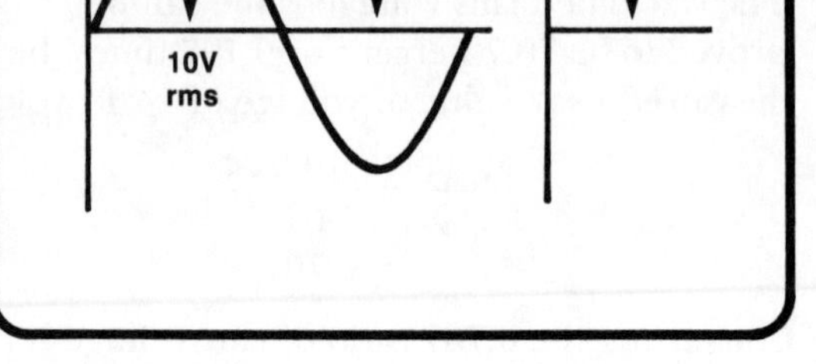

Figure 2.10 *10 Volts DC Is as Effective as 14.14 Peak Volts AC*

RMS, Peak, and Peak-to-Peak Conversion Examples

The following examples are typical conversion problems that you will encounter when working with ac circuits.

Peak and Peak-to-Peak From RMS

Calculate E_{peak} and $E_{peak\text{-}to\text{-}peak}$ of 120VAC. Use the conversion equations *2–19* and *2–3*. The VAC designation indicates that the voltage is an rms value. From now on E_{pk} will designate peak voltage instead of E_{peak} and E_{PP} will designate peak-to-peak voltage instead of $E_{peak\text{-}to\text{-}peak}$.

$$E_{pk} = 1.414E_{rms} = (1.414)(120VAC) = 169.68V$$

and

$$E_{PP} = 2E_{pk} = (2)(169.68) = 339.36V$$

This is summarized in *Figure 2.11*.

Peak and RMS From Peak-to-Peak

If a peak-to-peak value is given, the conversion will be to peak and rms values using equations *2–4* and *2–7*. For example, E_{pk} and E_{rms} of a 60 volt peak-to-peak sine wave is:

$$\begin{aligned} E_{pk} &= 0.5E_{pp} \\ &= (0.5)(60V_{pp}) \\ &= 30V \\ E_{rms} &= 0.707E_{pk} \\ &= (0.707)(30V) \\ &= 21.21VAC \end{aligned}$$

This is summarized in *Figure 2.12*. Remember then that ac voltage amplitudes can be specified in one of three ways: peak, peak-to-peak, or rms, and that the same specifications can be used for current amplitudes.

THE SINE WAVE AND SINE TRIGONOMETRIC FUNCTION

The term sinusoidal has been used to describe a waveform produced by an ac generator. The term sinusoidal comes from a trigonometric function called the *sine* function.

Right-Triangle Side and Angle Relationships

As you may know, trigonometry is the study of triangles and their relationships. The basic triangle studied in trigonometry is a *right triangle* which is a triangle that has a 90° angle as one of its three angles. A 90° triangle has a unique set of relationships from which the rules for trigonometry are derived.

Figure 2.13 shows a right triangle and identifies the 90° angle. In order to study the relationships of a right triangle one of the remaining two angles will be labeled with the Greek letter theta, θ.

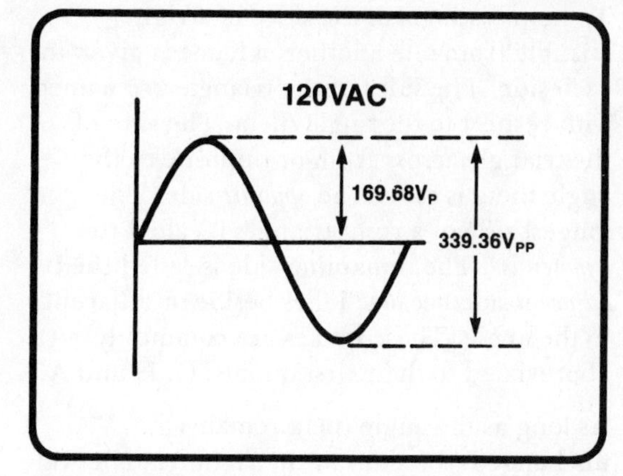

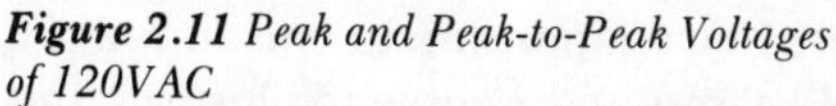
***Figure 2.11** Peak and Peak-to-Peak Voltages of 120VAC*

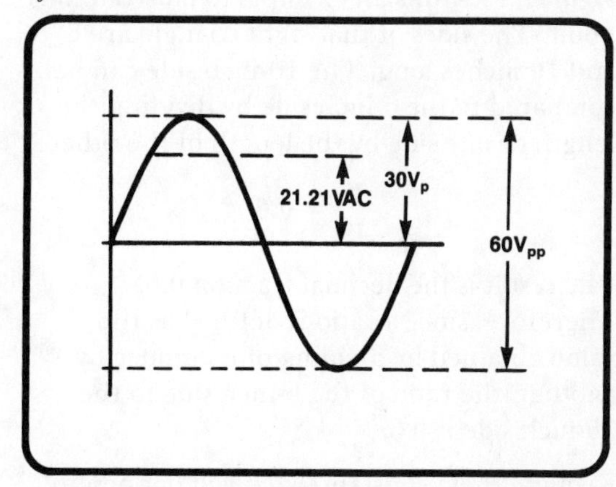

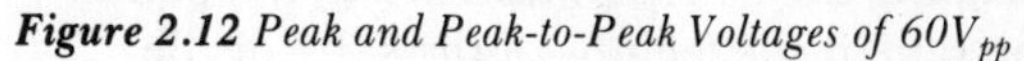
***Figure 2.12** Peak and Peak-to-Peak Voltages of 60V_{pp}*

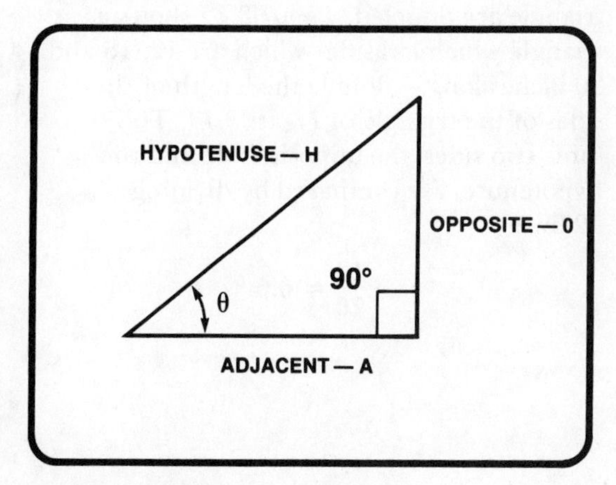

***Figure 2.13** A 90° Right Triangle*

■ Right-Triangle Side and Angle Relationships

To help distinguish the sides of a right triangle from one another, a name is given to each side. The sides of the triangle are named with respect to the angle theta. The side of the triangle across from or opposite to the angle theta is called the *opposite* side. The longest side of a right triangle is called the *hypotenuse*. The remaining side is called the *adjacent* side because it lies beside or adjacent to the angle. These names are commonly abbreviated to their first initials: O, H and A.

As long as the angle theta remains unchanged, the sides of the right triangle will retain the same relationship to one another. *Figure 2.14* shows an example to illustrate this point. The sides of that right triangle are 6, 8, and 10 inches long. The 10-inch side can be compared to the 6-inch side by dividing the length of one side by the length of the other:

$$\frac{6}{10} = 0.6$$

The result is the decimal fraction 0.6. Therefore, since a ratio is defined as the value obtained by dividing one number by another, the ratio of the 6-inch side to the 10-inch side is 0.6.

Keeping theta constant, what will happen to the ratio if the lengths of the sides of the triangle are doubled? *Figure 2.15* shows a triangle which has sides which are 12, 16 and 20 inches long—double the length of the sides of the triangle of *Figure 2.14*. The same two sides, the opposite side and the hypotenuse, are compared by dividing 12 by 20:

$$\frac{12}{20} = 0.6$$

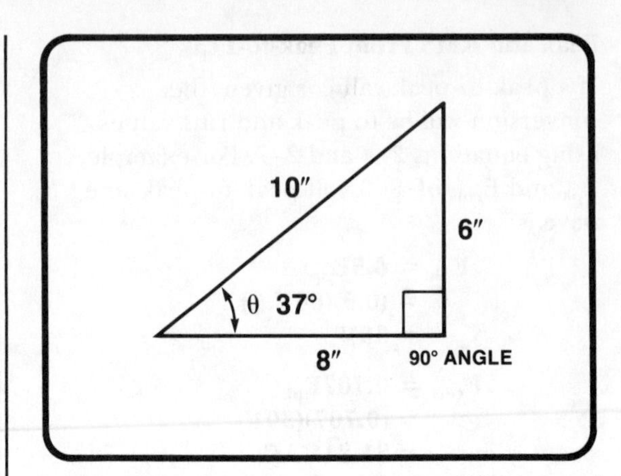

Figure 2.14 Triangle With Length of Its Sides 6, 8, and 10 Inches Long

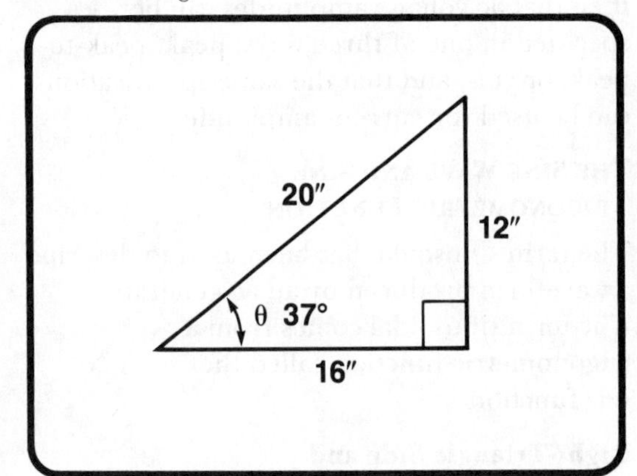

Figure 2.15 Triangle With Length of Its Sides 12, 16, and 20 Inches Long

Therefore, the ratio is 0.6, the same ratio as the first triangle. The ratio of the length of the sides remained constant because the angle theta remained constant. The ratios of the sides of right triangles will remain constant no matter what the lengths of their sides as long as the angle theta is not changed. However, if the angle theta is changed, then the ratio of the length of the sides will also change.

■ Basic Trigonometric Functions
■ The Sine Function Related to an AC Waveform

Basic Trigonometric Functions

In trigonometry, these ratios have specific names. The three most commonly-used ratios in the study of right triangles are called *sine*, *cosine*, and *tangent*. The sine of the angle theta is equal to the ratio formed by the length of the opposite side divided by the length of the hypotenuse:

$$\text{sine } \theta = \frac{\text{opposite}}{\text{hypotenuse}} \qquad (2\text{-}21)$$

The cosine of the angle theta is equal to the ratio formed by length of the adjacent side divided by the length of the hypotenuse:

$$\text{cosine } \theta = \frac{\text{adjacent}}{\text{hypotenuse}} \qquad (2\text{-}22)$$

The tangent of the angle theta is equal to the ratio formed by length of the opposite side divided by the length of the adjacent side:

$$\text{tangent } \theta = \frac{\text{opposite}}{\text{adjacent}} \qquad (2\text{-}23)$$

Remember that the sine, cosine and tangent represent a *ratio* of sides of triangles. Specifically, these ratios compare the lengths of the sides of a triangle.

The names sine, cosine and tangent are often abbreviated *sin* for sine, *cos* for cosine and *tan* for tangent.

The cosine and tangent functions will be used in later lessons. In this lesson the sine function will be discussed.

The Sine Function Related to an AC Waveform

In the previous two examples, the lengths of two sides of two triangles were compared. In both examples the ratio of the opposite side to the hypotenuse was determined. In fact, what was determined was the conditions for the angle theta. For both triangles theta is 37°. The sine of theta for both triangles is 0.60. It was stated that no matter how large the triangle might become or how long the sides, as long as the angle theta does not change, the ratio of the sides will not change and the sine of 37° will always be 0.60181502 (rounded off to two places this is 0.60).

Every angle theta has its own specific sine value. This is true for all angles between 0° and 360°. The sine for each angle between 0° and 360° could be calculated by dividing the length of the opposite side of the triangle by the length of the hypotenuse. Fortunately, this has already been done and published as a set of trigonometric tables. *Figure 2.16* shows a part of a trigonometric table.

Moreover, the values of these common trigonometric functions are provided as functions available on scientific calculators. Therefore, such calculators can be used to find the sine, cosine or tangent of an angle instead of a trigonometric table. Specific examples are included at the end of this lesson.

■ Rotating AC Generator

Rotating AC Generator

You may be wondering why all of this emphasis is being placed on the sine function. It is for one important reason: the sine of an angle theta is an exact mathematical statement describing the *voltage produced by an ac generator*. Therefore, the table which tabulates the sine of theta for any angle between 0° and 360° enables ac voltages to be expressed at any point in a cycle. To explain these statements, the relationship of the rotation of an ac generator, its sinusoidal output, the resulting vector right triangles, and the sine function will be discussed.

To demonstrate, a circle with a rotating arrow as shown in *Figure 2.17* will be used to represent an ac generator producing one cycle of an ac waveform. The arrow in *Figure 2.17* has been rotated 30 degrees, and a line perpendicular to the horizontal axis has been drawn downward from the tip of the arrow. As you can see, a right triangle has been formed. The arrow is the hypotenuse. The perpendicular line is the opposite side. The horizontal axis is the adjacent side. When the generator is rotated to different angles, the length of the hypotenuse (arrow) will never change. Only the opposite and adjacent sides will change length.

Recall from equation *2–21* that the sine theta is equal to the length of the opposite side divided by the length of the hypotenuse:

$$\sin \theta = \frac{\text{opposite}}{\text{hypotenuse}}$$

ANGLE	SIN	COS	TAN
0 °	0.0000	1.000	0.0000
1	.0175	.9998	.0175
2	.0349	.9994	.0349
3	.0523	.9986	.0524
4	.0698	.9976	.0699
5	.0872	.9962	.0875
6	.1045	.9945	.1051
7	.1219	.9925	.1228
8	.1392	.9903	.1405
9	.1564	.9877	.1584
10	.1736	.9848	.1763

Figure 2.16 A Part of a Trigonometric Table

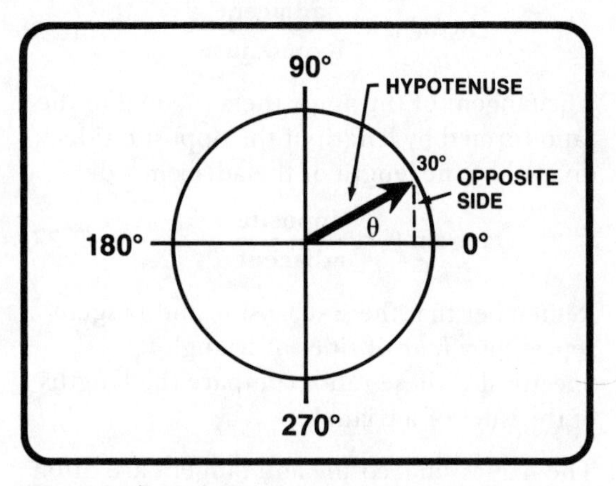

Figure 2.17 Circle Representation of an AC Generator

In *Figure 2.17* the length of the hypotenuse never changes, it is a constant. By setting the constant length of the hypotenuse equal to 1 or unity, it remains constant at 1 as the generator is rotated. This simplifies the equation for sine theta as shown in equation *2–24* because now the opposite side will be divided by 1:

$$\sin \theta = \frac{\text{opposite}}{1} \qquad (2\text{-}24)$$

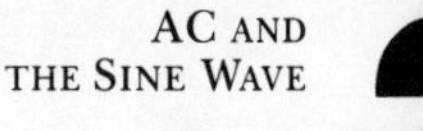

So in this instance the sine of theta is equal to the length of the opposite side itself:

$$\sin \theta = \text{opposite} \qquad (2\text{-}25)$$

Therefore, examining the length of the opposite side as theta varies will help demonstrate how the sine function varies.

During the following discussion the length of the opposite side of the resulting right triangle at different degree points around the circle will be compared to the unity value of the hypotenuse. The resulting values of these ratios will be the value of the sine function for different angles. The values will be plotted on the graph shown in *Figure 2.18*.

Let's begin when the rotating arrow in *Figure 2.17* is at 0°. In this case, the length of opposite side of the right triangle is zero units long. Therefore,

$$\sin 0° = \frac{0}{1} = 0.00$$

The original position of the arrow in *Figure 2.17* is 30°. If you were to measure the opposite side you would find that it would be one-half the length of the hypotenuse, or 0.5 units long. Therefore,

$$\sin 30° = \frac{0.5}{1} = 0.5$$

This point is plotted at 30° on the graph shown in *Figure 2.18*. Remember this point indicates the voltage amplitude when the hypotenuse has been rotated 30 degrees from the 0° reference.

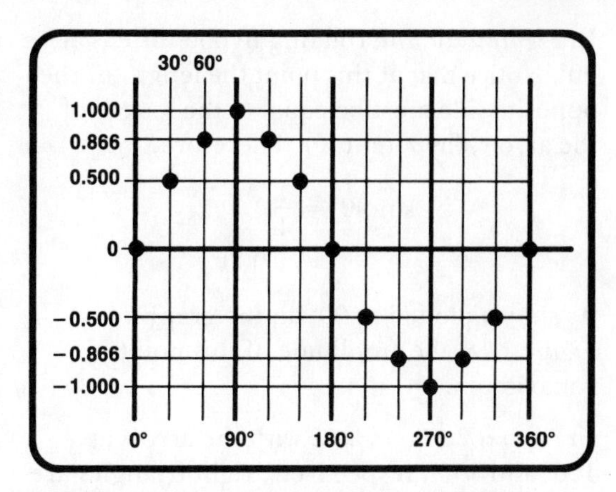

Figure 2.18 *Sine Values Between 0° and 360° at 30-Degree Intervals*

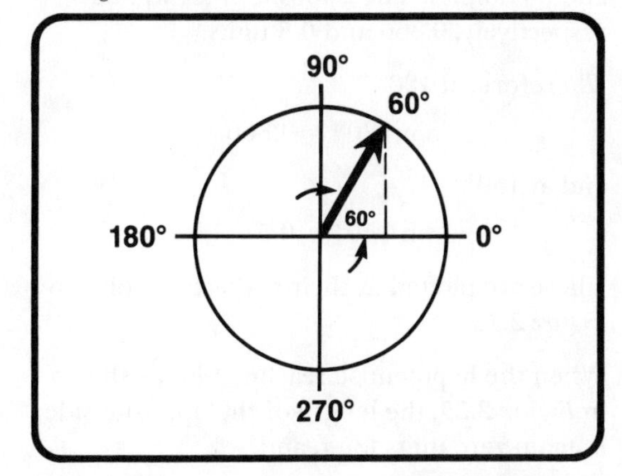

Figure 2.19 *Generator Has Been Rotated 60 Degrees*

In *Figure 2.19*, the generator has been rotated to 60°. The length of the opposite side is 86.6 percent of the length of the hypotenuse, or 0.866 units long. Therefore,

$$\sin 60° = \frac{0.866}{1} = 0.866$$

This point is plotted at 60° on the graph shown in *Figure 2.18*.

■ Rotating AC Generator

In *Figure 2.20*, the rotating hypotenuse is at 90°. Notice that at this point the length of the opposite side must be equal to the length of the arrow, (hypotenuse). Therefore,

$$\sin 90° = \frac{1}{1} = 1.000$$

As shown plotted at 90° on the graph of *Figure 2.18*, the amplitude of this point is a maximum.

In *Figures 2.21 and 2.22* with the arrow at 120° and 150°, respectively, right triangles are produced with angles of theta equal to the 60° and 30° angles. The opposite sides will be respectively, 0.866 and 0.5 units.

Therefore, at 120°,

$$\text{Sin } 120° = 0.866$$

and at 150°,

$$\text{Sin } 150° = 0.5$$

These are plotted at their respective points on *Figure 2.18*.

When the hypotenuse reaches 180° as shown in *Figure 2.23*, the length of the opposite side is again zero units long, and,

$$\sin 180° = \frac{0}{1} = 0.000$$

This is plotted at 180° on *Figure 2.18*. Note that all the values plotted are positive values. In *Figure 2.24* all of the points are connected with a curve indicating the continuously changing values as the generator rotates.

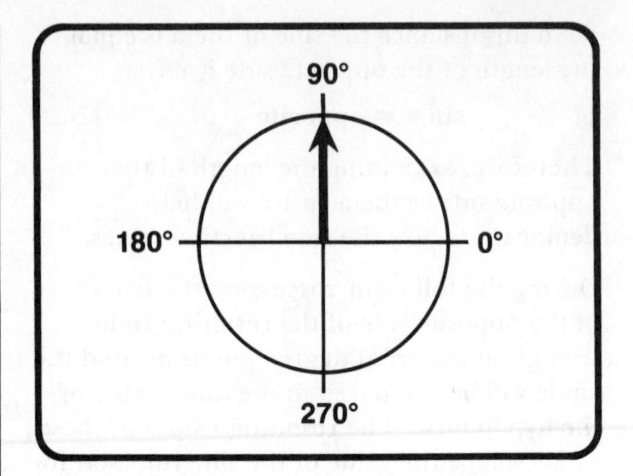

Figure 2.20 *Generator Has Been Rotated 90 Degrees*

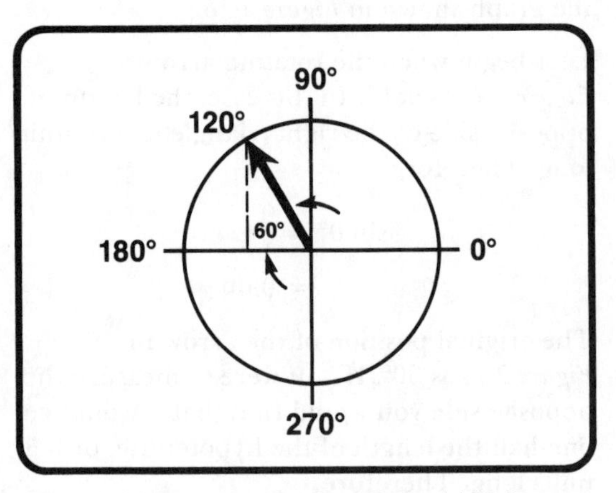

Figure 2.21 *Generator Has Been Rotated 120 Degrees*

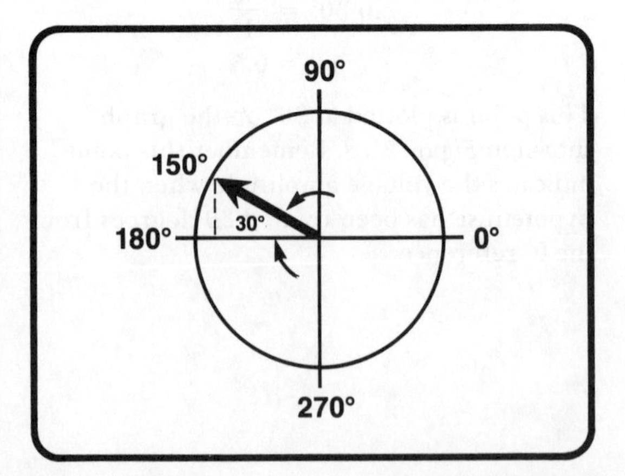

Figure 2.22 *Generator Has Been Rotated 150 Degrees*

Note that from 0° to 90° of rotation the amplitude of the sine wave goes from 0 to 1, and from 90° to 180° that the amplitude comes down from 1 to 0 following a similarly-shaped curve. Therefore, the amplitude values of a sine wave repeat themselves during the second 90 degrees of rotation in reverse order from the amplitude variations during the first 90 degrees of rotation.

Thus far, all of the points have been in the positive region of the graph because the perpendicular line has extended in an upward, or positive, direction. If the hypotenuse is rotated to the bottom half of the circle as shown in *Figure 2.25*, right triangles are produced which duplicate the right triangles in the top part of the circle. However, the opposite side of the right triangle is now extending in a downward direction which gives the sine of the angles negative values. The points are plotted on *Figure 2.18* and connected by a curve in *Figure 2.24*. Note that the negative portion of the curve is an exact replica of the positive portion; both sides are symmetrical.

In this discussion and demonstration, you should have become aware of the fact that the sine wave is an exact mathematical statement of the output of an ac generator. Also, *the sine wave is merely a plot of the sines of angles between 0° and 360°*. Therefore the voltage or current output of an ac generator at any angle of rotation, theta, depends directly on the sine of the angle theta. This dependence is a very important point, and, in a later lesson, more explicit use will be made of the sine function in relation to the ac sine wave.

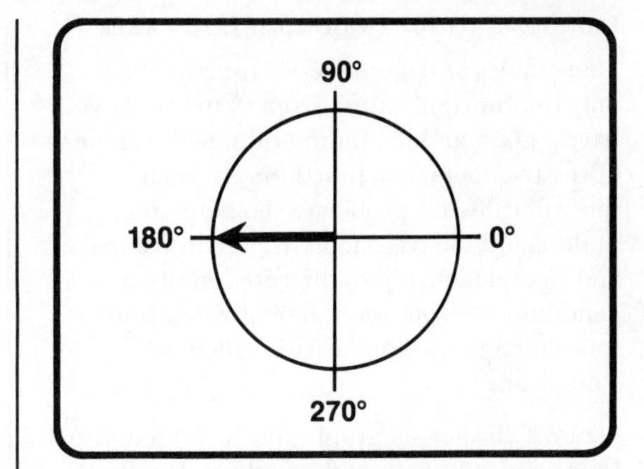

Figure 2.23 *Generator Has Been Rotated 180 Degrees*

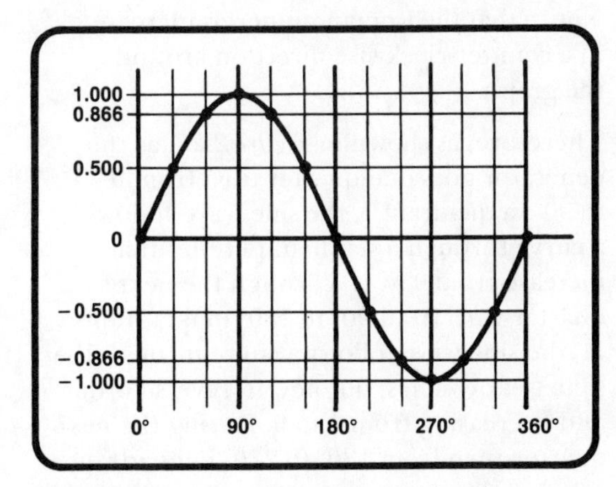

Figure 2.24 *Plot Points Connected With a Line*

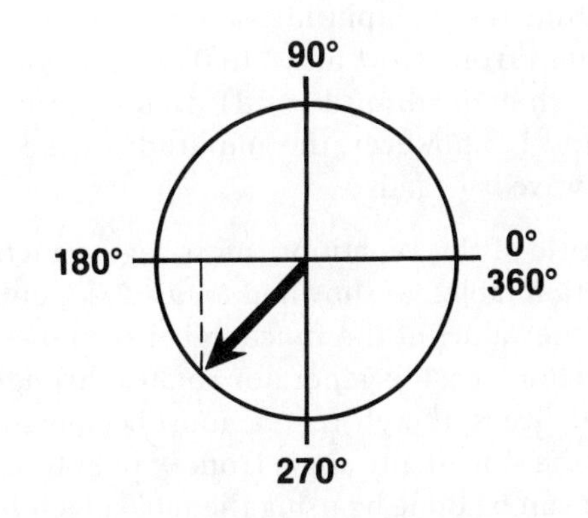

Figure 2.25 *Theta Between 180° and 360°*

■ **Rotating AC Generator**

USING A 0° TO 90° TRIGONOMETRIC TABLE

Many tables of trigonometric functions list only the function values from 0° to 90°. If you have a good understanding of how the angles of the trigonometric functions are related, this will cause no problems. To help you understand the trigonometric relationship and the relationship to the rotation of an ac generator the rotation will be divided into four equal parts. Each part is called a quadrant.

Figure 2.26 shows a graph that is divided into four quadrants with each quadrant labeled with a Roman numeral I, II, III, or IV. Notice that the Roman numerals increase in a counter-clockwise direction around the graph.

Therefore, as shown in *Figure 2.27*, as the generator goes one-quarter turn from 0° to 90° in quadrant I, the sine wave follows a curve through a set of amplitudes that increase from 0 to 1. Through the next quarter-turn from 90° to 180° in quadrant II, the sine wave follows a curve through the same set of values, but now in reverse order and decreasing from 1 to 0. During the next quarter-turn from 180° to 270° in quadrant III, and the next quarter-turn from 270° to 360° in quadrant IV, the same curves through the same set of amplitude values are gone through from 0 to 1 and 1 to 0 as were gone through in quadrant I and II. In quadrants III and IV, however, the amplitude of the sine wave is negative.

Because of this repetition, most trigonometric function tables, as shown in *Figure 2.28*, only give the values of the function between 0° and 90°. However, the generator rotates through 360 degrees; therefore, one must be able to find the sine of any angle from 0° to 360°. That can be done by using the table given for quadrant I to also determine the sine of angles of quadrant II, III, and IV, as well.

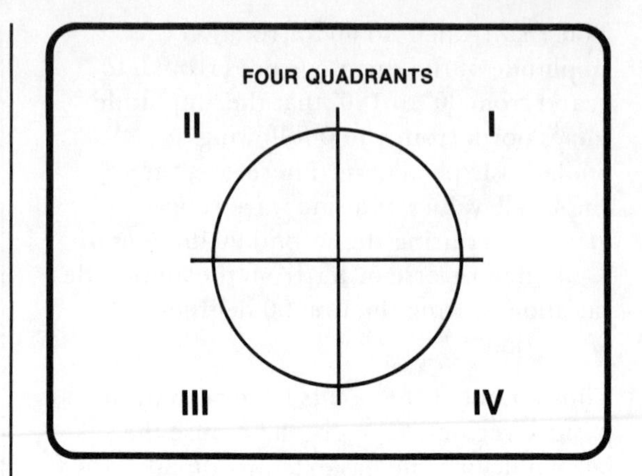

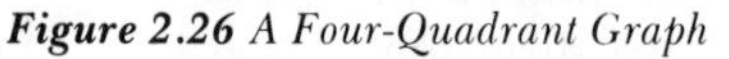

Figure 2.26 *A Four-Quadrant Graph*

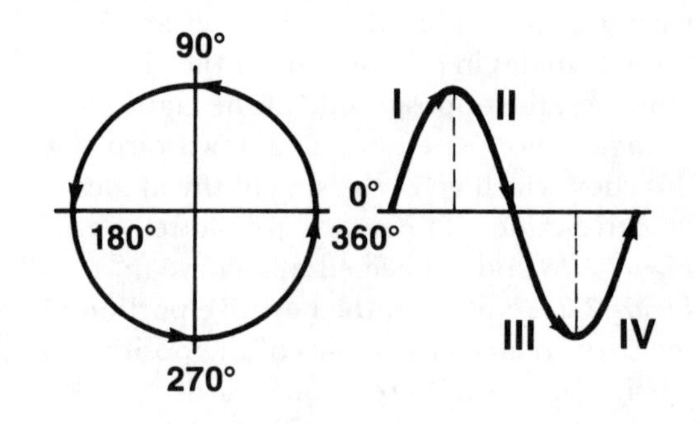

Figure 2.27 *Relationship of the Quadrants and a Sine Wave*

ANGLE	SIN	COS	TAN
0°	0.0000	1.000	0.0000
5	.0872	.9962	.0875
10	.1736	.9848	.1763
15	.2588	.9659	.2679
20	.3420	.9397	.3640
25	.4226	.9063	.4663
30	.5000	.8660	.5774
35	.5736	.8192	.7002
40	.6428	.7660	.8391
45	.7071	.7071	1.0000
50	.7660	.6428	1.1918
55	.8192	.5736	1.4281
60	.8660	.5000	1.7321
65	.9063	.4226	2.1445
70	.9397	.3420	2.7475
75	.9659	.2588	3.7321
80	.9848	.1736	5.6713
85	.9962	.0872	11.43
90	1.0000	.0000	∞

Figure 2.28 *A Part of a Trigonometric Table*

■ **Rotating AC Generator**

To find the sine of an angle less than 90° as shown in *Figure 2.29* the value is read directly from the table. For example, suppose you want to find the sine of 30°. You would look in the table, such as the one shown in *Figure 2.28*, under the angle column for 30°. Then you would look to the right of the angle in the sine column to locate the value of the sine of this angle. You can see in the table that it is 0.5. Thus, the sine of 30° is 0.5.

The sine of quadrant II angles, that is, angles between 90° and 180°, can be determined using a 0°-90° trig table by first converting the angle in the second quadrant to what is called a first-quadrant equivalent angle. For angles in the second quadrant, theta$_{II}$, the first-quadrant equivalent, theta$_{I}$, is determined by subtracting theta$_{II}$ from 180°:

$$\theta_I = 180° - \theta_{II} \qquad (2\text{–}26)$$

For example, suppose that you want to determine the sine of the second quadrant angle theta$_{II}$ = 120°, as shown in *Figure 2.30*. The first-quadrant equivalent angle of 120° is

$$\theta_I = 180° - \theta_{II}$$
$$= 180° - 120°$$
$$= 60°$$

Now, using the 0°-90° table of *Figure 2.28*, you would look up the sine of 60°. This is 0.8660 and which, therefore, also equals the sine of 120°.

Angles in quadrant III and quadrant IV also must be converted to quadrant I equivalents to be able to use a 0°-90° table to find the sine values of these angles. Use equation *2–27* to determine first-quadrant equivalents for angles in quadrant III.

$$\theta_I = \theta_{III} - 180° \qquad (2\text{-}27)$$

Use equation *2–28* for angles in Quadrant IV.

$$\theta_I = 360° - \theta_{IV} \qquad (2\text{-}28)$$

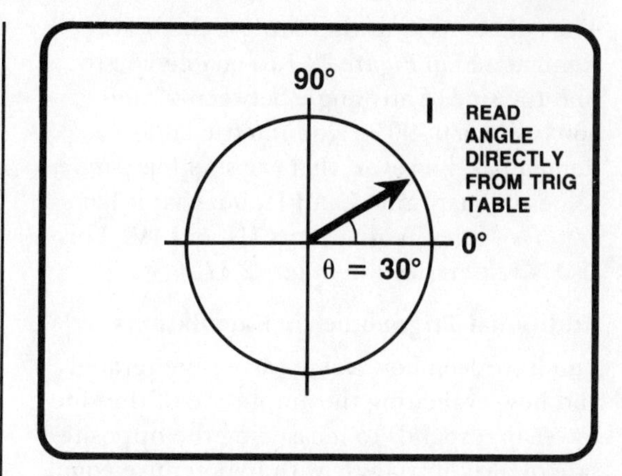

Figure 2.29 *Any Angle in Quadrant I Can Be Read Directly From Trig Table*

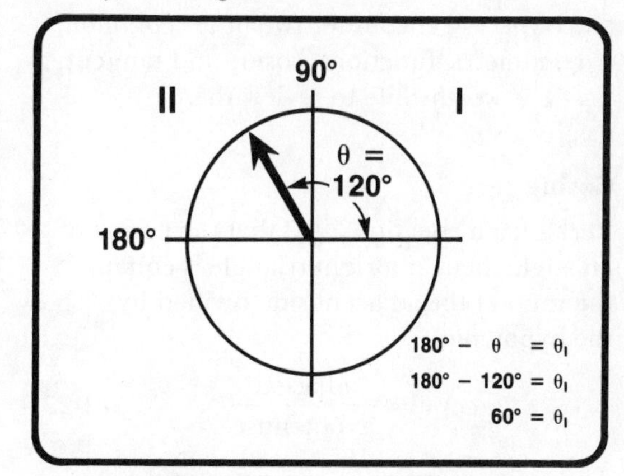

Figure 2.30 *Angle of 120° on Quadrant Graph*

The equations and where to use them are summarized in *Figure 2.31* to enable you to find the sine of any angle between 0° and 360° using a 0°-90° trigonometric table. Remember, however, that the sine has *positive* values in quadrants I and II, but that it has *negative* values in quadrants III and IV. This also is summarized in *Figure 2.31*.

Additional Trigonometric Calculations

You have seen how a sine wave is generated and how evaluating the amplitude of the sine wave corresponds to measuring the opposite side of a right triangle with hypotenuse equal to 1, as the hypotenuse rotates from 0 to 360°.

Since you will encounter two other common trigonometric functions, cosine and tangent, it will be worthwhile to review them briefly also.

Cosine

Recall from equation *2–22* that the cosine of an angle theta of a right triangle is equal to the ratio of the adjacent side divided by the hypotenuse.

$$\cos\theta = \frac{\text{adjacent}}{\text{hypotenuse}} \qquad (2\text{-}22)$$

The cosine function is generated in the same way as the sine function except that now the amplitude of the cosine waveform corresponds to measuring the adjacent side of a right triangle with hypotenuse equal to 1.

In *Figure 2.32a* is shown the same hypotenuse that was rotated for the sine wave. Its amplitude is kept constant at 1 (equal to unity). The hypotenuse is positioned at a rotation of 15° in solid lines and at 75° in dotted lines. Note that at 15° the adjacent side is almost 1 so that the ratio of the adjacent

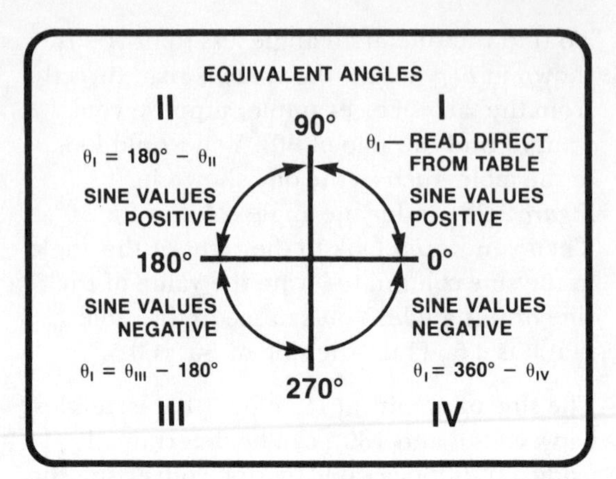

Figure 2.31 Quadrant Graph With Equivalent-Angle Equations

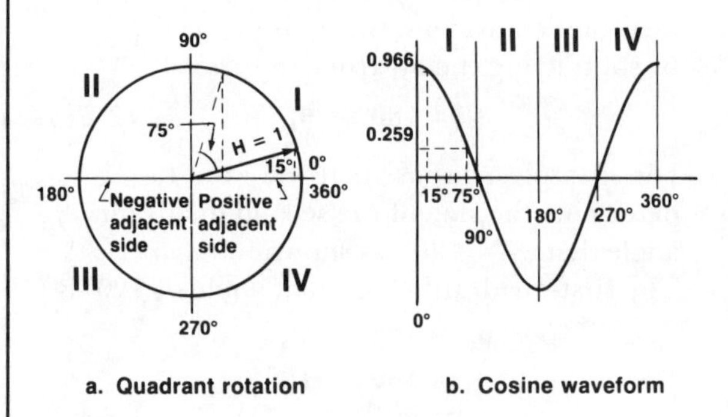

Figure 2.32 Variation of Cosine

side to the hypotenuse is 0.966, as plotted on the cosine curve of *Figure 2.32b* at 15°. If the hypotenuse were rotated back to 0°, the adjacent side equals the hypotenuse and the cosine is equal to 1.

At a rotation of 75°, as shown in *Figure 2.32a*, the adjacent side is reduced to 0.259 of the hypotenuse and therefore, the cosine is 0.259. This is plotted at 75° in *Figure 2.32b*. Therefore, as the hypotenuse rotates through the first quadrant from 0° to 90°, the cosine decreases from 1 to 0.

■ Tangent

You can go through the exercise of rotating the hypotenuse through quadrants II, III and IV, calculate the equivalent first–quadrant angle, and verify that the cosine varies from 0 to −1 in quadrant II, from −1 to 0 in quadrant III, and from 0 to 1 in quadrant IV. It is particularly important to recognize that the adjacent side is a *negative value* for quadrant II and III.

Tangent

The tangent of the angle theta of a right triangle from equation *2–23* is equal to the ratio of the side opposite to the side adjacent.

$$\tan\theta = \frac{\text{opposite}}{\text{adjacent}} \qquad (2\text{-}23)$$

The hypotenuse equal to unity is again rotated as shown in *Figure 2.33a*. Now the opposite side of the formed right triangle must be divided by the adjacent side in order to form the tangent ratio. The ratio does not correspond to the measurement of just one side as it did for the sine and cosine functions. For this reason the tangent waveform has a much different shape than the sine and cosine waveform.

Recall that as the sine function is formed it varies from 0 to 1 as the rotation goes from 0° to 90°. In other words the opposite side is increasing from 0 to 1. From *Figure 2.32* for the cosine, the adjacent side decreases from 1 to 0 through this same rotation. Therefore, the tangent has the numerator (the opposite side) increasing from 0 to 1 and the denominator decreasing from 1 to 0,

$$\tan\theta = \frac{0 \rightarrow 1}{1 \leftarrow 0} \qquad (2\text{-}29)$$

as shown in equation *2-29*, for theta from 0° to 90°.

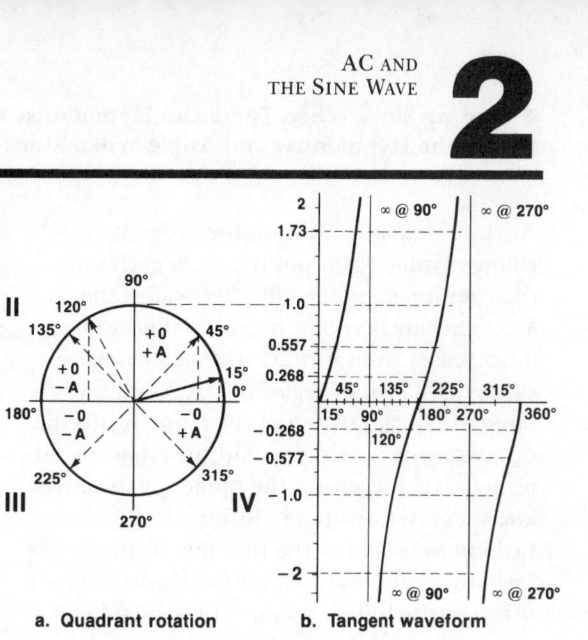

Figure 2.33 Variation of Tangent

Let's look at *Figure 2.33a*. When the rotation is at 0°, the adjacent side is 1 and the opposite side is 0. Therefore, the tangent is 0 as plotted on *Figure 2.33b*. At 15° rotation the opposite side has taken on a small value and the adjacent side is still near 1 (0.966 as shown on *Figure 2.32b*). The tangent is 0.268 and is plotted on *Figure 2.33b* at 15°. At 45° both the opposite side and the adjacent side are equal to the same value so that the tangent is 1 as plotted on *Figure 2.33b*. As the rotation increases above 45° the opposite side is larger than the adjacent side and the tangent is increasing and greater than 1 (specifically 1.73 at 60°). At 90° the opposite side is 1 and the adjacent side is 0. Division by 0 results in the tangent having a value of infinity at 90° and again at 270° as shown in *Figure 2.33b*. The opposite side is positive and the adjacent side is positive so that the tangent is positive in the first quadrant.

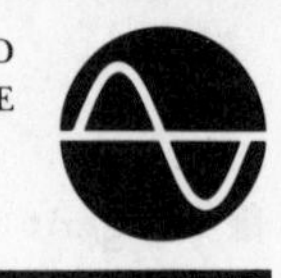

As the rotation continues, the tangent changes value from positive to negative as the rotation goes beyond 90° because the adjacent side becomes negative. Therefore, it decreases from infinity to smaller negative values as shown in *Figure 2.33b*. At 180° the tangent is again 0. In the third quadrant, the opposite side is negative and the adjacent side is negative, therefore, the tangent is positive and increases, from 0 to infinity in the same fashion as it did for the first quadrant. In like fashion, as shown in *Figure 2.33b*, the rotation through the fourth quadrant repeats the same tangent values as for the second quadrant. Each time the equivalent first quadrant angle is 45 degrees in any of the quadrants, the tangent is equal to 1. These points are plotted at 45°, 135°, 225° and 315° in *Figure 2.33b*.

Finding Sides When Theta and Hypotenuse are Given

One more important extension of right triangle mathematics and trigonometry that is important concerns calculating or evaluating the sides of a right triangle when the angle theta is given.

Look at *Figure 2.34a*. What is known about the sides in relationship to the angle theta? First of all, the ratio of the lengths of the sides is known by the equations *2–21*, *2–22* and *2–23*. (Abbreviations are inserted for convenience, H for hypotenuse, O opposite, and A for adjacent.)

$$\sin \theta = \frac{O}{H} \qquad (2\text{-}21)$$

$$\cos \theta = \frac{A}{H} \qquad (2\text{-}22)$$

$$\tan \theta = \frac{O}{A} \qquad (2\text{-}23)$$

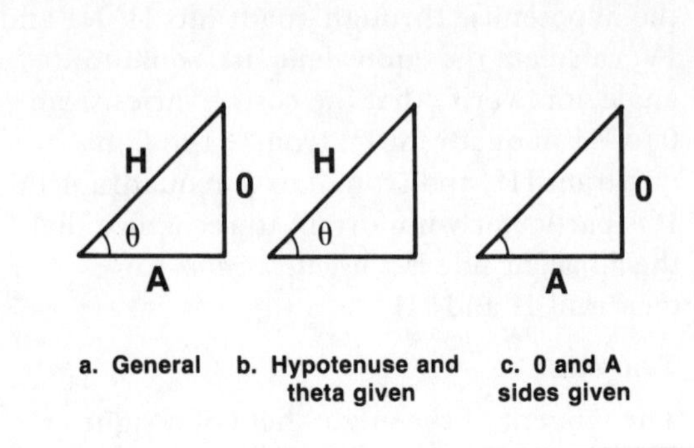

Figure 2.34 Determining Sides When Theta is Given

Secondly, when the hypotenuse and the angle are given as shown in *Figure 2.34b*, the opposite side and adjacent side can be determined easily. Rearranging equation *2–21* into equation *2–30*.

$$O = H \sin \theta \qquad (2\text{-}30)$$

allows the opposite side to be calculated if the hypotenuse and angle theta are given.

Rearranging equation *2–22* into equation *2–31*,

$$A = H \cos \theta \qquad (2\text{-}31)$$

allows the adjacent side to be calculated if the hypotenuse and angle theta are given.

Find the Hypotenuse and Angle When Sides are Given

Thirdly, if the opposite and adjacent sides are given, as shown in *Figure 2.34c*, the hypotenuse and angle theta can be calculated easily.

From the given opposite and adjacent sides the tangent of theta can be determined by using equation *2–23*. Knowing that theta is an angle that has a tangent equal to the calculated ratio, the angle can be determined from the trigonometric tables or by using a calculator that has the trigonometric functions such as the TI-35 shown in *Figure 2.35*.

When the angle theta is known the hypotenuse can be determined by using equation *2–30* or *2–31*, rearranged.

$$H = \frac{O}{\sin \theta} \qquad (2\text{-}32)$$

or

$$H = \frac{A}{\cos \theta} \qquad (2\text{-}33)$$

First Example

Here's an example of *how to determine unknowns*. *Figure 2.36a* shows a right triangle with an angle, theta, of 39° and a hypotenuse of 100.

Using equation *2–30*,

$$\begin{aligned} O &= H \sin \theta \\ &= 100 \sin 39° \\ &= (100)(0.629) \\ &= 62.9 \end{aligned}$$

The opposite side equals 62.9.

You can use a trigonometric table or a calculator. If a trigonometric table is used, you enter the table in the degrees column and scan across to the sine column and read the sine of 39° and use it in the equation.

Figure 2.35 *Typical Calculator with Trigonometric Functions*

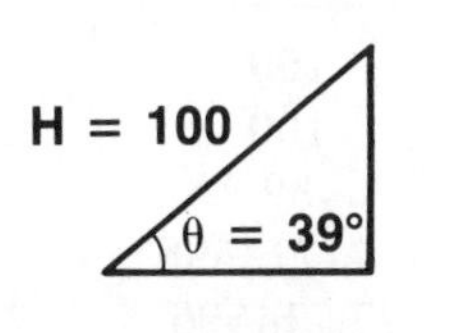

a. Hypotenuse and theta given

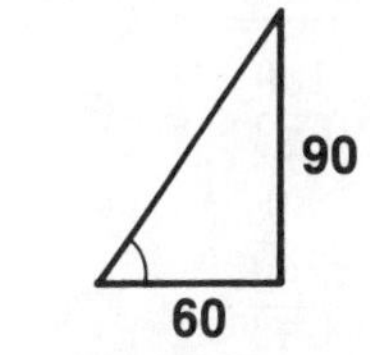

b. Opposite and adjacent sides given

Figure 2.36 *Example Problem*

■ Second Example

When a calculator is used then the result can be obtained almost immediately by following these keystrokes:

Press	*Display/Comments*
100	100
[×]	100
39	39
[sin]	0.62932039
[=]	62.932039

Using equation *2–31*,

$$\begin{aligned} A &= H\cos\theta \\ &= 100\cos 39^\circ \\ &= (100)(0.777) \\ &= 77.7 \end{aligned}$$

The adjacent side equals 77.7.
The calculator keystrokes are:

Press	*Display/Comments*
100	100
[×]	100
39	39
[cos]	0.77714596
[=]	77.714596

Second Example

In *Figure 2.36b* shows a right triangle which has the opposite side equal to 90 and the adjacent side equal to 60. Using equation *2–23*

$$\begin{aligned} \tan\theta &= \frac{O}{A} \\ &= \frac{90}{60} \\ &= 1.5 \end{aligned}$$

Locating 1.5 in the tangent column of a trigonometric function table and scanning across to the degree column you find that the angle is 56.31 degrees.

If a calculator is used, the following keystrokes are followed to find the angle:

Press	*Display/Comments*
90	90
[÷]	90
60	60
[=]	1.5
* [INV]	1.5
* [tan]	56.309932

*might be $\tan^{-1}$ key

The hypotenuse can be determined by using equation *2–32*,

$$\begin{aligned} H &= \frac{O}{\sin\theta} \\ &= \frac{90}{\sin 56.31^\circ} \\ &= \frac{90}{0.832} \\ &= 108.17 \end{aligned}$$

The hypotenuse is 108.17.

This is easily calculated using a calculator with the following keystrokes:

Press	*Display/Comments*
90	90
[÷]	90
56.31	56.31
[sin]	0.83205095
[=]	108.16645

SUMMARY

In this lesson the relationship of period to frequency was reviewed and you were shown how to specify the amplitude of a sine wave in terms of its peak, peak-to-peak, and rms values. The definitions of the various trigonometric functions were presented and related to right triangles and a quadrant graph. You were shown how to determine the value of the sine of any angle between 0° and 360° and how to use it, the cosine, and tangent to solve right triangle problems.

In the next lesson, you will learn about an instrument that provides an electronic picture of an ac sine wave, and you will learn how to use it to study further the characteristics of sine waves.

Worked-Out Examples

1. Given the frequency of waveform, determine the period, or time of one cycle.

f = 16kHz, T = ____________

Solution:

$$T_{(s)} = \frac{1}{f(Hz)} = \frac{1}{16kHz} = \frac{1}{16 \times 10^3} = 0.0625 \times 10^{-3} = \mathbf{62.5\mu s}$$

2. Given the time of one cycle, determine the frequency of the waveform.

T = 265μs, f = ____________

Solution:

$$f_{(Hz)} = \frac{1}{T_{(s)}} = \frac{1}{265\mu s}$$

$$265 \times 10 = \frac{1}{265 \times 10^{-6}} = 0.00377 \times 10^6 = \mathbf{3.77kHz}$$

3. Given the rms amplitude of a sine waveform, determine the peak and peak-peak values.

E_{rms} = 8.3V

E_{pk} = ____________

E_{pp} = ____________

Solution:

$E_{pk} = 1.414 \times E_{rms} = 1.414 \times 8.3V = \mathbf{11.7V}$

$E_{pp} = 2 \times E_{pk} = 2 \times 11.7V = \mathbf{23.4V}$

4. Given the peak amplitude of a sine waveform, determine the rms and peak-to-peak values.

E_{pk} = 8mV

E_{rms} = ____________

E_{pp} = ____________

Solution:

$E_{rms} = 0.707 \times E_{pk} = 0.707 \times 8mV = \mathbf{5.66mV}$

$E_{pp} = 2 \times E_{pk} = 2 \times 8mV = \mathbf{16mV}$

■ Worked-Out Examples

5. Given the peak-to-peak amplitude of a sine waveform, determine the peak and rms values.

I_{pp} = 16mA

I_{pk} = ______________

I_{rms} = ______________

Solution:

$I_{pk} = 0.5 \times I_{pp} = 0.5 \times 16\text{mA} = \mathbf{8mA}$

$I_{rms} = 0.707 \times I_{pk} = 0.707 \times 8\text{mA} = \mathbf{5.66mA}$

6. a. Given the following right triangle, determine the sine of the angle and with a calculator or using tables find the angle.

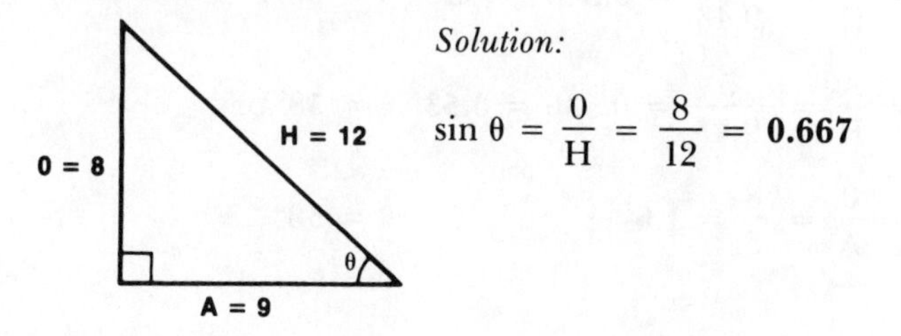

Solution:

$$\sin\theta = \frac{O}{H} = \frac{8}{12} = \mathbf{0.667}$$

Calculator Solution for Angle:

Enter the ratio into the calculator so it is displayed. Press inv and sin keys or $\sin^{-1}$ key and the calculator will display the angle in decimal degrees.

Table Solution for Angle:

Enter the table of natural trigonometric functions by locating the ratio in the sine column. Read across to the angle in decimal degrees in the degree column.

$$\text{arc}\sin 0.667 = \mathbf{41.8°}$$

b. Given the following right triangle, determine the cosine of the angle and with a calculator or using tables find the angle.

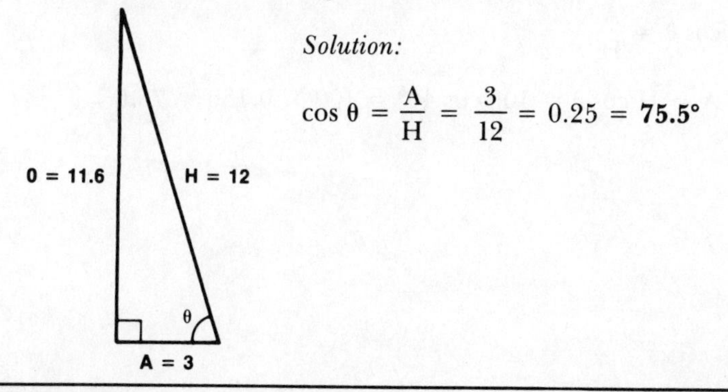

Solution:

$$\cos\theta = \frac{A}{H} = \frac{3}{12} = 0.25 = \mathbf{75.5°}$$

Worked-Out Examples

Calculator Solution for Angle:

Enter the ratio into the calculator so it is displayed. Press inv and cos keys or the $\cos^{-1}$ key and the calculator will display the angle in decimal degrees.

Table Solution for Angle:

Enter the table of natural trigonometric functions by locating the ratio in the cosine column. Read across to the angle in decimal degrees in the degree column.

7. Given the following right triangle, determine the sine, cosine, and tangent of the angle and the angle itself using a calculator or table as in problem 6.

Solution:

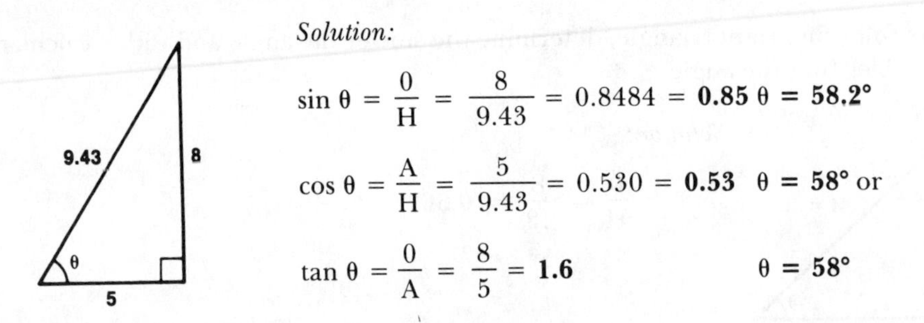

$$\sin\theta = \frac{0}{H} = \frac{8}{9.43} = 0.8484 = \mathbf{0.85}\ \theta = \mathbf{58.2°}$$

$$\cos\theta = \frac{A}{H} = \frac{5}{9.43} = 0.530 = \mathbf{0.53}\ \ \theta = \mathbf{58°}\text{ or}$$

$$\tan\theta = \frac{0}{A} = \frac{8}{5} = \mathbf{1.6}\qquad \theta = \mathbf{58°}$$

There may be as much as 0.2 to 0.3 of a degree variation in the angle calculated from the sine, cosine or tangent depending on the rounding off of the sine, cosine or tangent ratio as it is calculated.

8. Given the following right triangle, determine the remaining sides.

a.

Solution:

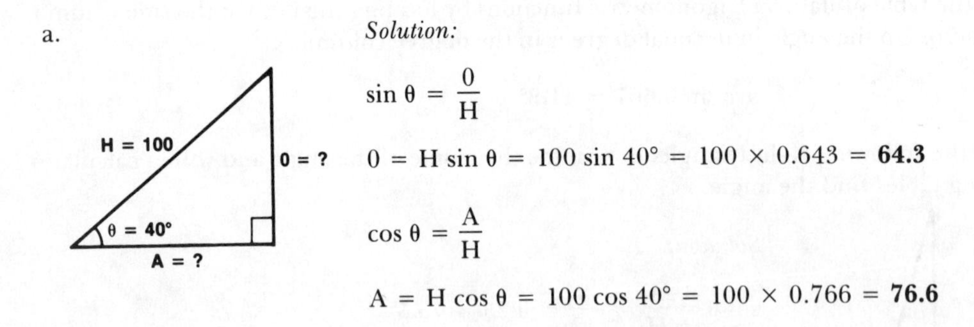

$$\sin\theta = \frac{0}{H}$$

$$0 = H\sin\theta = 100\sin 40° = 100 \times 0.643 = \mathbf{64.3}$$

$$\cos\theta = \frac{A}{H}$$

$$A = H\cos\theta = 100\cos 40° = 100 \times 0.766 = \mathbf{76.6}$$

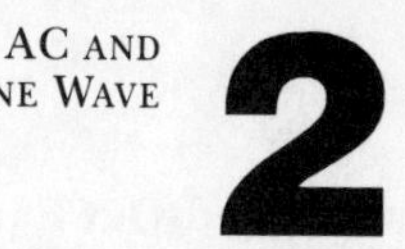

■ Worked-Out Examples

b.

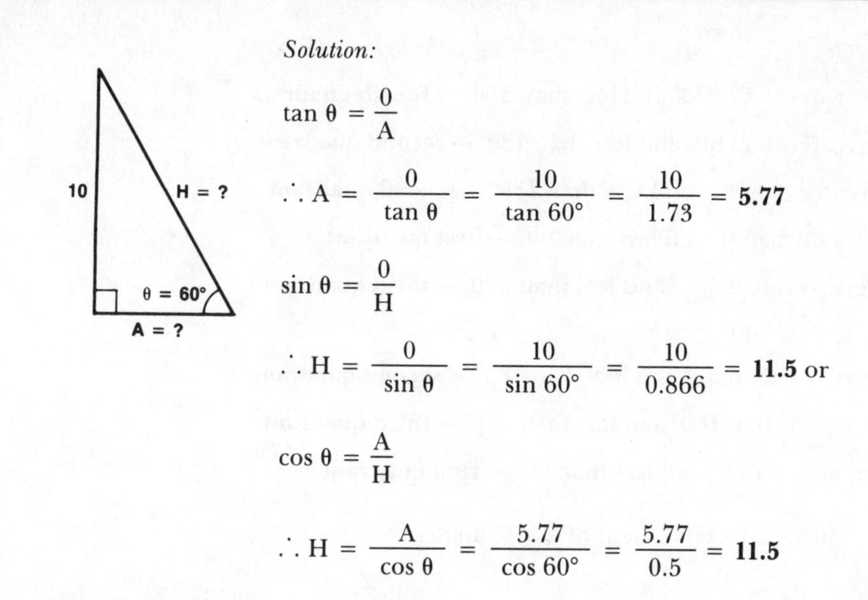

Solution:

$$\tan\theta = \frac{0}{A}$$

$$\therefore A = \frac{0}{\tan\theta} = \frac{10}{\tan 60°} = \frac{10}{1.73} = \mathbf{5.77}$$

$$\sin\theta = \frac{0}{H}$$

$$\therefore H = \frac{0}{\sin\theta} = \frac{10}{\sin 60°} = \frac{10}{0.866} = \mathbf{11.5} \text{ or}$$

$$\cos\theta = \frac{A}{H}$$

$$\therefore H = \frac{A}{\cos\theta} = \frac{5.77}{\cos 60°} = \frac{5.77}{0.5} = \mathbf{11.5}$$

Calculator Solution for Sine, Cosine or Tangent:

Enter the angle in the calculator so it is displayed. Press the sine, cosine or tangent key for the respective function required and the value will read out directly on the display.

Table Solution for Sine, Cosine or Tangent:
Enter the table of natural trigonometric functions by locating the angle in the degrees column. Scan across to the respective sine, cosine or tangent column and read the value in whole numbers and/or decimals.

9. If a hypotenuse of length equal to unity (length = 1) is rotated from the zero-degree position by the following angles, identify the quadrant in which the hypotenuse is located.

a.	278°	e.	229°
b.	153°	f.	460°
c.	92°	g.	191°
d.	10°	h.	89°

■ Worked-Out Examples

Solution:

a. 278° is greater than 270° and less than 360°—**fourth quadrant**

b. 153° is greater than 90° and less than 180°—**second quadrant**

c. 94° is greater than 90° and less than 180°—**second quadrant**

d. 10° is greater than 0° and less than 90°—**first quadrant**

e. 229° is greater than 180° and less than 270°—**third quadrant**

f. 460° − 360° = 100°

100° is greater than 90° and less than 180°—**second quadrant**

g. 191° is greater than 180° and less than 270°—**third quadrant**

h. 89° is greater than 0° and less than 90°—**first quadrant**

10. a. Find the quadrant I equivalent of a 136° angle.

Solution:

$\theta_I = 180° - \theta_{II} = 180° - 136° = \mathbf{44°}$

b. Find the quadrant I equivalent of a 197° angle.

Solution:

$\theta_I = \theta_{III} - 180° = 197° - 180° = \mathbf{17°}$

c. Find the quadrant I equivalent of a 338° angle.

Solution:

$\theta_I = 360° - \theta_{IV} = 360° - 338° = \mathbf{22°}$

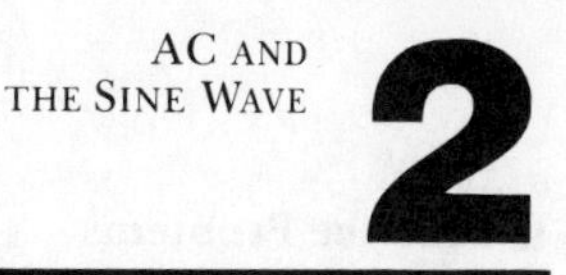

■ Practice Problems

1. Given the following frequencies, calculate the period of the waveform.

a. f = 6.3kHz, T = ____________

b. f = 276Hz, T = ____________

c. f = 37.8kHz, T = ____________

d. f = 7.6MHz, T = ____________

2. Given the following times of one cycle of a waveform, calculate the frequency of each waveform.

a. T = 50ms, f = ____________

b. T = 0.0057s, f = ____________

c. T = 33.8μs, f = ____________

d. T = 4.7μs, f = ____________

3. Given the following peak-to-peak amplitudes, calculate the peak and rms amplitude values.

a. E_{pp} = 370V

(1.) E_{pk} = ____________

(2.) E_{rms} = ____________

b. E_{pp} = 17.8V

(1.) E_{pk} = ____________

(2.) E_{rms} = ____________

c. E_{pp} = 7.5mV

(1.) E_{pk} = ____________

(2.) E_{rms} = ____________

d. I_{pp} = 32mA

(1.) I_{pk} = ____________

(2.) I_{rms} = ____________

4. Given the following rms amplitudes, calculate the peak and peak-to-peak amplitude values.

a. E_{rms} = 10V

(1.) E_{pk} = ____________

(2.) E_{pp} = ____________

b. E_{rms} = 120V

(1.) E_{pk} = ____________

(2.) E_{pp} = ____________

c. E_{rms} = 7.8mV

(1.) E_{pk} = ____________

(2.) E_{pp} = ____________

d. I_{rms} = 2.5mA

(1.) I_{pk} = ____________

(2.) I_{pp} = ____________

■ Practice Problems

5. Given the following right triangles, calculate the sine, cosine, and tangent of the angles.

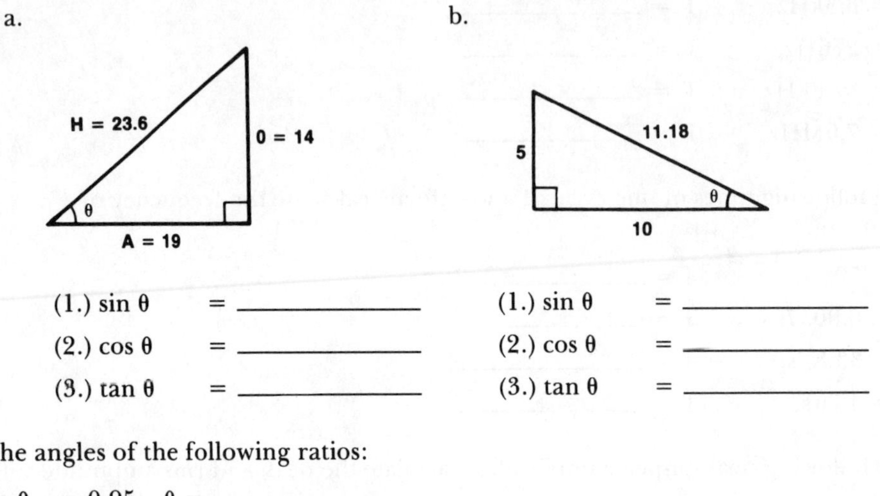

a.

(1.) sin θ = ____________

(2.) cos θ = ____________

(3.) tan θ = ____________

b.

(1.) sin θ = ____________

(2.) cos θ = ____________

(3.) tan θ = ____________

6. Find the angles of the following ratios:

a. sin θ = 0.95 θ = ____________

b. cos θ = 0.5 θ = ____________

c. tan θ = 1.0 θ = ____________

d. sin θ = 0.5 θ = ____________

e. cos θ = 0.12 θ = ____________

f. tan θ = 0.07 θ = ____________

7. Determine the value of the hypotenuse H of the following right triangles with opposite side 0 and adjacent side A and the angle θ.

	θ	0	A	H
a.	30°	10		____________
b.	60°		10	____________
c.	45°		10	____________
d.	75°	100		____________

Practice Problems

8. Determine the missing sides for the following right triangles with the hypotenuse and angle θ given.

	θ	H	A	O
a.	20°	100	____________	____________
b.	40°	100	____________	____________
c.	60°	100	____________	____________
d.	80°	100	____________	____________

9. In which quadrant are the following angles?

		I	II	III	IV
a.	91°				
b.	362°				
c.	200°				
d.	145°				
e.	300°				

10. Find the quadrant I equivalent angle of the following angles.

	θ
a.	46°
b.	100°
c.	205°
d.	420°

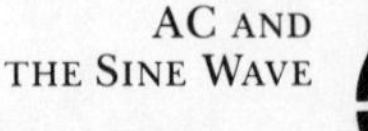

■ Quiz

1. Given the following periods or frequencies, determine the frequency or time of one cycle of the waveform.

a. f = 5.3MHz, T = 189 ns
b. f = 163kHz, T = 6.14 μs
c. T = 37ms, f = 27.0 Hz
d. T = 2.8μs, f = 357 kHz

2. Given the following peak amplitudes, determine the peak-to-peak and rms values.

a. E_{pk} = 37V
(1.) E_{pp} = 74.0 Vpp
(2.) E_{rms} = 26.2 Vac
b. E_{pk} = 75mV
(1.) E_{pp} = 150 mVpp
(2.) E_{rms} = 53.0 mVac
c. I_{pk} = 540μA
(1.) I_{pp} = 1.08 ma pp
(2.) I_{rms} = 382 μa ac

3. Given the following rms amplitude values, determine the peak and peak-to-peak values.

a. E_{rms} = 60V
(1.) E_{pk} = 84.9 Vpk
(2.) E_{pp} = 170 Vpp
b. I_{rms} = 6.7mA
(1.) I_{pk} = 9.50 mapk
(2.) I_{pp} = 19.0 mapp
c. I_{rms} = 3.15A
(1.) I_{pk} = 4.50 apk
(2.) I_{pp} = 8.91 app

4. Given the following right triangles, calculate the sine, cosine, and tangent of the angles.

a.

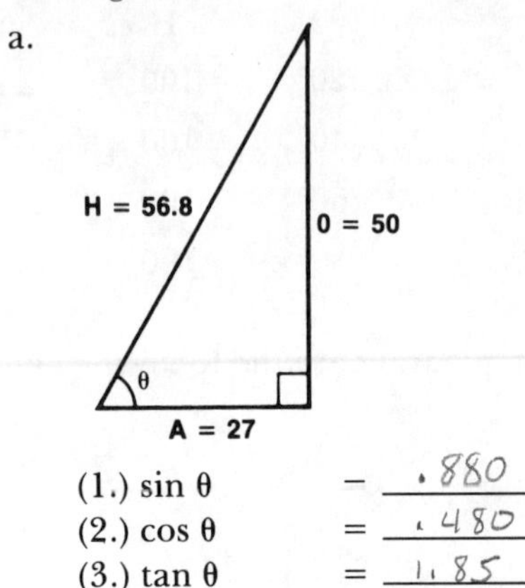

(1.) sin θ = .880
(2.) cos θ = .480
(3.) tan θ = 1.85

b.

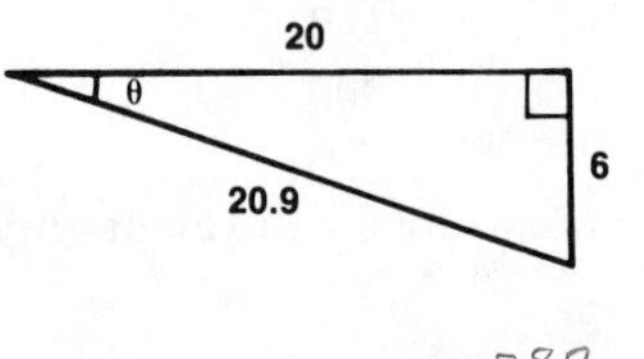

(1.) sin θ = .287
(2.) cos θ = .957
(3.) tan θ = .300

5. Using the data given, determine if the opposite side of a right triangle is greater than, equal to, or less than the adjacent side when the angle theta is:

θ	sin	cos	tan
30°	0.5	0.866	0.577
45°	0.707	0.707	1.0
60°	0.866	0.5	1.73

a. 30° $<$
b. 45° =
c. 60° $>$

■ Quiz

6. The opposite side of a right triangle

a. is the one larger than the hypotenuse.
b. remains constant.for changing hypotenuse.
c. plots the sine wave functions as the angle theta changes.
d. is always equal to the adjacent side.

7. The tangent trigonometric function of the angle theta for a right triangle

a. always has the opposite side and adjacent side equal.
b. is the ratio of the opposite side to the adjacent side.
c. is equal to the sine plus the cosine.
d. is equal to the hypotenuse times the sine of theta.

8. An angle in the second quadrant

a. is a negative angle.
b. is between 270° and 360°.
c. is one for which the sine function varies from 1 to 0.
d. is the same as an angle in the third quadrant.

9. If the hypotenuse has rotated 361°, it is in quadrant

a. I
b. II
c. III
d. IV

10. Find the quadrant I equivalent angle of the following angles.

a. 137° 43°
b. 232° 52°
c. 307° 53°

LESSON 3

The Oscilloscope and its Use

In this lesson one of the most important electronic measuring instruments for ac voltages and currents is discussed—the oscilloscope. It is particularly important because it allows one to "take" an electronic picture of ac waveforms. The construction of an oscilloscope is explained. The function and adjustment of an oscilloscope's controls are described. And you are told how to use an oscilloscope to perform some electrical measurements.

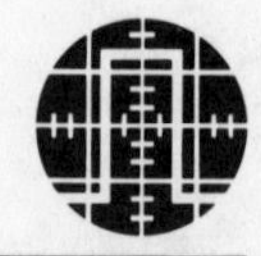

■ Objectives

At the end of this lesson you should be able to:

1. Describe the major sections of an oscilloscope.
2. Explain how the vertical and horizontal deflection systems of an oscilloscope function.
3. Adjust the basic controls of a typical oscilloscope.
4. Use an oscilloscope to measure the peak and peak-to-peak voltages of a waveform, and use these to calculate rms values.
5. Use an oscilloscope to determine the period of a sinusoidal waveform and calculate the waveform's frequency.

■ Introduction

INTRODUCTION

In the first two lessons, discussion concerned how ac is generated and how it is specified in terms of amplitude and frequency. You learned how to determine the instantaneous amplitude at various points in the cycle, and how the sine wave derives its name from the sine function.

Before continuing with an investigation of actual ac circuits, however, one of the instruments which is very important for making electrical measurements in the circuits which you will learn about throughout this book is the oscilloscope (or scope for short). A typical oscilloscope is shown in *Figure 3.1*.

In earlier lessons it was explained how dc and ac voltages versus time are graphed as shown in *Figure 3.2*. The oscilloscope is capable of *automatically* displaying an ac or dc voltage graphed versus time as shown in *Figure 3.3*. The oscilloscope, then, can be called a "visual voltmeter". But, in fact, it is more than just a voltmeter; the scope actually displays waveforms so that the intricacies of waveforms can be observed clearly. It is an instrument which converts electrical signals to visual waveforms on a screen.

An oscilloscope performs three basic functions: One of these is *waveform observation*. The scope allows the size and shape and type of waveform to be observed. A second function is *amplitude measurement*. The oscilloscope vertical deflection is calibrated on the screen so that actual voltage amplitudes can be measured. The third function is *a measurement of time*. The oscilloscope sweep across the screen horizontally is calibrated in time increments which allows the measurement of time periods or time duration.

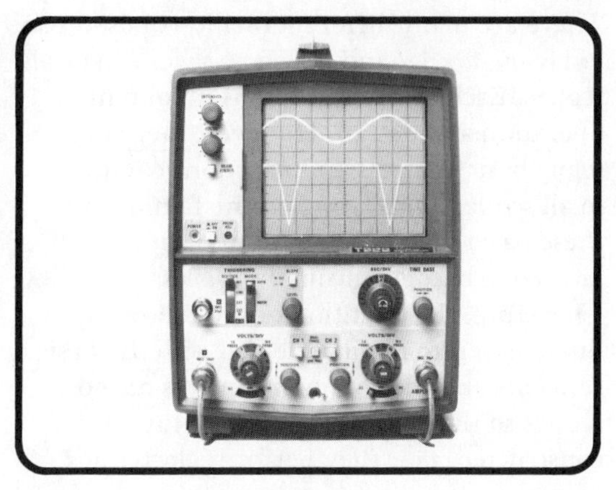

Figure 3.1 *A Typical Oscilloscope*

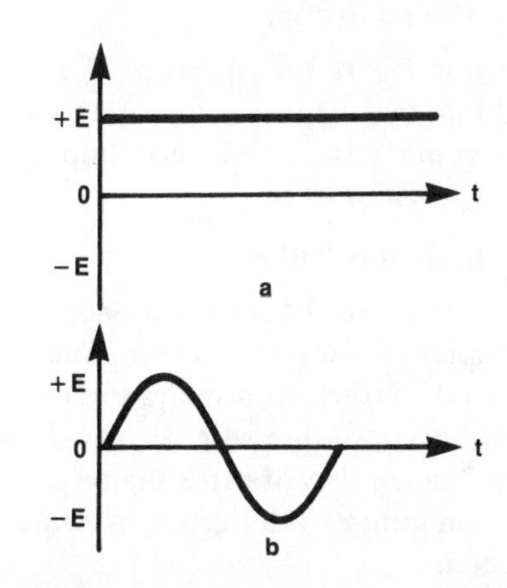

Figure 3.2 *Plots of: a. DC Voltage; b. AC Voltage*

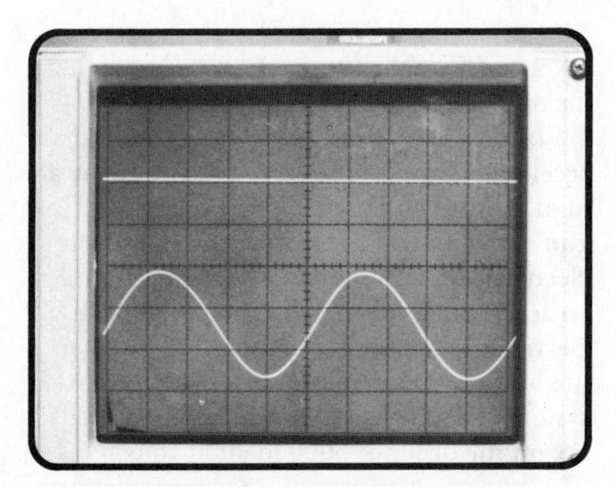

Figure 3.3 *Scope Face with AC and DC Waveforms Displayed*

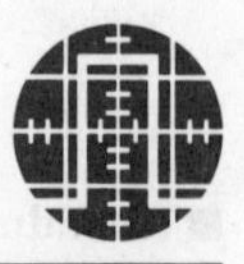

There are many different oscilloscopes in use today. *Figure 3.4* and *3.5* show several typical scopes. Each of these scopes is unique in operation and design. However, there are many basic controls and functions common to all scopes. Once you become familiar with these basic controls, operation of practically any scope becomes a simple process. Therefore, the intention of the following discussion is to familiarize you with the basic concepts and the common controls of most scopes so that you will learn how they are constructed, how they perform electrical measurements, and how they are used.

A BASIC OSCILLOSCOPE

As shown in *Figure 3.6*, the scope can be divided into two major segments: 1) the *cathode-ray tube* (also called a crt), and 2) the *controlling circuits.*

The Cathode-Ray Tube

The crt is the heart of the oscilloscope. Its 'face' displays the waveform being measured, similar to the crt which displays the picture received by a television receiver. As shown in *Figure 3.7*, it consists of three major parts: 1) an electron gun; 2) a deflection system; and 3) a screen.

Electron Gun

The electron gun is located at the rear of the crt, away from the screen. Its job is to emit a narrow stream of electrons. These electrons are (accelerated), focused into a beam and accelerated toward the screen of the crt by a high positive potential applied to the electron gun anode placed near the front end of the electron gun. Note: oscilloscopes may employ an acceleration system that accelerates the electrons in the beam either before or after the beam passes through the deflection system. The pre-deflection anode potentials are on the order of 3000 to 4000 volts.

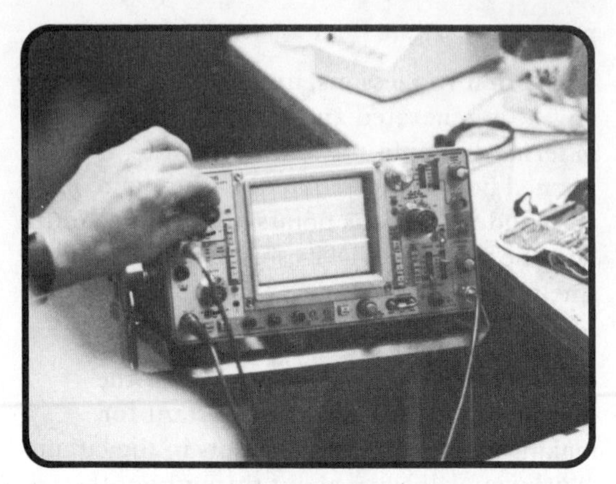

Figure 3.4 *Typical Oscilloscope*

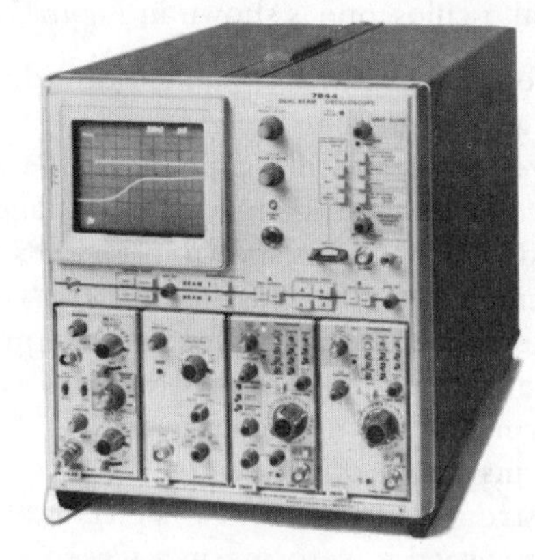

Figure 3.5 *Typical Oscilloscope*

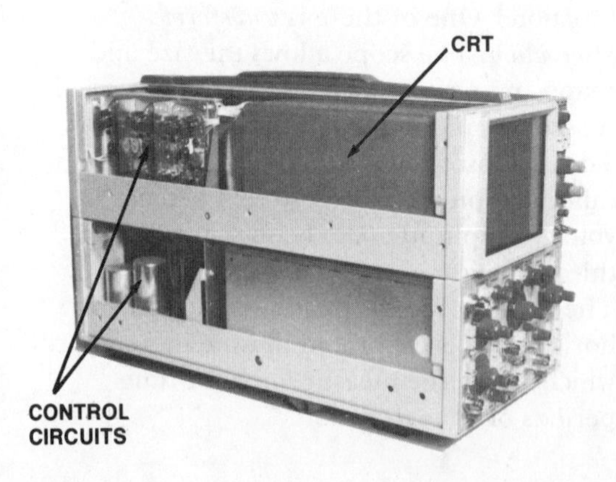

Figure 3.6 *Interior of a Typical Oscilloscope*

■ Electron Gun

Post-deflection anode potentials may be several thousand volts higher. Some may reach 25,000 volts.

The electron beam travelling toward the screen passes through the deflection system. The deflection system consists of four deflection plates as shown in *Figure 3.8*. Two plates—one on the top and one on the bottom—are called vertical deflection plates. The two plates on the sides are the horizontal deflection plates. By applying positive or negative potentials to these deflection plates, the electron beam is caused to deflect up or down and right or left as it passes through them.

The more potential applied to the plates, the more the electron beam deflects. Therefore, the amount of this deflection is actually a measure of the voltage, or potential difference, applied to the plates. The force bending the beam is electrostatic force and it follows the first law of electrostatics: like charges repel and unlike charges attract.

If a voltage is applied across the vertical deflection plates as shown in *Figure 3.9a*, the electron beam moves upwards. If the polarity applied to the plates is reversed, as shown in *Figure 3.9b*, the beam moves downwards.

If a voltage is applied across the horizontal deflection plates as shown in *Figure 3.10a*, the electron beam will travel from left to right. If the polarity applied to the deflection plates is reversed as shown in *Figure 3.10b*, the beam moves from right to left.

If voltages are applied to the vertical and horizontal deflection plates simultaneously, the beam moves vertically and horizontally at the same time, diagonally. This is shown in *Figure 3.11*.

If no potential is applied to the plates, the beam returns to the center of the tube. This was the original position shown in *Figure 3.8*.

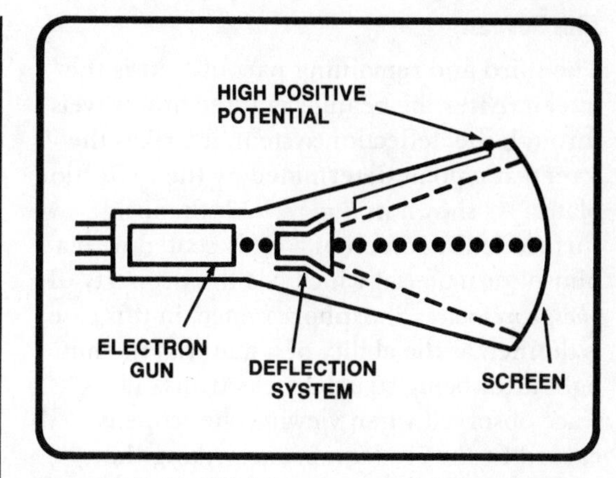

Figure 3.7 *Simplified Construction of a Typical CRT*

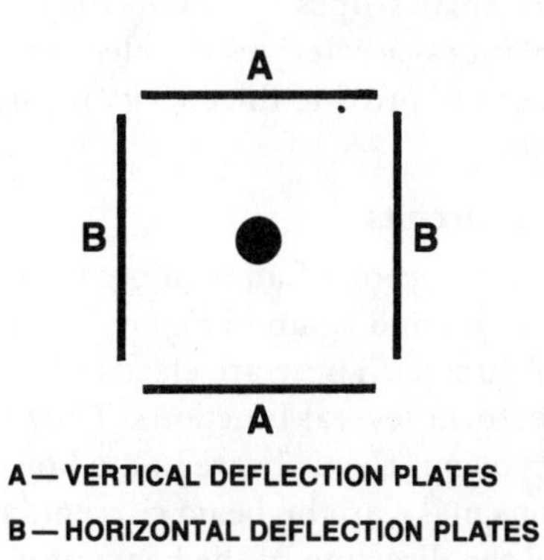

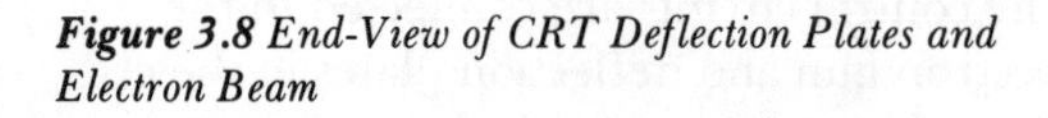

Figure 3.8 *End-View of CRT Deflection Plates and Electron Beam*

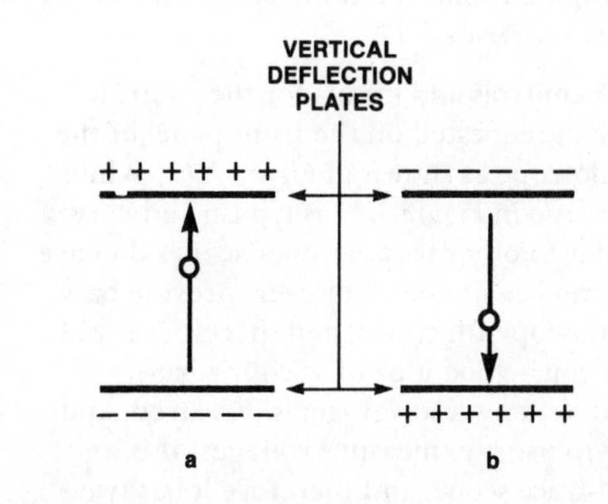

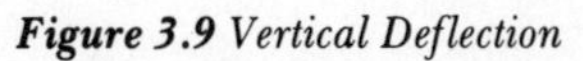
Figure 3.9 *Vertical Deflection*

The Screen

The third and remaining part of a crt is the screen. After the beam is emitted and travels through the deflection system, it strikes the screen at a point determined by the deflection plates. As shown in *Figure 3.12*, the inside surface of the screen of a crt is coated with a phosphor material which has the property of *phosphorescence*. Phosphorescence, in this case, is defined as the ability of a material to emit light after being struck by electrons. The trace observed when viewing the scope is caused by the electron beam striking the phosphor material of the screen. The very high positive potential (typically from 3,000 volts for small scopes to 25,000 volts for tv picture tubes) accelerates the electrons to the screen to provide the energy for light emission.

Control Circuits

The other portion of an oscilloscope shown in *Figure 3.6* and again in *Figure 3.12* is its control circuits. These are electronic circuits that perform several functions. They cause the crt to emit electrons, regulate how many electrons make up the beam current, and control the direction of the beam of electrons. The control circuits are connected to the electron gun and deflection plates in the crt through a connector at the base of the crt as shown in *Figure 3.12*.

Most controls and inputs for the control circuit are located on the front panel of the oscilloscope as shown in *Figure 3.13*. While the scope in *Figure 3.13* is typical and very similar to other scopes, other scopes do have differences. It does, however, provide basic oscilloscope functions, and therefore it will give you a good idea of a scope's basic controls, how a waveform is displayed, and how to use it to measure voltages. It is a dual-trace scope, and therefore it has two

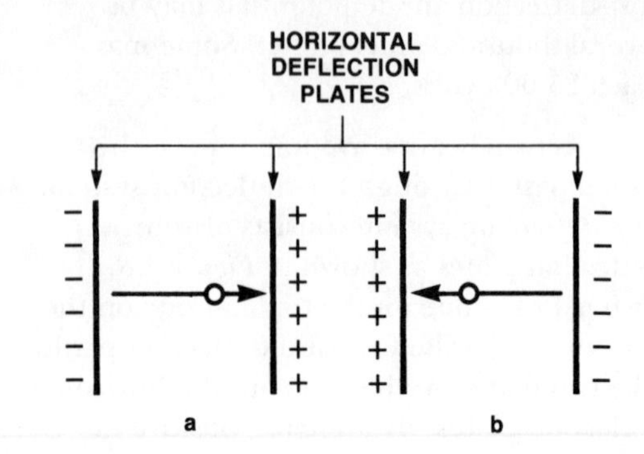

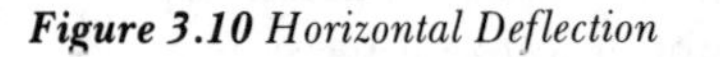

Figure 3.10 Horizontal Deflection

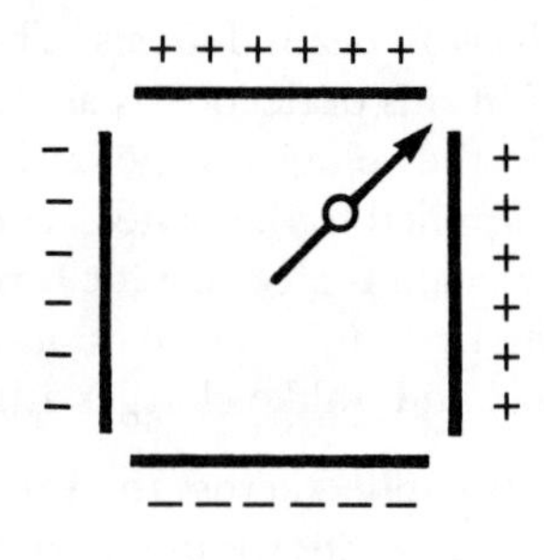

Figure 3.11 Diagonal Deflection

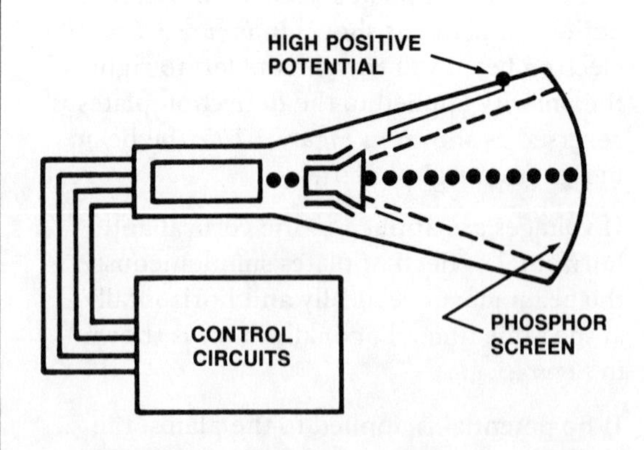

Figure 3.12 CRT and Control Circuits

identical vertical input controls so that it can display the traces of two different input signals at the same time.

The controls can be divided into three major groups: 1) the *mainframe group*, 2) the *vertical control group*, and 3) the *horizontal control group*. This is shown in *Figure 3.14*.

MAINFRAME CONTROLS

An enlarged view of the mainframe controls is shown in *Figure 3.15*. Once the scope is connected to a standard wall receptacle, power can be applied to *all* circuits in the scope with the *power* switch. After the power switch is activated, one to two minutes are required before the scope circuits are operational. After this "warm up" period, a trace may or may not be observed on the screen of the crt.

Beam Control – Intensity

If a trace is not observed, *the intensity control* should be turned clockwise. The intensity control regulates the number of electrons in the beam striking the screen. As the intensity control knob nears mid-position, a spot or a trace should be visible. If the intensity control is set too high as shown in *Figure 3.16*, the spot or trace (horizontal line across screen) on the screen will not be crisp and sharp; it will instead appear overbright and be fuzzy. This is called "blooming". If a high intensity like this is allowed to continue, the phosphor can be burned off the inside of the screen where the beam strikes it, and that portion of the screen will be damaged. It is advisable to increase the intensity only to the point where the trace is barely visible. You should remember to turn the intensity down before turning the scope off to keep the crt from blooming the next time the scope is turned on. In addition, it is not advisable to leave the intensity control turned up when there is a single spot on the screen rather than a trace. It could burn the screen.

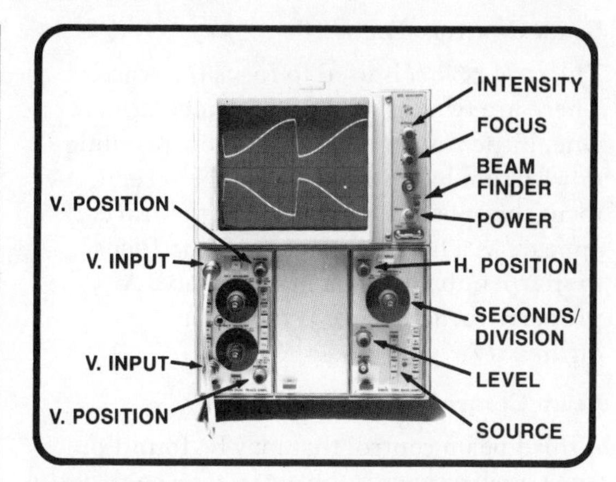

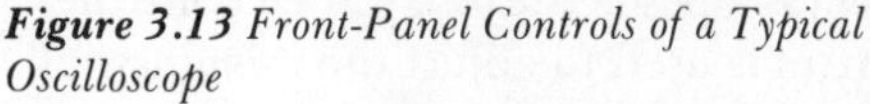
Figure 3.13 Front-Panel Controls of a Typical Oscilloscope

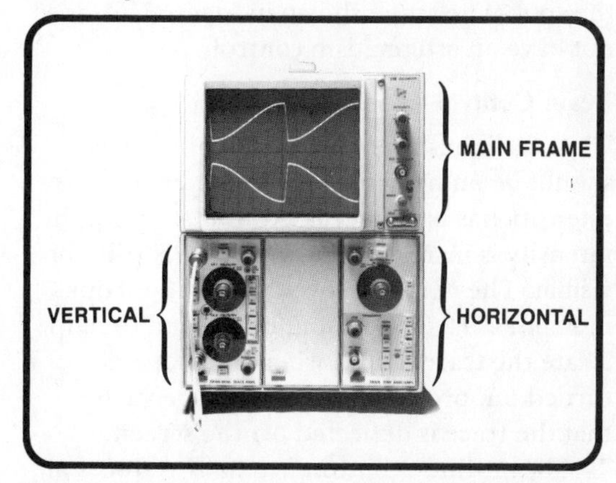

Figure 3.14 The Three Major Control Groups of a Typical Scope

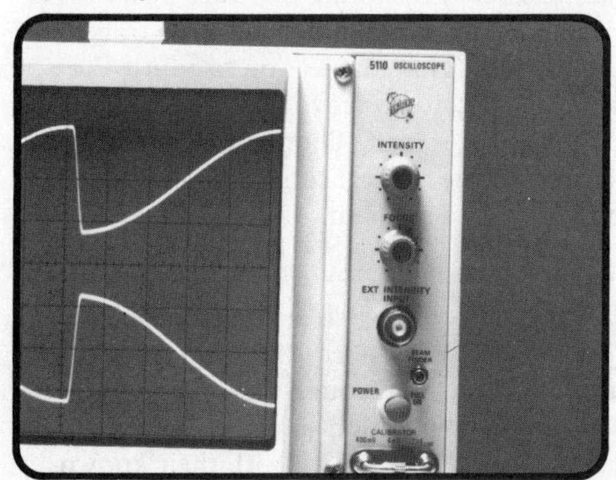

Figure 3.15 The Mainframe Controls

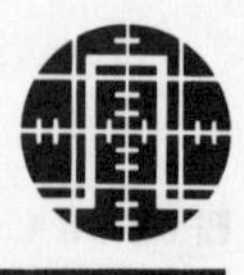

Beam Control – Focus

The *focus control* is used to focus the trace. There are two basic ways this can be done. One, if the trace appears as a spot travelling slowly from left to right across the screen, focus the spot to a *point*. Two, if the trace appears as a horizontal line, use the focus to sharpen the line to a narrow trace. A properly-focused trace is shown in *Figure 3.17*.

Beam Control – Astigmatism

A third beam control that may be found on some oscilloscopes is the *astigmatism* control. This control is used to adjust the *roundness* of the spot. The scope shown in *Figure 3.15* does not have an astigmatism control.

Beam Control – Position

Thus far it has been assumed that the trace should be immediately evident after the intensity has been increased. But suppose the intensity is increased and the trace is still not visible. The *beam-finder* button on the scope in *Figure 3.15* then should be pushed to help locate the trace. Often when the scope is turned on, previous settings may be such that the trace is deflected off the screen. The beam-finder button normally reduces all deflection potentials so that the trace appears on the screen, By noting the trace position, you know the direction in which the trace is deflected. This enables you to adjust the position controls to move the trace back to the center of the screen. The position controls, located on the vertical control panel, will be described in a moment.

Calibration

Below the power switch is the *calibrator loop*. A signal of a specific frequency and amplitude passes through this loop. A typical squarewave calibration signal is shown in *Figure 3.18*. These signals with their known specifications can be used to check the

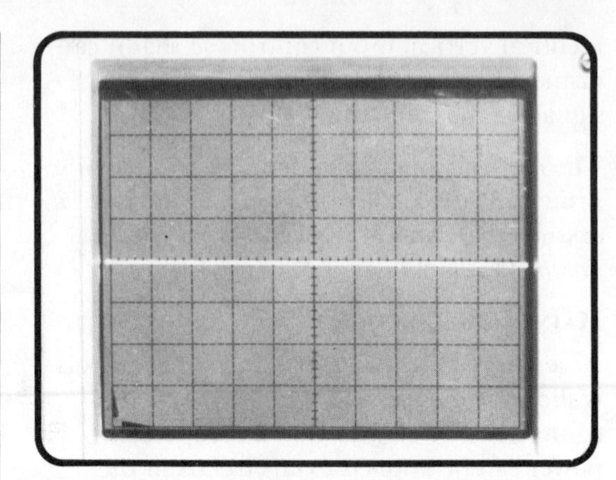

Figure 3.16 *Excessive Brightness Causes "Blooming"*

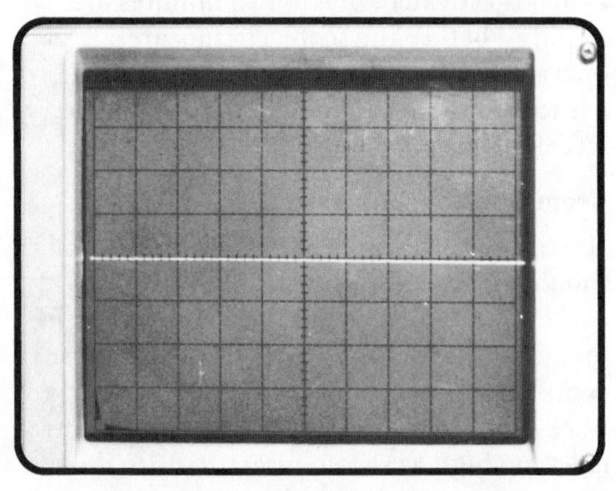

Figure 3.17 *A Well-focused Trace*

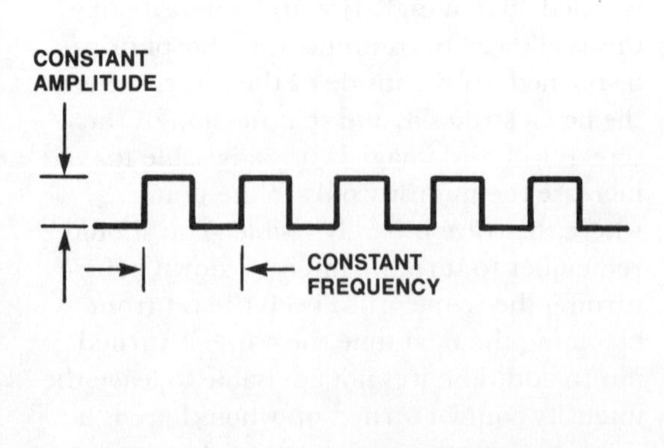

Figure 3.18 *A Typical Square Wave Calibration Signal*

accuracy of the vertical and horizontal deflection systems of the scope. The procedure, involving the use of a known signal, is called *calibration*.

Screen Graticule

To the left of all of these mainframe controls, of the scope in *Figure 3.13* is the crt. *Figure 3.19* shows an enlarged view of the front of the screen. Notice that the screen has been marked off into eight vertical divisions and ten horizontal divisions. Each division has been further marked off into five equal increments. Each increment represents two-tenths of one division. This scale is called a *graticule*.

Equipped with a graticule, the oscilloscope provides an electronic graph of voltage against time. It essentially is a calibrated scale with the vertical divisions of the graticule representing *voltage* values and the horizontal divisions representing *time* increments.

Some scopes have an illumination control which provides *lighting* for the graticule. The scope shown in *Figure 3.13* does not have that function.

Vertical Control

The vertical controls shown in *Figure 3.20* adjust circuits that are connected to the vertical deflection plates and performs two control functions: 1) they control the trace's vertical position, and 2) they select the amount of voltage each vertical division represents.

The vertical *position* control, is used to position the electron beam which forms the spot or trace on the graticule vertically. Any initial vertical reference level can be selected. It is usually positioned without any signal applied to the scope input. The most common level is on the center horizontal line of the graticule as shown in *Figure 3.21*. The spot can be moved vertically by applying a voltage

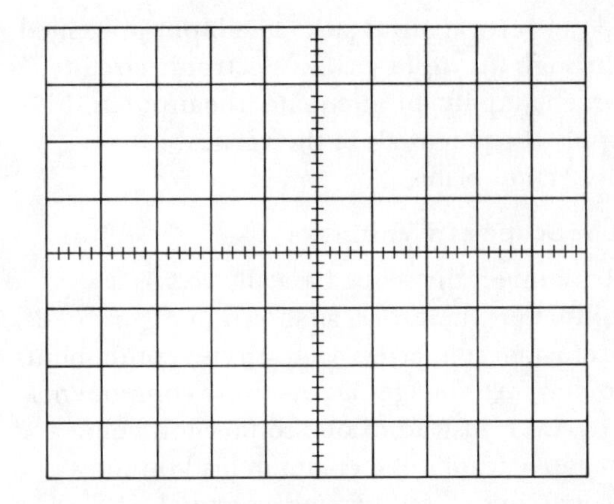

Figure 3.19 *A Typical Graticule*

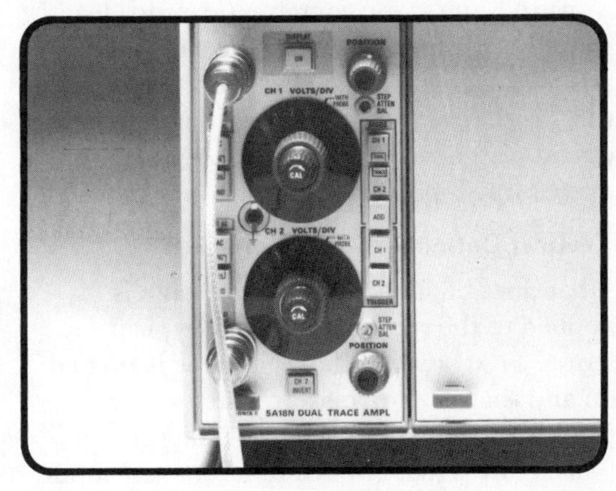

Figure 3.20 *Vertical Control Group*

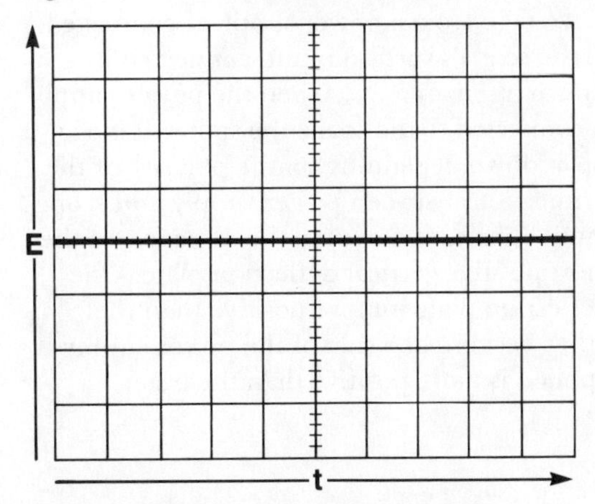

Figure 3.21 *Vertical Trace Positioned on Zero-Reference Line*

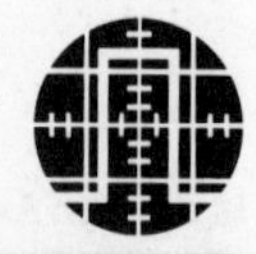

to the vertical input jack. The input is applied through the input jack to electronic circuits which amplify, or attenuate, the input and apply the potentials to the vertical deflection plates.

The Scope as a Voltmeter

Remember, the scope basically acts as a voltmeter; therefore, as shown in *Figure 3.22*, it must be connected *across* circuit components to measure voltages across those components. However, instead of one connection being a negative lead, it is a common lead on most scopes, and it is at ground potential. If measurements are to be made in a circuit with a ground, the scope common (ground) lead should be connected *only* to the circuit ground as shown in *Figure 3.23*. Failure to practice this procedure will often cause damage to the circuit or the scope, or it could expose the operator to a possible electrical shock.

Vertical Deflection – DC Voltage

What does happen when a dc voltage is applied to the vertical input of a typical scope? First, assume that the scope is turned on and that its controls have been set to produce a spot at the center of the screen *without* any signal applied to its vertical input. The screen is as shown in *Figure 3.24*. Now a dc voltage from a power supply is connected to the scope's vertical input, connected as shown in *Figure 3.25*. When the power supply is connected to the scope the spot will move up *or* down depending on the polarity of the connections between power supply and scope input. The spot moves in the direction of the most positive vertical deflection plate. One deflection plate is more positive than the other because one side of the power supply applied is more positive than the other.

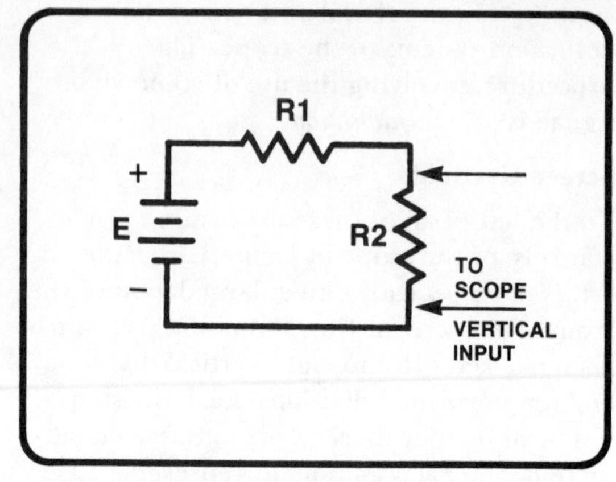

Figure 3.22 *A Scope Must Connected Across Components to Measure Voltages in Those Components*

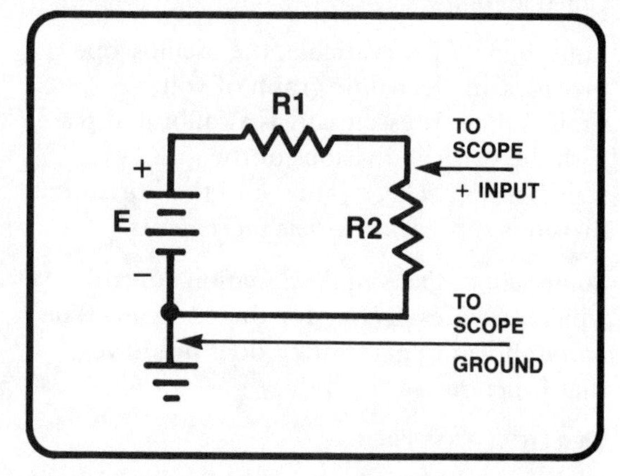

Figure 3.23 *Scope's Ground Lead Should be Connected Only to Circuit Ground*

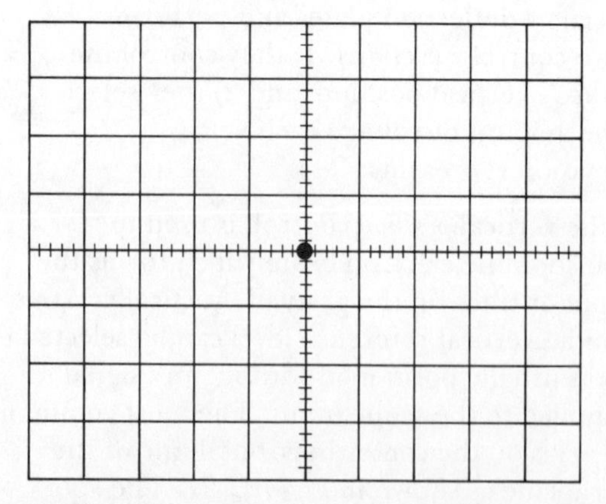

Figure 3.24 *Spot at Center of CRT Without Any Signal Applied to Vertical Input*

■ Vertical Deflection — AC Voltage

Therefore, the spot — which is actually a beam of negatively-charged electrons — is attracted to the more positive of the deflection plates. If the connections between the power supply and scope input are reversed as shown in *Figure 3.26*, the spot will move in the opposite direction it moved before the connections were reversed. This is because the opposite vertical deflection plate is now more positive than the other. The important fact is that the spot can be made to move *up* or *down* from a reference position by applying a voltage of the proper polarity to the vertical input. The examples here were for constant dc voltage inputs.

Vertical Deflection – AC Voltage

If an ac voltage is applied to a scope's vertical input, the spot will deflect up and down periodically from the center reference point of *Figure 3.24*. It does this because, as you know, ac is sinusoidal, constantly changing polarity from positive to negative and back again with time in a sine wave fashion. Therefore, a straight vertical line will appear on the scope face as shown in *Figure 3.27a*. The line is being created by the spot moving up and down very rapidly between the rapidly-changing positive and negative potentials on the vertical deflection plates. Even though a sinusoidal ac voltage is being input to the scope as shown in *Figure 3.27b*, it doesn't look like a sinewave because the voltage is being applied only to the vertical deflection plates, and therefore, there is no horizontal deflection. The line is simply an indication of the voltage amplitude changes of the applied ac voltage, as shown in *Figure 3.27b*. The length of the line will become longer or shorter as the ac voltage amplitude is increased or decreased. It obviously is not the standard graph of a sinusoidal waveform.

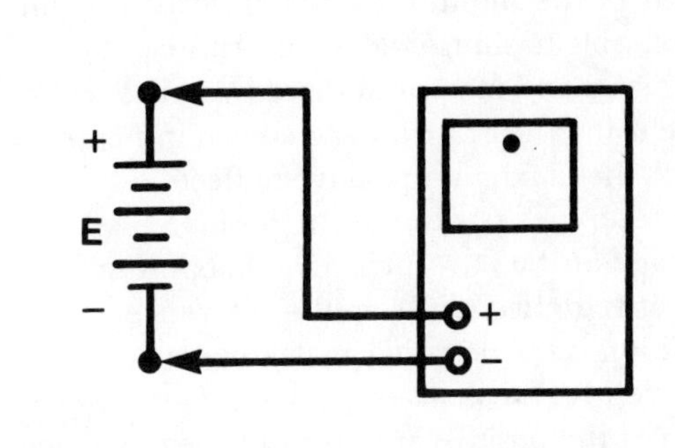

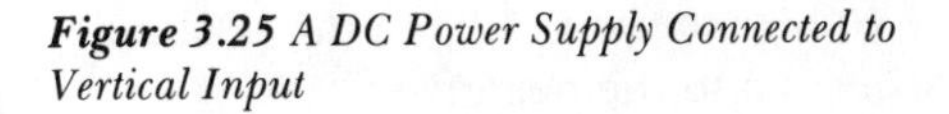
Figure 3.25 *A DC Power Supply Connected to Vertical Input*

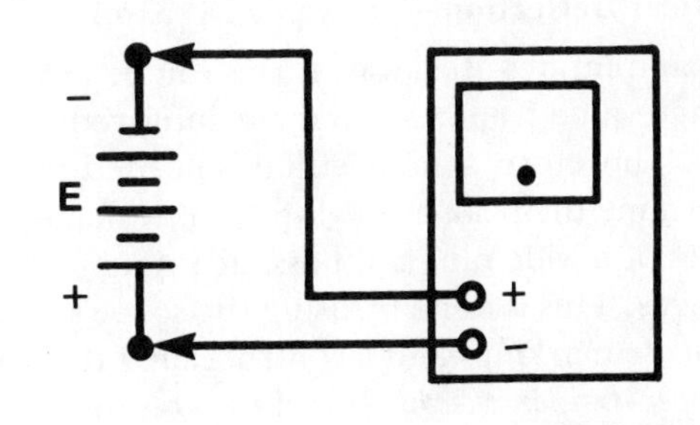

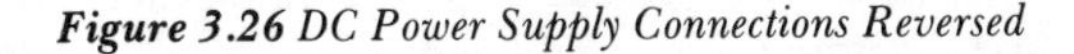
Figure 3.26 *DC Power Supply Connections Reversed*

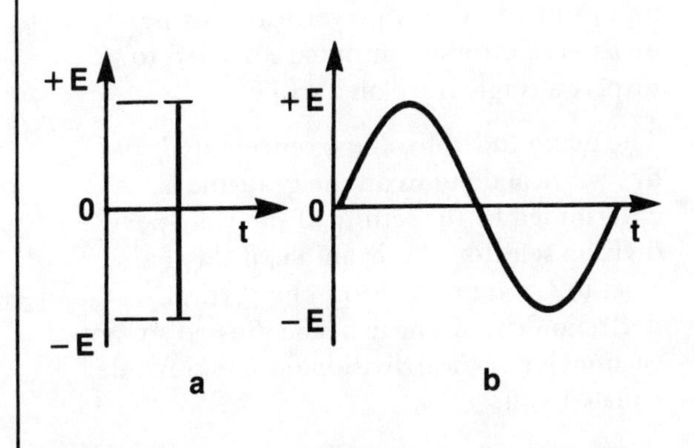

Figure 3.27 *a. Voltage Display on CRT; b. Sinusoidal Voltage Input to Scope*

■ **Voltage Measurements**
■ **Vertical Deflection — Volts per Division**

Voltage Measurements

Some of the amplitude units of measurement applicable to sinusoidal waveforms can be observed and measured directly on the scope face using the graticule. As shown in *Figure 3.28*, the maximum positive deflection from the reference position is the positive peak voltage of the waveform. The maximum negative deflection from the reference position is the negative peak voltage. Therefore, the peak-to-peak voltage is the sum of the positive and negative peak voltage values. Or, the peak-to-peak voltage is the value of voltage measured from the positive peak deflection to the negative peak deflection.

Vertical Deflection – Volts per Division

The amplitudes of ac waveforms can be any voltage value from zero to many hundreds of volts. Therefore, some method is needed to determine the peak or peak-to-peak voltage values of a wide range of possible input voltages. This is done by using the scope's graticule markings and a control called the *volts-per-division selector*. It is shown on the scope's vertical control panel in *Figure 3.20*. Two vertical volts-per-division selectors are shown on the panel because it is a dual-trace scope designed to provide two separate waveform traces from two separate vertical input sources. Only one vertical volts-per-division selector and input are needed to display a single trace on a scope.

The magnitude of voltage represented by a vertical division on the graticule is determined by the setting of the volts-per-division selector. For example, if the dial is set to 1 volt per division, the vertical deflection circuits have been adjusted so that each major vertical division on the graticule equals 1 volt.

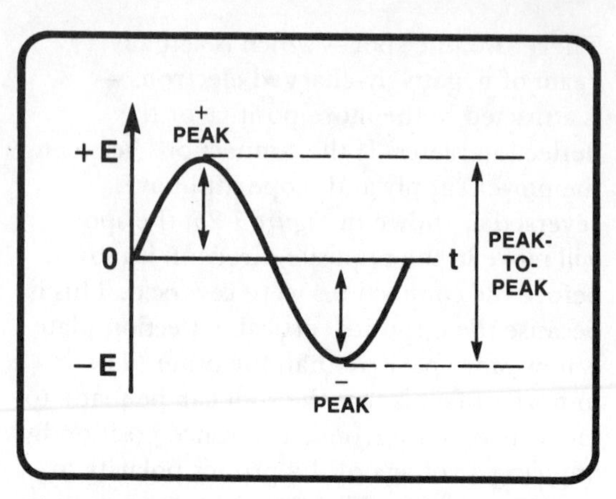

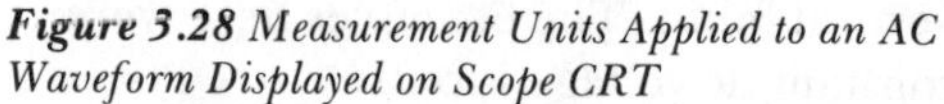
Figure 3.28 *Measurement Units Applied to an AC Waveform Displayed on Scope CRT*

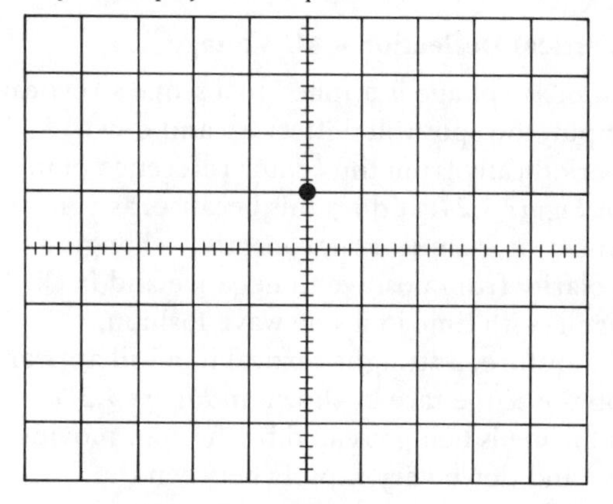

Figure 3.29 *Spot Deflected +1 Volt*

Now, a dc voltage is applied to the scope's vertical input. If the spot deflects up from a reference position exactly one division, as shown in *Figure 3.29*, a potential of +1 volt is being measured. If the spot deflects down exactly one division as shown in *Figure 3.30*, a potential of −1 volt is being measured.

Each major vertical division is further divided into five subdivisions. The voltage interval represented by each subdivision is two-tenths of the voltage of one major division.

■ An Example

If the spot deflects up one major division from a reference position plus four-tenths of the next major division (two subdivisions) as shown in *Figure 3.31*, then the number of divisions is 1.4. The potential being measured is calculated:

(1.4 divisions)(1 volt-per-division) = 1.4 volts

The voltage being measured is 1.4 volts. Other applied voltages are calculated the same way. The value of the voltage is determined by multiplying the number of graticule divisions by the setting on the volts-per-division selector.

$$E = (\text{number of divisions}) \times (\text{number of volts-per-division}) \quad (3\text{–}1)$$

An Example

For example, assume that an ac voltage whose voltage level is unknown is applied to a scope's vertical input. The waveform on the scope's crt is as shown in *Figure 3.32*. The scope's vertical selector is set to 2 volts-per-division. What are the peak and peak-to-peak voltages of the input voltage?

To evaluate the scope measurement, count the number of graticule divisions from the zero (center) horizontal graticule reference to the top of the waveform. In this example that is two and one-half divisions. This is the peak voltage of the ac voltage. It can be calculated using equation *3–1*.

$$\begin{aligned} \mathbf{E_{pk}} &= \textbf{(number of divisions)} \times \textbf{(volts-per-division)} \\ &= \mathbf{(2.5)(2V)} \\ &= \mathbf{5V} \end{aligned}$$

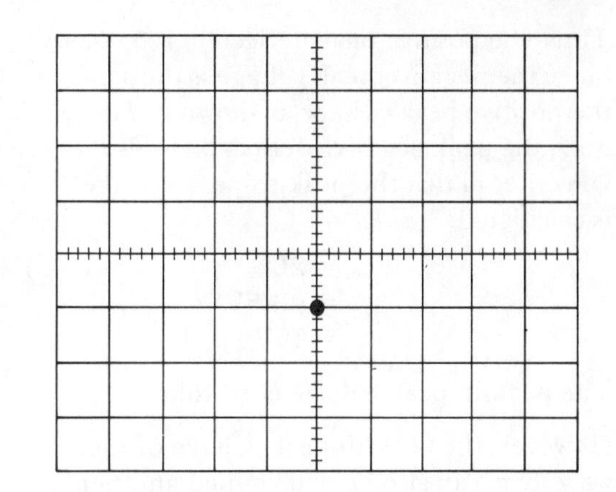

Figure 3.30 *Spot Deflected −1 Volt*

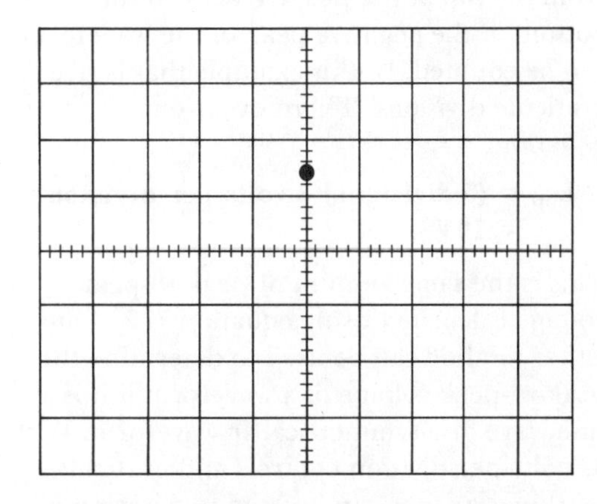

Figure 3.31 *Spot Deflected +1.4 Volts*

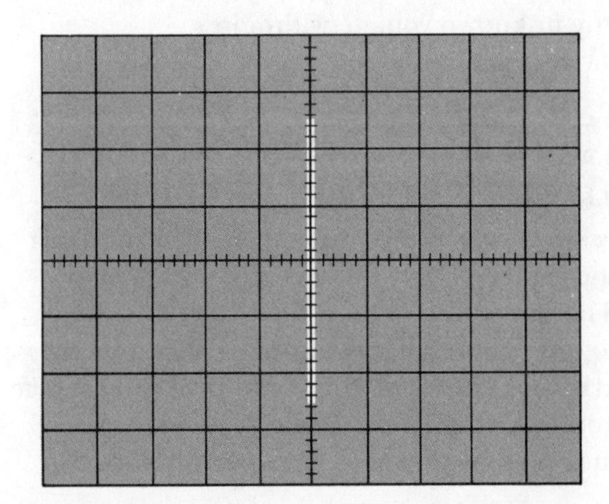

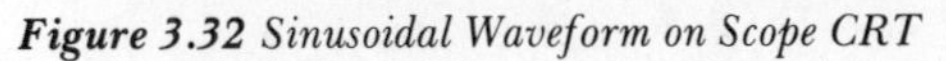

Figure 3.32 *Sinusoidal Waveform on Scope CRT*

Thus, the positive peak deflection is 5 volts. Since the negative peak voltage is the same as the positive peak voltage as shown in *Figure 3.33*, the peak negative deflection is also 5 volts. Recall that the peak-to-peak voltage is calculated:

$$E_{pp} = 2E_{pk} \quad (3\text{–}2)$$
$$= (2)(5V)$$
$$= 10V$$

The peak-to-peak voltage is 10 volts.

However, the peak-to-peak voltage of the waveform can also be determined another way. The total number of graticule divisions from the top of the positive peak to the bottom of the negative peak of the waveform can be counted. In this example that is five graticule divisions. Therefore, using equation *3–1*,

$$E_{pp} = (5 \text{ divisions})(2 \text{ volts-per-division})$$
$$= 10V$$

This is the same amount of peak-to-peak voltage calculated using equation *3–2*. Thus either method can be used to determine the peak-to-peak voltage of a waveform if it is a sine wave or a symmetrical ac waveform. With its volts-per-division controls and graticule markings a scope can be used to determine the peak and peak-to-peak values of any unknown voltage within its measurement range.

Vertical Deflection – Variable Volts-per-Division

The scope shown in *Figure 3.20* has a *variable* volts-per-division control located in the center of the volts-per-division selector. This control must be turned fully clockwise to the calibrated position as indicated by the arrow on the inner control to read the correct amplitudes from the scope graticule as indicated by the main volts-per-division selector. (This is a detent position called CAL, or there may even be a light indicator on latest model scope.) If the variable control is taken out of the calibrated position, the number of volts for each graticule division would become more than the setting on the main volts-per-division selector. This control should remain in the *CAL* position except when making special-purpose measurements which are outlined in the scope manual.

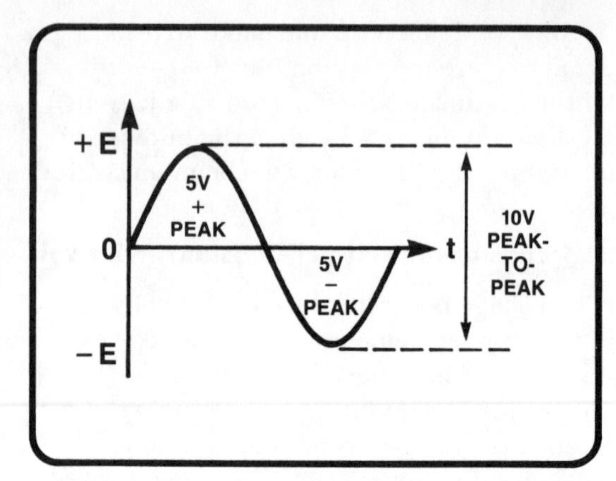

Figure 3.33 Voltage Waveform for Calculation Example

Vertical Coupling Control

Beside the vertical input connector on the scope in *Figure 3.20* there is a pushbutton control that is labeled *vertical coupling control.* It can be set to AC or DC. In general, the vertical coupling control should be set to ac when measuring ac voltages such as in *Figure 3.27b*. The control should be set to dc to measure dc voltages. If a particular waveform has both ac and dc voltage components as shown in *Figure 3.34*, set the control on the scope to the dc position to examine both components and it will appear as shown. If you wish to measure only the ac component of a waveform with both dc and ac components, change the vertical coupling control to the ac position. The dc component will be eliminated as shown in *Figure 3.35*.

■ Horizontal Sweep

Another vertical input control pushbutton in *Figure 3.20* labeled GND is used to set the trace to zero or reference position before taking measurements. When this control is pushed in, the input signal is removed, and zero volts, or ground potential is applied to the scope input circuit. The trace can then be moved to a point on the graticule representing the zero point. This zero reference point is normally set at the center horizontal line of the graticule, but it can be moved to any vertical position on the screen using the vertical deflection positioning control.

HORIZONTAL CONTROLS

The horizontal control section, identified in *Figure 3.14* and shown magnified in *Figure 3.36* determines the potentials on the right and left horizontal deflection plates causing the spot to move from side to side on the scope face. This side-to-side movement of the spot is called *sweep*. The spot is always swept from left to right across the screen of the scope as shown in *Figure 3.37*.

Horizontal Sweep

Recall that the vertical deflection of the spot is proportional to the *voltage amplitude* of the waveform. The horizontal sweep of the spot on the other hand, is proportional to an *amount of time* it takes time for the spot to move from one side of the scope's crt face to the other. Thus, as stated previously, the vertical deflection of the scope plots voltage and the horizontal deflection plots time.

If the spot is swept with the vertical input grounded, there will be no vertical deflection and only the effect of the horizontal controls on the spot will be viewed.

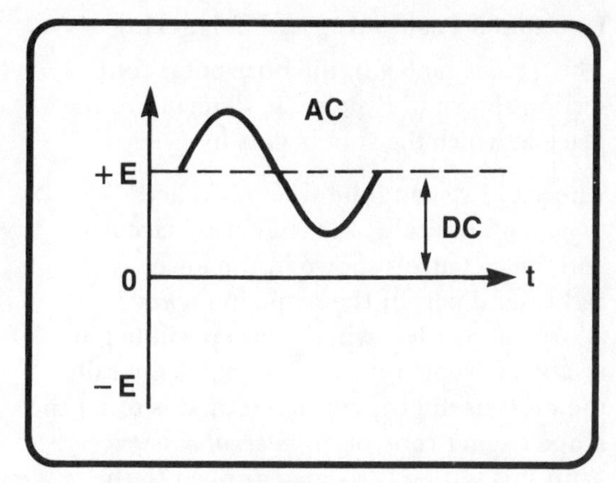

Figure 3.34 *Waveform Containing AC and DC Components Displayed Using DC Coupling*

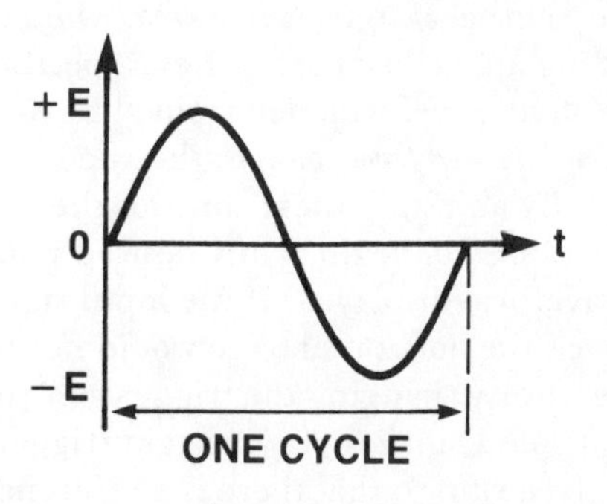

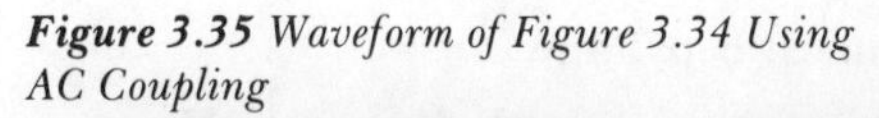

Figure 3.35 *Waveform of Figure 3.34 Using AC Coupling*

Figure 3.36 *Horizontal Control Group*

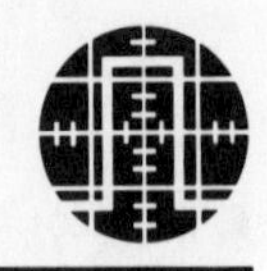

Horizontal Positioning and Triggering

The *position control* in the horizontal control section shown in *Figure 3.36* determines the place at which the spot begins its sweep.

The act of starting the sweep is called *triggering*. Basically, the triggering circuits and their controls shown in the lower right-hand part of the scope in *Figure 3.36* allow you to select when a sweep will begin relative to some reference signal. Generally, the most useful triggering method is to set the scope trigger controls to *internal auto trigger*. With this setting the signal applied to the vertical plates, and therefore observed on the screen, actually acts to trigger its own sweep.

In the internal auto trigger mode, when the incoming applied signal reaches a specific point on its waveform, determined by the scope's *slope and level controls*, the sweep begins. By adjusting these controls the sweep can be caused to begin at any desired point of the waveform displayed. If the input signal is removed the horizontal trace would not be triggered. By changing the trigger controls to the *auto* mode the scope will begin triggering itself (free run) so that there is a reference trace on the crt face.

Horizontal Sweep Time

The spot is swept across the screen by the horizontal deflection circuits changing the electrical potentials on the horizontal deflection plates. The time interval required for the trace to travel across the screen is controlled by the *horizontal deflection time selector*. The selector is calibrated in units of *seconds per division*. This is comparable to the volts-per-division units of the vertical deflection selector. The sweep is measured in units of time, usually in seconds or some fraction of a second.

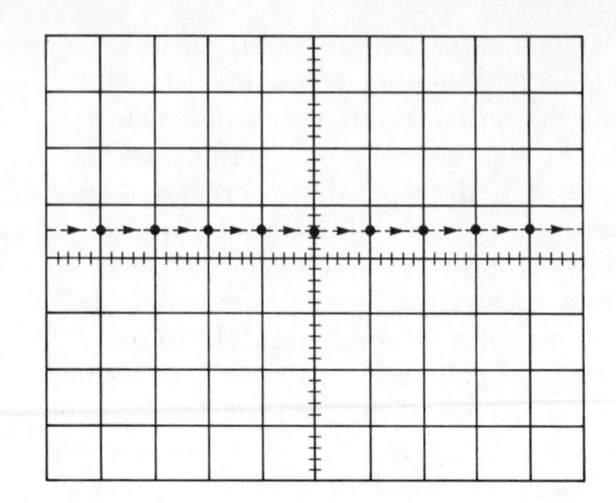

Figure 3.37 *Spot Moves From Left to Right Across Scope Face*

When a time in seconds or decimal fractions of a second is selected by a horizontal time selector setting, the spot then sweeps one division on the graticule horizontal scale in the selected time. For example, if the horizontal time (sweep) selector is set to one second per division, the spot moves from left to right on the scope face at a rate of one major graticule division per second.

In *Figure 3.37* for instance, it would take one second for the spot to move across the scope face from one major graticule mark to another. There are ten major divisions across the graticule. Therefore, with a horizontal setting of one second per division, it would take ten seconds for the spot to move from the left-hand edge of the scope face to its right-hand edge.

Also note in *Figure 3.37* that each major horizontal division is divided into five subdivisions. This is further clarified in *Figure 3.38*. The time interval represented by each subdivision represents two-tenths of the time of one major division. For example, if the horizontal time (sweep) selector is set to one second per division, each subdivision on the horizontal axis represents two-tenths of one second. If, however, the selector is set to two microseconds per division then the spot moves across the screen at the rate of one major graticule division in two microseconds, and each subdivision represents 0.4 microseconds as shown in *Figure 3.39*.

As the sweep time interval per division is decreased by changing the setting of the horizontal time control, the sweep moves more rapidly across the scope face. Eventually, the sweep (the spot) is moving so rapidly that it appears to be a continuous line. Actually, the sweep ends and starts just like on the larger time per division sweep settings, but the motion is too fast for the eye to see.

The selection of horizontal time units may be different on different scopes. With the scope shown in *Figure 3.36*, the trace can be made to move across the screen from a rate of five seconds per division to a rate up to one microsecond per division. The common name for the horizontal time control is horizontal sweep selector, therefore, that name will be used from now on.

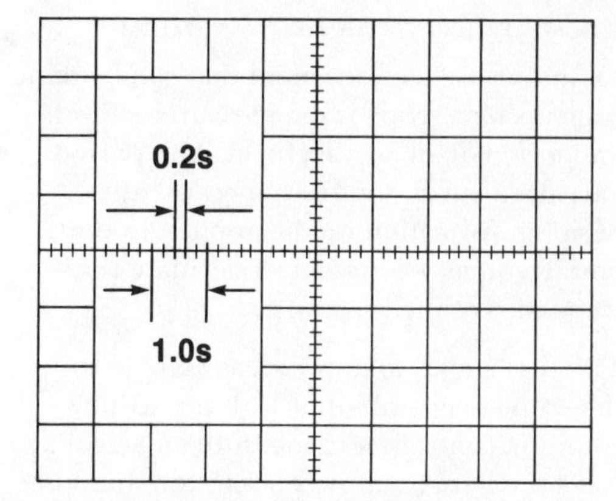

***Figure 3.38** Time Required to Move Spot Across Graticule When Horizontal Control is Set to 1 Second per Division*

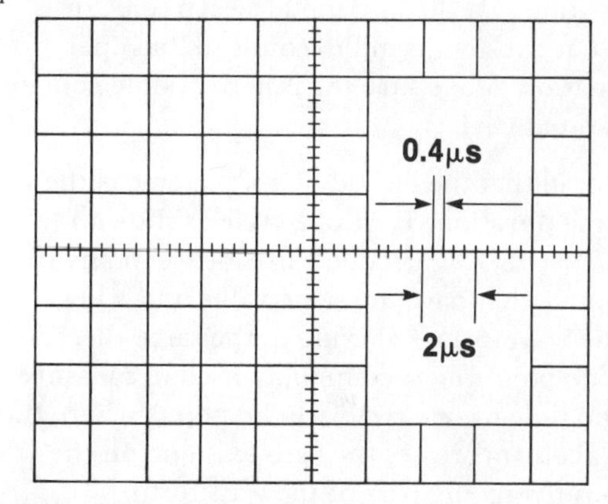

***Figure 3.39** Time Required to Move Spot Across' Graticule When Horizontal Control is Set to 2 Microseconds per Division*

Variable Sweep Time

In the center of the horizontal sweep control is a variable-control knob. This knob is similar to the variable control of the vertical deflection control. For the seconds-per-division control setting to indicate the correct sweep time it must be in the calibrated position, turned fully clockwise as indicated by the arrow. This is the CAL position. This control is in the *CAL* position except when making special-purpose measurements which are outlined in the scope manual.

MEASUREMENT OF AN AC WAVEFORM

An important measurement that can be done easily using a scope is the measurement of the period of an ac waveform. The period measurement is significant because the period information can be applied to the time-frequency equation to calculate the frequency of the waveform.

The first step is to apply an ac voltage to the scope's vertical input with the scope's coupling control set to measure an ac voltage. Next the horizontal sweep selector should be adjusted to provide a display of one or two cycles of the waveform. For this example, assume that the horizontal sweep selector is set at two-tenths milliseconds (0.2ms) per division. Make sure the inner variable control is set at CAL.

Recall that the period of a waveform is the time duration, T, of one cycle as shown in *Figure 3.35*. One cycle exists between any two points that have the same value and where the waveform is varying in the same direction. The points most commonly used to measure the time of one cycle are the points where the waveform crosses the zero axis line on the positive-going part of the waveform.

For this example, the waveform displayed on the face of the scope is as shown in *Figure 3.40*. Two hashmarks have been drawn to indicate one cycle of the waveform. To determine the amount of time between the two hashmarks, the number of graticule divisions between them is counted. There are exactly five divisions between them.

Recall that the scope's horizontal sweep selector is set to two-tenths milliseconds (0.2ms) per division. The period of the waveform is then calculated by multiplying the time per division by the number of divisions:

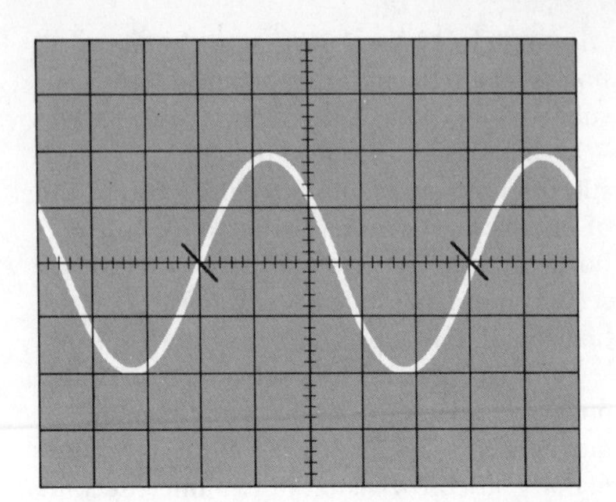

Figure 3.40 *Waveform for Period Calculation Example*

$$\begin{aligned} \mathbf{T} &= \mathbf{(seconds\ per\ division)} \times \\ &\quad \mathbf{(number\ of\ divisions)} \\ &= \mathbf{(0.2ms/division)(5\ divisions)} \\ &= \mathbf{1ms} \end{aligned} \qquad (3\text{–}3)$$

The period of the waveform is one millisecond.

The period of any sinusoidal waveform can be determined in the same way.

Now that the period of the waveform is known, the frequency of the waveform can be calculated easily using this equation:

$$\mathbf{f = \frac{1}{T}} \qquad (3\text{–}4)$$

Since T is one millisecond;

$$\begin{aligned} \mathbf{f} &= \mathbf{\frac{1}{1ms}} \\ &= \mathbf{\frac{1}{0.001}} \\ &= \mathbf{1 \times 10^3} \\ &= \mathbf{1kHz} \end{aligned}$$

The frequency of the waveform is 1 kilohertz.

You have now learned how to use a scope to determine the amplitude and frequency of an ac waveform. Another example will help you practice the technique. We will start with an

unknown waveform and use the scope to determine its amplitude and frequency.

Figure 3.41 shows the unknown waveform as it would appear displayed on the face of a scope when the trigger control is set for internal. The scope's vertical output control is set to 2 volts per division. The scope's horizontal sweep selector is set to 2 milliseconds per division.

The waveform's peak-to-peak amplitude can be determined, as stated previously, one of two ways. In this example, the number of graticule divisions between the top of the positive peak and the bottom of the negative peak will be counted. Then that number and equation *3–1* will be used to calculate the peak-to-peak voltage of the waveform. Thus, since the waveform amplitude covers six divisions, the peak-to-peak voltage is:

$$\begin{aligned} E_{pp} &= (6)(2V) \\ &= 12V_{pp} \end{aligned}$$

The unknown waveform is 12 volts peak-to-peak.

The frequency is calculated by determining the period of the waveform by counting horizontal graticule divisions. In this case, the number of graticule divisions for one period of the waveform is six. The time per division is 2 milliseconds. Using equations *3–3* and *3–4*, the period and the frequency of the waveform can be calculated:

$$\begin{aligned} T &= (6)(2 \times 10^{-3}) \\ &= 12 \times 10^{-3} \text{ seconds} \\ &= 12\text{ms} \end{aligned}$$

and

$$\begin{aligned} f &= \frac{1}{T} \\ &= \frac{1}{12 \times 10^{-3}} \\ &= 0.0833 \times 10^{3} \\ &= 83.3\text{Hz} \end{aligned}$$

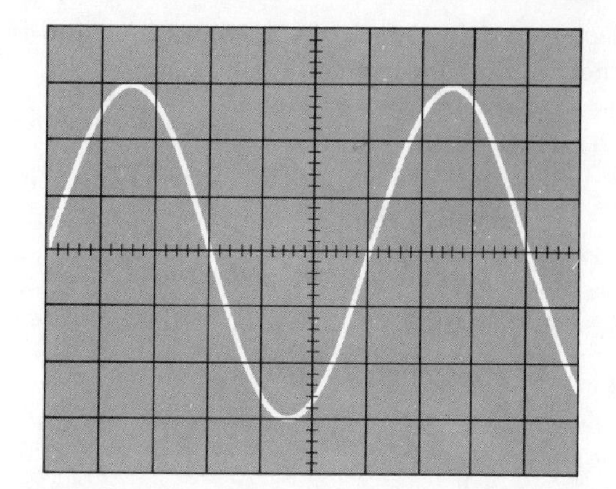

Figure 3.41 Waveform for Amplitude and Frequency Calculation Example

The waveform's frequency is 83.3 hertz.

Therefore, the unknown waveform is a 12 volt peak-to-peak ac voltage with a frequency of 83.3 hertz.

SUMMARY

Discussion in this lesson has tried to familiarize you with one of the most important pieces of electronic measuring equipment — the oscilloscope. The construction of a scope was described. It was explained why the scope is capable of measuring the voltage and time of electrical signals very accurately. The function of a scope's controls and how they must be adjusted, were described, and you were shown how to use a scope to measure and determine amplitude, period and frequency of ac waveforms. In following lessons these qualities will be required of many circuits that will be studied. The scope is a very useful instrument for these measurements.

One other point: there are a great number of different types of scopes in use. A scope is supplied, generally, with an operational instruction manual. If you need any additional information regarding any control on a particular scope you are using, you should refer to the manual for that scope.

Worked-Out Examples

Determine the values of voltage, period, and frequency for the following waveforms shown as they would appear on an oscilloscope.

1. The waveform shown is a dc voltage. Since it is below the OV-reference level, it is a negative voltage:

$$E = (2 \text{ divisions})(5 \text{ V/divisions})$$

$$E = 10V$$

But remember, this is a negative voltage; therefore,

E is a −10V.

A constant dc voltage has no frequency.

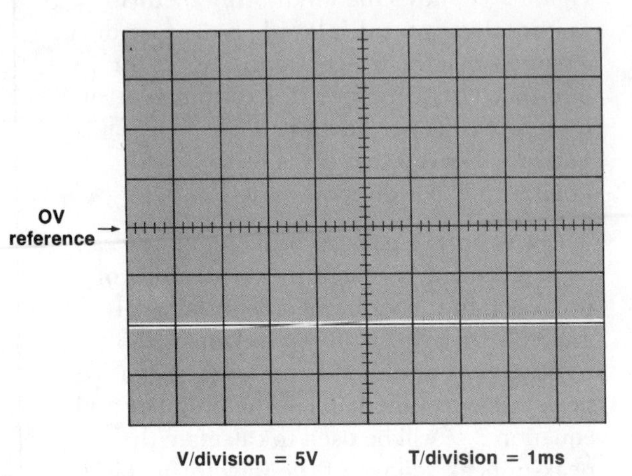

2. The waveform is an ac sine wave. Therefore its amplitude can be specified in terms of peak, peak-to-peak, or rms voltage. From its positive peak value to its negative peak value it spans 6 vertical divisions. Thus its peak-to-peak voltage is

$$E_{pp} = (6 \text{ divisions})(0.2V/\text{divisions})$$

$$E_{pp} = \mathbf{1.2V}$$

Its peak voltage is one-half its peak-to-peak value, so

$$E_{pk} = \tfrac{1}{2}(1.2V) = \mathbf{0.6V}$$

Its rms voltage is 0.707 times its peak voltage value. Therefore,

$$E_{rms} = 0.707(0.6V)$$

$$E_{rms} = \mathbf{0.42V}$$

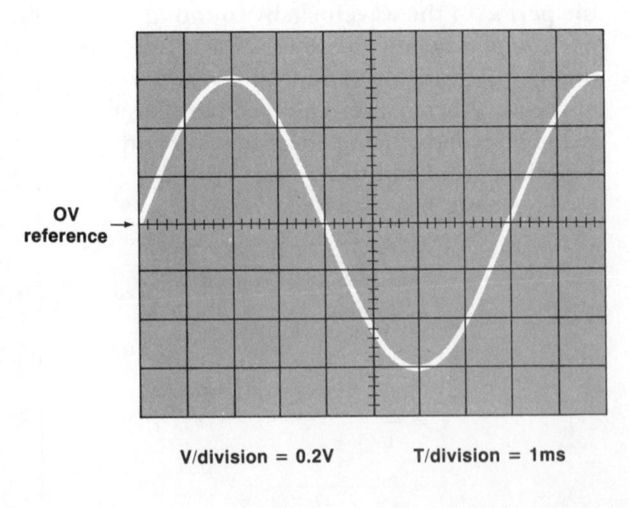

■ Worked-Out Examples

The period of the waveform is contained within 8 horizontal divisions. Thus

$$T = (8 \text{ divisions})(1\text{ms/division})$$

$$T = \mathbf{8\ ms}$$

Knowing the period of the waveform, the frequency can be calculated:

$$f = \frac{1}{T} = \frac{1}{8\text{ms}} = \mathbf{125Hz}$$

3. The waveform is called a square wave. It is like the waveform that is usually used as a calibration waveform for most oscilloscopes. Note that it alternates between a dc voltage level two divisions above the OV-reference level and two divisions below the OV-reference level. Its amplitude is the difference between these two levels:

$$E = (4 \text{ divisions})(2\text{V/division})$$

$$E = \mathbf{8V}$$

This square wave also has a frequency. Its frequency is the rate at which it alternates between its +4-volt and −4-volt levels. The cycle of the waveform exists between the points x and y shown on the graticule. The number of divisions which the cycle spans is four. Therefore, the period of the waveform is

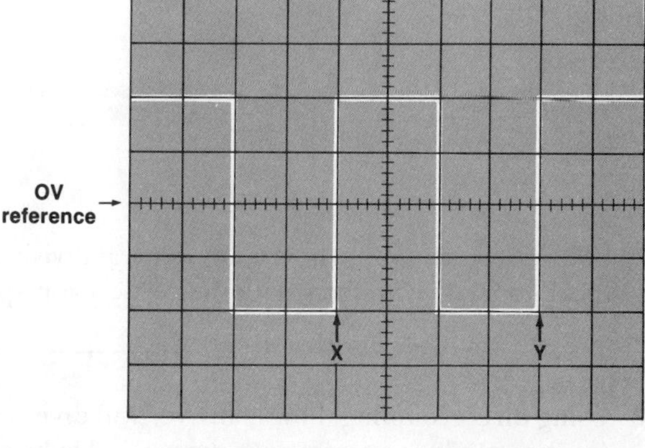

$$T = (4 \text{ divisions})(10\mu\text{s/division})$$

$$T = \mathbf{40\mu s}$$

The frequency of the square wave can now be determined as

$$f = \frac{1}{T} = \frac{1}{40\mu\text{s}} = \mathbf{25kHz}$$

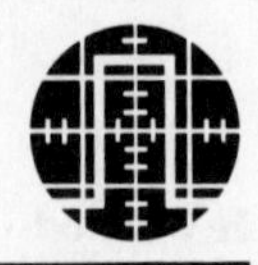

Practice Problems

1. Use the oscilloscope diagram on the following page to identify the following controls and specify the section of the oscilloscope controls in which they are located.

Sections: M = mainframe control group
V = vertical control group
H = horizontal control group (time base)

_______ a. scope ground jack
_______ b. level control
_______ c. external input jack
_______ d. volts/div control
_______ e. intensity control
_______ f. line trigger source control
_______ g. ac/dc coupling control (for signal measured/displayed)
_______ h. input ground control
_______ i. beam finder
_______ j. power switch
_______ k. automatic trigger mode control
_______ l. seconds/division control
_______ m. +/− slope control
_______ n. external trigger source control
_______ o. single sweep/reset mode control
_______ p. graticule
_______ q. focus control
_______ r. input jack (for signal measured/displayed)
_______ s. calibrator
_______ t. up/down position control for trace
_______ u. left/right position control for trace

2. Using direct coupling, how many vertical divisions are required to display an 8-volt (peak-to-peak) waveform with the vertical volts-per-division control set to 2V/division?

number of divisions = ___________

3. Using direct coupling, how many vertical divisions are required to display a 2.5-volt (peak-to-peak) waveform with the vertical volts-per-division control set to 0.2V/division?

number of divisions = ___________

■ Practice Problems

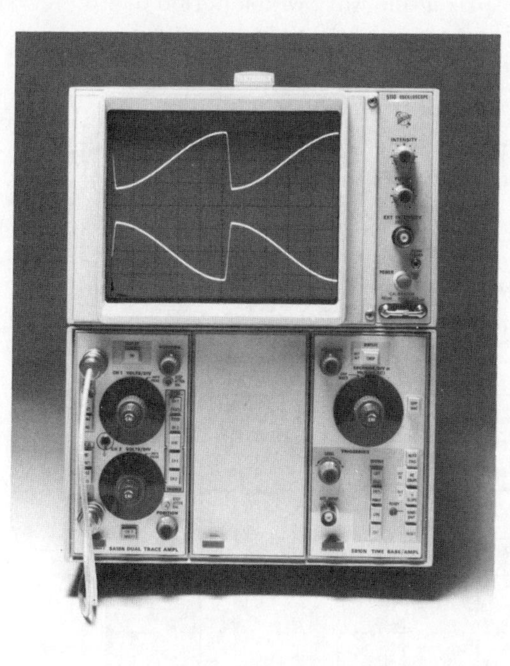

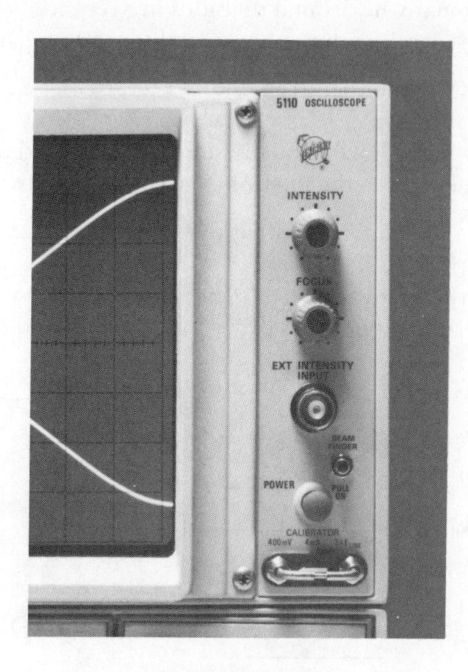

MAINFRAME

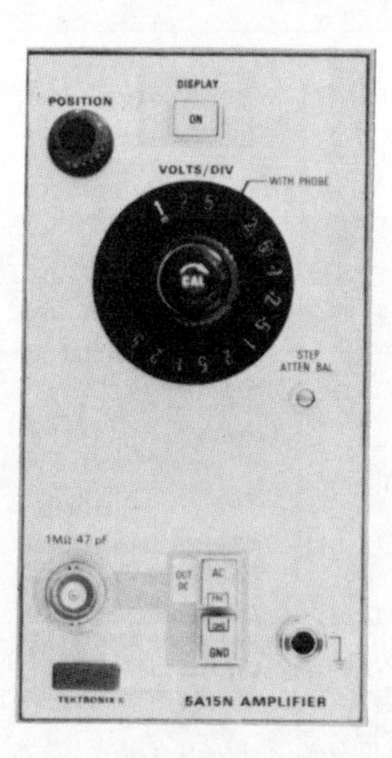

VERTICAL

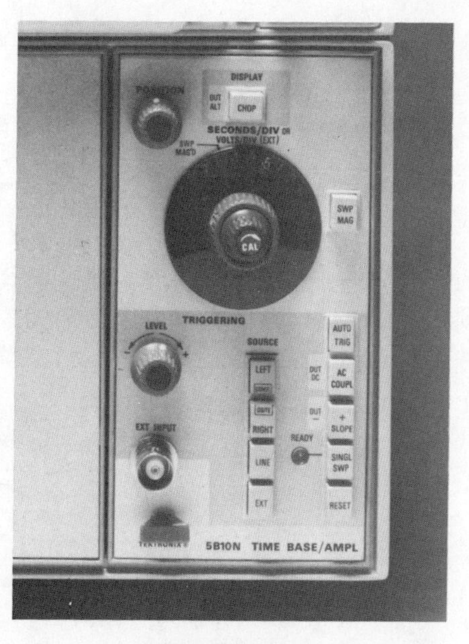

HORIZONTAL

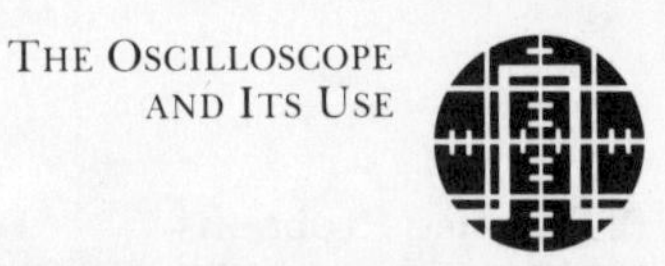

Practice Problems

4. How many horizontal divisions are required to display a sine wave with a period of 1.6 milliseconds with the horizontal seconds-per-division control set to 0.5ms/division?

number of divisions = ______________

5. How many horizontal divisions are required to display a sine wave with a frequency of 5 kilohertz with the horizontal seconds-per-division control set to 0.1ms/division?

number of divisions = ______________

For the following waveforms shown as they would appear on the graticule of an oscilloscope, determine the values of voltage, period, and frequency as specified:

6. V/div = 0.2V

T/div = 2ms

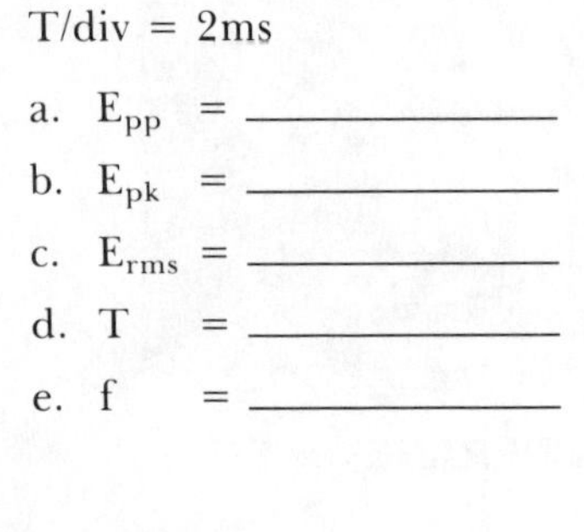

a. E_{pp} = ______________

b. E_{pk} = ______________

c. E_{rms} = ______________

d. T = ______________

e. f = ______________

7. V/div = 5V

T/div = 0.01μs

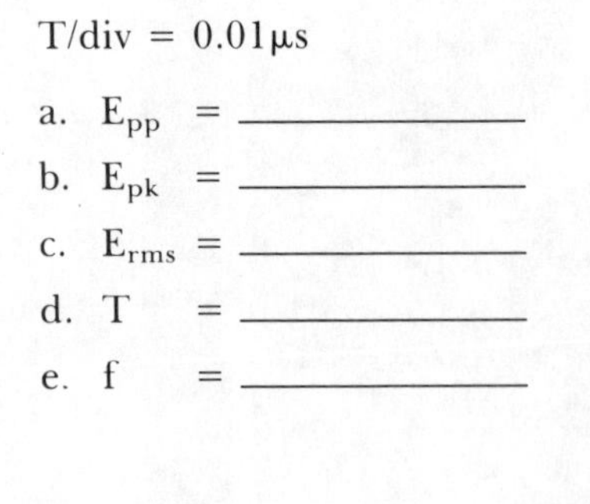

a. E_{pp} = ______________

b. E_{pk} = ______________

c. E_{rms} = ______________

d. T = ______________

e. f = ______________

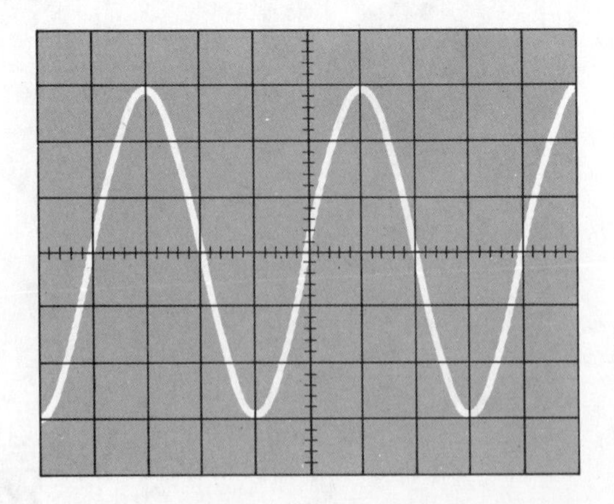

■ Practice Problems

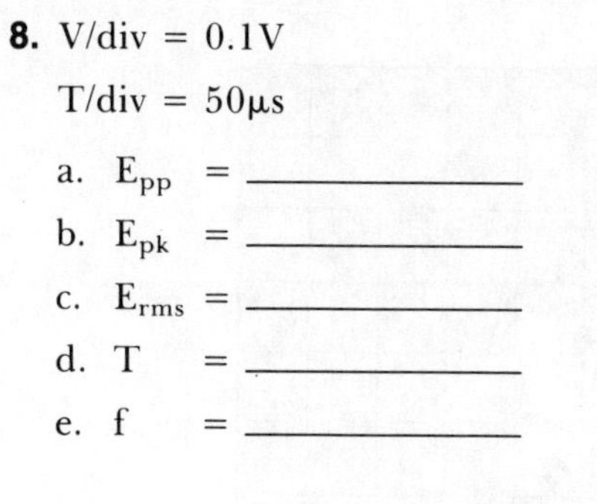

8. V/div = 0.1V

T/div = 50μs

a. E_{pp} = ________________

b. E_{pk} = ________________

c. E_{rms} = ________________

d. T = ________________

e. f = ________________

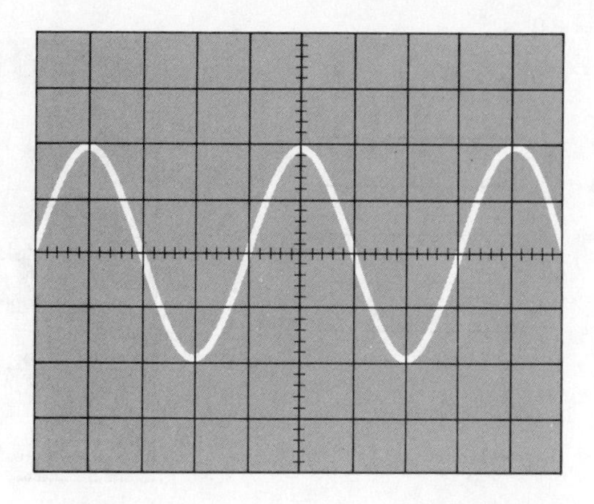

9. V/div = 0.01

T/div = 20μs

a. E_{pp} = ________________

b. E_{pk} = ________________

c. E_{rms} = ________________

d. T = ________________

e. f = ________________

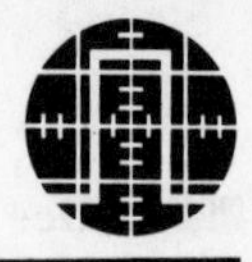

■ Practice Problems

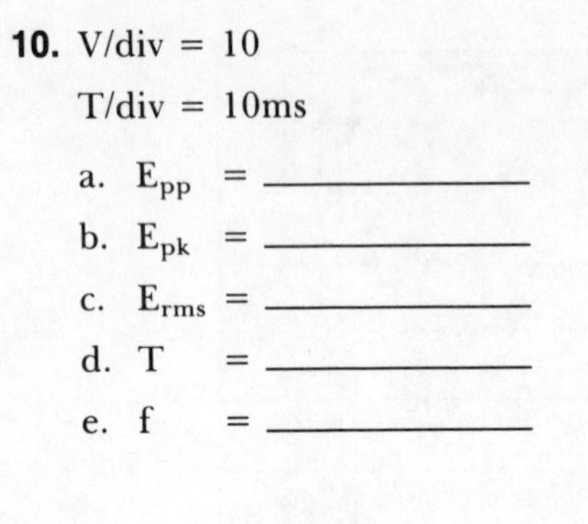

10. V/div = 10

T/div = 10ms

a. E_{pp} = ____________

b. E_{pk} = ____________

c. E_{rms} = ____________

d. T = ____________

e. f = ____________

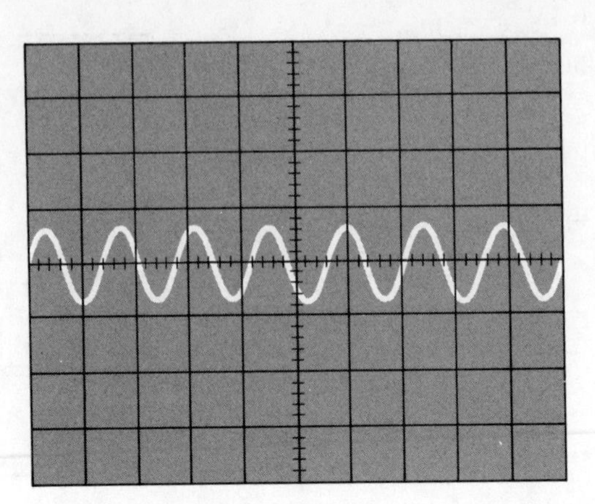

■ Quiz

1. Using the oscilloscope diagram on the following page, identify the location of each of the controls listed below by placing the letter designator of the control in the space beside the name of the control:

__P__ a. level
__A__ b. intensity
__G__ c. vertical position control
__N__ d. horizontal position control
__H__ e. vertical volts/division control
__O__ f. horizontal seconds/division control
__Q__ g. external input jack
__M__ h. vertical input ground control
__K__ i. vertical input ac/dc coupling control
__J__ j. variable volts/division control
__C__ k. beam finder
__E__ l. calibrator loop
__S__ m. variable seconds/division control
__R__ n. line triggering source control
__X__ o. external triggering source control
__F__ p. graticule
__D__ q. power switch
__T__ r. automatic triggering mode control
__B__ s. focus control
__U__ t. +/− slope control
__V__ u. single sweep triggering mode control
__W__ v. reset (for single sweep) control
__I__ w. vertical input jack
__L__ x. oscilloscope ground jack

The following 20 questions are multiple choice; circle the letter of the most-correct answer.

2. The three basic functions that an oscilloscope performs are

a. waveform observation, current measurement, frequency measurement
b. frequency measurement, voltage measurement, period measurement
c. waveform observation, voltage measurement, time measurement
d. period measurement, power measurement, current measurement

3. The two major segments of the oscilloscope are the

a. CRT and electron gun
b. screen and vertical section
c. CRT and controlling circuits
d. vertical and mainframe circuits

4. The CRT is composed of three major parts:

a. screen, deflection system, and screen coating
b. electron gun, deflection system, and screen
c. electron gun, cathode, and grid
d. anode, cathode, and pentode

5. The controlling circuits are divided into three major groups:

a. mainframe, underframe, and overframe control groups
b. mainframe, vertical, and CRT
c. mainframe, vertical, and horizontal
d. front, back, and side

6. The type of deflection employed by most oscilloscopes is

a. magnetic
b. electrostatic
c. yoke
d. both a and c

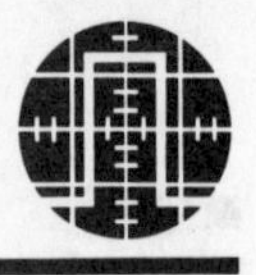

■ **Quiz**

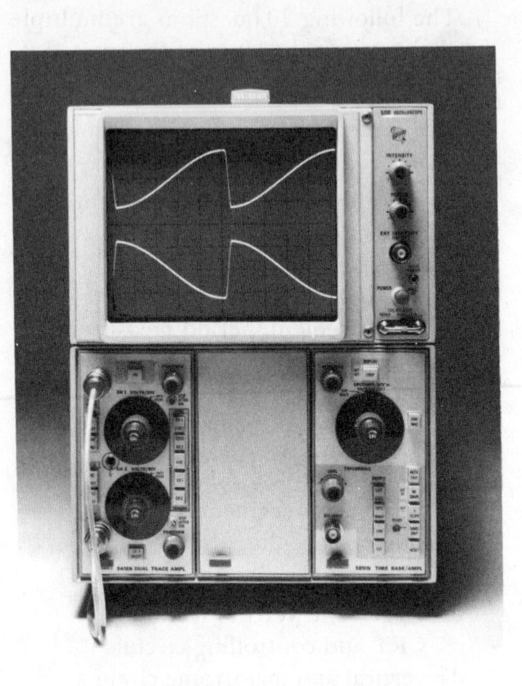

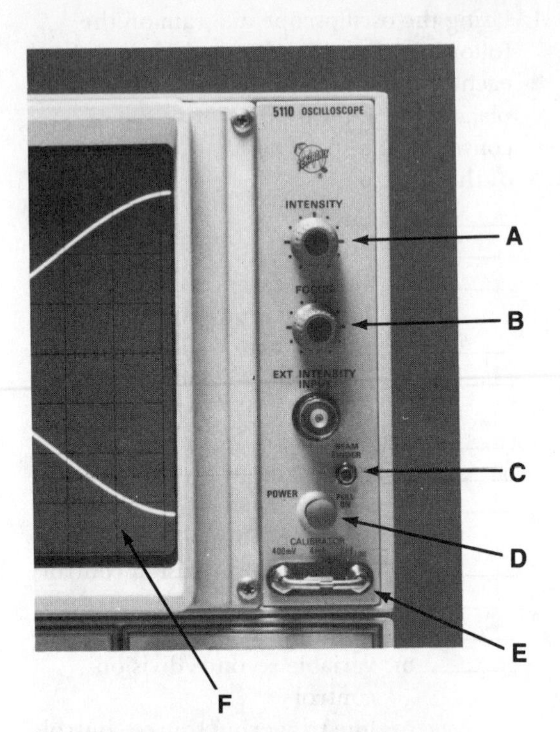
5110 OSCILLOSCOPE
INTENSITY
FOCUS
EXT INTENSITY INPUT
POWER
CALIBRATOR
A
B
C
D
E
F

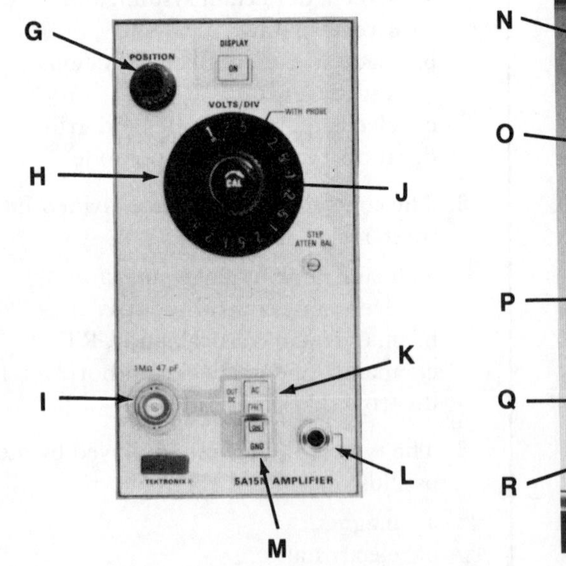
G
POSITION
DISPLAY
ON
VOLTS/DIV
WITH PROBE
CAL
STEP ATTEN BAL
H
J
1MΩ 47 pF
I
OUT DC
AC
GND
K
L
M
TEKTRONIX
5A15N AMPLIFIER

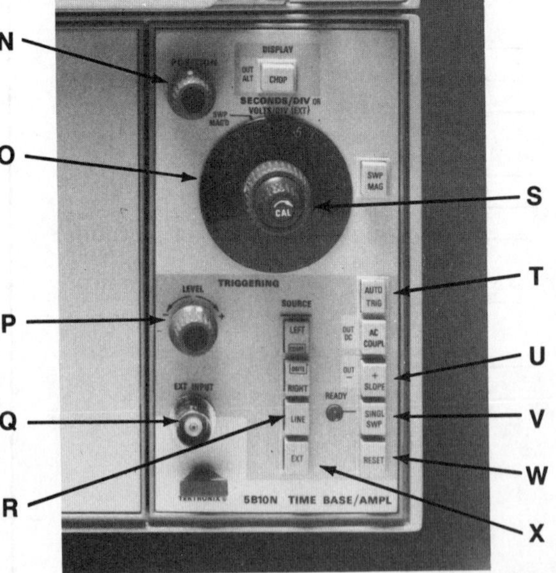
N
POSITION
DISPLAY
OUT ALT
CHOP
SECONDS/DIV OR VOLTS/DIV (EXT)
SWP MAG'D
O
S
SWP MAG
CAL
TRIGGERING
LEVEL
SOURCE
LEFT
RIGHT
LINE
EXT
EXT INPUT
AUTO TRIG
OUT DC
AC COUPL
OUT
+ SLOPE
READY
SINGL SWP
RESET
P
Q
R
T
U
V
W
X
TEKTRONIX
5B10N TIME BASE/AMPL

■ Quiz

7. The ability of a material to emit light after being struck by electrons thus enabling you to see the location of the electron beam on the screen of the oscilloscope is called

a. phosphorescence
b. luminance
c. lasing
d. ionization

8. The purpose of the power switch is to control power

a. to the circuit under test
b. only to the CRT
c. to all scope circuits
d. none of the above

9. The purpose of the intensity control is to adjust the

a. graticule illumination
b. brilliance of the electron trace
c. current to the circuit under test
d. calibrator output

10. The purpose of the focus control is to focus the

a. scale illumination
b. electron trace to a spot or line
c. graticule markings
d. output pulses of the calibrator

11. The purpose of the beam finder control is to

a. automatically reposition the trace
b. turn off the scope power until the beam is found
c. reduce deflection potentials so that the trace appears
d. none of the above

12. The purpose of the calibrator is to

a. supply a square wave signal of known amplitude and frequency for checking the scope's accuracy (i.e. calibration)
b. supply an appropriate signal used to self-calibrate the scope
c. automatically calibrate the scope each time it is turned on
d. b and c

13. Each small mark on the center vertical and horizontal graticule lines is

a. 1/10 of a major division
b. 1/2 of a major division
c. 1/4 of a major division
d. 2/10 of a major division

14. The purpose of the vertical position control is to control the

a. vertical position of the trace
b. vertical attitude of the scope
c. position of the vertical attenuator knob
d. none of the above

15. The purpose of the vertical attenuator (volts/division control) is to

a. control the level of the signal from the oscilloscope vertical circuits to the circuit under test
b. attenuate the signal to the triggering circuits
c. select the number of volts each vertical graticule major division represents
d. none of the above

■ Quiz

16. The purpose of the ac/dc vertical input coupling control is to

a. turn on an ac or dc voltage when switched to the corresponding position
(b.) eliminate the dc component of a waveform when in the ac position
c. eliminate the ac component of a waveform when in the dc position
d. none of the above

17. The purpose of the ground vertical input (pushbutton) control is to

a. ground the chassis of the scope to earth ground
b. remove the input signal from the vertical deflection circuits
(c.) apply ground potential (zero volts) to the vertical deflection circuits in order to set the trace to zero or reference position before taking measurements
→ d. both b and c

18. The purpose of the horizontal position control is to control the horizontal position of the

a. time/division knob
(b.) trace
c. scope
d. graticule markings

19. The purpose of the horizontal seconds/division control is to

(a.) set the number of seconds/division each major horizontal graticule mark represents
b. set the speed at which the trace (electron beam) moves from right to left across the face (graticule) of the scope
c. assist in determining the period and hence frequency of a measured waveform
→ d. all of the above

20. The purpose of the triggering controls is to select when

a. the scope will reset
b. the trace will be blanked (disappear)
(c.) the sweep will begin
d. all of the above

21. For most general-purpose measurements, the triggering method to use is to set the scope controls on

a. non-automatic, line triggering
b. automatic, external triggering
(c.) internal, automatic triggering
d. external, line triggering

22. Using direct coupling, how many divisions (vertically and horizontally) are required to display one complete cycle of a 3-volt (peak-to-peak), 40 kilohertz waveform with the vertical volts/division control set to 0.5V/division and the horizontal seconds/division control set to 5μs division?

number of vertical divisions = 6
number of horizontal divisions = 5

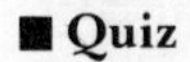

■ Quiz

23. V/div = 0.5
T/div = 10ms

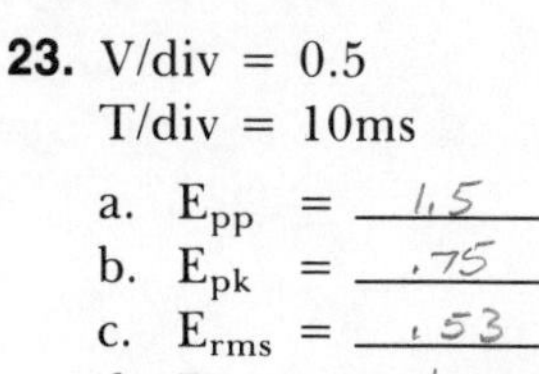

a. E_{pp} = 1.5
b. E_{pk} = .75
c. E_{rms} = .53
d. T = 40 ms
e. f = 25 Hz

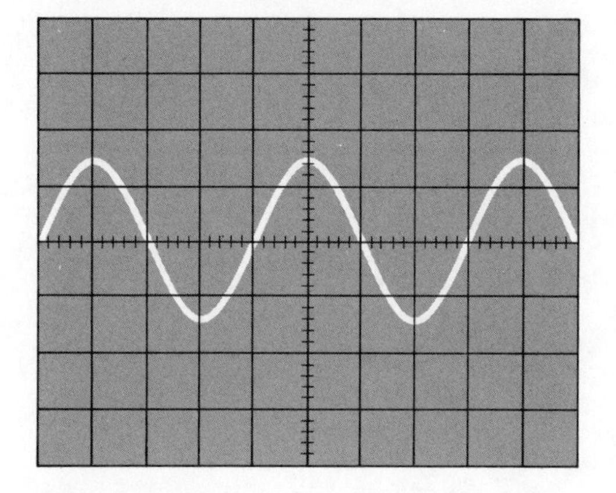

24. V/div = 0.2
T/div = 2μs

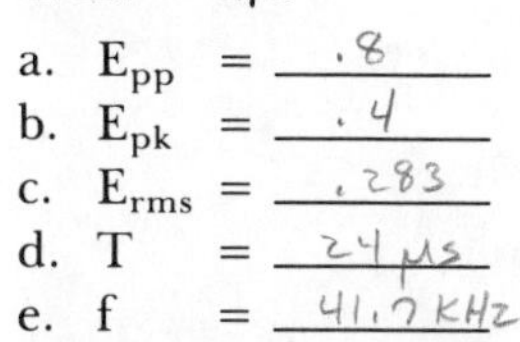

a. E_{pp} = .8
b. E_{pk} = .4
c. E_{rms} = .283
d. T = 24 μs
e. f = 41.7 KHz

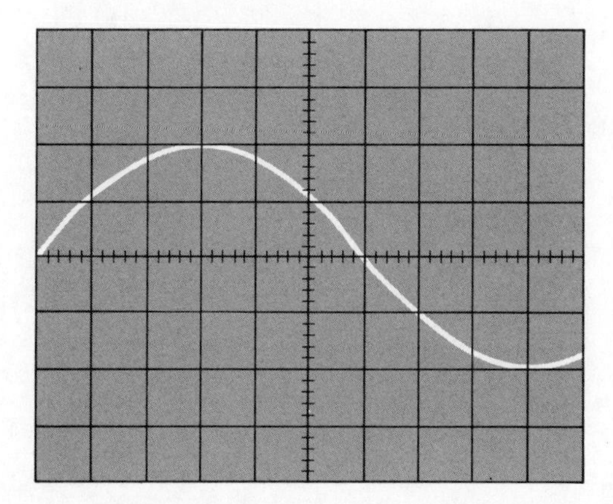

25. V/div = 2
T/div = 0.2ms

a. E_{pp} = 4.8
b. E_{pk} = 2.4
c. E_{rms} = 1.7
d. T = 560 μs
e. f = 1.79 kHz

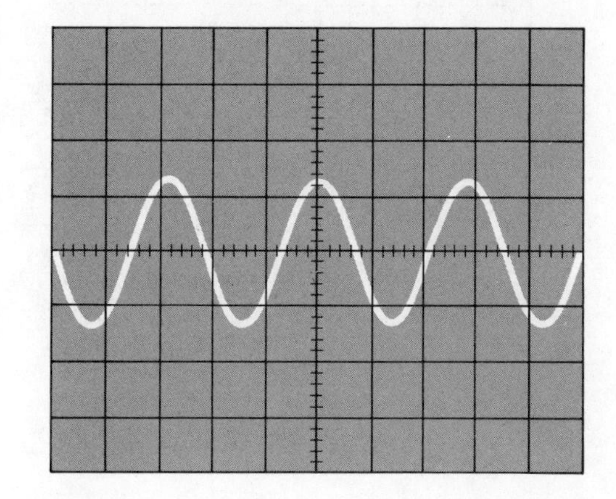

26. V/div = 5
T/div = 1ms

a. E_{pp} = 10
b. E_{pk} = 5
c. E_{rms} = 3.54
d. T = 8.6 ms
e. f = 116 Hz

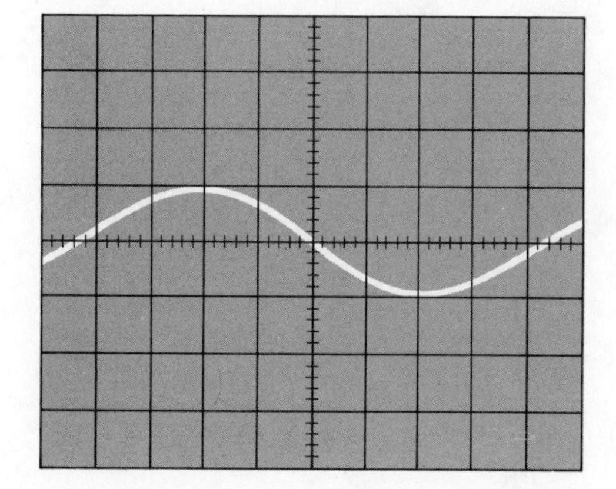

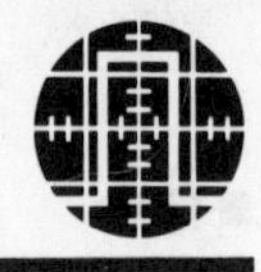

■ Quiz

27. V/div = 10
T/div = 0.5ms
E = 25

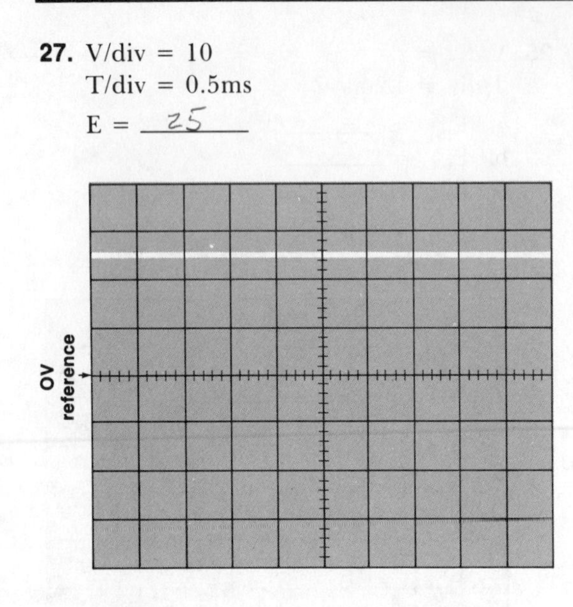

LESSON 4

The Sine Wave and Phase

In this lesson, the use of the sine function to determine instantaneous voltage and current values of a sine wave is discussed. The relationship between the period of a sinusoidal waveform and the electrical degrees of a cycle is described. The radian as a unit of angular measure is introduced. It is explained how to determine and specify the phase relationships of sinusoidal waveforms and the use of phasor notation is described.

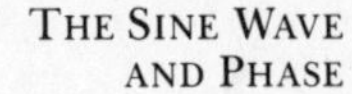

■ Objectives

At the end of this lesson you should be able to:

1. Determine the instantaneous value of a waveform at a specified angle using the equation $e_i = E_{pk}\sin\theta$ when given the amplitude of a sinusoidal waveform and a number of degrees into the waveform cycle.

2. Determine the instantaneous value of a waveform at a specified time using the equation $e_i = E_{pk}\sin(360ft)$ when given the amplitude and frequency of a sinusoidal waveform and an elapsed time into the waveform.

3. Plot a sine wave having a specified amplitude by using the equation $e_i = E_{pk}\sin\theta$ using 10° or 15° intervals.

4. Define the term sinusoidal.

5. Define the term non-sinusoidal.

6. Identify waveforms as being either sinusoidal or non-sinusoidal.

7. Express an angle in radians when given its value in degrees.

8. Express an angle in degrees when given its value in radians.

9. Specify the phase relationship of two sinusoidal waveforms by stating the lead or lag difference of one waveform from the other and the angular difference between the two waveforms.

10. Given a pair of sinusoidal waveforms, represent each pair by an equivalent phasor diagram showing both phase relationships and magnitudes.

11. Sketch the sinusoidal waveform representation of two sinusoidal waveforms showing correct amplitudes and phase relationships when given a phasor diagram of the two waveforms.

12. Specify the lead or lag difference and the angular difference of two sinusoidal waveforms when given a phasor diagram of the two waveforms.

■ Sine Wave Instantaneous Values

INTRODUCTION

In the last lesson discussion departed from the sine function and its relation to the sinusoidal ac waveform to investigate the practical operation and use of the oscilloscope. In Lesson 2 the relationship between the period and frequency of a periodic ac waveform was discussed and some practical examples were analyzed. Three ways in which a sinusoidal ac voltage's amplitude may be specified were introduced: in terms of its peak, peak-to-peak or RMS amplitude. The concept of trigonometric functions and specifically, the sine function, was introduced. The use of the table of trigonometric functions and how to determine the value of the sine function for any angle 0° to 360° were explained.

INSTANTANEOUS VALUES OF ELECTRICAL DEGREE POINTS

In Lesson 2 the concept of *instantaneous voltage* was introduced briefly during an explanation of the 0.707 constant. The concept of instantaneous voltage is vital to an understanding of an ac waveform. Therefore, this section describes in detail exactly what is meant by the instantaneous voltage of a sinusoidal waveform.

Note that the sine wave shown in *Figure 4.1* has a peak amplitude of 10 volts. If the voltage is measured at various instantaneous points throughout the cycle, however, it will be found that the voltage is not always 10 volts. At the beginning of the cycle, the voltage is zero volts. As the cycle progresses, the value of the voltage increases until one-quarter of the way through the cycle, at 90°, the voltage is at its maximum positive value, at its peak of 10 volts.

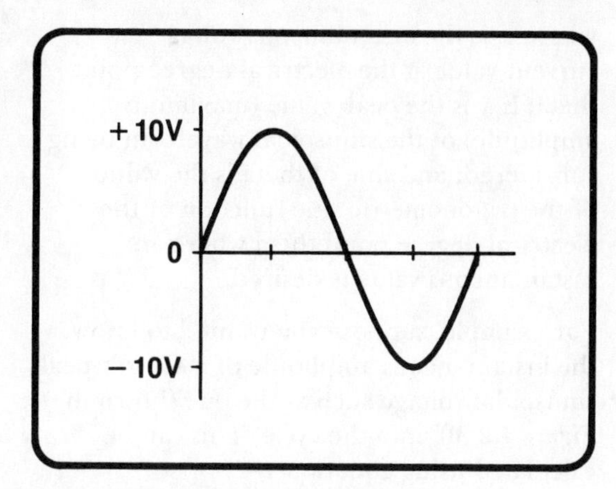

Figure 4.1 *A 10V_{pk} AC Waveform*

Obviously, the voltage when going from zero volts to 10 volts had to increase through many instantaneous values. They are called instantaneous because they are not constant. They are only momentary. They are that particular value only for an *instant.*

Past 90°, the instantaneous value of the voltage decreases until at 180° through the cycle, it is zero volts again. The negative alternation or half-cycle is simply a mirror image of the positive half-cycle. Recall that instantaneous values of the negative half-cycle, however, are said to be negative voltages since the polarity of this cycle is opposite that of the first half-cycle.

Sine Wave Instantaneous Values

The relationship of the sinusoidal waveform to the trigonometric sine function is useful in determining the instantaneous value of a sinusoidal voltage or current waveform at any electrical degree point. The relationship of instantaneous voltage values to the sine function is expressed mathematically by equation *4–1,*

$$\mathbf{e_i = E_{pk} sin\theta} \qquad (4\text{–}1)$$

where e_i is the instantaneous voltage or current value at the electrical degree point theta; E_{pk} is the peak value (maximum amplitude) of the sinusoidal waveform being considered; and sine of theta is the value of the trigonometric sine function of the electrical degree point theta where an instantaneous value is desired.

For example, suppose you wanted to know the instantaneous amplitude of a 40-volt peak sinusoidal voltage such as the one shown in *Figure 4.2* 30° into the cycle. This can be calculated using equation *4–1:*

$$\begin{aligned} e_i &= E_{pk}\sin\theta \\ &= 40V\sin 30° \\ &= 40V(0.5) \\ &= 20V \end{aligned}$$

The instantaneous voltage at 30° is 20 volts.

This calculation was comparatively simple since theta was between 0° and 90°, and therefore, the sine of theta could easily be determined from almost any trigonometric table.

Angles Greater Than 90°

Suppose, however, the angle theta at which you want to determine the instantaneous voltage is an angle greater than 90°. To find the sine of theta then, you must first determine the first quadrant equivalent angle as discussed in Lesson 2. The chart for determining equivalent angles from Lesson 2 is shown again in *Figure 4.3*.

For example, suppose you wanted to know the instantaneous voltage amplitude of a 50-volt peak waveform at 200° into the cycle as illustrated in *Figure 4.4*. It is calculated:

$$\begin{aligned} e_i &= E_{pk}\sin\theta \\ &= 50V\sin 200° \end{aligned}$$

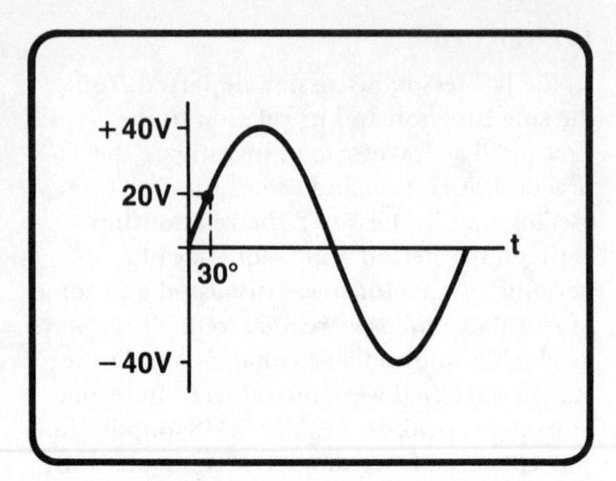

Figure 4.2 *A 40V$_{pk}$ AC Waveform*

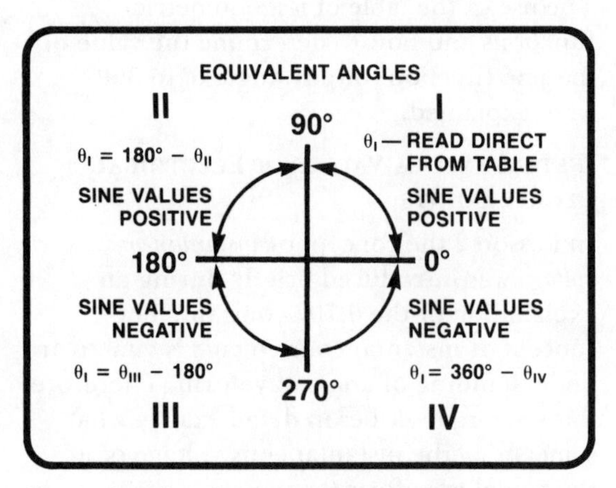

Figure 4.3 *Chart for Determining Equivalent Angles*

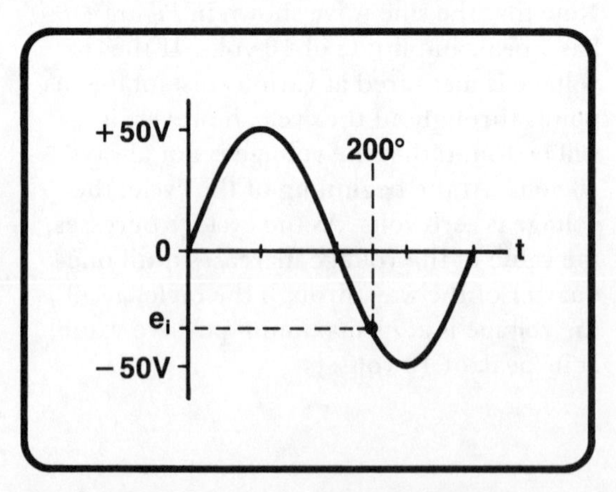

Figure 4.4 *200° Point of a 50V$_{pk}$ AC Waveform*

200° is in the third quadrant. It is in the first-half of the negative half-cycle of the waveform. From *Figure 4.3*, to find the first quadrant equivalent (or first-half of the *positive* half-cycle equivalent) angle of 200°, 180° must be subtracted from the angle 200°. This yields a result of 20° as follows:

$$\theta_I = \theta_{III} - 180° = 200° - 180° = 20°$$

This 20° is the angle of which you must find the sine.

$$e_i = E_{pk}\sin\theta = 50V\sin 20° = 50V(0.3420) = 17.1V$$

But this value of voltage lies in the negative half-cycle, and therefore, it is a negative value of voltage. Thus, the instantaneous voltage of the 50-volt peak sinusoidal waveform 200° into the cycle is −17.1 volts. Remember, instantaneous values for angles from 180° to 360° will be negative as shown in *Figure 4.5*.

Plotting a Sine Wave—Electrical Degrees

Since equation *4-1* provides instantaneous values of a sinusoidal waveform, the entire waveform can be plotted using this equation. For example, suppose you wanted to plot a sinusoidal waveform of 20-volts peak amplitude versus time. Using the equation *4–1*,

$$e_i = 20V\sin\theta$$

the instantaneous value of voltage at any value of theta into the cycle can be determined.

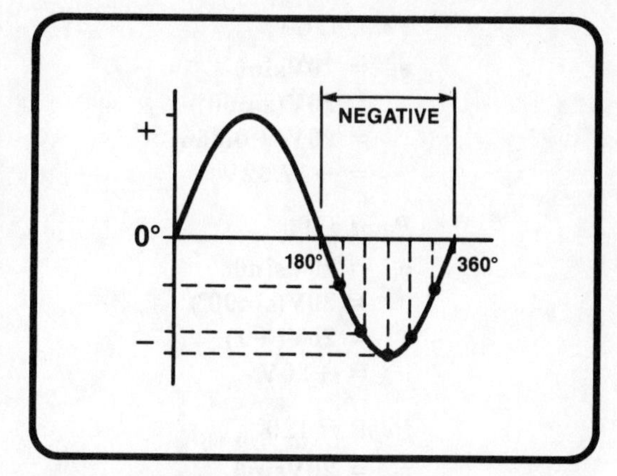

Figure 4.5 *Instantaneous Values for Angles from 180° to 360° are Negative*

To help you understand this concept, the instantaneous values for a 20-volt peak waveform will be calculated at 30° waveform increments throughout the cycle beginning at 0°. This will be very similar to the way the sine function was plotted in Lesson 2. When the electrical degrees of the cycle progress beyond 90° then equivalent quadrant I angles must be calculated as shown in the following calculations. Either a 0° - 90° trigonometric table or a calculator with trigonometric functions can be used to determine sine theta when performing the calculations.

Point 1: 0°

$$e_i = 20V\sin\theta = 20V(\sin 0°) = 20V(0) = 0V$$

Point 2: 30°

$$e_i = 20V\sin\theta = 20V(\sin 30°) = 20V(+0.5) = 10V$$

■ Plotting a Sine Wave—Electrical Degrees

Point 3: 60°

$$\mathbf{e_i = 20V\sin\theta = 20V(\sin 60°) = 20V(+0.866) = +17.32V}$$

Point 4: 90°

$$\mathbf{e_i = 20V\sin\theta = 20V(\sin 90°) = 20V(+1) = +20V}$$

Point 5: 120°

$$\mathbf{e_i = 20V\sin\theta = 20V(\sin 120°) = 20V(\sin 60°)^* = 20V(+0.866) = +17.32V}$$

$$\mathbf{^*\theta_I = 180° - \theta_{II} = 180° - 120° = 60°}$$

Point 6: 150°

$$\mathbf{e_i = 20V\sin\theta = 20V(\sin 150°) = 20V(\sin 30°)^* = 20V(+0.5) = +10V}$$

$$^*\theta_I = 180° - \theta_{II} = 180° - 150° = 30°$$

Point 7: 180°

$$\mathbf{e_i = 20V\sin\theta = 20V(\sin 180°) = 20V(\sin 0°)^* = 20V(0) = 0V}$$

$$^*\theta_I = 180° - \theta_{II} = 180° - 180° = 0°$$

Point 8: 210°

$$e_i = 20V\sin\theta = 20V(\sin 210°) = 20V(\sin - 30°)^* = 20V(-0.5)^{**} = -10V$$

$$^*\theta_I = -(\theta_{III} - 180°) = -(210° - 180°) = -30°$$

**Note:

$$\sin(-\theta) = -\sin\theta$$

i.e.

$$\sin(-30°) = -(\sin 30°) = -0.5$$

Point 9: 240°

$$e_i = 20V\sin\theta = 20V(\sin 240°) = 20V(\sin - 60°)^* = 20V(-0.866) = -17.3V$$

$$^*\theta_I = -(\theta_{III} - 180°) = -(240° - 180°) = -60°$$

Point 10: 270°

$$e_i = 20V\sin\theta = 20V(\sin 270°) = 20V(\sin - 90°)^* = 20V(-1) = -20V$$

$$^*\theta_I = (\theta_{III} - 180°) = -(270° - 180°) = -90°$$

*First quadrant equivalent angle.

Point 11: 300°

$$\begin{aligned} e_i &= 20V\sin\theta \\ &= 20V(\sin 300°) \\ &= 20V(\sin -60°)^* \\ &= 20V(-0.866) \\ &= -17.3V \end{aligned}$$

$$\begin{aligned} ^*\theta_I &= -(360° - \theta_{IV}) \\ &= -(360° - 300°) \\ &= -60° \end{aligned}$$

Point 12: 330°

$$\begin{aligned} e_i &= 20V\sin\theta \\ &= 20V(\sin 330°) \\ &= 20V(\sin - 30°)^* \\ &= 20V(-0.5) \\ &= -10V \end{aligned}$$

$$\begin{aligned} ^*\theta_I &= -(360° - \theta_{IV}) \\ &= -(360° - 330°) \\ &= -30° \end{aligned}$$

Point 13: 360°

$$\begin{aligned} e_i &= 20V\sin\theta \\ &= 20V(\sin 360°) \\ &= 20V(\sin 0°)^* \\ &= 20V(0) \\ &= 0V \end{aligned}$$

$$\begin{aligned} \theta_I &= -(360° - \theta_{IV}) \\ &= -(360° - 360°) \\ &= -(0°) \\ &= 0° \end{aligned}$$

*First quadrant equivalent angle.

All of these calculated instantaneous voltage values can be plotted in graph-form as shown in *Figure 4.6*. The electrical degrees are plotted on the X axis. If these points are connected with a smooth curve, the result is a sinusoidal waveform with a peak amplitude of 20 volts. Obviously, this is the same waveform as was plotted for the sine function in Lesson 2 modified, of course, by the 20 volts amplitude.

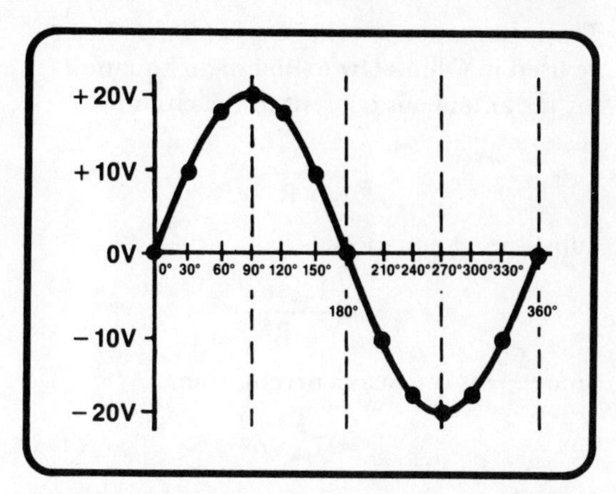

Figure 4.6 *Instantaneous Voltage Values of a 20-V_{pk} Sine Wave at 30° Increments*

Instantaneous Current

The current that flows as a result of this applied sinusoidal voltage will also be a sinusoidal waveform. The peak value of the current can be calculated. The peak value of current in a resistive circuit is simply the peak value of voltage divided by the resistance in the circuit:

$$I_{pk} = \frac{E_{pk}}{R} \qquad (4\text{–}2)$$

Using equation *4–2*, if the 20-volt peak waveform is applied to a resistance of 10-ohms as shown in *Figure 4.7a*, a peak current of 2 amperes will be produced. The calculation is:

$$\begin{aligned} I_{pk} &= \frac{E_{pk}}{R} \\ &= \frac{20V}{10\Omega} \\ &= 2A \end{aligned}$$

The waveform of this current is shown graphically in *Figure 4.7b*.

Sinusoidal Versus Non-Sinusoidal Waveform

The instantaneous voltage equation *4–1* can be used in Ohm's law to obtain an equation for instantaneous current in the circuit.

$$i_i = \frac{e_i}{R}$$

Substituting equation *4–1,*

$$i_i = \frac{E_{pk}\sin\theta}{R}$$

Since $\frac{E_{pk}}{R}$ is the peak current, then

$$\mathbf{i_i = I_{pk}\sin\theta} \qquad (4\text{–}3)$$

Note this equation is similar to the equation used to calculate the value of voltage. Therefore, the instantaneous value of current at any electrical degree point of the current waveform can be calculated in the same way an instantaneous voltage value is calculated.

For example, if the instantaneous current value of this 2-ampere waveform at 30° is desired as shown in *Figure 4.8,* the known values can be substituted in equation *4–3* to calculate i_i:

$$\begin{aligned} \mathbf{i_i} &= \mathbf{I_{pk}\sin\theta} \\ &= \mathbf{2A\sin 30°} \\ &= \mathbf{2(0.5)} \\ &= \mathbf{1A} \end{aligned}$$

The instantaneous current value at the 30° point of the current waveform is 1 ampere.

Sinusoidal Versus Non-Sinusoidal Waveform

As expressed previously in Lesson 2, a *sinusoidal waveform* can be defined as any waveform that may be expressed mathematically by using the sine function. The sinusoidal waveform always has the same general appearance shown in *Figure 4.9.*

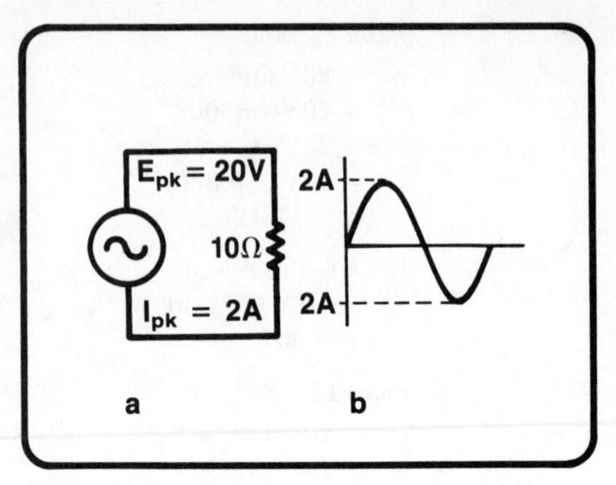

Figure 4.7. a. AC Circuit; b. Current Waveform

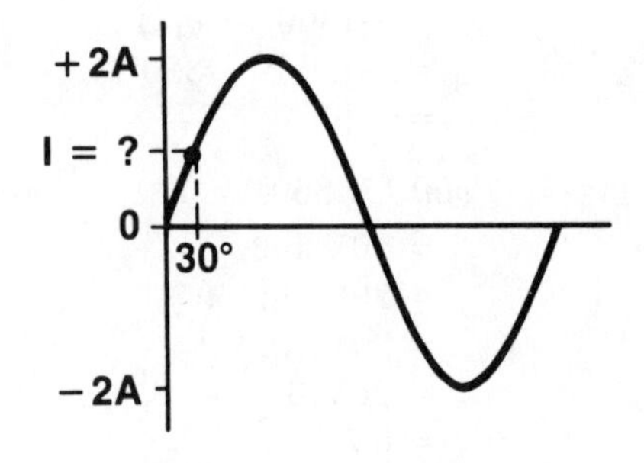

Figure 4.8 Instantaneous Value of Current 30° Point of Current Waveform

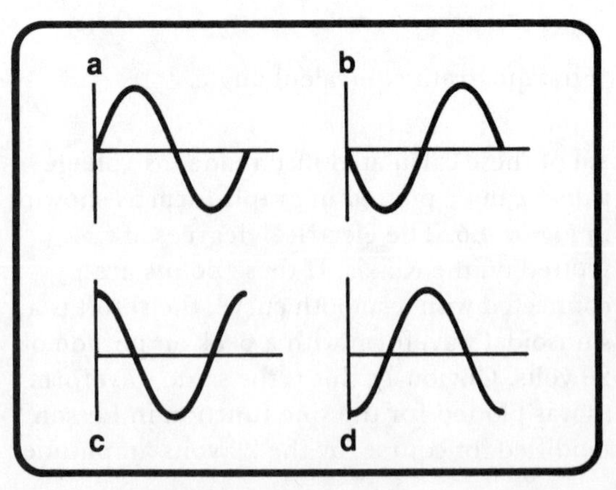

Figure 4.9 Sinusoidal Waveforms

There are also, however, *non-sinusoidal waveforms*. As its name implies, a non-sinusoidal waveform is any waveform that cannot be expressed mathematically by the sine function. For example, the waveforms shown in *Figure 4.10* are non-sinusoidal. Each of those waveforms has a distinctly different shape than that of the sine wave.

INSTANTANEOUS VALUES AT TIME INTERVALS

So far in this lesson, you have seen how an instantaneous voltage or current value of a sine wave can be determined at any degree point in its cycle using equations *4–1* and *4–3*. Therefore, if you know E_{pk}, you can determine e_i at any number of degrees, θ, into the cycle.

Now, suppose, however, it is necessary to calculate e_i at some *time interval* into the cycle. That too, can be done. Previously, you were told that every periodic waveform such as the one shown in *Figure 4.11* has a period which is related to its frequency. The equation which states this relationship is:

$$f = \frac{1}{T} \qquad (4\text{–}4)$$

Modifying the $e_i = E_{pk}\sin\theta$ Equation

Recall that the period, T, of a sinusoidal waveform is the time duration of one cycle, and that a cycle is the result of a conductor traveling in a circular path through 360 electrical degrees. Therefore, 360 electrical degrees is equivalent to the period of a cycle as shown in *Figure 4.11*. Moreover, as illustrated in *Figure 4.12* an amount of time, t, less than the period of a cycle, T, can be equated with a number of electrical degrees, θ, less than 360 degrees.

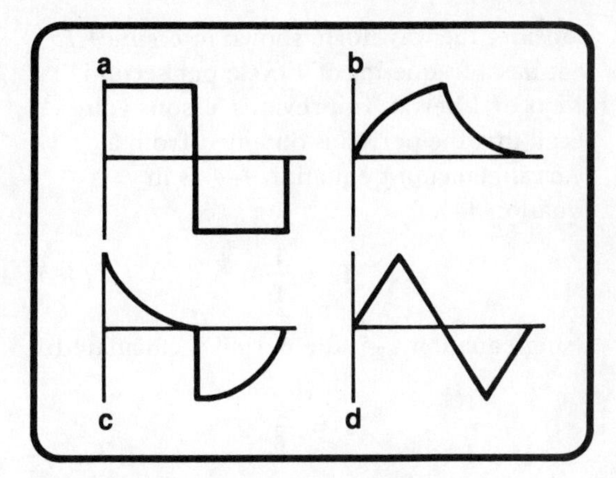

Figure 4.10 *Non-sinusoidal Waveforms*

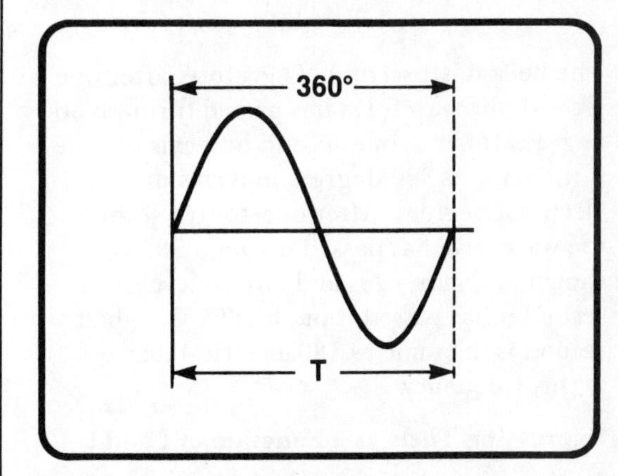

Figure 4.11 *Period of a Sine Wave*

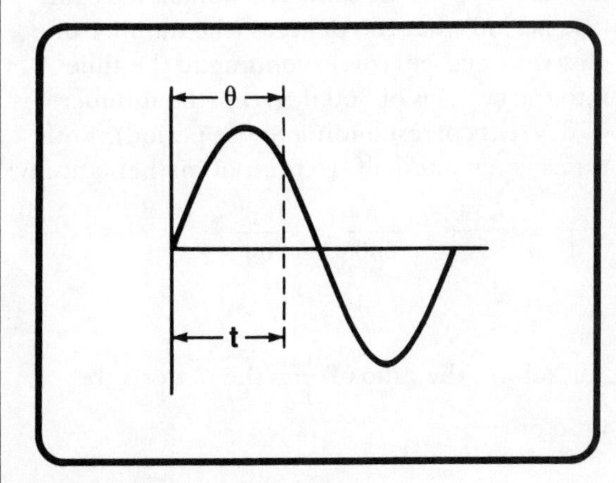

Figure 4.12 *Relationship of t to θ*

■ Expressing Theta as a Function of f and t

Consider the waveform shown in *Figure 4.13* that has a frequency of 1 cycle per second (1 cps or 1 hertz). In previous lessons you recall that the period is obtained from a rearrangement of equation *4–4* as in equation *4–5*.

$$T = \frac{1}{f} \qquad (4\text{–}5)$$

Using equation *4–5*, the period is calculated:

$$T = \frac{1}{f}$$

$$= \frac{1}{1\text{Hz}}$$

$$= 1\text{s}$$

The period is 1 second. Therefore, after one second, the waveform has passed through 360 degrees. That is, one second in terms of time is the same as 360 degrees in terms of electrical degrees. After one-fourth second the waveform has passed through 90° as shown in *Figure 4.14*, and after one-half second it has passed though 180°. One-half second is the same as 180 electrical degrees at this frequency.

Expressing Theta as a Function of f and t

Note that one-half second (an amount of time into the cycle) is the same fraction of 1 second (the period) that 180 degrees (the number of electrical degrees corresponding to the time into the cycle) is of 360 degrees (the number of degrees corresponding to the period). Both ratios equal one-half. Expressed mathematically:

$$\frac{t}{T} = \frac{0.5\text{s}}{1\text{s}} = \frac{1}{2} \qquad \frac{\theta}{360°} = \frac{180°}{360°} = \frac{1}{2}$$

Therefore, the ratio of $\frac{t}{T}$ is the same as the ratio of $\frac{\theta}{360°}$.

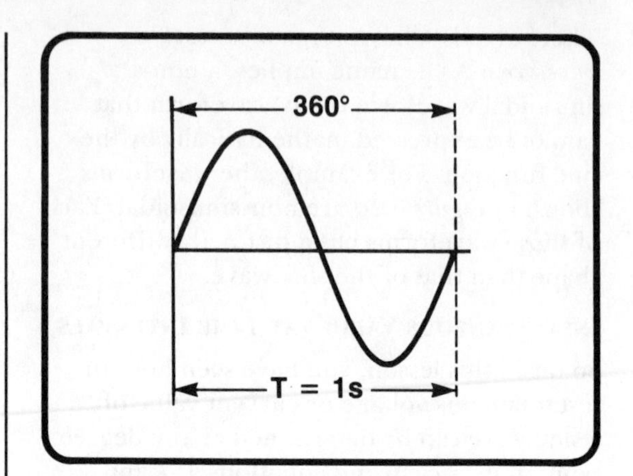

Figure 4.13 Sine Wave With a Period of 1 Second

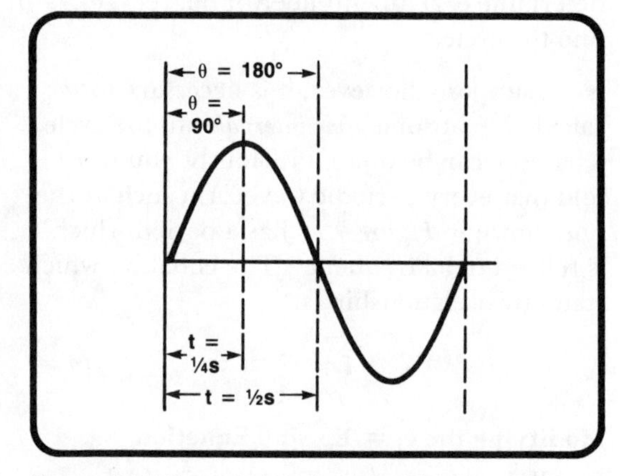

Figure 4.14 ¼s and ½s Points of a Sine Wave With T = 1s

If the frequency of a waveform changes, the electrical *degrees* of the waveform will correspond to *different times* (absolute times) into the cycle because the period of the cycle has changed. However if an instantaneous voltage is required at a particular elapsed time, t, into the cycle the ratio of the time elapsed into the period divided by the period will be the same as the electrical degrees at time, t, into the cycle divided by 360 degrees. The relationship of T, t, θ, and 360 electrical degrees is shown graphically in *Figure 4.15*.

In terms of a ratio:

$$\frac{t}{T} = \frac{\theta}{360°} \qquad (4\text{–}6)$$

If both sides of this equation are multiplied by 360°,

$$360°\left(\frac{t}{T}\right) = \left(\frac{\theta}{360°}\right) 360°$$

$$360°\left(\frac{t}{T}\right) = \theta$$

Since frequency equals 1/T, the equation can be rewritten:

$$\theta = 360°\left(\frac{t}{T}\right) \qquad (4\text{–}7)$$

$$\theta = 360°\left(\frac{1}{T}\right) t$$

$$\theta = 360° \text{ (f) t} \qquad (4\text{–}8)$$

Now, when determining the instantaneous voltage of an ac waveform, theta can be replaced in the expression by 360 ft because theta equals 360 times f times t. Thus, equation *4–1* for the instantaneous value of the voltage can be rewritten:

$$e_i = E_{pk}\sin\theta$$
$$= E_{pk}\sin(360ft) \qquad (4\text{–}9)$$

At first glance 360 ft may not appear to yield θ in degrees. But remember, as long as you express the frequency (f) in hertz, and the time elapsed into the cycle (t) in seconds, that 360 ft will yield θ in degrees.

Instantaneous Voltages at Specific Times

For example, consider a 50-volt peak, 4-kilohertz waveform such as the one in *Figure 4.15*. The period is 250 microseconds. Suppose you want to know the instantaneous amplitude of the waveform 100 microseconds into the cycle. Using equation *4–9*, the instantaneous voltage can be calculated:

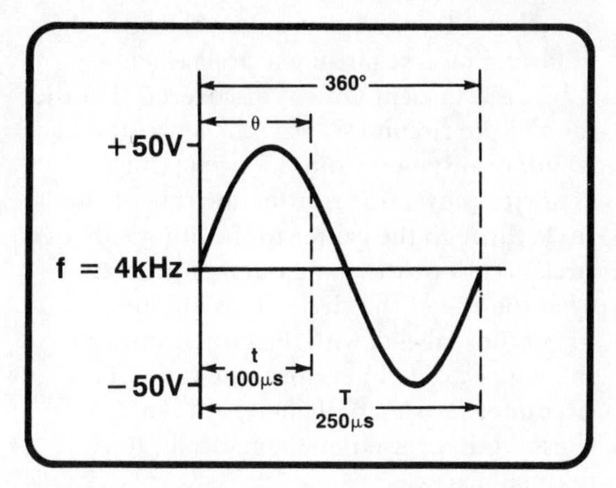

***Figure 4.15** Relationship of T, t, θ, and 360 Electrical Degrees*

$$e_i = E_{pk}\sin(360ft)$$
$$= 50V\sin(360ft)$$
$$= 50V\sin(360 \times 4000Hz \times 100\mu s)$$
$$= 50V\sin(360 \times 4 \times 10^{+3} \times 0.1 \times 10^{-3})$$
$$= 50V\sin(144°)$$
$$= 50V(0.5878)$$
$$= 29.39V$$

29.39 volts is the instantaneous voltage 100 microseconds into a cycle of an ac voltage of 50 volts peak and a frequency of 4 kilohertz.

RADIANS

An alternate unit of angular measure used when dealing with sinusoidal waveforms is the radian (abbreviated RAD). A radian is defined as the angle included within an arc equal to the radius of a circle. That is, if you measure off the radius of a circle on its edge as shown in *Figure 4.16,* the value of the angle ρ (greek letter rho) defined by the arc R is equal to one radian.

Radians

To help understand the radian, consider the following facts regarding a circle (*Figure 4.17*). The ancient Greeks discovered that the ratio of the circumference of a circle (the length or distance around a circle) to its diameter (how far it is from one side of the circle through the center to the other side of a circle) is always the same number no matter what the size of the circle. This number, which they labeled with the Greek letter pi (π), equals 3.14. Therefore, the ratio of circumference (c) to diameter (d) is a constant. Expressed mathematically, to two decimal places,

$$\frac{c}{d} = \pi = 3.14 \qquad (4\text{–}10)$$

This means that the circumference of a circle is 3.14 times longer than its diameter or said another way, the diameter divides into the circumference 3.14 times.

The radius of a circle is one-half its diameter. Therefore, as shown in *Figure 4.18*, the radius will divide the circumference twice as many times as the diameter, 6.28 times ($2 \times 3.14 = 6.28$). The circumference of a circle (c) is 6.28 or 2π times longer than its radius. Expressed mathematically,

$$c = 2\pi R = 6.28R \qquad (4\text{–}11)$$

Since a radian is the number of degrees included within the arc marked off by the radius on the circumference, you should see that there are 6.28, or 2π, radians in 360° (the number of degrees in a circle). Therefore, one radian equals 360° divided by 6.28 or 57.3° as shown in *Figure 4.19*. Expressed mathematically,

$$6.28 \text{ RAD} = 360° \qquad (4\text{–}12)$$

$$1\text{RAD}(\rho) = \frac{360°}{6.28} = 57.3° \qquad (4\text{–}13)$$

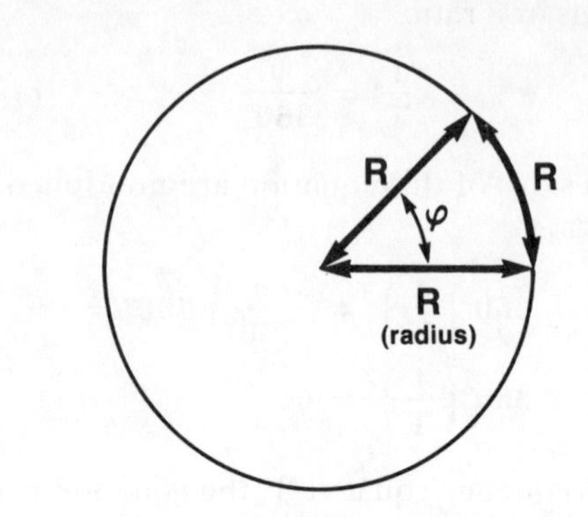

Figure 4.16 *The Unit Rho (ρ) is Equal to One Radian*

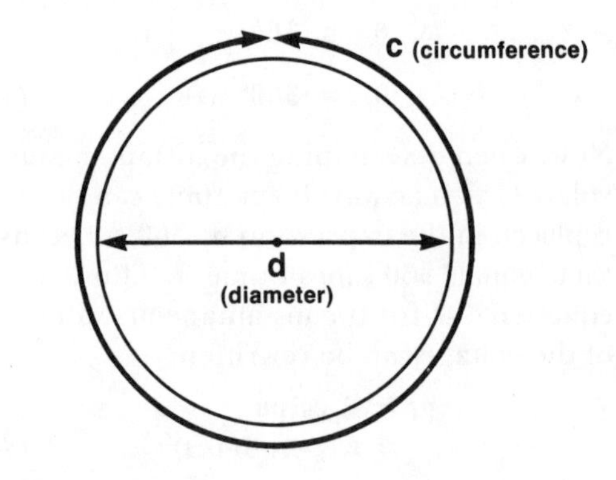

Figure 4.17 *Diameter and Circumference of a Circle*

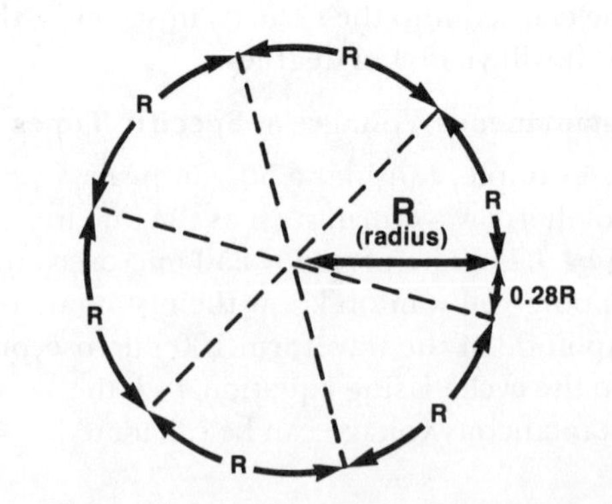

Figure 4.18 *The Radius Will Divide a Circle Two Times More Than the Diameter*

Instantaneous Voltage at Radian Rotation

The instantaneous voltage equation *4–9*,

$$e_i = E_{pk}\sin(360ft) \qquad (4\text{–}9)$$

can be rewritten substituting 2π (the number of radians in 360°) for 360. The resulting expression is:

$$e_i = E_{pk}\sin 2\pi ft \qquad (4\text{–}14)$$

Expressed in this $2\pi ft$ form of the angle theta has the dimensions of radians of angular rotation. The $2\pi f$ portion of this equation is given the name ω (Greek letter omega). It is not the capital omega used for resistance measurements (Ω), but the lower-case omega that looks like a small w. This $2\pi f$ form will be used at times in equations in the following lessons in this book.

PHASE RELATIONSHIPS

An important basic concept included in the subject of ac is that of the phase relationship —called thc phase angle—of two or more instantaneous waveforms.

Recall that in discussions in a previous lesson that when the single-loop generator began its rotation through the magnetic field its rotation began at a point where the induced voltage in the loop was a minimum. Thus, the sinusodial waveform it produced was like the one in *Figure 4.20*. In the following discussion it will be called waveform A.

Some time after this generator A has started, suppose that you decided to start another identical generator, generator B, and you want to compare its output voltage with that of the first generator. You wait until generator A has gone one-fourth turn before you start generator B and then observe the output voltage. B's output would appear as shown in *Figure 4.21*. Note that waveform B begins at the time that generator A is 90 degrees into its cycle because it began one-fourth turn (90 degrees) later.

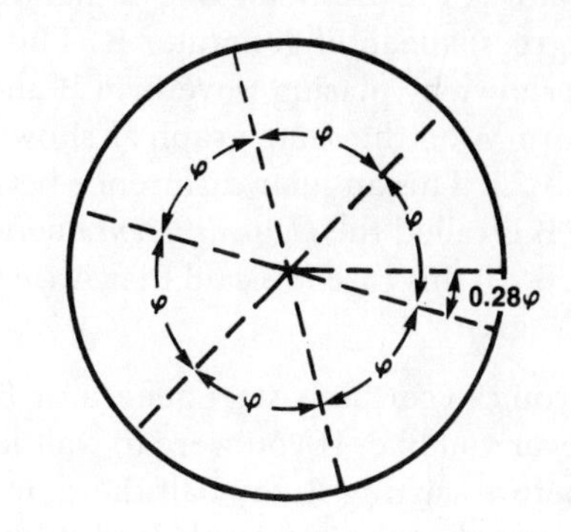

Figure 4.19 *One Radian (ρ) Is Equivalent to 57.3°*

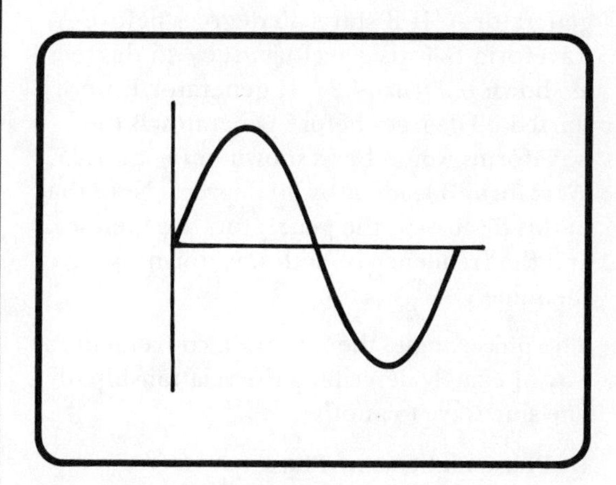

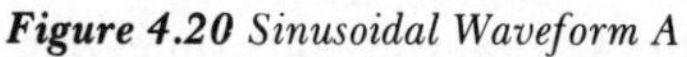

Figure 4.20 *Sinusoidal Waveform A*

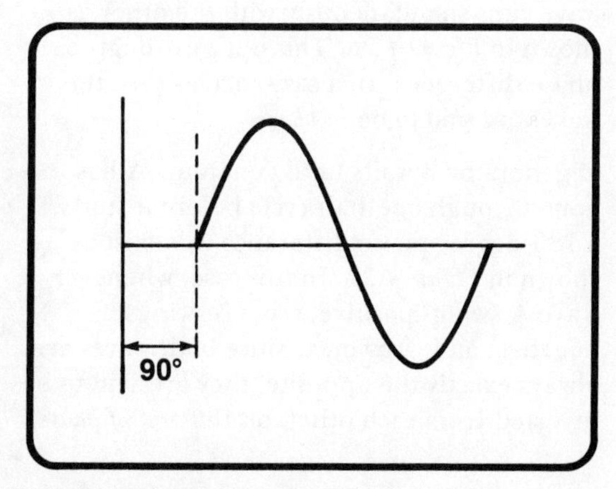

Figure 4.21 *Sinusoidal Waveform B*

Generator A will always be one-fourth turn (90 degrees) ahead of generator B. This can be shown by placing waveform B and waveform A on the same graph as shown in *Figure 4.22*. This angular difference between A and B is called the *phase difference* between A and B. In this case it is said that A *leads* B by 90 degrees.

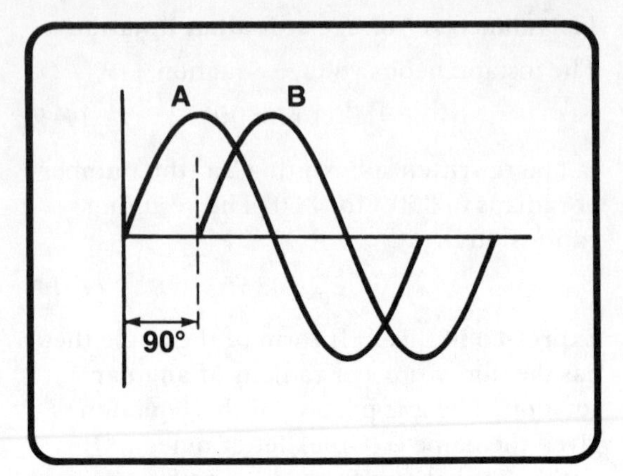

Figure 4.22 *Waveform A Leads Waveform B by 90 Degrees*

Now, you can certainly start generator B whenever you like. If you were to wait less time before starting B, say half the time previously, then A would only lead B by 45 degrees, as seen in *Figure 4.23*. Or generator B could just as easily be started before generator A. If B starts 45 degrees before A, waveform B leads waveform A by 45 degrees as shown in *Figure 4.24*. If generator B were started 90 degrees before generator B the waveforms would be as shown in *Figure 4.25*. Waveform B leads A by 90 degrees. Note that in this discussion the generators are identical and the frequency of both waveforms is the same.

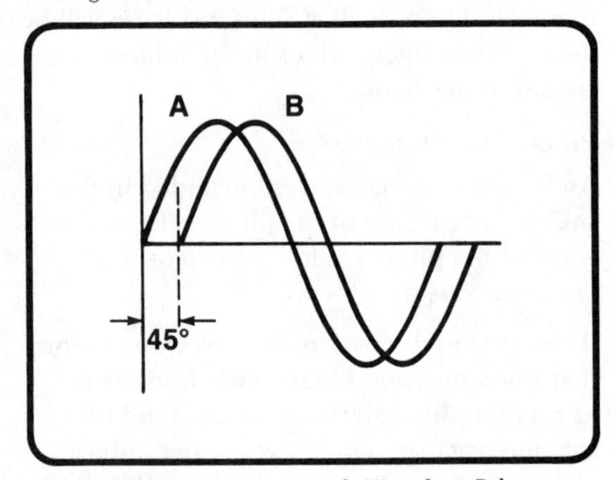

Figure 4.23 *Waveform A Leads Waveform B by 45 Degrees*

The phase angle, then, is just a convenient way of exactly describing the relationship of one sine wave to another.

In-Phase and Out-of-Phase

If both generators start at the same time, one wave runs simultaneously with the other as shown in *Figure 4.26*. This is a zero-degree phase difference. In a case such as this, the waves are said to be *in phase*.

If generator B waits until generator A has gone through one-half cycle before it starts, a 180-degree phase difference will exist as shown in *Figure 4.27*. In this case, whenever wave A swings positive, wave B swings negative, and vice versa. Since both waves are always exactly the opposite, they are said to be inverted from each other, or 180° *out of phase*.

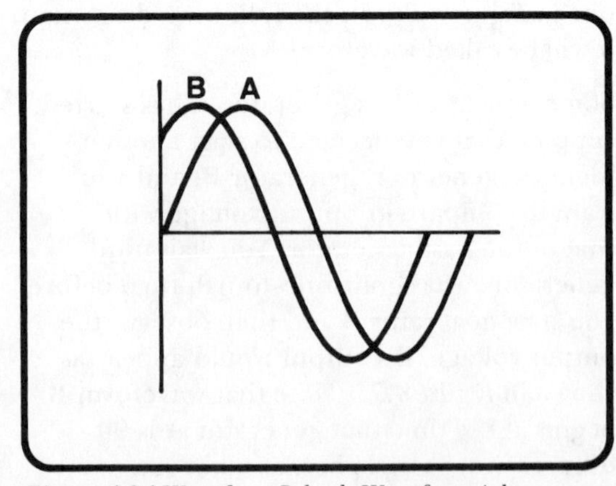

Figure 4.24 *Waveform B leads Waveform A by 45 Degrees*

■ Determining Phase Difference for Partial Waveforms

Now, up to this point, we've only specified that one wave *leads* another. This situation is much like two runners in a race as shown in *Figure 4.28*. You can say runner A is in the lead, but you could just as well say that runner B is lagging behind. Both mean the same thing; both are correct. Similarly, when describing the phase relationship of the two waveforms shown in *Figure 4.22* you can say A *leads* B by 90 degrees or you can just as correctly say, B *lags* A by 90 degrees. Both statements mean the same thing. The 90-degree phase difference is a common one, and it will be studied in more detail in later lessons. Therefore, it is important that you are able to recognize phase differences.

Determining Phase Difference for Partial Waveforms

In oscilloscope patterns, you may often have two waveforms shown and not really see the beginning of either wave. One of the simplest methods you can use to determine the phase difference is to first choose a point at which both waveforms are of the same instantaneous value of voltage or current. A convenient level to choose is the zero reference level. In *Figure 4.29*, waveform A and waveform B are the same value at all points where they cross the zero reference: zero volts. But, you must choose two points which are *side by side* and where waveform A is moving in the *same direction* as waveform B. That is, both must be either increasing or decreasing in amplitude. Points X and Y are side by side, and both waveforms are changing from their designated positive polarity to negative polarity, or decreasing. Between these two points, then, you can measure the phase angle. In this case that is 90 degrees.

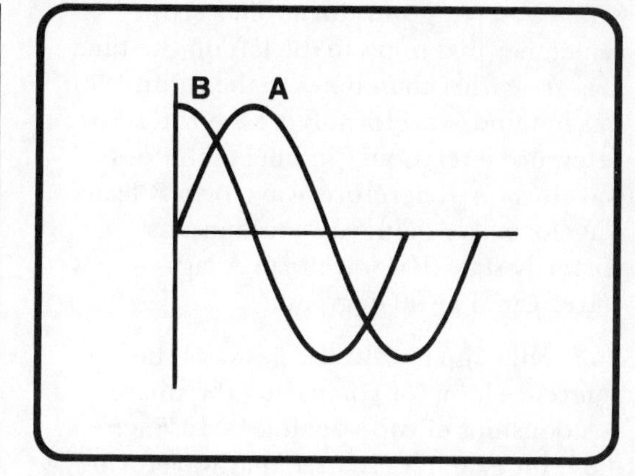

Figure 4.25 *Waveform B Leads Waveform A by 90 Degrees*

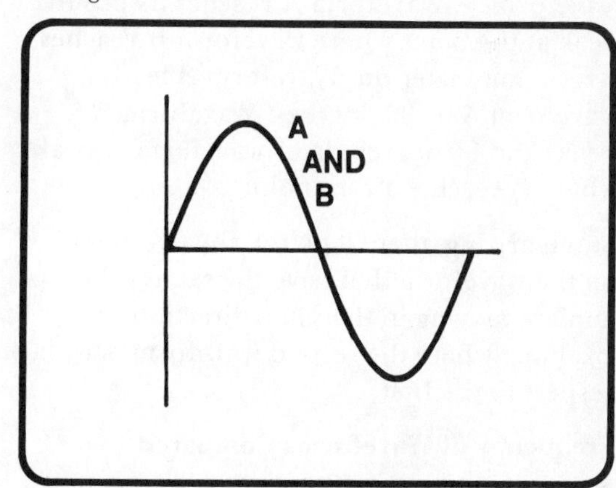

Figure 4.26 *Waveforms A and B In Phase*

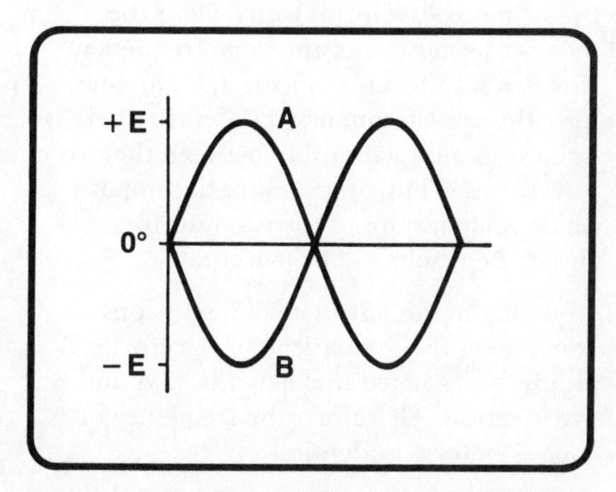

Figure 4.27 *Waveforms A and B 180° Out of Phase*

■ Frequency of Waveforms Compared

When observing waveforms on a scope, remember that times to the left on the time axis are earlier than times to the right. With this in mind, waveform B crosses the zero reference level sooner in time than does waveform A. Therefore, waveform B leads waveform A. Again, you could just as correctly state that waveform A lags waveform B by 90 degrees.

You could also look at the peaks as the reference level for comparing the phase relationships of two waveforms. In *Figure 4.29*, for example, you see that waveform B reaches its positive peak to the left of waveform A. Waveform A reaches its positive peak at the point where waveform B reaches a zero point later on. Waveform B leads waveform A by 90 degrees. Waveform B peaks and 90 degrees later waveform A peaks where B reaches a zero point.

Summarizing, then, look for any two points on the waveform that have the same value and are moving in the same direction. Examine where the second waveform is with respect to the first.

Frequency of Waveforms Compared

An important fact that you must remember when comparing the phase differences of two or more waveforms is that all of the waveforms must be of the same frequency. The two waveforms of *Figure 4.30* do not have the same frequency. Therefore, there is not a constant relationship between the two waveforms. Thus, any attempt to compare phase relationships of the two waveforms is futile; the results will be inaccurate.

Note that in this lesson in all discussions concerning the comparison of waveforms A and B, it was stated that generators A and B are identical. Therefore, the frequency of each waveform is identical.

Figure 4.28 *Runners A and B are Similar to Waveforms A and B*

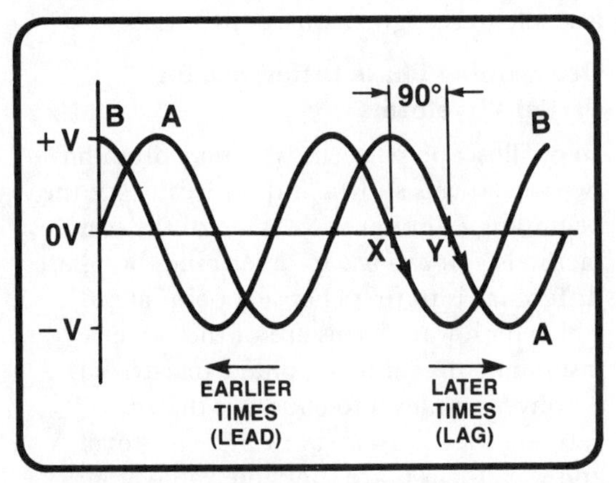

Figure 4.29 *Measuring Phase Difference Using Waveforms*

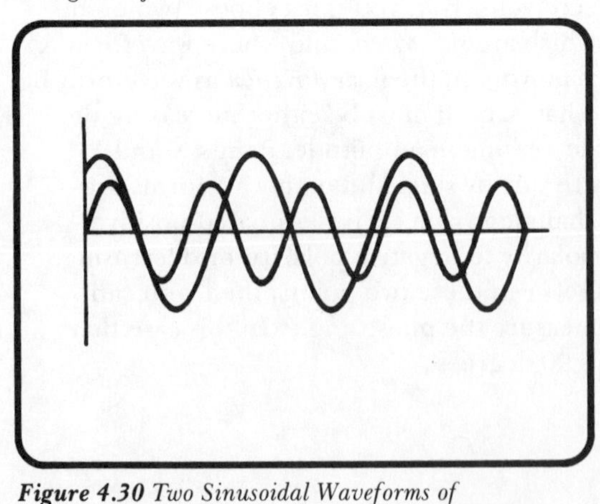

Figure 4.30 *Two Sinusoidal Waveforms of Different Frequency*

Conventional Phase Difference Specification

There is another interesting point about the phase of waveforms. In *Figure 4.31*, waveform B leads wave A by 90 degrees. Waveform A crosses the zero reference axis at point W. Note that if you look to the right, waveform B reaches a similar point (point Z) 270 degrees later on. Focusing on this relationship, it can be stated that waveform A leads waveform B by 270 degrees. Originally it was stated that waveform B leads A by 90 degrees. By comparing the phase at points X and Y this is verified.

These different phase relationships illustrate a key point—there are many ways of correctly expressing the phase between two waveforms. By convention, however, usually an angle of less than 180° is used to specify the phase. In other words, if A leads B by 225°, it is more common to say that B lags A by 135°. Other equivalent specifications can be found because of the repetitive nature of the sine function, and these are just as correct.

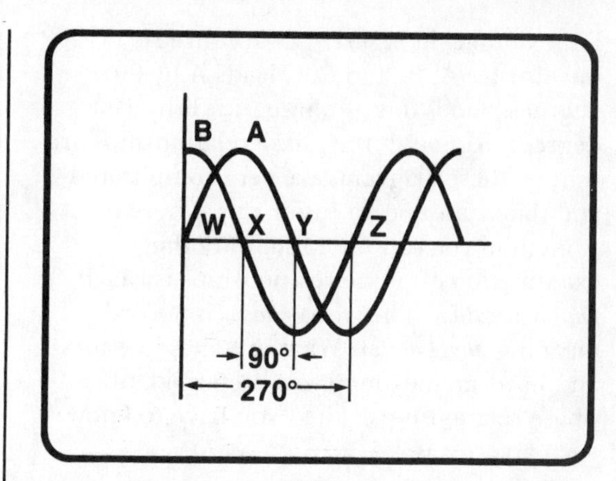

Figure 4.31 *Waveform A Leads Waveform B By 270°*

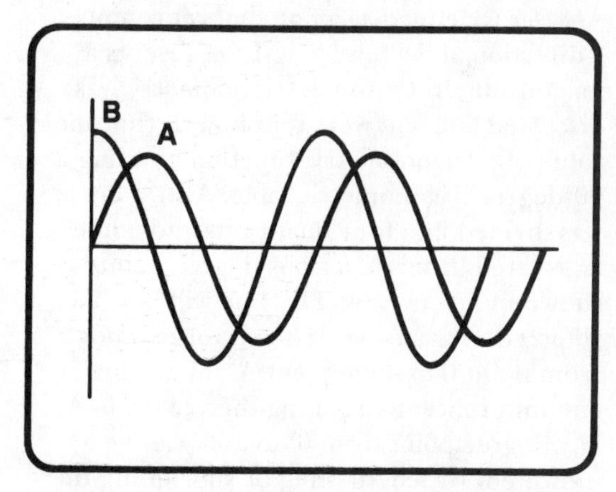

Figure 4.32 *Two Waveforms of Different Amplitudes*

Amplitude Versus Phase

Another important point about phase relationships is that the amplitude of a sine wave has no bearing on its phase. The phase relationship of two sine waves of different amplitude is determined in the same way as two sine waves of the same amplitude. For example in *Figure 4.32* waveform B leads A by 90 degrees. B has a greater amplitude than A, but the frequency of both is the same. Therefore, it is possible to determine accurate phase relationships between the two waveforms.

Phasor Notation

Thus far in this discussion of phase, sine wave plots of both waveforms have been used to determine and describe the phase. You probably have realized that if you wanted to decipher the phase relationships of three or more waveforms using these diagrams, it could become very tedious if not virtually impossible.

For example, in *Figure 4.33* are three waveforms, A, B, and C. C leads A by 90 degrees, and is out of phase with B by 100 degrees. Although the phase relationships are simple, these diagrams are very complicated and there must be an easier way. There is. A method you can use to simplify the examination of phase relationships is called *phasor notation.* The word *phasor* is a word meaning *phase vector.* What are phase vectors, and how can they help you keep track of phase relationships? First, you have to know what a vector is.

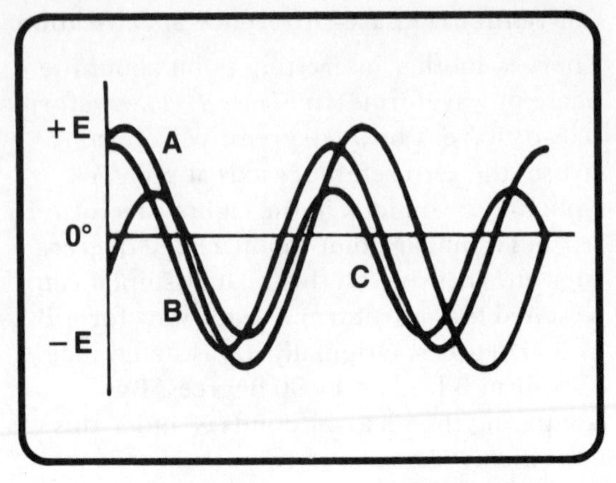

Figure 4.33 *Three Different Sinusoidal Waveforms*

Vectors

A *vector* is defined as a line that represents a direction and whose length represents a magnitude. In Lesson 2, trigonometry was discussed and you were told how to find the value of a trigonometric function by using a 90-degree trigonometric table. Also, a circle was divided into four quadrants and a line, A, rotated through various degree points as shown in *Figure 4.34.* This line will now be referred to as a *vector.* The vector extends from point 0 to some point A out on the circumference. By rotating the vector, 0–A, any degree point from 0° to 360° can be indicated. By lengthening or shortening the vector line, any magnitude can be indicated. Vectors are used in this fashion to represent voltage and current phase relationships, and when they do they are called *phasors.*

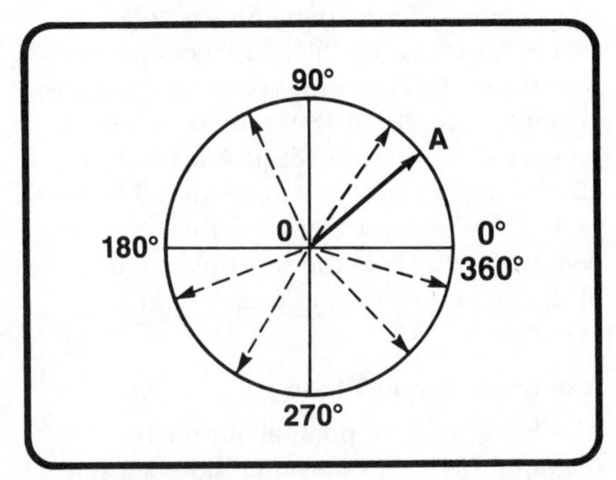

Figure 4.34 *Vector 0-A Can Represent Any Degree From 0° to 360°*

Using Phasor Notation

Phasor notation can be used to represent the phase relationship between two sinusoidal waveforms such as those shown in *Figure 4.35.*

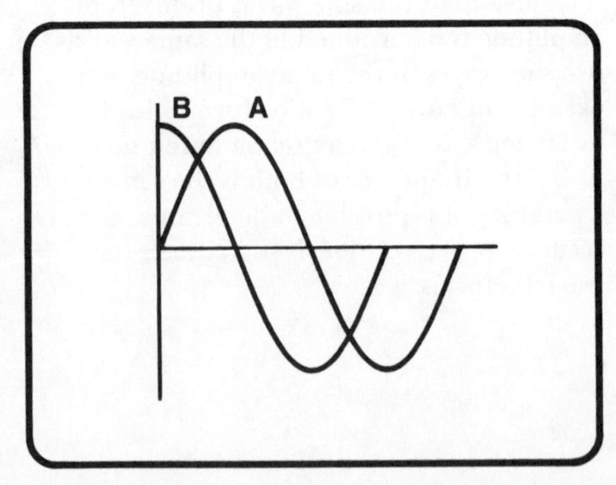

Figure 4.35 *Example Waveforms for Phasor Notation*

■ Using Phasor Notation

When using phasor notation, first one waveform must be chosen as the reference. In this example, the reference will be waveform A. The reference waveform phasor, E_A, is then positioned along the X axis, as shown in *Figure 4.36,* at the zero-degree rotational reference. This phasor is a vector representing the voltage of an ac generator as its conductors are rotated through a magnetic field. By convention, this rotation will always be rotated in a counterclockwise (ccw) direction.

If you look at the waveform diagram of *Figure 4.35*, you see that waveform B leads waveform A by 90 degrees. Waveform B, then, can be represented by another phasor, E_B, placed so that it leads phasor E_A by 90 degrees. Since E_A is at the 0° point, and the phasors are to be rotated counter-clockwise, this means that phasor E_B should be placed at the 90° point as shown in *Figure 4.37*.

Since the two phasors, E_A and E_B, represent voltages generated in conductors, adding E_B is like adding another conductor to the ac generator. Therefore, when the two conductors are rotated in the magnetic field of the ac generator, the voltage from conductor E_B will lead conductor E_A by 90 degrees throughout the cycle as shown in *Figure 4.38*, and the waveform E_B will lead the waveform E_A by 90 degrees throughout the entire cycle. Thus, since the phasors represent voltages generated by conductors, the phasor representation of the generator action graphically shows the phase relationship of two or more waveforms throughout an entire cycle.

Remember that phasors that are positioned counter-clockwise from E_A will *lead* E_A, and phasors positioned clockwise from E_A will *lag* E_A.

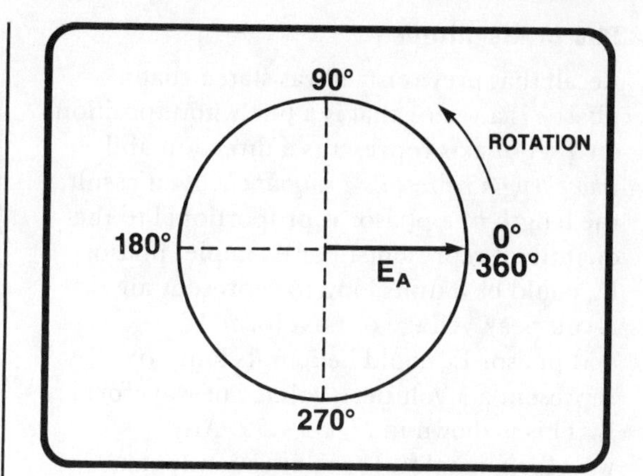

Figure 4.36 *Phasor E_A Is at the 0° Point*

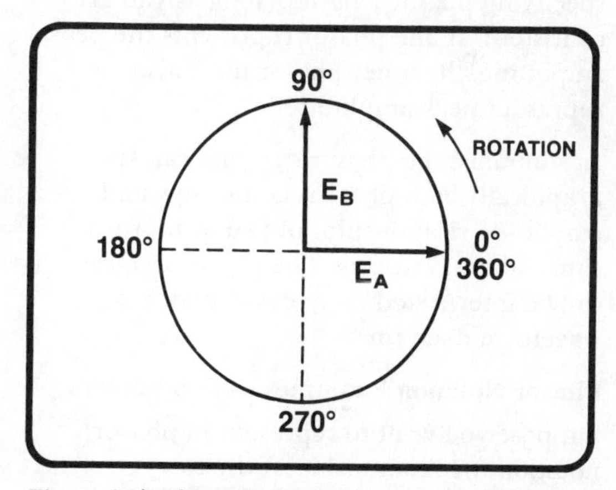

Figure 4.37 *Phasor E_B Leads Phasor E_A by 90 Degrees*

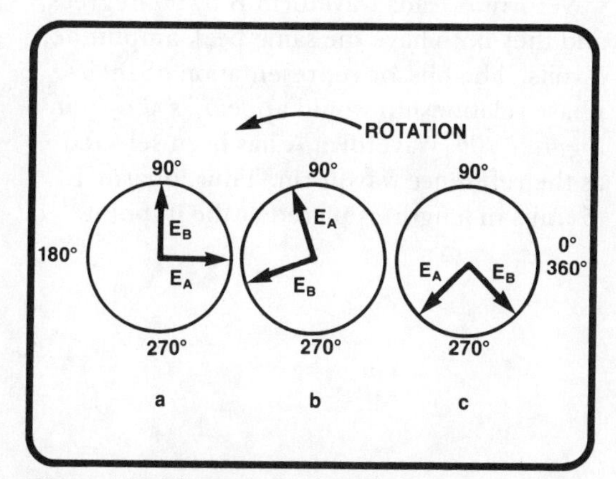

Figure 4.38 *Phasor E_B Leads E_A by 90 Degrees Throughout the Entire Cycle*

■ Phasor Magnitude
■ Phasor Notation Examples — Two Vectors

Phasor Magnitude

Recall that previously it was stated that a phasor is a vector that is a line whose position on a set of axis represents a direction and *whose length represents a magnitude.* As a result, the length of a phasor is proportional to the quantity it represents. For example, phasor E_A could be 8 units long to represent an 8-volt peak voltage of waveform A, and phasor E_B could be 5 units long to represent a 5-volt peak voltage of waveform B. This is shown in *Figure 4.39*. Any waveform amplitude specification (peak, peak-to-peak, rms) can be used when specifying phasor length as long as you are consistent. If one phasor represents the peak amplitude, the other phasor must also represent peak amplitude.

In summary, the phasor diagram can show graphically both phase relationships and amplitude relationships of two or more sinusoidal waveforms. The phasor diagram can be interpreted more easily than a waveform diagram.

Phasor Notation Examples — Two Vectors

Suppose you want to represent in phasor notation the relationship of the two waveforms shown in *Figure 4.40a*. Note that waveform A leads waveform B by 90 degrees, and they both have the same peak amplitude, 5 volts. The phasor representation of their phase relationship would appear as shown in *Figure 4.40b*. Waveform A has been selected as the reference waveform. Thus, phasor E_A (5 units in length) is placed at the 0° point.

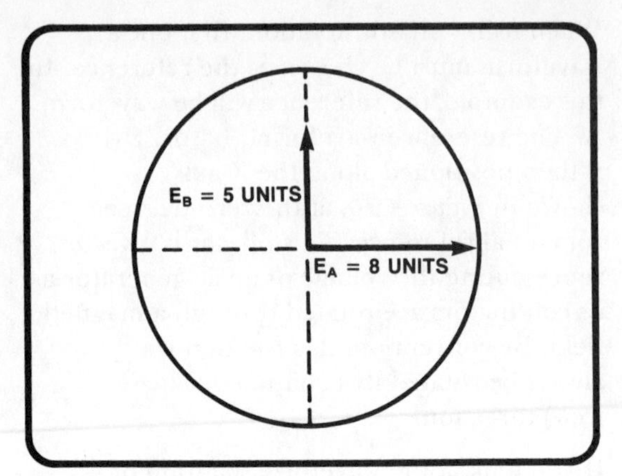

Figure 4.39 *Phasor Representation of Waveforms A and B with Different Amplitudes*

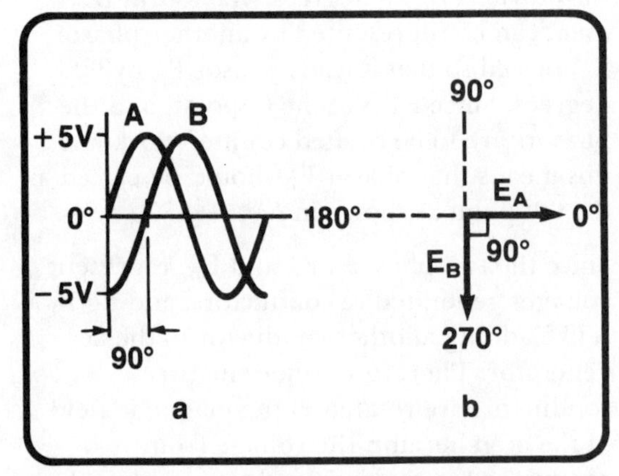

Figure 4.40 *Two Waveforms 90° Out of Phase, E_A at 0°: a. Waveforms; b. Phasor Diagram*

Since waveform A leads waveform B, phasor E_B (5 units in length) is placed at 270°, 90 degrees out of phase with E_A. Both phasors have the same length since both represent the same peak amplitude, 5 volts. From the phasor diagram, then, you can interpret that E_A *leads* E_B by 90 degrees or that E_B *lags* E_A by 90 degrees.

■ Phasor Notation Examples — Two Vectors

As another example, suppose that you are shown the phasor diagram illustrated in *Figure 4.41a*. What is represented? That diagram shows that E_A leads E_B by 90 degrees. It also shows that E_A and E_B are the same amplitude because phasor E_A and E_B are the same length. What would the waveforms represented by the phasor diagram look like?

The sinusoidal waveform representation of these waveforms is shown in *Figure 4.41b*. The phasor diagram of *Figure 4.41a* represents the voltages at the instantaneous degree point of point X showing E_A at the 90° point and E_B at the 0° point in *Figure 4.41b*.

Figure 4.42a shows a phasor diagram of two voltages which are the same amplitude, but 180 degrees out of phase. The sinusoidal waveforms showing this phase relationship are as shown in *Figure 4.42b*. Note that the waveform E_A is passing through zero amplitude at the 0° point and is increasing in a positive direction. To correspond to this, the phasor E_A is positioned at the 0° reference on the phasor diagram. The E_B waveform also is passing through zero amplitude but is increasing negatively, so the axis crossing point is at 180 electrical degrees. To represent this, the phasor is postioned at 180° on the phasor diagram.

Figure 4.43 shows two voltages of different amplitude that are in phase. Note that E_A is greater in amplitude than E_B, since it is longer than E_B on the phasor diagram.

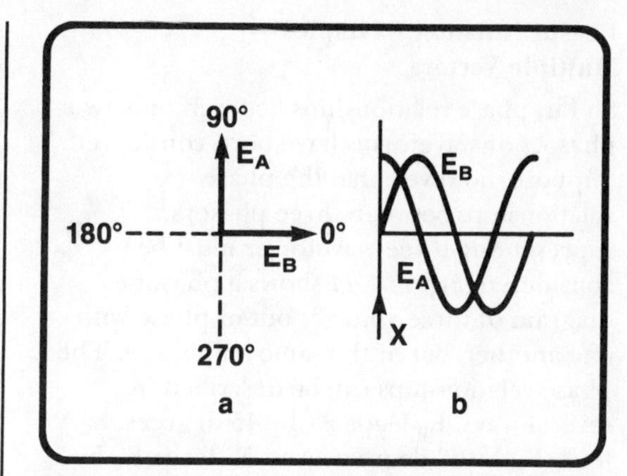

Figure 4.41 *Two Waveforms 90° Out of Phase, E_A at 90°: a. Phasor Diagram; b. Waveforms*

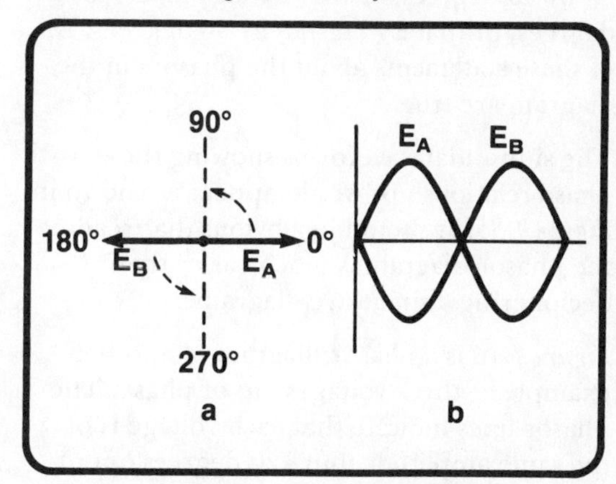

Figure 4.42 *Two Waveforms 180° Out of Phase: a. Phasor Diagram; b. Waveforms*

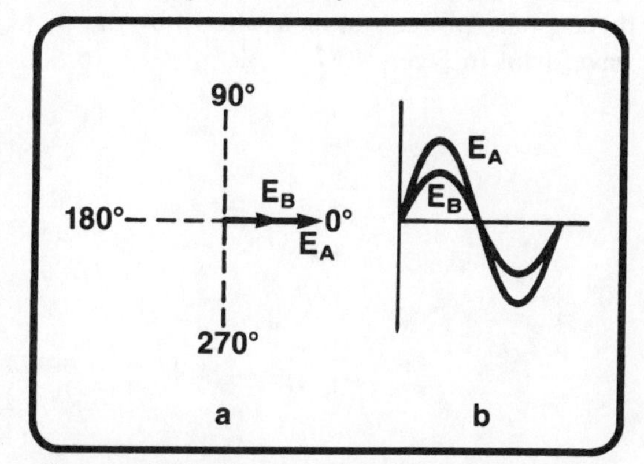

Figure 4.43 *Two In-Phase Waveforms with Different Amplitudes: a. Phasor Diagram b. Waveforms Represented by Phasor Diagram*

■ Phasor Notation Examples—Multiple Vectors

Phasor Notation Examples—Multiple Vectors

So far, phase relationships between only two phasors or waveforms have been considered. Suppose, however, that the phase relationships between three phasors, representing three waveforms must be considered. *Figure 4.44* shows a phasor diagram of three voltages out of phase with one another, but of the same amplitude. The phase relationships can be described in several ways. E_B leads E_A by 45 degrees. E_C leads E_A by 90 degrees. And E_C leads E_B by 45 degrees. It could also be said that E_B lags E_C by 45 degrees, or that E_A lags E_B by 45 degrees, or that E_A lags E_C by 90 degrees. All of these statements about the phasors in the diagram are true.

The sinusoidal waveforms showing these phase relationships would appear as shown in *Figure 4.45*. It should be obvious that reading the phasor diagram is much easier than deciphering a sine wave diagram.

Figure 4.46 is a phasor diagram of another example of three voltages out of phase. The phasor lines indicate that each voltage is of the same amplitude, but 120 degrees out of phase with each other. This type of phase relationship exists between the voltages in a three-phase power system as shown by the waveforms in *Figure 4.47*.

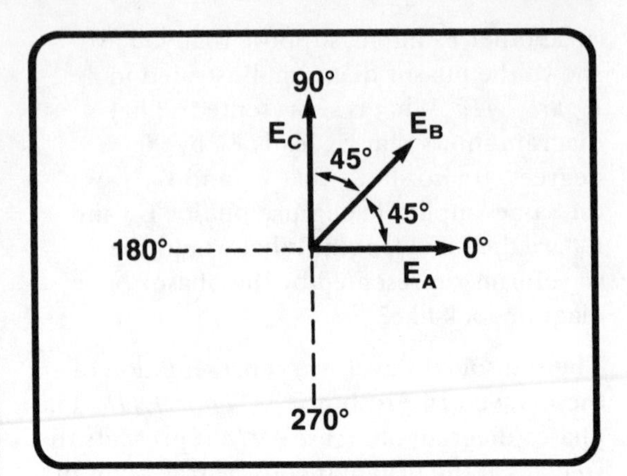

Figure 4.44 *Phasor Diagram of Three Voltages Out of Phase*

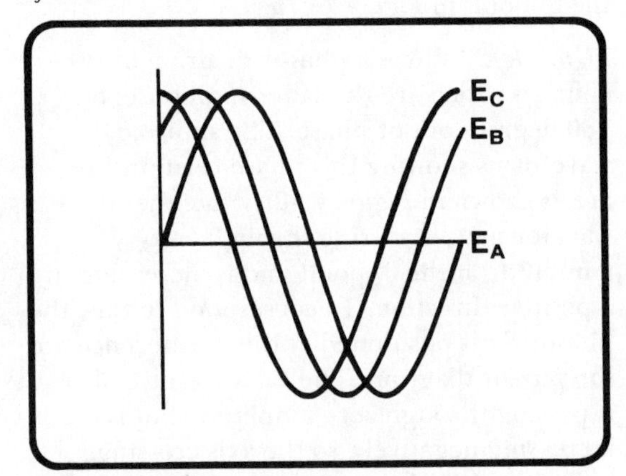

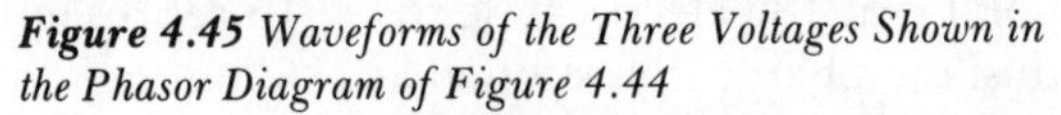

Figure 4.45 *Waveforms of the Three Voltages Shown in the Phasor Diagram of Figure 4.44*

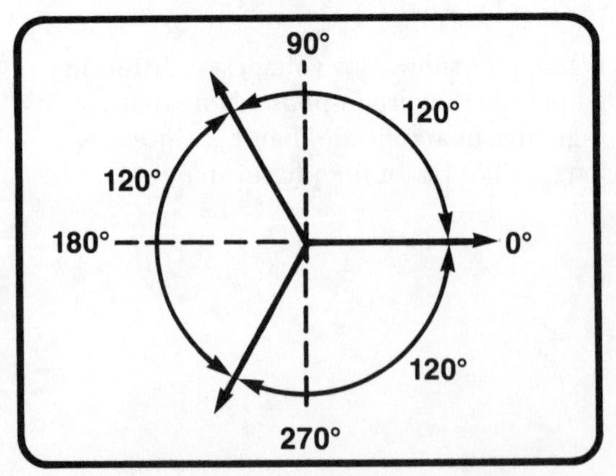

Figure 4.46 *Phasor Diagram of Three Voltages Out of Phase*

It should be obvious that phasor notation greatly simplifies the visual description of the phase relationships of voltages. Phasors can also be used to show current relationships, or voltage-current relationships, or even power relationships. In fact, phasor diagrams showing voltage-current phase relationships are used extensively when working with ac. A typical example is the phasor diagram of *Figure 4.48* which shows a voltage leading a current by 90 degrees. In fact, as you will see in later lessons, one of the quantities in an ac circuit that is important is the phase angle— the phase difference in electrical degrees between the voltage applied to a circuit and the total circuit current.

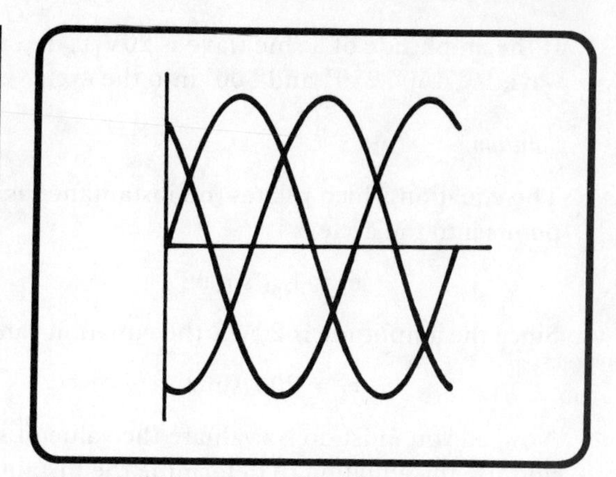

Figure 4.47 *Waveforms of the Three Voltages of Figure 4.46*

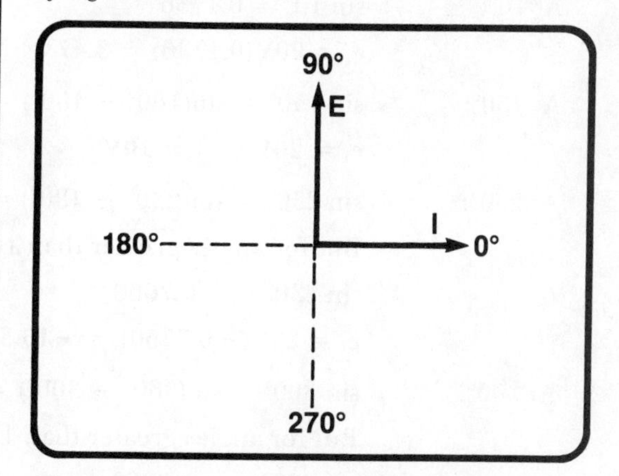

Figure 4.48 *Voltage-Current Phasor Diagram*

SUMMARY

In this lesson you were shown how to calculate the instantaneous voltage or current at any point in a sinusoidal waveform by using the sine function. The difference between sinusoidal waveforms was described and examples of each were shown. The relationship of the electrical degrees of a cycle and the period of a cycle was explained. The phase relationship of sinusoidal waveforms was described and you were shown how to simplify the visualization of those phase relationships with vector notation by using phasors.

Worked-Out Examples

1. If the amplitude of a sine wave is $20V_{pk}$, determine the instantaneous amplitude of the sine wave 10°, 150°, 230°, and 300° into the cycle.

Solution:

The equation which relates the instantaneous amplitude of a sine wave at various degree points into the cycle is

$$e_i = E_{pk} \sin \theta$$

Since the amplitude is $20V_{pk}$ the equation can be rewritten as

$$e_i = 20V \sin \theta$$

Now, all you must do is evaluate the value of sin θ at 10°, 150°, 230° and 300° into the cycle and use the equation to determine the instantaneous amplitudes at these points:

At 10°: $\sin 10° = 0.1736$

$e_i = 20V(0.1736) = \mathbf{3.47V}$

At 150°: $\sin 150° = \sin(180° - 150°) = \sin 30° = 0.5$

$e_i = 20V(0.5) = \mathbf{10V}$

At 230°: $\sin 230° = \sin(230° - 180°) = \sin 50° = 0.7660$

But for angles greater than 180°, the sine wave is negative; therefore,

$\sin 230° = -0.7660$

$e_i = 20V(-0.7760) = \mathbf{-15.32V}$

At 300°: $\sin 300° = \sin(360° - 300°) = \sin 60° = 0.8660$

But for angles greater than 180°, the sine wave is negative; therefore,

$\sin 230° = -0.8660$

$e_i = 20V(-0.8660) = \mathbf{-17.32V}$

■ Worked-Out Examples

For clarity, these values are plotted on the sine wave below:

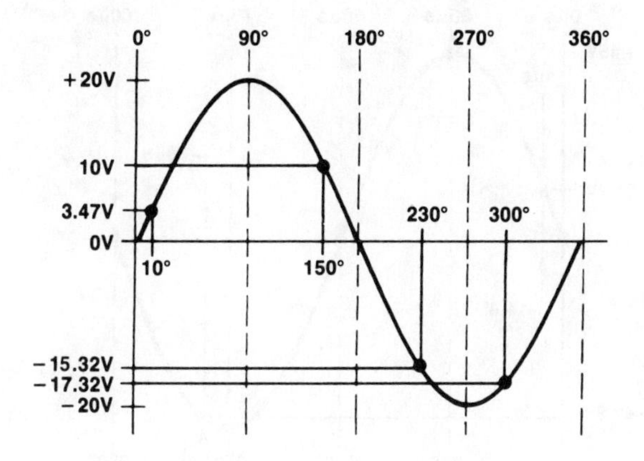

2. If the amplitude of a sine wave is $45V_{pk}$, and has a frequency of 10kHz, determine the instantaneous value of the sine wave at 4μs and 80μs into the cycle.

Solution:

The equation which relates the time, frequency, and amplitude of a sine wave to an instantaneous value of voltage at that point is

$$e_i = E_{pk} \sin(360ft)$$

Since the amplitude and frequency are specified, the equation can be rewritten:

$$e_i = 45V \sin(360 \times 10kHz \times t)$$

Now, all that must be done to evaluate the instantaneous voltage at times into the cycle is to use (360 × 10kHz × t) to determine the number of degrees that specified times into the cycle (t) represent and solve for the sine of these angles.

The equation can then be used to determine the instantaneous value of voltage as in example 1.

4μs:

$$360ft = (360)(10kHz)(4\mu s)$$
$$= (360)(10 \times 10^3 Hz)(4 \times 10^{-6} s) = 14.4°$$
$$e_i = 45V \sin 14.4° = 45V(0.2487) = \mathbf{11.2V}$$

80μs:

$$360ft = (360)(10kHz)(80\mu s)$$
$$= (360)(10 \times 10^3 Hz)(80 \times 10^{-6} s) = 288°$$
$$e_i = 45V \sin 288° = 45V(-0.9511) = \mathbf{-42.8V}$$

■ Worked-Out Examples

For clarity, these values are plotted on the sine wave below:

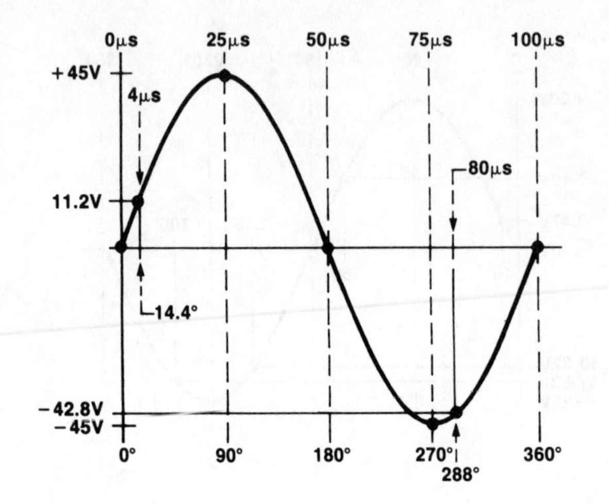

3. Plot a sine wave having an amplitude of 12 VAC. Tabulate values for each 20° between 0° and 360°, then plot the sine wave. Use $e_i = E_{pk} \sin \theta$. Also plot values at 0°, 90°, 180°, and 270°.

The table of values for angles between 0° and 360° is shown below. Note that a 12 VAC sine wave has a peak value of $17V_{pk}$. ($E_{pk} = 1.414 \times E_{rms} = 1.414 \times 12V = 17V$) Thus, the equation used is $e_i = 17V \sin \theta$.

θ	sin θ	E_{pk}	θ	sin θ	E_{pk}
0	0	0	200	−0.3420	−5.8V
20	0.3420	5.8V	220	−0.6428	−10.9V
40	0.6428	10.9V	240	−0.8660	−14.7V
60	0.8660	14.7V	260	−0.9848	−16.7V
80	0.9848	16.7V	270	−1	−17V
90	1	17V	280	−0.9848	−16.7V
100	0.9848	16.7V	300	−0.8660	−14.7V
120	0.8660	14.7V	320	−0.6428	−10.9V
140	0.6428	10.9V	340	−0.3420	−5.8V
160	0.3420	5.8V	360	0	0
180	0	0			

Worked-Out Examples

Solution:

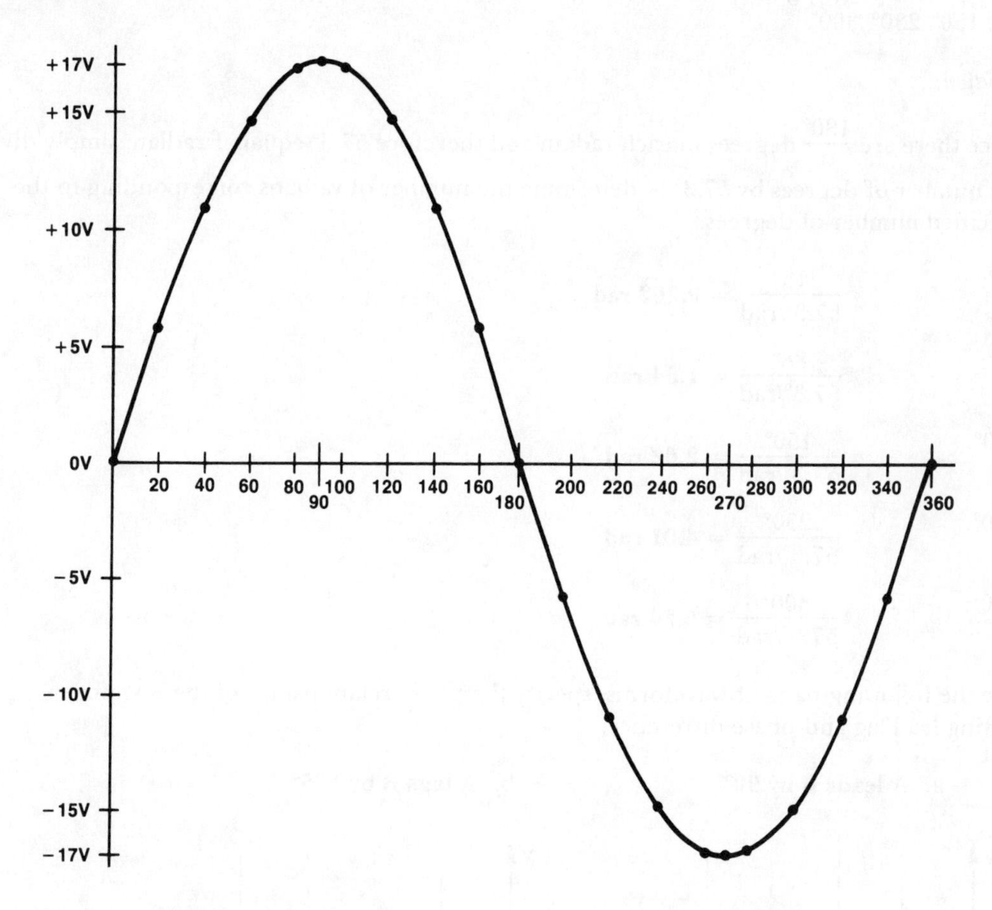

4. For the following angles specified in radians, determine the equivalent angle in degrees: 1 rad, 3.6 rad, 4.7 rad, and 5.9 rad.

Solution:

Since there are 2π radians in 360°, there must be π radians in 180°. Therefore, there are $\frac{180°}{\pi}$ degrees in each radian. Thus, 57.29578° or 57.3° equals 1 radian. To determine the number of degrees a certain number of radians represents, simply multiply the number of radians by 57.3°.

1 rad:	1 rad × 57.3°/rad = **57.3°**
3.6 rad:	3.6 rad × 57.3°/rad = **206.3°**
4.7 rad:	4.7 rad × 57.3°/rad = **269.3°**
5.9 rad:	5.9 rad × 57.3°/rad = **338°**

■ Worked-Out Examples

5. For the following angles specified in degrees, determine the equivalent angle in radians: 15°, 88°, 150°, 230°, 300°.

Solution:

Since there are $\frac{180°}{\pi}$ degrees in each radian and therefore 57.3° equals 1 radian, simply divide the number of degrees by 57.3° to determine the number of radians corresponding to the specified number of degrees.

15°: $\frac{15°}{57.3°/\text{rad}} = \mathbf{0.262\ rad}$

88°: $\frac{88°}{57.3°/\text{rad}} = \mathbf{1.54\ rad}$

150°: $\frac{150°}{57.3°/\text{rad}} = \mathbf{2.62\ rad}$

230°: $\frac{230°}{57.3°/\text{rad}} = \mathbf{4.01\ rad}$

300°: $\frac{300°}{57.3°/\text{rad}} = \mathbf{5.24\ rad}$

6. For the following pairs of waveforms, specify the phase relationships of the waveforms stating lead/lag and phase difference.

a. A **leads** B by **90°**

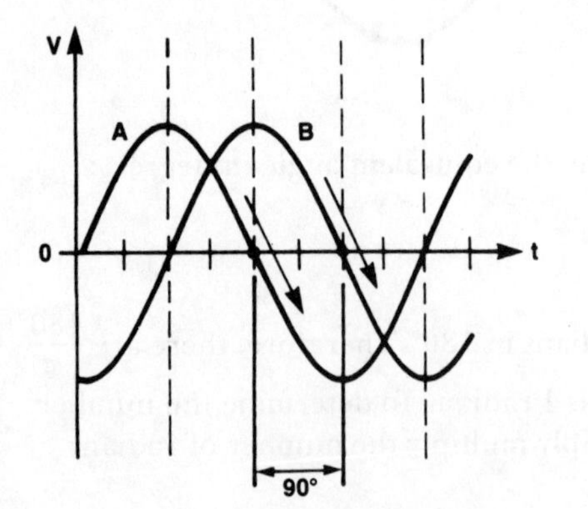

A crosses the zero level first, 90° ahead of B.

b. A **lags** B by **135°**

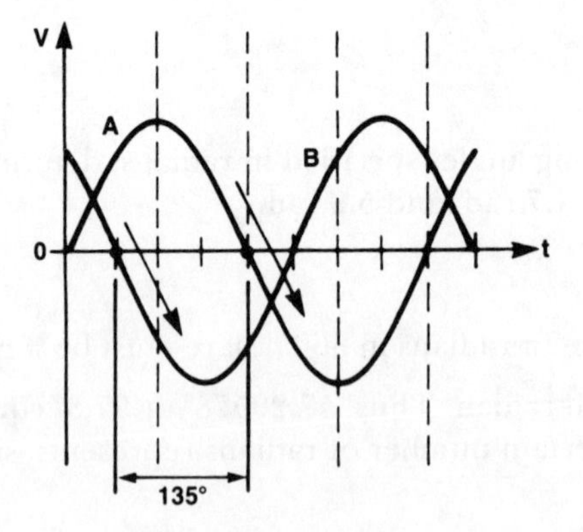

A crosses the zero level 135° after B crosses it.

■ Worked-Out Examples

7. Sketch the equivalent phasor diagram for the waveforms in question 6. Use waveform A as reference at 0° position.

a. b.

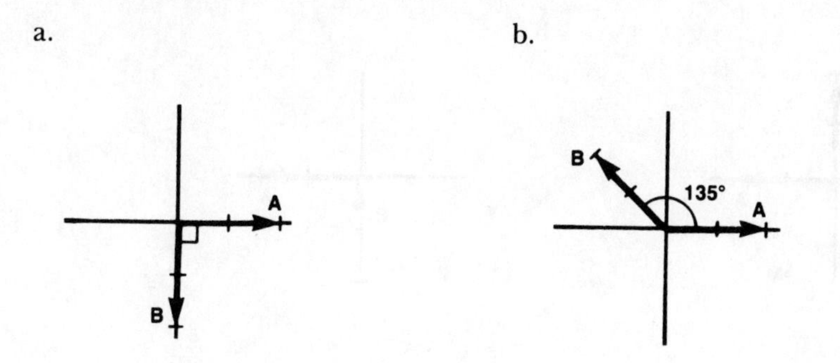

8. For the two following phasor diagrams, state the phase relationship of the phasors stating lead/lag and phase difference.

a. A **lags** B by **45°**

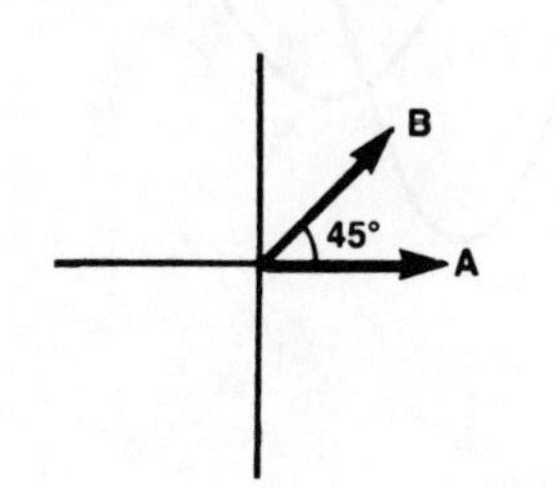

b. A **leads** B by **135°**

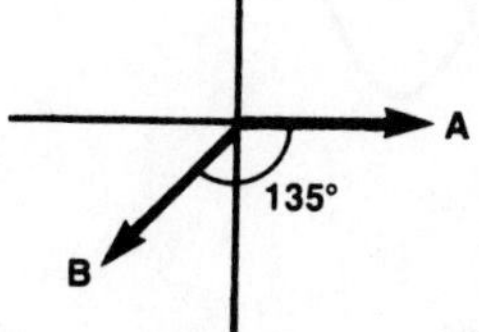

■ Worked-Out Examples

9. For the following phasor diagrams shown below, sketch the sinusoidal waveform representation of the pair showing appropriate amplitudes and phase relationships.

a. b.

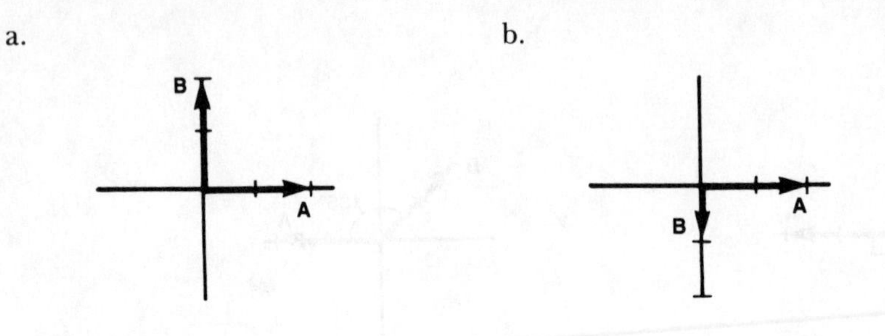

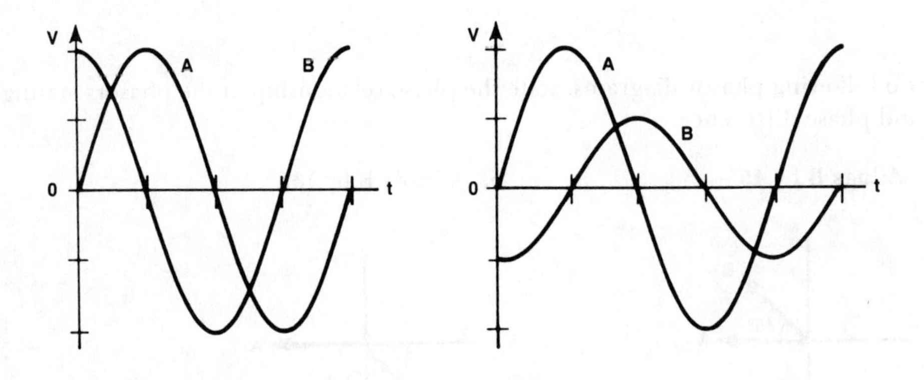

B leads A by **90°**

A is the same amplitude as B since both phasors are the same length.

A **leads** B by **90°**

B is one-half the amplitude of A since B's phasor length is one-half A's phasor length.

■ Practice Problems

1. Determine the instantaneous amplitude of the following sine waves at the specified number of degrees into the cycle:

   a. $40V_{pk}$, 38° into the cycle: ____________
   b. $28V_{pk}$, 156° into the cycle: ____________
   c. 25VAC, 65° into the cycle: ____________
   d. 58VAC, 271° into the cycle: ____________
   e. $87V_{pk}$, 322° into the cycle: ____________
   f. 36VAC, 94° into the cycle: ____________

2. Determine the instantaneous amplitude of the following sine waves at the specified time elapsed into the cycle:

   a. $35V_{pk}$, 30kHz, 20μs into the cycle: ____________
   b. 15VAC, 400Hz, 800μs into the cycle: ____________
   c. 220VAC, 60kHz, 3μs into the cycle: ____________
   d. $50V_{pk}$, 4MHz, 44ns into the cycle: ____________
   e. $30V_{pk}$, 16kHz, 51μs into the cycle: ____________
   f. 120VAC, 455kHz, 540ns into the cycle: ____________

3. Plot a sine wave having an amplitude of $48V_{pp}$ using 18° increments, from 0° to 360°.

4. For the following angles expressed in radians, determine the equivalent angle in degrees:

   a. 3 rad = ____________ °
   b. 5.2 rad = ____________ °
   c. 2.3 rad = ____________ °
   d. 4.1 rad = ____________ °
   e. 1.8 rad = ____________ °
   f. 0.6 rad = ____________ °

5. For the following angles expressed in degrees, determine the equivalent angle in radians:

   a. 38° = ____________ rad
   b. 100° = ____________ rad
   c. 231° = ____________ rad
   d. 285° = ____________ rad
   e. 344° = ____________ rad
   f. 84° = ____________ rad

■ Practice Problems

6. For the following pairs of waveforms, specify the phase relationships of the waveforms stating lead/lag and phase difference:

a. A ________ B by ________°

b. A ________ B by ________°

c. A ________ B by ________°

d. A ________ B by ________°

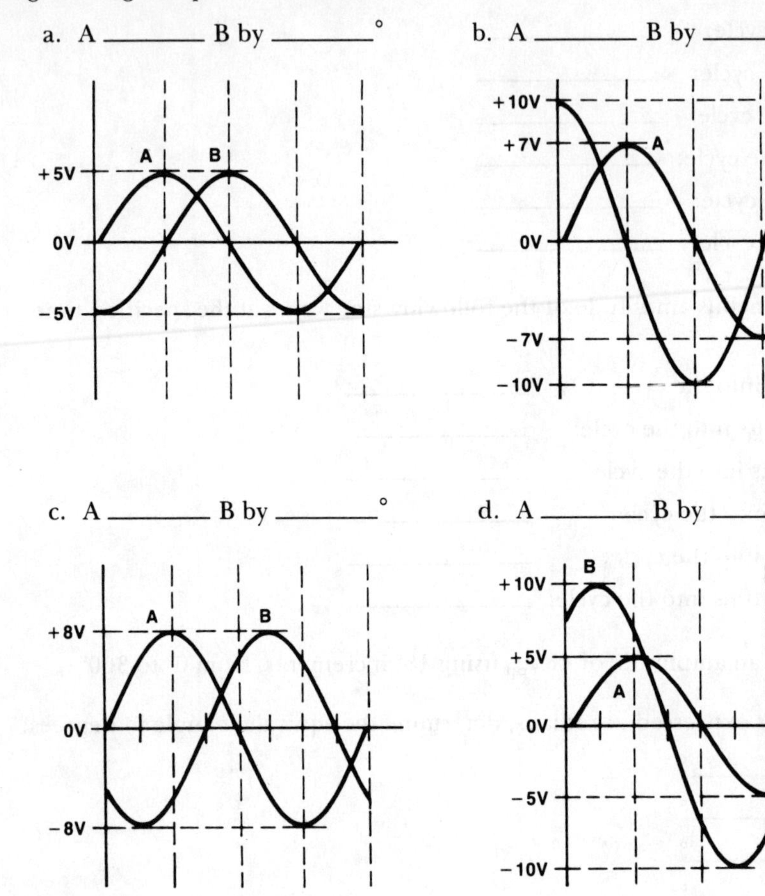

■ Practice Problems

7. Sketch the equivalent phasor diagram for the waveforms in question 6. Use waveform A as reference at the 0° position.

a.

b.

c.

d.

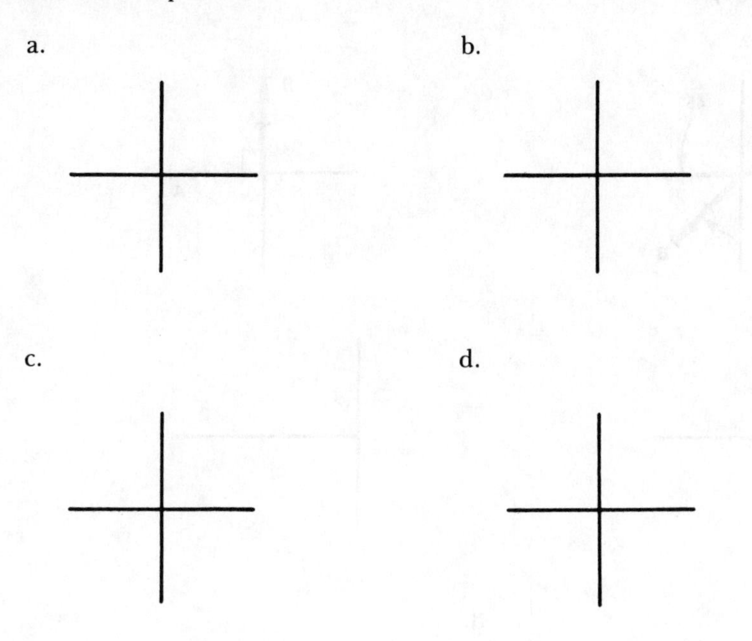

8. For the following phasor diagrams, state the phase relationship of the phasors stating lead/lag and phase difference:

a. A ________ B by ________°

b. A ________ B by ________°

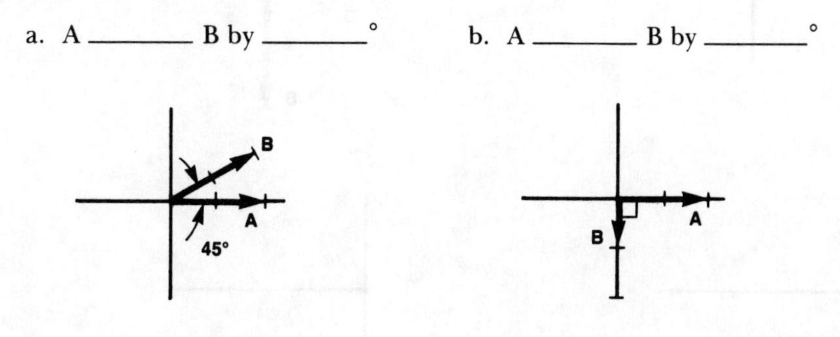

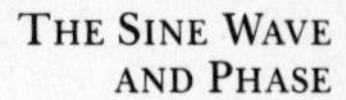

■ Practice Problems

9. For the following phasor diagrams, sketch the sinusoidal waveform representation of the pair showing appropriate amplitudes and phase relationships:

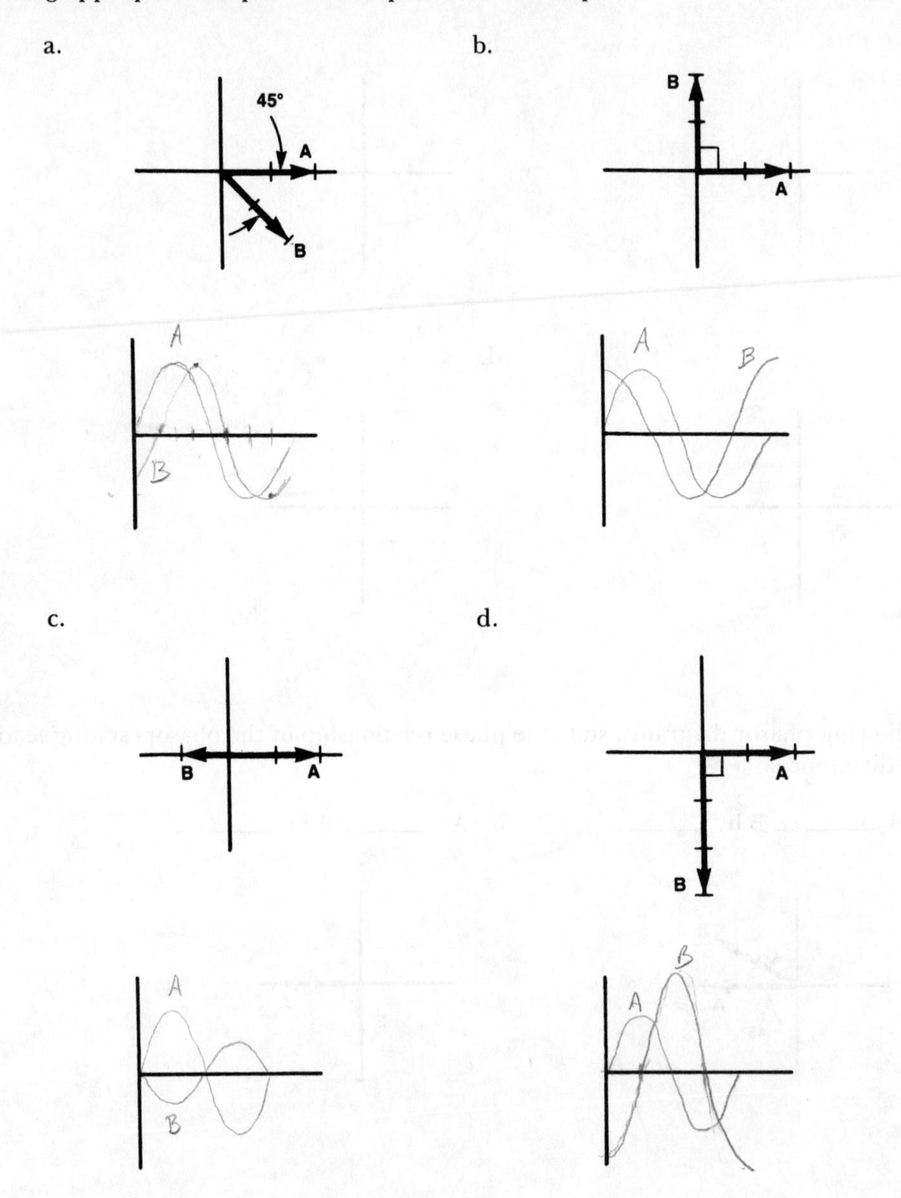

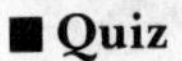

Quiz

1. Determine the instantaneous amplitude of the following sine waves at the specified number of degrees into the cycle:

a. $45V_{pk}$ at 54° into the cycle: 36.4 V

b. 18VAC at 220° into the cycle: −16.4 V

$E_{pk} = \frac{18v.ac}{.707} \approx 25.5V_{pk}$

2. Determine the instantaneous amplitude of the following sine waves at the specified time elapsed into the cycle:

a. $33V_{pk}$, 25kHz, 4μs into the cycle: 19.4 V

$e_i = E_{pk}\sin(360ft)$

b. 10VAC, 50Hz, 3ms into the cycle: 11.4 V

$E_{pk} = \frac{10vac}{.707} = 14.1$

3. For the following angles expressed in radians, determine the equivalent angle in degrees:

a. 3.9 rad = 223.5 °

b. 5.5 rad = 315.2 °

$\frac{rad \times 360°}{2\pi\, rad}$

4. For the following angles expressed in degrees, determine the equivalent angle in radians:

a. 44° = 0.770 rad

b. 248° = 4.33 rad

$\frac{° \times 2\pi\, rad}{360°}$

5. For the following pairs of waveforms, specify the phase relationships of the waveforms stating lead/lag and phase difference, then sketch the equivalent phasor diagram using waveform A as reference at the 0° position.

a.

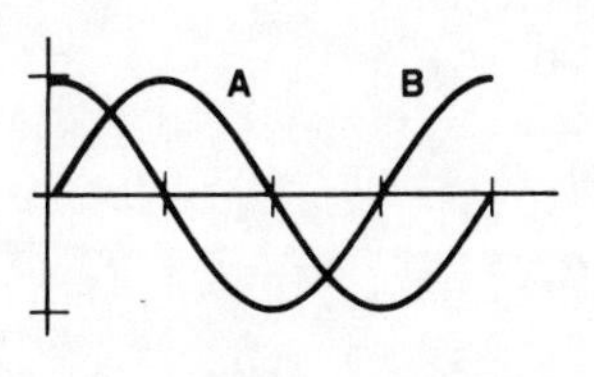

A lags B by 90 °

b.

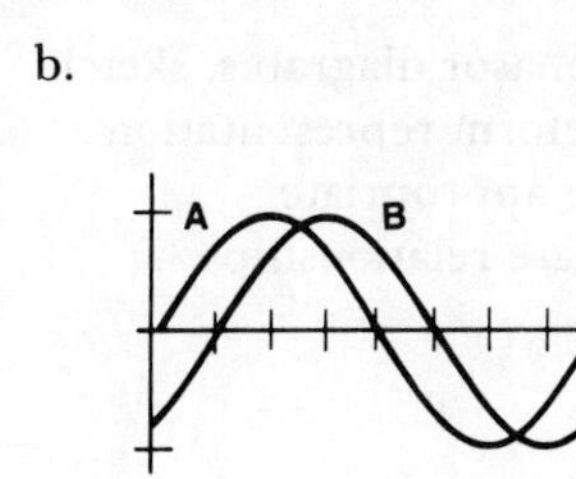

A leads B by 45 °

c. (for a.)

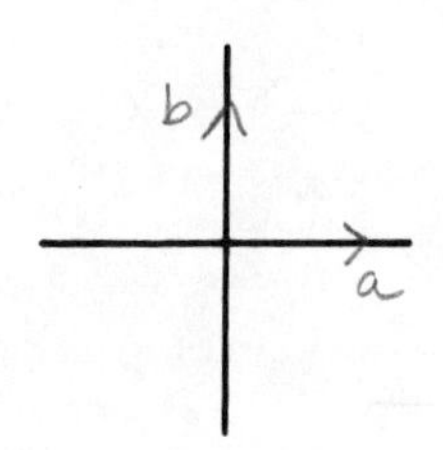

d. (for b.)

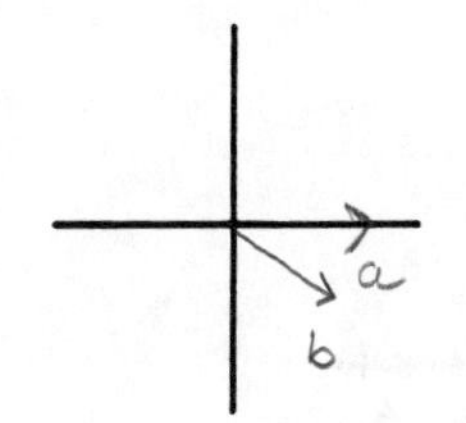

■ Quiz

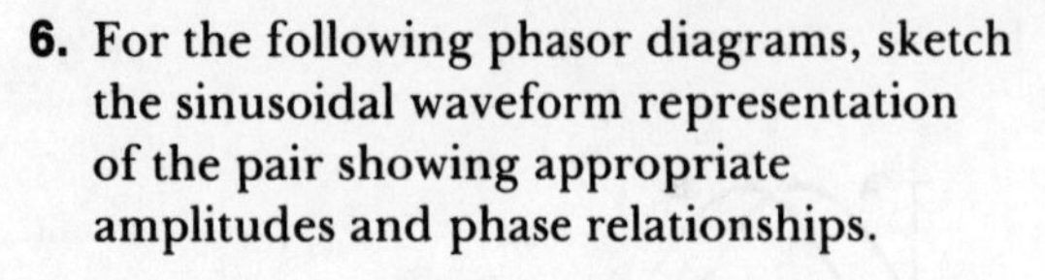

6. For the following phasor diagrams, sketch the sinusoidal waveform representation of the pair showing appropriate amplitudes and phase relationships.

a.

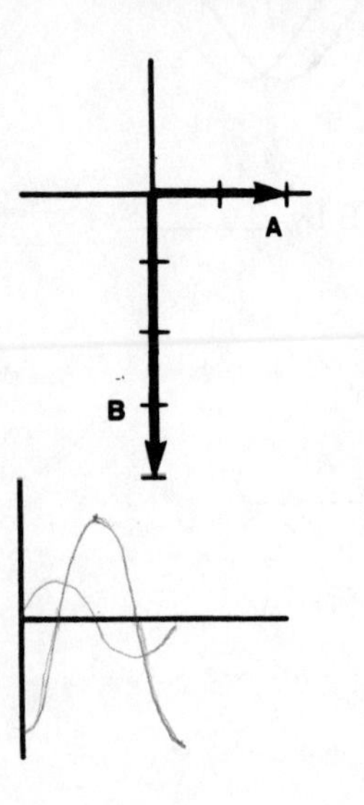

b.

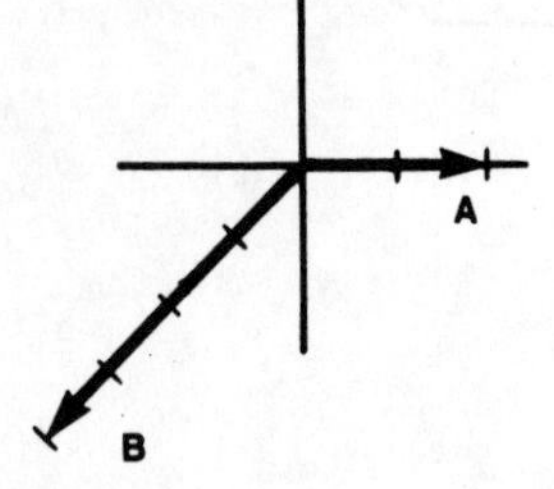

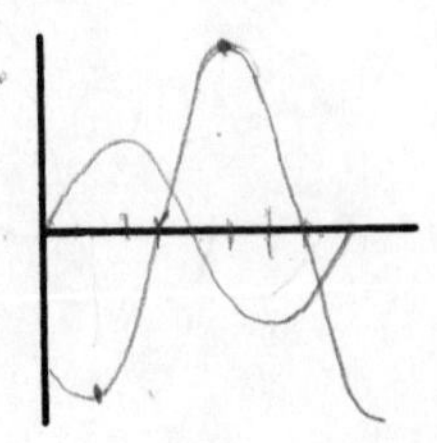

c.

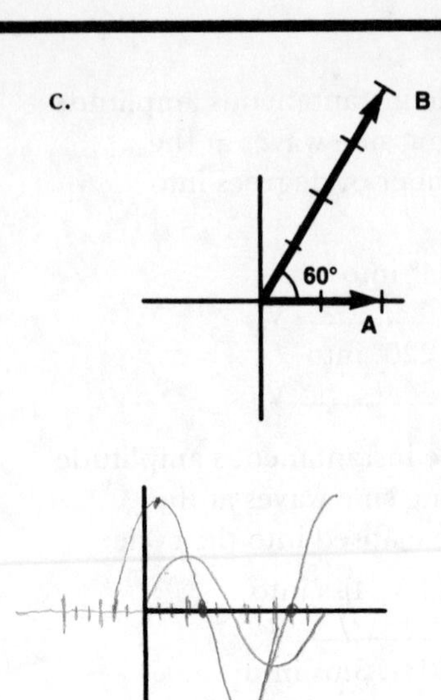

LESSON 5

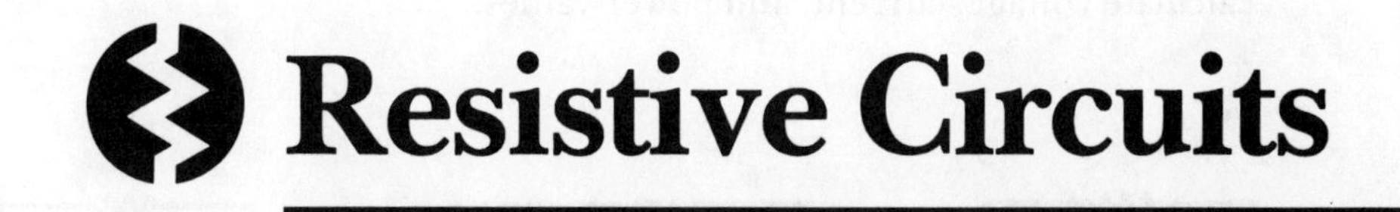

Resistive Circuits

In previous lessons, ac was discussed in general terms. With this lesson, a more detailed investigation of actual circuits with an ac voltage applied will begin. This lesson describes how to determine voltages and current in series, parallel, and series-parallel resistive circuits when an ac voltage is applied. Peak, peak-to-peak, and rms voltage, current and power values are calculated. The phase and frequency of circuit current and component voltages in ac resistive circuits are analyzed.

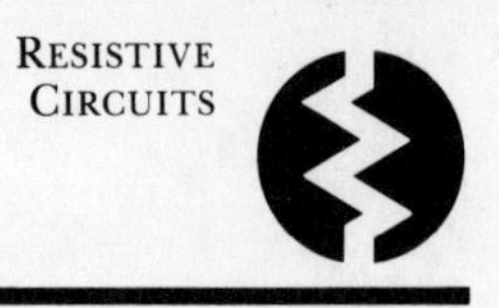

■ Objectives

At the end of this lesson, you should be able to:

1. Analyze the phase and frequency relationships of current and voltage in an ac resistive circuit.

2. Explain the relationship of peak, peak-to-peak, and rms values for an ac resistive circuit and be able to convert from one value to another.

3. Calculate the instantaneous voltage and current values in an ac resistive circuit and plot the results.

4. Analyze series, parallel, and series-parallel ac circuits as shown and calculate voltage, current, and power values.

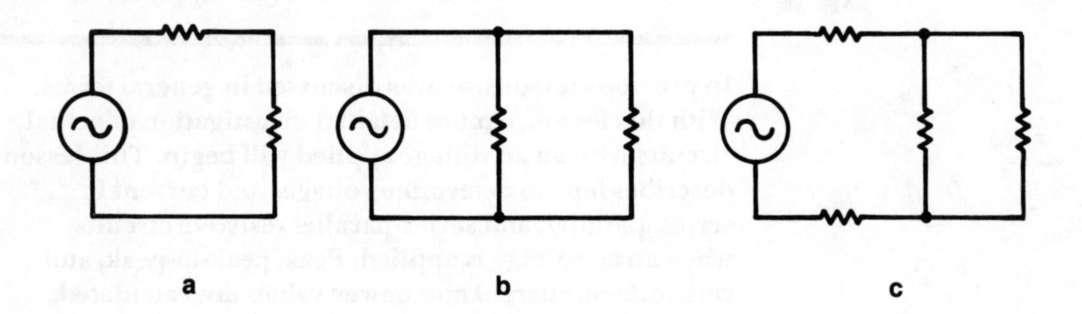

■ **AC Applied to a Resistor**
■ **Calculating Instantaneous Voltages and Currents**

REVIEW OF A FEW BASICS

AC Applied to a Resistor

Recall that when a fixed dc voltage is applied to a resistor as shown in *Figure 5.1,* current flows in only *one* direction and its value is determined by the value of the voltage and the resistor in the circuit. According to Ohm's law, the current is equal to the voltage divided by the resistance.

$$I = \frac{E}{R} \qquad (5\text{–}1)$$

$$= \frac{10V}{1000\Omega}$$

$$= 10mA$$

Therefore, the current in the circuit of *Figure 5.1* is 10 milliamperes.

If the connections to the battery are reversed as shown in *Figure 5.2,* the direction of the current reverses, but the current value is the same. The current is still 10 milliamperes.

If, now, a sinusoidal voltage is applied to the same circuit with the resistor as shown in *Figure 5.3,* the voltage in the circuit will be changing constantly. This is shown in *Figure 5.4.* Ohm's law still applies and can be used at each *instant of time* to calculate the current just as it was used to calculate the current when a constant dc voltage was applied. This is an important point of this lesson and bears repeating. When a known ac voltage is applied to a purely resistive circuit Ohm's law can be used at any instant of time to calculate the current at that same instant of time.

Calculating Instantaneous Voltages and Currents

Recall that the instantaneous value of the voltage for a sinusoidal waveform can be calculated using this equation:

$$e_i = E_{pk}\sin\theta \qquad (5\text{–}2)$$

Figure 5.1 *A Typical Resistive DC Circuit*

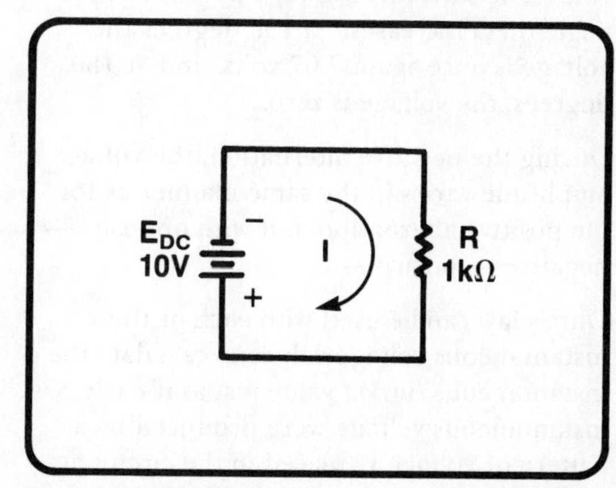

Figure 5.2 *The Resistive DC Circuit of Figure 5.1 With Battery Connections Reversed*

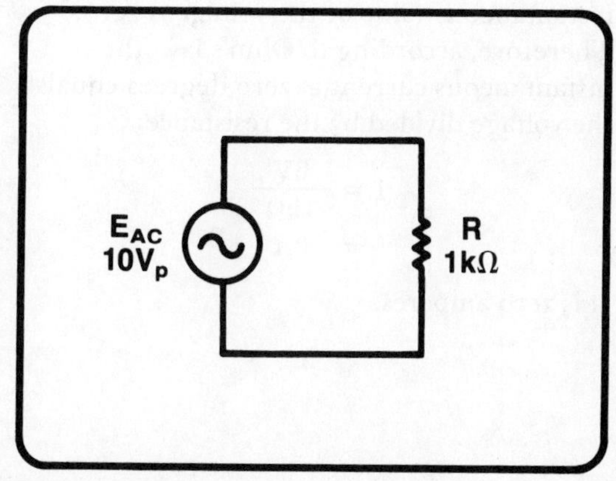

Figure 5.3 *The Resistive Circuit With AC Voltage Applied*

■ Calculating Instantaneous Voltages and Currents

The instantaneous voltage is equal to the peak voltage times the sine theta.

Figure 5.5 shows a sinusoidal waveform that can be described with equation *5–2*. The peak amplitude of that waveform is 10 volts. Since the sine of zero degrees is zero, at zero degrees the voltage is zero. At 45 degrees the voltage has risen to 7.07 volts calculated as follows:

$$\begin{aligned} e_i &= E_{pk}\sin\theta \\ &= (10)\,(0.707) \\ &= 7.07V \end{aligned}$$

At 90 degrees the voltage has risen to 10 volts. The waveform voltage declines as the angle theta increases. At 135 degrees the voltage is once again 7.07 volts, and at 180 degrees, the voltage is zero.

During the negative alternation, the voltage amplitude varies in the same manner as for the positive alternation, but with opposite— negative— polarities.

Ohm's law can be used with each of these instantaneous voltage values to calculate the instantaneous *current* value just as if each instantaneous voltage were produced by a battery of voltage E_i placed in the circuit at the right moment as shown in *Figure 5.6*. At zero degrees (point A of *Figure 5.5*) the instantaneous value of the voltage is zero. Therefore, according to Ohm's law, the instantaneous current at zero degrees equals the voltage divided by the resistance:

$$\begin{aligned} i &= \frac{0V}{1k\Omega} \\ &= 0A \end{aligned}$$

It is zero amperes.

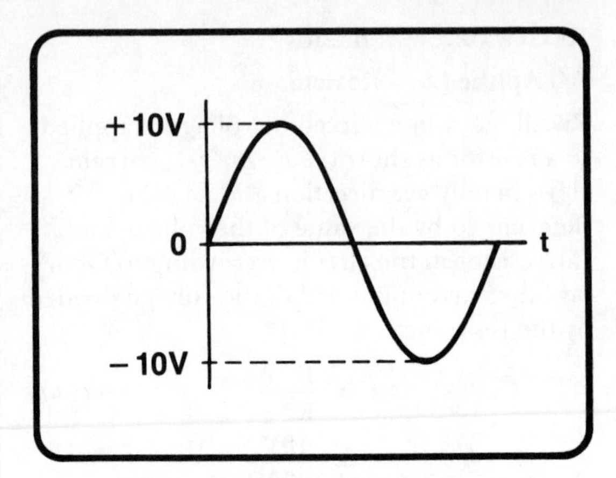

Figure 5.4 *An AC Voltage is Changing Constantly*

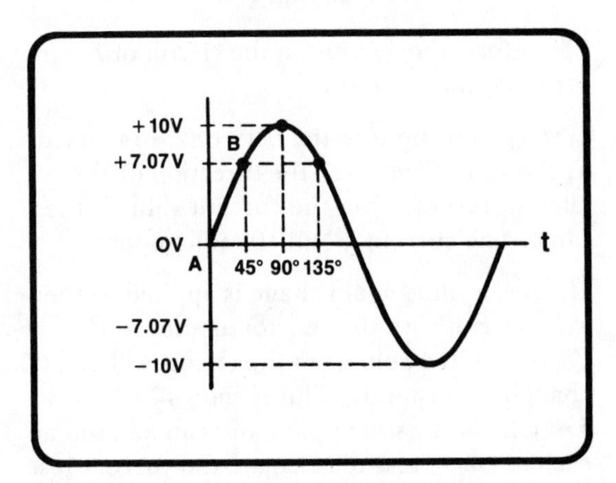

Figure 5.5 *A 10-Volt Peak Sinusoid Waveform*

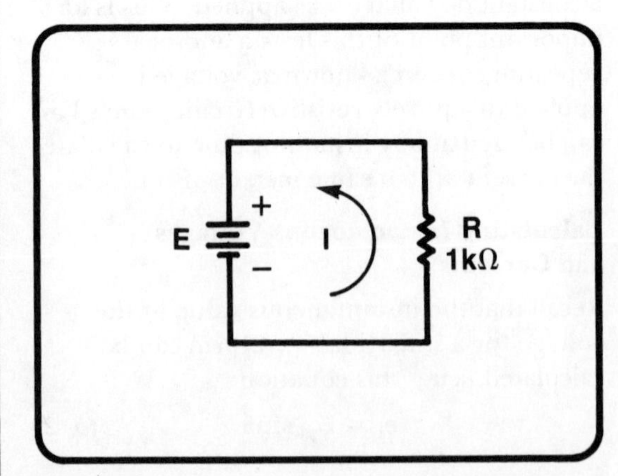

Figure 5.6 *DC Equivalent Circuit Used to Calculate Instantaneous Current Values*

At 45 degrees (point B of *Figure 5.5)* the instantaneous voltage, e_i, is 7.07 volts. This can be represented by a 7.07 volt battery as shown in the circuit in *Figure 5.7*. Using Ohm's law as before, the current at 45 degrees is:

$$i = \frac{7.07V}{1k\Omega} = 7.07mA$$

The current flowing at that instant in the cycle is 7.07 milliamperes.

Instantaneous current values continue to vary this way throughout the entire cycle as shown in *Figure 5.8*. In the negative alternation, however, the current is plotted as negative because its direction of flow is opposite to the direction called positive. If all of these plotted points are connected with a smooth curve, as shown in *Figure 5.9*, it is found that the current is also sinusoidal.

Relationship of Peak, Peak-to-Peak, and RMS Values

Recall that the amplitude of a sinusoidal voltage can be specified in three ways: peak, peak-to-peak, and rms. The voltage waveform in the previous example has a peak amplitude of 10 volts as shown in *Figure 5.10*. Thus, it has a peak-to-peak amplitude of 20 volts.

$$E_{pp} = 2(E_{pk}) = (2)(10) = 20V$$

It also has an rms amplitude of

$$E_{rms} = 0.707(E_{pk}) = (0.707)(10) = 7.07V$$

Thus, its rms voltage is 7.07 volts.

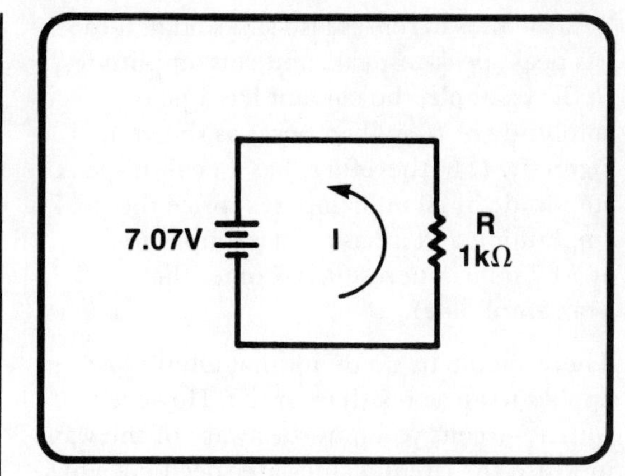

Figure 5.7 DC Equivalent Circuit When AC Instantaneous Voltage, e_i, is 7.07 Volts

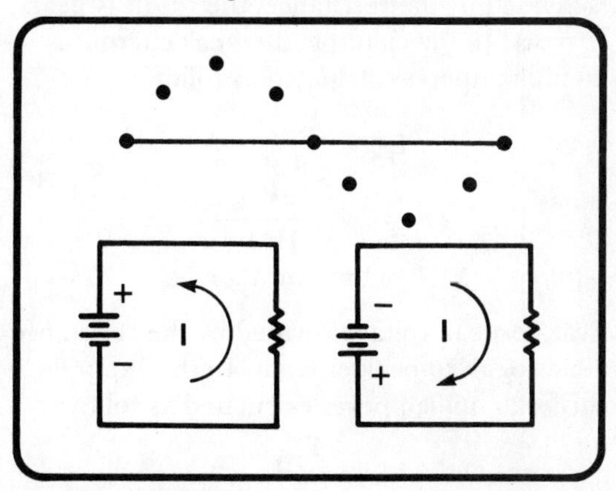

Figure 5.8 Instantaneous Current Values During One Cycle

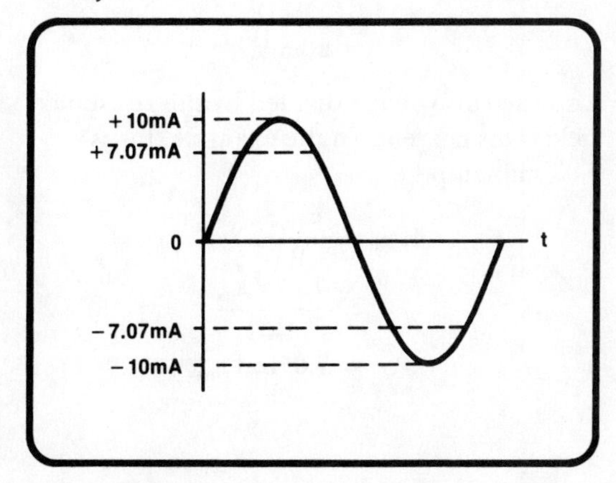

Figure 5.9 The Current Waveform is Sinusoidal

Because the current is also sinusoidal, it too has peak, peak-to-peak, and rms amplitudes. In the example, the current has a peak amplitude of 10 milliamperes as shown in *Figure 5.11*. It, therefore, has a peak-to-peak amplitude of 20 milliamperes (twice the peak amplitude); and it has an rms amplitude of 7.07 milliamperes (0.707 times the peak amplitude).

There should be no doubt that Ohm's law applies to an ac resistive circuit. However, with ac circuits you must be aware of the way in which the circuit values are specified. For example, if the peak voltage of the waveform is divided by the resistance, the result is peak current. In the example the peak current is 10 milliamperes calculated as follows:

$$I_{pk} = \frac{E_{pk}}{R} = \frac{10V_{pk}}{1k\Omega} = 10mA_{pk}$$

Peak-to-peak voltage divided by the resistance yields peak-to-peak current. In the example this is 20 milliamperes calculated as follows:

$$I_{pp} = \frac{E_{pp}}{R} = \frac{20V_{pp}}{1k\Omega} = 20mA_{pp}$$

And the rms voltage divided by the resistance yields rms current. In the example this is 7.07 milliamperes.

$$I_{rms} = \frac{E_{rms}}{R} = \frac{7.07V_{rms}}{1k\Omega} = 7.07mA_{rms}$$

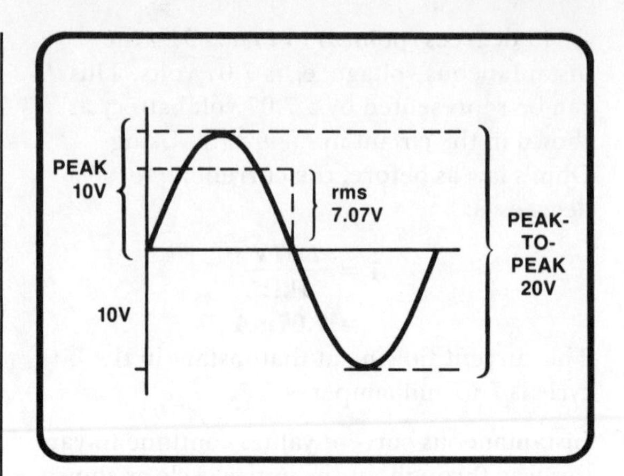

Figure 5.10 *Voltage Waveform for Example Circuit*

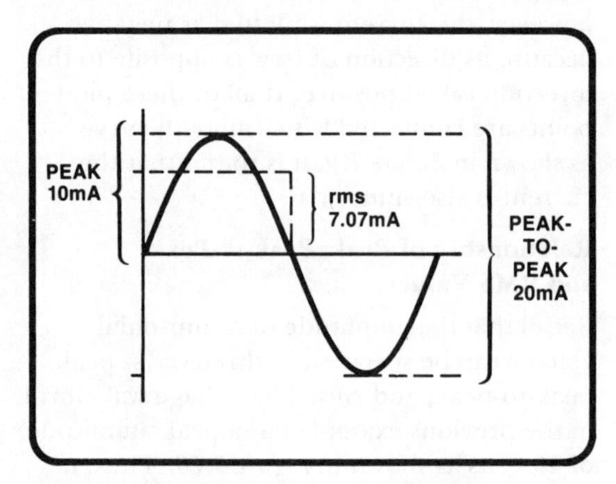

Figure 5.11 *Current Waveform for Example Circuit*

Converting Values

An applied voltage or current can by specified in one of these three ways. But suppose a current or voltage value is specified as peak and its rms amplitude is needed. As you may recall from previous discussions, it is easy to convert the peak specification to its rms value by calculation. RMS voltage or current is equal to 0.707 times the peak voltage or peak current.

$$E_{rms} = 0.707\ E_{pk} \quad (5\text{–}3)$$

$$I_{rms} = 0.707\ I_{pk} \quad (5\text{–}4)$$

Or it may be necessary to convert from any one of these peak, peak-to-peak, and rms specifications to another. A table, Table 5.1, of conversion equations is included to enable you to convert from any one of these specifications to another for either voltage or current.

Voltage and Current Phase Relationship

If the voltage waveform and the current waveform are examined together as shown in *Figure 5.12*, it can be seen that when the voltage is at a maximum or peak, the current is also at a maximum or peak. The current flow is less as the value of the instantaneous voltage decreases. When the voltage is at zero, the current is at zero. Also note that the current is flowing in a direction determined by the polarity of the voltage. That is, in the example shown, a *positive* 10 milliamperes flows when the voltage is a *positive* 10 volts and a *negative* 10 milliamperes flows when the voltage is a *negative* 10 volts.

As shown in *Figure 5.13*, the voltage and current peak at the same time. Because they both peak at the same time, and are zero at the same time, and move in the same direction, the voltage and current are *in phase*. This means there is *no phase* difference between them. Their amplitudes and polarities are the same for each instant of time. Therefore, the phase angle between them is zero.

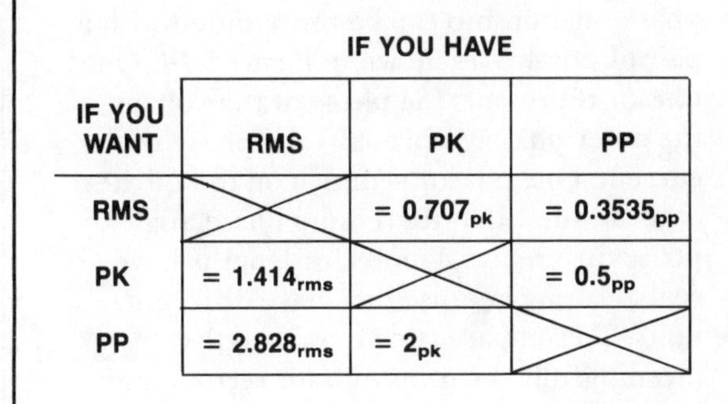

IF YOU WANT	IF YOU HAVE RMS	PK	PP
RMS		$= 0.707_{pk}$	$= 0.3535_{pp}$
PK	$= 1.414_{rms}$		$= 0.5_{pp}$
PP	$= 2.828_{rms}$	$= 2_{pk}$	

Table 5.1 *Conversions*

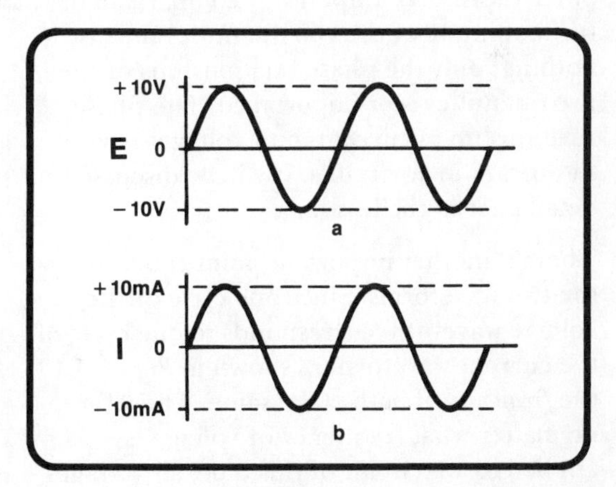

Figure 5.12 *a. Voltage Waveform; b. Current Waveform*

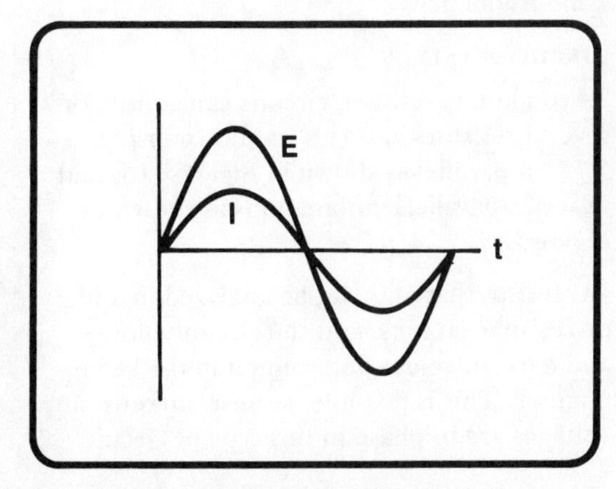

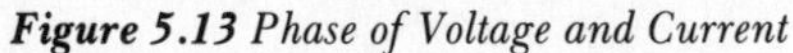

Figure 5.13 *Phase of Voltage and Current*

■ Voltage and Current Phase Relationship

As you recall from previous sessions, this phase relationship can be represented with a pair of phasors as shown in *Figure 5.14*. One phasor represents the phase of the voltage; the other phasor represents the phase of the current. One phasor is drawn on top of the other to show that there is no phase angle or phase difference. The vector lengths have no real meaning because they have different units. To compare two vector lengths meaningfully the units of both vectors must be the same—both must represent volts, amperes, ohms, milliamperes, etc. In *Figure 5.14* one vector represents volts while the other represents amperes. A comparison of the amplitude of the two quantities means nothing; only the phase relationships of the two quantities is of importance. The phase relationship of currents and voltages is very significant in ac circuits. It will be discussed in detail later in the lesson.

There's another important point concerning the two waveforms. Since one cycle of the voltage waveform corresponds to one cycle of the current waveform as shown in *Figure 5.13* the *frequency* of both is the same. Therefore, no matter what frequency of voltage is applied to *any* circuit of this type, all voltages and currents in that circuit will have the same frequency.

CIRCUIT ANALYSIS

As you know, resistive circuits can consist of several resistors in series, as shown in *Figure 5.15*, in parallel as shown in *Figure 5.16*, and in series-parallel combinations as shown in *Figure 5.17*.

AC resistive circuits can be analyzed just like dc resistive circuits, and the circuit voltage and current values determined in the same manner. This is possible because currents and voltages are in phase in this type of circuit.

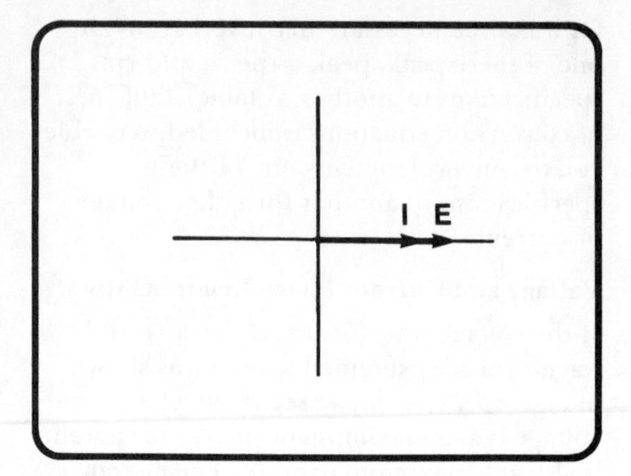

Figure 5.14 *Voltage, E, and Current, I, Vectors*

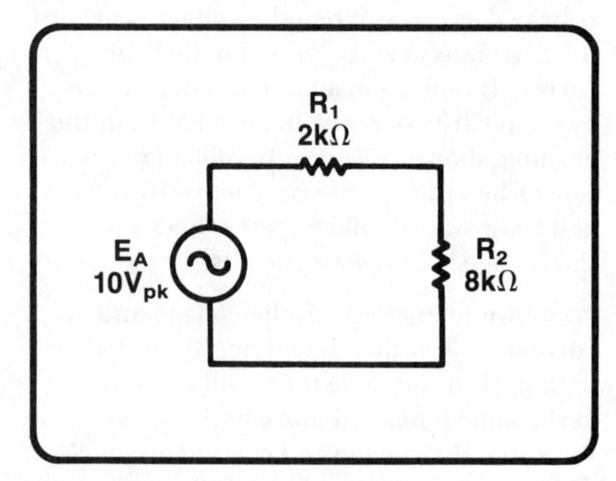

Figure 5.15 *Series Resistive AC Circuit*

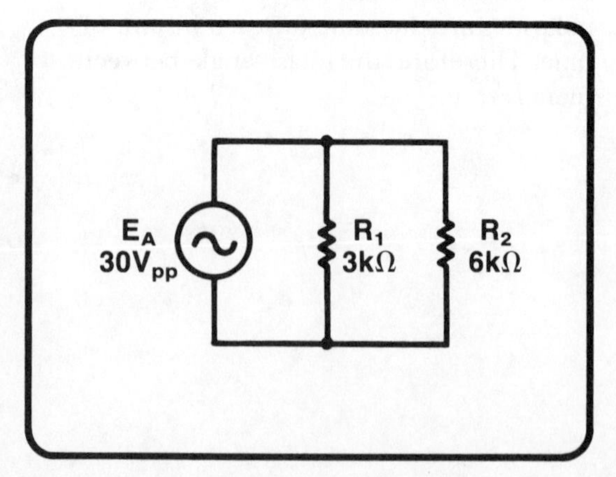

Figure 5.16 *Parallel Resistive AC Circuit*

■ Series Resistive AC Circuit

The voltage or current values can be described in peak, peak-to-peak, or rms values. You must make the necessary conversions to obtain your answers in the desired units.

Series Resistive AC Circuit

The first circuit to be analyzed is the series circuit shown in *Figure 5.15*. There are two resistors in this circuit in series with an applied ac voltage of 10 volts peak. To determine the voltages and currents in this circuit, first the total resistance must be calculated. The total resistance in the circuit is the sum of the individual resistances,

$$\begin{aligned} R_T &= R_1 + R_2 \\ &= 2k\Omega + 8k\Omega \\ &= 10k\Omega \end{aligned}$$

The total resistance of the circuit is 10 kilohms. Once R_T has been determined the voltages and currents in the circuit can be calculated.

Since the total resistance we calculated is 10 kilohms, to find the total current in the circuit, divide the applied voltage by the total resistance:

$$\begin{aligned} I_T &= \frac{E_A}{R_T} \\ &= \frac{10V_{pk}}{10k\Omega} \\ &= 1mA_{pk} \end{aligned}$$

The total current in the circuit is 1 milliampere peak. Since this is a series circuit, this current is the only current in the circuit. It follows only one path—through both resistors in the circuit.

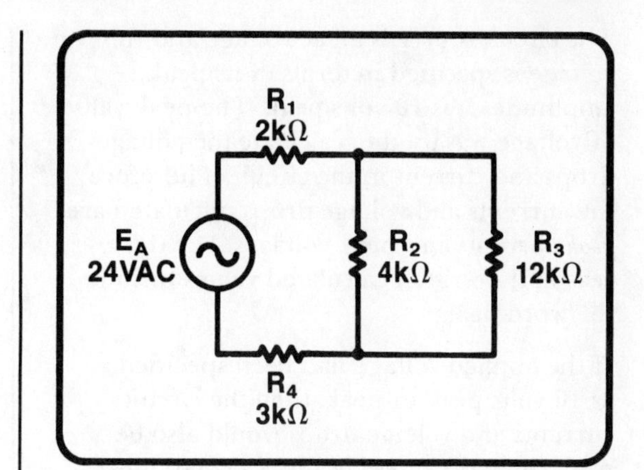

Figure 5.17 Series-Parallel Resistive AC Circuit

Next, the voltage drops across each resistor can be found using Ohm's law. The voltage drop across R_1 is equal to the current through R_1, which is the total current, times the value of R_1:

$$\begin{aligned} E_{R1} &= I_T R_1 \\ &= (1mA_{pk})(2k\Omega) \\ &= 2V_{pk} \end{aligned}$$

The drop across R_2 is calculated similarly:

$$\begin{aligned} E_{R2} &= I_T R_2 \\ &= (1mA_{pk})(8k\Omega) \\ &= 8V_{pk} \end{aligned}$$

Notice that Kirchhoff's voltage law is satisfied because the voltage drop across R_1, 2 volts peak, plus the voltage drop across R_2, 8 volts peak, equals the applied voltage, 10 volts peak:

$$\begin{aligned} E_T &= E_{R1} + E_{R2} \\ 10_{pk} &= 2V_{pk} + 8V_{pk} \end{aligned}$$

■ Series Resistive AC Circuit

The circuit source is an ac voltage and this voltage is specified in terms of its peak amplitude. It is 10 volts peak. The peak value of voltage was used to calculate the voltage drops and current in the circuit. Therefore, the currents and voltage drops calculated are *peak* currents and peak voltages. For this reason the written calculated values include the word peak.

If the applied voltage had been specified as 10 volts peak-to-peak, then the circuit currents and voltage drops would also be peak-to-peak values because calculations would be performed using the 10 volts peak-peak. The specification peak-to-peak would be written along with the calculated value as shown in *Figure 5.18*. Similarly, if the applied voltage had been specified as rms, then the voltage drops and currents calculated using this rms value would also be rms values.

In summary then, to determine the voltages and currents in a resistive circuit with an ac voltage source, calculate the voltages and currents in the circuit the same as these factors are calculated in a dc series circuit. The circuit voltages and currents are specified the same as the ac applied voltage: peak, peak-to-peak or rms.

Once all voltages and currents in a circuit have been calculated using one amplitude specification, the other two types of amplitude values of these same circuit currents and voltages can be determined readily by directly converting each value using the appropriate conversion equation.

Using the series circuit problem we just solved as an example, a table of values can be created as shown in *Figure 5.19*. In the column labeled peak are all the peak values calculated for the circuit. The peak-to-peak value is twice their peak value. The rms

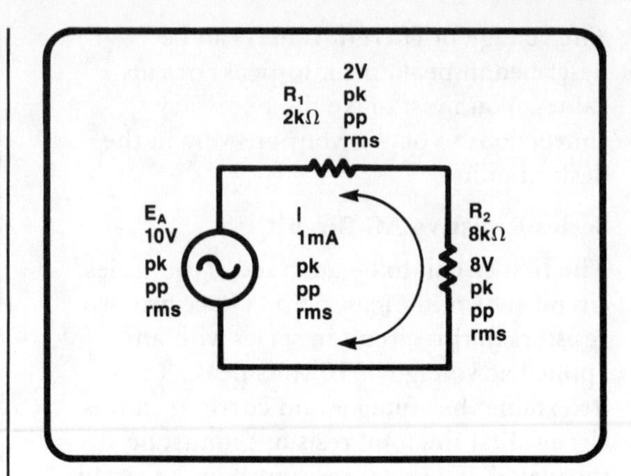

Figure 5.18 *Specification of E_A as Peak, Peak-to-Peak, or RMS, Determines Specification of Other Circuit Values*

	PEAK	PEAK-TO-PEAK	RMS
E_A	$10V_p$	$20V_{pp}$	7.07V
E_{R1}	$2V_p$	$4V_{pp}$	1.414V
E_{R2}	$8V_p$	$16V_{pp}$	5.656V
I_T	$1mA_p$	$2mA_{pp}$	707μA

Figure 5.19 *Calculated Values for Series AC Circuit Example*

■ **Series Resistive AC Circuit**

values are 0.707 of the peak values. For example, the rms values have been calculated using equations *5–3* and *5–4* as follows:

$$E_{rms} = 0.707E_{pk}$$
$$E_{Arms} = (0.707)(10V) = 7.07V$$
$$E_{R1rms} = (0.707)(2V) = 1.414V$$
$$E_{R2rms} = (0.707)(8V) = 5.656V$$
$$I_{rms} = 0.707I_{pk}$$
$$I_{Trms} = (0.707)(1mA) = 0.707mA = 707\mu A$$

In other words, in the circuit, the applied voltage is 10 volts peak, which equals 20 volts peak-to-peak or 7.07 volts rms. These values are shown graphically in *Figure 5.20*.

The 2 volt peak voltage drop across R_1 equals 4 volts peak-to-peak or 1.414 volts rms as shown in *Figure 5.21*. The three values, 8 volts peak, 16 volts peak-to-peak and 5.656 volts rms are shown in *Figure 5.22* for E_{R2}.

The circuit current shown graphically in *Figure 5.23*, is 1 milliampere peak. This equals 2 milliamperes peak-to-peak or 707 microamperes rms.

Remember that all voltages and currents in this purely resistive circuit are *in phase* and of the *same frequency*. Also note in *Figure 5.19* that in each column the value of E_{R1} and E_{R2} add up to the applied voltage in each column.

If a circuit has the source voltage specified in a peak-to-peak value, but answers are required in rms values, the voltages and currents can be determined one of two ways: either by using the peak-to-peak values and then converting all answers to rms values or by converting the source voltage to an rms value before performing any other calculations. Then all of the following calculations will be in rms values.

Also an important point to remember is that whenever the designation VAC is used, it is understood, by convention, to mean an rms value.

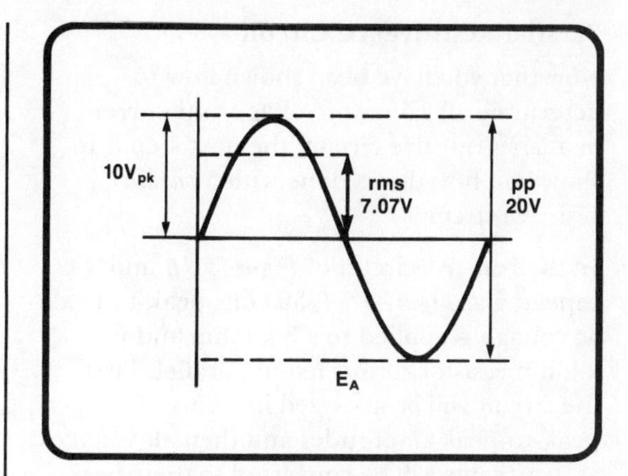

Figure 5.20 *E_A for Series AC Circuit Example*

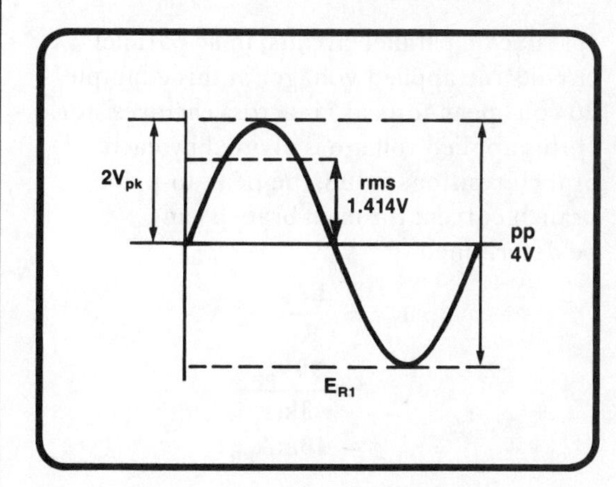

Figure 5.21 *E_{R1} for Series Circuit Example*

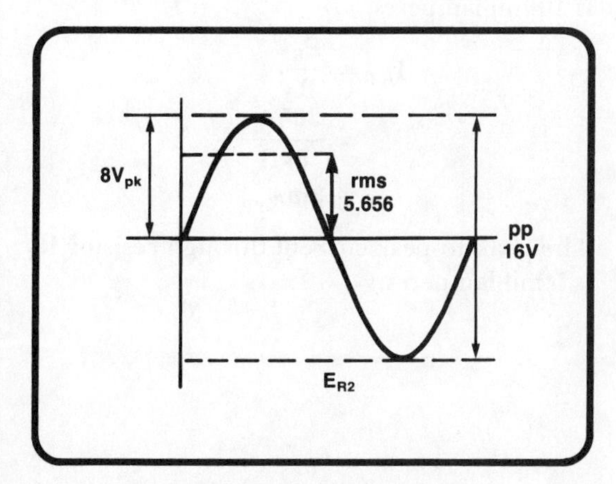

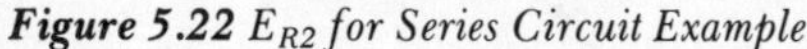

Figure 5.22 *E_{R2} for Series Circuit Example*

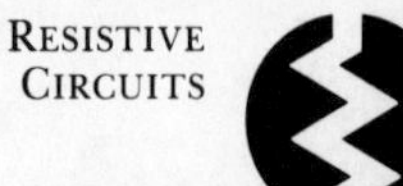

■ Parallel Resistive AC Circuit

Parallel Resistive AC Circuit

Now that you have been shown how to determine all values of voltage and current in a *series* resistive circuit, the next step is to show you how this is done with a *parallel* resistive circuit.

In the resistive circuit of *Figure 5.16* and repeated as *Figure 5.24*, 30 volts peak-to-peak ac voltage is applied to a 3 kilohm and 6 kilohm resistor connected in parallel. First the circuit will be analyzed in terms of peak-to-peak amplitudes and then all voltages and currents will be converted to their peak, and rms amplitudes.

Just like dc parallel circuits, in ac parallel circuits the applied voltage (in this example 30 volts peak-to-peak) is across each resistor. If this applied voltage is divided by each branch resistor's value, the peak-to-peak branch current for each branch can be determined.

$$I_{R1} = \frac{E_{R1}}{R_1} = \frac{30V_{pp}}{3k\Omega} = 10mA_{pp}$$

The peak-to-peak current through resistor R_1 is 10 milliamperes.

$$I_{R2} = \frac{E_{R2}}{R_2} = \frac{30V_{pp}}{6k\Omega} = 5mA_{pp}$$

The peak-to-peak current through resistor R_2 is 5 milliamperes.

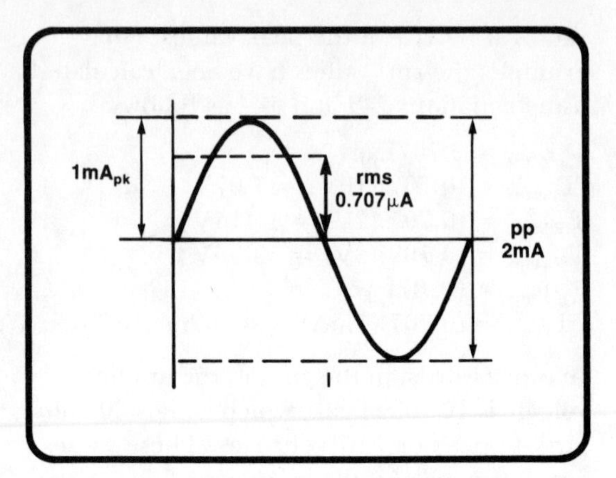

Figure 5.23 *I for Series Circuit Example*

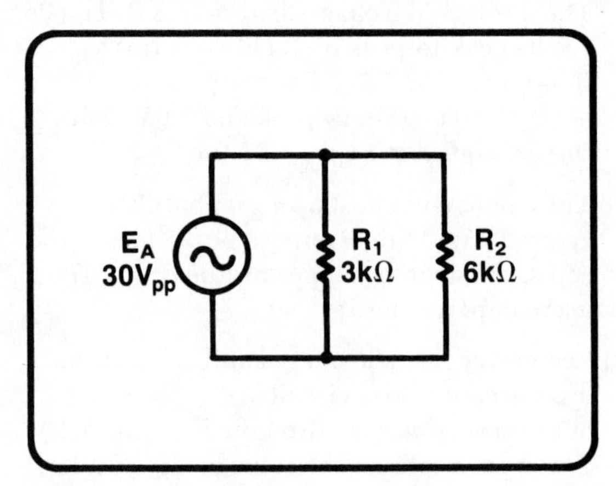

Figure 5.24 *Parallel Resistive AC Circuit*

The total current is the sum of the branch currents. Since the branch currents are known, they can be added to determine the total current.

$$I_T = I_{R1} + I_{R2} = 10mA_{pp} + 5mA_{pp} = 15mA_{pp}$$

The total current is 15 milliamperes peak-to-peak.

■ Parallel Resistive AC Circuit

Therefore, as in a dc parallel circuit, the *current* through each branch of an ac parallel circuit may be different depending on the resistor values, while the voltage drop across each resistor is the same. Also, the algebraic sum of the branch currents equals the total current.

To verify that this is the correct total current the applied voltage can be divided by the total resistance in the circuit. Recall that the total resistance for two parallel resistors is their product divided by their sum. A special symbol is used to indicate the parallel connection. Therefore,

$$\mathbf{R_1 \| R_2}$$

indicates that R_1 is in parallel with R_2. Thus,

$$\mathbf{R_1 \| R_2 = \frac{R_1 \times R_2}{R_1 + R_2}} \qquad (5\text{–}5)$$

is the expression to calculate the total resistance of two resistors connected in parallel.

In the example circuit of *Figure 5.24*, the total resistance, calculated by using equation *5–5*, is:

$$\begin{aligned} \mathbf{R_T} &= \mathbf{\frac{R_1 \times R_2}{R_1 + R_2}} \\ &= \mathbf{\frac{(3k)(6k)}{3k + 6k}} \\ &= \mathbf{\frac{18k}{9k}} \\ &= \mathbf{2k\Omega} \end{aligned}$$

The total current in the circuit is calculated using Ohm's law:

$$\begin{aligned} \mathbf{I_T} &= \mathbf{\frac{E_T}{R_T}} \\ &= \mathbf{\frac{30V_{pp}}{2k\Omega}} \\ &= \mathbf{15mA_{pp}} \end{aligned}$$

This calculation shows that 15 milliamperes peak-to-peak current is the total current and verifies previous calculations. Using these peak-to-peak values for the circuit, the peak and rms values can be calculated. Peak values are one-half the peak-to-peak values. The rms values are 0.707 times the peak value or 0.3535 times the peak-to-peak value. The calculations for the peak values are as follows:

$$E_{Apk} = \frac{30V_{pp}}{2} = 15V$$

$$I_{R1pk} = \frac{10mA_{pp}}{2} = 5mA$$

$$I_{R2pk} = \frac{5mA_{pp}}{2} = 2.5mA$$

$$I_{Tpk} = \frac{15mA_{pp}}{2} = 7.5mA_{pp}$$

The calculations for the rms values are as follows:

$$\begin{aligned} E_{Arms} &= (0.3535)(30V_{pp}) &&= 10.6V \\ I_{R1rms} &= (0.3535)(10mA_{pp}) &&= 3.54mA \\ I_{R2rms} &= (0.3535)(5mA_{pp}) &&= 1.77mA \\ I_{Trms} &= (0.3535)(15mA_{pp}) &&= 5.30mA \end{aligned}$$

Once calculated they can be written in tabular form as shown in *Figure 5.25*. Remember that all voltages and currents are in phase, and that they all have the same frequency. Also, the branch currents sum to the total current in each column.

■ Series-Parallel Resistive AC Circuit
■ Simplified Circuit

Series-Parallel Resistive AC Circuit

Now that a series resistive circuit and a parallel-resistive circuit have been analyzed, the next circuit to be studied is a series-parallel resistive circuit shown in *Figure 5.26*. This is the same circuit as *Figure 5.17*.

In the circuit, 24 volts ac is applied to a series-parallel combination of four resistors. Recall that if an ac voltage is specified as VAC then the value is an rms amplitude. That is, the applied voltage is 24 volts rms.

Simplified Circuit

The series-parallel combination of resistors consists of R_1 and R_4 in series with the parallel combination of R_2 and R_3. To begin determining the voltages and currents in this circuit, as with dc circuits, first the circuit is simplified by finding the equivalent resistance of R_2 and R_3 in parallel. It is calculated using the form of equation 5–5.

$$R_{23} = R_2 \| R_3 = \frac{R^2 \times R^3}{R^2 + R^3}$$

$$= \frac{(4k)(12k)}{4k + 12k}$$

$$= \frac{48k}{16k}$$

$$= 3k\Omega$$

Thus, the parallel combination of R_2 and $R_3(R_{23})$ can be thought of as a 3 kilohm resistor substituted in the circuit as shown in *Figure 5.27*. As you probably realize, the circuit is now a simple series circuit. Therefore, its total resistance is equal to the sum of the individual resistances.

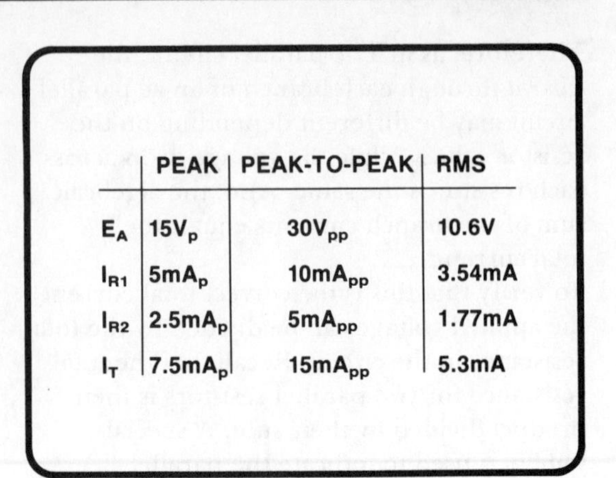

	PEAK	PEAK-TO-PEAK	RMS
E_A	$15V_p$	$30V_{pp}$	10.6V
I_{R1}	$5mA_p$	$10mA_{pp}$	3.54mA
I_{R2}	$2.5mA_p$	$5mA_{pp}$	1.77mA
I_T	$7.5mA_p$	$15mA_{pp}$	5.3mA

Figure 5.25 *Calculated Values of the Circuit of Figure 5.24*

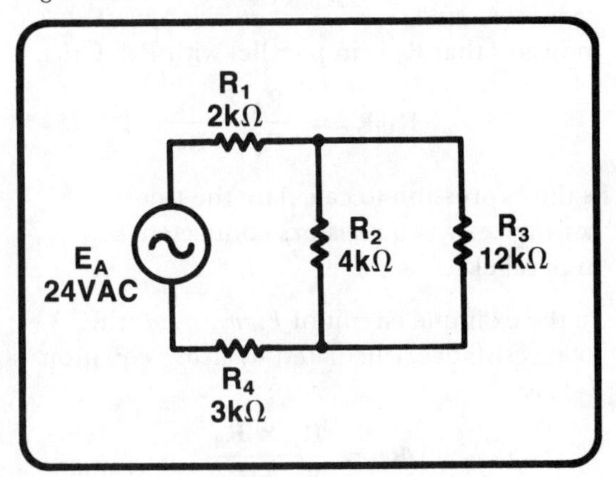

Figure 5.26 *Series-Parallel Resistive AC Circuit*

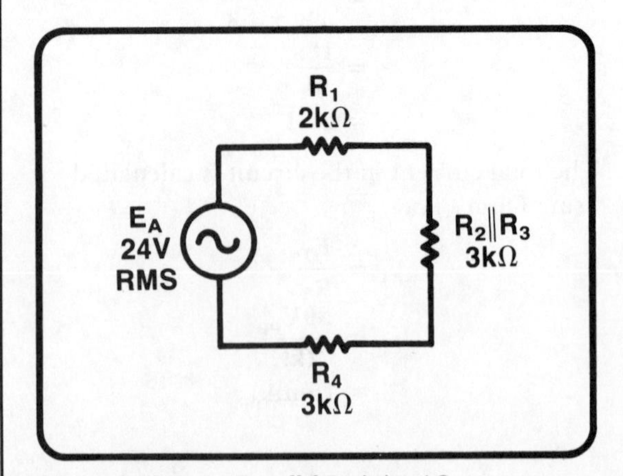

Figure 5.27 *Series-Parallel Resistive AC Circuit Simplified*

Total Resistance

The total resistance equals 2 kilohms for R_1, plus 3 kilohms for the parallel combination of R_2 and R_3, plus 3 kilohms for R_4:

$$\begin{aligned} R_T &= R_1 + (R_2 \| R_3) + R_4 \\ &= 2K\Omega + 3K\Omega + 3k\Omega \\ &= 8k\Omega \end{aligned}$$

The total resistance is 8 kilohms.

Total Current

To determine the total current supplied by the ac voltage generator to the circuit the applied voltage is divided by the total resistance. 24 volts divided by 8 kilohms is:

$$\begin{aligned} I_T &= \frac{E_A}{R_T} \\ &= \frac{24V_{rms}}{8k\Omega} \\ &= 3mA_{rms} \end{aligned}$$

The total current is 3mA rms. This total current flows through R_1, through the *parallel combination* of R_2 and R_3, and through R_4. Therefore, I_{R1} and I_{R4} are $3mA_{rms}$.

Voltage Drops

Ohm's law is now used to determine the *voltage drop* across each resistance. It is the value of the resistance multiplied by the current through it. Thus,

$$\begin{aligned} E_{R1} &= I_T R_1 \\ &= (3mA_{rms})(2k\Omega) \\ &= 6V_{rms} \end{aligned}$$

The voltage drop across R_1 is 6 volts rms.

The voltage drop across R_2 in parallel with R_3 is equal to the total current times the value of the equivalent parallel resistance R_{23}:

$$\begin{aligned} E_{R23} &= I_T R_{23} \\ &= (3mA_{rms})(3K\Omega) \\ &= 9V_{rms} \end{aligned}$$

The voltage drop across R_{23} is 9 volts rms.

The voltage drop across R_4 is calculated in a similar manner as:

$$\begin{aligned} E_{R4} &= I_T R_4 \\ &= (3mA_{rms})(3k\Omega) \\ &= 9V_{rms} \end{aligned}$$

The voltage drop across R_4 is also 9 volts rms.

Branch Currents

Now the current through R_2 and R_3 can be calculated. Since there are 9 volts across R_{23}, the parallel combination of R_2 and R_3, there are 9 volts across R_2 and 9 volts across R_3. Thus, you see that this parallel combination is treated just like the separate parallel circuit. Using Ohm's law the value of the current through R_2 and through R_3 (these are like the branch circuits in the separate parallel circuit) can be determined by dividing the voltage across either resistor by the value of the resistor.

For R_2, the value of the current is:

$$\begin{aligned} I_{R2} &= \frac{E_{R2}}{R_2} \\ &= \frac{9V_{rms}}{4k\Omega} \\ &= 2.25mA_{rms} \end{aligned}$$

I_{R2} of *Figure 5.26* is 2.25 milliamperes rms.

The current through R_3 is calculated similarly:

$$I_{R3} = \frac{E_{R3}}{R_3}$$

$$= \frac{9V_{rms}}{12k\Omega}$$

$$= 0.75mA_{rms}$$

I_{R3} of *Figure 5.26* is 0.75 milliamperes rms.

Voltage Chart

Figure 5.28 shows a tabulation of the voltages in this series-parallel circuit. These values are in the rms column since the applied voltages and all the voltage drops calculated are rms. These rms values can be converted directly to both peak, and peak-to-peak values. The peak values are simply 1.414 times the rms values. To determine the peak-to-peak amplitude, either the peak values can be doubled, or the rms values can be multiplied by 2.828. Note that in each column the voltage across R_2 and R_3 is the same since they are in parallel, and this voltage along with the other series circuit voltages E_{R1} and E_{R4}, add up to the applied voltage.

Current Chart

Figure 5.29 shows a tabulation of the values of the currents for the circuit for *Figure 5.26*. The rms current values are the original ones calculated and these values are used to convert to peak and peak-to-peak current values. The peak current calculations are as follows:

$$I_{R1pk} = (1.414)(3mA) = 4.24mA$$
$$I_{R2pk} = (1.414)(2.25mA) = 3.18mA$$
$$I_{R3pk} = (1.414)(0.75mA) = 1.06mA$$
$$I_{R4pk} = (1.414)(3mA) = 4.24mA$$

	PEAK	PEAK-TO-PEAK	RMS
E_A	$33.9V_p$	$67.8V_{pp}$	24V
E_{R1}	$8.5V_p$	$17V_{pp}$	6V
E_{R2}	$12.7V_p$	$25.4V_{pp}$	9V
E_{R3}	$12.7V_p$	$25.4V_{pp}$	9V
E_{R4}	$12.7V_p$	$25.4V_{pp}$	9V

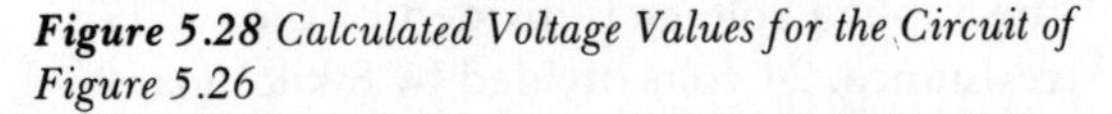

Figure 5.28 *Calculated Voltage Values for the Circuit of Figure 5.26*

	PEAK	PEAK-TO-PEAK	RMS
I_{R1}	4.24mA	8.48mA	3mA
I_{R2}	3.18mA	6.36mA	2.25mA
I_{R3}	1.06mA	2.12mA	0.75mA
I_{R4}	4.24mA	8.48mA	3mA

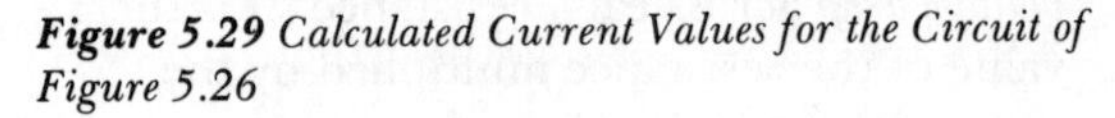

Figure 5.29 *Calculated Current Values for the Circuit of Figure 5.26*

■ Current Chart

The peak-to-peak current values are as follows:

$$I_{R1pp} = (2)(4.24mA) = 8.48mA$$
$$I_{R2pp} = (2)(3.18mA) = 6.36mA$$
$$I_{R3pp} = (2)(1.06mA) = 2.12mA$$
$$I_{R4pp} = (2)(4.24mA) = 8.48mA$$

The results are a bit different through the voltage chart. Now the two branch currents through the parallel combination of R_2 and R_3 add to the total series current, which is the same through R_1 and R_4.

The effects of ac voltages and currents in series, parallel, and series-parallel resistive circuits have now been discussed thoroughly. However, there is another important consideration in these circuits — the *power dissipated.*

POWER DISSIPATION

In ac circuits there are three kinds of power: *real, reactive* and *apparent* power. The kind of power that exists in purely resistive circuits is what is called *real* power. *Real* power is power that is dissipated in the form of heat and is measured in *watts* as illustrated in *Figure 5.30*.

Recall, as shown in *Figure 5.31*, that in dc circuits the amount of power dissipated in watts can be calculated by multiplying the voltage across a resistor by the current passing through it.

$$P = EI \qquad (5\text{–}6)$$

The power, P, is in watts, when the voltage, E, is in volts, and the current, I, is in amperes. It turns out that equation *5–6* is true whether a dc voltage or an ac voltage is applied. This is shown in *Figures 5.30* and *5.31*.

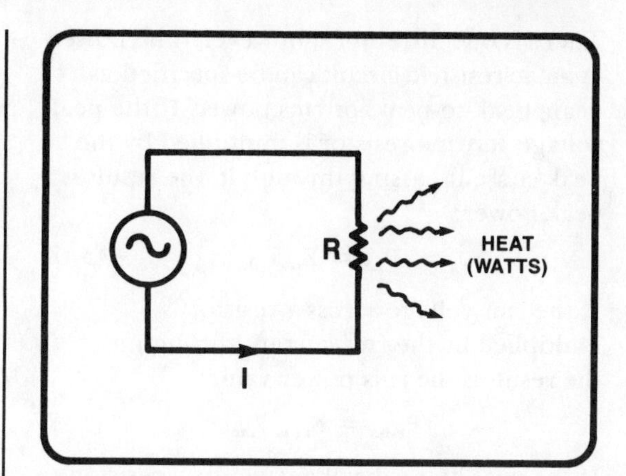

Figure 5.30 *Power Dissipation in an AC Circuit*

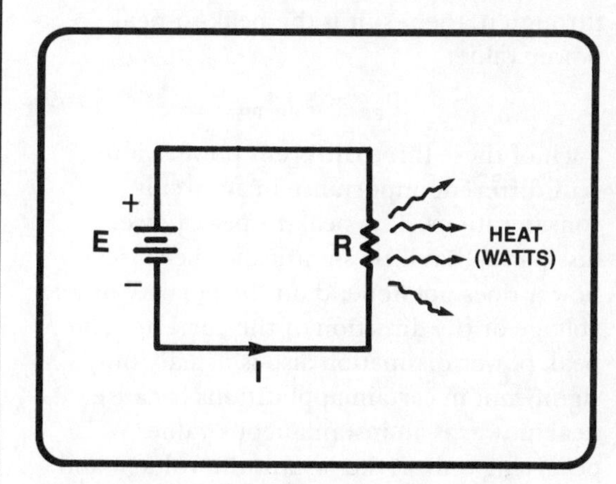

Figure 5.31 *Power Dissipation in a DC Circuit*

■ Power Dissipation

There is one difference, however. The power in an ac resistive circuit can be specified as peak, peak-to-peak, or rms power. If the peak voltage across a resistor is multiplied by the peak current passing through it, the result is peak power:

$$P_{pk} = E_{pk}I_{pk} \quad (5\text{–}7)$$

If the rms voltage across a resistor is multiplied by the rms current through it, the result is the rms power value:

$$P_{rms} = E_{rms}\,I_{rms} \quad (5\text{–}8)$$

If the peak-to-peak voltage across a resistor is multiplied by the peak-to-peak current through it, the result is the peak-to-peak power value:

$$P_{pp} = E_{pp}I_{pp} \quad (5\text{–}9)$$

Each of these three different power values is of different importance in ac circuit considerations. The peak-to-peak power dissipation has little significance because power does not depend on the polarity of a voltage or the direction of the current. The peak power dissipation also is usually only significant in certain applications because peak power is an instantaneous value occurring only at the instant the voltage and current are at their peak values. The most significant power specification in an ac circuit is the rms power. It is more significant because it is the type of ac power that can be equated with dc power. (Remember that rms was defined to provide the same equivalent heating effect as dc).

Using equation *5–6* the power dissipated by a resistor is determined by multiplying the voltage across the resistor by the current flowing through it.

$$P_R = E_RI_R \quad (5\text{–}10)$$

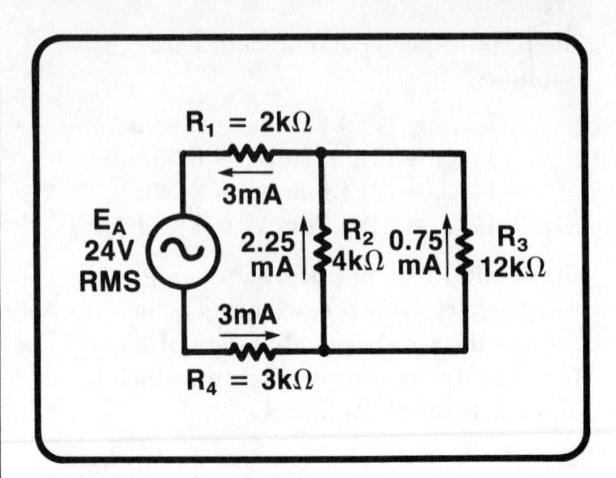

Figure 5.32 *Series-Parallel Resistive AC Circuit Power Calculation Example*

Thus, this equation can be used to calculate the rms power dissipated in series, parallel, and series-parallel ac resistive circuits. For example, it can be used to calculate the power dissipated in each of the four resistors in our previous series-parallel circuit shown again in *Figure 5.32*. This is the same circuit as *Figure 5.26* with the rms current calculated values added. Keep in mind that it is a series-parallel circuit and that the principles discussed can be applied to determine the rms power dissipated in simple series or simple parallel circuits as well.

The voltage drop across R1 of 6 volts rms was previously calculated. It is used along with the current through R1 to calculate the power dissipated in R1 as follows:

$$\begin{aligned} P_{R1} &= E_{R1}I_{R1} \\ &= (6V)(3mA) \\ &= 18mW \end{aligned}$$

■ **Summary**

The rms power dissipated by R_1 is 18 milliwatts.

The power being dissipated by resistors R_2, R_3, and R_4 can be calculated similarly. Recall that the voltage drop across the parallel combination of R2 and R3 was calculated as 9 volts and across R4 as 9 volts. Therefore,

$$\begin{aligned} P_{R2} &= E_{R2}I_2 \\ &= (9V)(2.25mA) \\ &= 20.25mW \end{aligned}$$

$$\begin{aligned} P_{R3} &= E_{R3}I_{R3} \\ &= (9V)(0.75mA) \\ &= 6.75mW \end{aligned}$$

$$\begin{aligned} P_{R4} &= E_{R4}I_{R4} \\ &= (9V)(3mA) \\ &= 27mW \end{aligned}$$

Therefore, the rms power dissipated by each of these resistors is: R_2 = 20.25 milliwatts; R_3 = 6.75 milliwatts; and R_4 = 27 milliwatts.

The total power being dissipated within the entire circuit equals the total applied voltage times the total current.

$$\begin{aligned} P_T &= E_AI_T \\ &= (24V)(3mA) \\ &= 72mW \end{aligned}$$

The total power in the circuit is 72 milliwatts.

All individual power dissipation values should add up to the total power calculated:

$$P_{R1} + P_{R2} + P_{R3} + P_{R4} = P_T$$
$$18mW + 20.25mW + 6.75mW + 27mW = 72mW$$

The total is 72 milliwatts which verifies the previous calculation for total power.

Remember that in all of these power calculation examples, calculations have been rms power.

SUMMARY

In this lesson purely resistive circuits with an ac voltage applied were analyzed. Calculations were performed to show you how to determine the amplitudes of the voltages, currents, and power dissipations in series, parallel, and series-parallel ac resistive circuits. You were also shown how to convert between the peak, peak-to-peak, and rms amplitude specifications of these quantities.

It should be remembered that all voltages and currents in these circuits have the same period and therefore, the same frequency. In pure resistive ac circuits there is no phase difference between the voltages and currents, so the voltage and current are said to be "in phase".

■ Worked-Out Examples

1. Given the peak-to-peak circuit values shown in the table, complete the table by converting to peak and rms values.

	peak-to-peak	peak	rms
E_{R1}	50V	a. ______	e. ______
E_{R2}	30V	b. ______	f. ______
E_{R3}	27V	c. ______	g. ______
E_{R4}	14V	d. ______	h. ______

Solution:

a. $E_{R1pk} = 0.5 \times E_{R1pp} = 0.5 \times 50V = \mathbf{25v}$

b. $E_{R2pk} = 0.5 \times E_{R2pp} = 0.5 \times 30 = \mathbf{15V}$

c. $E_{R3pk} = 0.5 \times E_{R3pp} = 0.5 \times 27V = \mathbf{13.5V}$

d. $E_{R4pk} = 0.5 \times E_{R4pp} = 0.5 \times 14V = \mathbf{7V}$

e. $E_{R1rms} = 0.707 \times E_{R1pk} = 0.707 \times 25V = \mathbf{17.7V}$

f. $E_{R2rms} = 0.707 \times E_{R2pk} = 0.707 \times 15V = \mathbf{10.6V}$

g. $E_{R3rms} = 0.707 \times E_{R3pk} = 0.707 \times 13.5V = \mathbf{9.54V}$

h. $E_{R4rms} = 0.707 \times E_{R4pk} = 0.707 \times 7V = \mathbf{4.95V}$

2. For the circuit below, calculate the instantaneous voltage and current values specified by the table:

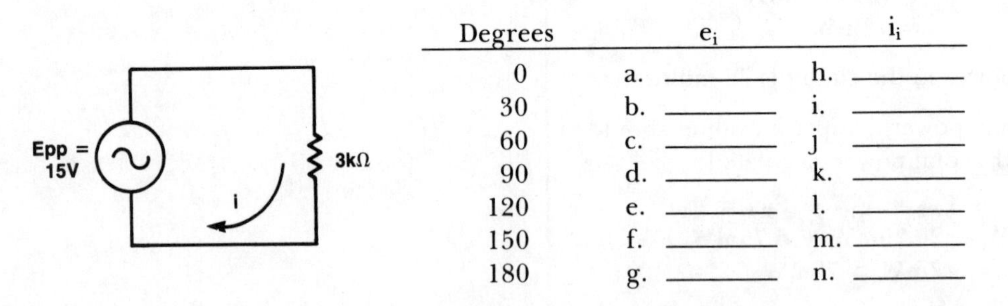

Degrees	e_i	i_i
0	a. ______	h. ______
30	b. ______	i. ______
60	c. ______	j ______
90	d. ______	k. ______
120	e. ______	l. ______
150	f. ______	m. ______
180	g. ______	n. ______

■ Worked-Out Examples

Solution:

First, E_{pk} is determined:

$E_{pk} = 0.5 \times E_{pp} = 0.5 \times 15V_{pp} = \mathbf{7.5V}$

Then the instantaneous voltage and current values can be calculated.

Finding $e_{i(x°)}$:

a. $e_{i(0°)} = \sin 0° \times E_{pk} = 0.00 \times 7.5V = \mathbf{0V}$

b. $e_{i(30°)} = \sin 30° \times E_{pk} = 0.5 \times 7.5V = \mathbf{3.75V}$

c. $e_{i(60°)} = \sin 60° \times E_{pk} = 0.866 \times 7.5V = \mathbf{6.5V}$

d. $e_{i(90°)} = \sin 90° \times E_{pk} = 1 \times 7.5V = \mathbf{7.5V}$

e. $e_{i(120°)} = e_{i(60°)} = \mathbf{6.5V}$

f. $e_{i(150°)} = e_{i(30°)} = \mathbf{3.75V}$

g. $e_{i(180°)} = e_{i(0°)} = \mathbf{0V}$

Finding $i_{(x°0)}$:

h. $i_{(0°)} = \dfrac{e_{i(0°)}}{R_T} = \dfrac{0V}{3k\Omega} = \mathbf{0.0A}$

i. $i_{(30°)} = \dfrac{e_{i(30°)}}{R_T} = \dfrac{3.75V}{3k\Omega} = \mathbf{1.25mA}$

j. $i_{(60°)} = \dfrac{e_{i(60°)}}{R_T} = \dfrac{6.5V}{3k\Omega} = \mathbf{2.17mA}$

k. $i_{(90°)} = \dfrac{e_{i(90°)}}{R_T} = \dfrac{7.5V}{3k\Omega} = \mathbf{2.5mA}$

l. $i_{(120°)} = i_{(60°)} = \mathbf{2.17mA}$

m. $i_{(150°)} = i_{(30°)} = \mathbf{1.25mA}$

n. $i_{(180°)} = i_{(0°)} = \mathbf{0.0A}$

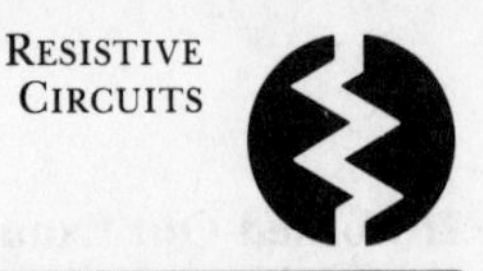

■ Worked-Out Examples

3. For the circuit in Example 2, plot the current and voltage waveforms based on the calculated instantaneous values.

Solution:

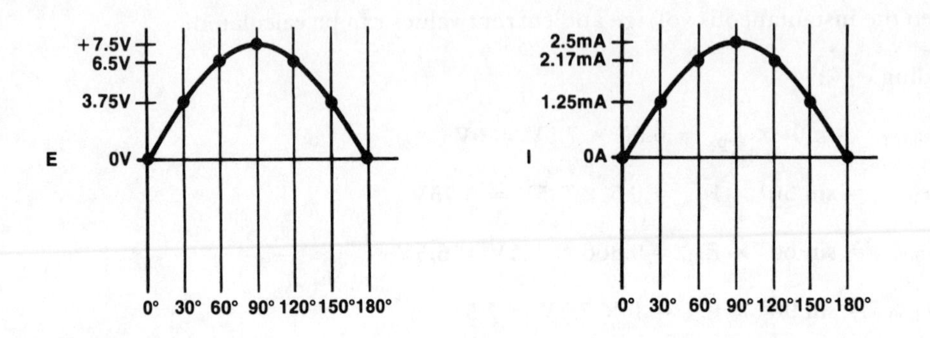

4. For the following circuit and the typical circuit values shown, determine the circuit values specified in rms values.

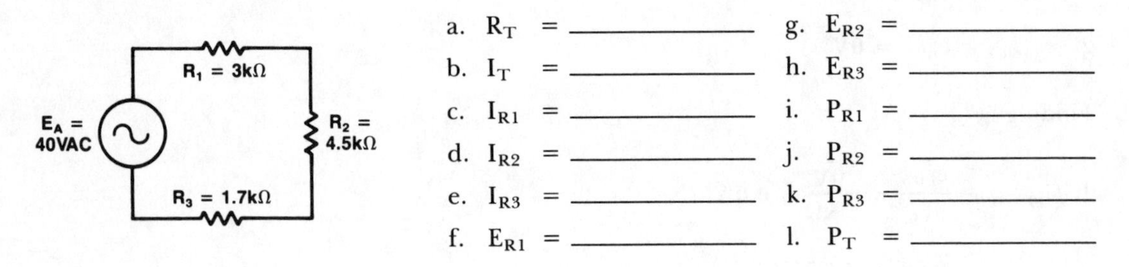

a. R_T = ____________ g. E_{R2} = ____________
b. I_T = ____________ h. E_{R3} = ____________
c. I_{R1} = ____________ i. P_{R1} = ____________
d. I_{R2} = ____________ j. P_{R2} = ____________
e. I_{R3} = ____________ k. P_{R3} = ____________
f. E_{R1} = ____________ l. P_T = ____________

Solutions:

a. $R_T = R_1 + R_2 + R_3 = 3k\Omega + 4.5k\Omega + 1.7k\Omega = \mathbf{9.2k\Omega}$

b. $I_T = \frac{E_T}{R_T} = \frac{40VAC}{9.2k\Omega} = \mathbf{4.35mA_{rms}}$

c.,d.,e. $I_{R1} = I_{R2} = I_{R3} = I_T = \mathbf{4.35mA_{rms}}$

f. $E_{R1} = I_{R1} \times R_1 = 4.35mA_{rms} \times 3k\Omega = \mathbf{13.0VAC}$

g. $E_{R2} = I_{R2} \times R_2 = 4.35mA_{rms} \times 4.5k\Omega = \mathbf{19.6VAC}$

h. $E_{R3} = I_{R3} \times R_3 = 4.35mA_{rms} \times 1.7k\Omega = \mathbf{7.4VAC}$

i. $P_{R1} = I_{R1} \times E_{R1} = 4.35mA_{rms} \times 13VAC = \mathbf{56.5mW_{rms}}$

j. $P_{R2} = I_{R2} \times E_{R2} = 4.35mA_{rms} \times 19.6VAC = \mathbf{85.2mW_{rms}}$

■ Worked-Out Examples

k. $P_{R3} = I_{R3} \times E_{R3} = 4.35mA_{rms} \times 7.4VAC = \mathbf{32.2mW_{rms}}$

l. $P_T = I_T \times E_A = 4.35mA_{rms} \times 40VAC = \mathbf{174mW_{rms}}$

5. For the following circuit and typical circuit values shown, determine the circuit values specified in peak values.

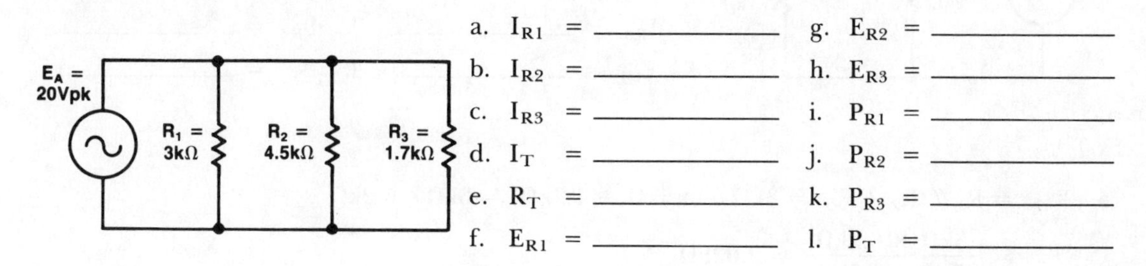

Solutions:

a. $I_{R1} = \frac{E_{R1}}{R_1} = \frac{20V_{pk}}{3k\Omega} = \mathbf{6.67mA_{pk}}$

b. $I_{R2} = \frac{E_{R2}}{R_2} = \frac{20V_{pk}}{4.5k\Omega} = \mathbf{4.44mA_{pk}}$

c. $I_{R3} = \frac{E_{R3}}{R_3} = \frac{20V_{pk}}{1.7k\Omega} = \mathbf{11.8mA_{pk}}$

d. $I_T = I_{R1} + I_{R2} + I_{R3} = 6.67mA_{pk} + 4.44mA_{pk} + 11.8mA_{pk} = \mathbf{22.9mA_{pk}}$

e. $R_T = \frac{E_A}{I_T} = \frac{20V_{pk}}{22.9mA_{pk}} = \mathbf{873\Omega}$

f.,g.,h. $E_{R1} = E_{R2} = E_{R3} = E_A = \mathbf{20V_{pk}}$

i. $P_{R1} = I_{R1} \times E_{R1} = 6.67mA_{pk} \times 20V_{pk} = \mathbf{133mW_{pk}}$

j. $P_{R2} = I_{R2} \times E_{R2} = 4.44mA_{pk} \times 20V_{pk} = \mathbf{88.8mW_{pk}}$

k. $P_{R3} = I_{R3} \times E_{R3} = 11.8mA_{pk} \times 20V_{pk} = \mathbf{236mW_{pk}}$

l. $P_T = I_T \times E_A = 22.9mA_{pk} \times 20V_{pk} = \mathbf{458mW_{pk}}$

Worked-Out Examples

6. For the following circuit and typical circuit values shown, determine the circuit values specified in peak-to-peak values.

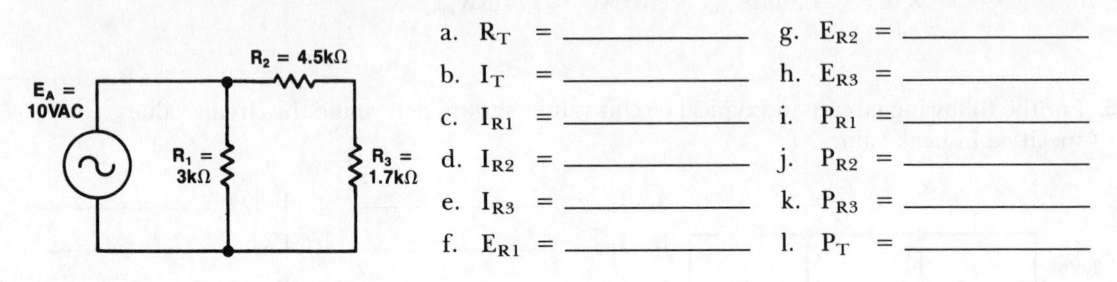

a. R_T = ____________ g. E_{R2} = ____________

b. I_T = ____________ h. E_{R3} = ____________

c. I_{R1} = ____________ i. P_{R1} = ____________

d. I_{R2} = ____________ j. P_{R2} = ____________

e. I_{R3} = ____________ k. P_{R3} = ____________

f. E_{R1} = ____________ l. P_T = ____________

Solution:

a. $R_T = R_1 \,//\, (R_2 + R_3) = 3\text{k}\Omega \,//\, (4.5\text{k}\Omega + 1.7\text{k}\Omega) = 3\text{k}\Omega \,//\, 6.2\text{k}\Omega$

$$= \frac{3\text{k}\Omega \times 6.2\text{k}\Omega}{3\text{k}\Omega + 6.2\text{k}\Omega} = \mathbf{2.02k\Omega}$$

b. $$I_T = \frac{E_A}{R_T} = \frac{10\text{VAC}}{2.02\text{k}\Omega} = \mathbf{4.95mA_{rms}}$$

c. $$I_{R1} = \frac{E_{R1}}{R_1} = \frac{E_A}{R_1} = \frac{10\text{VAC}}{3\text{k}\Omega} = \mathbf{3.33mA_{rms}}$$

d.,e. $$I_{R2} = I_{R3} = \frac{E_{R2} + E_{R3}}{R_2 + R_3} = \frac{E_A}{R_2 + R_3} = \frac{10\text{VAC}}{6.2\text{k}\Omega} = \mathbf{1.61mA_{rms}}$$

f. $E_{R1} = I_{R1} \times R_1 = 3.33\text{mA}_{rms} \times 3\text{k}\Omega = 10\text{VAC} = \mathbf{E_A}$

g. $E_{R2} = I_{R2} \times R_2 = 1.61\text{mA}_{rms} \times 4.5\text{k}\Omega = \mathbf{7.25VAC}$

h. $E_{R3} = I_{R3} \times R_3 = 1.61\text{mA}_{rms} \times 1.7\text{k}\Omega = \mathbf{2.74VAC}$

i. $P_{R1} = I_{R1} \times E_{R1} = 3.33\text{mA}_{rms} \times 10\text{VAC} = \mathbf{33.3mW_{rms}}$

j. $P_{R2} = I_{R2} \times E_{R2} = 1.61\text{mA}_{rms} \times 7.25\text{VAC} = \mathbf{11.7mW_{rms}}$

k. $P_{R3} = I_{R3} \times E_{R3} = 1.61\text{mA}_{rms} \times 2.74\text{VAC} = \mathbf{4.41mW_{rms}}$

l. $P_T = I_T \times E_A = 4.95\text{mA}_{rms} \times 10\text{VAC} = \mathbf{49.5mW_{rms}}$

■ Practice Problems

1. Given the rms circuit values below, complete the table by converting to peak and peak-to-peak values.

	rms	peak	peak-to-peak
I_{R1}	75mA	a. ______	e. ______
I_{R2}	32.7mA	b. ______	f. ______
I_{R3}	7.35mA	c. ______	g. ______
I_{R4}	248.2mA	d. ______	h. ______

2. For the following circuit, calculate the instantaneous voltage and current values specified by the table:

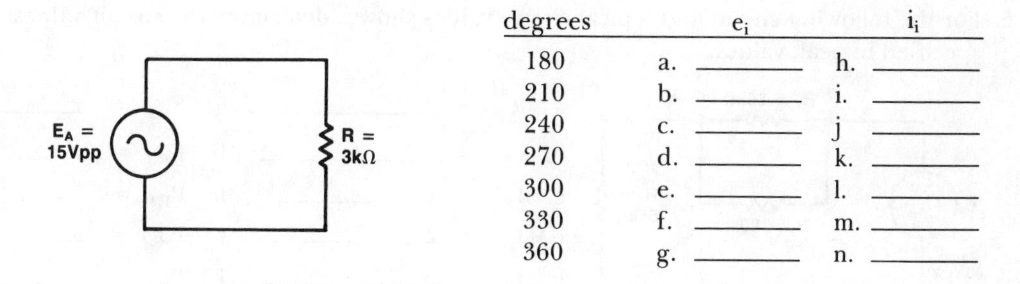

degrees	e_i	i_i
180	a. ______	h. ______
210	b. ______	i. ______
240	c. ______	j ______
270	d. ______	k. ______
300	e. ______	l. ______
330	f. ______	m. ______
360	g. ______	n. ______

3. For the circuit in Problem 2, plot the current and voltage waveforms based on the calculated instantaneous values.

4. For the following circuit and typical circuit values shown, determine the circuit values specified in peak-to-peak values.

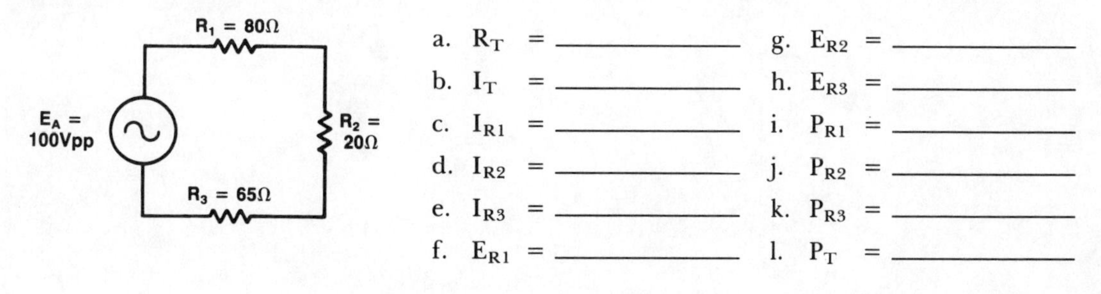

a. R_T = ______ g. E_{R2} = ______

b. I_T = ______ h. E_{R3} = ______

c. I_{R1} = ______ i. P_{R1} = ______

d. I_{R2} = ______ j. P_{R2} = ______

e. I_{R3} = ______ k. P_{R3} = ______

f. E_{R1} = ______ l. P_T = ______

■ Practice Problems

5. For the following circuit and typical circuit values shown, determine the circuit values specified in rms values.

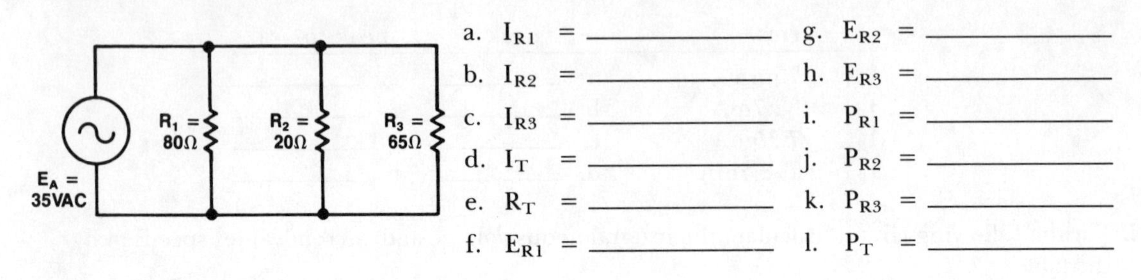

6. For the following circuit and typical circuit values shown, determine the circuit values specified in peak values.

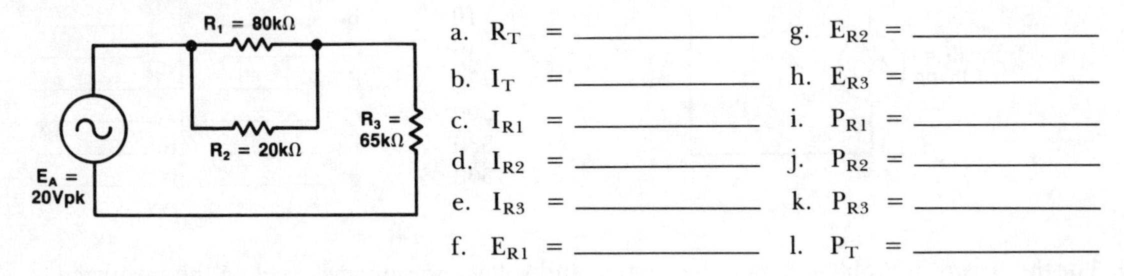

■ Quiz

1. Given the peak circuit values shown in the table, complete the table by converting to peak-to-peak and rms values.

	peak	peak-to-peak	rms
I_{R1}	44mA	a. ____	e. ____
I_{R2}	573μA	b. ____	f. ____
E_{R1}	16.3V	c. ____	g. ____
E_{R2}	109.7V	d. ____	h. ____

2. For the following circuit, calculate the instantaneous voltage and current values specified by the table.

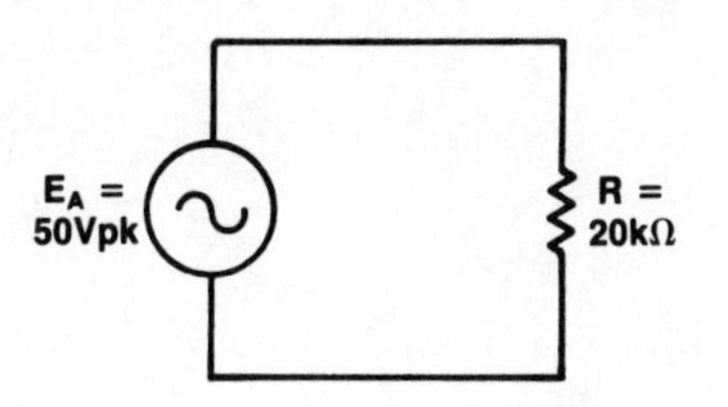

degrees	e_i	i_i
0	a. ______	j. ______
45	b. ______	k. ______
90	c. ______	l. ______
135	d. ______	m. ______
180	e. ______	n. ______
225	f. ______	o. ______
270	g. ______	p. ______
315	h. ______	q. ______
370	i. ______	r. ______

3. For the circuit in problem 2, plot the current and voltage waveforms based on the calculated instantaneous values.

4. For the following circuit and the typical circuit values, determine the circuit values specified in peak values.

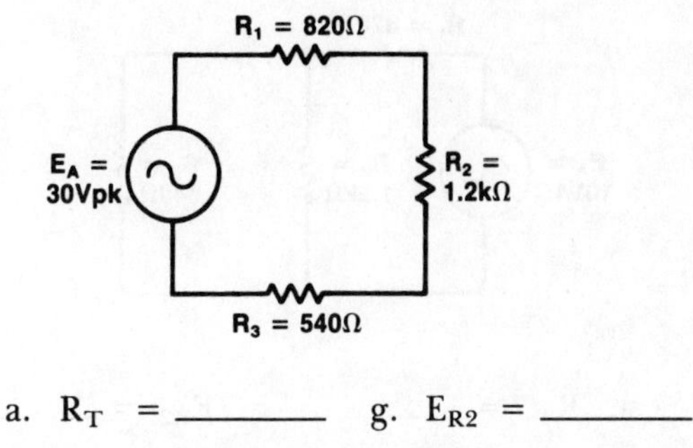

a. R_T = ______ g. E_{R2} = ______
b. I_T = ______ h. E_{R3} = ______
c. I_{R1} = ______ i. P_{R1} = ______
d. I_{R2} = ______ j. P_{R2} = ______
e. I_{R3} = ______ k. P_{R3} = ______
f. E_{R1} = ______ l. P_T = ______

5. For the following circuit and typical circuit values shown, determine the circuit values specified in peak-to-peak values.

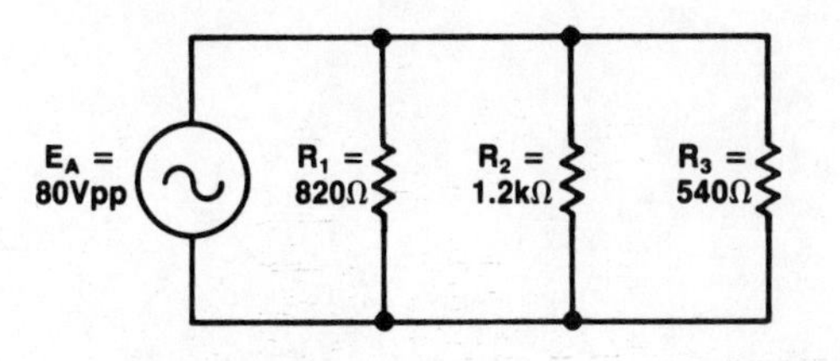

a. I_{R1} = ______ g. E_{R2} = ______
b. I_{R2} = ______ h. E_{R3} = ______
c. I_{R3} = ______ i. P_{R1} = ______
d. I_T = ______ j. P_{R2} = ______
e. R_T = ______ k. P_{R3} = ______
f. E_{R1} = ______ l. P_T = ______

■ Quiz

6. For the following circuit and typical values shown, determine the circuit values specified in rms values.

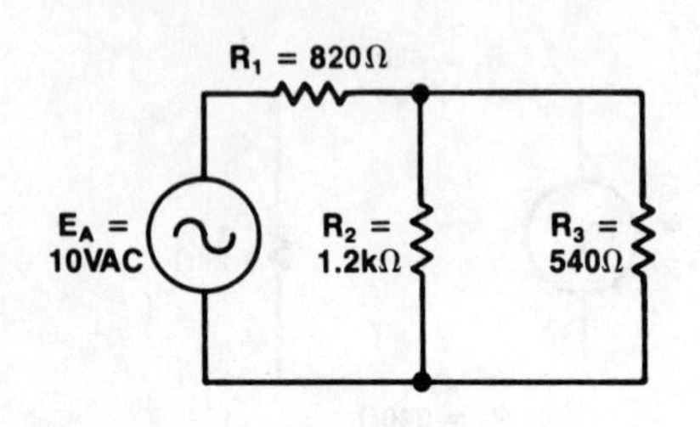

a. R_T = ________

b. I_T = ________

c. I_{R1} = ________

d. I_{R2} = ________

e. I_{R3} = ________

f. E_{R1} = ________

g. E_{R2} = ________

h. E_{R3} = ________

i. P_{R1} = ________

j. P_{R2} = ________

k. P_{R3} = ________

l. P_T = ________

LESSON 6

Capacitance

In this lesson, the electrical property of capacitance and another electronic circuit component, the capacitor, are introduced. The physical and electrical properties of capacitors are discussed. Series and parallel capacitive ac circuits are described and analyzed. The capacitor and the effects of capacitive action in ac series and parallel circuits are described.

■ Objectives

At the end of this lesson, you should be able to:

1. List the names of common types of capacitors and discuss the characteristics of each.
2. Define capacitance and the capacitive property.
3. Describe two effects of capacitors in ac circuits.
4. Describe the result of using different types of dielectrics, different sizes of plates, and increasing the distance between the plates of capacitors.
5. Given a schematic diagram of the types shown and typical circuit values, you should be able to calculate capacitance, capacitive reactance, current, voltage, and power for each capacitor and for the circuit total.

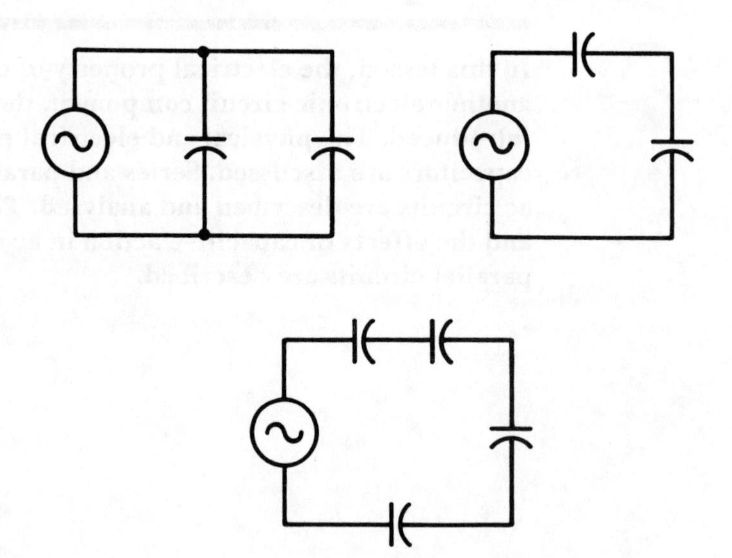

■ Definition of Capacitance
■ Basic Schematic Symbol

INTRODUCTION

In the previous lesson, discussion concerned the action of alternating current in resistive circuits. As you probably noted, this concept was very similar to an analysis of direct current in resistive circuits. Resistor series, parallel, and series-parallel circuits were analyzed. Total and individual values of current, voltage, resistance, and power were determined for each of these types of circuits.

In this lesson, capacitive properties and the effects of capacitance in ac circuits will be described and circuits containing capacitors connected in series and parallel will be analyzed.

Capacitors, like resistors, do not amplify or increase voltage or current. However, they perform a unique function by storing an electrical charge. This has the effect of opposing any changes in voltage.

PROPERTIES OF CAPACITANCE

Definition of Capacitance

The ability of a nonconductor to store an electrical charge is called *capacitance.* This capability is particularly important for applications that require short bursts of relatively large amounts of current. For example, a flash tube for a camera uses a burst of current to produce light for a fraction of a second. An electronic ignition system in a car has a capacitor in it to help provide interrupted current conditions that produce the high voltage for the spark discharge. Spot-welding equipment uses a capacitor to obtain peak current almost instantaneously. Capacitors are also used in medical equipment. For example, in a defibrillator used to revive a patient's heartbeat, a capacitor is discharged between two paddles which are applied to the patient's body. The paddles, thus, deliver a controlled amount of electrical energy to the patient.

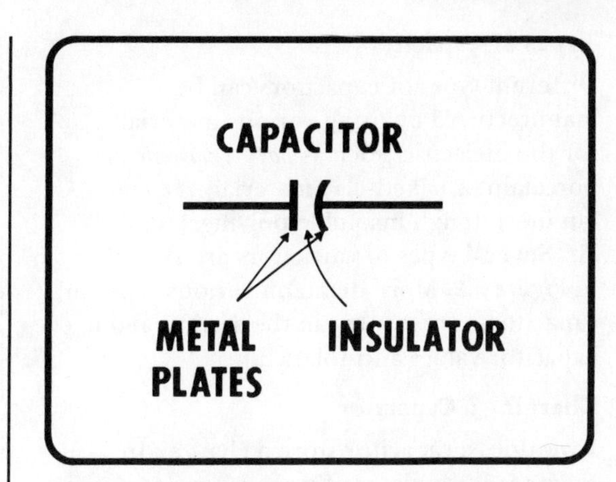

Figure 6.1 *A Capacitor's Schematic Symbol*

PHYSICAL DESCRIPTION OF A CAPACITOR

Basic Schematic Symbol

Essentially, a capacitor is two conductive metal plates separated by an insulator. *Figure 6.1* shows a capacitor's schematic symbol. The straight and curved lines represent the two conducting plates of the capacitor. The space between the two lines is the *dielectric* of the capacitor. A dielectric is a nonconductor of electric current.

■ Types of Capacitors
■ Charging a Capacitor

Types of Capacitors

Different types of capacitors can be manufactured by using various materials for the dielectric, such as *paper, ceramic* (a porcelain or baked-clay material), *Teflon** (an inert, tough insoluble polymer), and air. Several types of capacitors are shown in *Figure 6.2*. More detail on various types of capacitors is presented in the section about capacitor values and following.

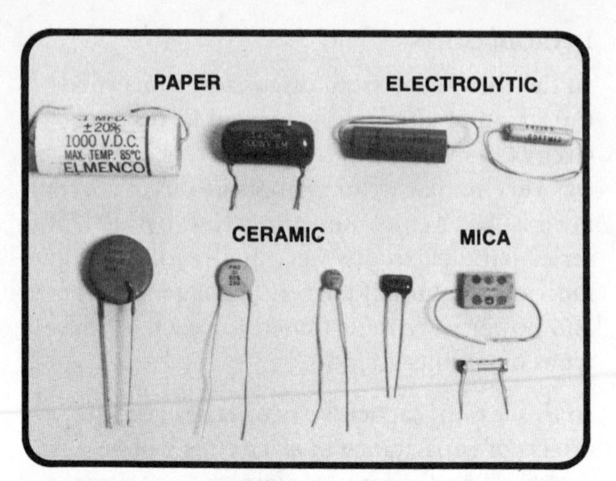

Figure 6.2 Several Types of Capacitors

Charging a Capacitor

How does a capacitor store a charge? In *Figure 6.3* two plates of a capacitor are shown in a circuit with a battery. Switch 1 is open. Therefore, the plates contain equal numbers of neutral atoms, positive ions, and free electrons that are associated with all conductive materials. This equality makes the plates electrically neutral. When a battery is connected between the two plates by closing switch 1, as shown in *Figure 6.4,* current flows. (As in all other discussions in this book, current is *electron* current.) Current flows because the positive potential applied to plate A attracts free electrons from the neutral atoms of the plate. This action creates an excess of positive ions. The free electrons are pulled to the *positive* terminal of the battery. They are deposited on the *negative* terminal by the chemical action within the battery and repelled by the negative battery potential and then they collect on plate B of the capacitor. The creation of positive ions on plate A constitutes a net positive charge on the plate. The excess electrons on plate B creates a net negative charge on that plate. The difference between the charges on the two plates is a potential difference which is defined as voltage. If there is no restriction of current flow in the circuit, the voltage, or potential difference, between the plates of the capacitor becomes equal to the battery voltage very quickly.

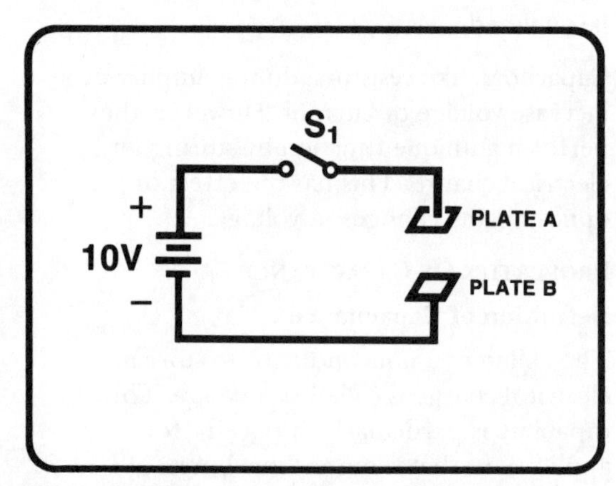

Figure 6.3 Plates of a Capacitor Normally Are Electrically Neutral

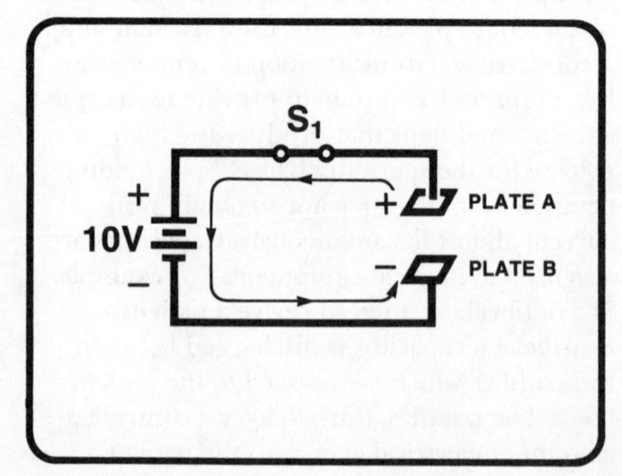

Figure 6.4 Capacitor Connected to a Battery and Charging

*Trademark of DuPont Company

- Storing a Charge
- Discharging a Capacitor

In *Figure 6.5*, the positive potential on plate A is equal to the positive potential on the battery. Therefore, no more electrons are removed from the positive plate, A. The negative potential on plate B is equal to the negative potential on the battery. Therefore, the B plate cannot collect any more electrons. At this point the current in the circuit stops. The charges have been distributed on the positive and negative plates until the voltage across the capacitor is equal to the battery voltage, and the capacitor is fully charged. The amount of time necessary for the capacitor to charge and the values of voltage across the capacitor after different time periods are discussed in Lesson 10 of this book and in Lesson 13 of the TI Learning Center textbook *Basic Electricity and DC Circuits.*

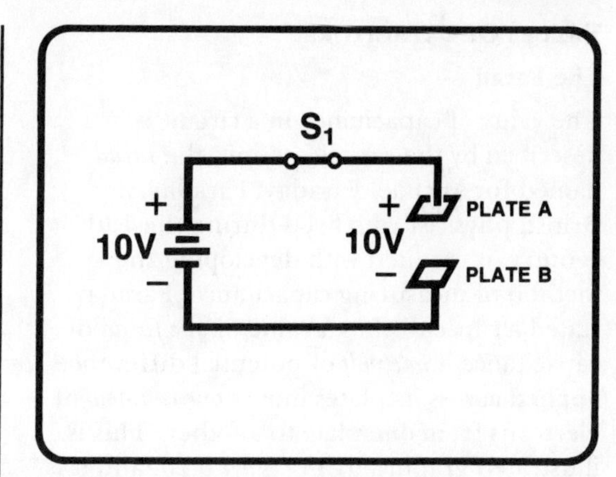

Figure 6.5 *Capacitor is Fully Charged*

Storing a Charge

If the capacitor is disconnected from the battery by opening the switch as shown in *Figure 6.6,* the attraction of the charges on the two plates maintains the potential difference between the two plates. Also, the dielectric (insulator) material prevents electrons from moving from plate B to neutralize the positive ions on plate A. Therefore, the charge is stored. By storing a charge equal to the applied battery voltage, the capacitor opposes any change in voltage which normally occurs when the battery is disconnected from the circuit.

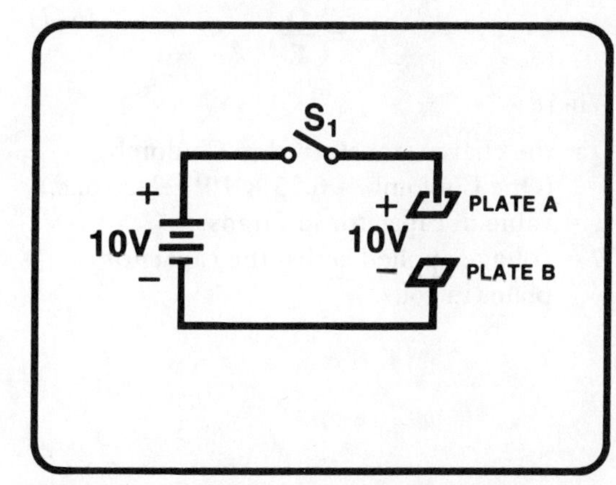

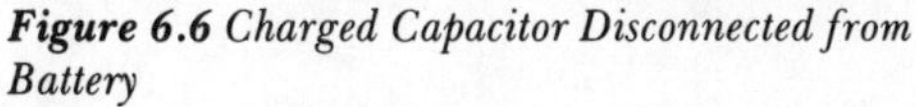

Figure 6.6 *Charged Capacitor Disconnected from Battery*

Discharging a Capacitor

To discharge the capacitor, a wire can be connected between the two plates, as shown in *Figure 6.8*. Current flows during a period of time, the excess electrons from plate B neutralize the positive ions on plate A, and both plates return to their original neutral condition and current stops as shown in *Figure 6.9*. The discharge time is related to the charge time. This also is discussed in Lesson 10.

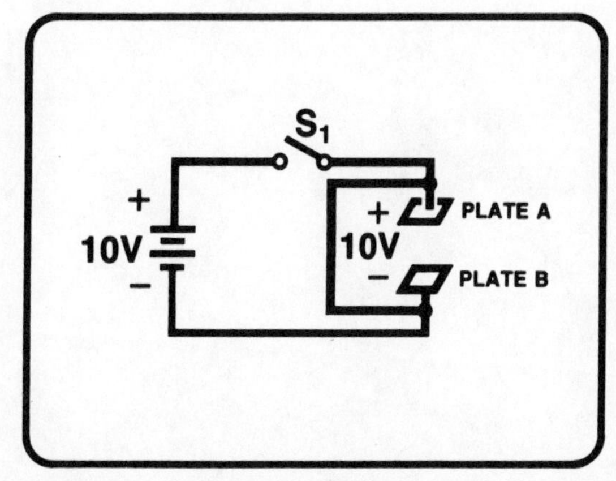

Figure 6.7 *A Wire Connected Between the Two Plates*

VALUES OF CAPACITORS

The Farad

The value of capacitance in a circuit is described by the capacitive unit, the *farad*, named for Michael Faraday. Faraday, a British physicist who lived during the 19th century, is credited with developing the method of measuring capacitance. Faraday stated a capacitor has a value of *one farad* of capacitance if *one volt* of potential difference applied across its plates moves *one coulomb* of electrons from one plate to another. This is illustrated graphically in *Figure 6.10*, and it is expressed in this equation:

$$C = \frac{Q}{E} \qquad (6\text{–}1)$$

Where

Q = the charge transferred in Coulombs. (One Coulomb = 6.25×10^{18} electrons.)
C = value of capacitor in farads.
E = voltage applied across the capacitor plates in volts.

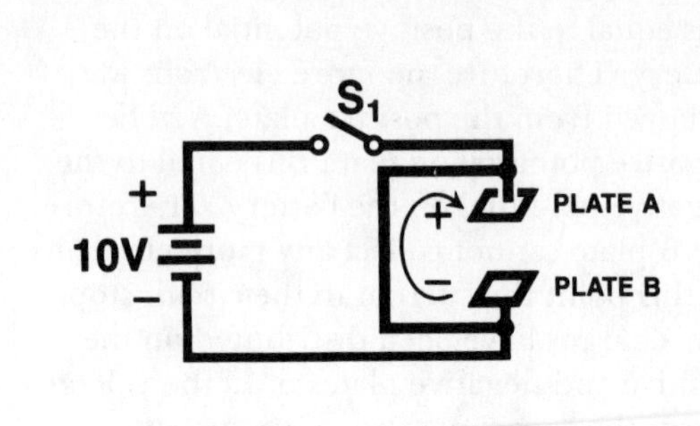

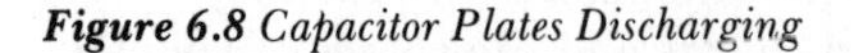

Figure 6.8 *Capacitor Plates Discharging*

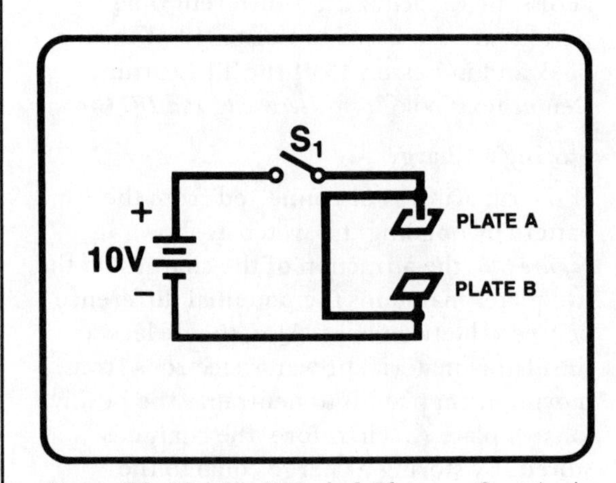

Figure 6.9 *After Discharge both Plates are Once Again Electrically Neutral*

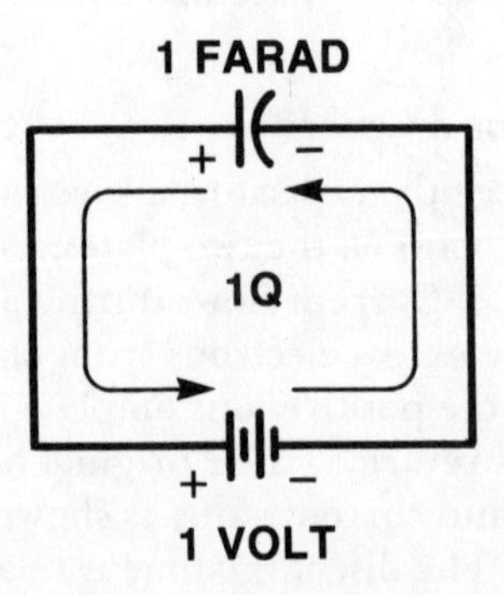

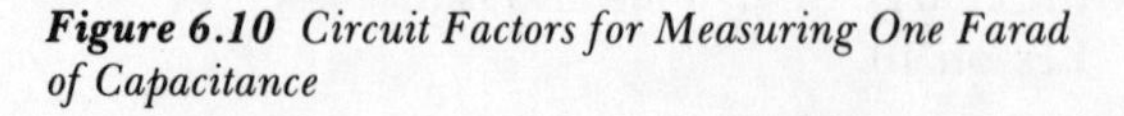

Figure 6.10 *Circuit Factors for Measuring One Farad of Capacitance*

Standard Values of Capacitance

6.25×10^{18} electrons is a very large number of electrons; therefore, a one farad capacitor is very large electrically and physically. Practical units of capacitance vary from a small capacitor of one picofarad (1 pF = 0.000000000001 farads = 10^{-12} farads) to 1000 microfarads (1000 μF = 0.001 farads = 1×10^{-3} farads), for a large capacitor. Typical capacitors of these values are shown in *Figure 6.11*.

Abbreviations of Capacitive Unit

The capacitance value of most capacitors is in either microfarads or picofarads. There are many ways that these units can be expressed and abbreviated. For example, the value of a 0.001 microfarad capacitor can be written:

0.001×10^{-6} F or 1000×10^{-12} F
or 1000 picofarads or 1000 pF or 1 kpF
or 1000 micromicrofarads or 1000 μμF,
or $1000 \times 10^{-6} \times 10^{-6}$F.
or 1 nanofarad or 1×10^{-9} F.

All other capacitance values can be expressed and abbreviated in as many ways. Although the capacitance value of a capacitor is usually printed on it, it can be in any one of many different abbeviations because all manufacturers have not established a standard abbreviation code for capacitance units. This can pose some deciphering problems. However, as a rule of thumb, remember that practical values of capacitance are fractions of one farad, and they typically range from one picofarad to several thousand microfarads.

***Figure 6.11** Typical Small-Value and Large-Value Capacitors*

Equation for Calculating Capacitive Values Based on Physical Parameters

Capacitors of different values are manufactured by varying several factors, such as the area of the plates, the distance between the plates, and the type of dielectric material used. An equation which relates these factors is

$$\mathbf{C} = \mathbf{k_e}\left(\frac{\mathbf{A}}{\mathbf{d}}\right)\epsilon_o \qquad (6\text{–}2)$$

Where
- C = the value of the capacitor in farads
- k_e = the dielectric constant of the dielectric material (no units)
- A = the area of either plate (square meters)
- d = the distance between the plates (meters)
- ϵ_o = the permittivity of dry air (8.85×10^{-12} farads per meter)

Area of the Plates

If the area of the plates is increased, the capacitor will be able to store more charge for every volt applied across its plates. Thus, if the dielectric and the distance between the plates remain constant, capacitance will be increased if the plate area is increased.

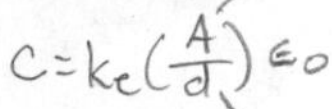

Distance Between Plates

If the distance between the plates of a capacitor is increased, then the capacitor stores less charge for every volt applied across its plates. Therefore, if the dielectric and the plate area remain constant, capacitance will be decreased if the distance between plates is increased.

Dielectric Constant

The dielectric constant of a dielectric material is a ratio of the permittivity of a dielectric material to the permittivity of a vacuum, or more practically, dry air. The permittivity of a material is its ability to concentrate the flux density of an electric field. For example, the dielectric constant of glass is 8. Therefore, glass has a permittivity of 8 times that of air. The ratio of the permittivity of glass to the permittivity of air is 8:1, the air having an assigned constant of 1.

The larger the value of dielectric constant, the greater the concentration of the electric field between the two charged plates. Therefore, assuming that the area of the plates and the spacing between plates remain constant, the value of a capacitor will be increased if the dielectric constant is increased, and it will be decreased if the dielectric constant is decreased.

A table of typical dielectric constants of various materials used in the construction of capacitors is shown in *Figure 6.12*.

DIELECTRIC MATERIAL	DIELECTRIC CONSTANT
AIR	1
CERAMICS	80-1200 VARIES WITH TYPE
GLASS	8
MICA	3-8 VARIES WITH TYPE
TEFLON	2.1
OIL	2-5 VARIES WITH TYPE
PAPER	2-6 VARIES WITH TYPE

Figure 6.12 *Dielectric Constant of Various Materials*

Voltage Ratings of Capacitors

In addition to their capacitance values, all capacitors are also rated as to the maximum allowable dc voltage that may be applied across the plates without arc-over and subsequent damage to it and the circuit. This rating is called the working voltage dc, usually abbreviated WVDC or simply VDC when printed on a capacitor. For example, if a capacitor is rated at 100 WVDC, no voltage greater than 100 volts dc should be applied across its plates.

If a capacitor is used in an ac circuit, the peak value of the ac voltage should be compared to WVDC rating of the capacitor to be sure that the voltage will not exceed that rating and cause the capacitor to arc between its plates. In a circuit in which the voltage across the capacitor is 40 volts peak to peak, the peak value of this voltage is 20 volts and therefore, the WVDC rating of the capacitor should be a minimum of 20 volts.

Dielectric Strength

The WVDC rating of a capacitor is related to the dielectric strength (breakdown voltage) of the type of dielectric material used in the capacitor's construction. The dielectric strength is different for different materials. The dielectric strength of a material is typically rated in volts per mil (V/mil). One mil is 0.001 inches. The V/mil rating of a dielectric material indicates how many volts a one-mil thickness of the material can withstand without breaking down. *Figure 6.13* lists some V/mil ratings of several dielectric materials commonly used in the construction of capacitors.

It is also important to remember that the working voltage of any capacitor decreases as temperature increases. Therefore, if the circuit in which a capacitor of a particular WVDC rating is to be used will be subjected to high temperatures, the capacitor should be chosen so that its working voltage will exceed expected circuit voltages at those temperatures.

MATERIAL	DIELECTRIC STRENGTH (VOLTS/MIL)
AIR	20
CERAMICS	600-1250 VARIES WITH TYPE
PYREX GLASS	330
MICA	600-1500 VARIES WITH TYPE
TEFLON	1525
OIL	375
PAPER	400-1250 VARIES WITH TYPE

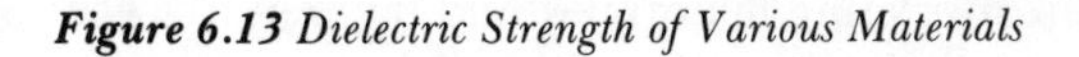

Figure 6.13 *Dielectric Strength of Various Materials*

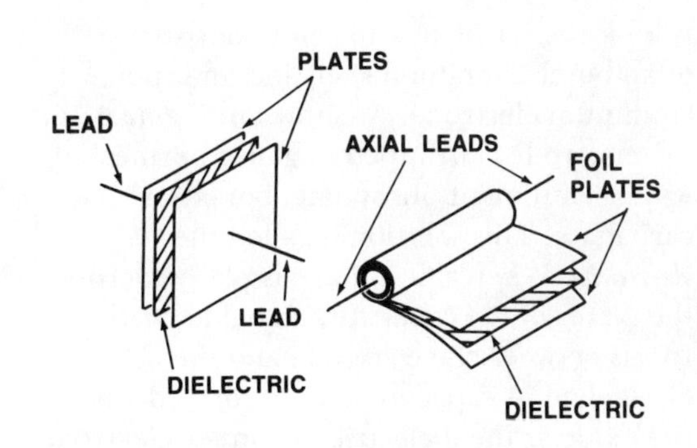

Figure 6.14 *Typical Paper Capacitor*

TYPES OF CAPACITORS

As stated previously, the dielectric material used in the construction of a capacitor primarily determines the type of capacitor. Some of the common types are ceramic, paper, air, and electrolytic capacitors.

Ceramic Disc Capacitors

Ceramic disc capacitors usally consist of two conductive discs on each side of a piece of ceramic insulator, one lead attached to each plate, and coated with some type of inert, waterproof coating, often made of some type of ceramic composition. Ceramic capacitors typically are manufactured in values from 1 picofarad up to thousands of microfarads and dc working volts from 10 volts up to 5000 volts. Several ceramic disc capacitors are shown in *Figure 6.2*.

Paper Capacitors

A paper capacitor typically consists of a sheet of paper insulator (dielectric) sandwiched between two sheets of foil, rolled in cylindrical form as shown in *Figure 6.14*. Paper capacitors are typically manufactured in capacitive values from about 1000 picofarads to one microfarad.

■ Electrolytic Capacitors

Paper, mylar, and other tubular film-type capacitors typically have a band at one end of the capacitor as shown in *Figure 6.15*. The band identifies the outside foil of the capacitor. The lead closest to the band should be connected to the ground or lowest potential point in the circuit. This provides rf shielding in high frequency circuits. It also provides a safety factor for the technician if the insulating material would happen to wear away and expose the capacitor's outside plate.

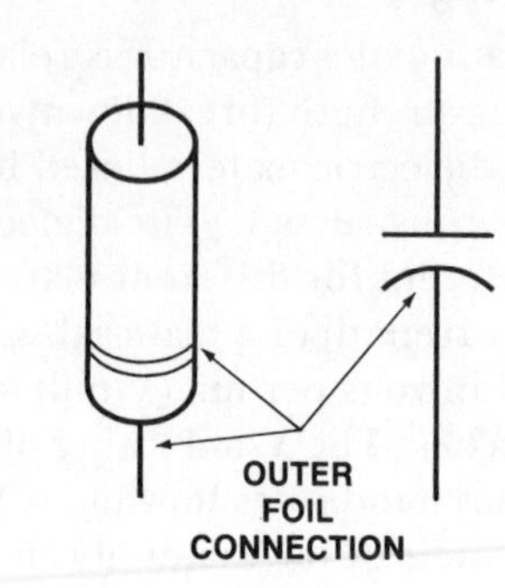

Figure 6.15 *Capacitor with Outer Foil Connection Indicated with Band*

Electrolytic Capacitors

A typical electrolytic capacitor consists of an outer aluminum shell and an inner aluminum electrode. As shown in *Figure 6.16*, the electrode is wrapped in gauze permeated with a solution of phosphate, borax, or carbonate. This solution is called the *electrolyte.* When a dc voltage is placed across the plates of the capacitor, an oxide coating forms between the electrode and the electrolyte. A capacitor is then formed with the oxide as the dielectric, the inner electrode as the positive plate (anode), and the outer shell and electrolyte as the negative plate (cathode).

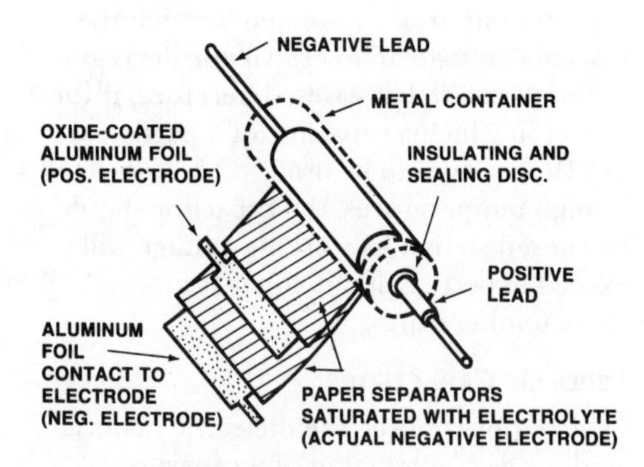

Figure 6.16 *Construction of an Electrolytic Capacitor*

Several capacitors of different values can be formed within one outer shell by using several electrodes and applying different potentials to the electrodes. *Figure 6.17* shows two views of a typical multisection electrolytic capacitor. The side view shows a typical way the capacitance and WVDC ratings of each section of the capacitor are marked on its side. Note that a small geometric symbol is marked beside each of the section ratings. These symbols are used to identify each of the capacitor's sections. The bottom view shows several lugs extending outward from the bottom of the capacitor. Each of the inner lugs is connected to an inner (usually positive) electrode of each individual capacitor section.

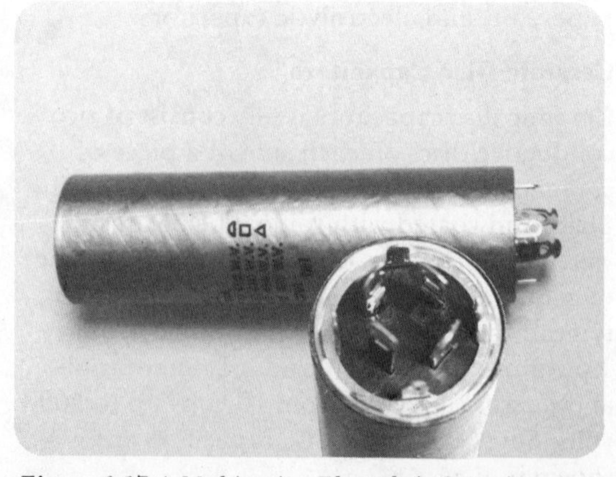

Figure 6.17 *A Multisection Electrolytic Capacitor*

Note that the small geometric symbol marked beside each of these lugs match the symbols marked on the side. Therefore, the capacitance value and WVDC rating of each section (lug) can be determined easily. Note also that there are several lugs around the outer edge of the capacitor. These lugs are part of the outer shell of the capacitor which is usually the negative (ground) side of the capacitor, common to each section. These lugs are used to mount the capacitor and provide a common negative connection in the circuit.

Some electrolytic capacitors use special metals called *tantalum* or *niobium* as the anode and are called *tantalum* or *niobium* capacitors. These capacitors have a larger capacitance in a much smaller size than other types of standard electrolytics. Because of their small physical size, they are ideal for use in miniaturized electronic circuits where space considerations are important.

Since an electrolytic capacitor utilizes a *chemical* process for its capacitive ability, it has a designated *shelf life*. That is, an electrolytic capacitor can be stored only for a specified length of time without use before it changes value.

The schematic symbol for an electrolytic capacitor has the added notation of the plus and minus signs. Electrolytic capacitors must be used in circuits with the end marked positive always at a more positive potential than its negative end. Thus, they are normally used only where dc or pulsating dc voltages are present. Special types of non-polarized electrolytic capacitors are available for use in ac circuits. Specified symbols for electrolytic capacitors are shown in *Figure 6.18*.

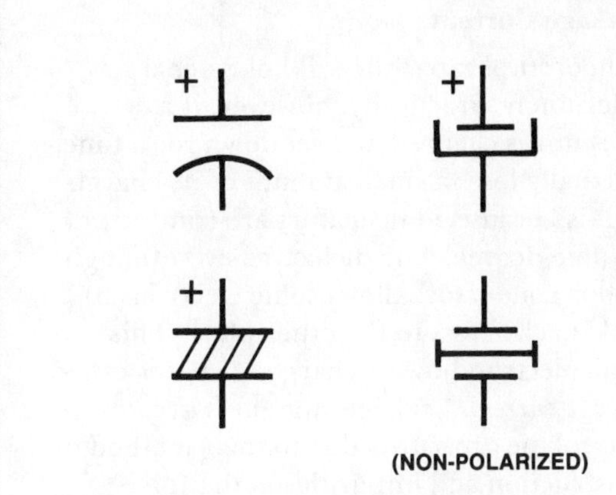

Figure 6.18 *Electrolytic Capacitor Symbols*

Electrolytic capacitors are available from about one microfarad to several thousand microfarads, with working voltage ranging from several volts to several hundred volts. The applied voltage to an electrolytic capacitor should be approximately equal to the voltage rating of the electrolytic. This will help insure that the proper value of capacitance will be present in the circuit.

Air Capacitors

With air capacitors, capacitance is varied by meshing and unmeshing the capacitor plate, with the air between the plates serving as the dielectric. As shown in *Figure 6.19,* air capacitors are commonly used as a variable tuning capacitor in receiver circuits.

The variable capacitor is designated schematically by adding an arrow to the standard schematic symbol also shown in *Figure 6.19*. Air capacitors are available usually in variable values from a few picofarads to several hundred picofarads.

■ Leakage Current

Leakage Current

A theoretical capacitor will hold a charge indefinitely. Practically, however, if a capacitor is charged and set down for a time, it actually loses a small amount of its charge. This is because all insulators are conductors to some degree. The dielectric, even though a poor conductor, allows some electrons to 'leak' back across to the other plate. This small electron flow discharge is what is called *leakage current.* Leakage current is greatest in electrolytic capacitors due to their method of construction and impurities in the foil and electrolyte.

A table showing leakage current for common types of capacitors for typical ranges of values and WVDC ratings is shown in *Figure 6.20*.

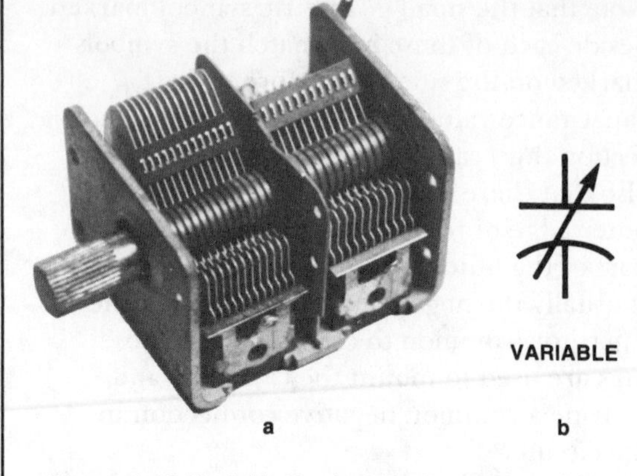

Figure 6.19 *Variable Tuning Air Capacitor and its Schematic Symbol*

Capacitor Type	Range of Capacitance	Range of WVDC (in volts)	Range of Temperature (°C)	R_I = Insulation Resistance I_L = Leakage Current (at 25°C)	Comments
Ceramic	1 pF – 2.5 μF	20 – 200	−55 +125	R_I = 100 GΩ/μF	Small size Low cost
Paper	0.001 – 2 μF	50 – 2000	−55 +105	R_I = 3 – 20 GΩ/μF	Low cost
Electrolytic	0.5 – 1,000,000 μF	2.5 – 700	−80 +125	I_L = 0.1 μA or more	Very small size Very low cost
Mylar	0.001 – 20 μF	50 – 1000	−55 +150	R_I = 50 GΩ/μF	Small size Relatively high cost
Air	10 – 400 pF	200/0.01 in air gap	—	—	Variable
Mica	1 pF – 1 μF	50 – 100,000	−55 +150	R_I = 10 – 100 GΩ/μF	Cap. change with age very small
Oil Filled	0.001 – 15 μF	100 – 12,500	−55 +85	R_I = 2 – 100 GΩ/μF	Low cost

Figure 6.20. *Typical Capacitor Parameters*

- Resistors in DC and AC Circuits
- Capacitors in DC Circuits
- Capacitors in AC Circuits

COMPARISON OF RESISTANCE AND CAPACITANCE

A comparison of resistive and capacitive properties should help you understand better the effects of capacitors in circuits. A resistor limits, and therefore, controls the flow of current in a circuit. A capacitor stores a charge and as a result, opposes a change in voltage in the circuit.

Resistors in DC and AC Circuits

Recall from previous lessons that in both dc and ac circuits, as shown in *Figure 6.21*, a resistor places a fixed amount of resistance in the circuit, and as a result, governed by Ohm's law, determines the amount of current flow. A capacitor, however, stores a charge and as a result, opposes any change in voltage in a circuit.

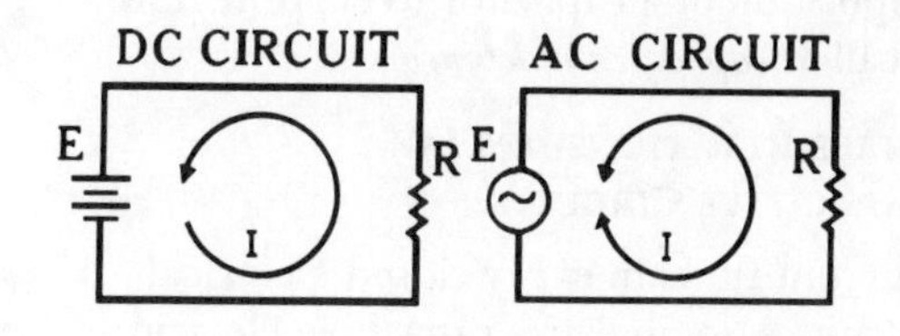

Figure 6.21 *Resistors in DC and AC Circuits*

Capacitors in DC Circuits

In dc circuits, a capacitor with no charge instantly acts as a short circuit by allowing a maximum value of current to flow, as shown in *Figure 6.22a.* As the capacitor charges, and the voltage across the capacitor increases in a polarity which opposes the battery or voltage source, the amount of current flowing decreases. When the capacitor is fully charged, current no longer flows in the circuit, and the capacitor then acts as an open circuit, as shown in *Figure 6.22b*. The time required for a capacitor to become fully charged is almost instantaneous, and is determined by RC time constants which will be discussed in detail in Lesson 10.

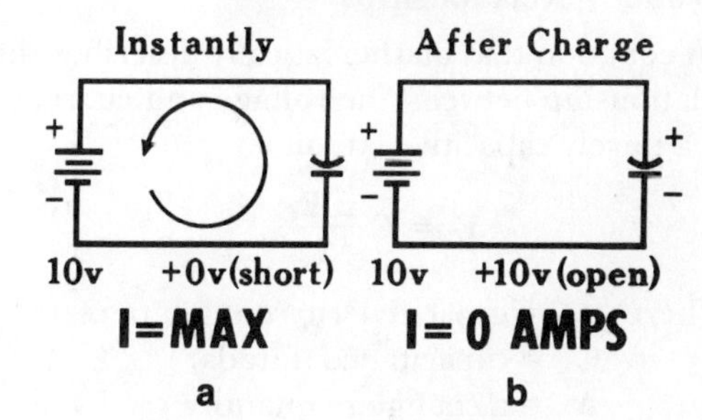

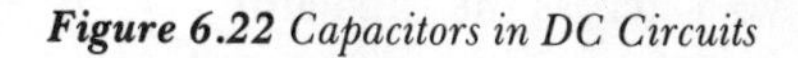

Figure 6.22 *Capacitors in DC Circuits*

Capacitors in AC Circuits

In ac circuits, the capacitive property of capacitors is observed in two related ways. First, the voltage across the capacitor lags the current through the capacitor by 90 degrees, as shown in *Figure 6.23*. Second, also shown in *Figure 6.23*, a capacitor represents a varying

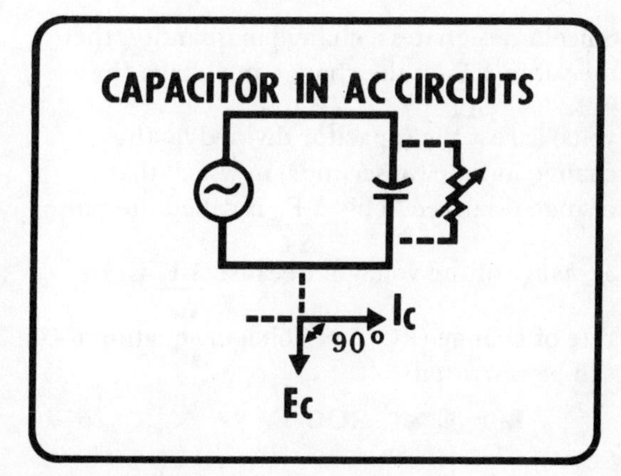

Figure 6.23 *Capacitor in AC Circuits*

opposition to current flow that is inversely related to the frequency of the ac source causing the current flow. This changing opposition of a capacitor to current flow is called *capacitive reactance.*

PHASE RELATIONSHIPS IN CAPACITIVE CIRCUITS

Current flows in every closed electrical circuit. Also, in every capacitive electrical circuit there are certain predictable and precise relationships between current, voltage, and capacitance. These relationships can be precisely described in terms of *phase relationships*. And a good way to begin a discussion and understanding of capacitive circuits is by analyzing phase relationships of the factors in such circuits.

E_C and I_C Relationships

An equation that mathematically describes the relationship between the voltage and current in a purely capacitive circuit is

$$I_C = C \frac{\Delta E_C}{\Delta t} \qquad (6\text{–}3)$$

Where I_C = capacitive current (amperes)
C = capacitance (farads)
Δ = a change in quantity
E_C = volts across the capacitor
t = time (seconds)

Since Δ designates a change in quantity, then the value $\frac{\Delta E_C}{\Delta t}$ is the change in voltage (in volts) across the capacitor divided by the change in time (in seconds) in which that change occurred. This $\frac{\Delta E_C}{\Delta t}$ is called the *rate of change* of the voltage. Because $\frac{\Delta E_c}{\Delta t}$ is the rate of change (ROC) of voltage, equation *6–3* can be rewritten:

$$I_C = C \times [\text{ROC } E_C] \qquad (6\text{–}4)$$

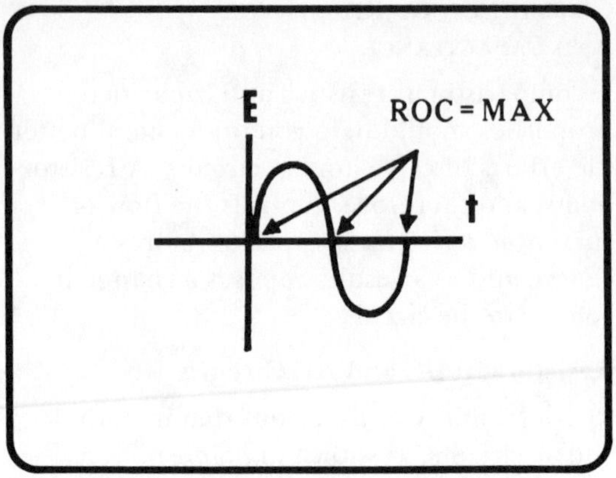

Figure 6.24 Maximum Rate of Change of Voltage

Rate of Change of E_C

Since there are rates of change of voltage in a capacitive ac circuit, it is important that you know where the maximum and minimum rates of change occur. Using equation *6–4*, certain characteristics of the current can be determined. A good example is to examine the rate of change of voltage at various points in a sine wave. A fixed amount of time, Δ t, will be moved along a time axis and examined to see how much change in voltage occurs at various intervals of the sine wave. As shown in *Figure 6.24,* the maximum change in voltage occurs as the sine wave crosses the axis at the zero voltage level. Thus, the rate of change of voltage is *maximum* as the sine wave crosses the zero voltage level.

On the other hand, the change in voltage is minimum or zero as the voltage peaks. Thus, the rate of change of voltage is zero as the sine wave peaks as shown in *Figure 6.25*. This should be obvious because when the sine wave peaks, the voltage *stops* increasing and begins decreasing.

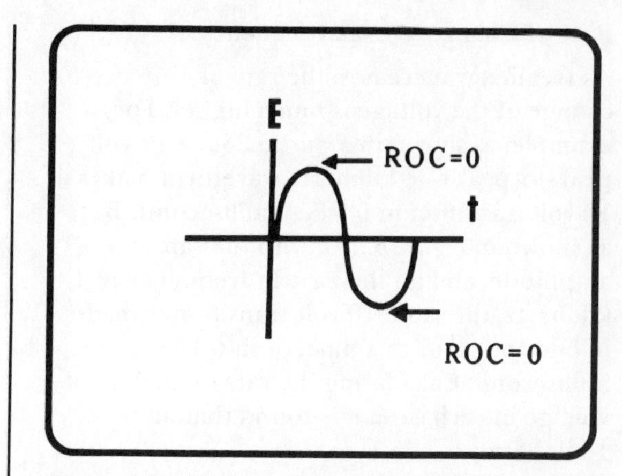

Figure 6.25 *Minimum Rate of Change of Voltage*

E_C Versus I_C

Remember that the current is equal to the value of the capacitor times the rate of change of the voltage. When the rate of change of voltage is zero (E_C is at either peak), the value of the current is zero. When the rate of change of voltage is maximum (as E_C crosses the zero voltage level), the value of the current is a maximum. The zero and maximum points of I_C are indicated by the Xs in *Figure 6.26*. Since the voltage changes sinusoidally, the rate of change of voltage changes in a sinusoidal shape and therefore, so does the current. Connecting these points with a sinusoidal waveform, as shown in *Figure 6.27*, the current waveform can be plotted. Note in *Figure 6.27* that the current leads the voltage by 90 degrees as was described earlier.

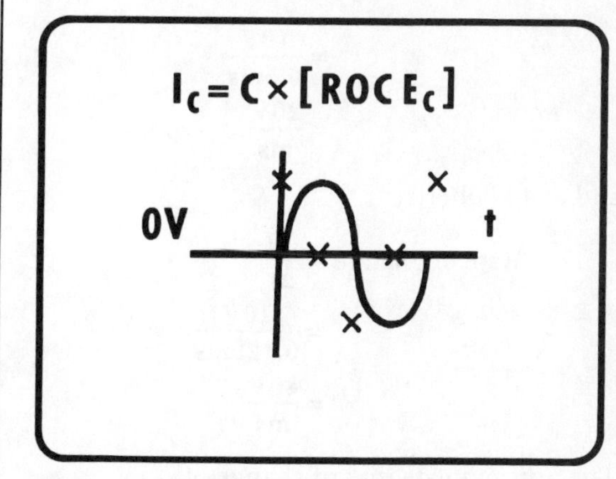

Figure 6.26 *Zero and Maximum Points of I_C*

FREQUENCY RELATED TO CAPACITIVE IMPEDANCE

In a circuit, a capacitor impedes—opposes—the flow of current. The opposition is called capacitive reactance. The capacitive reactance depends on the frequency of the source. The rate-of-change equation can be used to relate the capacitive opposition to the frequency of the voltage in a capacitive circuit.

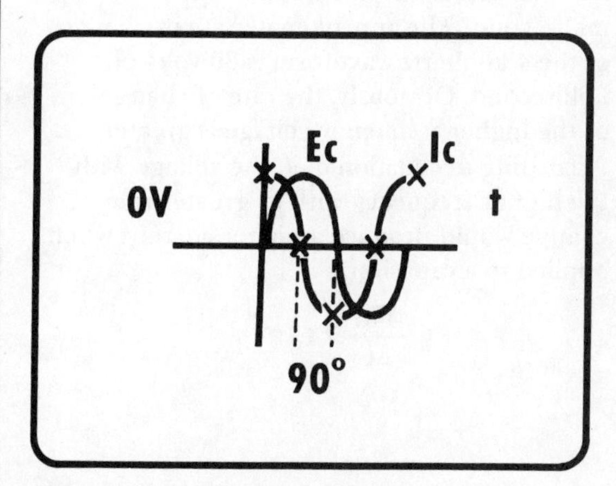

Figure 6.27 *Relationship of I_C to E_C*

- **Rate of Change Versus Frequency**
- **Equation for X_C**

Rate of Change Versus Frequency

As frequency increases, the rate of change of the voltage is much higher. For example, as shown in *Figure 6.28a*, a 10 volt peak-to-peak one kilohertz waveform makes a 10 volt transition in 0.5 (½) millisecond. But, as shown in *Figure 6.28b*, with the same amplitude, and an increase in frequency to 4 kilohertz, the same 10 volt transition is made in one-fourth of this time, or in 0.125 millisecond. Calculating the rate of change of voltage in each case, it is found that at 1 kilohertz:

$$\text{Rate of change} = \frac{\Delta E}{\Delta t} = \frac{10V}{0.5ms} = \frac{20V}{ms}$$

and at 4 kilohertz:

$$\text{Rate of change} = \frac{\Delta E}{\Delta t} = \frac{10V}{0.125ms} = \frac{80V}{ms}$$

The approximate rate of change of the 1 kilohertz waveform is 20 volts per millisecond. The approximate rate of change of the 4 kilohertz waveform is 80 volts per millisecond. Obviously, the rate of change of the higher frequency voltage is greater. According to equation *6–4*, the voltage with the higher frequency with its greater rate of change would produce a higher current when applied to a capacitor.

$$f\uparrow \ \frac{\Delta E_C}{\Delta t} \uparrow I_C \uparrow$$

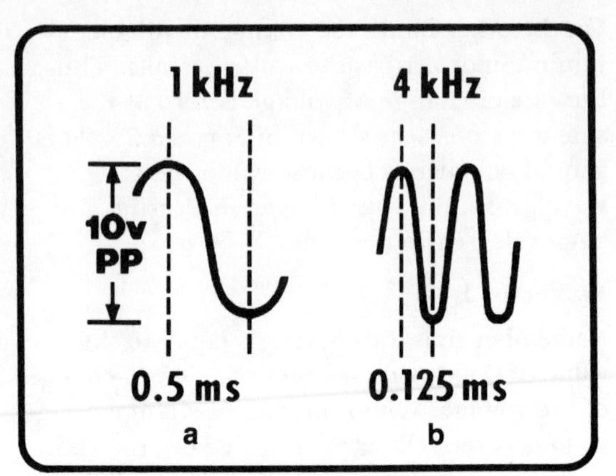

Figure 6.28 *Rate of Change of Voltage Versus Frequency*

According to Ohm's law, if voltage is constant and current increases then the opposition to current flow must have decreased.

$$I\uparrow = \frac{E}{X_C\downarrow}$$

On the other hand, if source frequency is decreased, the current will also decrease, which means an increase in opposition must have occurred. These relationships can be expressed as:

$$f\downarrow \ \frac{\Delta E_C}{\Delta t} \downarrow I_C \downarrow \text{ and } X_C \uparrow$$

Capacitive Reactance

As stated earlier, this changing opposition of a capacitor is called *capacitive reactance* and is inversely related to the source frequency.

Equation for X_C

Capacitive reactance is measured in *ohms* of reactance like resistance, and depends on the frequency of the *applied voltage* and the *value* of the *capacitor.*

$$X_C\,(\Omega) = \frac{1}{2\pi f(Hz)C(F)} \qquad (6\text{–}5)$$

where $2\pi = 6.28$.

■ Analysis of an AC Capacitive Circuit

The symbol for reactance is X. To specify a specific type of reactance, a subscript is used. In this case, since it's capacitive reactance, the subscript C is used. The constant 2π comes from the number of radians in one cycle of a sinusoidal ac waveform. Therefore, this equation is valid only for calculating the capacitive reactance of a capacitor to sinusoidal alternating current.

Analysis of an AC Capacitive Circuit

The circuit in *Figure 6.29* will be used to determine the capacitive reactance using the capacitive reactance equation. That circuit contains a 10 microfarad capacitor with an applied voltage with a frequency of 4 kilohertz. The capacitive reactance is calculated:

$$X_C = \frac{1}{2\pi fC}$$

$$= \frac{1}{2\pi(4 \times 10^3\text{Hz})\,(10 \times 10^{-6}\text{F})}$$

$$= \frac{1}{251.2 \times 10^{-3}}$$

$$X_C = 3.98\Omega$$

If the applied voltage is 10 volts, as shown in *Figure 6.30*, the current in the circuit will be the value of the applied voltage divided by the value of the capacitive reactance.

$$I_C = \frac{E_{AC}}{X_C}$$

$$= \frac{10\text{V}}{3.98\Omega}$$

$$= 2.51\text{A}$$

The current in the circuit is 2.51 amperes. Remember, the voltage and current values are rms values since they have not been otherwise specified.

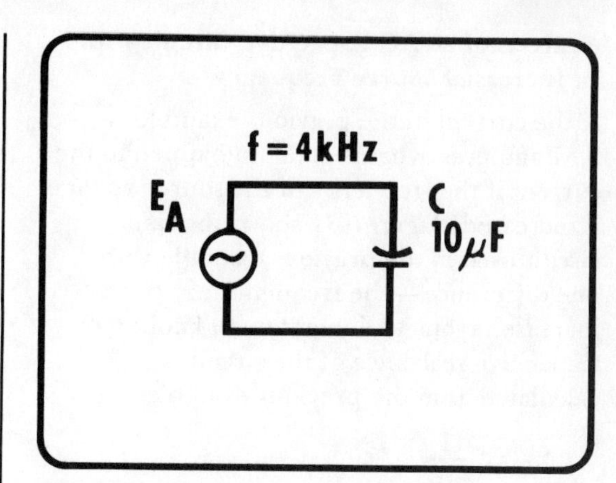

Figure 6.29 Example Circuit to Calculate X_C

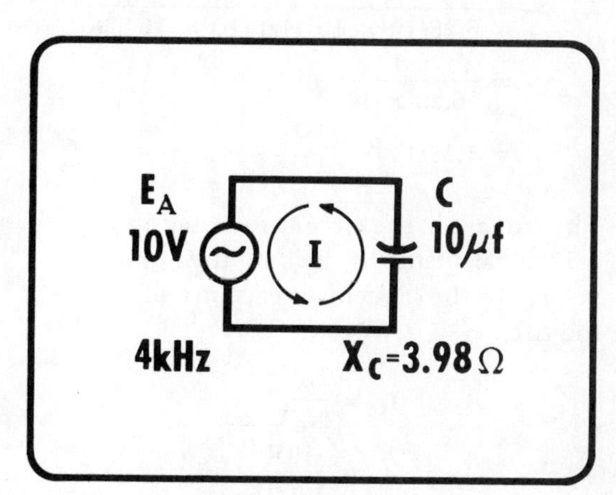

Figure 6.30 Example Circuit to Calculate I_C

■ **Analysis of an AC Capacitive Circuit with an Increased Source Frequency**

Analysis of an AC Capacitive Circuit with an Increased Source Frequency

If the current in the previous example is 2.51 amperes, what is going to happen to the current if the frequency of the source voltage is increased? *Figure 6.31* shows the same circuit used in the previous example with one difference—the frequency has been increased from 4 kilohertz to 10 kilohertz. Capacitive reactance of the circuit is calculated as in the previous example.

$$X_C = \frac{1}{2\pi fC}$$

$$= \frac{1}{6.28(10 \times 10^3 Hz)\ (10 \times 10^{-6} F)}$$

$$= \frac{1}{6.28 \times 10^{-1}}$$

$$= 1.59\Omega$$

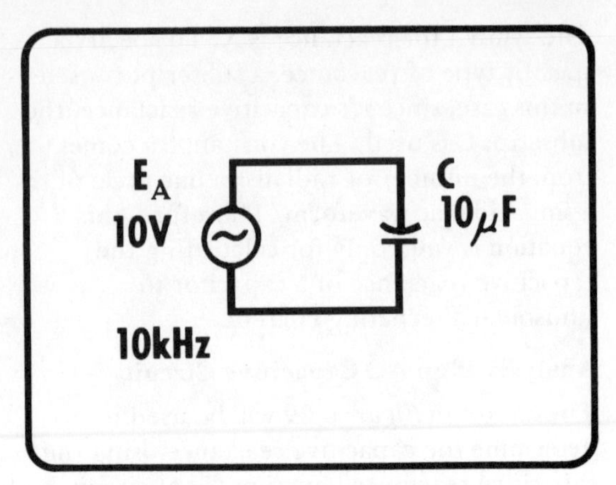

Figure 6.31 Example Circuit to Calculate X_C

The current in the circuit, repeated in *Figure 6.32*, is calculated by dividing the applied voltage by the capacitive reactance of the circuit:

$$I_C = \frac{E_A}{X_C}$$

$$= \frac{10V}{1.59\Omega}$$

$$= 6.28A$$

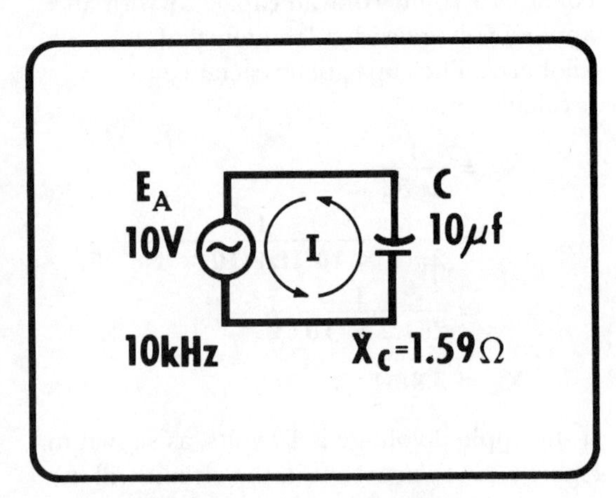

Figure 6.32 Example Circuit to Calculate I_C

Obviously, this *current* is *greater* than the current in the same circuit when the applied frequency was lower.

Figure 6.33 summarizes the relationships between capacitive reactance, X_C, circuit current, I_C, and frequency, f, of the source voltage for the two example circuits just discussed. From this summary, it is apparent that as frequency of source voltage increases, capacitive reactance decreases and current increases.

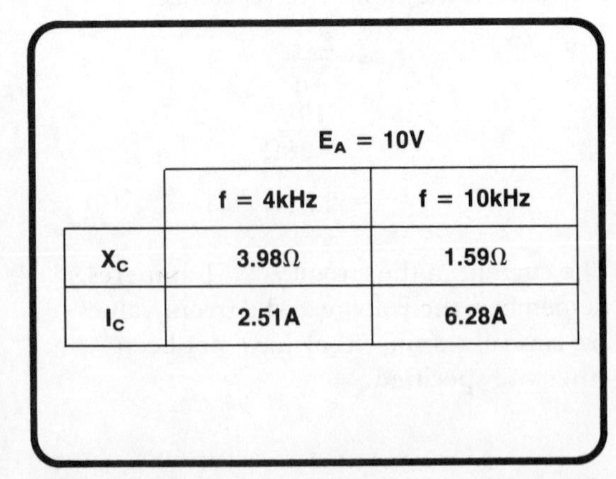

E_A = 10V

	f = 4kHz	f = 10kHz
X_C	3.98Ω	1.59Ω
I_C	2.51A	6.28A

Figure 6.33 X_C *Versus* I_C *at Two Different Frequencies of Source Voltage*

The Capacitor as a Variable Resistance

A capacitor can be thought of as a variable resistor whose value is controlled by the applied frequency. As frequency increases, its *opposition to current* or its capacitive *reactance* decreases, as shown in *Figure 6.34*. *Figure 6.35* shows the same concept graphically.

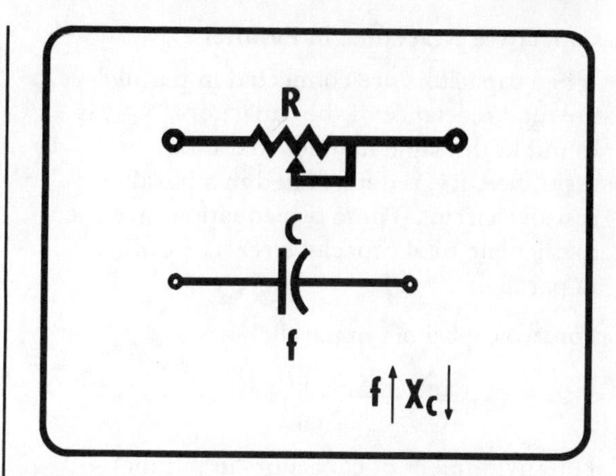

Figure 6.34 *A Capacitor as a Variable Resistor*

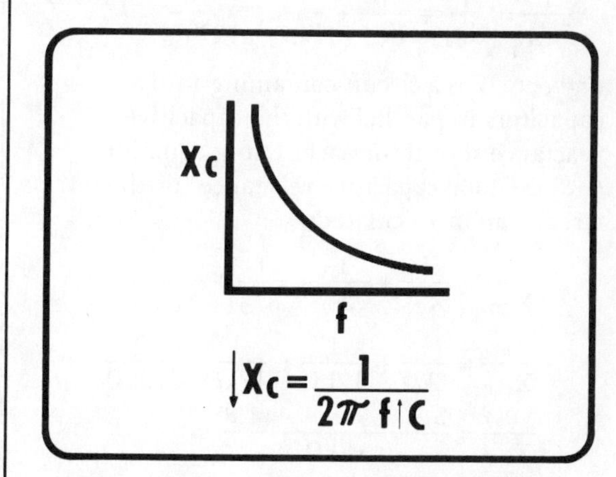

Figure 6.35 *Capacitive Reactance Versus Frequency*

CAPACITIVE REACTANCE IN SERIES AND PARALLEL

As observed in the preceding section, capacitors in ac circuits present opposition to current flow based on the capacitive reactance equation, equation *6–5*. As mentioned, it is similar to the resistance of a resistor. The X_C designation differentiates it from a resistor R. In a purely capacitive circuit where no resistance is present, totals of capacitive reactance can be calculated the same way that totals of resistance are calculated in purely resistive circuits.

Capacitive Reactance in Series

When capacitors are connected in series, the total reactance of the capacitors (X_{CT}) is simply a sum of the capacitive reactance of the capacitors present. This procedure is identical to the way total resistance (R_T) is determined in a series resistive circuit. Equation *6–6* is used to calculate total capacitive reactance (X_{CT}) in series:

$$X_{CT} = X_{C1} + X_{C2} + X_{C3} \ldots + X_{CN} \quad (6\text{–}6)$$

Figure 6.36 is a circuit containing four capacitors in series with the capacitive reactance shown for each. Using equation *6–6*, the total capacitive reactance for the circuit can be calculated:

$$X_{CT} = X_{C1} + X_{C2} + X_{C3} + X_{C4}$$
$$X_{CT} = 1\text{k}\Omega + 2.5\text{k}\Omega + 4\text{k}\Omega + 3.3\text{k}\Omega$$
$$X_{CT} = 10.8\text{k}\Omega$$

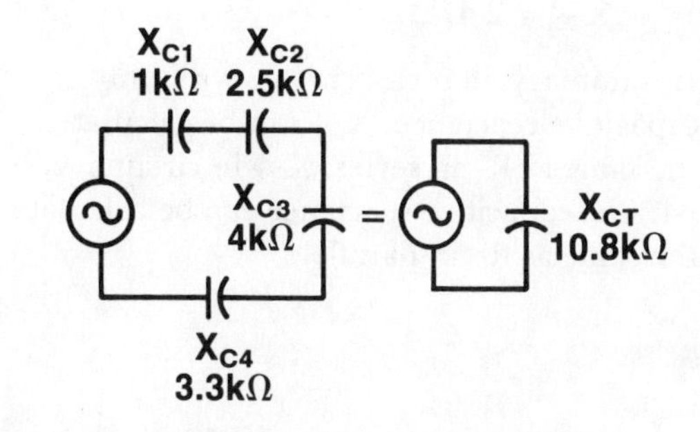

Figure 6.36 *Example Circuit for Calculating X_{CT} in Series*

Capacitive Reactance in Parallel

Capacitive Reactance in Parallel

When capacitors are connected in parallel, the total reactance of the capacitors, X_{CT}, is found in the same manner that total resistance, R_T, is determined in a parallel resistive circuit. These two equations are used to calculate total capacitive reactance, X_{CT}, in parallel:

For two capacitors in parallel—

$$X_{CT} = \frac{X_{CA} \times X_{CB}}{X_{CA} + X_{CB}} \qquad (6\text{–}7)$$

For any number of capacitors in parallel—

$$\frac{1}{X_{CT}} = \frac{1}{X_{C1}} + \frac{1}{X_{C2}} + \frac{1}{X_{C3}} \ldots + \frac{1}{X_{CN}} \qquad (6\text{–}8)$$

Figure 6.37 is a circuit containing four capacitors in parallel with the capacitive reactance shown for each. Using equation *6–8*, the total capacitive reactance for the circuit can be calculated:

$$\frac{1}{X_{CT}} = \frac{1}{X_{C1}} + \frac{1}{X_{C2}} + \frac{1}{X_{C3}} + \frac{1}{X_{C4}}$$

$$\frac{1}{X_{CT}} = \frac{1}{6k\Omega} + \frac{1}{12k\Omega} + \frac{1}{18k\Omega} + \frac{1}{10k\Omega}$$

$$\frac{1}{X_{CT}} = \frac{3 + 1.5 + 1 + 1.8}{18k\Omega}$$

$$\frac{1}{X_{CT}} = \frac{7.3}{18k\Omega}$$

$$\frac{X_{CT}}{1} = \frac{18k\Omega}{7.3}$$

$$X_{CT} = 2.47k\Omega$$

In summary, in series circuits with only capacitive reactance, X_{CT} can be calculated the same as R_T in series. X_{CT} in circuits with parallel capacitive reactance can be calculated the same as R_T in parallel.

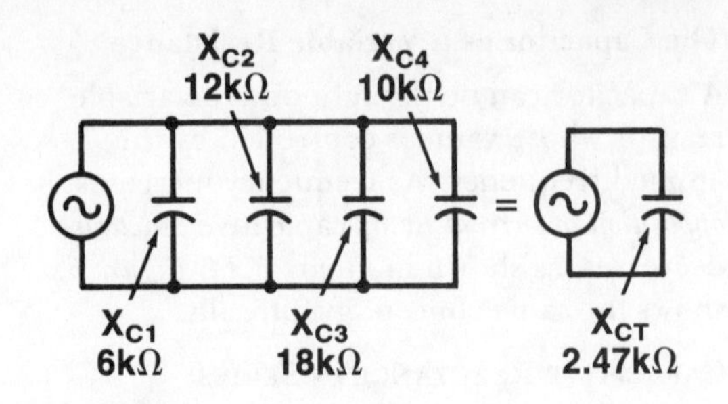

Figure 6.37 *Example Circuit for Calculating X_{CT} in Parallel*

Capacitance in Series and Parallel

Thus far, discussion has concerned capacitive reactance connected in series or in parallel in circuits. What about capacitance? What happens to total capacitance when several capacitors are combined in series or parallel? To best understand this concept, the variables that determine the value of a capacitor should first be examined.

Equation *6–2* explains the value of a capacitor, and it is repeated here:

$$\mathbf{C = k_e\left(\frac{A}{d}\right)} \qquad (6\text{–}9)$$

Notice that the value of a capacitor is determined by dividing the area of the plates by the distance between the plates and multiplying by a constant which is characteristic of the insulating materials. In this case the constant k_e is formed by multiplying the dielectric constant and ε_o, the permittivity of air. It is a simplified equation *6–2* which was

$$C = k_e\left(\frac{A}{d}\right)\epsilon_o \qquad (6\text{–}2)$$

Equation *6–9* indicates that the value of a capacitor increases if the plate area of the capacitor increases:

$$C\uparrow = k_e\left(\frac{A\uparrow}{d}\right)$$

Capacitance in Series

It also indicates that the value of a capacitor decreases if the distance separating the two plates increases.

$$C\downarrow = k_e\left(\frac{A}{d\uparrow}\right)$$

Capacitance in Series

With equation *6–9* in mind, look at *Figure 6.38* in which two capacitors of equal value are connected in series. Note that the total effective distance between the plates connected directly to the battery is doubled. Since an increase in distance between the plates decreases capacitance, it can be concluded that the total capacitance of the two equal capacitors connected in series will be only one-half of the capacitance of one capacitor. This reciprocal relationship between total capacitance and capacitors in series is described by these two equations.

For two capacitors in series—

$$C_T = \frac{C_1 \times C_2}{C_1 + C_2} \qquad (6\text{–}10)$$

For any number of capacitors in series—

$$\frac{1}{C_T} = \frac{1}{C_1} + \frac{1}{C_2} + \frac{1}{C_3} \dots + \frac{1}{C_N} \qquad (6\text{–}11)$$

These equations probably look familiar to you because they are similar to the equations for total resistance when resistors are connected in *parallel*.

Figure 6.39 is a circuit containing four capacitors in series with the capacitance value for each shown. Using equation *6–11*, the total capacitance for the circuit can be calculated:

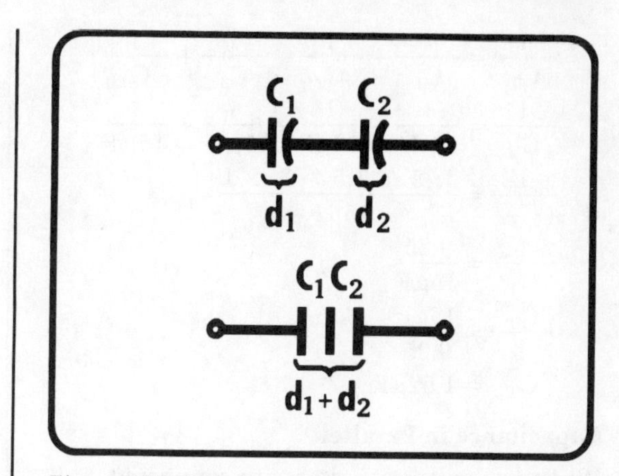

Figure 6.38 *Capacitances Connected in Series Decrease Total Capacitance*

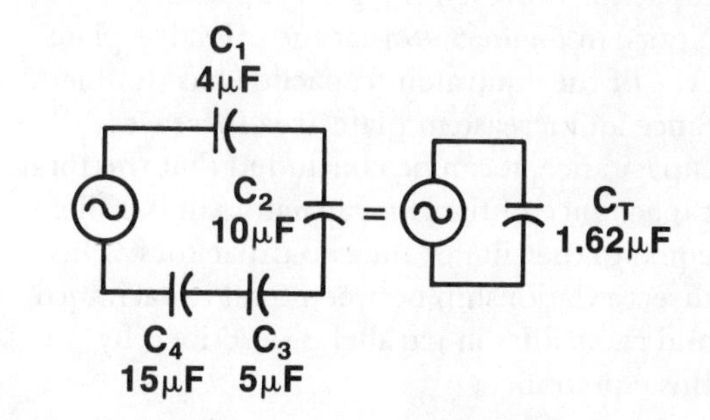

Figure 6.39 *Example Circuit for Calculating C_T*

Capacitance in Parallel

$$\frac{1}{C_T} = \frac{1}{C_1} + \frac{1}{C_2} + \frac{1}{C_3} + \frac{1}{C_4}$$

$$\frac{1}{C_T} = \frac{1}{4\mu F} + \frac{1}{10\mu F} + \frac{1}{5\mu F} + \frac{1}{15\mu F}$$

$$\frac{1}{C_T} = \frac{3.75 + 1.5 + 3 + 1}{15\mu F}$$

$$\frac{1}{C_T} = \frac{9.25}{15\mu F}$$

$$\frac{C_T}{1} = \frac{15\mu F}{9.25}$$

$$C_T = 1.62\mu F$$

Capacitance in Parallel

When two equal capacitors are connected in parallel, the plates of the individual capacitors, in effect, combine to form one capacitor representing total capacitance. Notice in *Figure 6.40* that the effective plate area of the equivalent capacitor has doubled. Since an increase in plate area increases capacitance, it can be concluded that the total capacitance of the two capacitors in parallel is equal to the sum of the two capacitors. This direct relationship between total capacitance and capacitors in parallel is described by this equation:

$$\mathbf{C_T = C_1 + C_2 + C_3 + \ldots C_N} \quad (6\text{–}12)$$

This equation should look familiar to you. It is similar to the equation used for calculating the total resistance of resistors connected in *series*.

Figure 6.41 is a circuit containing four capacitors connected in parallel with the capacitance value for each shown. Using equation *6–12*, the total capacitance for the circuit can be calculated:

$$C_T = C_1 + C_2 + C_3 + C_4$$
$$C_T = 2\mu F + 3\mu F + 6\mu F + 5\mu F$$
$$C_T = 16\mu F$$

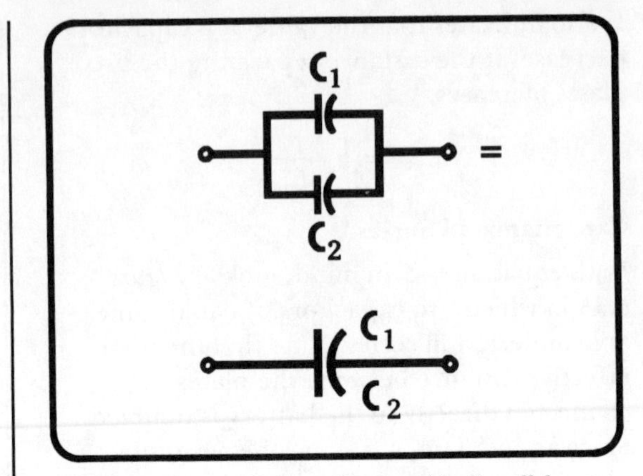

Figure 6.40 *Capacitances Connected in Parallel Increase Total Capacitance*

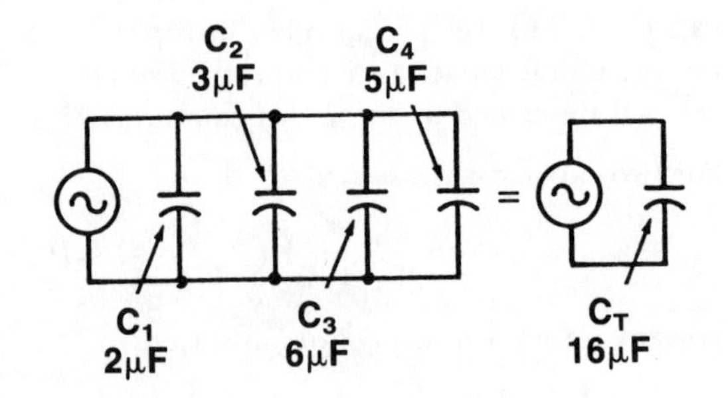

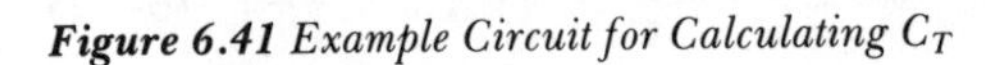
Figure 6.41 *Example Circuit for Calculating* C_T

In summary, the total capacitance of capacitors in series can be described mathematically with an equation (*6-11*) similar to that for calculating the total resistance of resistors connected in parallel. Similarly, total capacitance of capacitors connected in parallel can be calculated (equation *6–12)* like the total resistance of resistors connected in series.

ANALYSIS OF SERIES AND PARALLEL CAPACITIVE CIRCUITS

Thus far in this lesson, capacitive circuits have been analyzed that contain only a known voltage source and a single capacitor. In this next discussion, circuits with a known voltage source and multiple capacitors connected in series or parallel will be analyzed.

Series Capacitive Circuit Analysis

In the circuit of *Figure 6.42* are two capacitors, 4 microfarads and 12 microfarads, connected in series. Applied voltage is 10 volts, 3 kilohertz. Capacitive reactance is calculated first using equation *6–5*. The capacitive reactance of C_1 = 4 microfarads is:

$$X_{C1} = \frac{1}{2\pi fC_1} = 13.2\Omega$$

As shown, inserting values and calculating, it is found that X_{C1} = 13.2 ohms. Similar calculations are performed for the 12 microfarad capacitor, C_2.

$$X_{C2} = \frac{1}{2\pi fC_2} = 4.4\Omega$$

And we find that X_{C2} = 4.4 ohms. Notice that as capacitance increases, the capacitive reactance decreases:

$$\downarrow X_C = \frac{1}{2\pi fC \uparrow}$$

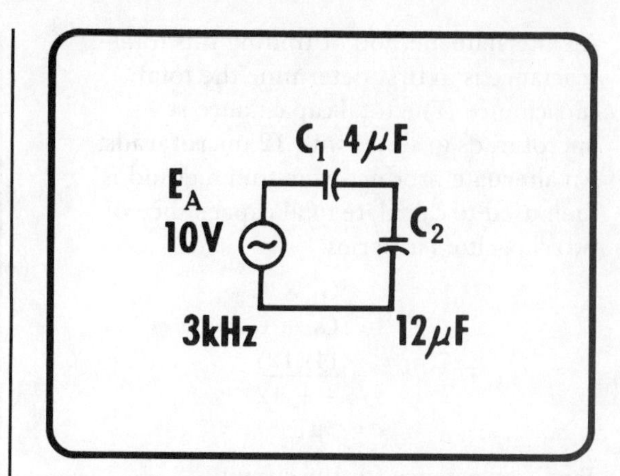

Figure 6.42 Example Series Capacitive Circuit

There is a definite mathematical relationship between increase and decrease of the values. In the circuit of *Figure 6.42* the capacitance of C_2 is three times the capacitance of C_1:

$$\frac{12\mu F}{4\mu F} = \frac{3}{1}$$

The capacitive reactance of C2 is one-third of the capacitive reactance of C1.

$$\frac{4.4\Omega}{13.2\Omega} = \frac{1}{3}$$

Now that the individual reactances in the circuit have been determined, the reactance ohms are treated as resistive ohms. That is, all series reactances are added.

$$X_{CT} = X_{C1} + X_{C2} = 13.2 + 4.4 = 17.6\Omega$$

The total capacitive reactance is 17.6 ohms. This *total capacitive reactance* is the total opposition that the circuit presents to current flow at the applied frequency.

■ Series Capacitive Circuit Analysis

An alternate method of finding this total reactance is to first determine the total capacitance. The total capacitance is 4 microfarads in series with 12 microfarads. An alternate product-over-sum method is then used to calculate total capacitance of two capacitors in series.

$$C_T = \frac{C_1 \times C_2}{C_1 + C_2} = \frac{(4)(12)}{4 + 12} = 3\mu F$$

Total capacitance for the circuit is 3 microfarads. Using this, the total capacitive reactance of the circuit can be calculated using equation *6–5* as a total capacitive reactance equation.

$$X_{CT} = \frac{1}{2\pi f C_T}$$

Substituting circuit values,

$$X_{CT} = \frac{1}{6.28(3kHz)(3\mu F)} = 17.6\Omega$$

Total capacitive reactance is 17.6 ohms,the same total capacitive reactance calculated earlier by adding individual reactances.

Once the total capacitive reactance has been determined, the total current in the circuit can be calculated. The total current is equal to:

$$\mathbf{I_T = \frac{E_A}{X_{CT}} = \frac{10V}{17.6\Omega} = 568mA}$$

The applied voltage divided by the total capacitive reactance equals 0.568 amperes—568 milliamperes. Therefore, as shown in *Figure 6.43*, most information about the circuit has been calculated.

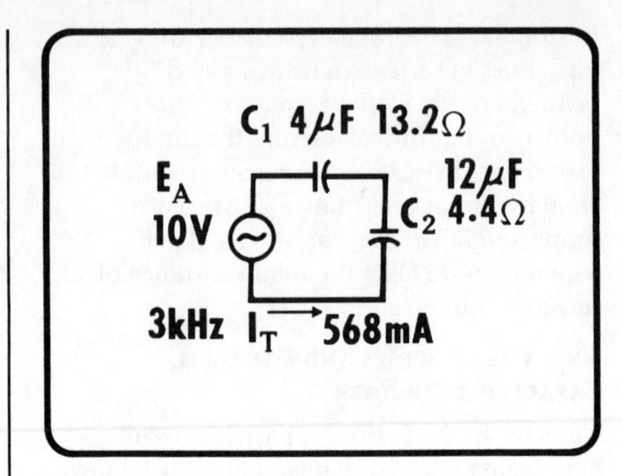

Figure 6.43 *Summation of Calculated Circuit Values*

Now the rules of series circuits are applied to find the voltage drops across C_1 and C_2. Since it is a series circuit, the current is the total current and it is the same throughout the circuit. The voltage drop across C_1 is equal to the value of its capacitive current times its opposition to that current:

$$\mathbf{E_{C1} = I_{C1}X_{C1} = I_T X_{C1} = (568mA)(13.2\Omega) = 7.5V}$$

The voltage drop across C_2 is calculated similarly.

$$\mathbf{E_{C2} = I_{C2}X_{C2} = I_T X_{C2} = (568mA)(4.4\Omega) = 2.5V}$$

In a purely capacitive circuit, the voltage drops add to the total applied voltage as in a series resistive circuit:

$$\mathbf{E_A = E_{C1} + E_{C2} = 7.5V + 2.5V = 10V}$$

■ Parallel Capacitive Circuit Analysis

The circuit with these calculated voltages indicated is shown in *Figure 6.44*.

Thus, you can see that knowing just a few factors about a circuit can allow you to calculate most all other factors.

Parallel Capacitive Circuit Analysis

Now, a parallel combination of two capacitors, as shown in *Figure 6.45*, will be analyzed. The same voltage, frequency and two capacitors used in the circuit of *Figure 6.42* will be used in this analysis. The capacitive reactance of the respective capacitors will be the same— 13.2 ohms for X_{C1} and 4.4 ohms for X_{C2}— since the values of the capacitors and the applied frequency are the same. The branch current for the C_1 branch may be determined by dividing the voltage across the branch by the opposition in the branch.

$$I_{C1} = \frac{E_{C1}}{X_{C1}} = \frac{E_A}{X_{C1}} = \frac{10V}{13.21\Omega} = 757mA$$

The current for the C_2 branch is calculated similarly.

$$I_{C2} = \frac{E_{C2}}{X_{C2}} = \frac{E_A}{X_{C2}} = \frac{10V}{4.4\Omega} = 2.273A$$

The total current in the circuit is the sum of the branch currents:

$$I_T = I_{C1} + I_{C2} = 0.757A + 2.273A = 3.03A$$

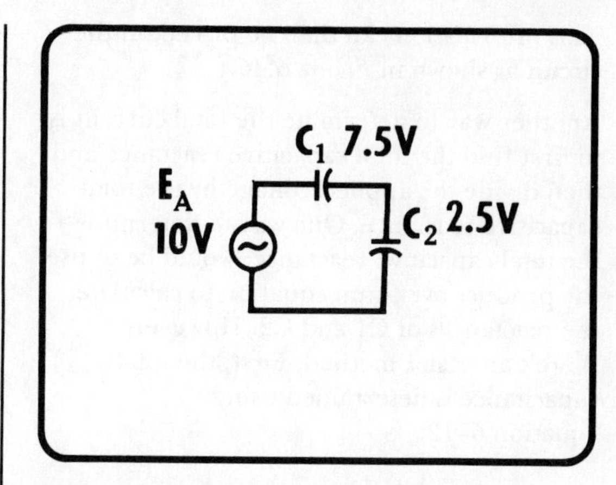

Figure 6.44 Example Circuit Calculated Voltages

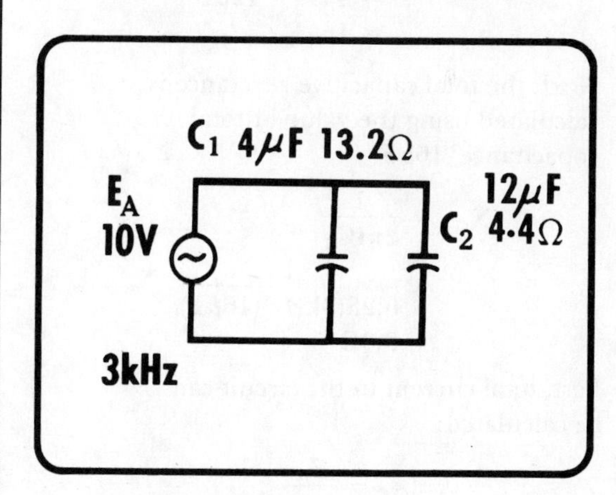

Figure 6.45 Example Parallel Circuit

■ Calculations of Reactive Power

This information can then be placed on the circuit as shown in *Figure 6.46*.

Another way to determine the total current is to first find the total capacitive reactance and then divide the applied voltage by the total capacitive reactance. One way to determine the total capacitive reactance, would be to use the product-over-sum equation to calculate the reactances of C1 and C2. However, there's an easier method. First, the total capacitance is determined using equation *6–12*.

$$C_T = C_1 + C_2$$
$$= 4\mu F + 12\mu F$$
$$= 16\mu F$$

Next, the total capacitive reactance is calculated using the value of total capacitance, 16µF.

$$X_{CT} = \frac{1}{2\pi f C_T}$$
$$= \frac{1}{6.28(3kHz)(16\mu F)}$$
$$= 3.3\Omega$$

Last, total current in the circuit can be calculated:

$$I_{CT} = \frac{E_A}{X_{CT}}$$
$$= \frac{10V}{3.3\Omega}$$
$$= 3.03A$$

The total current calculated is 3.03 amperes, which is the identical answer calculated previously using a different equation. Therefore, performing this last calculation proved that the earlier calculation was correct.

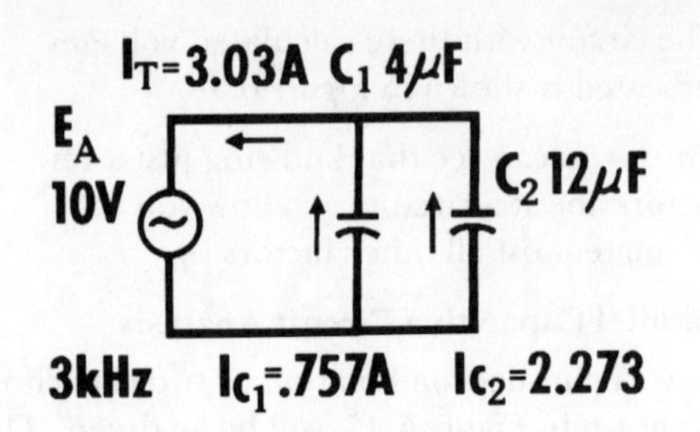

Figure 6.46 *Example Parallel Circuit with Calculated Values*

Calculations of Reactive Power

Power in capacitive circuits can be calculated similar to the way power is calculated in resistive circuits.

Recall that power in resistive circuits is converted to heat and dissipated. That is, the electrical energy is converted into heat energy. This is not true with capacitors. The electrical energy in capacitive circuits is stored temporarily on the plates of the capacitors in what is called an electrostatic field. The energy is then returned to the circuit.

Recall that the basic power equation is

$$P = EI \qquad (6\text{–}13)$$

This same equation is used to calculate power in capacitive circuits. However, the power in capacitive circuits is measured in units called VAR, not watts as in resistive circuit. VAR stands for Volts-Amperes-Reactive. The power in a capacitive circuit is called reactive power since the opposition to current in the circuit is totally reactive. Therefore, the basic equation to calculate power in a capacitive circuit (with appropriate units) is:

$$\underset{(VAR)}{P} = \underset{(volts)}{E} \times \underset{(amperes)}{I} \qquad (6\text{–}14)$$

■ Power in Series Capacitive Circuit
■ Power in Parallel Capacitive Circuit

Power in Series Capacitive Circuit

The series capacitive circuit of *Figure 6.47* (which is the circuit that was discussed earlier) will be used for the example of calculating power in such a circuit. In that circuit, E_{C1} is 7.5 volts and the current through C1 is 568 milliamperes (0.568A).

The power equation can be used to calculate the reactive power of C_1. P_{C1} equals E_{C1} times I_{C1}.

$$\begin{aligned} P_{C1} &= E_{C1}I_{C1} \\ &= (7.5V)(0.568A) \\ &= 4.26VAR \end{aligned}$$

As shown this equals 7.5 volts times 0.568 amperes which equals 4.26 VAR.

Similarly, the reactive power of C_2 can be calculated.

$$\begin{aligned} P_{C2} &= E_{C2}I_{C2} \\ &= (2.5V)(0.568A) \\ &= 1.42VAR \end{aligned}$$

By adding these two values of reactive power the total reactive power of the circuit is obtained:

$$\begin{aligned} P_{XT} &= P_{C1} + P_{C2} \\ &= 4.26VAR + 1.42VAR \\ &= 5.68VAR \end{aligned}$$

The total reactive power of the circuit could also have been determined by multiplying the total applied voltage times the total current.

$$\begin{aligned} P_{XT} &= E_AI_T \\ &= (10V)(0.568A) \\ &= 5.68VAR \end{aligned}$$

10 volts times 0.568 amperes equals 5.68 VAR which is the same power calculated previously, and proves that calculation was correct. Thus, using the basic power equation as the reactive power equation, the individual and total reactive powers have been calculated.

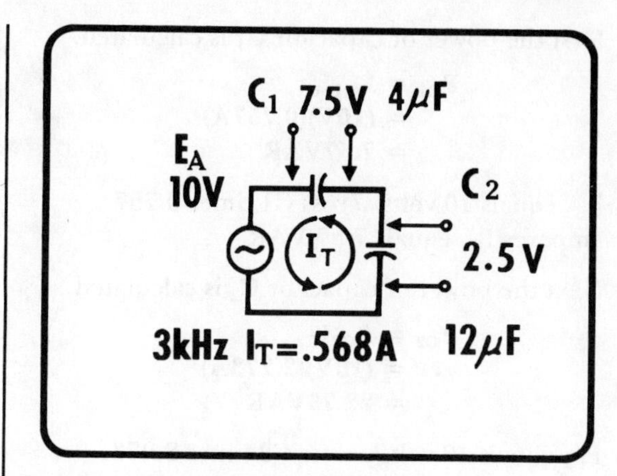

Figure 6.47 Example Series Circuit

Power in Parallel Capacitive Circuit

The same type of calculations can be performed to obtain the reactive power quantities for a parallel capacitive circuit. The parallel capacitive circuit of *Figure 6.45* used earlier will be used again, and is repeated in *Figure 6.48*.

■ **Summary**

First the power of capacitor C_1 is calculated.

$$\begin{aligned} P_{C1} &= E_{C1}I_{C1} \\ &= (10V)(0.757A) \\ &= 7.57VAR \end{aligned}$$

P_{C1} equals 10 volts across C1 times 0.757 amperes I_{C1} equals 7.57 VAR.

Next the power of capacitor C_2 is calculated.

$$\begin{aligned} P_{C2} &= E_{C2}I_{C2} \\ &= (10V)(2.273A) \\ &= 22.73VAR \end{aligned}$$

P_{C2} equals 10 volts across C2 times 2.273 amperes, I_{C2}, equals 22.73 VAR.

Last, the total reactive power of the circuit is calculated.

$$\begin{aligned} P_{XT} &= P_{C1} + P_{C2} \\ &= 7.57VAR + 22.73VAR \\ &= 30.3VAR \end{aligned}$$

Total reactive power equals 30.3 VAR. Also, the total reactive power equals the total applied voltage times the total current. In this case, that would be 10 volts times 3.03 amperes which equals 30.3 VAR. This is the same answer calculated previously and proves the previous calculations are correct.

Using the equations discussed you should be able to calculate similar information about any other series or parallel capacitive circuit.

SUMMARY

This lesson has provided an introduction to capacitive circuits. The physical properties of different types of capacitors and the concept of capacitance and charge were discussed. How a capacitor functions when it is charged and discharged relating capacitive properties to a stored charge and an opposition to voltage changes, was described. The concepts of capacitive reactance as an opposition to current flow and as a function of frequency were covered. Current/voltage relationships were explained using the rate of change equation. Series and parallel capacitive circuits were analyzed determining reactance voltage drops, currents, and power. All of these concepts will be used in following lessons to solve more complex circuits.

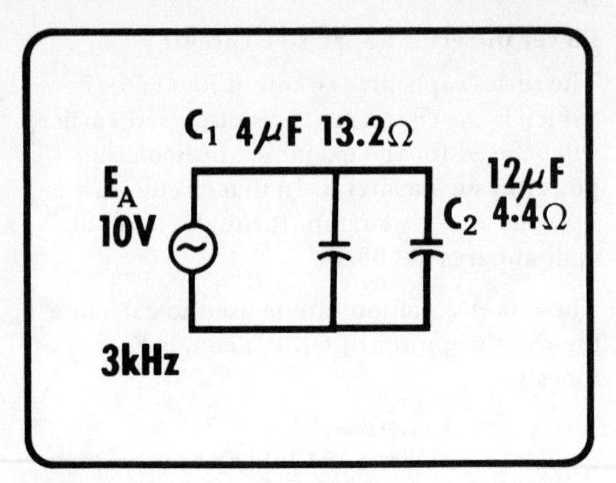

Figure 6.48 *Example Parallel Circuit*

■ Worked-Out Examples

1. Describe the characteristics of a ceramic disc capacitor.

Solution: The ceramic disc capacitor is available in small values of capacitance from approximately 1 picofarad to 2.5 microfarads. Typical working voltages are 20 volts up to about 200 volts. It is a good insulator, and it has a permittivity of almost 1000 times more than that of dry air.

2. Define capacitance.

Solution: Capacitance is the ability of a nonconductor to store a charge. If a voltage of 50 volts is applied to a capacitor for a period of time, the capacitor will charge to the 50-volt value and retain that difference in potential for a period of time after that voltage is removed.

3. Describe one effect of the capacitive property in ac circuits.

Solution: In ac circuits, the current through the capacitive branch of the circuit leads the voltage across the capacitor by 90 degrees. This is explained by the equation $I_C = C \times ROC$.

4. A mylar capacitor with a dielectric 2 mils thick has a value of 0.01 microfarads. What would be the result of increasing the dielectric 2 mils thickness to 4 mils if all of the other variables remain the same?

Solution: The equation describing the physical parameters of a capacitor is:

$$C = K_e \left(\frac{A}{d}\right) \epsilon_o$$

The d variable represents the distance between the plates, and the equation shows that distance is indirectly related to capacitance. If the distance between the plates increases as proposed in the problem, capacitance will decrease. Because the distance has doubled, the capacitance value will be halved, and the value of capacitance will be 0.005 microfarads.

5. Calculate the area of either plate of a 1 farad paper capacitor with a dielectric 1 millimeter thick. Assume $K_e = 4$.

Solution:

$$C = K_e \left(\frac{A}{d}\right) \epsilon_O$$

Solving for A:

$$A = \frac{C \times d}{K_e \epsilon_O} = \frac{1F \times 0.001m}{4 \times 8.85 \times 10^{-12}} = \frac{0.001}{3.54 \times 10^{-11}} = \mathbf{28.2 \times 10^6 \text{ square meters}}$$

or about 12 square miles.

■ Worked-Out Examples

6. Solve for X_{CT} in this circuit:

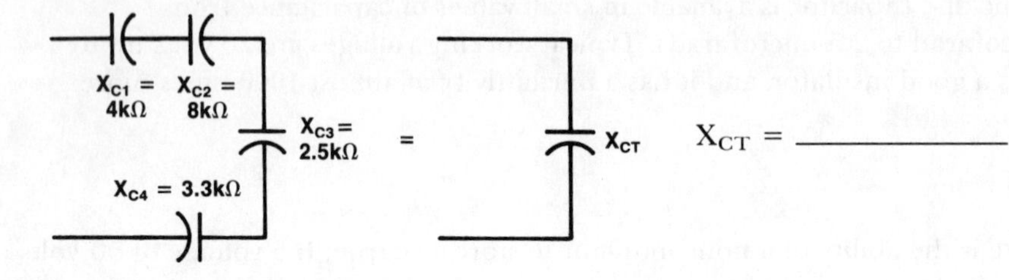

Solution: In a series capacitive circuit, total capacitive reactance is calculated the same as total resistance in series is calculated.

$$X_{CT} = X_{C1} + X_{C2} + X_{C3} + X_{C4} = 4\text{k}\Omega + 8\text{k}\Omega + 2.5\text{k}\Omega + 3.3\text{k}\Omega = \mathbf{17.8k\Omega}$$

7. Solve for C_T in this circuit:

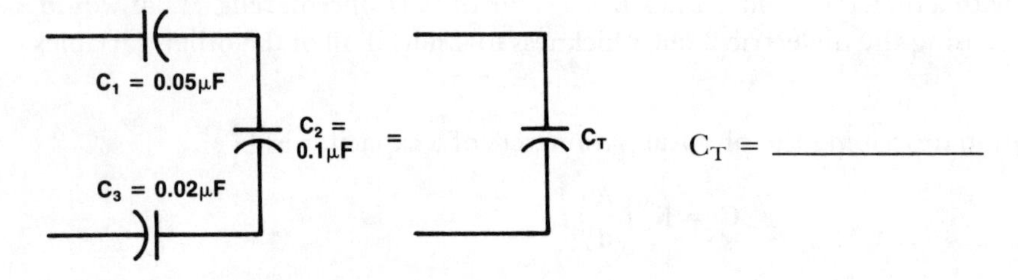

Solution: In a series capacitive circuit, total capacitance is calculated similar to the way total resistance is calculated in a parallel resistive circuit.

$$\frac{1}{C_T} = \frac{1}{C_1} + \frac{1}{C_2} + \frac{1}{C_3} = \frac{1}{0.05\mu F} + \frac{1}{0.1\mu F} + \frac{1}{0.02\mu F}$$

$$= \frac{1}{0.05 \times 10^{-6}} + \frac{1}{0.1 \times 10^{-6}} + \frac{1}{0.02 \times 10^{-6}} = 20 \times 10^6 + 10 \times 10^6 + 50 \times 10^6 F$$

$$= 80 \times 10^6 F$$

$$C_T = \frac{1}{80 \times 10^6} F = \mathbf{0.0125\mu F}$$

■ **Worked-Out Examples**

8. Solve for X_C in this circuit:

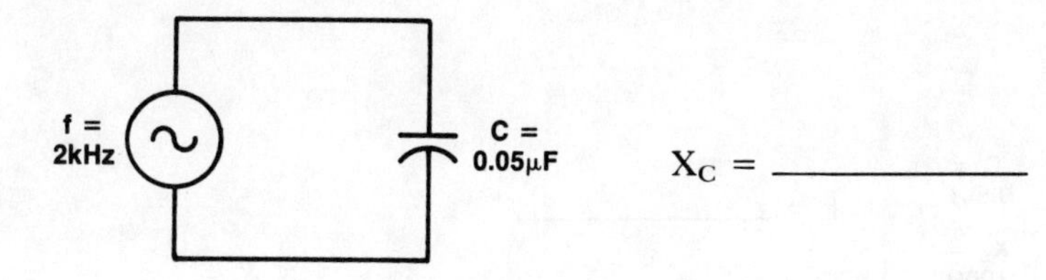

X_C = ____________

Solution:

$$X_C = \frac{1}{2\pi fC} = \frac{1}{628 \times 2 \times 10^3 \times 0.05 \times 10^{-6}} = \frac{1}{6.28 \times 10^{-6}\Omega} = 1.59 \times 10^3 = \mathbf{1.59k\Omega}$$

9. Solve for C_T in this circuit:

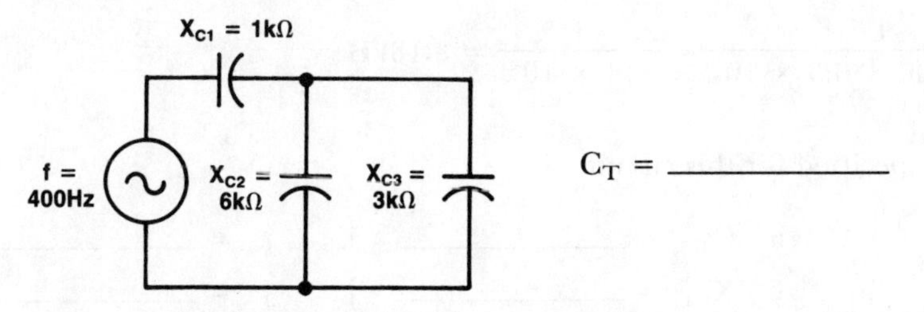

C_T = ____________

Solution: Solve for X_{CT} by circuit simplication.

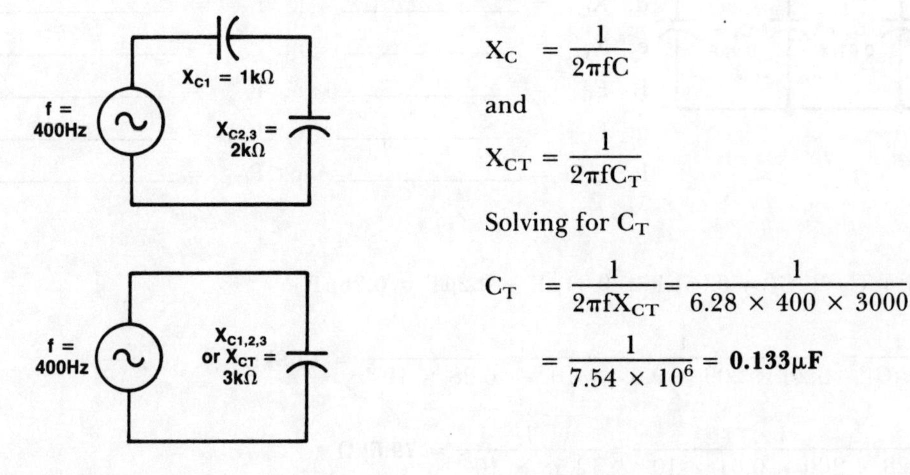

$$X_C = \frac{1}{2\pi fC}$$

and

$$X_{CT} = \frac{1}{2\pi fC_T}$$

Solving for C_T

$$C_T = \frac{1}{2\pi fX_{CT}} = \frac{1}{6.28 \times 400 \times 3000}$$

$$= \frac{1}{7.54 \times 10^6} = \mathbf{0.133\mu F}$$

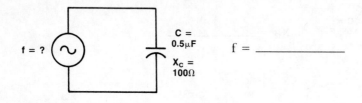

■ Worked-Out Examples

10. What source frequency must be applied to a 0.5 microfarad capacitor so that the capacitor will have a capacitive reactance of 100 ohms?

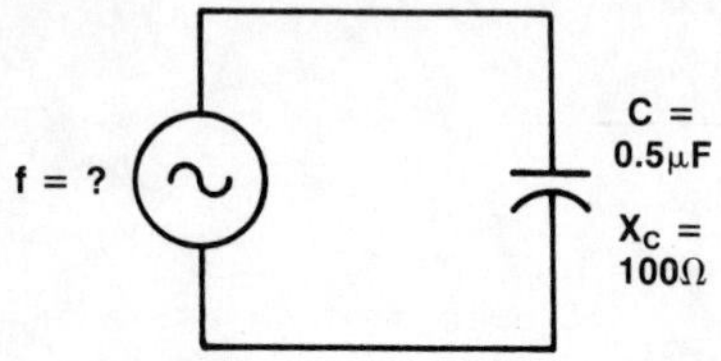

f = ____________

Solution:

$$X_C = \frac{1}{2\pi fC}$$

Solving for f

$$f = \frac{1}{2\pi X_C C} = \frac{1}{6.28 \times 100 \times 0.5 \times 10^{-6}} = \frac{1}{314 \times 10^{-6}} = \mathbf{3.18kHz}$$

11. Solve for the circuit values specified for this circuit:

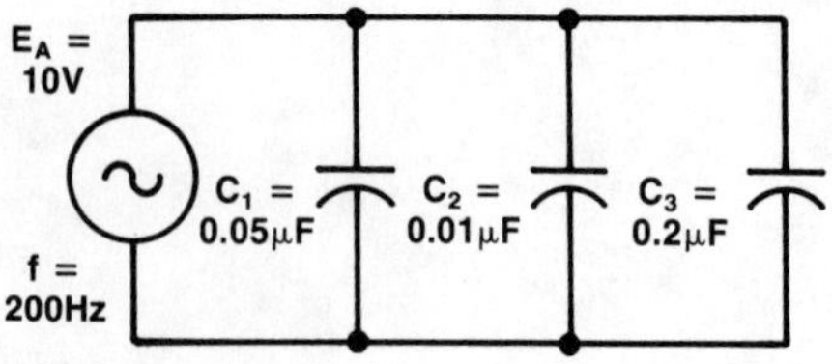

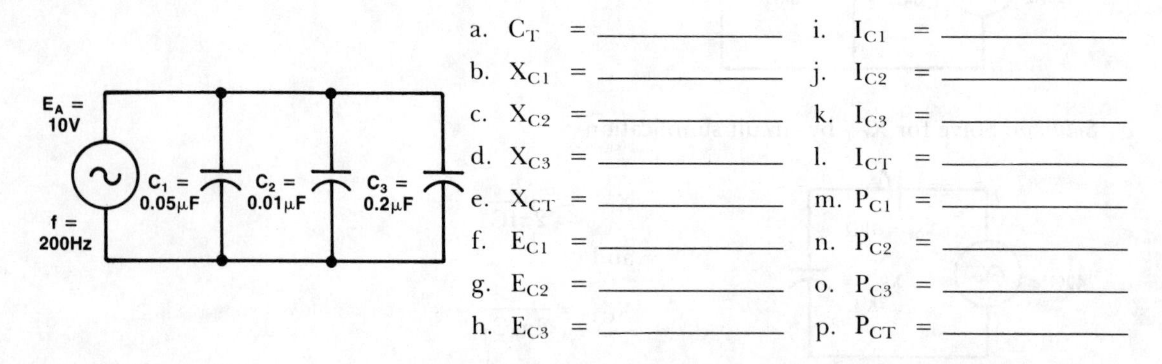

a. C_T = ____________
b. X_{C1} = ____________
c. X_{C2} = ____________
d. X_{C3} = ____________
e. X_{CT} = ____________
f. E_{C1} = ____________
g. E_{C2} = ____________
h. E_{C3} = ____________
i. I_{C1} = ____________
j. I_{C2} = ____________
k. I_{C3} = ____________
l. I_{CT} = ____________
m. P_{C1} = ____________
n. P_{C2} = ____________
o. P_{C3} = ____________
p. P_{CT} = ____________

Solution:

a. $C_T = C_1 + C_2 + C_3 = 0.05\mu F + 0.01\mu F + 0.2\mu F = \mathbf{0.26\mu F}$

b. $X_{C1} = \frac{1}{2\pi fC_1} = \frac{1}{6.28 \times 200 \times 0.05 \times 10^{-6}} = \frac{1}{6.28 \times 10^{-5}} = \mathbf{15.9k\Omega}$

c. $X_{C2} = \frac{1}{6.28 \times 200 \times 0.01 \times 10^{-6}} = \frac{1}{12.58 \times 10^{-6}} = \mathbf{79.6k\Omega}$

■ Worked-Out Examples

d. $X_{C3} = \dfrac{1}{6.28 \times 200 \times 0.2 \times 10^{-6}} = \dfrac{1}{251 \times 10^{-6}} = \mathbf{3.98k\Omega}$

e. $\dfrac{1}{X_{CT}} = \dfrac{1}{X_{C1}} + \dfrac{1}{X_{C2}} + \dfrac{1}{X_{C3}} = \dfrac{1}{15.9k} + \dfrac{1}{79.6k} + \dfrac{1}{3.98k}$

$= 62.9 \times 10^{-3} + 12.6 \times 10^{-3} + 2.51 \times 10^{-3} = 327 \times 10^{-3} = \dfrac{1}{327 \times 10^{-3}} = \mathbf{3.06k\Omega}$

or $X_{CT} = \dfrac{1}{2\pi FC_T} = \dfrac{1}{6.28 \times 200 \times 0.26 \times 10^{-6}} = \dfrac{1}{3.27 \times 10^{-4}}$

$= 0.306 \times 10^4 = \mathbf{3.06k\Omega}$

f,g,h. **In parallel E_A, 10VAC, is across all components.**

i. $I_{C1} = \dfrac{E_{C1}}{X_{C1}} = \dfrac{10V}{15.9k\Omega} = \mathbf{630\mu A_{rms}}$

j. $I_{C2} = \dfrac{E_{C2}}{X_{C2}} = \dfrac{10V}{79.6k\Omega} = \mathbf{126\mu A}$

k. $I_{C3} = \dfrac{E_{C3}}{X_{C3}} = \dfrac{10V}{3.98k\Omega} = \mathbf{2.51mA}$

l. $I_{CT} = I_{C1} + I_{C2} + I_{C3} = 630\mu A + 126\mu A + 2.51mA_{rms} = \mathbf{3.27mA}$ or

$I_{CT} = \dfrac{E_A}{X_{CT}} = \dfrac{10V}{3.06} = \mathbf{3.27mA}$

m. $P_{C1} = I_{C1} \times E_{C1} = 630\mu A \times 10V = \mathbf{6.3mVAR}$

n. $P_{C2} = I_{C2} \times E_{C2} = 126\mu A \times 10V = \mathbf{126mVAR}$

o. $P_{C3} = I_{C3} \times E_{C3} = 2.51mA \times 10V = \mathbf{25.1mVAR}$

p. $P_{CT} = P_{C1} + P_{C2} + P_{C3} = 6.3mVAR + 1.26mVAR + 25.1mVAR = \mathbf{32.7mVAR}$ or

$P_{CT} = I_{CT}E_A = (3.27mA)\,(10V) = \mathbf{32.7mVAR}$

■ Worked-Out Examples

12. Solve for the circuit values specified for this circuit:

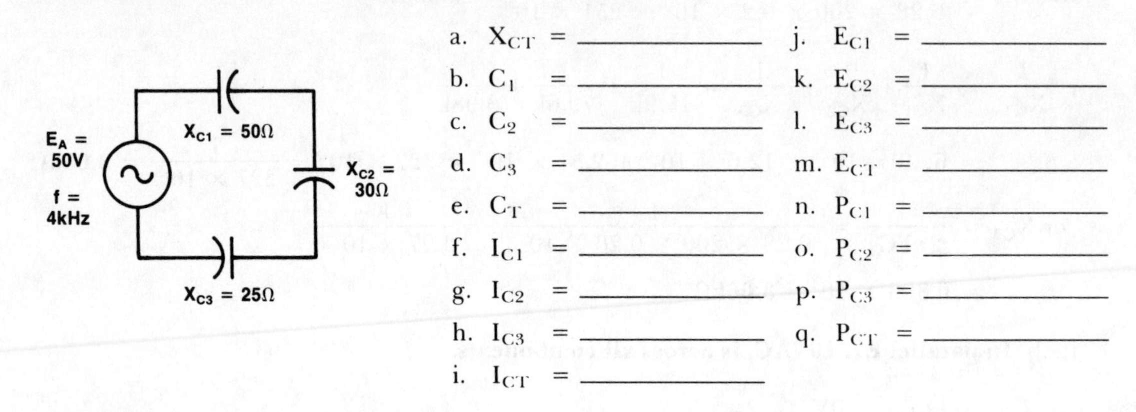

Solution:

a. $X_{CT} = X_{C1} + X_{C2} + X_{C3} = 50\Omega + 30\Omega + 25\Omega = \mathbf{105\Omega}$

b. $C_1 = \frac{1}{2\pi f X_{C1}} = \frac{1}{6.28 \times 4k \times 50} = \frac{1}{1.256 \times 10^6} = \mathbf{0.796\mu F}$

c. $C_2 = \frac{1}{2\pi f X_{C2}} = \frac{1}{6.28 \times 4k \times 30} = \frac{1}{754 \times 10^3} = \mathbf{1.33\mu F}$

d. $C_3 = \frac{1}{2\pi f X_{C3}} = \frac{1}{6.28 \times 4k \times 25} = \frac{1}{628 \times 10^3} = \mathbf{1.59\mu F}$

e. $C_T = \frac{1}{2\pi f X_{CT}} = \frac{1}{6.28 \times 4k \times 105} = \frac{1}{2.64 \times 10^6} = \mathbf{0.379\mu F}$

f,g,h. **In series, total current flows through all components, See I_{CT}.**

i. $I_{CT} = \frac{E_A}{X_{CT}} = \frac{50V}{105\Omega} = \mathbf{0.476A(rms)}$

j. $E_{C1} = I_{C1} \times X_{C1} = 0.476A \times 50\Omega = \mathbf{23.8V}$

k. $E_{C2} = I_{C2} \times X_{C2} = 0.476A \times 30\Omega = \mathbf{14.3V}$

l. $E_{C3} = I_{C3} \times X_{C3} = 0.476A \times 25\Omega = \mathbf{11.9V}$

m. $E_{CT} = E_A = E_{C1} + E_{C2} + E_{C3} = 23.8V + 14.3V + 11.9V = \mathbf{50V}$

n. $P_{C1} = I_{C1} \times E_{C1} = 0.476A \times 23.8V =$ **11.3VAR**

o. $P_{C2} = I_{C2} \times E_{C2} = 0.476A \times 14.3V =$ **6.81VAR**

p. $P_{C3} = I_{C3} \times E_{C3} = 0.476A \times 11.9V =$ **5.66VAR**

q. $P_{CT} = P_{C1} + P_{C2} + P_{C3} = 11.3 + 6.81 + 5.66 =$ **23.8VAR** or

$P_{CT} = I_{CT} \times E_A = 0.476A \times 50V =$ **23.8VAR**

13. Solve for the circuit values specified for this circuit:

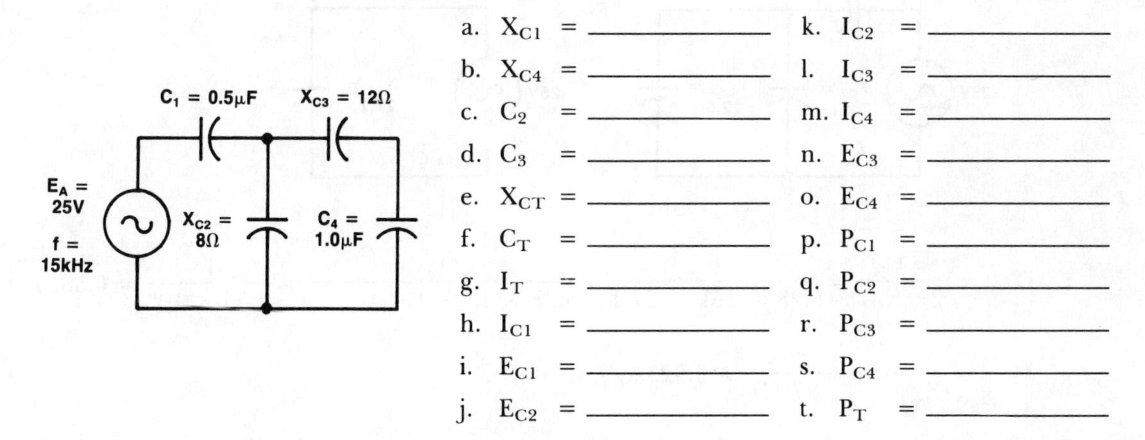

a. X_{C1} = ____ k. I_{C2} = ____
b. X_{C4} = ____ l. I_{C3} = ____
c. C_2 = ____ m. I_{C4} = ____
d. C_3 = ____ n. E_{C3} = ____
e. X_{CT} = ____ o. E_{C4} = ____
f. C_T = ____ p. P_{C1} = ____
g. I_T = ____ q. P_{C2} = ____
h. I_{C1} = ____ r. P_{C3} = ____
i. E_{C1} = ____ s. P_{C4} = ____
j. E_{C2} = ____ t. P_T = ____

Solution:

a. $X_{C1} = \frac{1}{2\pi f C_1} = \frac{1}{6.28 \times 15k \times 0.5 \times 10^{-6}} = \mathbf{21.2\Omega}$

b. $X_{C4} = \frac{1}{6.28 \times 15k \times 1 \times 10^{-6}} = \mathbf{10.6\Omega}$

c. $C_2 = \frac{1}{2\pi f X} = \frac{1}{6.28 \times 15k \times 8} = \mathbf{1.33\mu F}$

d. $C_3 = \frac{1}{6.28 \times 15k \times 12} = \mathbf{0.885\mu F}$

Worked-Out Examples

e. X_{CT} = **by simplification**

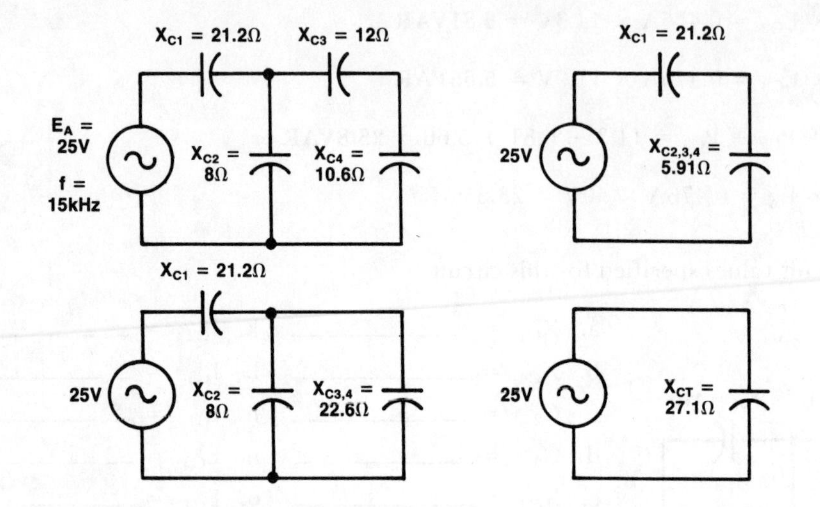

f. $C_T = \frac{1}{2\pi f X_{CT}} = \frac{1}{6.28 \times 15k \times 27.1} = \frac{1}{6.28 \times 15 \times 10^3 \times 27.1} = \frac{1}{2.55 \times 10^6} = \mathbf{0.39\mu F}$

g. $I_T = \frac{E_A}{X_{CT}} = \frac{25V}{27.1\Omega} = \mathbf{0.923A_{rms}}$

h. $I_{C1} = I_T = \mathbf{0.923A}$

i. $E_{C1} = I_{C1} \times X_{C1} = 0.923A \times 21.2\Omega = \mathbf{19.6V}$

j. $E_{C2} = E_{C2,\,3,\,4} = I_T \times X_{C2,\,3,\,4} = I_T \left(\frac{8\Omega\,(12\Omega + 10.6\Omega)}{8\Omega + 12\Omega + 10.6\Omega}\right) = 0.923A \times 5.91\Omega = \mathbf{5.45V}$

k. $I_{C2} = \frac{E_{C2}}{X_{C2}} = \frac{5.45V}{8\Omega} = \mathbf{0.681A}$

l. m. $I_{C3} = I_{C4} = \frac{E_{C3,\,4}}{X_{C3,\,4}} = \frac{E_{C2,\,3,\,4}}{X_{C3,\,4}} = \frac{5.45V}{12\Omega + 10.6\Omega} = \frac{5.45V}{22.6\Omega} = \mathbf{0.241A}$

n. $E_{C3} = I_{C3} \times X_{C3} = 0.24A \times 12\Omega = \mathbf{2.88V}$

■ Worked-Out Examples

o. $E_{C4} = I_{C4} \times X_{C4} = 0.24\text{A} \times 10.6\Omega =$ **2.54V**

p. $P_{C1} = I_{C1} \times E_{C1} = 0.923\text{A} \times 19.6\text{V} =$ **18.1VAR**

q. $P_{C2} = I_{C2} \times E_{C2} = 0.681\text{A} \times 5.45\text{V} =$ **3.71VAR**

r. $P_{C3} = I_{C3} \times E_{C3} = 1.241\text{A} \times 2.88\text{V} =$ **0.695VAR**

s. $P_{C4} = I_{C4} \times E_{C4} = 0.241\text{A} \times 2.54\text{V} =$ **0.612VAR**

t. $P_T = P_{C1} + P_{C2} + P_{C3} + P_{C4} =$ **23.1VAR** or

$P_T = I_{CT} \times E_A = 0.923\text{A} \times 25\text{V} =$ **23.1VAR**

■ Practice Problems

1. Describe the characteristics of a paper capacitor. Use the tables provided in the text.

2. Predict what will happen in the circuits below: a. with the switch in position 1; b. with the switch moved from position 1 to position 2; c. with the switch moved from position 2 to position 3.

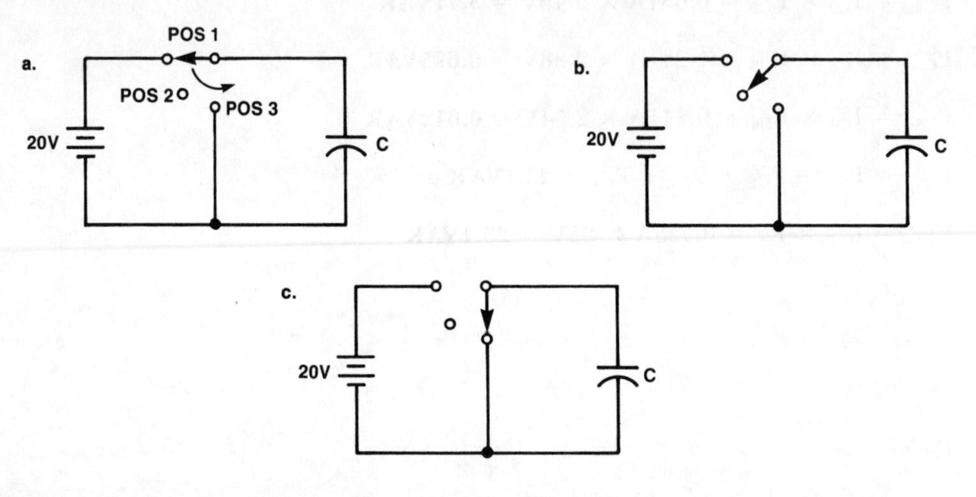

3. In Example 3 of the Worked-Out Examples, one effect of the capacitive property was discussed. Discuss briefly the other effect.

4. Relate the plate area of a capacitor to capacitors connected in parallel.

5. Solve for X_{CT} in this circuit:

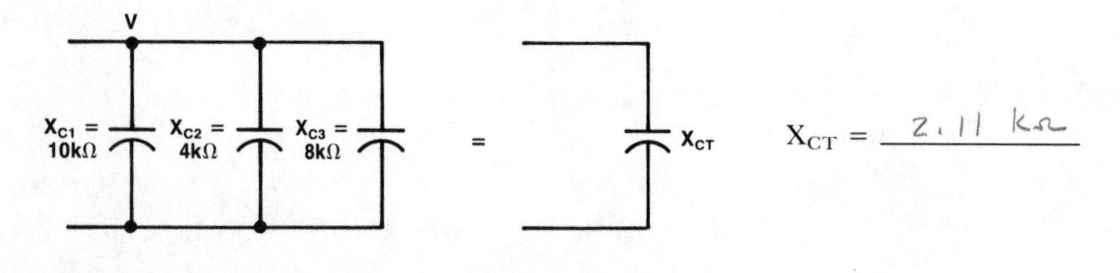

6. Solve for C_T in this circuit:

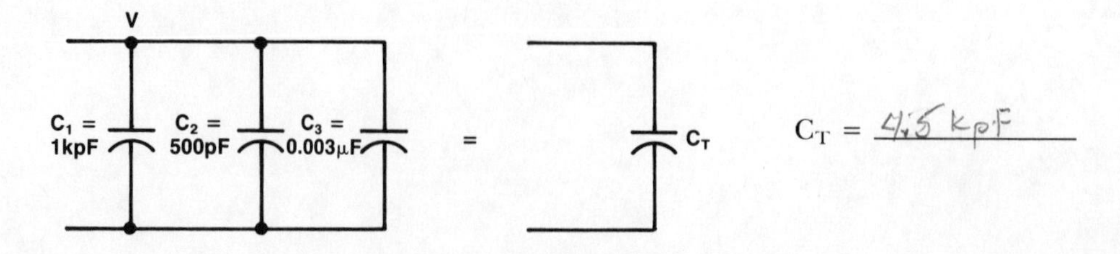

■ Practice Problems

7. What value of capacitor will have a resistance of 2.5kΩ in this circuit:

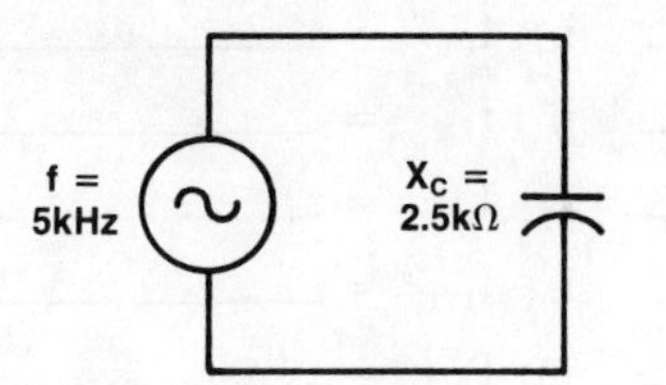

C = 12.7 kpF

8. Solve for X_{CT} and C_T in this circuit:

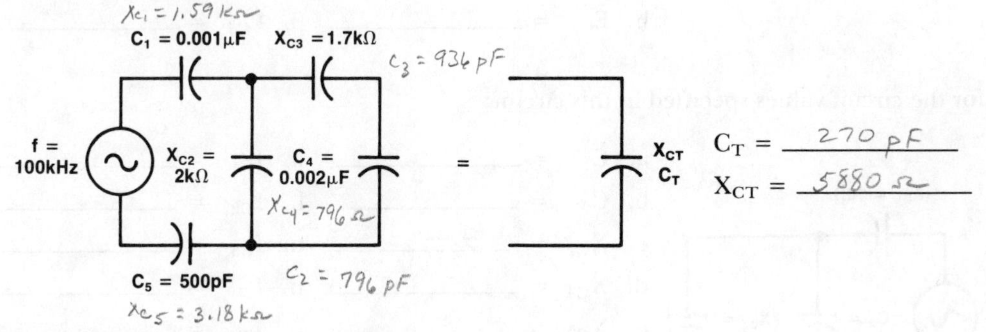

9. Solve for the circuit values specified in this circuit:

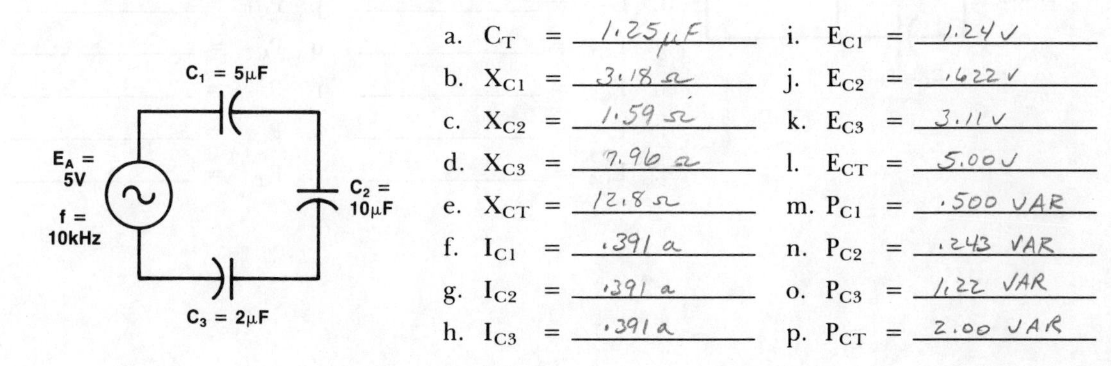

a. C_T = 1.25µF
b. X_{C1} = 3.18 Ω
c. X_{C2} = 1.59 Ω
d. X_{C3} = 7.96 Ω
e. X_{CT} = 12.8 Ω
f. I_{C1} = .391 a
g. I_{C2} = .391 a
h. I_{C3} = .391 a
i. E_{C1} = 1.24 V
j. E_{C2} = .622 V
k. E_{C3} = 3.11 V
l. E_{CT} = 5.00 V
m. P_{C1} = .500 VAR
n. P_{C2} = .243 VAR
o. P_{C3} = 1.22 VAR
p. P_{CT} = 2.00 VAR

■ Practice Problems

10. Solve for the circuit values specified in this circuit:

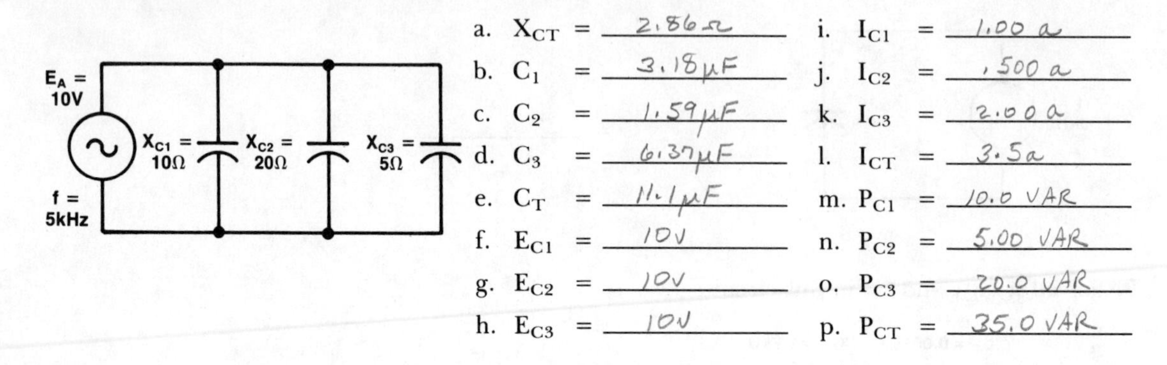

E_A = 10V
X_{C1} = 10Ω
X_{C2} = 20Ω
X_{C3} = 5Ω
f = 5kHz
a. X_{CT} = 2.86 Ω
b. C_1 = 3.18 μF
c. C_2 = 1.59 μF
d. C_3 = 6.37 μF
e. C_T = 11.1 μF
f. E_{C1} = 10V
g. E_{C2} = 10V
h. E_{C3} = 10V
i. I_{C1} = 1.00 a
j. I_{C2} = .500 a
k. I_{C3} = 2.00 a
l. I_{CT} = 3.5 a
m. P_{C1} = 10.0 VAR
n. P_{C2} = 5.00 VAR
o. P_{C3} = 20.0 VAR
p. P_{CT} = 35.0 VAR

11. Solve for the circuit values specified in this circuit:

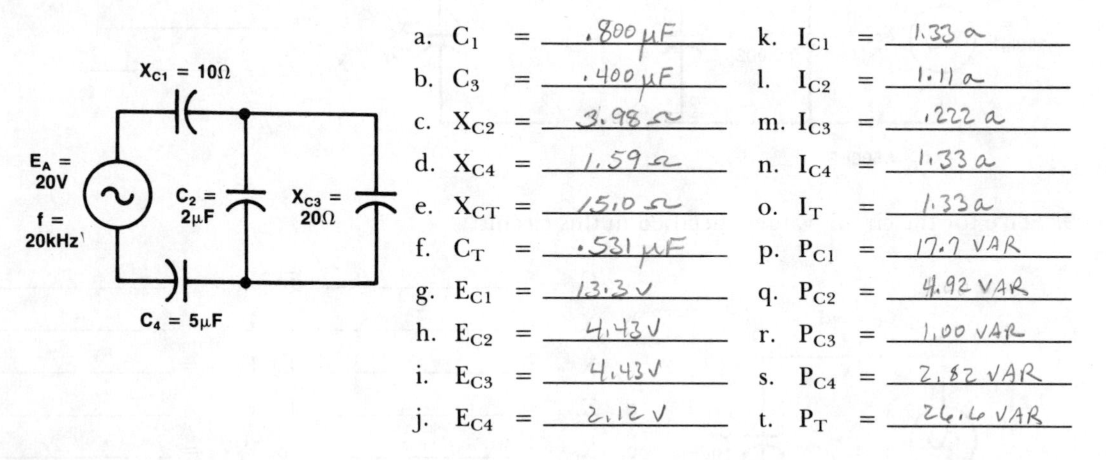

X_{C1} = 10Ω
E_A = 20V
f = 20kHz
C_2 = 2μF
X_{C3} = 20Ω
C_4 = 5μF
a. C_1 = .800 μF
b. C_3 = .400 μF
c. X_{C2} = 3.98 Ω
d. X_{C4} = 1.59 Ω
e. X_{CT} = 15.0 Ω
f. C_T = .531 μF
g. E_{C1} = 13.3 V
h. E_{C2} = 4.43 V
i. E_{C3} = 4.43 V
j. E_{C4} = 2.12 V
k. I_{C1} = 1.33 a
l. I_{C2} = 1.11 a
m. I_{C3} = .222 a
n. I_{C4} = 1.33 a
o. I_T = 1.33 a
p. P_{C1} = 17.7 VAR
q. P_{C2} = 4.92 VAR
r. P_{C3} = 1.00 VAR
s. P_{C4} = 2.82 VAR
t. P_T = 26.6 VAR

$I_T = \frac{E_A}{X_{CT}} = \frac{20V}{15Ω} = 1.33a$

$X_C = 3.33Ω$ $E = (3.33)(1.33) = 4.43V$

$I_T = 1.33a$

Quiz

1. Using the tables provided in the text, answer the following questions: Which type of capacitor has:

a. The greatest dielectric strength?
b. The largest dielectric constant?
c. The largest capacitance value?
d. The largest value of working voltage?

2. a. If a neutrally-charged capacitor is connected to a 50-volt source, what will happen to the capacitor?
b. If the source is then removed from the capacitor, what happens?
c. If a charged capacitor is shorted with a piece of wire, what occurs in the resulting circuit?

3. Draw a phasor diagram showing the relationship of I_C and E_C in an ac capacitive circuit.

4. Draw a sine wave graph showing the relationship of I_C and E_C in an ac capacitive circuit.

5. A capacitor in an ac circuit may be best defined as:

a. An open circuit
b. A short circuit
c. A fixed resistance
d. A variable resistance
e. A short circuit instantly, an open circuit after the capacitor is charged.

6. a. What is the source frequency? Note the given values of C_1 and X_{C1}.
b. X_{CT} = ________
c. C_T = ________

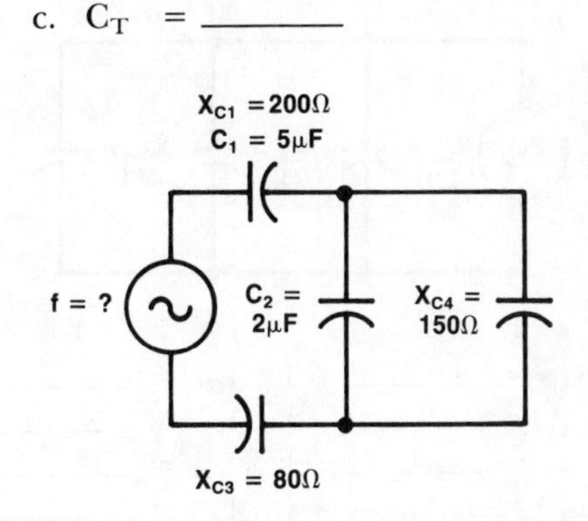

7. Solve for the circuit values specified for this circuit:

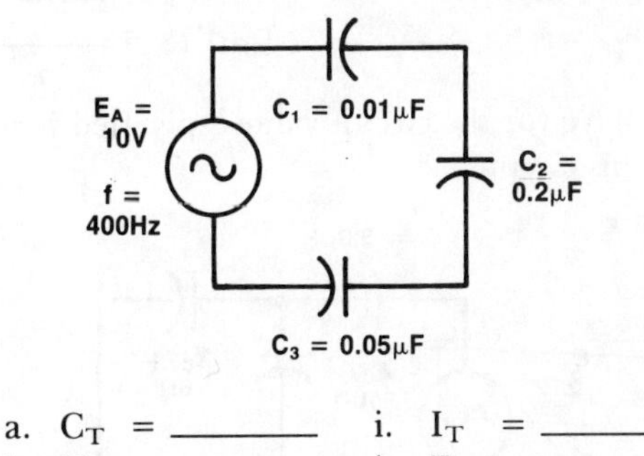

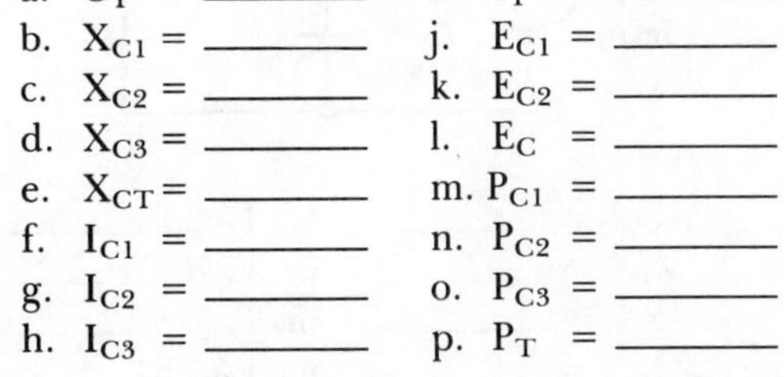

a. C_T = ________
b. X_{C1} = ________
c. X_{C2} = ________
d. X_{C3} = ________
e. X_{CT} = ________
f. I_{C1} = ________
g. I_{C2} = ________
h. I_{C3} = ________
i. I_T = ________
j. E_{C1} = ________
k. E_{C2} = ________
l. E_C = ________
m. P_{C1} = ________
n. P_{C2} = ________
o. P_{C3} = ________
p. P_T = ________

Quiz

8. Solve for the circuit values specified for this circuit:

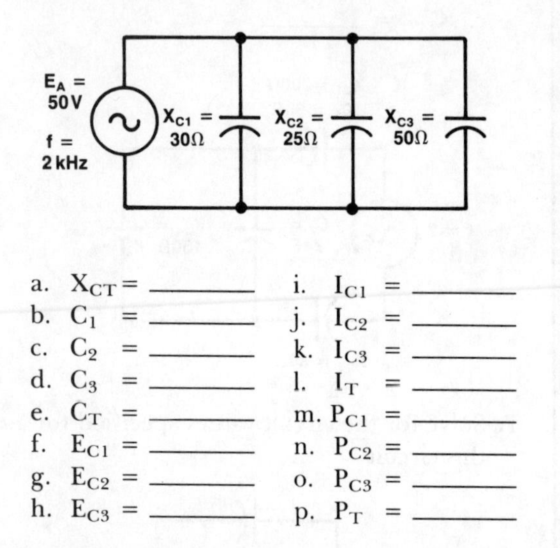

a. X_{CT} = ________
b. C_1 = ________
c. C_2 = ________
d. C_3 = ________
e. C_T = ________
f. E_{C1} = ________
g. E_{C2} = ________
h. E_{C3} = ________
i. I_{C1} = ________
j. I_{C2} = ________
k. I_{C3} = ________
l. I_T = ________
m. P_{C1} = ________
n. P_{C2} = ________
o. P_{C3} = ________
p. P_T = ________

9. Solve for the circuit values specified for this circuit:

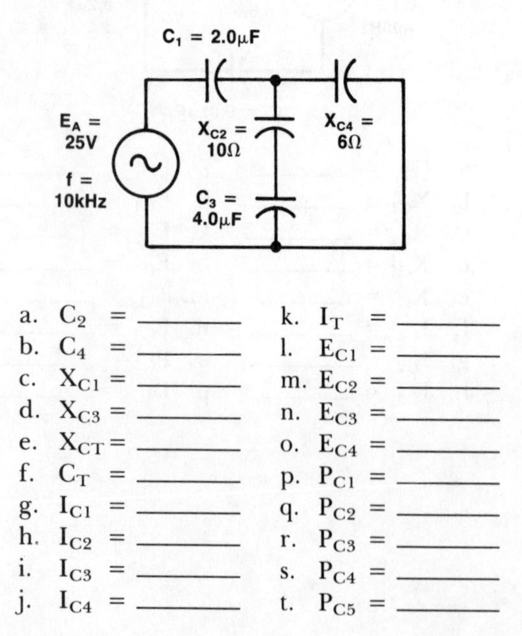

a. C_2 = ________
b. C_4 = ________
c. X_{C1} = ________
d. X_{C3} = ________
e. X_{CT} = ________
f. C_T = ________
g. I_{C1} = ________
h. I_{C2} = ________
i. I_{C3} = ________
j. I_{C4} = ________
k. I_T = ________
l. E_{C1} = ________
m. E_{C2} = ________
n. E_{C3} = ________
o. E_{C4} = ________
p. P_{C1} = ________
q. P_{C2} = ________
r. P_{C3} = ________
s. P_{C4} = ________
t. P_{C5} = ________

10. Are the voltage, current, and power values calculated in Problems 7, 8, and 9 rms, peak, or peak to peak values?

LESSON 7

RC Circuit Analysis

In this lesson, discussion concerns ac circuits in which capacitors and resistors are combined in series and parallel. Phasor diagrams are developed for both types of circuits. Ohm's law and the Pythagorean theorem are used to calculate the various circuit values.

■ Objectives

This lesson describes series and parallel RC ac circuits and the various methods used to analyze these types of circuits. At the end of this lesson, you should be able to:

1. Draw phasor diagrams which show the phase relationships of the various circuit values in series and parallel RC circuits.

2. Identify the circuit values in series and parallel circuits that are described by Pythagorean theorem relationships.

3. Use the tangent trigonometric function to solve for phase angles.

4. Define the term *phase angle*, and describe positive and negative phase angles in series and parallel RC ac circuits.

5. Analyze series and parallel RC ac circuits as shown and calculate voltage, current, and power values.

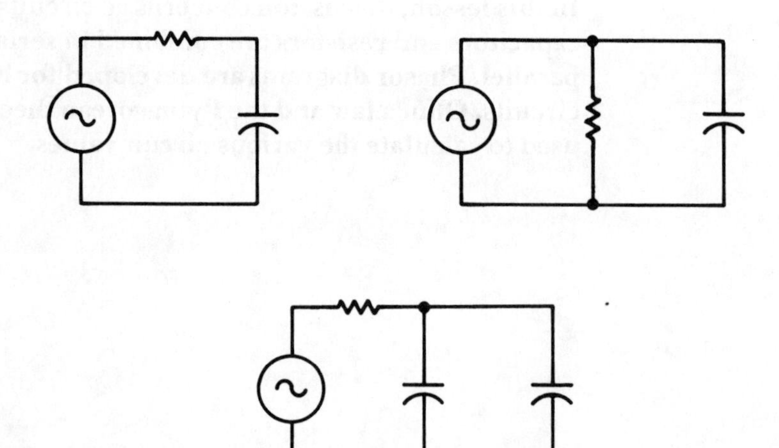

■ Phase Relationships in Series RC Circuits

INTRODUCTION

In the previous lesson, the capacitor and its properties were discussed and ac capacitive circuits were analyzed. A more common circuit, however, is one in which capacitors are combined in series or parallel with resistors as shown in *Figure 7.1*. In this lesson, discussion will concern how to solve various ac circuit problems for total resistance, total capacitance, power, and other electrical values using series and parallel RC circuits.

PHASE RELATIONSHIPS IN SERIES RC CIRCUITS

A good way to begin to learn about RC circuits is by analyzing a simple circuit in which a single resistor is connected in series with a single capacitor. The circuit, shown in *Figure 7.2*, is called a series RC circuit. In this circuit, like in any series circuit, the current flowing through all components is the same value.

$$I_T = I_C = I_R \quad (7\text{–}1)$$

However, the algebraic sum of the voltage drop across the resistor and the voltage drop across the capacitor does not equal the applied voltage as it would in either a purely resistive or purely capacitive circuit. Therefore,

$$E_T \text{ is not } E_R + E_C \quad (7\text{–}2)$$

This can be demonstrated with a specific example. Suppose the resistance of *Figure 7.2* is 40 ohms and the capacitive reactance, X_C, is 30 ohms. With an applied voltage of 10 volts, the voltage drops across R and C are calculated to be 8 volts and 6 volts respectively. The algebraic sum of the voltage drops is 14 volts and does not equal the applied voltage of 10 volts.

This is true because the phase relationship between the voltage across and the current through each component is different.

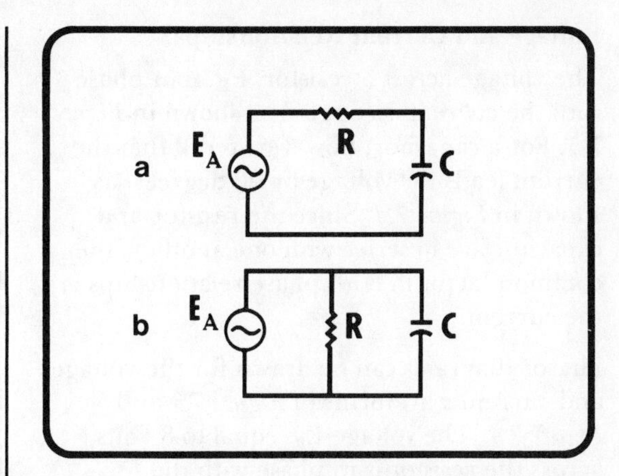

***Figure 7.1** Typical Series and Parallel RC AC Circuits*

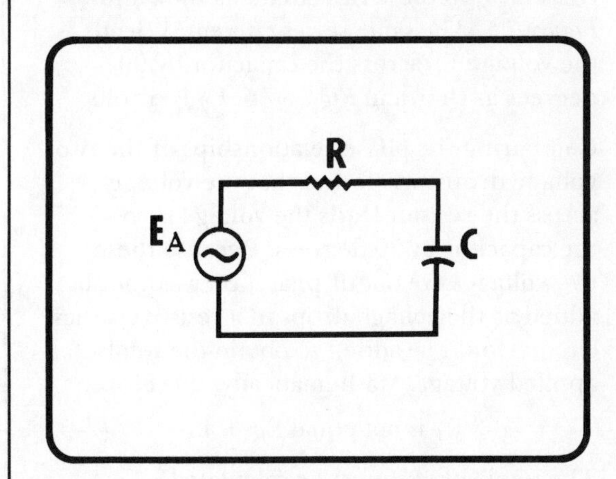

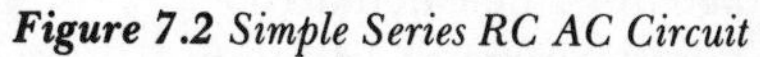

***Figure 7.2** Simple Series RC AC Circuit*

Voltage and Current Relationships

The voltage across a resistor, E_R, is in phase with the current through it as shown in *Figure 7.3*. For a capacitor, however, recall that the current leads the voltage by 90 degrees, as shown in *Figure 7.4*. Since the resistor and capacitor are in series with one another, the common factor in both phase relationships is the current.

Phasor diagrams can be drawn for the voltage and current waveforms of *Figure 7.3* and *Figure 7.4*. The voltage E_R, equal to 8 volts, across the resistor is in phase with the current, I, in the series circuit as shown in *Figure 7.5*. The same series current, I, leads the voltage E_C across the capacitor by 90 degrees as shown in *Figure 7.6*. E_C is 6 volts.

Comparing the phase relationships of the two voltage drops, it is found that the voltage across the resistor leads the voltage across the capacitor by 90 degrees. Because these two voltages are out of phase, they cannot be added as the voltage drops in a resistive series circuit would be added to obtain the total applied voltage. Mathematically, therefore:

$$E_T \text{ is not equal } E_R + E_C \qquad (7\text{–}3)$$

The total voltage must be calculated, therefore, by another type of addition known as *vector addition.*

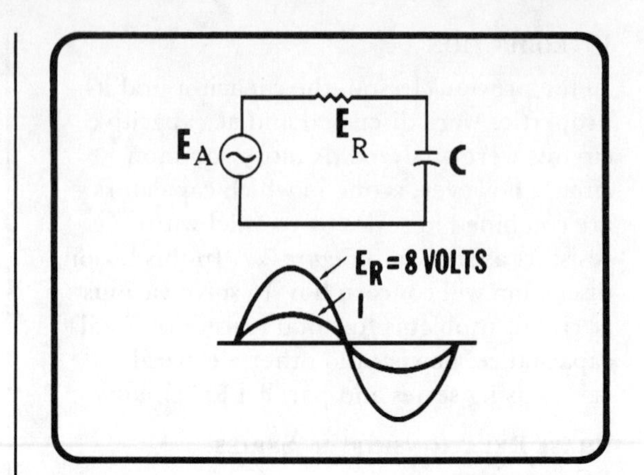

Figure 7.3 *Phase Relationship of Resistor Voltage and Current*

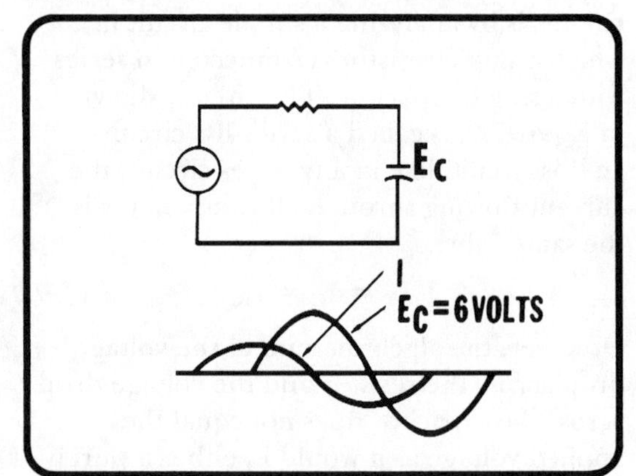

Figure 7.4 *Phase Relationship of Capacitor Voltage and Current*

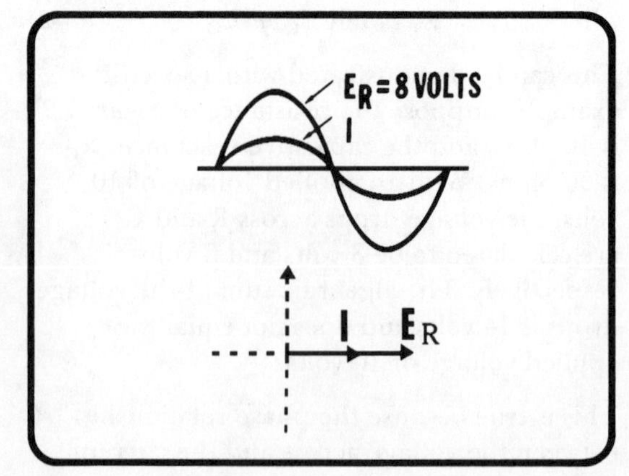

Figure 7.5 *Phasor Diagram for E_R and I*

Vector Addition

Common Phasor Diagram

Since I is the same in both of the phasor diagrams of *Figure 7.5* and *Figure 7.6*, a common phasor diagram can be formed as shown in *Figure 7.7*. It is now very apparent that E_R and E_C are 90 degrees out of phase. More specifically, the voltage across the resistor leads the voltage across the capacitor by 90 degrees. In *Figure 7.7* E_R and E_C are two vectors. Their length represents their amplitude and their direction represents their phase.

Parallelogram Method

When two vectors are added, the *parallelogram method* is used. The sum of the two vectors, as shown in *Figure 7.8*, is represented by a third vector called the *resultant*. To add two vectors using the parallelogram method, the parallelogram begun by the two *original* vectors are merely completed. The two original vectors in *Figure 7.8* are the two solid lines with arrows at points E_R and E_C. Then a line parallel to each vector beginning at the tip of the other and shown as the two dotted lines in *Figure 7.8* are drawn.

The first line begins at the tip of E_C and is parallel to line E_R. The second line begins at the tip of E_R and is parallel to line E_C. The point where these two dotted lines intersect determines where the tip of the resultant vector will be located. The resultant vector is then drawn from the origin (as are all vectors) to the point of intersection as shown.

If the length and phase of E_R and E_C are drawn to scale, the length of the resultant vector obtained in this manner will indicate its magnitude. Therefore, using the diagram, the resultant vector's magnitude and phase relationship to the other two vectors can be measured. The resultant is the *vector sum* of E_R and E_C.

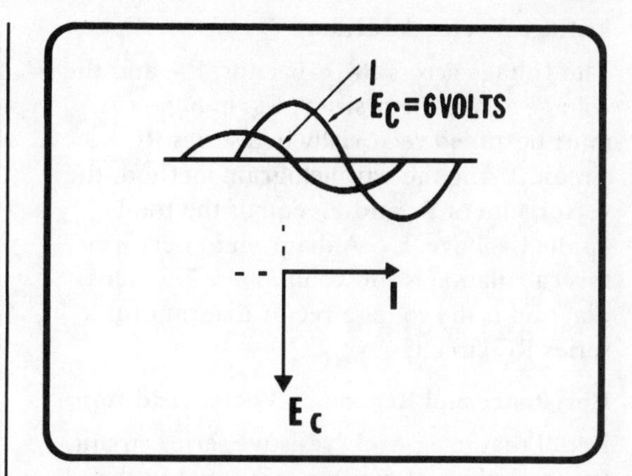

Figure 7.6 *Capacitor Voltage and Current Phase Relationships*

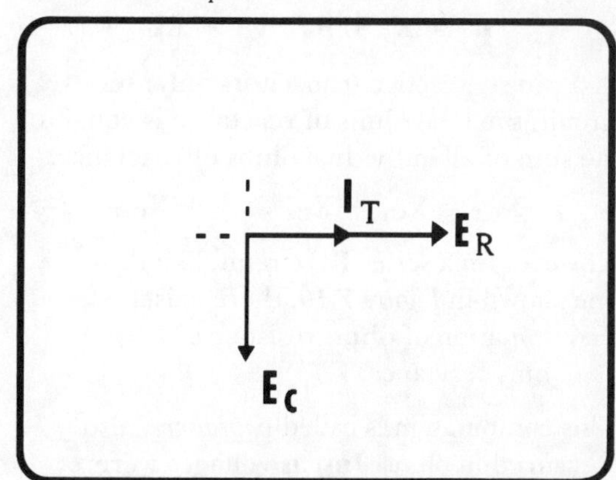

Figure 7.7 *Phase Relationships of the Resistor and Capacitor Voltage Drops*

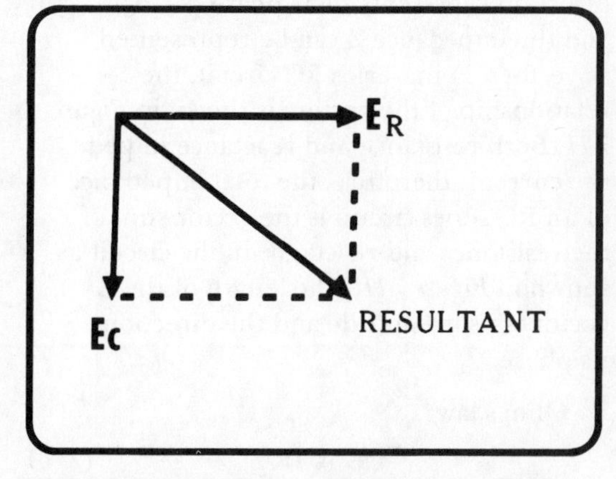

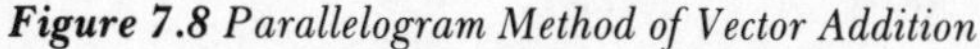

Figure 7.8 *Parallelogram Method of Vector Addition*

■ Voltage Vector Addition
■ Resistance and Reactance Vector Addition

Voltage Vector Addition

The voltage across the capacitor, E_C, and the voltage across the resistor, E_R, in *Figure 7.7* must be added vectorially in a series RC circuit. Using the parallelogram method, the vector sum of E_R and E_C equals the total applied voltage, E_A. A diagram to perform this calculation is shown in *Figure 7.9*. This diagram is the voltage vector diagram for a series RC circuit.

Resistance and Reactance Vector Addition

Recall that in a purely resistive series circuit, the total ohms of resistance is equal to the sum of all individual ohms of resistance.

$$R_T = R_1 + R_2 \ldots + R_N \qquad (7\text{–}4)$$

In a purely reactive (capacitors only) series circuit, the total ohms of reactance is equal to the sum of all individual ohms of reactance.

$$X_{CT} = X_{C1} + X_{C2} \ldots + X_{CN} \qquad (7\text{–}5)$$

However, in a series RC circuit, like the one shown in *Figure 7.10*, there exists a combination of ohms resistance *and* ohms reactance.

This combination is called *impedance*, also measured in ohms. Just as voltages were represented by vectors whose length represents their magnitude and whose direction represents their phase, R, X_C, and the impedance Z can be represented as vectors. For a series RC circuit, the relationship of the vectors is shown in *Figure 7.11*. Both resistance and reactance impede the current; therefore, the total impedance of an RC series circuit is the vector sum of the resistance and reactance in the circuit as shown in *Figure 7.11*. The length of the Z vector is its magnitude and the direction its phase.

By Ohm's law,

$$E = IR \qquad (7\text{–}6)$$

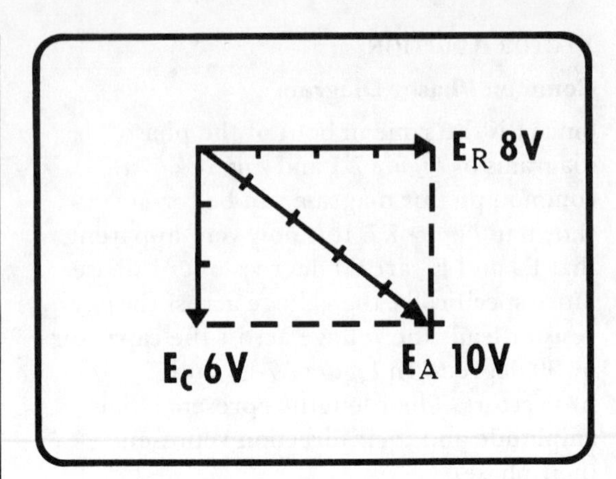

Figure 7.9 Voltage Vector Diagram for a Series RC AC Circuit

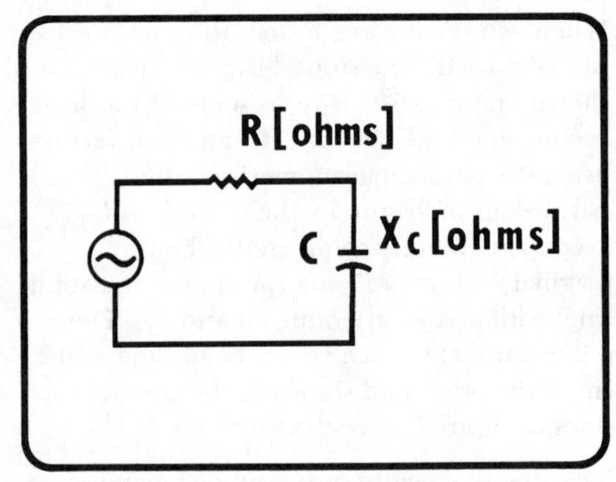

Figure 7.10 Typical Series RC Circuit

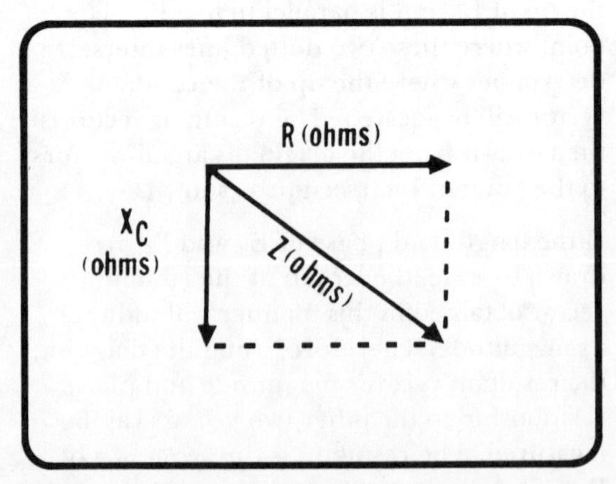

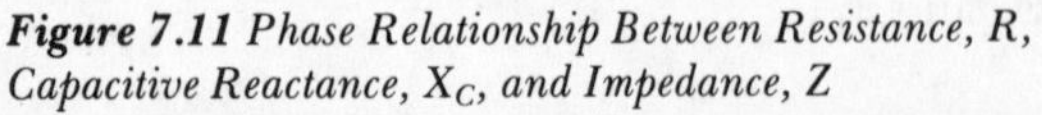

Figure 7.11 Phase Relationship Between Resistance, R, Capacitive Reactance, X_C, and Impedance, Z

■ Resistance and Reactance Vector Addition

The current, I, flowing in the series RC circuit of *Figure 7.10* is the same through each circuit element. Thus,

$$E_R = IR \qquad (7\text{–}7)$$

and the vector E_R in *Figure 7.12* is made equal to IR. The current is the same for the capacitor, therefore,

$$E_C = IX_C \qquad (7\text{–}8)$$

and the E_C vector of *Figure 7.12* is made equal to IX_C. E_C and E_R previously were plotted at right angles since E_R leads E_C by 90 degrees.

Also by Ohm's law, the total applied voltage is equal to the circuit current I times the total impedance so that,

$$E_A = IZ \qquad (7\text{–}9)$$

As a result, IZ is substituted for the applied voltage E_A in the voltage vector diagram as shown in *Figure 7.12*. In that diagram:

$$E_A = \text{vector sum of } E_R \text{ and } E_C \qquad (7\text{–}10)$$
$$IZ = \text{vector sum of IR and } IX_C \qquad (7\text{–}11)$$

Because the total current is the common factor, it can be factored out of equation 7–11.

$$\not{I}Z = \text{vector sum of } \not{I}R \text{ and } \not{I}X_C \qquad (7\text{–}12)$$

Equation 7–12 becomes, for a series RC Circuit,

$$Z = \text{vector sum of R and } X_C. \qquad (7\text{–}13)$$

This shows clearly that Z is the vector sum of R and X_C.

The resultant diagram is shown in *Figure 7.13*.

The parallelogram method is used to show the relationship of the resultant total impedance, Z, to the value of the resistance and capacitance in a series RC circuit. Its magnitude is its length and its direction its phase.

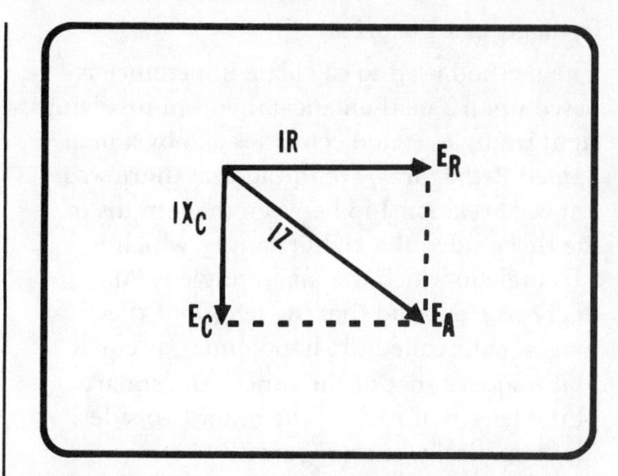

Figure 7.12 *Vector Factors for R, X_C, and Z in Series RC Circuit*

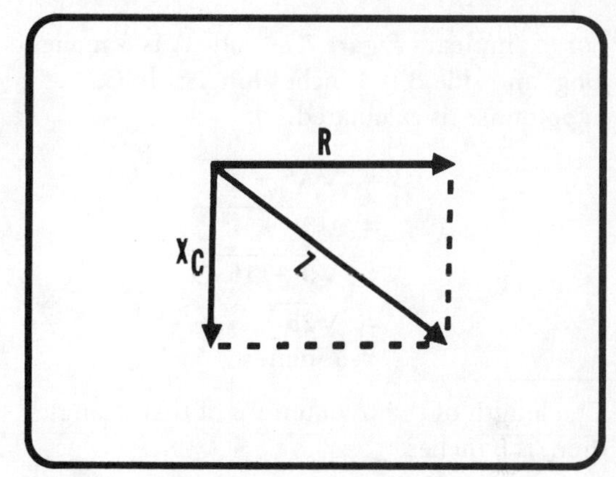

Figure 7.13 *Impedance as a Vector Sum in a Series RC Circuit*

VECTOR SOLUTIONS

It is possible, then, to determine the impedance by graphing the resistance and reactance vector to scale, using the parallelogram method to obtain the resultant impedance, and then measuring the length of the impedance vector to determine the value of the circuit impedance. This graphing method, however, can be tedious and somewhat inaccurate. Therefore, it is more reliable to use a method of calculation to determine the impedance.

■ Pythagorean Theorem
■ Pythagorean Theorem Applied to Impedance Solutions

Pythagorean Theorem

The method used to calculate impedance is based upon a mathematical theorem involving right triangles stated centuries ago by a man named Pythagoras. He found that there was a special relationship between the lengths of the three sides of a right triangle, which is a triangle in which the major angle is 90 degrees. He found that the length of the longest side, called the hypotenuse, is equal to the square root of the sum of the square of the length of each of the other two sides. Mathematically, from *Figure 7.14*.

$$\mathbf{C(Hypotenuse)} = \sqrt{A^2 + B^2} \quad (7\text{–}14)$$

For example, in *Figure 7.14,* side A is 3 inches long, and side B is 4 inches long. Side C, hypotenuse, is calculated:

$$\begin{aligned} C &= \sqrt{A^2 + B^2} \\ &= \sqrt{3^2 + 4^2} \\ &= \sqrt{9 + 16} \\ &= \sqrt{25} \\ &= 5 \text{ inches} \end{aligned}$$

The length of the hypotenuse of this triangle, then, is 5 inches.

Pythagorean Theorem Applied to Impedance Solutions

The Pythagorean theorem can be applied to circuit problems involving resistance and reactance. *Figure 7.15* shows the resistive-reactance phasor diagram for a series RC circuit. The total impedance of the circuit, Z, is the vector sum of the resistance, R, and reactance, X_C. If the values of the resistance and reactance are known, the impedance can be calculated using the Pythagorean theorem.

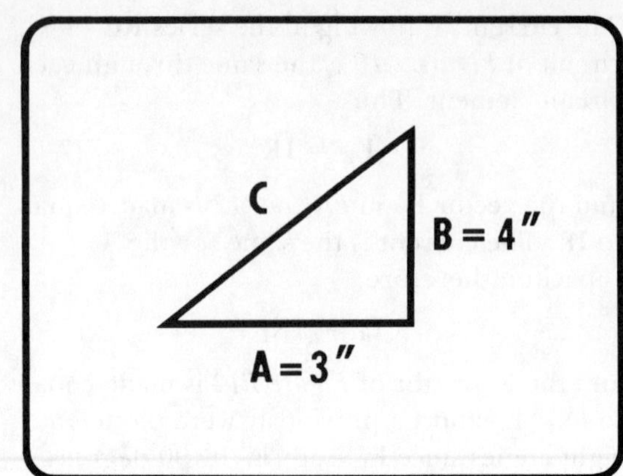

Figure 7.14 *Right Triangle Used in Example Problem*

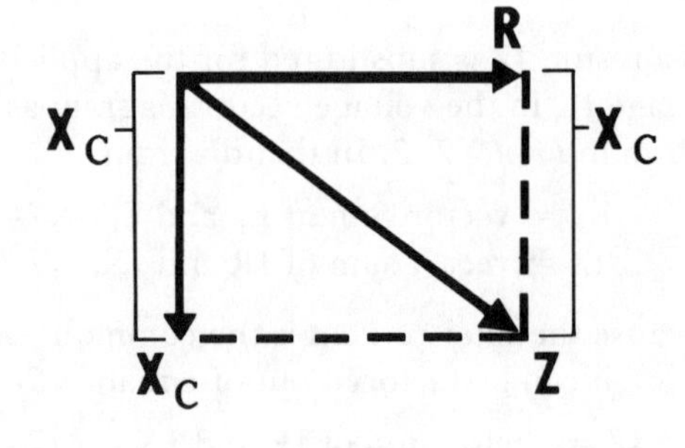

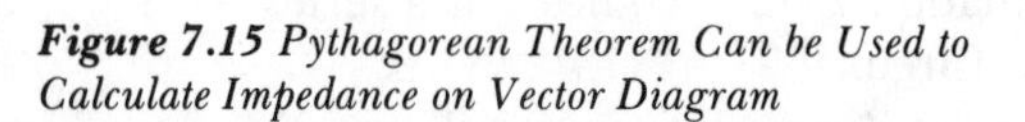

Figure 7.15 *Pythagorean Theorem Can be Used to Calculate Impedance on Vector Diagram*

Notice that the length of the reactance vector is the same as the length between the tip of the resistance vector and the tip of the impedance vector. In fact, the vector diagram can be drawn with the reactance vector placed as shown in *Figure 7.16*. Now, the right triangle of the Pythagorean theorem becomes evident. Applying the Pythagorean theorem, the total impedance (the hypotenuse) can be calculated,

$$Z = \sqrt{R^2 + X_C^2} \qquad (7\text{–}15)$$

Total impedance is equal to the square root of the resistance squared plus the reactance squared.

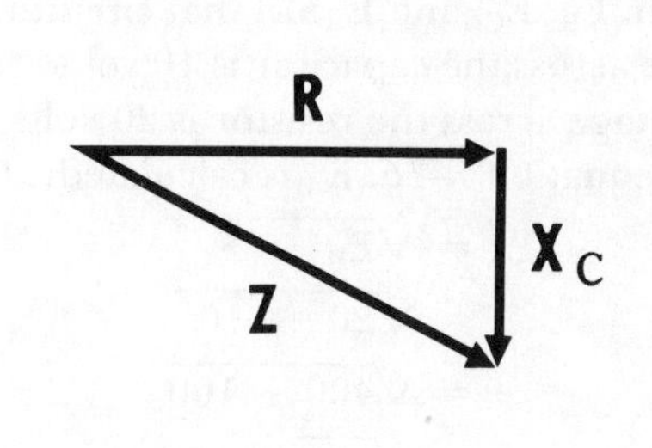

Figure 7.16 *Vector Diagram Redrawn with the Reactance Vector Obvious*

Figure 7.17 shows a typical series RC circuit. The impedance can be calculated using equation *7–15*. The resistance of this RC circuit is 12 ohms and the capacitive reactance is 16 ohms.

$$\begin{aligned} Z &= \sqrt{R^2 + X_C^2} \qquad (7\text{–}15) \\ &= \sqrt{12^2 + 16^2} \\ &= \sqrt{144 + 256} \\ &= \sqrt{400} \\ &= 20\Omega \end{aligned}$$

Thus, the total impedance of the circuit is 20 ohms.

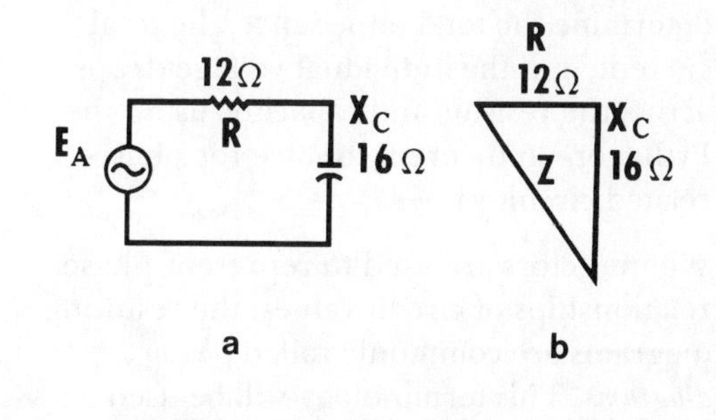

Figure 7.17 *Example Circuit for Calculating Impedance Using Equation 7–15*

Pythagorean Theorem Applied to Voltage Solutions

Similar mathematical calculations using the Pythagorean theorem can be performed to show the relationship between the voltage drops and applied voltage in a series RC circuit. In this case the phasor diagram is drawn as shown in *Figure 7.18*. Shifting the capacitive voltage vector, as shown in *Figure 7.19*, equation *7–15* can be rewritten to calculate easily the applied voltage, E_A.

$$E_A = \sqrt{E_R^2 + E_C^2} \qquad (7\text{–}16)$$

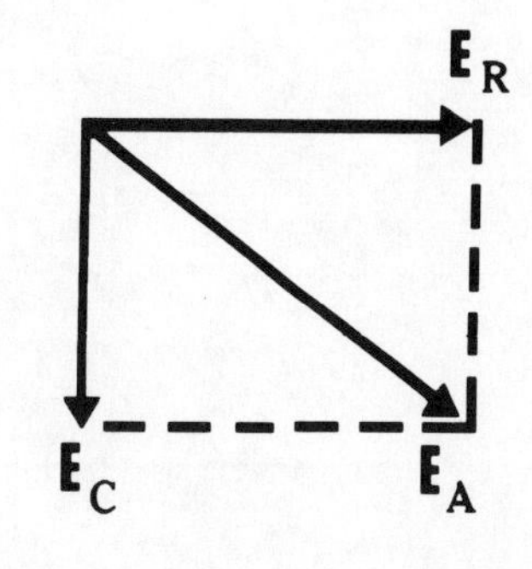

Figure 7.18 *Phasor Diagram of Voltage Relationship in a Series RC Circuit*

■ Pythagorean Theorem Applied to Voltage Solutions

An example circuit is shown in *Figure 7.20* which also shows the vector relationship between E_R, E_C, and E_A. In that circuit the voltage across the capacitor is 10 volts, rms; the voltage across the resistor is 20 volts, rms. Using equation *7–16*, E_A is calculated:

$$
\begin{aligned}
E_A &= \sqrt{E_R{}^2 + E_C{}^2} \\
&= \sqrt{20^2 + 10^2} \\
&= \sqrt{400 + 100} \\
&= \sqrt{500} \\
&= 22.36\text{V}
\end{aligned}
$$

Thus, the total applied voltage to the circuit is 22.36 volts(rms).

SERIES RC CIRCUIT ANALYSIS

It's now time to take a series RC circuit and determine the total impedance, the total current, and the individual voltage drops across the resistor and capacitor using the Pythagorean theorem to solve for phase-related circuit values.

When vectors are used to represent phase relationships of circuit values, the resulting diagrams are commonly called *phasor diagrams.* This terminology will be used during the remainder of this book.

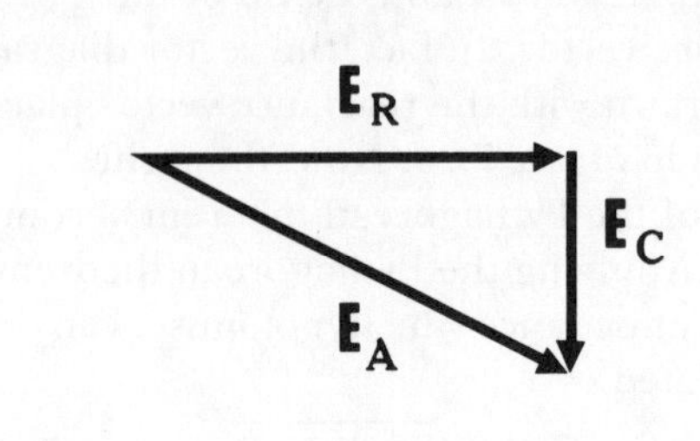

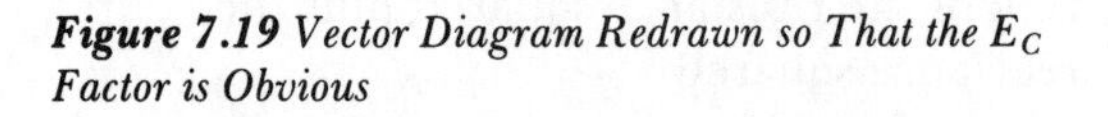

Figure 7.19 *Vector Diagram Redrawn so That the E_C Factor is Obvious*

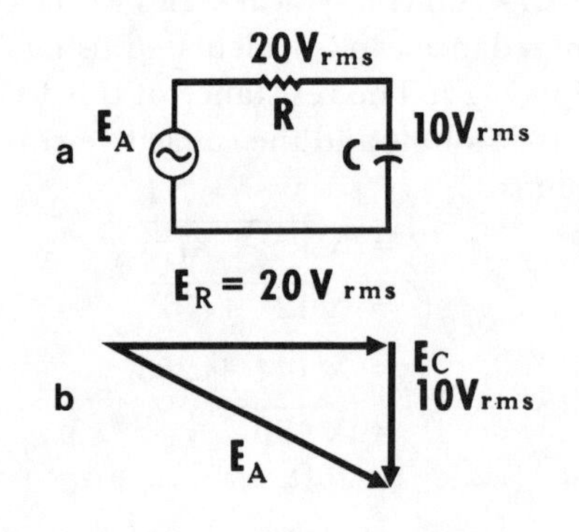

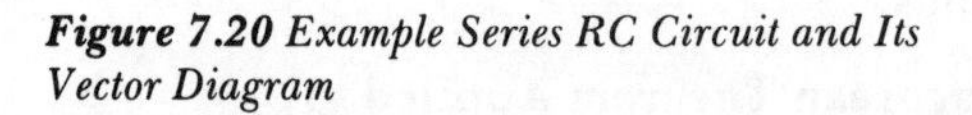

Figure 7.20 *Example Series RC Circuit and Its Vector Diagram*

Calculations of X_C

First, capacitive reactance, X_C, will be calculated in a typical series RC circuit shown in *Figure 7.21*. The value of the resistor is 30 ohms. The value of the capacitor is 4 microfarads. The applied voltage is 100 VAC with a frequency of 995 hertz. To determine the total impedance of this circuit, the value of the capacitive reactance must first be calculated.

$$X_C = \frac{1}{2\pi fC}$$
$$= \frac{1}{(6.28)\,(995\text{Hz})\,(4\mu\text{F})}$$
$$= 40\Omega$$

Therefore, X_C equals 40 ohms.

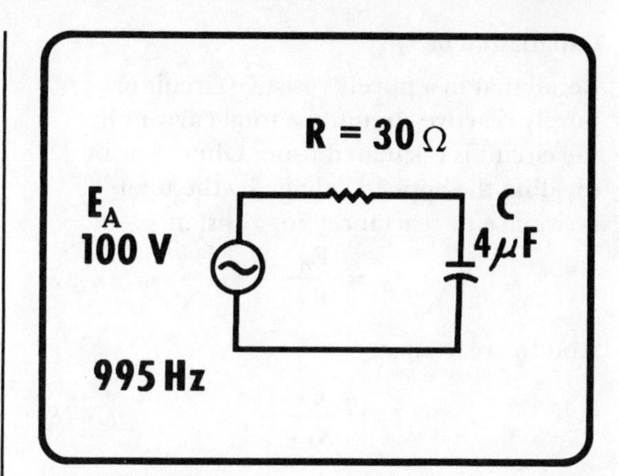

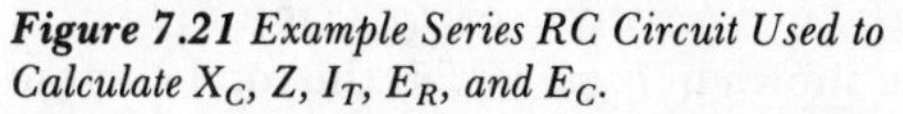
Figure 7.21 *Example Series RC Circuit Used to Calculate X_C, Z, I_T, E_R, and E_C.*

Calculation of Z

The impedance phasor diagram may now be drawn to show the relationship between the values of resistance, reactance, and impedance in the circuit as shown in *Figure 7.22*. The value of impedance is calculated by using the Pythagorean theorem (equation *7–15*). The vectors being plotted are the sides of a right triangle.

$$Z = \sqrt{R^2 + X_C{}^2}$$
$$= \sqrt{30^2 + 40^2}$$
$$= \sqrt{900 + 1600}$$
$$= \sqrt{2500}$$
$$= 50\Omega$$

Thus, the value of the total impedance of this circuit is 50 ohms. Impedance is defined as the total *opposition* of a resistive-reactance circuit to the flow of alternating current. It is something that impedes the current.

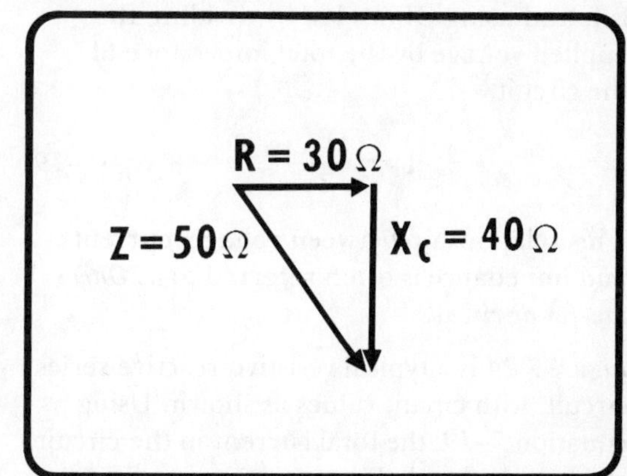

Figure 7.22 *Impedance Phasor Diagram of Circuit in Figure 7.21*

■ Calculation of I_T

Calculation of I_T

Recall that in a purely resistive circuit or purely reactive circuit, the total current in the circuit is calculated using Ohm's law by dividing the applied voltage by the total resistance or reactance. For resistance:

$$I_T = \frac{E_A}{R_T} \tag{7–17}$$

And for reactance:

$$I_T = \frac{E_A}{X_{CT}} \tag{7–18}$$

Similarly, in a resistive-reactive circuit, like the one shown in *Figure 7.23*, the total current is found using Ohm's law by dividing the applied voltage by the total impedance of the circuit.

$$I_T = \frac{E_A}{Z_T} \tag{7–19}$$

This relationship between voltage, current and impedance is often referred to as *Ohm's law for ac circuits*.

Figure 7.24 is a typical resistive-reactive series circuit, with circuit values as shown. Using equation *7–19*, the total current in the circuit can be calculated. That equation requires that Z_T be known. As calculated in the previous section, Z_T for the circuit is 50 ohms. Therefore,

$$\begin{aligned} I_T &= \frac{E_A}{Z_T} \\ &= \frac{100V}{50\Omega} \\ &= 2A \end{aligned}$$

The total circuit current equals 2 amperes.

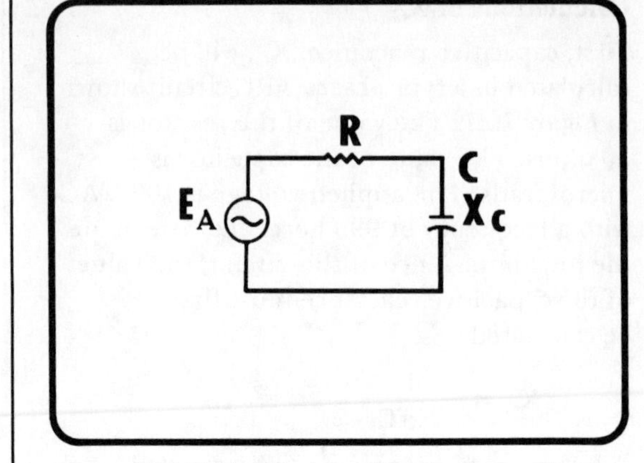

Figure 7.23 *Typical Resistive-Reactive Series Circuit*

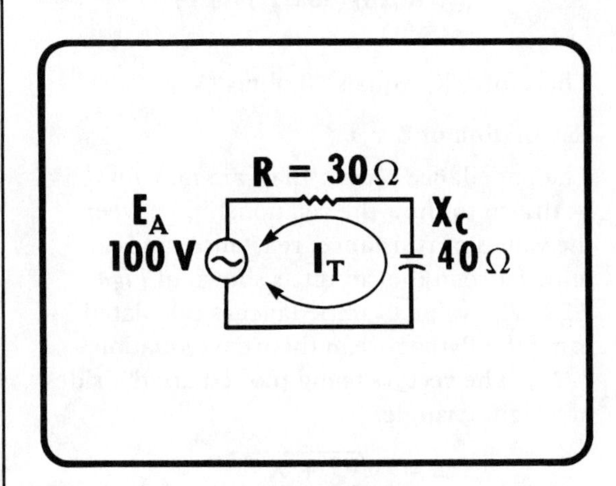

Figure 7.24 *Example Circuit for Calculating I_T*

■ Calculations of E_R and E_C
■ Relationship of E_R, E_C and E_A in Series RC Circuits

Calculations of E_R and E_C

Voltage drops across the resistor and capacitor are now calculated as they would be in any series circuit. The voltage drop across the resistor is calculated:

$$\begin{aligned} E_R &= I_T R \\ &= (2A)\,(30\Omega) \\ &= 60V \end{aligned}$$

The voltage drop across the capacitor is then calculated:

$$\begin{aligned} E_C &= I_T X_C \\ &= (2A)\,(40\Omega) \\ &= 80V \end{aligned}$$

Relationship of E_R, E_C and E_A in Series RC Circuits

Notice that the voltage across the capacitor plus the voltage across the resistor add to 140 volts. This is more than the applied voltage and at first, seems to contradict Kirchhoff's voltage law. Recall that Kirchhoff's voltage law essentially states that the total voltage applied to any closed circuit path always equals the sum of the voltage drops across the individual parts of the path. Also remember, however, that the voltage across the resistor and the voltage across the capacitor are out of phase by 90 degrees as shown in *Figure 7.25*. Therefore, they must be added vectorially like this:

$$\begin{aligned} E_A &= \sqrt{E_R^2 + E_C^2} \\ &= \sqrt{60^2 + 80^2} \\ &= \sqrt{3600 + 6400} \\ &= \sqrt{10,000} \\ &= 100VAC \end{aligned}$$

The applied voltage in the example circuit is 100 VAC. Thus, the *vector sum* of the circuit voltage drops equals the applied voltage. This type of calculation can be performed to check the calculated voltage drops in the circuit.

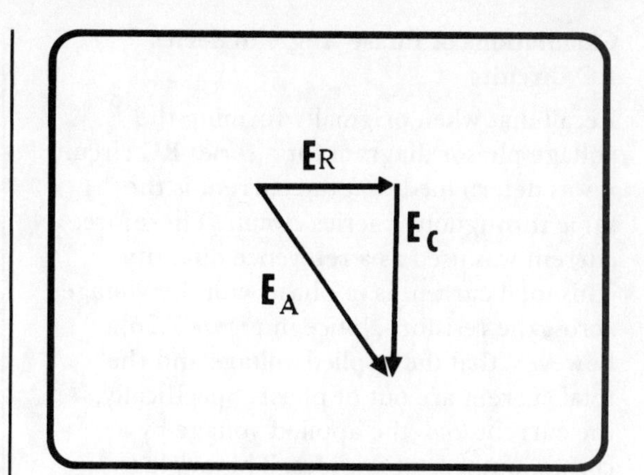

Figure 7.25 Phase Relationship of E_R and E_C

Calculations of Phase Angle in Series RC Circuits

Recall that when originally forming the voltage phasor diagram for a series RC circuit it was determined that the current is the same throughout a series circuit. Therefore, current was used as a reference quantity. This total current is in phase with the voltage across the resistor. Notice in *Figure 7.26,* however, that the applied voltage and the total current are out of phase. Specifically, the current *leads* the applied voltage by a certain number of degrees. This angle by which the total applied voltage and total current are out-of-phase is called the *phase angle* of the circuit. The phase angle, by definition, is the number of degrees the total current and the total applied voltage are out of phase. The phase angle is usually denoted by the Greek letter theta, θ.

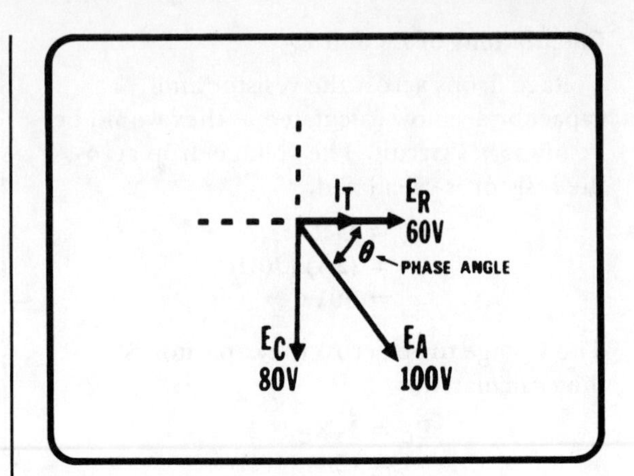

Figure 7.26 *Circuit's Phase Angle*

The phase angle in a series RC circuit can also be recognized as the angle between the voltage across the resistor and applied voltage as shown in *Figure 7.27. Figure 7.12* showed that the impedance phasor diagram is proportional to the voltage phasor diagram because

$$E_R = IR$$
$$E_C = IX_C$$
$$E_A = IZ$$

and the fact that I cancels because it is common to each term. Therefore, as shown in *Figure 7.27,* the voltage across the resistor E_R is in phase with the series circuit current. The phase angle in a series RC circuit is also the angle between the resistance vector and the impedance vector as shown in *Figure 7.28.*

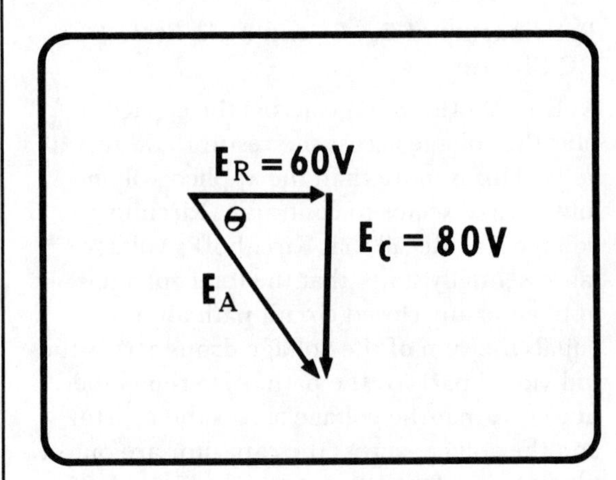

Figure 7.27 *Phase Angle — Voltage Phasor Diagram*

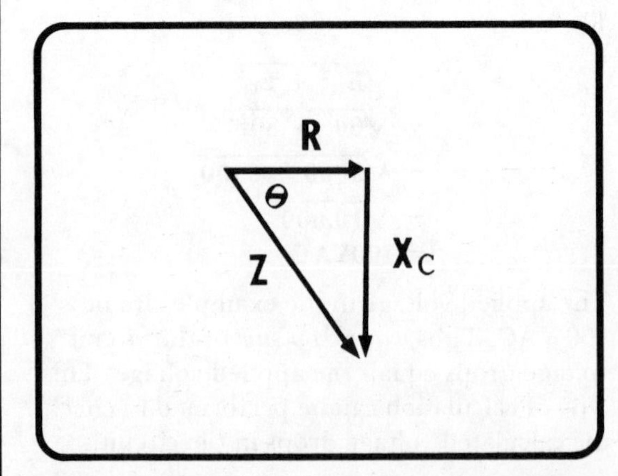

Figure 7.28 *Phase Angle — Impedance Phasor Diagram*

Definition of Tangent Function

The value of the phase angle can be calculated by using a trigonometric function called the *tangent function.* Recall from Lesson 2 that the tangent, abbreviated *tan,* of an angle in a right triangle, is equal to the ratio of the length of the side opposite the angle divided by the length of the side adjacent to the angle:

$$\tan\theta = \frac{\textbf{opposite side}}{\textbf{adjacent side}} \qquad (7\text{–}20)$$

For example, as shown in *Figure 7.29*, the side of a right triangle opposite from the angle θ is 10 units long and the side adjacent to the angle θ is 5 units long. Then,

$$\tan\theta = \frac{\text{opposite}}{\text{adjacent}} = \frac{10}{5} = 2$$

Thus, the tangent of θ in this right triangle is equal to 2. A calculator or trigonometric function table is then used to determine the angle which has a tangent of 2. The angle which has a tangent of 2 is about 64°.

Relationship of Tangent and Arctangent

Recall from equation *7–20* that the value obtained by dividing the lengths of the sides of a right triangle is called the *ratio* of the two sides. It will be a ratio that has a value from 0 to infinity. If given the ratio, the angle can be found using a calculator or trigonometric table.

The inverse process of determining the angle when the ratio is known is called finding the *arctangent*, abbreviated arctan.

$$\arctan\frac{\text{opposite}}{\text{adjacent}} = \theta \qquad (7\text{–}21)$$

This equation reads: θ is an angle of a right triangle whose tangent is the ratio called out in the arctan function.

For example, the *arctangent* of the ratio of 2 is about 64°:

$$\arctan(2) = 64°$$

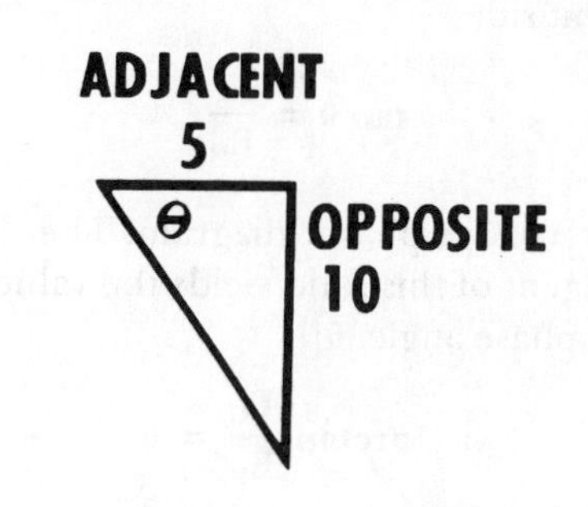

***Figure 7.29** Triangle Used to Calculate θ in an Example*

Said inversely, the *tangent* of 64° is approximately 2:

$$\tan 64° \cong 2$$

Taking a second example, the tangent of 35° = 0.7. Therefore, the arctangent of 0.7 = 35°. Sometimes, arctangent is abbreviated by writing: $\tan^{-1}$ (tan to the minus one). Both notations indicate the arctangent function. Therefore, the *arctangent* of a *ratio* indicates an *angle* whose tangent is the ratio.

$$\arctan\frac{\text{opposite}}{\text{adjacent}} = \theta$$

$$\tan^{-1}\frac{\text{opposite}}{\text{adjacent}} = \theta$$

Application of Tangent Function in Solving for Phase Angles

Using the tangent function to find the phase angle of the RC circuit in *Figure 7.24* the side opposite the phase angle in the voltage phasor diagram of *Figure 7.27* is the vector E_C. The length of this vector represents the value of the voltage drop across the capacitor. The side adjacent to the phase angle in the voltage phasor diagram is the vector, E_R, whose length represents the value of the voltage drop across the resistor.

■ Application of Tangent Function in Solving for Phase Angles

The tangent of the phase angle is equal to the ratio of the opposite side divided by the adjacent side,

$$\tan\theta = \frac{E_C}{E_R}$$

for the voltage phasor diagram. The arctangent of this ratio yields the value of the phase angle, θ;

$$\arctan\frac{E_C}{E_R} = \theta$$

In the series RC circuit shown in *Figure 7.30*, the voltage across the capacitor is 80 volts and the voltage across the resistor is 60 volts. Therefore,

$$\arctan\left(\frac{80V}{60V}\right) = \theta$$
$$\arctan(1.33) = \theta$$
$$53° = \theta$$

Completing the calculations the arctangent of E_R divided by E_C equals the arctangent of 80 volts divided by 60 volts which equals the arctangent of 1.33. Using your calculator or trigonometric table, you can determine that the tangent function has a value of 1.33 for an angle of about 53°, or the arctangent of 1.33 is 53°. Therefore, the phase angle of this RC series circuit is 53°.

Recall that earlier it was shown that the impedance phasor diagram of a series circuit is proportional to the voltage phasor diagram. Since these two phasor diagrams are proportional, the phase angle also can be determined using an impedance phasor diagram. This will provide an identical value for the phase angle for the RC circuit of *Figure 7.24*.

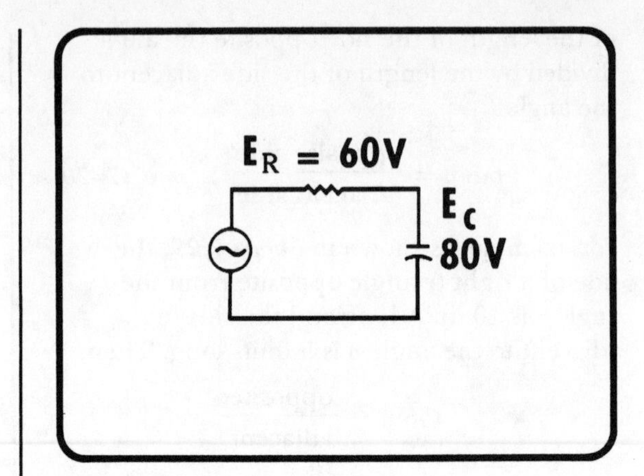

***Figure 7.30** Example Circuit Used to Calculate Phase Angle*

The side opposite to the phase angle in the impedance phasor diagram in *Figure 7.28* is the reactance vector, X_C. The length of this vector represents the value of the capacitive reactance. The side adjacent to the phase angle in the impedance phasor diagram is the resistance vector, R. Its length represents the value of the resistor. Recall that the tangent ratio is the opposite side divided by adjacent side, therefore,

$$\tan\theta = \frac{X_C}{R} \quad (7\text{-}22)$$

The arctangent of this ratio yields the value of the phase angle.

$$\arctan\left(\frac{X_C}{R}\right) = \theta \quad (7\text{-}23)$$

There is one other important point about calculating theta of a series RC circuit. In the series RC circuit of *Figure 7.24*, the value of the capacitive reactance is 40 ohms and the value of the resistor is 30 ohms. Therefore:

$$\arctan\left(\frac{40\Omega}{30\Omega}\right) = \theta$$
$$\arctan(1.33) = \theta$$
$$53° = \theta$$

40 divided by 30 results in a ratio of 1.33. The arctangent of 1.33 is 53°. This is the same phase angle derived using the voltage phasor diagram. Thus, it should be remembered that the value of the phase angle in a series RC circuit can be determined from either the voltage *or* impedance phasor diagram.

Negative Phase Angle in Series RC Circuits

The *value* of the phase angle in a *series RC circuit* is considered to be a *negative value*. That is, the phase angle in the example circuit of *Figure 7.24* is actually *negative 53° degrees* as shown in *Figure 7.31*. The reason the phase angle is considered to be negative is because, previously, the angle of rotation of phasors in a *counter-clockwise* direction was designated as *positive*. Thus, the angle of rotation of a phasor in a *clockwise* direction, in agreement with the convention, is said to be *negative*. This is illustrated graphically in *Figure 7.32*.

Remember that the total current in the series RC circuit was used as the reference, which is at zero degrees. Thus, as shown in *Figure 7.33*, the negative sign of the phase angle indicates that the applied voltage is rotated 53 degrees *clockwise* from the current vector direction.

CALCULATION OF POWER IN RESISTIVE AND REACTIVE CIRCUITS

In previous examples, power was calculated in purely resistive circuits (real power) and in purely reactive circuits (reactive power). In RC circuits there is a combination of real power, P_R, (in watts), and reactive power, P_C, (in VAR).

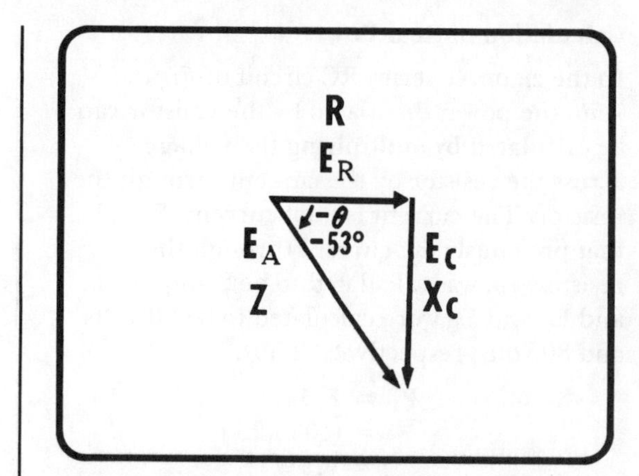

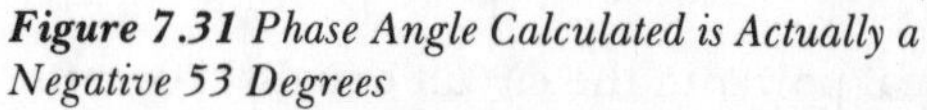

Figure 7.31 *Phase Angle Calculated is Actually a Negative 53 Degrees*

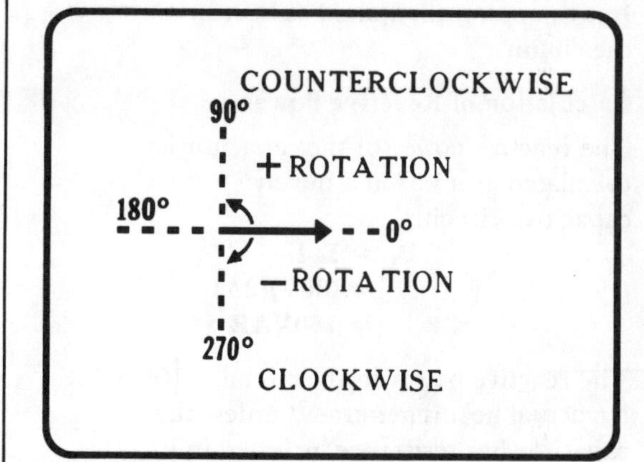

Figure 7.32 *Negative and Positive Phase Angles*

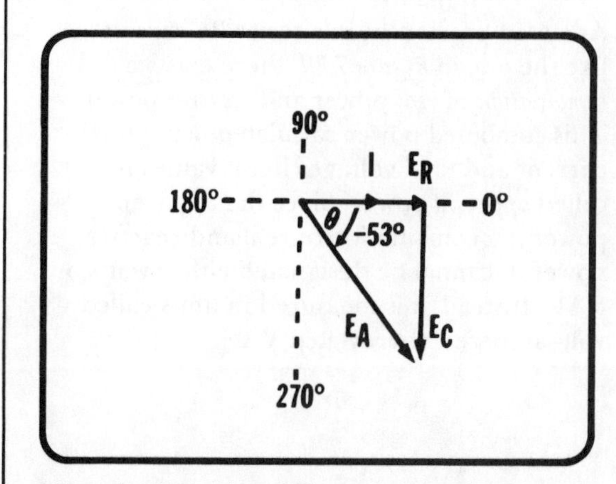

Figure 7.33 *Rotation of Phasors in Example Circuit is Clockwise, Therefore, Negative*

Calculation of Real Power

In the example series RC circuit of *Figure 7.30*, the power dissipated by the resistor can be calculated by multiplying the voltage across the resistor by the current through the resistor. The current is total current. Recall that previously the current through the resistor, I_R, was calculated to be 2 amperes, and E_R and E_C were calculated to be 60 volts and 80 volts, respectively. Thus,

$$\begin{aligned} \mathbf{P_R} &= \mathbf{E_R I_R} \\ &= \mathbf{(60V)(2A)} \\ &= \mathbf{120W} \end{aligned}$$

The real power in the circuit equals 120 watts. It is the power dissipated as heat in the circuit.

Calculation of Reactive Power

The reactive power of the capacitor is calculated as it was in a purely capacitive circuit.

$$\begin{aligned} \mathbf{P_C} &= \mathbf{E_C I_C} \\ &= \mathbf{(80V)(2A)} \\ &= \mathbf{160VAR} \end{aligned}$$

The reactive power in the circuit is 160 VAR. No actual heat is generated unless the capacitor has resistance included in it.

Calculation of Apparent Power

As stated previously, in a series RC circuit like the one in *Figure 7.30*, there exists a *combination* of *real* power and *reactive* power. This combined power calculated using total current and total voltage circuit values is called *apparent* power. Since the apparent power is a combination of real and reactive power, it cannot be designated either watts or VAR. Instead, it is measured in units called volt–amperes, abbreviated VA.

Power in a circuit is always calculated by multiplying a voltage times a current.

$$\mathbf{P_T = E_A I_T} \qquad (7\text{–}24)$$

Equation *7–24* is such an equation and it can be used to calculate total power in any resistive–reactive circuit. In the circuit being used (*Figure 7.30*) E_A = 100 volts and I_T = 2 amperes. Thus,

$$\begin{aligned} \mathbf{P_T} &= \mathbf{(100V)(2A)} \\ &= \mathbf{200VA} \end{aligned}$$

Recall that in a purely resistive circuit the *total real power* is the sum of all the individual real power values:

$$\mathbf{P_R = P_{R1} + P_{R2} \ldots + P_{RN}} \qquad (7\text{–}25)$$

and in a purely reactive circuit the *total reactive power* is the sum of all individual reactive power values:

$$\mathbf{P_T = P_{C1} + P_{C2} \ldots + P_{CN}} \qquad (7\text{–}26)$$

However, in resistive and reactive circuits, the simple sum of the real power and reactive power does not equal the apparent power.

$$\mathbf{P_T \text{ is not } P_R + P_C} \qquad (7\text{–}27)$$

This occurs because the phase relationship between the voltage across and the current through each component are different.

■ Calculation of Apparent Power

In series RC circuits, the phase relationships of the three power determinations are similar to the voltage phase relationships, as shown in *Figure 7.34*.

Since each power determination is made by multiplying the voltage shown times the current, the power phasor diagram is proportional to the voltage phasor diagram by a factor of the total current, as shown in *Figure 7.34*.

Mathematically:

$$E_A I_T = \text{phasor sum of } E_R I_T + E_C I_T \quad (7\text{–}28)$$

or

$$P_T(\text{apparent}) = \text{phasor sum of } P_{real} + P_{reactive} \quad (7\text{–}29)$$

The total apparent power is calculated using the Pythagorean theorem.

$$\mathbf{P_{T(a)} = \sqrt{P_R^2 + P_C^2}}$$

It is equal to the square root of the real power squared plus the reactive power squared. Thus, the total apparent power in the example circuit of *Figure 7.24* and *Figure 7.30* as calculated, 200 volt-amperes, should be equal to the square root of 120 watts squared plus 160 VAR squared.

$$\begin{aligned} \mathbf{P_T} &= \mathbf{\sqrt{P_R^2 + P_C^2}} \\ &= \mathbf{\sqrt{120^2 + 160^2}} \\ &= \mathbf{\sqrt{14{,}400 + 25{,}600}} \\ &= \mathbf{\sqrt{40{,}000}} \\ &= \mathbf{200VA} \end{aligned}$$

This is the relationship between the apparent power and the real and reactive power. Note in *Figure 7.35* that the angle between the real and apparent power is the phase angle. In this example the phase angle is negative 53 degrees.

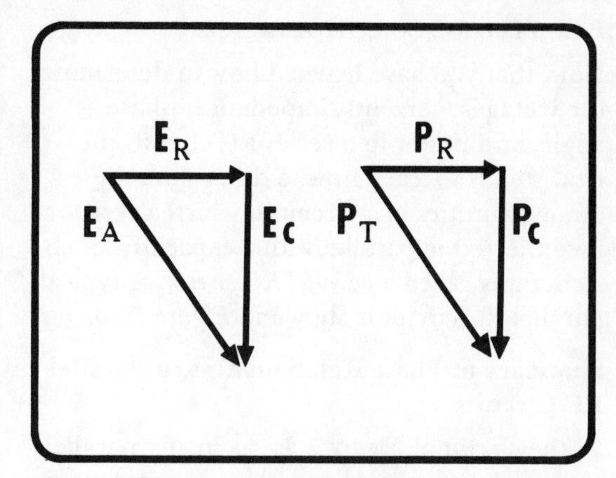

***Figure 7.34** Relationshiop of Voltage and Power Factors in an RC Series Circuit*

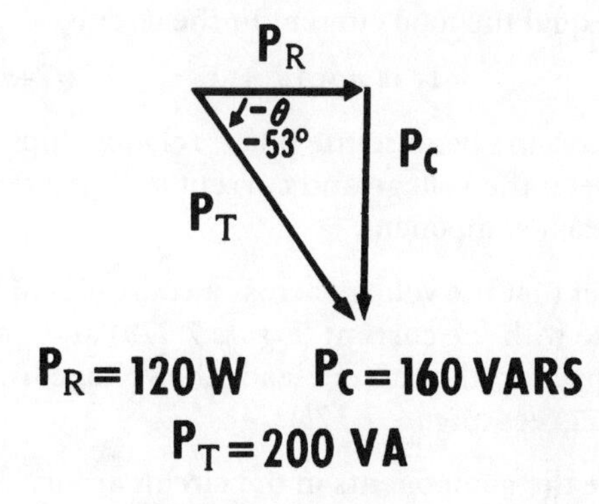

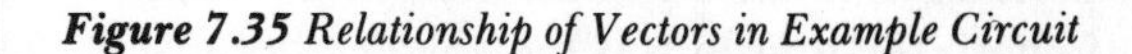

P_T = 200 VA

***Figure 7.35** Relationship of Vectors in Example Circuit*

PARALLEL RC CIRCUIT ANALYSIS

Now, that you have learned how to determine the voltages, currents, impedance, phase angle, and power in a series RC circuit, the next step is to learn how to determine these same quantities in a circuit in which a resistor is connected in parallel with a capacitor. Such a circuit is called a *parallel RC circuit.* A typical parallel RC circuit is shown in *Figure 7.36*.

Summary of Phase Relationships in Parallel RC Circuits

In the circuit of *Figure 7.36*, as in any parallel circuit, the voltage across all components is the same: $E_A = E_R = E_C$. However, the algebraic sum of the branch currents does not equal the total current in the circuit.

$$I_T \text{ is not } I_R + I_C \qquad (7\text{–}31)$$

This occurs because the phase relationships between the voltage and current are different for each component.

Recall that the voltage across a resistor is in phase with the current (*Figure 7.37a*) and, in a capacitor, the current leads the voltage by 90 degrees (*Figure 7.37b*).

Since the components in the circuit are in parallel with one another, the common factor in both relationships is the voltage, because the voltage across both components is the same. This voltage is the applied voltage, E_A.

For discussion purposes, the diagram of *Figure 7.37b* is rotated by 90 degrees to make E_C horizontal, as shown in *Figure 7.38a*. Because the components in the example circuit are parallel with one another, the common factor in both relationships is the voltage. The voltage across both components is the applied voltage, therefore, E_C is equal to E_A and E_R equals E_A.

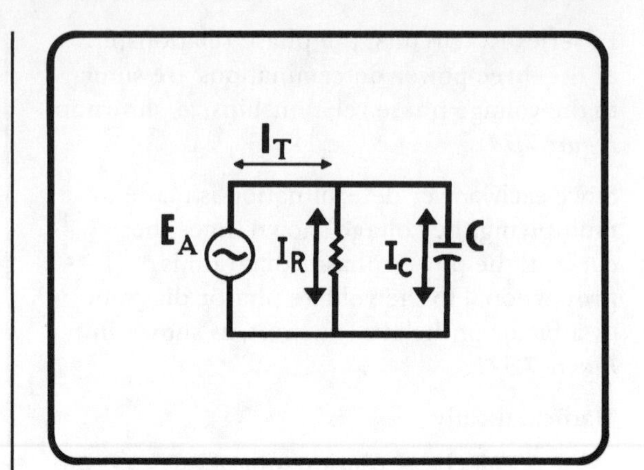

Figure 7.36 Typical Parallel RC Circuit

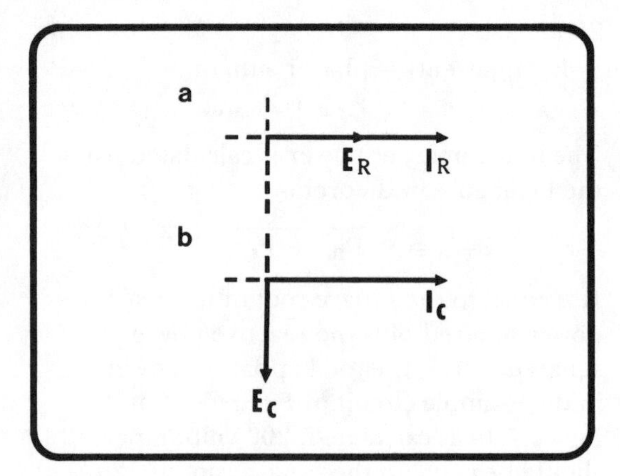

Figure 7.37 Phase Relationship of E_R, I_R, and E_C, I_C

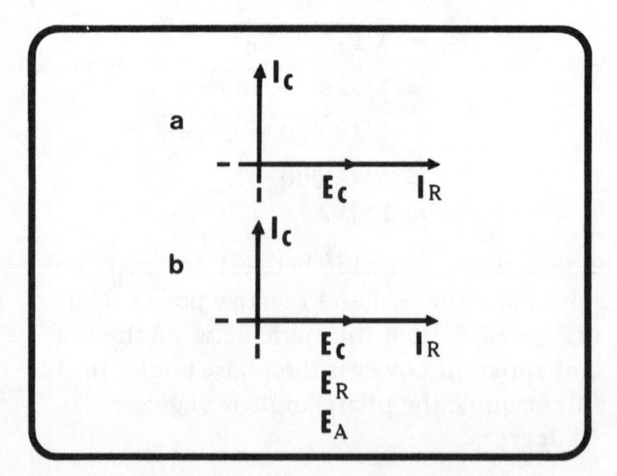

Figure 7.38 Phase Relationship of I_C, I_R, E_C, E_R, E_A

- Calculation of I_R and I_C
- Calculation of I_T

Using the voltage as a common reference (instead of current which was used in an analysis of a series RC circuit), the two individual phasor diagrams can be combined into one as shown in *Figure 7.38b*. However, in order to match common voltage, the E_C vector has been rotated 90 degrees to match the direction of the E_R and E_A vectors. Also note that the current through the resistor is shown in phase with the applied voltage across it, while holding to the theory that the capacitive current is leading the applied voltage by 90 degrees. Comparing the phase relationships of the two branch currents, the capacitive current leads the resistive current by 90 degrees.

Calculation of I_R and I_C

The individual branch currents in a parallel circuit can be calculated as they were calculated in a purely resistive or purely capacitive circuit. Simply divide the voltage across the branch by the opposition to current in that branch.

In the resistive branch, the opposition to the flow of current is measured in ohms of resistance. Thus, in the circuit, shown in *Figure 7.39*, the resistive current is determined by dividing the applied voltage by the value of the resistor.

$$I_R = \frac{E_R}{R} = \frac{E_A}{R} \qquad (7\text{–}32)$$

In the capacitive branch, the opposition to the flow of current is measured in ohms of reactance. Therefore the capacitive current is determined by dividing the applied voltage by the reactance of the capacitor.

$$I_C = \frac{E_C}{X_C} = \frac{E_A}{X_C} \qquad (7\text{–}33)$$

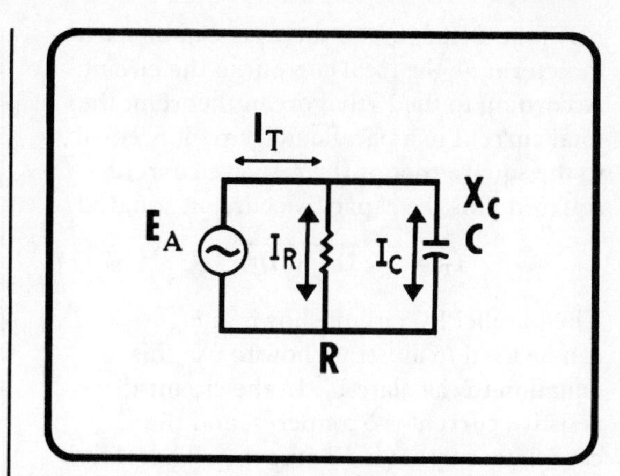

Figure 7.39 Example RC Parallel Circuit

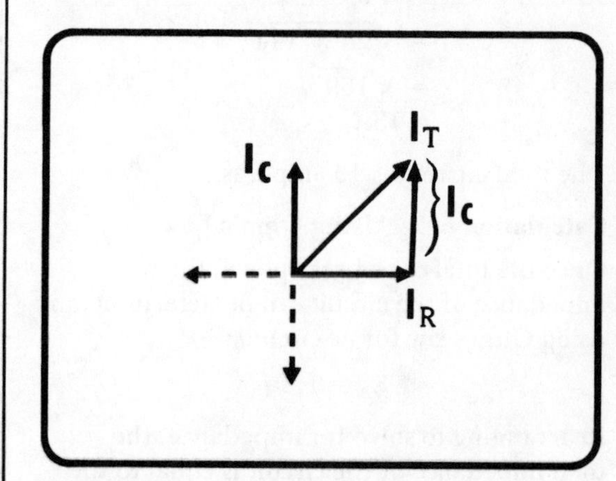

Figure 7.40 Currents Must be Added Vectorially

Calculation of I_T

To obtain the total current, I_T, the currents in a parallel RC circuit cannot simply be added as the branch currents in a parallel resistive circuit would be added. Instead, these currents must be added *vectorially*.

As shown in *Figure 7.40*, the length of the capacitive current vector is the same as the length between the tip of the total current vector and the tip of the resistive current vector. Thus, the capacitive current vector may be shifted over to form a right triangle.

■ Calculation of Z_T Using Ohm's Law
■ Calculation of Z_T Using Impedance Vectors

Now, the Pythagorean theorem can be used to determine the total current in the circuit. According to the Pythagorean theorem, the total current in a parallel RC circuit is equal to the square root of the resistive current squared plus the capacitive current squared.

$$I_T = \sqrt{I_R^2 + I_C^2} \qquad (7\text{–}34)$$

The parallel RC circuit shown in *Figure 7.41* can be used to illustrate how to use this equation to calculate I_T. In the circuit the resistive current is 5 amperes, and the capacitive current is 12 amperes. Therefore,

$$\begin{aligned} I_T &= \sqrt{5^2 + 12^2} \\ &= \sqrt{25 + 144} \\ &= \sqrt{169} \\ &= 13\text{A} \end{aligned}$$

The total current is 13 amperes.

Calculation of Z_T Using Ohm's Law

Once the total current is known, the total impedance of the circuit can be determined using Ohm's law for ac circuits:

$$E_A = I_T Z_T \qquad (7\text{–}35)$$

Rearranging to solve for impedance, the total impedance of the circuit is equal to the applied voltage divided by the total current:

$$Z_T = \frac{E_A}{I_T} \qquad (7\text{–}36)$$

For example, in the circuit in *Figure 7.41* the applied voltage is 26 volts. The circuit impedance is calculated:

$$\begin{aligned} Z_T &= \frac{E_A}{I_T} \\ &= \frac{26\text{V}}{13\text{A}} \\ &= 2\Omega \end{aligned}$$

The circuit's impedance is 2 ohms.

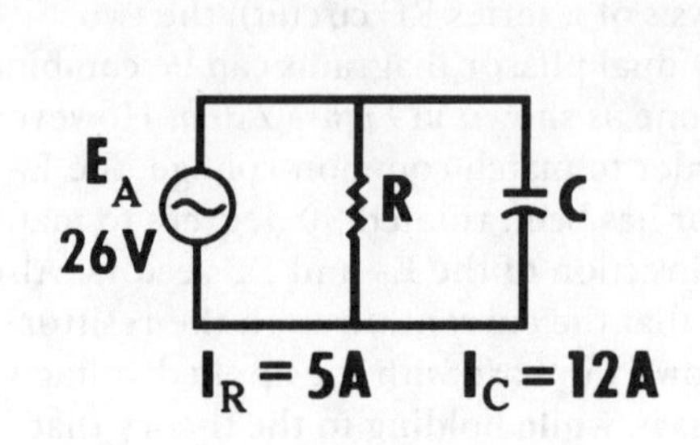

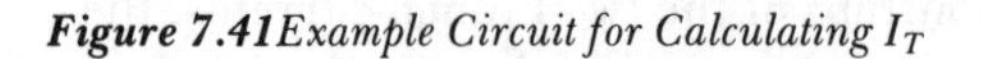

***Figure 7.41** Example Circuit for Calculating I_T*

Calculation of Z_T Using Impedance Vectors

The total impedance can also be determined directly. However, a more complex equation is involved. Note that in *Figure 7.42a* the total current is equal to the applied voltage divided by the total impedance. The resistive current is equal to the applied voltage divided by the resistance. And the capacitive current is equal to the applied voltage divided by the capacitive reactance. These E over Z, E over R, and E over X quantities can be substituted for the current they equal in the current phasor diagram as shown in *Figure 7.42b.*

■ Calculation of Z_T Using Impedance Vectors

Since the applied voltage is the common factor, it can be factored out. The result, as shown in *Figure 7.43,* is an impedance phasor diagram relating the reciprocals of resistance, reactance and impedance. Now, using the Pythagorean theorem, it can be written that:

$$\frac{1}{Z} = \sqrt{\left(\frac{1}{R}\right)^2 + \left(\frac{1}{X_C}\right)^2} \qquad (7\text{–}37)$$

Therefore,

$$Z = \frac{1}{\sqrt{\left(\frac{1}{R}\right)^2 + \left(\frac{1}{X_C}\right)^2}} \qquad (7\text{–}38)$$

The reciprocal of the impedance is equal to the square root of the reciprocal of the resistance squared plus the reciprocal of the reactance squared. The impedance, then, is equal to one divided by the square root of the reciprocal of the resistance squared plus the reciprocal of the reactance squared.

Calculations using

$$R = \frac{26V}{5A}$$

$$\frac{1}{R} = \frac{5A}{26V}$$

and

$$X_C = \frac{26V}{12A}$$

$$\frac{1}{X_C} = \frac{12A}{26V}$$

show that

$$Z_T = \frac{26}{\sqrt{5^2 + 12^2}}$$

$$= \frac{26}{13}$$

$$= 2\Omega$$

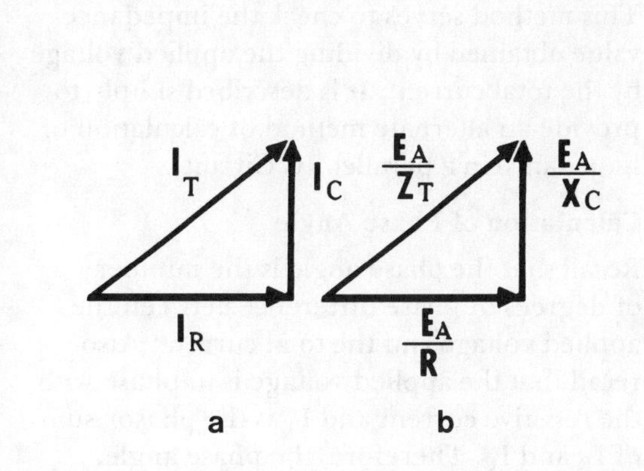

Figure 7.42 a. Current Phasor Diagram; b. Factors Substituted for Current They Equal

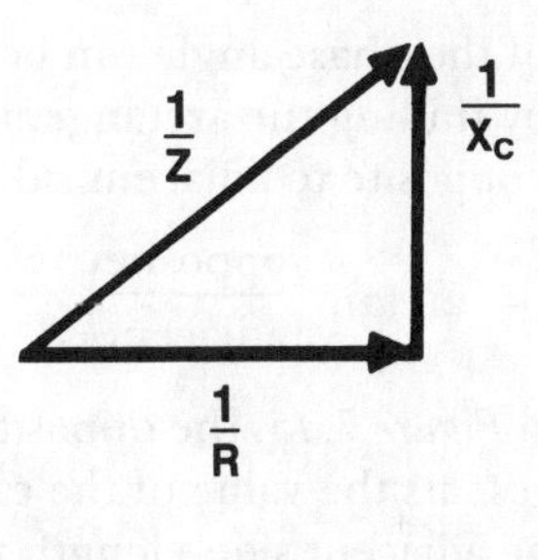

Figure 7.43 Impedance Phasor Diagram Relating the Reciprocals of Resistance Reactance, and Impedance

■ Calculation of Phase Angle

This method serves to check the impedance value obtained by dividing the applied voltage by the total current. It is described simply to provide an alternate method of calculation of impedance in a parallel RC Circuit.

Calculation of Phase Angle

Recall that the phase angle is the number of degrees of phase difference between the applied voltage and the total current. Also recall that the applied voltage is in phase with the resistive current and I_T is the phasor sum of I_R and I_C. Therefore, the phase angle, theta, is located between the E_A value and I_T value on the phasor diagram in *Figure 7.44*. The phase angle in a parallel RC circuit can also be recognized as the angle between the resistive current and the total current.

The value of the phase angle can be calculated by finding the arctangent of the ratio of the opposite to adjacent sides:

$$\theta = \arctan\left(\frac{\text{opposite}}{\text{adjacent}}\right) \qquad (7\text{–}39)$$

As shown in *Figure 7.45*, the opposite side's length represents the value of the capacitive current. The adjacent side's length represents the value of the resistive current. Thus, the tangent of the phase angle, θ, is equal to the ratio of the capacitive current divided by the resistive current as shown in equation *7–40*.

$$\tan\theta = \frac{I_C}{I_R} \qquad (7\text{–}40)$$

Therefore,

$$\theta = \arctan\left(\frac{I_C}{I_R}\right) \qquad (7\text{–}41)$$

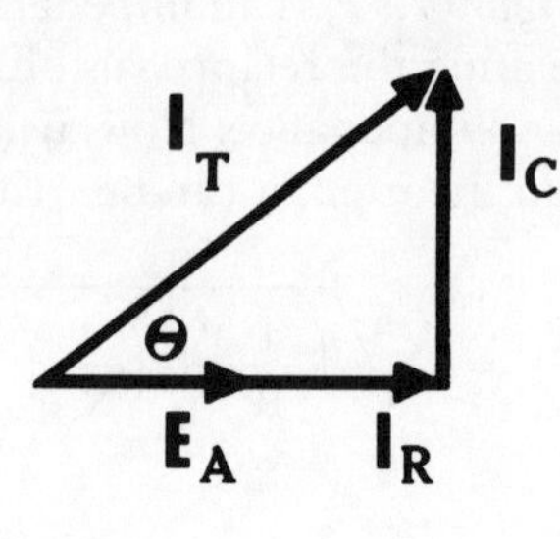

Figure 7.44 *Relationship of θ, E_A, and I_T Values on Phasor Diagram*

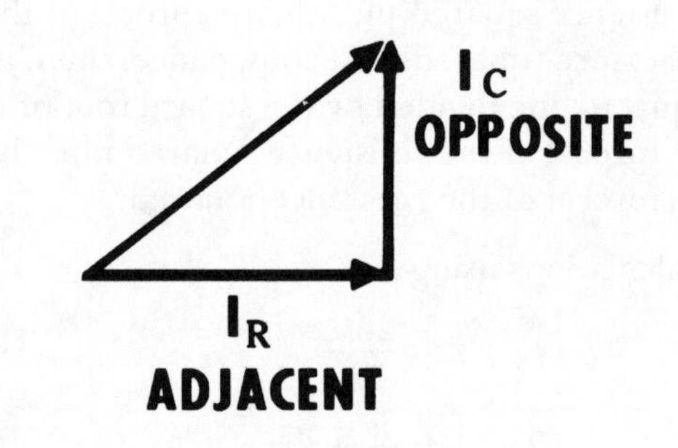

Figure 7.45 *Adjacent and Opposite Sides' Relationship*

The arctangent of the capacitive current divided by the resistive current equals the value of the phase angle. The phase angle, theta is an angle whose tangent is the ratio of capacitive current divided by resistive current.

For the example of *Figure 7.41*

$$\theta = \arctan\left(\frac{12}{5}\right)$$

$$\theta = \arctan 2.4$$

$$\theta \cong 67°$$

ANALYSIS OF PARALLEL RC CIRCUITS: EXAMPLE

With this background a parallel RC circuit will be analyzed further. The individual branch currents, the total current, the total impedance, and the phase angle will be determined. The Pythagorean theorem will be used to calculate phase-related circuit values.

Circuit Summary

The parallel RC circuit to be used in this example is shown in *Figure 7.46*. The value of the resistor is 15 ohms. The value of the capacitor is such that at the applied frequency, it has a capacitive reactance of 20 ohms. The applied voltage is 60 VAC.

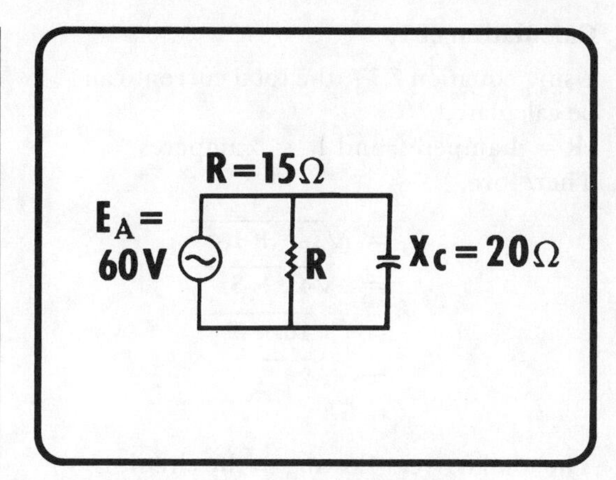

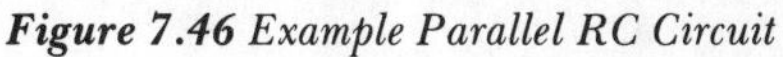
Figure 7.46 Example Parallel RC Circuit

Calculation of I_R and I_C

First the resistive branch current, I_R, is determined

$$I_R = \frac{E_R}{R} = \frac{60V}{15\Omega} = 4A$$

The capacitive branch current, I_C, is determined in a similar manner:

$$I_C = \frac{E_C}{X_C} = \frac{60V}{20\Omega} = 3A$$

The current phasor diagram can now be drawn to show the relationships between the resistive and capacitive branch currents and the total circuit current. This diagram is shown in *Figure 7.47*.

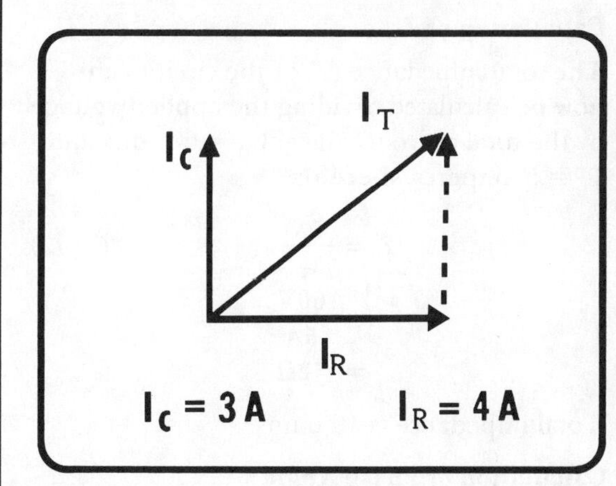

Figure 7.47 *Relationship Between I_R and I_C in Example Circuit*

Calculation of I_T

Using equation *7.34,* the total current can be calculated.
IR = 4 amperes, and I_C = 3 amperes.
Therefore,

$$\begin{aligned} I_T &= \sqrt{I_R^2 + I_C^2} \\ &= \sqrt{4^2 + 3^2} \\ &= \sqrt{16 + 9} \\ &= \sqrt{25} \\ &= 5A \end{aligned}$$

The total current flowing in the circuit is 5 amperes.

Calculation of Z

The total impedance (Z) of the circuit can now be calculated dividing the applied voltage by the total current. Since, E_A = 60 volts and I_T = 5 amperes, therefore,

$$\begin{aligned} Z &= \frac{E_A}{I_T} \quad (7\text{–}42) \\ &= \frac{60V}{5A} \\ &= 12\Omega \end{aligned}$$

Total impedance is 12 ohms.

Calculation of Phase Angle

The phase angle can be recognized as the angle between the resistive current and the total current on the current phasor diagram in *Figure 7.48*. The value of I_C = 3 amperes, and I_R = 4 amperes. The phase angle is calculated using equation *7–41:*

$$\begin{aligned} \theta &= \arctan\left(\frac{I_C}{I_R}\right) \\ &= \arctan\left(\frac{3A}{4A}\right) \\ &= \arctan(0.75) \\ &= 37° \end{aligned}$$

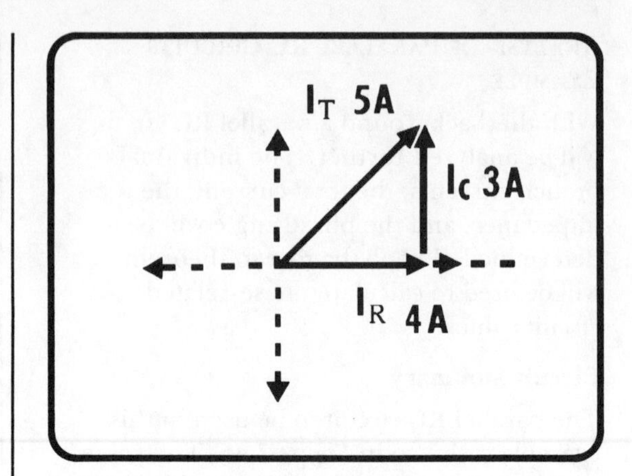

Figure 7.48 *Phasor Diagram of Example Circuit for Calculating θ*

The phase angle is equal to the arctangent of the ratio of the capacitive current to the resistive current. Thus, the phase angle, θ, is equal to the arctangent of 3 amperes divided by 4 amperes which equals the arctangent 0.75. Using a calculator or trigonometric table, the angle whose arctangent is 0.75 is approximately 37 degrees. Thus, the phase angle of this parallel RC circuit is approximately 37 degrees.

Positive Phase Angle in Parallel RC Circuit

The phase angle is said to be a positive 37 degrees, as shown in *Figure 7.49*. This is because the applied voltage in the parallel RC circuit was used as a reference at zero degrees. The positive sign is used to indicate that the total current phasor is rotated 37 degrees counter-clockwise from the applied voltage phasor.

In *Figure 7.50* it can be seen that in a *series* RC circuit, the phase angle is *negative*. But in a *parallel* circuit, as shown in *Figure 7.51*, the phase angle is *positive*. The sign of the phase angle is used simply to indicate in which direction of rotation *from the reference* that the phase angle is measured, therefore, if you know only that the circuit is an RC circuit,

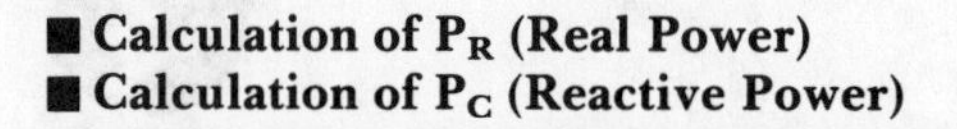

you can readily determine whether the resistor and capacitor are connected in series or parallel by the sign of the phase angle.

POWER CALCULATIONS IN PARALLEL RC CIRCUITS

In the example parallel RC circuit, shown in *Figure 7.52*, the power relationships are similar to those of a series RC circuit.

Calculation of P_R (Real Power)

The real power, P_R, is equal to the voltage across the resistor times the value of the current flowing through it.

$$P_R = E_R I_R$$
$$= (60V)(4A)$$
$$= 240W$$

Real power is 240 watts.

Calculation of P_C (Reactive Power)

The reactive power, P_C, is equal to the voltage across the capacitor times the value of the current through it.

$$P_C = E_C I_C \qquad (7\text{–}43)$$

In the example circuit, E_C = 60 volts; I_C = 3 amperes. Therefore:

$$P_C = E_C I_C$$
$$= (60V)(3A)$$
$$= 180VAR$$

Reactive power is 180 VAR.

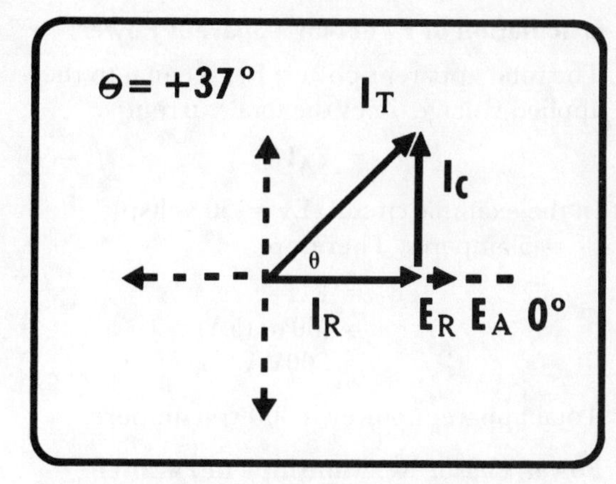

Figure 7.49 The Phase Angle for Example Circuit is Positive

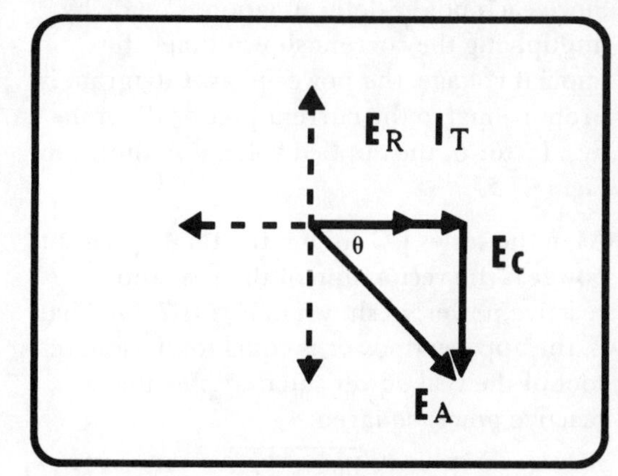

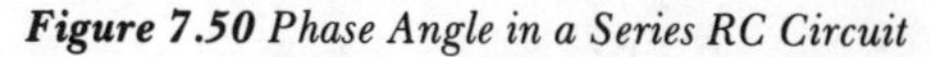

Figure 7.50 Phase Angle in a Series RC Circuit

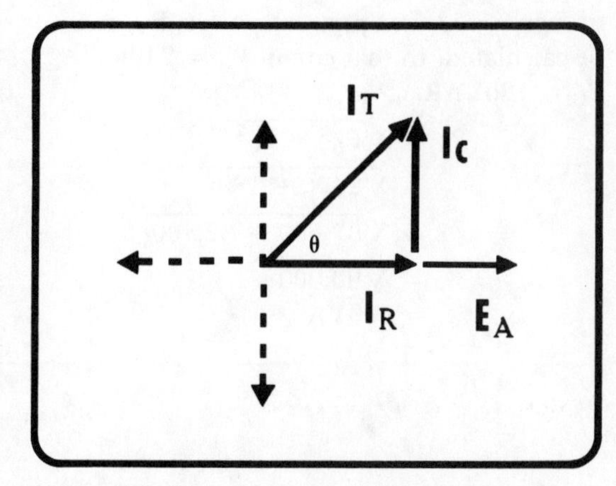

Figure 7.51 Phase Angle in a Parallel RC Circuit

■ **Calculation of P_T (Total Apparent Power)**
■ **Power Phasor Relationships in Parallel RC Circuits**

Calculation of P_T (Total Apparent Power)

The total apparent power, P_T, is equal to the applied voltage times the total current:

$$P_T = E_A I_T \qquad (7\text{–}44)$$

In the example circuit, $E_A = 60$ volts; $I_T = 5$ amperes. Therefore:

$$\begin{aligned} P_T &= E_A I_T \\ &= (60V)\,(5A) \\ &= 300VA \end{aligned}$$

Total apparent power is 300 volt amperes.

Power Phasor Relationships in Parallel RC Circuits

Since each power determination is made by multiplying the current shown times the applied voltage, the power phasor diagram is proportional to the current phasor diagram by a factor of the applied voltage as shown in *Figure 7.53*.

As in the series RC circuit, the total apparent power is the vector sum of the real and reactive power, as shown in *Figure 7.54*. That is, the apparent power is equal to the square root of the real power squared plus the reactive power squared.

$$P_T = \sqrt{P_R{}^2 + P_C{}^2} \qquad (7\text{–}45)$$

Using the parallel RC circuit values as shown in *Figure 7.52* the apparent power, P_T, can be calculated. In that circuit $P_R = 240$w; $P_C = 180$VAR.

$$\begin{aligned} P_T &= \sqrt{P_R{}^2 + P_C{}^2} \\ &= \sqrt{240^2 + 180^2} \\ &= \sqrt{57{,}600 + 32{,}400} \\ &= \sqrt{90{,}000} \\ &= 300VA \end{aligned}$$

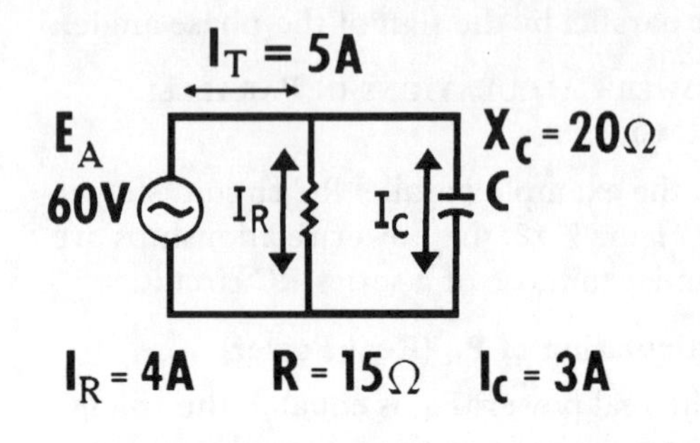

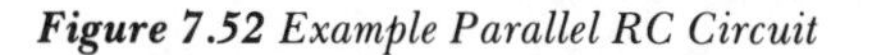

***Figure 7.52** Example Parallel RC Circuit*

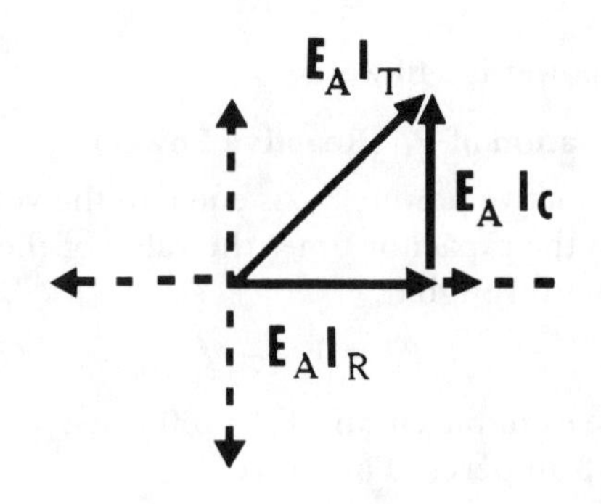

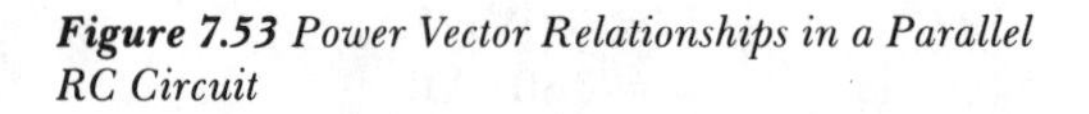

***Figure 7.53** Power Vector Relationships in a Parallel RC Circuit*

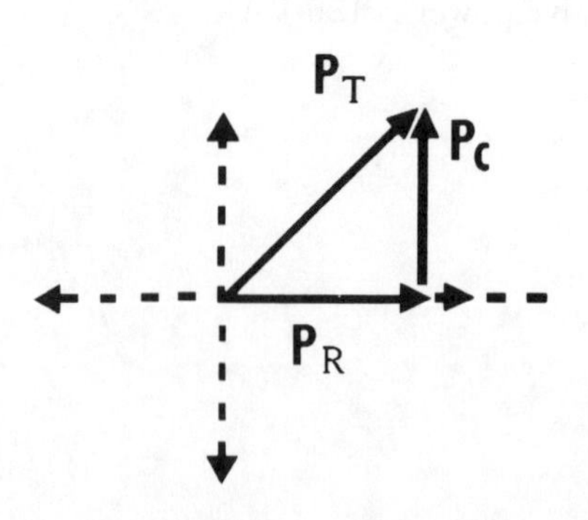

***Figure 7.54** Relationship Between P_T, P_R, and P_C in Parallel RC Circuit*

Therefore, the total apparent power calculated in this way is 300 volt-amperes which matches previous calculations and serves as a check that they were correct.

Summary

In this lesson, techniques were discussed which are used to analyze series and parallel RC circuit operation and calculate circuit values. The different phase relationships between voltage and current for each component and the effects upon the circuit value for each type of circuit were described. You were told how to determine the currents and voltages, impedance, phase angle, and power values for both a series and parallel RC circuit. You can use these methods to determine the same circuit values of any series or parallel RC circuit.

■ Worked-Out Examples

1. Draw the phasor diagrams for the voltage and power circuit values in the circuit below.

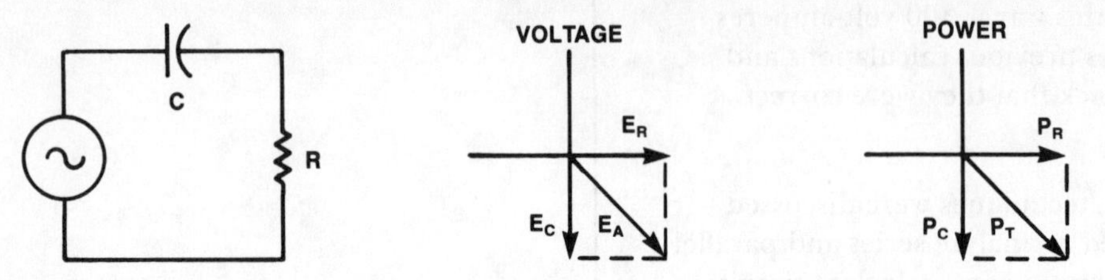

Solution: In a series circuit, current is common and used as the reference. E_R is plotted in phase with I_T, I_R and I_C. E_C lags I_C by 90 degrees. E_A is the phasor sum of E_R and E_C. Current is a common factor in the power equations, and the power values are plotted in phase with the voltage values.

2. Draw the phasor diagrams for the voltage and power circuit values in the circuit below.

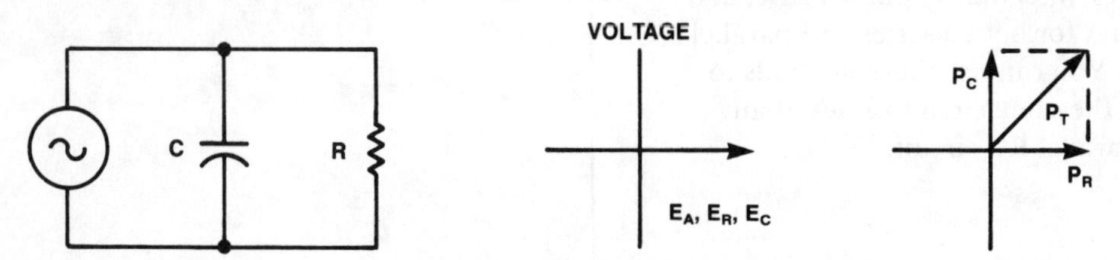

Solution: In a parallel circuit, voltage is common and used as the reference circuit value in the phasor diagram. Voltage is a common factor in the power equations, and the power values are plotted in phase with the current values. I_C leads E_C by 90°. I_R is in phase with E_R.

3. Draw the Pythagorean theorem circuit value relationships for R, X_C and Z_T for the circuit in problem 1.

Solution:

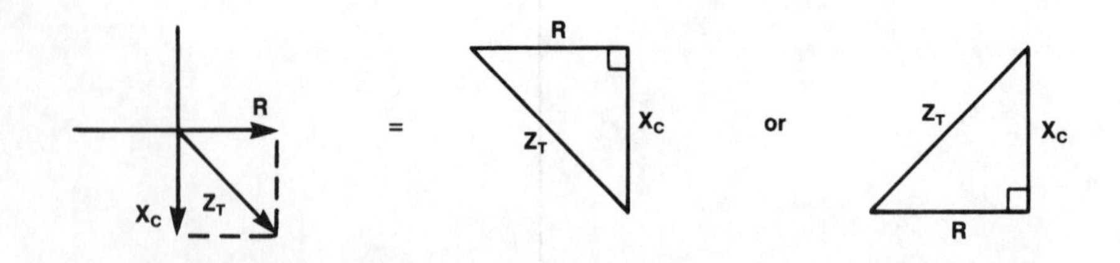

4. Given the following angles find the tangent of the angle.

a. 82°
b. 48°
c. 33.5°
d. 21.9°
e. 11°

Calculator Solution: Enter the angle in degrees into the calculator and press the tan key. The tangent of the angle will be on the calculator.

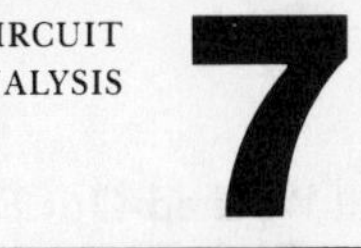

Table Solution: Obtain a table of natural trigonometric functions for angles in decimal degrees. Enter the table by locating the angle in the degree column. Scan across to the tangent column and read the tangent value.

Angle	*Tangent*
82°	**7.115**
48°	**1.111**
33.5°	**0.662**
21.9°	**0.402**
11°	**0.194**

5. Given the following arctangents find the angle.

a. 0.3
b. 0.477
c. 1.6
d. 3.00
e. 21.20

Calculator Solution: Enter the ratio into the calculator so it is displayed. Press the inv and tan keys or the $\tan^{-1}$ key and the calculator will display the angle in decimal degrees.

Table Solution: Enter the table of natural trigonometric functions by locating the ratio in the tangent column. Read across to the angle in decimal degrees in the degree column.

Ratio (tangent)	*Angle*
a. 0.3	**16.7°**
b. 0.477	**25.5°**
c. 1.6	**58.0°**
d. 3.00	**71.6°**
e. 21.20	**87.3°**

6. Fill in the blanks.

a. arctan 0.700 = ________
b. arctan 0.400 = ________
c. arctan 1.00 = ________
d. arctan ______ = 5°
e. arctan ______ = 85°

Solution: Use same calculator or table procedure as for question 4 or 5.

arctan 0.700 = **34.99° (35°)**
arctan 0.400 = **21.8°**
arctan 1.00 = **45°**
arctan **0.0875** = 5°
arctan **11.43** = 85°

■ Worked-Out Examples

7. In the phasor diagram for a parallel RC circuit shown, calculate the phase angle.

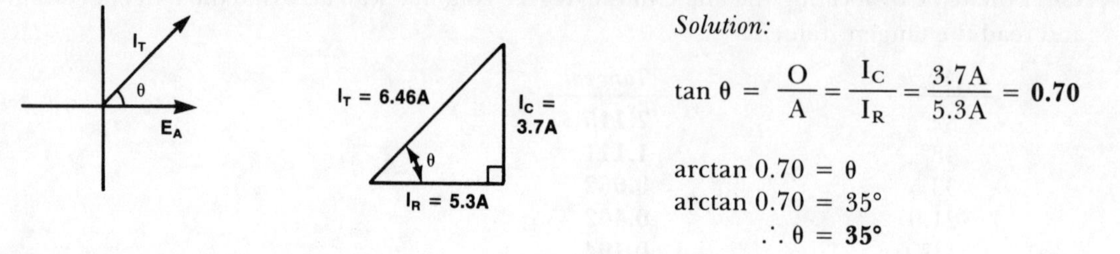

Solution:

$$\tan\theta = \frac{O}{A} = \frac{I_C}{I_R} = \frac{3.7A}{5.3A} = \mathbf{0.70}$$

$$\arctan 0.70 = \theta$$
$$\arctan 0.70 = 35°$$
$$\therefore \theta = \mathbf{35°}$$

8. Solve for the circuit values specified for the circuit below.

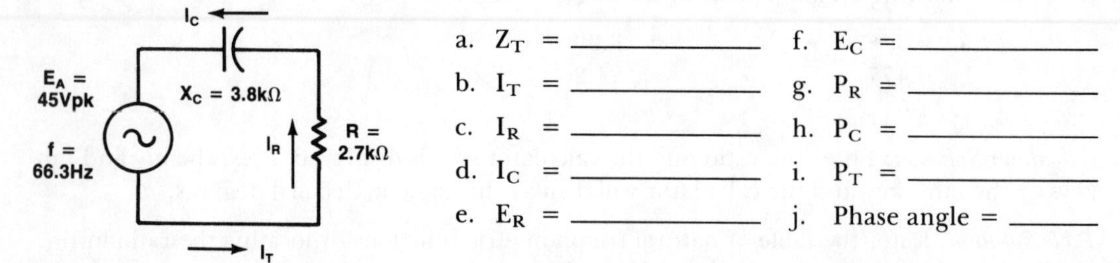

a. Z_T = ________
b. I_T = ________
c. I_R = ________
d. I_C = ________
e. E_R = ________
f. E_C = ________
g. P_R = ________
h. P_C = ________
i. P_T = ________
j. Phase angle = ________

Solution:

a. $Z_T = \sqrt{R^2 + X_C{}^2} = \sqrt{2.7k\Omega^2 + 3.8k\Omega^2} = \sqrt{21.73k\Omega^2} = \mathbf{4.66k\Omega}$

b. $I_T = \frac{E_A}{Z_T} = \frac{45V_{pk}}{4.66k\Omega} = \mathbf{9.66mA_{pk}}$

c., d. $I_T = I_R = I_C = \mathbf{9.66mA_{pk}}$

e. $E_R = I_R \times R = 9.66mA_{pk} \times 2.7k\Omega = \mathbf{26.1V_{pk}}$

f. $E_C = I_C \times X_C = 9.66mA_{pk} \times 3.8k\Omega = \mathbf{36.7V_{pk}}$

g. $P_R = I_R \times E_R = 9.66mA_{pk} \times 26.1V_{pk} = \mathbf{0.252W_{pk}}$

h. $P_C = I_C \times E_C = 9.66mA_{pk} \times 36.7V_{pk} = \mathbf{0.354VAR_{pk}}$

i. $P_T = I_T \times E_A = 9.66mA_{pk} \times 45V_{pk} = \mathbf{0.435VA_{pk}}$

j. $\tan\theta = \frac{X_C}{R} = \frac{3.8k\Omega}{2.7k\Omega} = \mathbf{1.41}$

arctan 1.41 = Phase angle
arctan 1.41 = 54.6°
$\therefore \theta = \mathbf{-54.6°}$ (series circuit)

■ Worked-Out Examples

9. Solve for the circuit values specified for the circuit below:

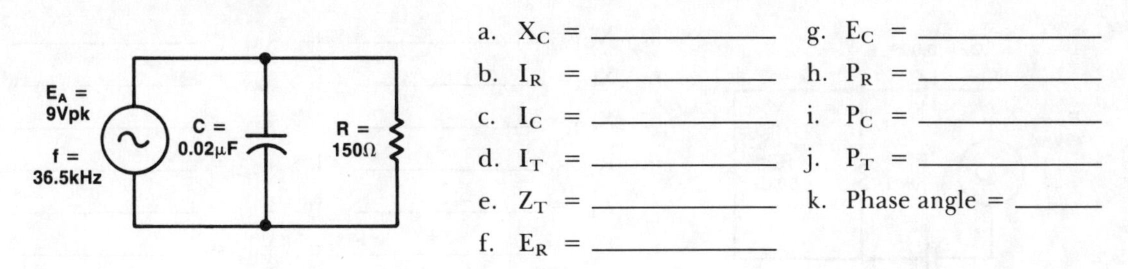

Solution:

a. $X_C = \dfrac{1}{2\pi fC} = \dfrac{1}{6.28 \times 36.5 \times 10^3 \times 0.02 \times 10^{-6}} = \dfrac{1}{4.58 \times 10^{-3}} = \mathbf{218\Omega}$

b. $I_R = \dfrac{E_R}{R} = \dfrac{9V_{pk}}{150\Omega} = \mathbf{0.06A_{pk}}$

c. $I_C = \dfrac{E_C}{X_C} = \dfrac{9V_{pk}}{218\Omega} = \mathbf{0.0413A_{pk}}$

d. $I_T = \sqrt{I_R{}^2 + I_C{}^2} = \sqrt{0.06A^2 + 0.0413A^2} = \sqrt{0.00531A^2} = \mathbf{0.0728A_{pk}}$

e. $Z_T = \dfrac{E_A}{I_T} = \dfrac{9V_{pk}}{0.0728A_{pk}} = \mathbf{124\Omega}$

f., g. $E_A = E_R = E_C = \mathbf{9V_{pk}}$

h. $P_R = I_R \times E_R = 0.06A_{pk} \times 9V_{pk} = \mathbf{0.54W_{pk}}$

i. $P_C = I_C \times E_C = 0.0413A_{pk} \times 9V_{pk} = \mathbf{0.372VAR_{pk}}$

j. $P_T = I_T \times E_A = 0.0728A \times 9V = \mathbf{0.655VA_{pk}}$

k. $\tan\theta = \dfrac{O}{A} = \dfrac{I_C}{I_R} = \dfrac{0.0413A_{pk}}{0.06A_{pk}} = \mathbf{0.688}$

$\arctan 0.688 = \theta$
$\arctan 0.688 = 34.5°$
$\therefore \theta = \mathbf{+34.5°}$ (parallel circuit)

■ Worked-Out Examples

10. Solve for the circuit values specified for the circuit below:

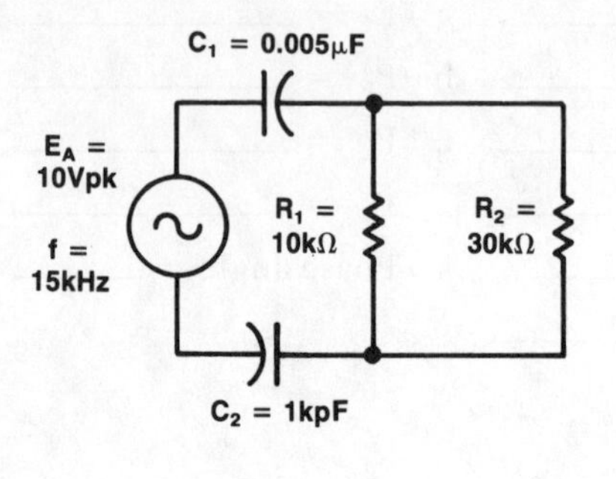

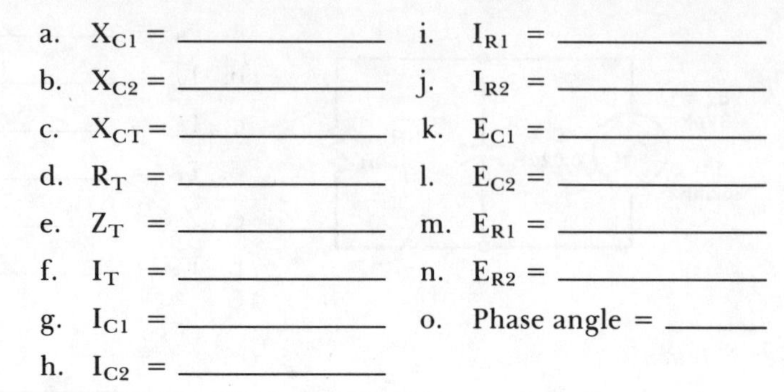

a. X_C1 = ______
b. X_C2 = ______
c. X_CT = ______
d. R_T = ______
e. Z_T = ______
f. I_T = ______
g. I_C1 = ______
h. I_C2 = ______
i. I_R1 = ______
j. I_R2 = ______
k. E_C1 = ______
l. E_C2 = ______
m. E_R1 = ______
n. E_R2 = ______
o. Phase angle = ______

Solution: Simplify the circuit.

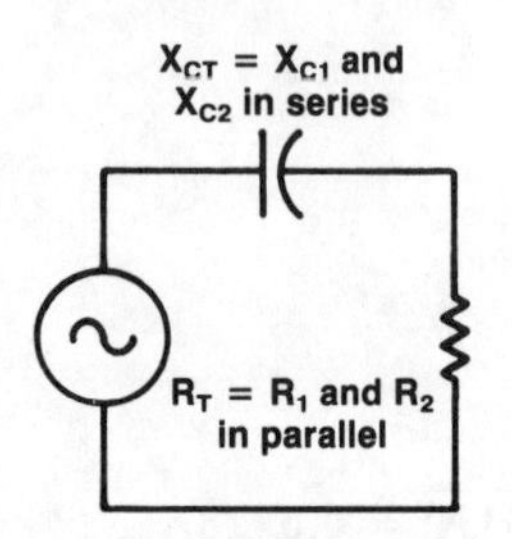

a. $X_{C1} = \dfrac{1}{2\pi f C_1} = \dfrac{1}{4.71 \times 10^{-4}} = \mathbf{2.12k\Omega}$

b. $X_{C2} = \dfrac{1}{2\pi f C_2} = \dfrac{1}{9.42 \times 10^{-5}} = \mathbf{10.6k\Omega}$

c. $X_{CT} = X_{C1}$ & X_{C2} in series
$= 2.12k\Omega + 10.06k\Omega = \mathbf{12.18k\Omega}$

d. $R_T = \dfrac{(10k\Omega)(30k\Omega)}{10k\Omega + 30k\Omega} = \mathbf{7.5k\Omega}$

e. $Z_T = \sqrt{R_T^{\,2} + X_{CT}^{\,2}}$
$= \sqrt{7.5k\Omega^2 + 12.18k\Omega^2}$
$= \sqrt{204.6k\Omega^2} = \mathbf{14.3k\Omega}$

f. $I_T = \dfrac{E_A}{Z_T} = \dfrac{10V_{pk}}{14.3k\Omega} = \mathbf{0.699mA_{pk}}$

g. $I_{C1} = I_T = \mathbf{0.699mA_{pk}}$

h. $I_{C2} = I_T = \mathbf{0.699mA_{pk}}$

i. $I_{R1} = \dfrac{E_{R1}}{R_1} = \dfrac{5.24V}{10k} = \mathbf{0.524mA_{pk}}$

j. $I_{R2} = \dfrac{E_{R2}}{R_2} = \dfrac{5.24V}{30k} = \mathbf{0.175mA_{pk}}$

k. $E_{C1} = I_T \times X_{C1}$
$= 0.699mA \times 2.12k\Omega = \mathbf{1.48V_{pk}}$

l. $E_{C2} = I_{C2} \times X_{C2}$
$= 0.699mA \times 10.6k\Omega = \mathbf{7.41V_{pk}}$

m. $E_{R1} = I_T \times R_T$
$= 0.699mA_{pk} \times 7.5k\Omega = \mathbf{5.24V_{pk}}$

n. $E_{R2} = E_{R1} = 5.24V_{pk}$

o. $\tan\theta = \dfrac{X_{CT}}{R_T} = \dfrac{12.18k\Omega}{7.5k\Omega} = 1.62$

$\arctan 1.62 = \theta$
$\arctan 1.62 = \mathbf{-58.4^\circ}$

Practice Problems

1. Given the following angles find the tangent of the angle.

a. 15° .268
b. 30° .577
c. 45° 1.00
d. 60° 1.73
e. 75° 3.73

2. Given the following tangent values, determine the angle.

a. 0.1 5.71°
b. 0.9 42.0°
c. 2.0 63.4°
d. 4.0 76.0°
e. 8.0 83.0°

3. What is the definition of arctangent?

4. In each of the diagrams a., b., c., and d., solve for the value of the hypotenuse of the right triangle.

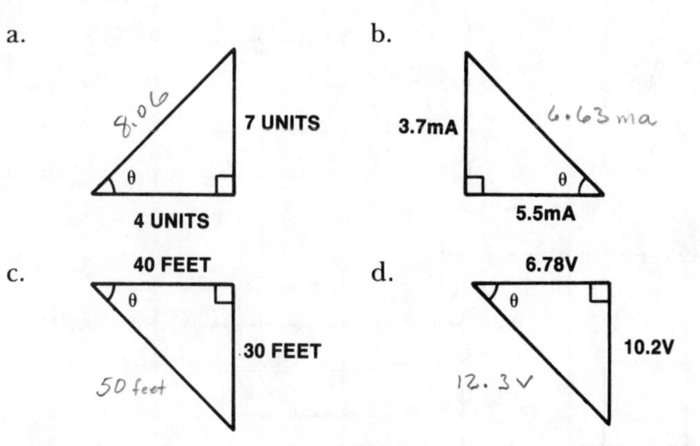

5. Solve for the phase angle in each of the circuits a., b., c., and d.

a.) θ = 60.3° c.) θ = 36.9°
b.) θ = 33.9° d.) θ = 56.4°

6. Draw a phasor diagram for the circuit shown. Show voltage, current and power phasors.

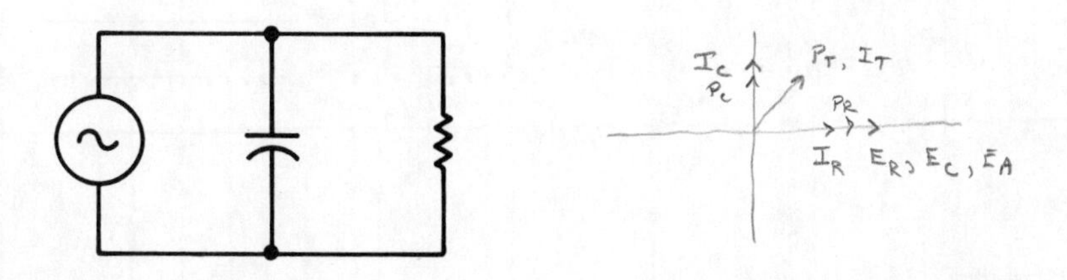

■ Practice Problems

7. a. Identify the type of circuit which fits the current phasor diagram shown. *parallel RC*

b. Calculate I_T. *19.6 ma rms*

c. Calculate phase angle. *23.4°*

I_C = 7.8mA(rms)

I_R = 18.0mA(rms)

8. Solve for the circuit values specified for this circuit.

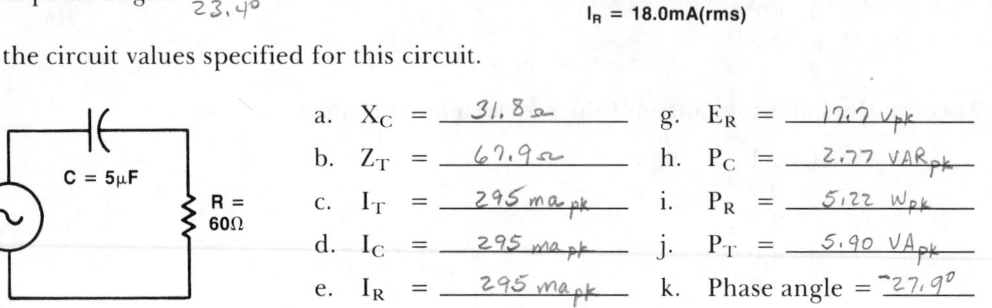

a. X_C = *31.8 Ω*

b. Z_T = *67.9 Ω*

c. I_T = *295 ma pk*

d. I_C = *295 ma pk*

e. I_R = *295 ma pk*

f. E_C = *9.38 Vpk*

g. E_R = *17.7 Vpk*

h. P_C = *2.77 VARpk*

i. P_R = *5.22 Wpk*

j. P_T = *5.90 VApk*

k. Phase angle = *−27.9°*

9. Solve for the circuit values specified for this circuit.

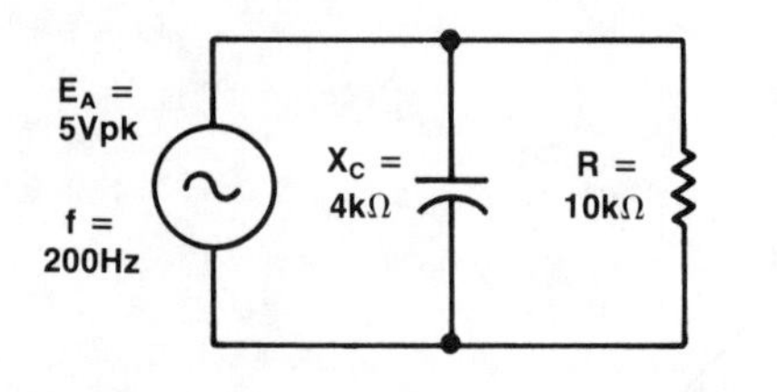

a. I_C = *1.25 ma pk*

b. I_R = *.500 ma pk*

c. I_T = *1.35 ma pk*

d. Z_T = *3.71 kΩ*

e. E_C = *5.00 Vpk*

f. E_R = *5.00 Vpk*

g. P_C = *6.25 m VARpk*

h. P_R = *2.50 mWpk*

i. P_T = *6.73 mVApk*

j. Phase angle = *68.2°*

10. Solve for the values specified for this circuit.

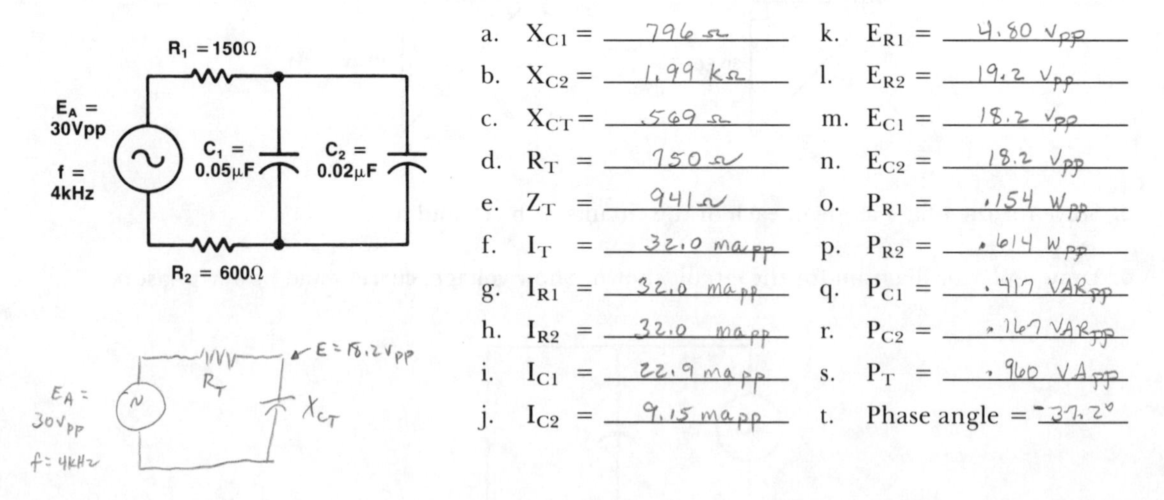

a. X_{C1} = *796 Ω*

b. X_{C2} = *1.99 kΩ*

c. X_{CT} = *569 Ω*

d. R_T = *750 Ω*

e. Z_T = *941 Ω*

f. I_T = *32.0 ma pp*

g. I_{R1} = *32.0 ma pp*

h. I_{R2} = *32.0 ma pp*

i. I_{C1} = *22.9 ma pp*

j. I_{C2} = *9.15 ma pp*

k. E_{R1} = *4.80 Vpp*

l. E_{R2} = *19.2 Vpp*

m. E_{C1} = *18.2 Vpp*

n. E_{C2} = *18.2 Vpp*

o. P_{R1} = *.154 Wpp*

p. P_{R2} = *.614 Wpp*

q. P_{C1} = *.417 VARpp*

r. P_{C2} = *.167 VARpp*

s. P_T = *.960 VApp*

t. Phase angle = *−37.2°*

■ Quiz

1. In each of the diagrams a., b., c., and d., solve for the value of the hypotenuse of the right triangle.

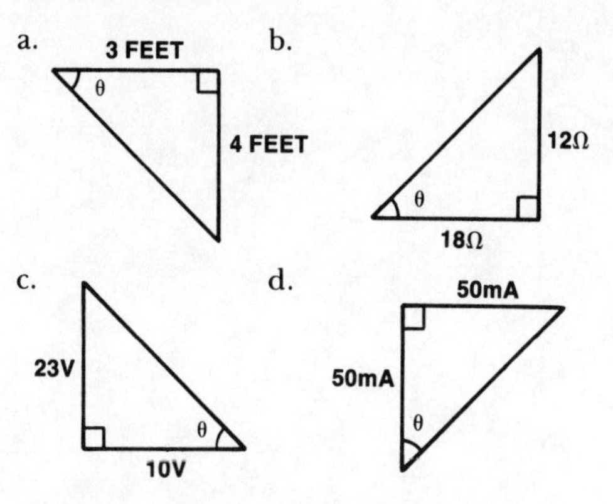

2. Solve for the phase angle in the circuits a., b., c., and d.

3. Draw a phasor diagram for this circuit. Show voltage, current, and power phasor.

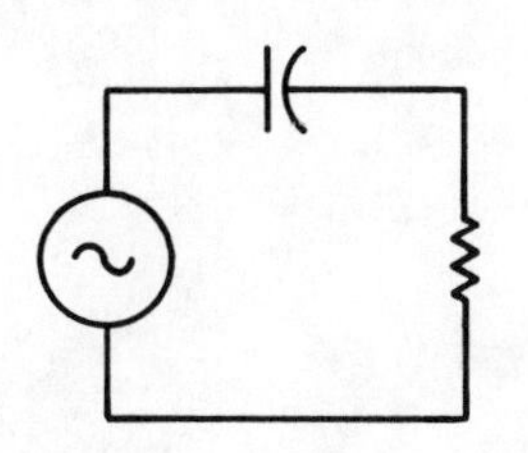

4. a. Identify the type of circuit which is the same as this voltage phasor diagram.
   b. Calculate its E_T.
   c. Calculate its phasor angle.

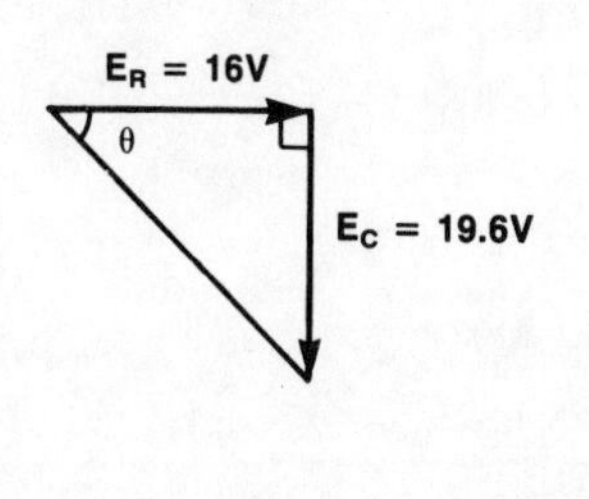

5. Solve for the circuit values specified for this circuit:

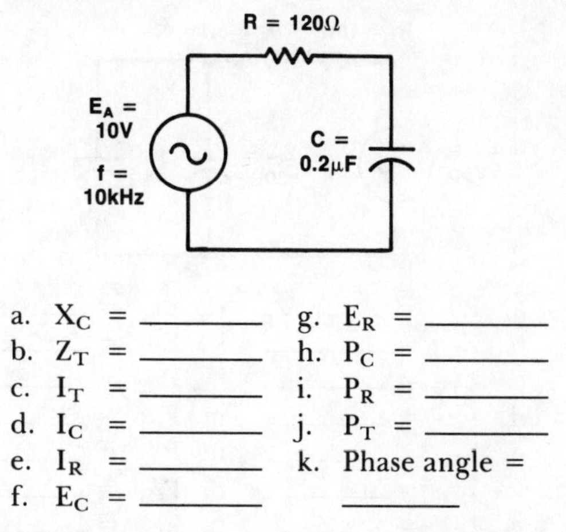

a. X_C = ________
b. Z_T = ________
c. I_T = ________
d. I_C = ________
e. I_R = ________
f. E_C = ________
g. E_R = ________
h. P_C = ________
i. P_R = ________
j. P_T = ________
k. Phase angle = ________

6. Solve for the circuit values specified for this circuit.

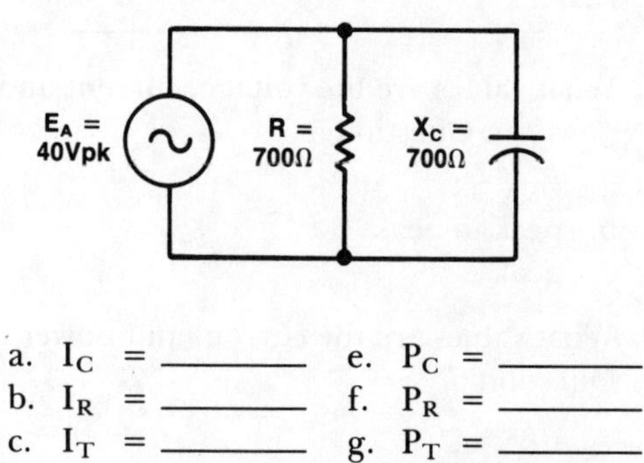

a. I_C = ________
b. I_R = ________
c. I_T = ________
d. Z_T = ________
e. P_C = ________
f. P_R = ________
g. P_T = ________
h. Phase angle = ________

■ Quiz

7. Solve for the circuit values specified for the circuit below.

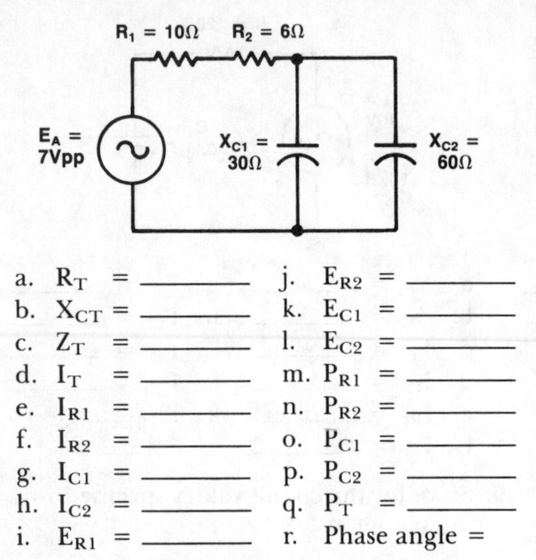

a. R_T = ________
b. X_{CT} = ________
c. Z_T = ________
d. I_T = ________
e. I_{R1} = ________
f. I_{R2} = ________
g. I_{C1} = ________
h. I_{C2} = ________
i. E_{R1} = ________
j. E_{R2} = ________
k. E_{C1} = ________
l. E_{C2} = ________
m. P_{R1} = ________
n. P_{R2} = ________
o. P_{C1} = ________
p. P_{C2} = ________
q. P_T = ________
r. Phase angle = ________

8. What values are the voltage, current and power in Question 5.

a. rms
b. peak to peak
c. peak

9. What values are the current and power in Question 6.

a. rms
b. peak to peak
c. peak

10. What values are the voltage, current and power in Question 7.

a. rms
b. peak to peak
c. peak

LESSON 8

Inductance and Transformers

This lesson introduces the inductor and transformer with a discussion of their physical and electrical properties. Circuits containing series and parallel-connected inductors are discussed and analyzed. The property of mutual inductance is introduced.

■ Objectives

At the end of this lesson, you should be able to:

1. Define inductance, state the basic unit of inductance, and be able to explain two concepts used to explain the counter EMF of an inductor.
2. Determine the effect of increasing or decreasing the permeability of the core material, the number of turns of conductor, and the cross-sectional area of the core or the length of the core on the value of an inductor.
3. State the relationship between the voltage across an inductor and the current through it.
4. Given a circuit of series-connected inductors or parallel-connected inductors, determine the total inductance of the circuit (no mutual inductance).
5. Given the value of two inductors connected in series or parallel aiding or opposing and their coefficient of coupling, determine the total inductance of their combination.
6. Given the turns ratio of a transformer, the input voltage, and the secondary load resistance, determine the secondary voltage, the primary and secondary current, the secondary frequency, and the primary and secondary power.
7. State the differences between step-up, step-down, and autotransformers, and their advantages and disadvantages.
8. Define inductive reactance and state the units in which it is measured.
9. Given a circuit of series-connected inductors, the applied voltage, and frequency of the applied voltage, determine the inductive reactance of each inductor, the total inductance, the total current, the voltage across each inductor, the reactive power of each inductor, and the total reactive power in the circuit.
10. Given a circuit of parallel-connected inductors, the applied voltage, and frequency of the applied voltage, determine the inductive reactance of each inductor, the total inductive reactance, the total inductance, each inductive branch current, the total current, the reactive power of each inductor, and the total reactive power in the circuit.

INTRODUCTION

In the last two lessons, discussion concerned the capacitor and how to analyze circuits composed of only capacitors or capacitors and resistors. This lesson is about the remaining passive circuit element— the inductor.

Figure 8.1 shows some types of inductors. Basically, any inductor is a coil of thin wire wrapped on a cylinder called the core. The core may be hollow, of laminated paper—an air core—or made of some type of iron—an iron core. Often an inductor is also called a choke or coil. The turns of wire of the inductor are electrically insulated from each other by a thin, non-conductive coating.

As shown in *Figure 8.2* the schematic symbol used to represent the inductor resembles what it is—wire wrapped on a core. The inductor's letter symbol is a capital L which represents linkages—flux linkages.

An inductor has magnetic properties. Therefore, a brief review of the subject of magnetism should help you understand better the electrical properties of an inductor.

ELECTROMAGNETIC PROPERTIES

Faraday's Discovery

Recall that in 1831, Michael Faraday showed that when a conductor connected in a closed circuit is moved through a magnetic field, an electron current flows as a result of a voltage induced in the conductor. (In this lesson, like in all other lessons in this book, current flow refers to *electron current flow*.) The amount of induced voltage is proportional to the rate of change of the magnetic field—the amount the magnetic flux changes in a specific time period:

$$\text{induced voltage} \approx \frac{\text{change in magnetic flux}}{\text{change in time}} \quad (8\text{–}1)$$

Figure 8.1 *Typical Inductors*

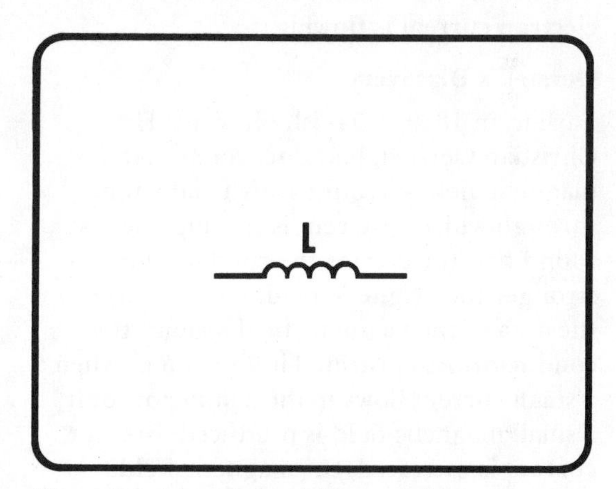

Figure 8.2 *Schematic Symbol for an Inductor*

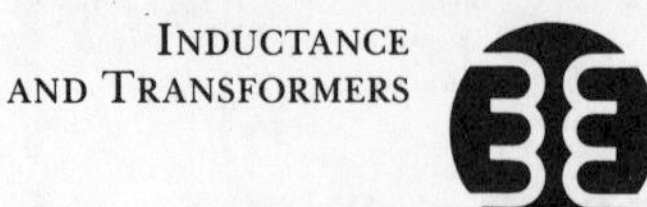

As shown in *Figures 8.3* and *8.4*, it makes no difference if the conductor moves through a stationary magnetic field, or if the conductor is stationary and the magnetic field moves. As long as the lines of flux are caused to cut though the conductor, a voltage is induced in the conductor.

Remember that the direction of the resulting electron current can be determined by the *left-hand rule for generators* which is shown in *Figure 8.5*. The thumb points in the direction of the conductor's motion. The forefinger points in the direction of the magnetic field. The index finger points in the direction the electron current is flowing.

Oersted's Discovery

Earlier, in 1819, a Danish physicist, Hans Christian Oersted, had discovered that a magnetic field surrounds any conductor through which a current is flowing. He also found that the greater the current, the stronger the magnetic field, and the larger the area of the magnetic field around the conductor. As illustrated in *Figure 8.6,* when a small current flows in the conductor, only a small magnetic field is produced. But as current is increased, the magnetic field expands. Conversely, if the current is decreased, the magnetic field decreases in size.

All lines of flux are considered to have a definite direction. The *direction* of the lines of flux surrounding a conductor can be determined by what is known as the *left-hand rule for conductors* which is shown in *Figure 8.7*. It states that if the conductor is grasped with the left-hand with the thumb pointing in the direction of the electron current, the fingers fold in the direction of the lines of flux.

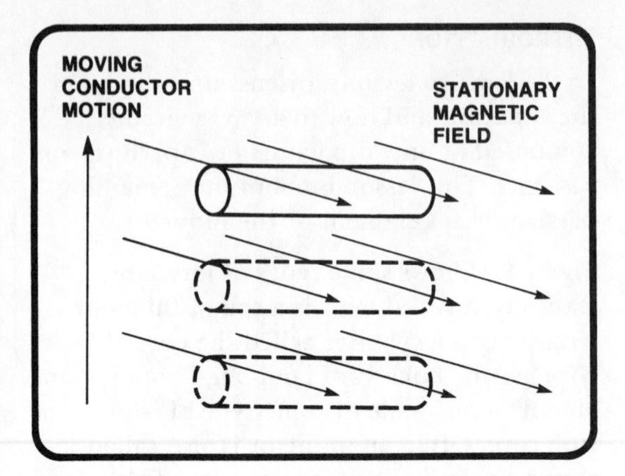

Figure 8.3 *Conductor Moving Through Magnetic Field*

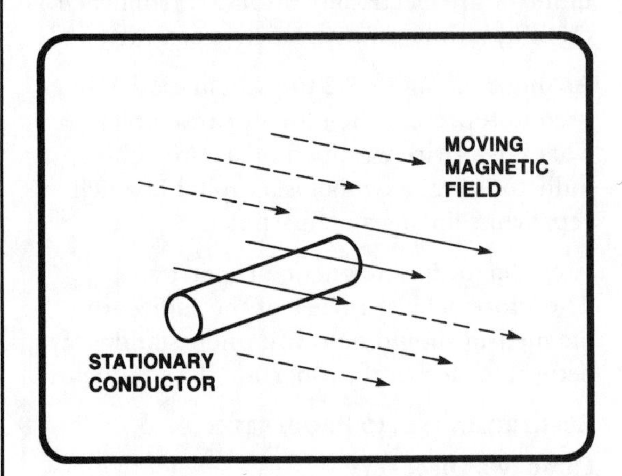

Figure 8.4 *Magnetic Field Moving Past Conductor*

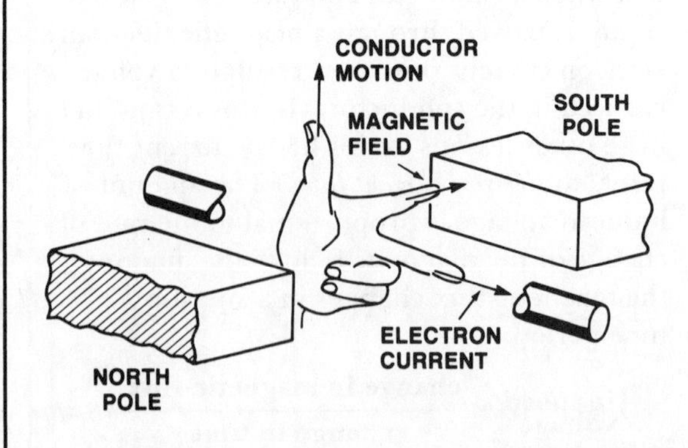

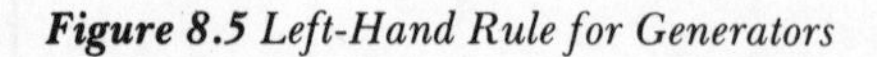

Figure 8.5 *Left-Hand Rule for Generators*

THE PROPERTY OF INDUCTANCE

The principle of inductance should be easier for you to understand if you realize that it is essentially the result of these two properties of electromagnetism: 1) That a voltage can be developed in a conductor which is in a changing magnetic field; 2) That a magnetic field is produced by current flowing in a conductor.

Inductance of a Wire

Consider a single piece of wire, as shown in *Figure 8.8a*, through which current is flowing in the direction indicated by the arrow. Suppose current is increasing in value from zero to some maximum, such as an alternating current would during a portion of its cycle. According to the left-hand rule, the changing magnetic field's lines of flux will have a clock-wise direction as indicated by the arrows in *Figure 8.8b,* and will expand as the current increases.

Imagine that the magnetic field starts at the center of the conductor and expands out as the current increases as shown in *Figure 8.8c*. As the magnetic field first develops and expands, it cuts through the body of the conductor itself and induces a voltage that opposes the voltage producing the initial current. In other words, the induced voltage produces a circuit current opposite in direction to the initial current.

This opposing current is called a counter-induced current. This inductive action occurs in any piece of wire through which an electrical current flows.

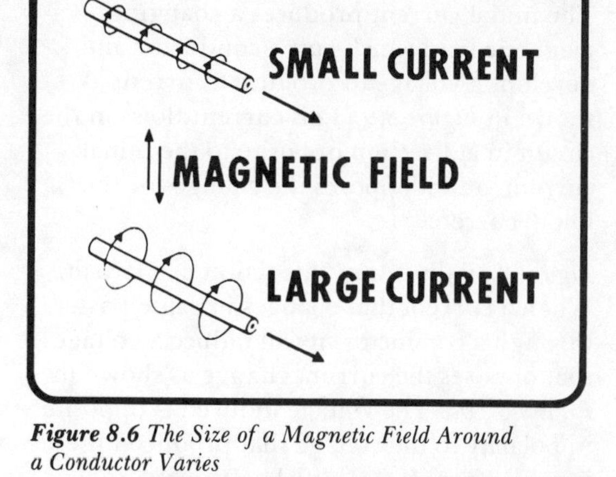

Figure 8.6 *The Size of a Magnetic Field Around a Conductor Varies*

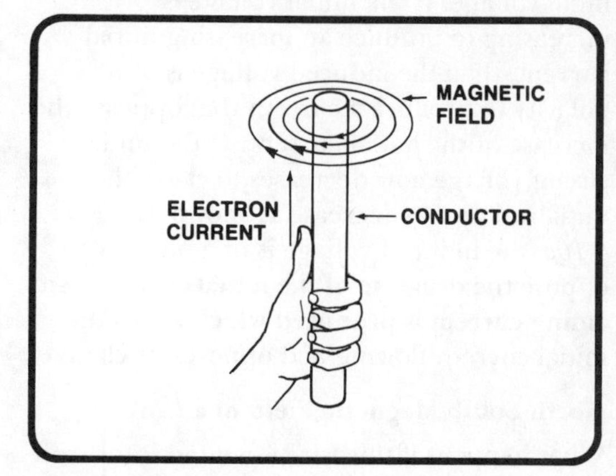

Figure 8.7 *Left-Hand Rule for Conductors*

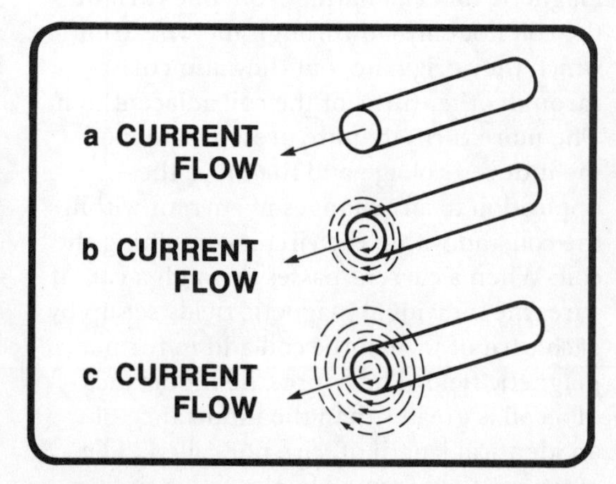

Figure 8.8 *A Length of Wire with an Electrical Current Flowing Through It*

So there are two properties working together. The initial current produces a changing magnetic field which cuts a conductor and develops a voltage to produce a current as shown in *Figure 8.9*. This current flows in the circuit in a direction opposite to the initial current, and it opposes the changes of the initial current.

Figure 8.10 illustrates this action in a circuit. When a current that changes in value passes through a conductor, it will induce a voltage that opposes the current change as shown in *Figure 8.10a.* The voltage induced is opposite in polarity to the voltage that produced the intial voltage. If the initial voltage is increasing to produce an increasing initial current, then the induced voltage is of a polarity to produce a current that opposes the increase of the initial current. If the initial circuit voltage now decreases to cause the initial current to decrease as shown in *Figure 8.10b,* the induced voltage is of a polarity to oppose the decrease of the initial current. An aiding current is produced which keeps the initial current flowing and opposes its change.

North-South Magnetic Field of a Coil

What happens if this wire is wound into a coil as shown in *Figure 8.11*? In a coil the lines of magnetic flux emanating from one turn of the coil not only cut through the wire from which they originate, but they also cut through other turns of the coil adjacent to it. The more turns that are present, the more the induced voltage and the more the opposition to any changes in current within the coil and within the circuit containing the coil. When a current passes through a coil of wire, the individual magnetic fields set up by each turn of wire in the coil add to form a magnetic field. Therefore, the inductance of a coil is greater than the inductance of an identical length of wire not coiled. The magnetic field formed by the coil is similar to that of a bar magnet.

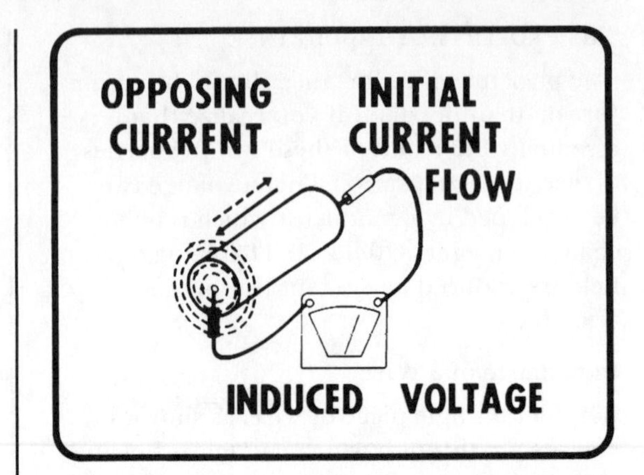

Figure 8.9 *Current Flow and Induced Voltage in a Wire*

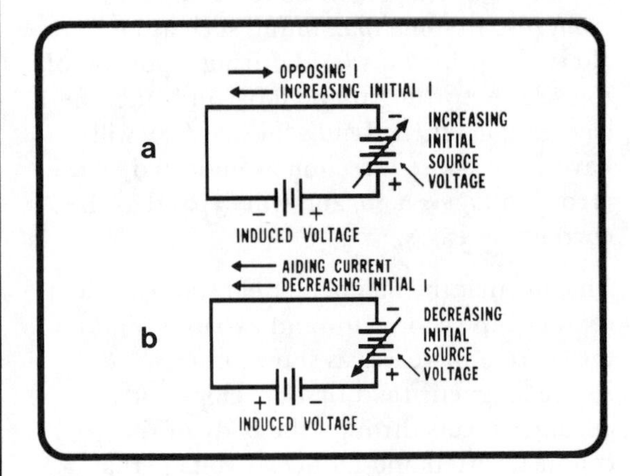

Figure 8.10 *Inductve Action in a Circuit*

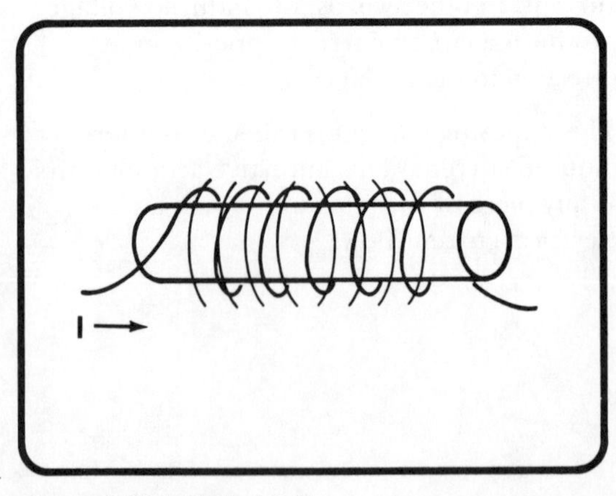

Figure 8.11 *A Typical Coil*

The orientation of the magnetic field around the coil can be determined by using the *left-hand rule for coils*. The rule states that if the fingers of the left hand are wrapped around the coil in the direction of current flow as shown in *Figure 8.12,* the thumb points toward the north pole end of the coil.

Inductance of a Coil

In a coil, the inductive action is multiplied because the lines of magnetic flux emanating from one section of the coil not only cut through the wire from which they originate, but they also cut through the sections of the coil adjacent to it. This is illustrated in *Figure 8.13a* and *8.13b.* The more wire that is present, the more the opposition will be to any changes in current within that wire. Thus, the number of turns of wire determines a coil's inductance value. The magnetic field of a coil is shown in *Figure 8.13c.* It is like the magnetic field of a bar magnet (*Figure 8.13d*).

Counter EMF

If an inductor is placed in a circuit as shown in *Figure 8.14* which has a changing voltage, E_A, applied to it, a current which changes in magnitude will attempt to flow in the circuit. The property of inductance that the inductor exhibits, however, will oppose the *change* of current through it. Thus, a voltage will be set up across the inductor due to the inductive action. This voltage has been given a special name It is called the counter EMF, abbreviated CEMF. This is the voltage that appears across inductors in ac circuits. It is due to the property called *self-inductance.*

Recall that earlier it was stated that a coil must be composed of *turns* of wire in which all turns are insulated one from the other. It should be clear now why this is true. If the turns of wire are not electrically separate, then the coil would not act as an inductor, but as a conductor.

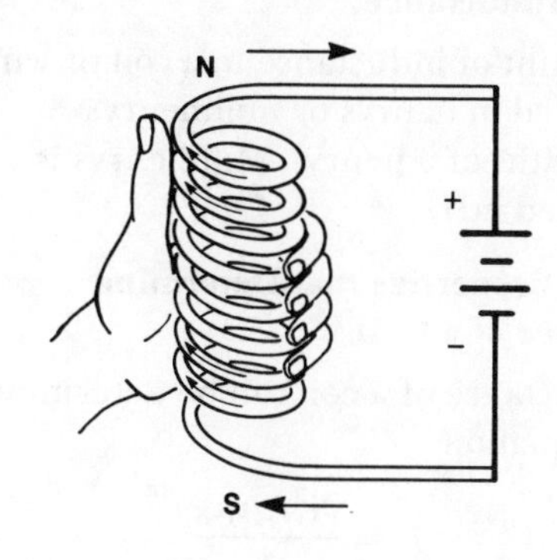

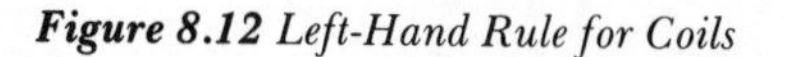
Figure 8.12 *Left-Hand Rule for Coils*

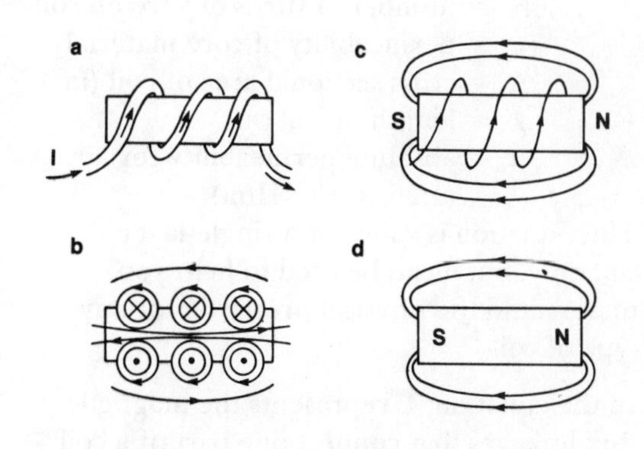

Figure 8.13 *Inductive Action of a Coil*

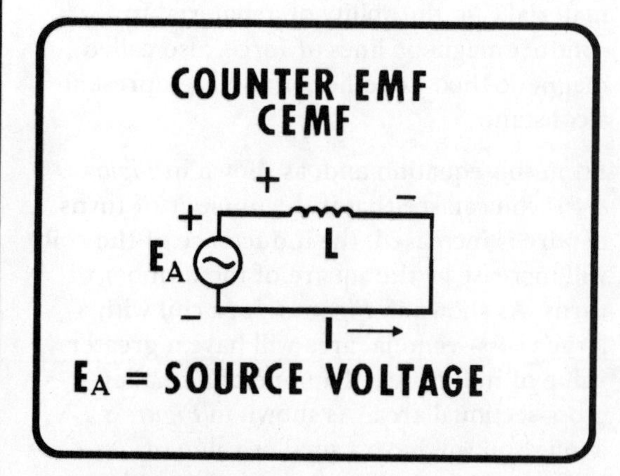

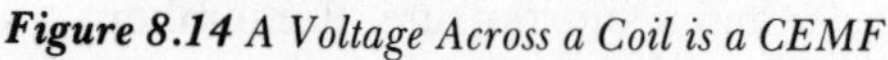
Figure 8.14 *A Voltage Across a Coil is a CEMF*

Units of Inductance

The amount of inductance in a coil of wire is measured in henrys or millihenrys (thousandths of a henry). Millihenrys is abbreviated mH.

Physical Properties that Determine Inductance of a Coil

The inductance of a coil can be determined by this equation:

$$L = \frac{N^2 A \mu \mu_o}{\ell} \qquad (8\text{–}2)$$

where

L = inductance of coil (henrys)
N = number of turns of wire on coil
μ = permeability of core material
A = cross sectional area of coil (m^2)
ℓ = length of coil (m)
μ_o = absolute permeability for air, (1.26×10^{-8} H/m)

This equation is valid for a single-layer coil only, but it can be used to help you understand the physical properties of any type of coil.

In the equation, L represents the magnetic flux linkages that connect one part of a coil to the next part, causing the property of inductance. The permeablity of the core material is μ, the ability of a material to conduct magnetic lines of force, also called magnetic flux. Together, μ and μ_0 represent a constant.

From this equation and, as shown in *Figure 8.15,* you can see that if the number of turns of wire is increased, the inductance of the coil will increase by the square of the number of turns. As shown in *Figure 8.16,* a coil with a large cross-sectional area will have a greater value of inductance than one of a smaller cross-sectional area. As shown in *Figure 8.17,* a long coil will have a smaller value of inductance than a shorter one. Remember that when determining the effect of varying

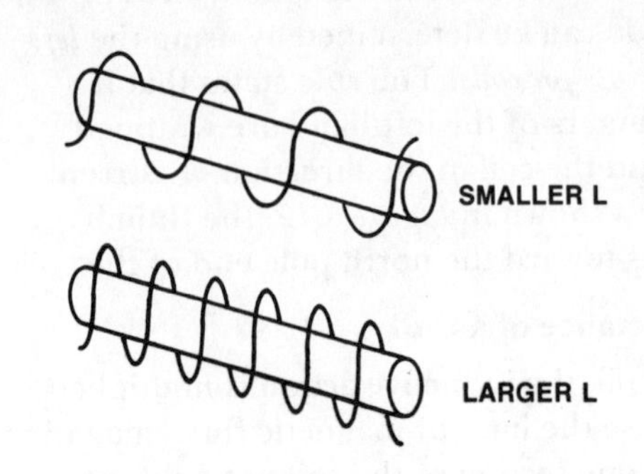

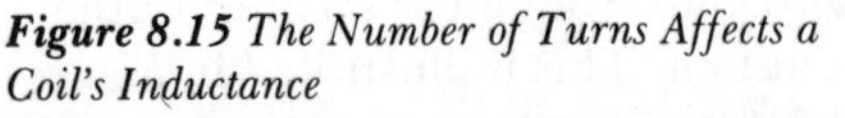

Figure 8.15 *The Number of Turns Affects a Coil's Inductance*

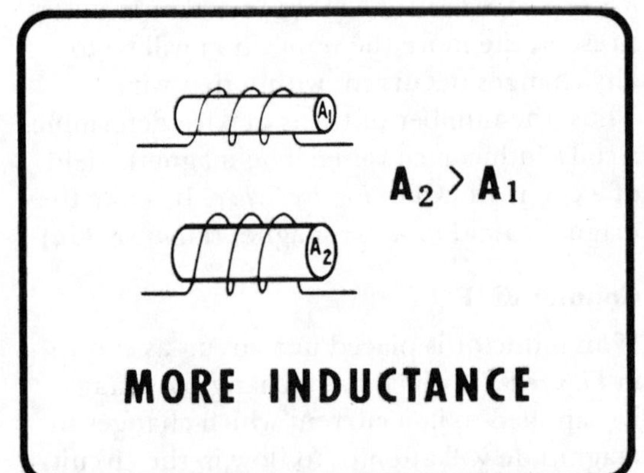

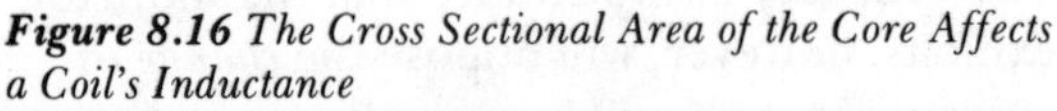

Figure 8.16 *The Cross Sectional Area of the Core Affects a Coil's Inductance*

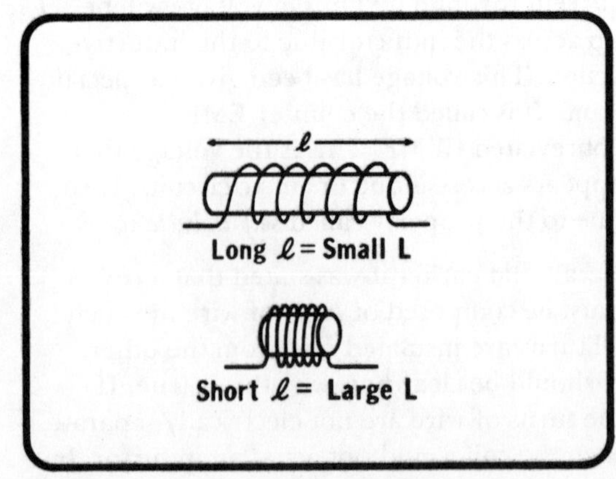

Figure 8.17 *Length Affects a Coil's Inductance*

one specific factor of a coil, that all other factors must be held constant. From equation *8–2*, the value of a manufactured inductor depends basically on these four factors: 1) the number of turns of wire used, 2) the cross-sectional area of the coil, 3) the overall length of the coil, and 4) the permeability of the core material.

Working an example using equation *8–2* should help you learn to use it to determine the inductance of a coil. In this example, as shown in *Figure 8.18,* a coil has 400 turns of wire wound on a round core 0.013 meter (0.5 inch) in diameter, a length of 0.076 meter (3 inches), and uses iron as a core material which has a permeability of 1000. What is the inductance of the coil?

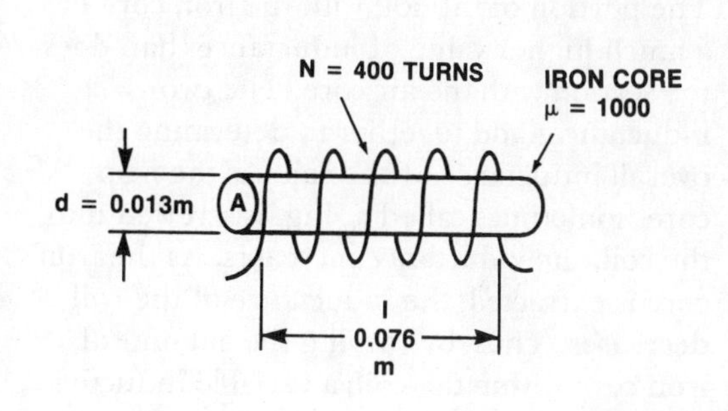

Figure 8.18 *Coil Used in Inductance Equation Example*

First of all, the cross sectional area of the coil is determined.

$$A = \pi r^2$$
$$r = \tfrac{1}{2}d$$
$$r = \tfrac{1}{2}(0.013\text{m})$$
$$= 0.0065\text{m}$$
$$A = \pi r^2$$
$$= (3.14)\,(.0065\text{m})^2$$
$$= 1.33 \times 10^{-4}\text{m}^2$$

Then the inductance can be calculated.

$$L = \frac{N^2 A \mu \mu_o}{\ell} \qquad (8\text{–}2)$$

$$= \frac{(400)^2(1.33 \times 10^{-4}\text{m}^2)(1000)(1.26 \times 10^{-8}\text{H/m})}{(0.076\text{m})}$$
$$= 3.53 \times 10^{-3}\text{H}$$
$$= 3.53\text{mH}$$

Variable Inductors

The fact that the permeability of the core material determines the value of a coil is put to use in manufacturing variable inductors. Iron is the most common core material used in the manufacture of inductors. Iron has a much higher permeability than air. Variable inductors are made by constructing an inductor with an iron core that can be inserted various lengths into the inductor.

Figure 8.19 shows a typical variable inductor. The portion of the coil with the iron core has a much higher value of inductance than does the section with the air core. The two inductances add together to determine the overall inductance of the coil. As the iron core, sometimes called a slug, is screwed into the coil, the inductance increases. As the iron core is extracted, the inductance of the coil decreases. Thus, by varying the amount of iron core within the coil, a variable inductor has been created which can be "tuned' to a specific inductance. This type of variable inductor is called a *permeability-tuned* inductor.

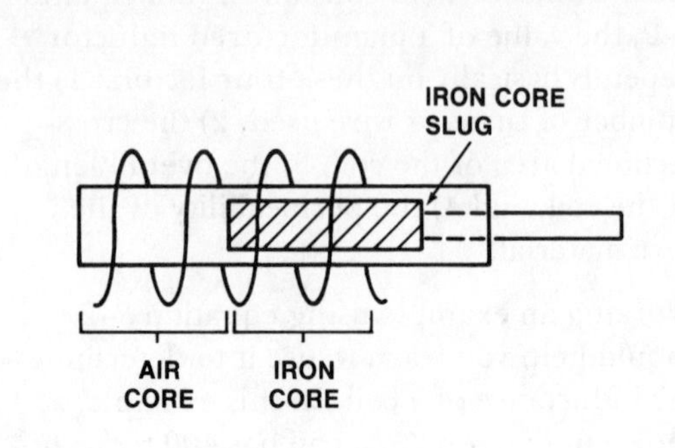

Figure 8.19 *Typical Variable Inductor*

INDUCTANCE IN A CIRCUIT

Effect of an Inductor on a Changing Current in a Circuit

When considering the electrical action of an inductor in a circuit it is helpful to compare its action to that of a resistor. Recall that in a circuit containing only resistors, as shown in *Figure 8.20,* that when a voltage is applied by closing the switch, the full value of current flows almost immediately, and it continues to flow through the resistors as long as the voltage is applied.

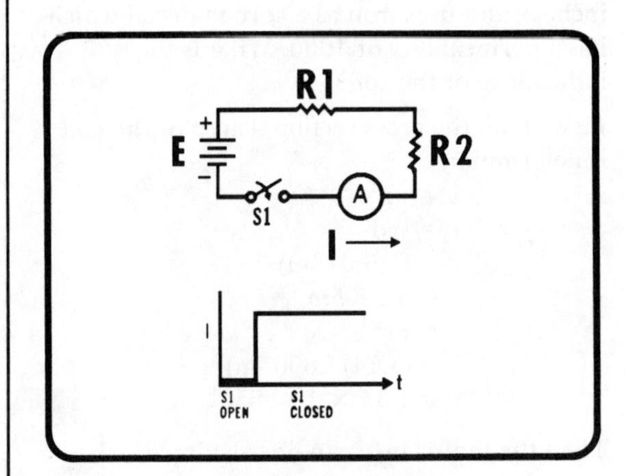

Figure 8.20 *A Typical Resistive Circuit*

If, however, the resistor R1 is replaced with an inductor, as shown in *Figure 8.21*, when the switch is closed the full value of current does not flow immediately. Instead, when the current first begins to increase in value from zero, the inductive action sets up a counter EMF that opposes the current change. When the switch is first closed, in fact, the rapid change in current that starts to flow causes a rapid change in magnetic flux that produces a high counter EMF. Therefore, very little current is allowed to flow initially. At this point in time, the inductance appears as an open circuit as illustrated in *Figure 8.22*.

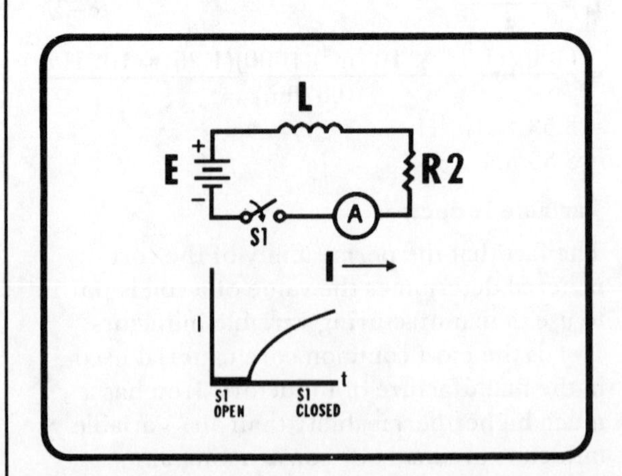

Figure 8.21 *A Typical Resistive-Inductive Circuit*

■ Internal (Winding) Resistance of a Coil
■ Exponential Change of Current

This can be compared to the physical property of inertia. Initially, it is difficult to get the current moving, but once moving, and if forces are not changed, it flows smoothly.

When the self-inductance has stopped the initial flow of current, the self-inductance of the coil ceases since it depends on a change of current. The current again attempts to increase. As it increases slightly, the self-inductance again hinders the flow of current in the circuit. Then, when the self-inductance of the inductor has stopped the increase in current, the self-inductance ceases and the current attempts to increase again. This back-and-forth action continues smoothly until the current in the circuit reaches its maximum value. The maximum current in this circuit is determined by the value of R1 and the internal resistance of the coil.

Internal (Winding) Resistance of a Coil

Every coil has some internal resistance because of the length of the wire used. The internal resistance of coils range from fractions of an ohm to several hundred ohms depending on the length and size (gauge) of wire used. You can measure this internal resistance easily with an ohmmeter.

Exponential Change of Current

Figure 8.23 shows graphically the action of a resistive-inductive (RL) circuit. At first, time t_1, the voltage across the inductor is at maximum since the inductor appears as an open circuit initially. Current is zero. All of the applied voltage appears across the inductor as counter EMF. Gradually, the current begins to increase until it reaches its maximum value. As the current increases, the counter EMF across the inductor decreases.

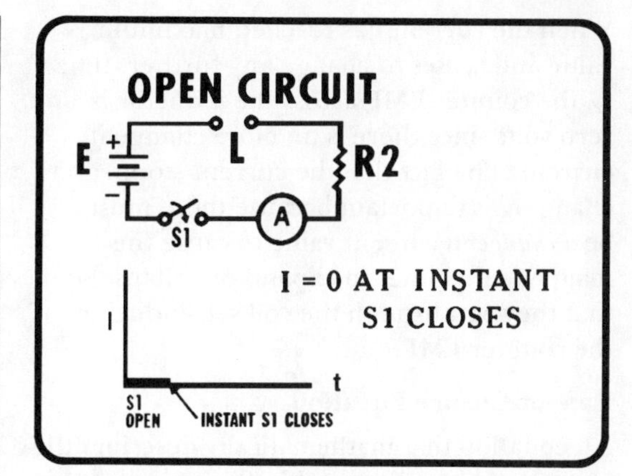

Figure 8.22 When S1 is First Closed the Circuit Appears as an Open Circuit

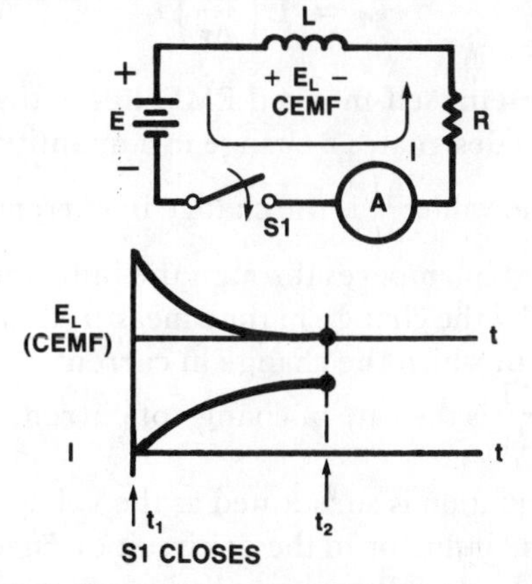

Figure 8.23 The Relationship of E and I in an RL Circuit

The increase of current and decrease of voltage is said to be *exponential*. However, if one examined the circuit action in very small periods of time, it would appear as small steps where current increases, counter EMF opposes it, current increases again, counter EMF opposes it again, and this action continues until the current attains its final value. As shown in *Figure 8.24*, it is, as stated, a step function. But the steps are infinitesimally small, and to us the action appears as a smooth exponential rise in current.

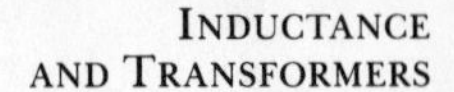

■ Rate-of-Change Equation

When the current has reached maximum value and ceases to change any further, time t_2, the counter EMF across the inductor is zero volts since there is no more change in current. The fact that the current stopped changing is important because there must be a *change* in current value to cause the magnetic flux lines to expand or collapse so that they cut through the coil self-inducing the counter EMF.

Rate-of-Change Equation

An equation that mathematically describes the current-voltage relationship for an inductor is

$$E_L = L\left(\frac{\Delta I}{\Delta t}\right) \qquad (8\text{-}3)$$

E_L represents self-induced EMF. Recall that a Δ symbol designates a change in a quantity. Thus, the value $\frac{\Delta I}{\Delta t}$ is the change in current measured in amperes through the inductor divided by the change in time measured in seconds in which the change in current occurs. $\frac{\Delta I}{\Delta t}$ is the rate of change of current.

If this equation is substituted as the voltage across the inductor in the dc circuit of *Figure 8.21*, and a dc voltage is applied when the switch is closed, then the solution for the circuit current will be the exponential curve shown in *Figure 8.21*.

But what will happen in the circuit of *Figure 8.21* if the dc voltage is replaced with an ac voltage source? *Figure 8.25* shows the circuit with an ac voltage source. In this circuit, the sinusoidal voltage is continually changing in magnitude and polarity, causing the magnetic field of the coil to continously expand and collapse.

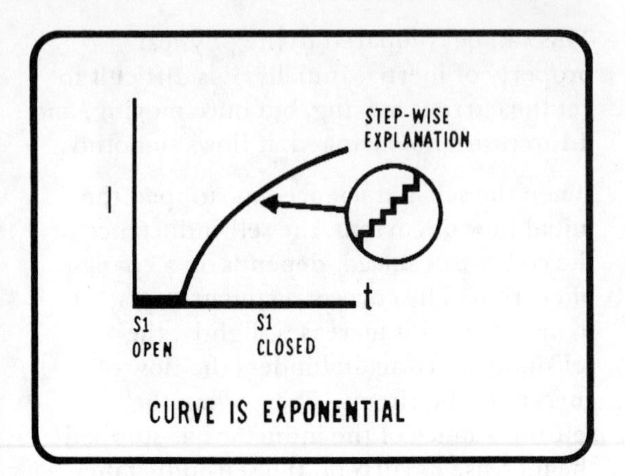

Figure 8.24 *Step-Function Action of Current*

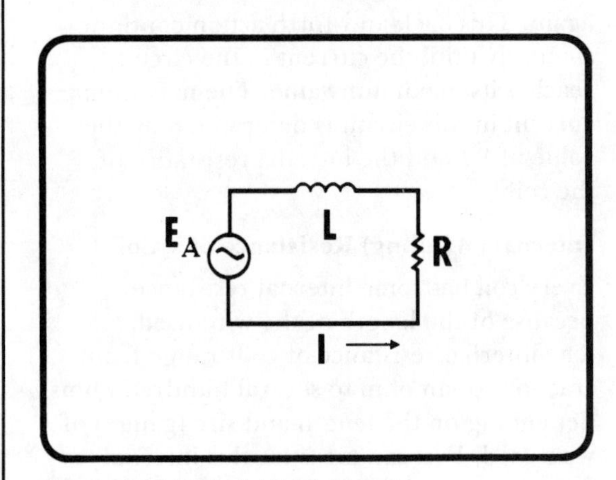

Figure 8.25 *Typical AC RL Circuit*

■ Rate-of-Change Equation

If the phase relationship of the voltage across the inductor, which is the counter EMF, is compared with the current in the inductor, it will be found that they are 90 degrees out of phase as shown in *Figure 8.26*. More specifically, the voltage leads the current by 90 degrees.

The voltage across the inductor, E_L, in the circuit of *Figure 8.25* can be calculated using equation *8–3*. But before this is done one thing should be noted. The rate-of-change of current $\frac{\Delta I}{\Delta t}$, may be abbreviated ROC.

Since any voltage that appears across the inductor is equal to the inductance times the rate of change of current through the inductor, equation *8–3* can be rewritten:

$$E_L = L(\text{ROC of I}). \qquad (8\text{-}4)$$

Discussion of an example will show how this equation describes the current-voltage relationship for an inductor. The sine wave of *Figure 8.27* represents the current passing through the inductor. Since E_L = L(rate of change of current), when the rate of change of current of the sinusoidal waveform is zero at its peak, the value of the voltage is zero. When the rate of change of current of the sinusoidal waveform is maximum as it crosses the zero axis either increasing positively or increasing negatively, the value of the voltage is a positive or negative maximum. Since the current changes sinusoidally, the rate-of-change of current changes sinusoidally and so does the voltage. By connecting these points with a smooth sinusoidal waveform as shown in *Figure 8.27*, the voltage waveform can be plotted. It can be seen that, as stated previously, the voltage leads the current by 90 degrees as shown in *Figure 8.28*. Therefore, it should give you a clear idea of the relationship of a current through and voltage across an inductance.

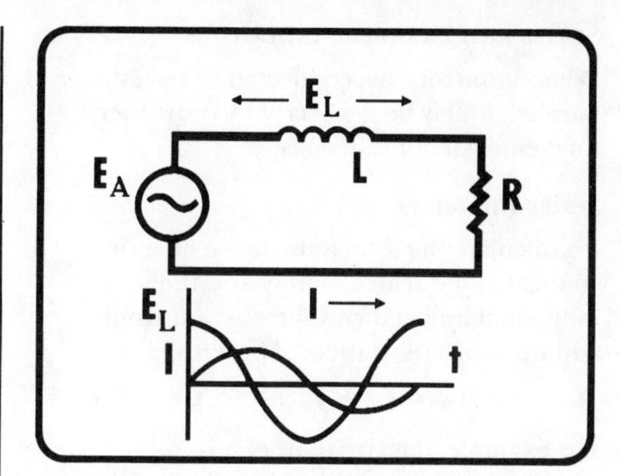

Figure 8.26 *Voltage Leads Current by 90 Degrees*

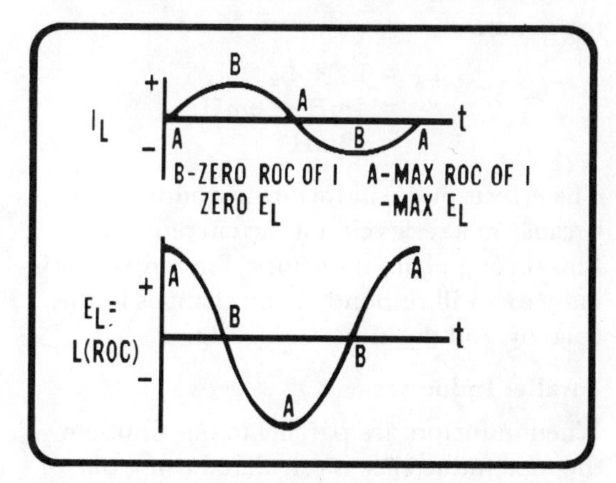

Figure 8.27 *Inductive Voltage and Current Relationship*

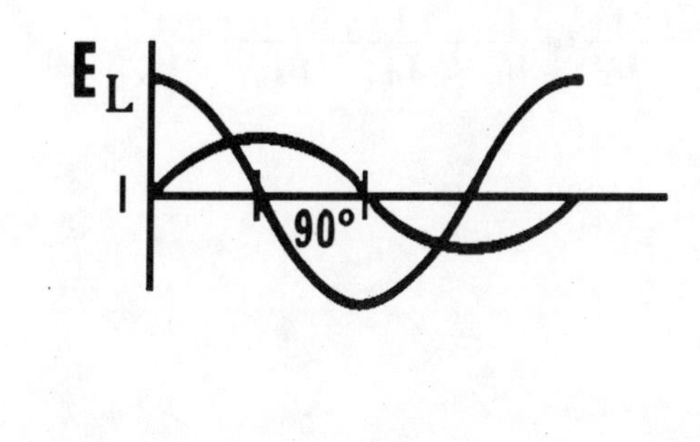

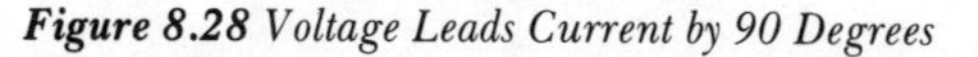

Figure 8.28 *Voltage Leads Current by 90 Degrees*

SERIES AND PARALLEL INDUCTORS

When inductors are connected in series or in parallel, it may be necessary to know their total equivalent inductance.

Series Inductors

To calculate the total inductance of series-connected inductors, simply add their individual inductance values as you would add up series resistances. Therefore,

$$L_T = L_1 + L_2 \ldots + L_N \qquad (8\text{–}5)$$

For example, the circuit of *Figure 8.29* contains two inductors, series-connected. Their total inductance value is simply their sum.

$$\begin{aligned} L_T &= L_1 + L_2 \\ &= 4\text{mH} + 8\text{mH} \\ &= 12\text{mH} \end{aligned}$$

The effects of the inductors are additive because in a series circuit the current must flow through both inductors. Therefore, both inductors will respond to any changes in this current, and the effects are additive.

Parallel Inductors

When inductors are parallel to one another the total inductance is calculated using one of two methods.

If three or more inductances are in parallel then the total inductance is calculated by this reciprocal equation.

$$\frac{1}{L_T} = \frac{1}{L_1} + \frac{1}{L_2} + \frac{1}{L_3} \cdots \frac{1}{L_n}$$

Figure 8.29 *Series Inductor Circuit*

For example, suppose three inductors of 3mH, 6mH and 12mH are in parallel. The total inductance is

$$\begin{aligned} \frac{1}{L_T} &= \frac{1}{3\text{mH}} + \frac{1}{6\text{mH}} + \frac{1}{12\text{mH}} \\ &= \frac{4 + 2 + 1}{12\text{mH}} \\ &= \frac{7}{12\text{mH}} \\ L_T &= \left(\frac{12}{7}\right)\text{mH} \\ &= 1.71\text{mH} \end{aligned}$$

■ Parallel Inductors

If two inductors are in parallel as shown in *Figure 8.30,* then total inductance is calculated by using the reciprocal addition process that was used to add parallel resistances. Like the parallel resistance equation for two resistors, the total inductance for two series inductors is calculated by using a product-over-sum equation.

$$L_T = \frac{L_1 \times L_2}{L_1 + L_2} \qquad (8\text{–}6)$$

If the two inductors are 3mH and 6mH, then, the total inductance of their combination is

$$\begin{aligned} L_T &= \frac{L_1 \times L_2}{L_1 + L_2} \\ &= \frac{(3mH)\,(6mH)}{3mH + 6mH} \\ &= \left(\frac{18}{9}\right) mH \\ &= 2mH \end{aligned}$$

Note that the combination of parallel inductances, L_T, is always less than the smallest individual inductance value. This result is similar to that obtained for a combination of two parallel resistances.

It is interesting to note that the solution for total inductance could be accomplished by taking the inductances "two at a time" and use the product over sum equation. For example, a total inductance of 2mH, as previously calculated, can be used for the combination of 3mH and 6mH in parallel. This resultant can then be used in a product over sum equation with the 12mH inductor to determine the final total inductance. Here is the calculation:

$$\begin{aligned} L_T &= \frac{(2mH)\,(12mH)}{(2mH + 12mH)} \\ &= \left(\frac{24}{14}\right) mH \\ &= \left(\frac{12}{7}\right) mH \\ &= 1.71mH \end{aligned}$$

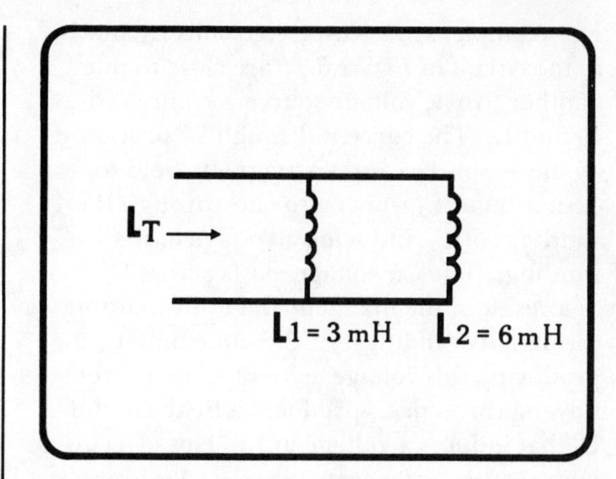

Figure 8.30 Parallel Inductor Circuit

Either method gives equal results.

MUTUAL INDUCTANCE

Although the total inductance of coils connected in series or parallel can be determined using the same technique used for determining the total resistance of resistors connected in series or parallel, inductors have a property not at all similar to resistive circuits. This property is called *mutual inductance.* When the current in an inductor increases or decreases, the magnetic flux field surrounding that inductor changes. The varying flux can cut across any other coil which may be nearby, causing induced voltage to exist in it.

■ Coefficient of Coupling

For example, series-connected coils L_1 and L_2 in the circuit of *Figure 8.31* are close to one another. An ac voltage source is connected to L_1 and L_2. The current through L_1 produced by the ac source causes a magnetic field to exist around L_1 which also cuts through the windings of L_2, inducing a voltage in its windings. Thus, a voltage exists across L_2 as a result of the magnetic field of L_1 cutting through its windings. At the same time L_1 is producing this voltage across L_2, the current flowing through L_2 produces a field around L_2 that induces a voltage in L_1. This effect is known as *mutual inductance.*

Mutual inductance is measured in henrys and is designated by the symbol L_M. The amount of mutual inductance between two coils, L_1 and L_2, can be calculated by this equation:

$$L_M = k\sqrt{L_1 \times L_2} \qquad (8\text{–}7)$$

Where

L_1 = value of L_1 in henrys
L_2 = value of L_2 in henrys
k = coefficient of coupling between the two coils
L_M = mutual inductance in henrys

Coefficient of Coupling

The coefficient of coupling is the fraction of total magnetic flux lines produced by both coils that is common to both coils. Stated as an equation:

$$k = \frac{\phi\text{common}}{\phi\text{total}} \qquad (8\text{–}8)$$

Where

ϕcommon = flux linkages common to L_1 and L_2
ϕtotal = total flux produced by L_1 and L_2

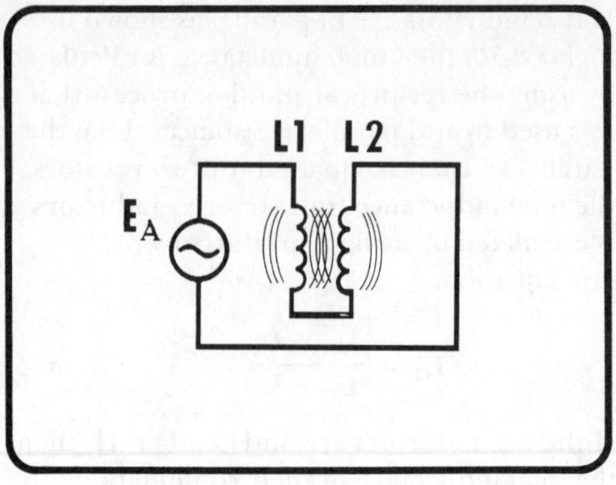

Figure 8.31 Mutual Inductance is Caused by Interaction of Adjacent Magnetic Field

For example, in a circuit there are two coils, L_1 and L_2, which produce 10,000 flux lines, all of which link both coils. Therefore, using equation *8–8* the coefficient of coupling can be calculated:

$$k = \frac{\phi\text{common}}{\phi\text{total}} = \frac{10{,}000}{10{,}000} = 1 \text{ (unity)}$$

In this case, there is unity coefficient of coupling and k = 1. However, if L_1 and L_2 produce 10,000 lines of flux and only 5,000 are common, the coefficient of coupling is

$$k = \frac{\phi\text{common}}{\phi\text{total}} = \frac{5{,}000}{10{,}000} = 0.5$$

Therefore, in this case, the coefficient of coupling is 0.5. There are no units for k since it is a ratio of two amounts of magnetic flux.

■ Calculating Mutual Inductance

A low value of k is called "loose coupling"; a high value of k is called "tight coupling". Coils wound on a common iron core can be considered to have a coefficient of coupling of one or "unity coupling". *Figure 8.32* summarizes the types of coupling and range of coefficients of coupling.

The value of k can be varied by factors relative to the coils. For example, as shown in *Figure 8.33*, k becomes larger (with a maximum of one) as the coils are placed closer together; are placed parallel to one another, as opposed to perpendicular; or are both wound on a common iron core. Conversely, k becomes smaller as the coils are placed farther apart; are placed perpendicular to one another; or are both wound on separate iron cores.

Calculating Mutual Inductance

Mutual inductance is calculated using equation 8–7. *Figure 8.34* shows a circuit with two inductors, L_1 and L_2. L_1 equals 20 henrys and L_2 equals 5 henrys. The coefficient of coupling is 0.4. The coils' mutual inductance is calculated as follows:

$$
\begin{aligned}
L_M &= k\sqrt{(L_1)(L_2)} \\
&= 0.4\sqrt{(20)\,(5)} \\
&= 0.4\sqrt{100} \\
&= 0.4\,(10) \\
&= 4\text{H}
\end{aligned}
$$

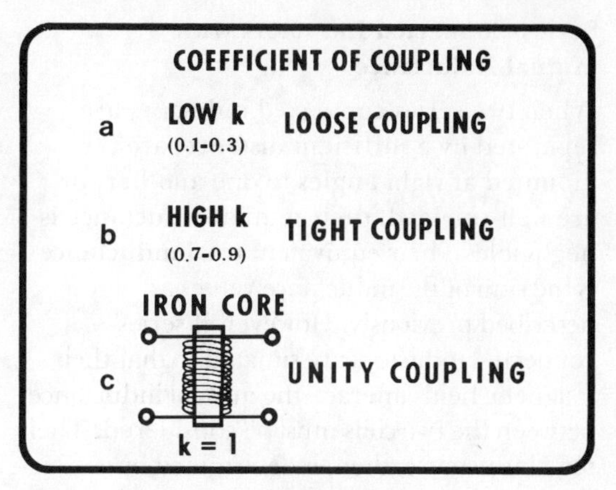

Figure 8.32 Coefficient of Coupling

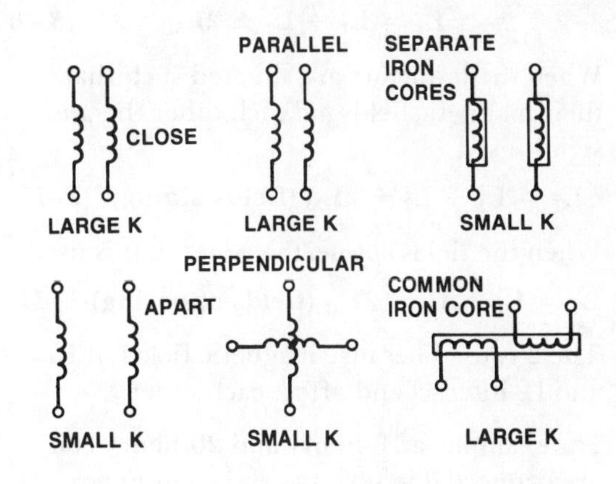

Figure 8.33 Techniques for Varying the Coefficient of Coupling

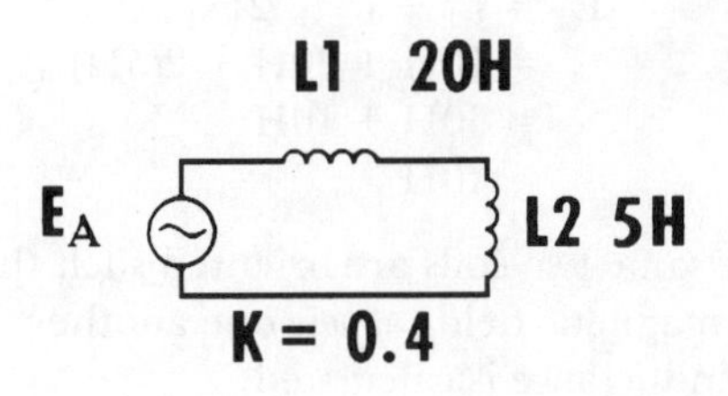

Figure 8.34 Example Circuit for Calculating Mutual Inductance

■ Series-Connected Inductors with Mutual Inductance
■ Parallel-Connected Inductors with Mutual Inductance

Series-Connected Inductors with Mutual Inductance

When two series-connected inductors are separated by a sufficient distance, are mounted at right angles to one another, or are well-shielded, their mutual inductance is negligible. Their equivalent total inductance is the sum of the inductance values as described previously. However, if series-connected inductors are situated so that their magnetic fields interact, the mutual inductance between the two coils must be considered. Their total inductance, then, is determined by this equation:

$$L_T = L_1 + L_2 \pm 2L_M \qquad (8\text{–}9)$$

When the inductors are situated such that their magnetic fields *aid* each other the *plus* sign is used.

$$L_T = L_1 + L_2 + 2L_M \text{ (fields aiding)} \qquad (8\text{–}10)$$

When the fields *oppose*, the *minus* sign is used.

$$L_T = L_1 + L_2 - 2L_M \text{ (fields opposing)} \qquad (8\text{–}11)$$

The 2 is used because magnetic fields of L_1 and L_2 interact and affect each other.

For example, a 10-henry and 20-henry coil are connected in series as shown in *Figure 8.35* such that their magnetic fields *aid* one another, and they have a mutual inductance of 5 henrys. Their total inductance is calculated:

$$\begin{aligned} L_T &= L_1 + L_2 + 2L_M \\ &= 10H + 20H + 2(5H) \\ &= 30H + 10H \\ &= 40H \end{aligned}$$

If the same two coils are oriented such that their magnetic fields *oppose* one another, their total inductance is calculated:

$$\begin{aligned} L_T &= L_1 + L_2 - 2L_M \\ &= 10H + 20H - 2(5H) \\ &= 30H - 10H \\ &= 20H \end{aligned}$$

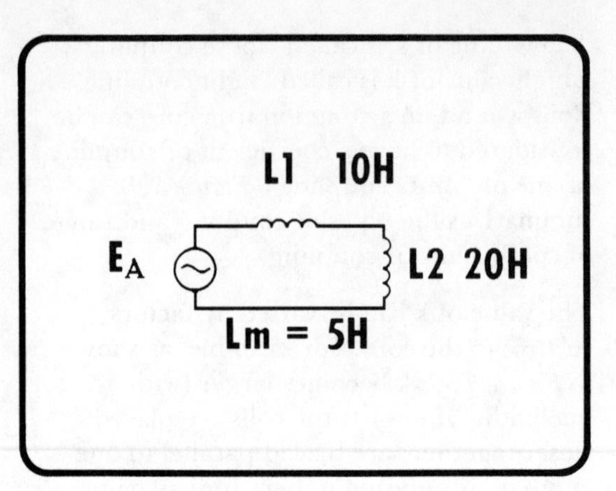

Figure 8.35 *Example Circuit for Calculating Total Inductance of Series-Aiding and Series-Opposing Inductors*

Parallel-Connected Inductors with Mutual Inductance

When two coils are parallel-connected without any mutual inductance, their equivalent total inductance is found by the reciprocal addition or product-over-sum method discussed earlier. However, if their magnetic fields interact, their equivalent total inductance is found by one of the following equations:

Parallel-aiding connection—
If the two coils are oriented such that their magnetic fields *aid* each other.

$$L_T = \frac{(L_1 + L_M)(L_2 + L_M)}{L_1 + L_2 + 2L_M} \qquad (8\text{–}12)$$

Parallel-opposing connection—
If the two coils are oriented such that their magnetic fields *oppose* each other.

$$L_T = \frac{(L_1 - L_M)(L_2 - L_M)}{L_1 + L_2 - 2L_M} \qquad (8\text{–}13)$$

Using the same 10 henry and 20 henry coils with a mutual inductance of 5 henrys due to an aiding field, but connected in parallel, the total inductance using equation *8–12* is

$$L_T = \frac{(10mH + 5mH)(20mH + 5mH)}{10mH + 20mH + 2(5mH)}$$

$$= \left(\frac{(15)(25)}{40}\right) mH$$

$$= 9.38mH$$

TRANSFORMERS

Basic Construction

A device in which the property of mutual inductance is put to practical use is the transformer. A typical transformer is shown in *Figure 8.36*. A typical standard transformer consists of two separate coils, wound on a common iron core as shown in the schematic of *Figure 8.37* and considered to have a coefficient of coupling of one. One coil is called the primary; the other is called the secondary. As a result of mutual inductance, a changing voltage across the primary will induce a changing voltage in the secondary. Thus, if the primary winding is connected to an ac source and the secondary to a load resistor, the transformer is able to transfer power from the primary to the secondary to the load resistance as illustrated in *Figure 8.38*. By having more or fewer turns in the secondary as compared to the primary, the primary voltage may be either stepped-up or stepped-down to provide the necessary operating voltage for the load.

Figure 8.36 *A Typical Transformer*

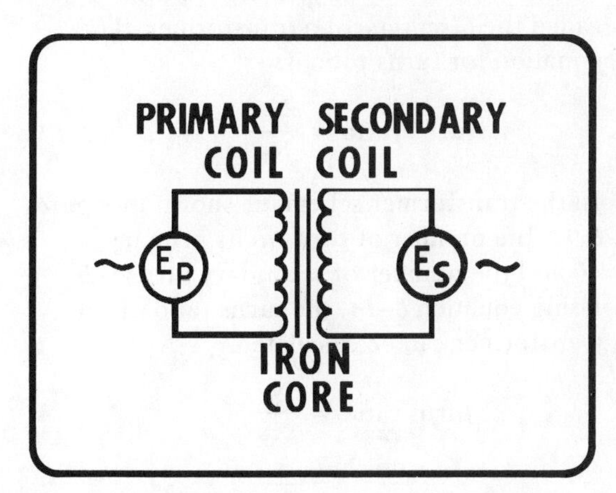

Figure 8.37 *Schematic Drawing of a Transformer*

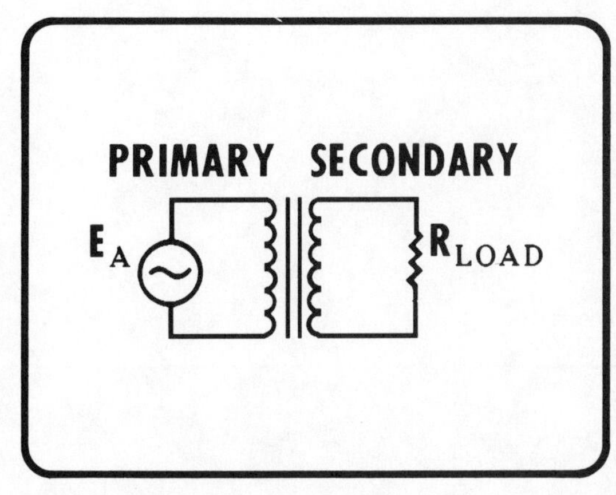

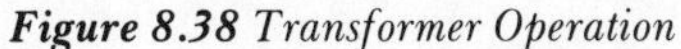

Figure 8.38 *Transformer Operation*

■ Turns Ratio versus Voltage

Turns Ratio versus Voltage

Recall that if a coil has a larger number of turns, a larger voltage is induced across the coil. With a smaller number of turns the voltage is less. Therefore it is easy to see that by having more or fewer turns in the secondary as compared to the primary, as shown in *Figures 8.39* and *8.40*, the voltage may either be stepped up or stepped down to provide the necessary operating voltage for the load.

The ratio of the number of turns in a transformer secondary winding to the number of turns in its primary winding is called the *turns ratio* of a transformer. The equation for turns ratio is:

$$\textbf{turns ratio} = \frac{N_S}{N_P} \qquad (8\text{–}14)$$

In the transformer schematic shown in *Figure 8.41*, the number of turns in its primary is 10 and the number of secondary turns is 5. Using equation *8–14*, the turns ratio of the transformer can be calculated.

$$\text{turns ratio} = \frac{N_S}{N_P} = \frac{5}{10} = \frac{1}{2}$$

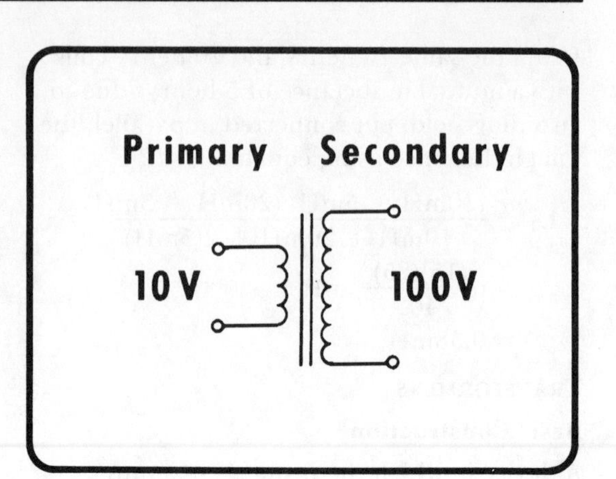

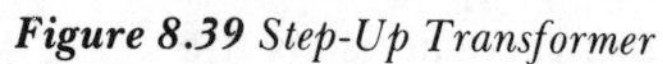
Figure 8.39 *Step-Up Transformer*

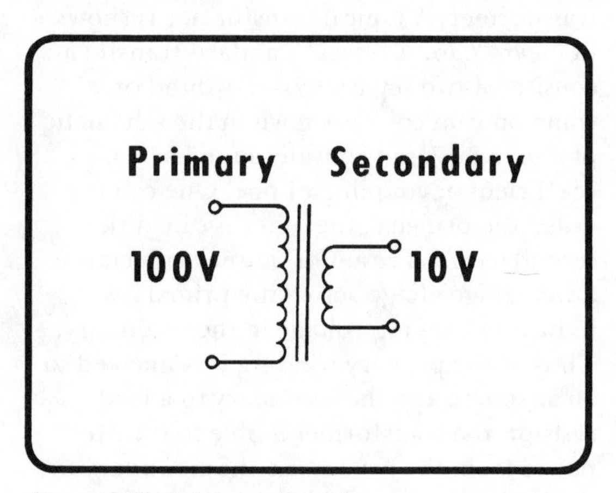

Figure 8.40 *Step-Down Transformer*

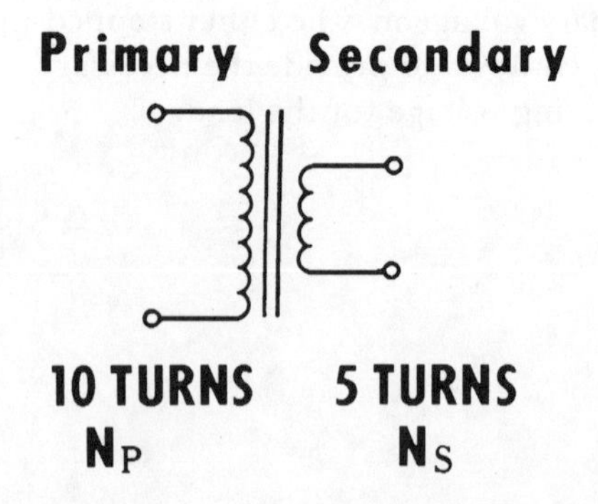

Figure 8.41 *Transformer Used to Calculate Turns Ratio*

■ Turns Ratio versus Voltage

Transformers have a unity coefficient of coupling. Therefore, the voltage induced in each turn of the secondary winding (E_{iS}) is the same as the voltage *self-induced* (E_{iP}) in each turn of the primary, as shown in *Figure 8.42*. The voltage self-induced in each turn of the primary equals the voltage applied to the primary divided by the number of turns in the primary. This can be written:

$$E_{iP} = \frac{E_P}{N_P} \qquad (8\text{–}15)$$

Figure 8.43 shows a schematic of a transformer in which there are 8 turns in the primary and 8 volts ac is applied to it. Using equation *8–15*, the voltage self-induced in each primary turn can be calculated.

$$E_{iP} = \frac{E_P}{N_P} = \frac{8}{8} = 1\text{V}$$

In this example, one volt is induced in each turn of the primary.

If each turn of the secondary has the same voltage induced in it, then the secondary voltage is equal to the number of secondary turns times the induced voltage. This can be written

$$E_S = N_S\left(\frac{E_P}{N_P}\right) \qquad (8\text{–}16)$$

Or rearranging,

$$E_S = E_P\left(\frac{N_S}{N_P}\right) \qquad (8\text{–}17)$$

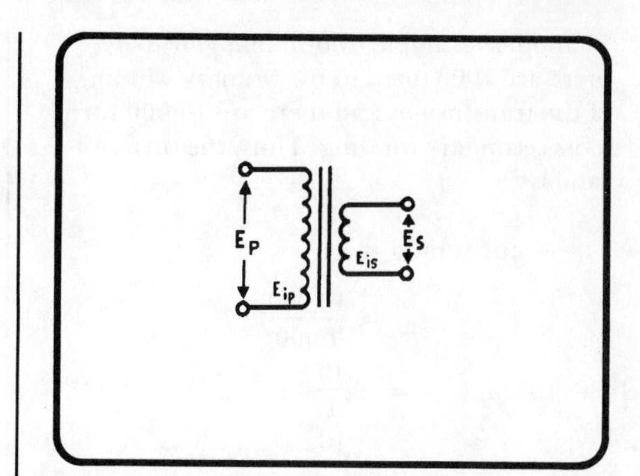

Figure 8.42 *Transformer Voltage Induction*

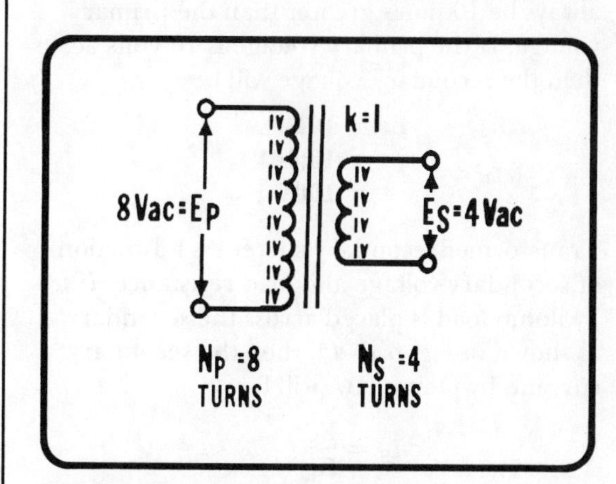

Figure 8.43. *Example Transformer Used to Calculate Self-Induced Voltage in Primary Turns*

The transformer shown in *Figure 8.43* has 4 turns in its secondary. Using equation *8–16*, the secondary voltage can be calculated.

$$E_S = N_S\left(\frac{E_P}{N_P}\right) = 4\left(\frac{8}{8}\right) = 4\text{V}$$

The transformer's secondary voltage is 4 volts — 4 turns times 1 volt per turn.

Turns Ratio versus Voltage

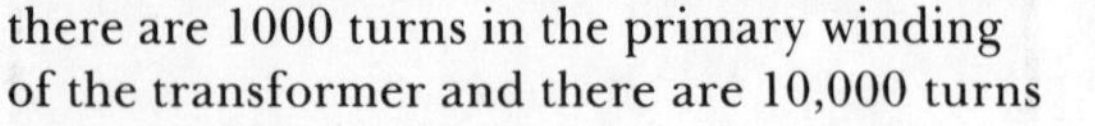

In another example, shown in *Figure 8–44*, there are 1000 turns in the primary winding of the transformer and there are 10,000 turns in its secondary winding. Thus, the turns ratio is

$$\begin{aligned}\text{turns ratio} &= \frac{N_S}{N_P}\\ &= \frac{10{,}000}{1{,}000}\\ &= \frac{10}{1}\\ &= 10\end{aligned}$$

Therefore, the secondary voltage would always be 10 times greater than the primary voltage. If the primary voltage is 10 volts ac, then the secondary voltage will be

$$\begin{aligned}\mathbf{E_S} &= \mathbf{10E_P}\\ &= \mathbf{10(10V)}\\ &= \mathbf{100V}\end{aligned}$$

Transformer secondary current is a function of secondary voltage and load resistance. If a 1 kilohm load is placed across the secondary as shown in *Figure 8–45*, then the secondary current, by Ohm's law, will be

$$\begin{aligned}\mathbf{I_S} &= \mathbf{\frac{E_S}{R_L}}\\ &= \mathbf{\frac{100V}{1k\Omega}}\\ &= \mathbf{0.1A}\\ &= \mathbf{100mA}\end{aligned}$$

The secondary current is 100mA. The transformer secondary acts as an ac voltage source to the load.

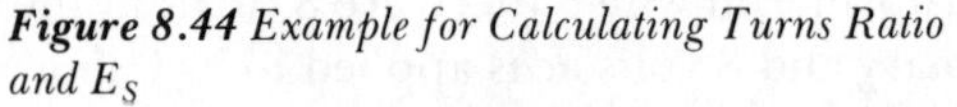

Figure 8.44 *Example for Calculating Turns Ratio and E_S*

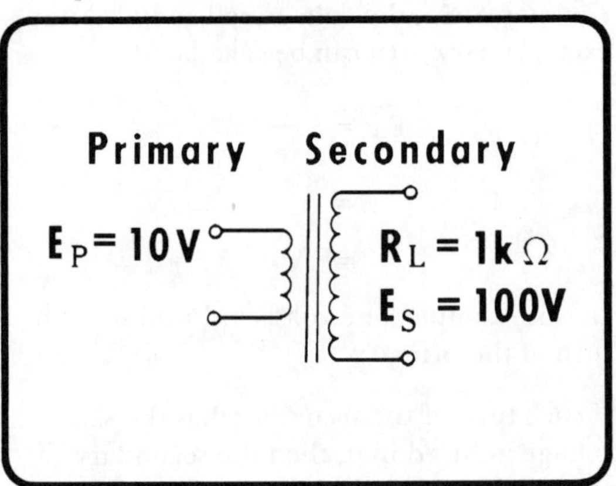

Figure 8.45 *Example for Calculating Transformer I_S*

Primary-to-Secondary Current Relationship

Modern transformers, with coefficient of coupling considered to be one, and with no real power consumed in the windings or the core can be considered to have no loss, as shown in *Figure 8.46*. Therefore, the power in the primary is considered to be the same as the power in the secondary, $P_P = P_S$. Since $P = EI$,

$$P_P = P_S$$
$$E_P I_P = E_S I_S$$

Rewriting this,

$$\frac{I_P}{I_S} = \frac{E_S}{E_P} \quad (8\text{–}18)$$

Note that the current relationship is the inverse of the voltage relationship. Thus, if the voltage is stepped up in a transformer by a factor of 10, the current must have been stepped down the same factor. This may be stated another way using equations *8–17* and *8–18*.

Since $$\frac{I_P}{I_S} = \frac{E_S}{E_P} = \frac{N_S}{N_P} \quad (8\text{–}19)$$

Then $$\frac{I_P}{I_S} = \frac{N_S}{N_P}$$

Or $$I_P = \left(\frac{N_S}{N_P}\right) I_S$$

Thus, in the example shown in *Figure 8.45*, if E_P is 10 volts, E_S is 100 volts, and if I_S, the secondary current, is 100 milliamperes, the primary current, I_P, is calculated as:

$$\begin{aligned} I_P &= \left(\frac{N_S}{N_P}\right) I_S \\ &= \left(\frac{100}{10}\right) 100\text{mA} \\ &= (10)\ 100\text{mA} \\ &= 1000\text{mA} \\ &= 1\text{A} \end{aligned}$$

The primary current in the transformer is one ampere.

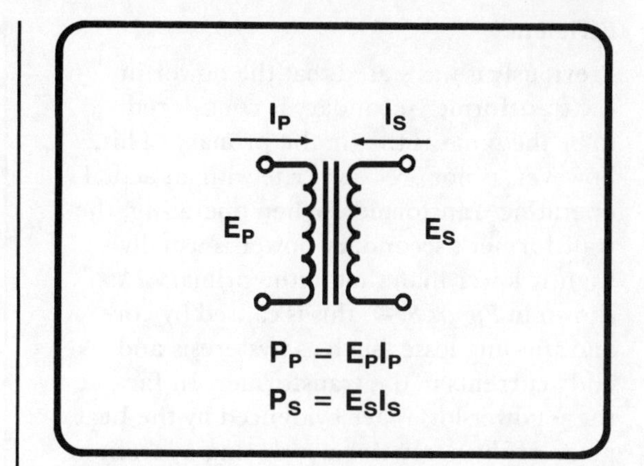

Figure 8.46 Relationship of Transformer Primary and Secondary Windings

Performing the following calculations it can be determined that both the primary and secondary power are equal; both are 10 watts.

$$\begin{aligned} P_P &= E_P I_P \\ &= (10\text{V})\ (1\text{A}) \\ &= 10\text{W} \\ P_S &= E_S I_S \\ &= (100\text{V})\ (100\text{mA}) \\ &= 10{,}000\text{mW} \\ &= 10\text{W} \end{aligned}$$

The transformer, then, either steps up or steps down the voltage and current, but conserves power from the primary to the secondary.

However, transformers do not affect the frequency of the ac voltage they act upon. If the frequency of the primary voltage and current is 60 hertz, then the secondary voltage and current will have a 60 hertz frequency.

Recall that a transformer will not operate with a dc voltage. That is because dc voltage is non-changing and cannot produce an expanding or collapsing magnetic field to cut the secondary windings to produce a secondary voltage.

■ Efficiency
■ Isolation

Efficiency

Previously it was stated that the power in the transformer secondary is considered to be the same as that in the primary. This, however, is not precisely true with an actual operating transformer. When operating, the transformer's secondary power is usually slightly lower than that in the primary. As shown in *Figure 8.47,* this is caused by core and winding losses such as hysteresis and eddy currents in the transformer. In fact, these power losses are evidenced by the heat given off by a transformer during operation.

How well a transformer delivers power from the primary to the secondary circuit is called its *efficiency*. Efficiency is defined as the ratio of the power in the secondary circuit divided by the power in the primary circuit. Because efficiency is usually stated as a percent, it can be written

$$\text{Percent Efficiency} = \left(\frac{P_S}{P_P}\right) 100 \qquad (8\text{–}20)$$

To determine the actual percent efficiency of a transformer, primary voltage and current and secondary voltage and current are measured, the primary and secondary power are calculated and using equation *8–20*, the percent efficiency is calculated. Typical transformer efficencies are about 80 to 90 percent.

Isolation

Another feature of the transformer is that it provides electrical *isolation* between primary and secondary windings, as shown in *Figure 8.48*. That is, there is no electrical connection between the primary and secondary windings of the transformer. (Only the electromagnetic field links the secondary with the primary). When a circuit is connected directly to the power source as shown in *Figure 8.49a,* no electrical isolation exists between the circuit

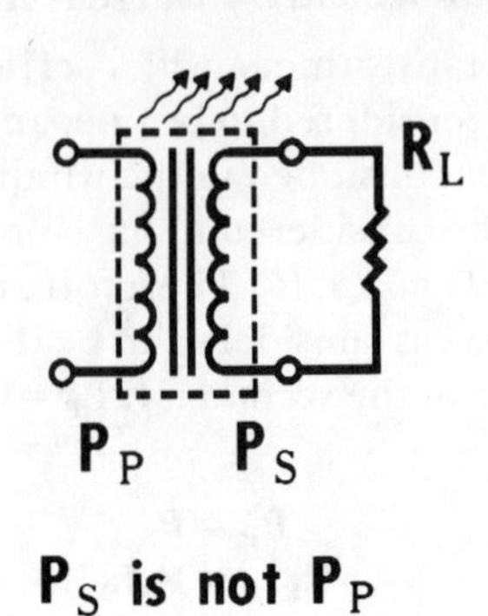

Figure 8.47 In Operation, A Transformer Loses Power in the Form of Heat

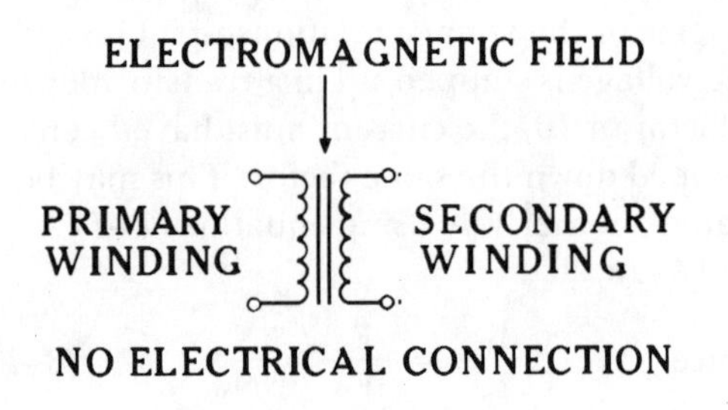

Figure 8.48 A Transformer's Secondary Winding is Isolated from its Primary Winding

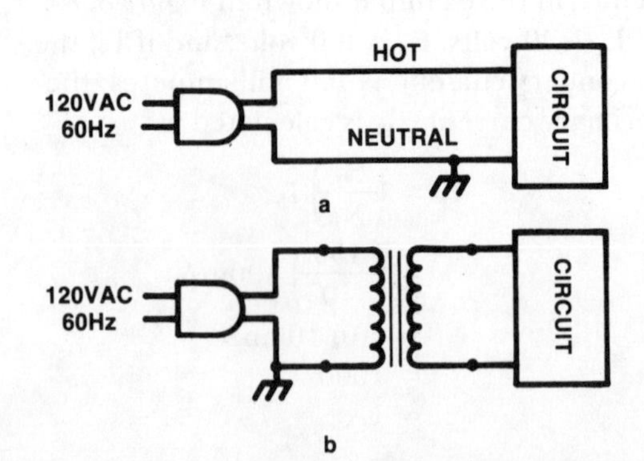

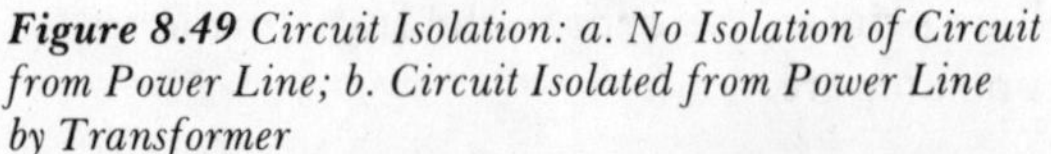

Figure 8.49 Circuit Isolation: a. No Isolation of Circuit from Power Line; b. Circuit Isolated from Power Line by Transformer

and power source, which can present a shock hazard. But any circuit connected to the secondary of a transformer, as shown in *Figure 8.49b,* is electrically isolated from the primary, and thus, the power source. This prevents a shock hazard from existing regardless of the connection of the primary to the electrical ac power line. In fact, transformers designed to simply provide isolation with no voltage change from primary to secondary (i.e. a 1:1 turns ratio) are referred to as "isolation transformers". A schematic of an isolation transformer is shown in *Figure 8.50.*

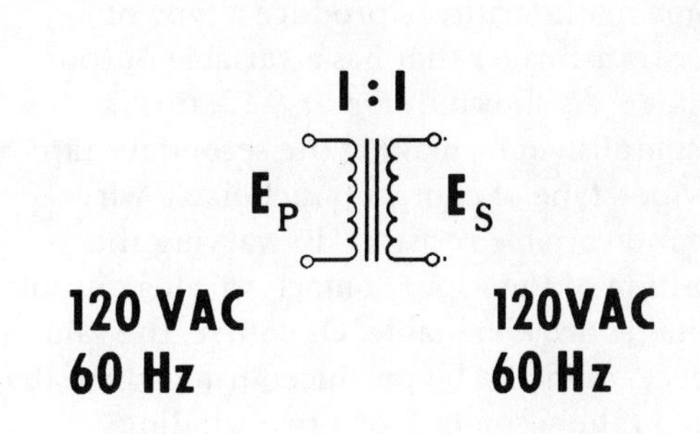

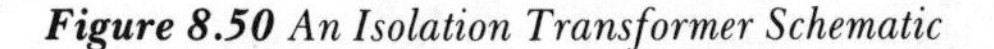

Figure 8.50 *An Isolation Transformer Schematic*

Autotransformers

Transformers do exist, however, which do not provide isolation. These use a single winding and are referred to as autotransformers. As shown in *Figure 8.51*, an autotransformer utilizes a single tapped winding with one lead common to both input and output sides of the transformer. Therefore, there is no isolation. The voltages present across the primary and secondary portions of the transformer are proportional to the number of turns across which they appear. These are calculated the same as in a transformer with isolated windings by using equation *8–17*.

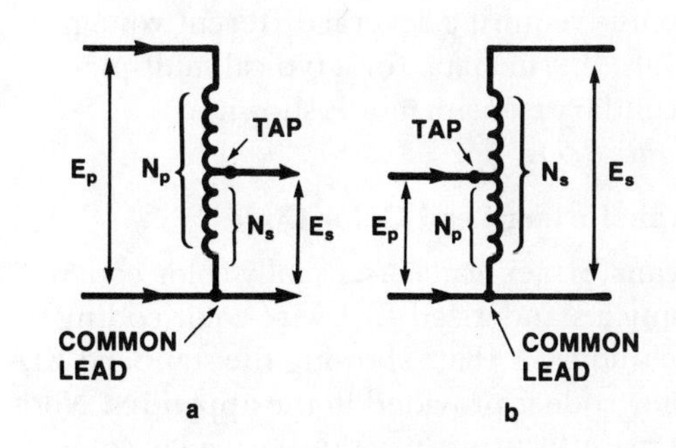

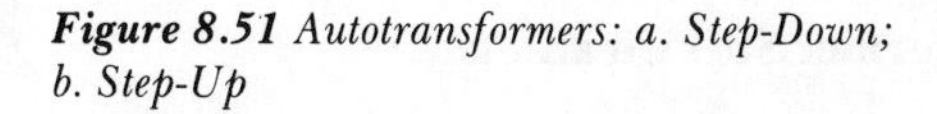

Figure 8.51 *Autotransformers: a. Step-Down; b. Step-Up*

The obvious advantage of this type of transformer is that only one winding is needed. But a less obvious advantage is that an autotransformer provides a more constant output voltage under varying load conditions than does the two-winding type of transformer. This is because the primary and secondary currents are 180 degrees out of phase and tend to cancel in the portion of the winding that is common to both primary and secondary.

Variable-output Transformers

Some manufacturers produce a type of autotransformer that has a variable output voltage. As shown in *Figure 8.52*, this is accomplished by making the secondary tap a wiper-type of contact (much like a wire-wound variable resistor). By varying the position of the wiper contact, various output voltages are obtainable. Of course, the same effect could also be produced using a variable tap on the secondary of a two-winding transformer as shown in *Figure 8.53*.

Multiple-secondary Transformers

Transformers are also produced which have multiple-secondary and center-tapped secondary windings in order to provide for circuits requiring several different voltage levels. A schematic for a typical multiple-secondary transformer is shown in *Figure 8.54*.

Transformer Lead Color Code

Transformer leads are usually color coded using a standarized EIA wire color coding technique. A chart showing the standard EIA color code is provided in the appendix. Not all manufacturers use this particular color code so there will be some variation.

Transformer Specifications

Manufacturers provide specifications for transformers. The specifications enable a user to select a transformer that best meets the requirement of the application. Transformer specifications usually include primary voltage and frequency, secondary voltage(s), impedance, dc winding resistances, and current capabilities. For example, the power transformer of *Figure 8.54* has the following specifications:

Primary voltage: 117V, 60Hz
High-voltage secondary: 240V-0-240V (center-tapped) 150mA
Low-voltage secondary: 6.3V, 2A
Low-voltage secondary: 5V, 3A

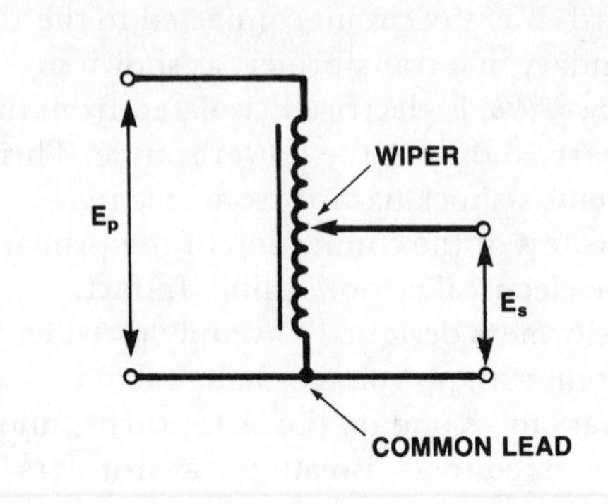

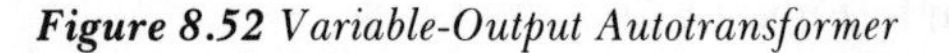

Figure 8.52 *Variable-Output Autotransformer*

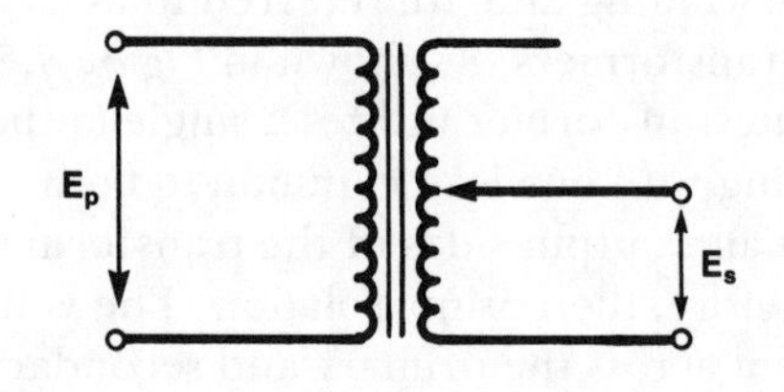

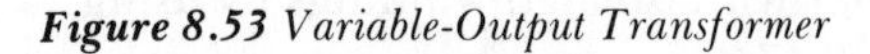

Figure 8.53 *Variable-Output Transformer*

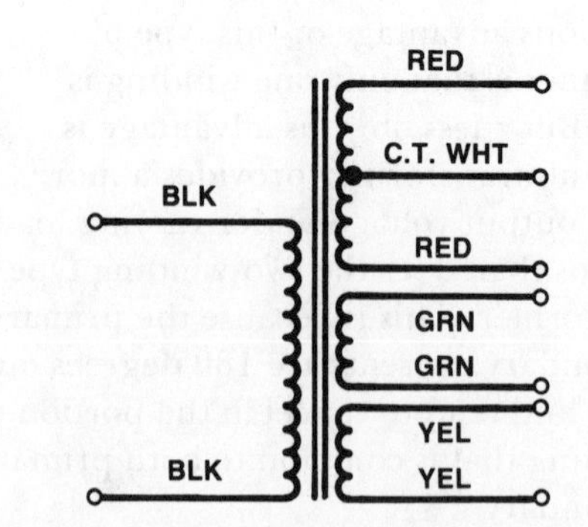

Figure 8.54 *Multiple-Secondary Power Transformer*

■ Transformer Specifications

Power transformers are multiple secondary winding transformers with both high and low voltage secondaries. Typical power, audio, and filament transformers are shown in *Figure 8.55*. Power transformers originally were developed for use with vacuum tube circuits in which high voltage for power supply levels and low voltage for vacuum tube filaments (heaters) were needed. The primary ratings specify the voltage and frequency at which the transformer is designed to be operated. The secondary ratings specify the voltages available from the various secondary windings as well as the maximum current which the secondaries can supply.

Audio transformers are designed for input/output audio applications and are rated according to their primary and secondary impedances, power capabilities (wattage), and turns ratio. They have only a single secondary winding.

Filament transformers are single secondary, low voltage, high current (several amperes, typically) transformers rated according to their primary voltage and frequency, secondary output voltage and maximum output current capabilities.

Figure 8.55 *Typical Power, Audio, and Filament Transformers*

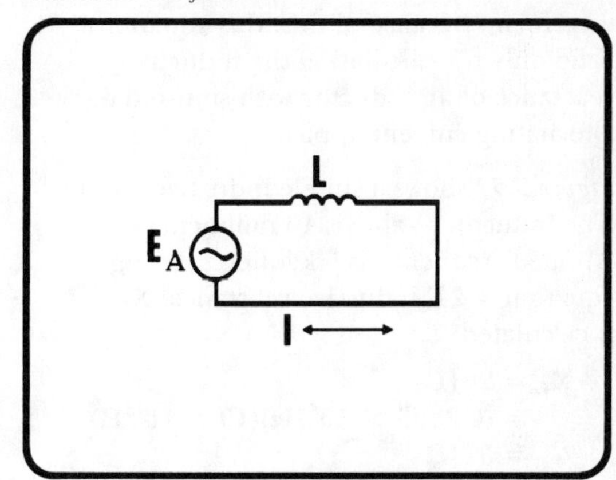

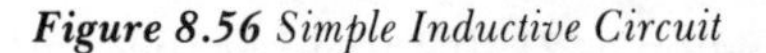

Figure 8.56 *Simple Inductive Circuit*

INDUCTIVE REACTANCE

Now that inductance, self-inductance, and transformer action have been discussed, the next step is a discussion of the effect of an inductor in an ac circuit.

Inductance is measured and inductors are rated in henrys. An inductor's effect in a circuit depends on the inductance and is expressed in a quantity called *inductive reactance*. Inductive reactance is a quantity that represents the opposition that a given inductance presents to an ac current in a circuit, such as is shown in *Figure 8.56*.

■ Inductive Reactance

Like capacitive reactance, it is measured in ohms and depends upon the frequency of the applied ac voltage and the value of the inductor. Inductive reactance can be expressed as follows:

$$X_L = 2\pi fL \qquad (8\text{–}21)$$

Where

X_L = inductive reactance (ohms)
2π = 6.28
f = frequency(Hz)
L = inductance(H)

The constant of 2π comes from the number of radians in one cycle of a sinusoidal ac waveform. Because of this, this equation is valid only for calculating the inductive reactance of an inductor with sinusoidal alternating current applied.

Figure 8.57 shows a simple inductive circuit. The inductor's value is 10 millihenrys. Applied frequency is 5 kilohertz. Using equation *8–21*, inductive reactance, X_L, is calculated:

$$\begin{aligned} X_L &= 2\pi fL \\ &= (6.28)(5 \times 10^3\text{Hz})(10 \times 10^{-3}\text{H}) \\ &= 314\Omega \end{aligned}$$

Note from equation *8–21* that if either the frequency or the inductance is increased the inductive reactance increases. *Figure 8.58* shows graphically how a change in either the frequency or inductance changes the inductive reactance, X_L. Note that the inductive reactance increases linearly with frequency and inductance. As the frequency or inductance increases, the inductor's opposition to the flow of current increases.

These plots of inductive reactance versus frequency and inductive reactance versus inductance shown in *Figures 8.59* and *8.60* will be examined more closely to help you understand these relationships more clearly.

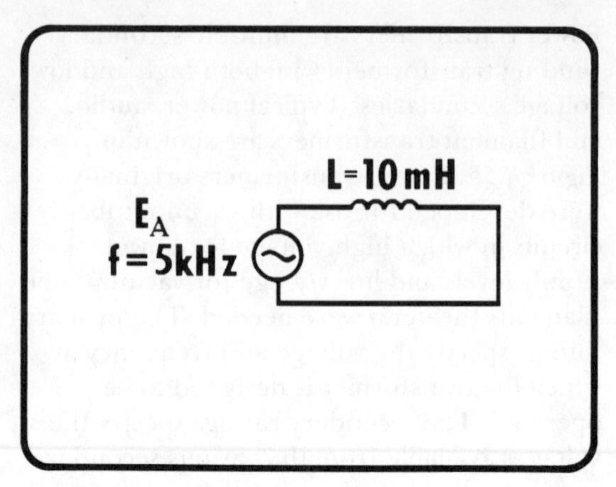

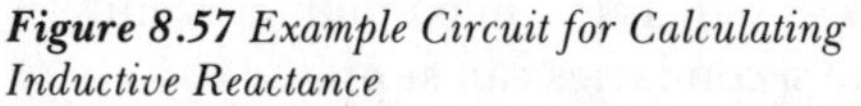

Figure 8.57 *Example Circuit for Calculating Inductive Reactance*

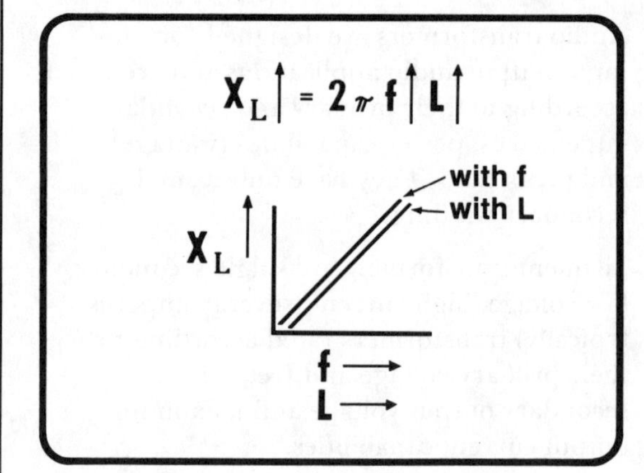

Figure 8.58 *Frequency and Inductance Versus Inductive Reactance*

■ Inductive Reactance

Figure 8.59 shows the inductive reactance versus frequency for an inductance of 10 millihenrys. It can be seen that as frequency increases so does the inductive reactance. For example, at a frequency of 159 hertz, the inductive reactance is 10 ohms. However, at a frequency of 1590 hertz the inductive reactance is now 100 ohms. Inductive reactance is directly proportional to frequency.

In *Figure 8.60*, which plots inductive reactance versus inductance at a frequency of 159 hertz, it can be seen that as inductance increases so does the inductive reactance. For example, with an inductance of 0.01 henrys (10 millihenrys), inductive reactance is 10 ohms. However, if the inductance is increased to 1 henry, the inductive reactance is now 1 kilohm. Inductive reactance also is directly proportional to inductance.

The basic equation for inductive reactance may be rewritten in two other forms:

$$f = \frac{X_L}{2\pi L} \qquad (8\text{–}22)$$

Or

$$L = \frac{X_L}{2\pi f} \qquad (8\text{–}23)$$

Equation *8–22* can be used to determine the frequency at which an inductance will produce a certain reactance. Equation *8–23* can be used to determine the inductance that will have a certain reactance at a certain frequency. For example, equation *8–22* can be used to determine the frequency at which an 8.5 henry inductor will have an inductive reactance of 5 kilohms.

$$f = \frac{X_L}{2\pi L} = \frac{5000\Omega}{(6.28)(8.5\text{H})} = 93.7\text{Hz}$$

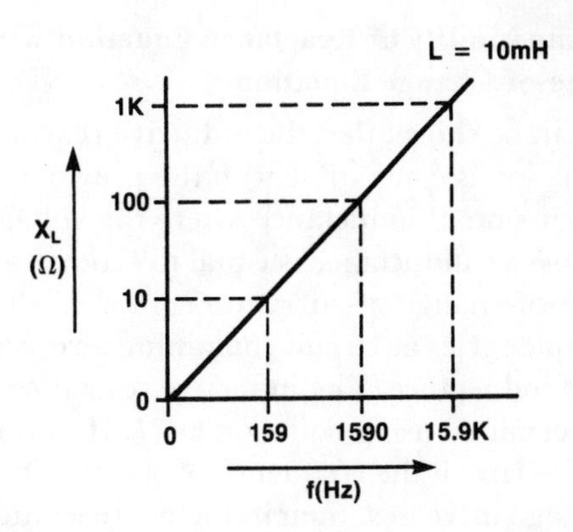

Figure 8.59 *Inductive Reactance Versus Frequency for an Inductance of 10 mH*

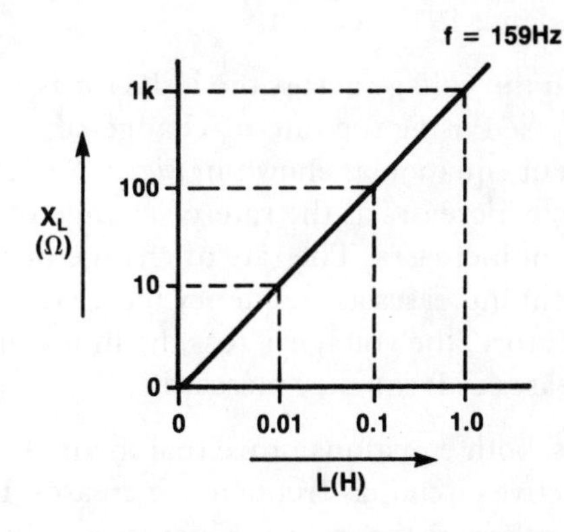

Figure 8.60 *Inductive Reactance Versus Inductance at a Frequency of 159 Hz*

Equation *8–23* can be used to determine the value of inductance needed to produce an inductive reactance of 10 kilohms at a frequency of 300 kilohertz.

$$L = \frac{X_L}{2\pi f} = \frac{10 \times 10^3\Omega}{(6.28)(300 \times 10^3\text{Hz})} = 0.531 \times 10^{-2}\text{H} = 5.31\text{mH}$$

Compatibility of Reactance Equation with Rate-of-Change Equation

It can be shown that the inductive reactance equation is compatible with the previous discussion of inductance where the voltage across an inductance is equal to L times a rate-of-change-of-current. *Figure 8.61* shows a simple series ac circuit containing a resistance and inductance. The inductive reactance is determined using equation *8–21*. If in this series circuit the frequency of the applied voltage increases, then inductive reactance increases and the voltage drop across the inductor increases because the inductive reactance is a larger portion of the total impedance in the circuit.

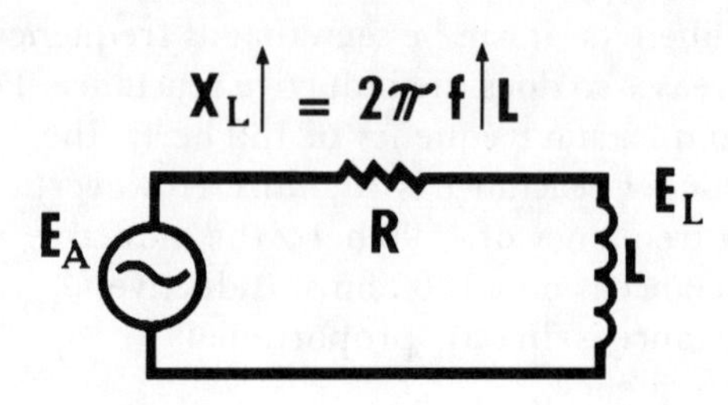

Figure 8.61 Inductive Reactance Equation can be used to Determine Circuit Operation

When the voltage across the inductor is expressed using the rate-of-change-of-current equation as shown in *Figure 8.62*, the voltage increases as the rate-of-change of current increases. The rate of change of current increases as frequency increases. Therefore, the voltage across the inductance increases as frequency increases.

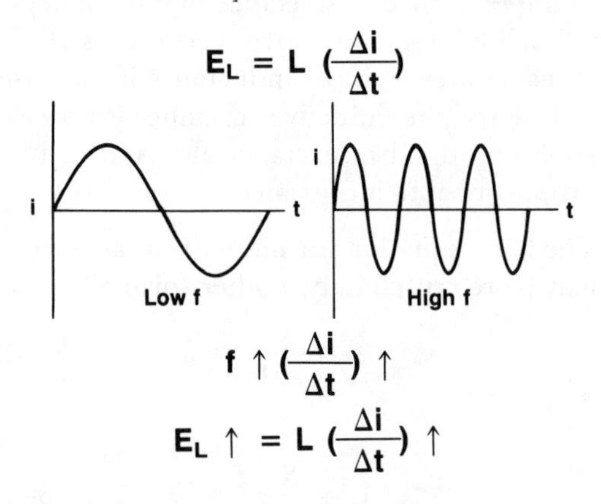

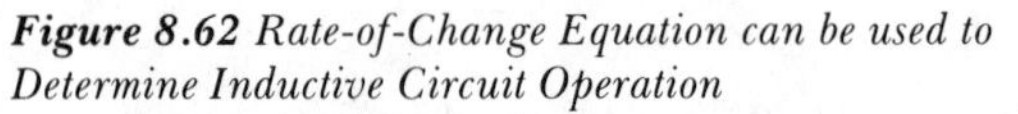

Figure 8.62 Rate-of-Change Equation can be used to Determine Inductive Circuit Operation

Thus, both equations prove that in an inductive circuit, as frequency increases, the voltage across the inductor increases. This analysis also proves that the inductive reactance and the rate-of-change equations state the same basic principle.

Like a capacitor, an inductor can be thought of as a variable resistor whose opposition to current flow (its inductive reactance) is controlled by the applied frequency. This is shown in *Figure 8.63*.

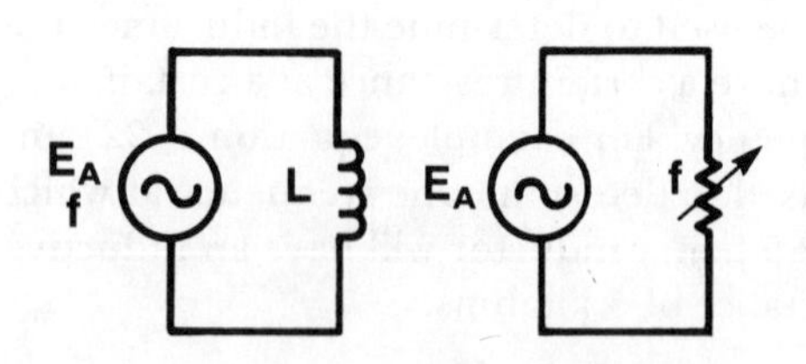

Figure 8.63 An Inductor can be Thought of as a Variable Resistor

SERIES AND PARALLEL INDUCTORS IN AC CIRCUITS

Thus far, only circuits with a single inductor have been discussed. More than one inductor can be used in a circuit. Therefore, an understanding of series and parallel combinations of inductors with applied voltage and frequency is important.

Series Inductors

Figure 8.64 shows a circuit in which two inductors, one 40 millihenrys, and the other 20 millihenrys, are connected in series. The applied voltage is 40 volts at 10 kilohertz.

First the inductive reactances of the two inductors will be calculated using the basic inductive reactance equation *8–21*.

$$\begin{aligned} \mathbf{X_{L1}} &= \mathbf{2\pi fL} \\ &= \mathbf{(6.28)\,(10 \times 10^3 Hz)\,(40 \times 10^{-3} H)} \\ &= \mathbf{2512\Omega} \\ \mathbf{X_{L2}} &= \mathbf{2\pi fL} \\ &= \mathbf{(6.28)\,(10 \times 10^3 Hz)\,(20 \times 10^{-3} H)} \\ &= \mathbf{1256\Omega} \end{aligned}$$

In this circuit the inductive reactance of L_1 is twice the inductive reactance of L_2. Thus, if the inductance increases by a factor of two, the inductive reactance increases by a factor of two.

Now that the individual reactances have been calculated, the reactance ohms are treated like resistive ohms, and series reactances are added.

$$\begin{aligned} \mathbf{X_{LT}} &= \mathbf{X_{L1} + X_{L2}} \\ &= \mathbf{2512\Omega + 1256\Omega} \\ &= \mathbf{3768\Omega} \end{aligned}$$

Therefore, the total reactance for the circuit is 3768 ohms. This total reactance is the total opposition that the circuit presents to alternating current flow at the applied frequency.

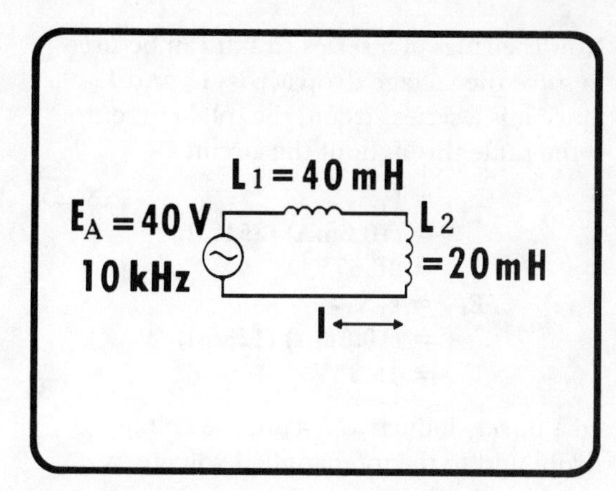

Figure 8.64 *Example Circuit for Determining Circuit Operation*

An alternate method of finding total reactance is to first calculate total inductance.

$$\begin{aligned} L_T &= L_1 + L_2 \\ &= 40mH + 20mH \\ &= 60mH \end{aligned}$$

Assuming no mutual inductance, the total inductance is 60 millihenrys. Now the total inductive reactance can be calculated using the basic inductive reactance equation:

$$\begin{aligned} X_{LT} &= 2\pi f L_T \\ &= (6.28)\,(10 \times 10^3 Hz)\,(60 \times 10^{-3} H) \\ &= 3768\Omega \end{aligned}$$

This is the same value of total inductive reactance calculated using the other method.

Because the total inductive reactance has been determined, it can be used to calculate the total current in the circuit. The total current, by Ohm's law, is equal to the applied voltage divided by the total inductive reactance.

$$\mathbf{I_T} = \frac{\mathbf{E_A}}{\mathbf{X_{LT}}} \qquad (8\text{–}24)$$

$$\begin{aligned} &= \frac{\mathbf{40V}}{\mathbf{3768\Omega}} \\ &= \mathbf{0.0106A} \\ &= \mathbf{10.6mA} \end{aligned}$$

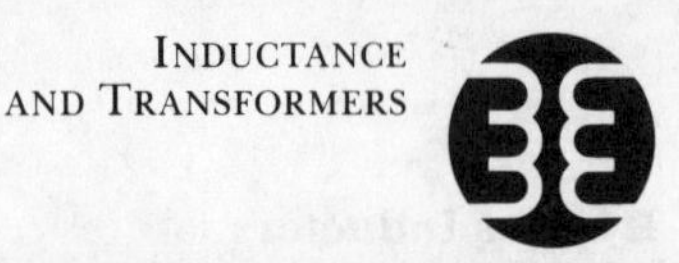

Now the rules of a series circuit can be used to solve the voltage drop across L_1 and L_2. Since it is a series circuit, the total current is the same throughout the circuit.

$$\begin{aligned} E_{L1} &= I_T X_{L1} \qquad (8\text{–}25)\\ &= (10.6\text{mA})\,(2512\Omega)\\ &= 26.67\text{V}\\ E_{L2} &= I_T X_{L2}\\ &= (10.6\text{mA})\,(1256\Omega)\\ &= 13.33\text{V} \end{aligned}$$

In a purely inductive circuit, the voltage drops add to the total applied voltage as in a series resistive circuit.

$$\begin{aligned} E_A &= E_{L1} + E_{L2} \qquad (8\text{–}26)\\ &= 26.67\text{V} + 13.33\text{V}\\ &= 40\text{V} \end{aligned}$$

Parallel Inductors

Figure 8.65 shows a circuit in which there is a parallel combination of two inductors. The voltage, frequency, and inductance are the same as those used in the previous discussion of series inductors. Keeping these the same should help you better understand the differences between series inductor and parallel inductor circuits. Since inductance of the coils is the same and applied frequency is the same, the inductive reactance of each of the inductors will be the same:

$$\begin{aligned} L_1 &= 2512\Omega\\ L_2 &= 1256\Omega \end{aligned}$$

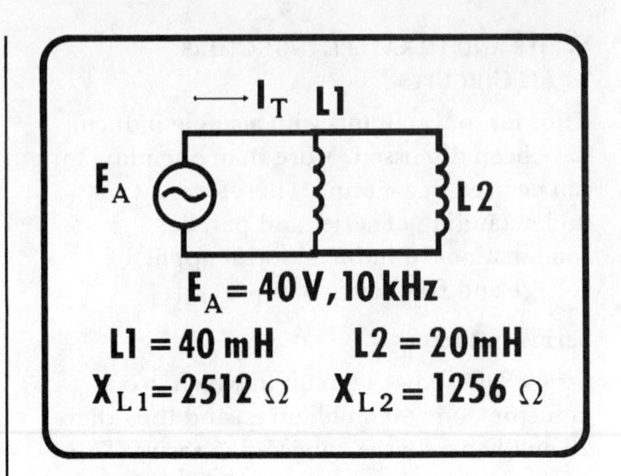

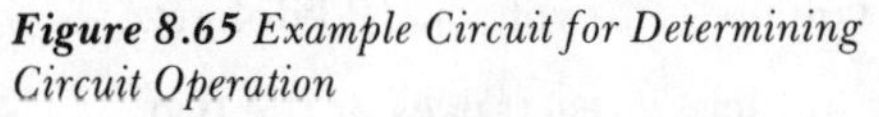

***Figure 8.65** Example Circuit for Determining Circuit Operation*

The branch current for each branch can be determined by dividing the voltage across the branch by the opposition in the branch.

$$\begin{aligned} I_{L1} &= \frac{E_{L1}}{X_{L1}}\\ &= \frac{E_A}{X_{L1}}\\ &= \frac{40\text{V}}{2512\Omega}\\ &= 15.9\text{mA}\\ I_{L2} &= \frac{E_{L2}}{X_{L2}}\\ &= \frac{E_A}{X_{L2}}\\ &= \frac{40\text{V}}{1256\Omega}\\ &= 31.8\text{mA} \end{aligned}$$

The total current is simply the sum of the branch currents.

$$\begin{aligned} I_T &= I_{L1} + I_{L2} \qquad (8\text{–}27)\\ &= 15.9\text{mA} + 31.8\text{mA}\\ &= 47.7\text{mA} \end{aligned}$$

To find the total inductive reactance, simply divide the applied voltage by the total current.

$$X_{LT} = \frac{E_A}{I_T} \qquad (8\text{–}28)$$
$$= \frac{40V}{47.7mA}$$
$$= 837\Omega$$

There are two alternate methods that can be used to find the total inductive reactance. The first method is to treat the inductive reactances as if they were parallel resistances and use the product-over-sum method of solution.

$$L_T = \frac{(X_{L1})(X_{L2})}{X_{L1} + X_{L2}} \qquad (8\text{–}29)$$
$$= \frac{(2512\Omega)(1256\Omega)}{2512\Omega + 1256\Omega}$$
$$= 837\Omega$$

The other method is to first determine the total inductance of the circuit, then calculate the total inductive reactance using the total inductance. Using this method,

$$L_T = \frac{(L_1)(L_2)}{L_1 + L_2} \qquad (8\text{–}30)$$
$$= \frac{(40mH)(20mH)}{40mH + 20mH}$$
$$= 13.3mH$$
$$X_{LT} = 2\pi fL_T$$
$$= (6.28)(10 \times 10^3 Hz)(13.3 \times 10^{-3} H)$$
$$= 837\Omega$$

Power in Inductive Circuits

In inductive circuits, power can be determined using calculations similar to the calculations used to determine power in resistive or capacitive circuits. Remember that in resistive circuits, electrical energy is converted into heat energy. However, in inductive circuits, the electrical energy is stored temporarily in the magnetic field around the inductor. This is similar to the storage of energy in an electrostatic field in a capacitor. Thus, inductors store *electromagnetic* energy temporarily; capacitors store *electrostatic* energy temporarily.

Recall that the basic power equation is

$$P = EI \qquad (8\text{–}31)$$

This power equation can be used to calculate the power in inductive circuits. The power in inductive circuits is measured in units called VAR as it is in capacitive circuits. VAR stands for Volts Amperes Reactive. The power in an inductive circuit is called reactive power since the opposition to current in the circuit is strictly reactive.

Power Calculations for Series-Inductive Circuits

In the series-inductive circuit discussed previously, and repeated in *Figure 8.66*, the voltage drop across L_1 is 26.67 volts. The current through L_1 and L_2 is 10.6 milliamperes. Reactive power for L_1 is calculated:

$$P_{L1} = (E_{L1})(I_{L1})$$
$$= (26.67V)(10.6mA)$$
$$= 282.7mVAR$$

Reactive power for L_2 is calculated:

$$P_{L2} = (E_{L2})(I_{L2})$$
$$= (13.33V)(10.6mA)$$
$$= 141.3mVAR$$

By adding these two values of reactive power, the total reactive power of the circuit can be determined. That is,

$$P_{XT} = P_{L1} + P_{L2} \quad (8\text{–}32)$$
$$= 282.7\text{mVAR} + 141.3\text{mVAR}$$
$$= 424\text{mVAR}$$

Total reactive power of the circuit could also have been calculated by multiplying the total applied voltage times the total current. In this case,

$$P_{XT} = E_A I_T$$
$$= (40\text{V})\,(10.6\text{mA})$$
$$= 424\text{mVAR}$$

Power Calculations for Parallel-Inductive Circuit

Similar calculations can be performed to obtain the reactive power for the parallel inductive circuit shown in *Figure 8.65*. Recall in that circuit I_{L1} = 15.9 milliamperes and I_{L2} = 31.8 milliamperes. Remember the voltage across each branch is the applied voltage. The reactive power of L_1 is:

$$P_{L1} = E_{L1} I_{L1}$$
$$= (40\text{V})\,(15.9\text{mA})$$
$$= 636\text{mVAR}$$

The reactive power of L_2 is:

$$P_{L2} = E_{L2} I_{L2}$$
$$= (40\text{V})\,(31.8\text{mA})$$
$$= 1272\text{mVAR}$$

The total reactive power is:

$$P_{XT} = P_{L1} + P_{L2}$$
$$= 636\text{mVAR} + 1272\text{mVAR}$$
$$= 1908\text{mVAR}$$

Also, the total reactive power in a parallel circuit equals the total applied voltage times the total current.

$$P_{XT} = E_A I_T$$
$$= (40\text{V})\,(47.7\text{mA})$$
$$= 1908\text{mVAR}$$

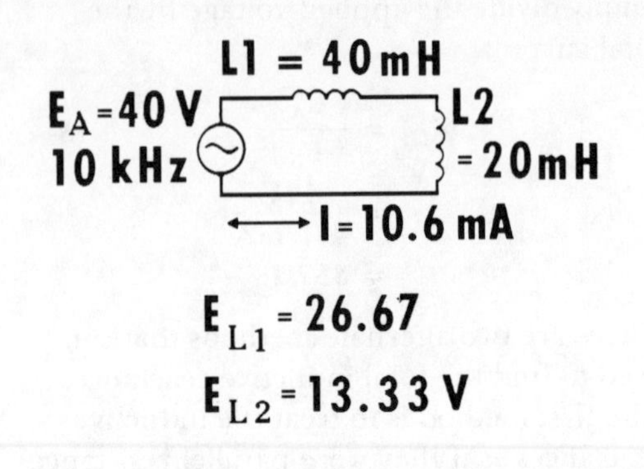

Figure 8.66 *Example Series Inductive Circuit*

SUMMARY

This lesson has been an introduction to the inductor, how it is structured, its schematic symbol, its typical units of inductance, and how it functions in typical circuits. The phase relationships of the voltage and current in an inductive circuit were discussed. Mutual inductance and how it is put to use in transformers, and how to make voltage and current calculations for transformer circuits were also discussed. Series and parallel inductive problems were solved, and reactive power calculations were described.

■ Worked-Out Examples

1. Describe the action of an inductor in a circuit.

Solution: A magnetic field surrounds any wire carrying current. As current increases through a wire, the magnetic field expands through the wire inducing a counter current which opposes the increase in the initial current. As current decreases in a wire, the magnetic field collapses through the wire inducing current in the same direction and aiding the current which is trying to decrease, thus opposing the decrease of current. When the wire is wound into a coil, the magnetic field produced by each turn of wire in the coil interacts with adjacent turns increasing this inductive effect. This coil of wire is called an inductor. If it is placed in a circuit such that a changing current passes through it, it will oppose the change (increase or decrease) of current.

2. Define inductance.

Solution: Inductance is the property of a circuit which opposes any change in current.

3. If the current through an 8 millihenry-coil is changing at the rate of 10 milliamperes every 5 seconds, determine the rate of change of the current in amperes per second, and the voltage (CEMF) induced across the coil.

a. Rate of change of current $= \frac{\Delta i}{\Delta t} = \frac{10\text{mA}}{5\text{ sec}} = \mathbf{2mA/sec}$

b. CEMF $= E_L = L\left(\frac{\Delta i}{\Delta t}\right) = (8\text{mH})(2\text{mA/sec}) = (8 \times 10^{-3}\text{H})(2 \times 10^{-3}\text{A/sec})$

$= 16 \times 10^{-6}\text{V} = \mathbf{16\mu V}$

4. If two coils are connected in series as shown, determine their total inductance with no mutual inductance, and their mutual inductance and total inductance considering mutual inductance (aiding and opposing) if $k = 0.4$.

Solution:

a. $L_T(\text{no } L_M) = L_1 + L_2 = 18\text{H} + 2\text{H} = \mathbf{20H}$

b. $L_T\text{ (aid)} = L_1 + L_2 + 2L_m = 18\text{H} + 2\text{H} + 2(2.4\text{H})$
$= 18\text{H} + 2\text{H} + 4.8\text{H} = \mathbf{24.8H}$

where $L_M = k\ \sqrt{L_1 \times L_2} = 0.4\ \sqrt{18\text{H} \times 2\text{H}}$
$= 0.4\ \sqrt{36\text{H}} = 0.4\ (6)\text{H} = \mathbf{2.4H}$

c. $L_T(\text{oppose}) = L_1 + L_2 - 2L_M = 18\text{H} + 2\text{H} - 2(2.4\text{H})$
$L_T(\text{oppose}) = 20\text{H} - 4.8\text{H} = \mathbf{15.2H}$

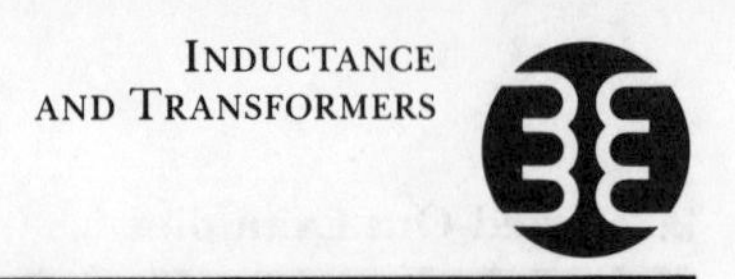

■ Worked-Out Examples

5. a. Given the circuit shown solve for the total inductance of the parallel-connected inductors if there is no mutual inductance.

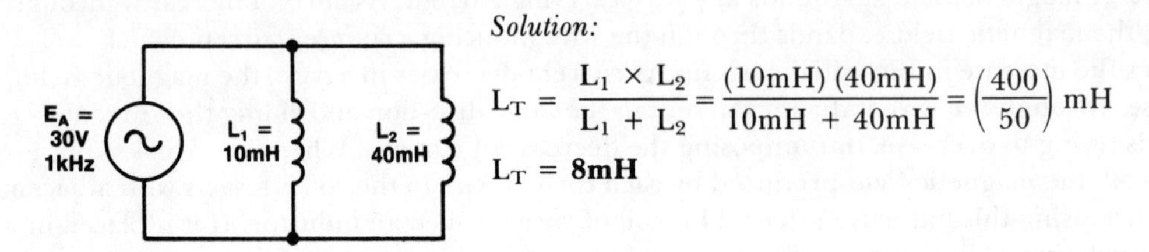

Solution:

$$L_T = \frac{L_1 \times L_2}{L_1 + L_2} = \frac{(10mH)(40mH)}{10mH + 40mH} = \left(\frac{400}{50}\right) mH$$

$$L_T = \mathbf{8mH}$$

b. Determine their mutual inductance and total inductance (aiding and opposing) if mutual inductance exists with a coefficient of 0.2.

Solution:

$$L_M = k\sqrt{L_1 \times L_2} = 0.2\sqrt{10mH \times 40mH} = 0.2\sqrt{400mH} = 0.2(20)mH$$

$$L_M = \mathbf{4mH}$$

$$L_T(aid) = \frac{(L_1 + L_M)(L_2 + L_M)}{L_1 + L_2 + 2L_M} = \frac{(10mH + 4mH)(40mH + 4mH)}{10mH + 40mH + 2(4mH)} = \frac{(14mH)(44mH)}{(58mH)}$$

$$L_T(aid) = \mathbf{10.62mH}$$

$$L_T(oppose) = \frac{(L_1 - L_M)(L_2 - L_M)}{L_1 + L_2 - 2L_M} = \frac{(10mH - 4mH)(40mH - 4mH)}{10mH + 40mH - 2(4mH)} = \mathbf{5.14mH}$$

6. If the primary voltage applied to a transformer is 120 VAC and the secondary voltage output is 480 VAC, determine the turns ratio for the transformer and state whether it is a step-up or step-down transformer.

Solution:

a. Turns ratio $= \frac{N_S}{N_P} = \frac{E_S}{E_P} = \frac{480V}{120V} = \frac{\mathbf{4}}{\mathbf{1}}$

or written in N_S:N_P form, **4:1**

b. **This is a step-up transformer since the secondary voltage is higher than the primary voltage.**

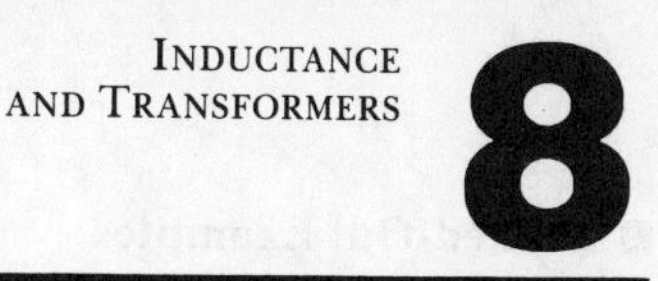

■ Worked-Out Examples

7. Given the transformer with turns-ratio and load-resistance specified, determine the following values: E_{sec}, I_{sec}, I_{pri}, P_{pri} and P_{sec}. (Assume 100 percent efficiency.)

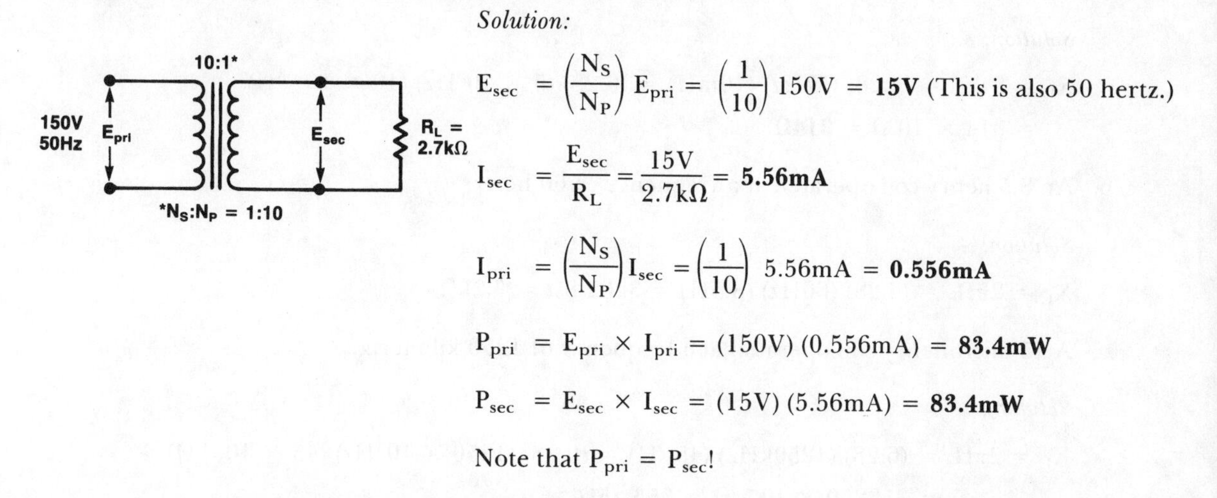

Solution:

$$E_{sec} = \left(\frac{N_S}{N_P}\right) E_{pri} = \left(\frac{1}{10}\right) 150\text{V} = \mathbf{15V} \text{ (This is also 50 hertz.)}$$

$$I_{sec} = \frac{E_{sec}}{R_L} = \frac{15\text{V}}{2.7\text{k}\Omega} = \mathbf{5.56mA}$$

$$I_{pri} = \left(\frac{N_S}{N_P}\right) I_{sec} = \left(\frac{1}{10}\right) 5.56\text{mA} = \mathbf{0.556mA}$$

$$P_{pri} = E_{pri} \times I_{pri} = (150\text{V})\,(0.556\text{mA}) = \mathbf{83.4mW}$$

$$P_{sec} = E_{sec} \times I_{sec} = (15\text{V})\,(5.56\text{mA}) = \mathbf{83.4mW}$$

Note that $P_{pri} = P_{sec}$!

8. If the primary voltage is 120 VAC with a primary current of 10mA and the secondary voltage is 12.6 VAC with a secondary current of 85 millamperes, determine the percent efficiency of this transformer. Explain the loss of power between primary and secondary.

Solution:

a. $P_{pri} = E_{pri} \times I_{pri} = (120\text{V})\,(10\text{mA}) = 1200\text{mW}$

$P_{sec} = E_{sec} \times I_{sec} = (12.6\text{V})\,(85\text{mA}) = 1071\text{mW}$

$$\%\ \text{Eff} = \frac{P_S}{P_P} \times 100\% = \left(\frac{1071\text{mW}}{1200\text{mW}}\right) \times 100\% = 0.893 \times 100\% = \mathbf{89.3\%}$$

b. **The power loss (10.7 percent of the primary power) between primary and secondary is due to eddy currents, hysteresis and winding resistance heat loss.**

Worked-Out Examples

9. Calculate the inductive reactance of the inductors at these specified frequencies:

a. 10 millihenry coil operated at a frequency of 5 kilohertz:

Solution:

$$X_L = 2\pi fL = (6.28)\ (5kHz)\ (10mH) = (6.28)\ (5 \times 10^3 Hz)\ (10 \times 10^{-3} H)$$
$$= 314 \times 10^0 \Omega = \mathbf{314\Omega}$$

b. An 8.5 henry coil operated at a frequency of 60 hertz:

Solution:

$$X_L = 2\pi fL = (6.28)\ (60Hz)\ (8.5H) = 3202.8\Omega = \mathbf{3.2k\Omega}$$

c. A 45 microhenry coil operated at a frequency of 1250 kilohertz:

Solution:

$$X_L = 2\pi fL = (6.28)\ (1250kHz)\ (45\mu H) = (6.28)\ (1250 \times 10^3 Hz)\ (45 \times 10^{-6} H)$$
$$= 353250 \times 10^{-3}\Omega = \mathbf{353.25\Omega}$$

10. Calculate the value of the inductor needed to produce the reactance specified at the given frequency:

a. A reactance of 1 megohm at a frequency of 40 kilohertz:

Solution:

$$L = \frac{X_L}{2\pi f} = \frac{1M\Omega}{(6.28)\ (40kHz)} = \frac{1 \times 10^6\Omega}{(6.28)\ (40 \times 10^3 Hz)} = \mathbf{3.9H}$$

b. A reactance of 47 kilohms at a frequency of 108 megahertz:

Solution:

$$L = \frac{X_L}{2\pi f} = \frac{47k\Omega}{(6.28)\ (108MHz)} = \frac{47 \times 10^3\Omega}{(6.28)\ (108 \times 10^6 Hz)} = \left(\frac{47 \times 10^3}{678.24 \times 10^6}\right) H$$
$$= 0.0693 \times 10^{-3} H = 0.0693mH = \mathbf{69.3\mu H}$$

■ Worked-Out Examples

11. Calculate the frequency at which the given inductors will have the specified reactance.

a. A reactance of 50 kilohms with a 4 millihenry inductor:

Solution:

$$f = \frac{X_L}{2\pi L} = \frac{50k\Omega}{(6.28)\ (4mH)} = \frac{50 \times 10^3\Omega}{(6.28)\ (4 \times 10^{-3}H)} = \left(\frac{50 \times 10^3}{25.12 \times 10^{-3}}\right) Hz$$

$$= 2 \times 10^6 Hz = \mathbf{2MHz}$$

b. A reactance of 25 ohms with a 5 millihenry inductor:

Solution:

$$f = \frac{X_L}{2\pi L} = \frac{25\Omega}{(6.28)\ (5mH)} = \frac{25\Omega}{(6.28)\ (5 \times 10^{-3}H)} = \frac{25}{0.0314} = \mathbf{796Hz}$$

12. Solve for the values indicated using the circuit shown. (Assume $L_M = 0$.)

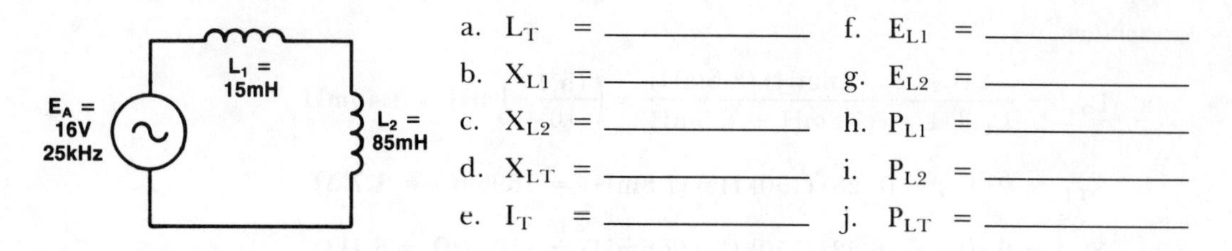

a. L_T = ______ f. E_{L1} = ______

b. X_{L1} = ______ g. E_{L2} = ______

c. X_{L2} = ______ h. P_{L1} = ______

d. X_{LT} = ______ i. P_{L2} = ______

e. I_T = ______ j. P_{LT} = ______

Solution:

a. $L_T = L_1 + L_2 = 15mH + 85mH = \mathbf{100mH}$

b. $X_{L1} = 2\pi f L_1 = (6.28)\ (25kHz)\ (15mH) = 2355\Omega = \mathbf{2.36k\Omega}$

c. $X_{L2} = 2\pi f L_2 = (6.28)\ (25kHz)\ (85mH) = 13345\Omega = \mathbf{13.35k\Omega}$

d. $X_{LT} = X_{L1} + X_{L2} = 2.36k\Omega + 13.35k\Omega = \mathbf{15.7k\Omega}$ or

$X_{LT} = 2\pi f L_T = (6.28)\ (25kHz)\ (100mH) = \mathbf{15.7k\Omega}$

e. $I_T = \frac{E_A}{X_{LT}} = \frac{16V}{15.7k\Omega} = \mathbf{1.02mA}$

f. $E_{L1} = I_{L1}X_{L1} = I_T X_{L1} = (1.02mA)\ (2.36k\Omega) = \mathbf{2.4V}$

g. $E_{L2} = I_{L2}X_{L2} = I_T X_{L2} = (1.02mA)\ (13.35k\Omega) = \mathbf{13.6V}$

■ **Worked-Out Examples**

h. $P_{L1} = E_{L1}I_{L1} = E_{L1}I_T = (2.4V)(1.02mA) = \mathbf{2.45mVAR}$

i. $P_{L2} = E_{L2}I_{L2} = E_{L2}I_T = (13.6V)(1.02mA) = \mathbf{13.87mVAR}$

j. $P_{LT} = P_{L1} + P_{L2} = 2.45mVAR + 13.87mVAR = \mathbf{16.32mVAR}$ or

$P_{LT} = E_AI_T = (16V)(1.02mA) = \mathbf{16.32mVAR}$

13. Solve for the values indicated using the circuit shown. (Assume $L_M = 0$.)

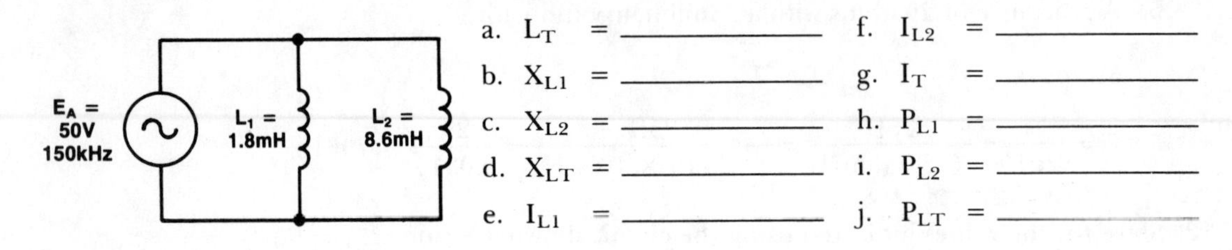

Solution:

a. $L_T = \dfrac{L_1L_2}{L_1 + L_2} = \dfrac{(1.8mH)(8.6mH)}{(1.8mH + 8.6mH)} = \left(\dfrac{15.48}{10.4}\right) mH = \mathbf{1.49mH}$

b. $X_{L1} = 2\pi fL_1 = (6.28)(150kHz)(1.8mH) = 1695.6\Omega = \mathbf{1.7k\Omega}$

c. $X_{L2} = 2\pi fL_2 = (6.28)(150kHz)(8.6mH) = 8101.2\Omega = \mathbf{8.1k\Omega}$

d. $X_{LT} = \dfrac{(X_{L1})(X_{L2})}{X_{L1} + X_{L2}} = \dfrac{(1.7k\Omega)(8.1k\Omega)}{(1.7k\Omega) + (8.1k\Omega)} = \left(\dfrac{13.77}{9.8}\right) k\Omega = \mathbf{1.4k\Omega}$ or

$X_{LT} = 2\pi fL_T = (6.28)(150kHz)(1.49mH) = 1403.6\Omega = \mathbf{1.4k\Omega}$

e. $I_{L1} = \dfrac{E_A}{X_{L1}} = \dfrac{50V}{1.7k\Omega} = \mathbf{29.4mA}$

f. $I_{L2} = \dfrac{E_A}{X_{L2}} = \dfrac{50V}{8.11k\Omega} = \mathbf{6.2mA}$

g. $I_T = I_{L1} + I_{L2} = 29.4mA + 6.2mA = \mathbf{35.6mA}$

h. $P_{L1} = E_{L1}I_{L1} = E_AI_{L1} = (50V)(29.4mA) = \mathbf{1470mVAR}$

i. $P_{L2} = E_{L2}I_{L2} = E_AI_{L2} = (50V)(6.2mA) = \mathbf{310mVAR}$

j. $P_{LT} = E_AI_T = (50V)(35.6mA) = \mathbf{1780mVAR}$ or

$P_{LT} = P_{L1} + P_{L2} = 1470mVAR + 310mVAR = \mathbf{1780mVAR}$

■ Practice Problems

1. State a short definition of inductance.

2. The concepts of two men are used to explain CEMF for inductors. Who are they?

3. If an iron core is extracted from a coil, will the coil's inductance increase or decrease? Why?

4. As the number of turns of wire used in a coil increases, does the value of its inductance increase or decrease?

5. If two coils are placed in proximity of one another and one coil produces 4000 lines of flux, 3500 of which cut the second coil, what is the coefficient of coupling of these two coils? k = ______ .

6. What is the range of values for the coefficient of coupling? ________ to ________ (upper and lower limits for k).

7. In the circuit shown, two coils are connected in series. Determine their total inductance with no mutual inductance. Then determine their mutual inductance and their combined inductance considering mutual inductance (aiding and opposing). k = 0.4, L_1 = 4 henrys, and L_2 = 9 henrys.

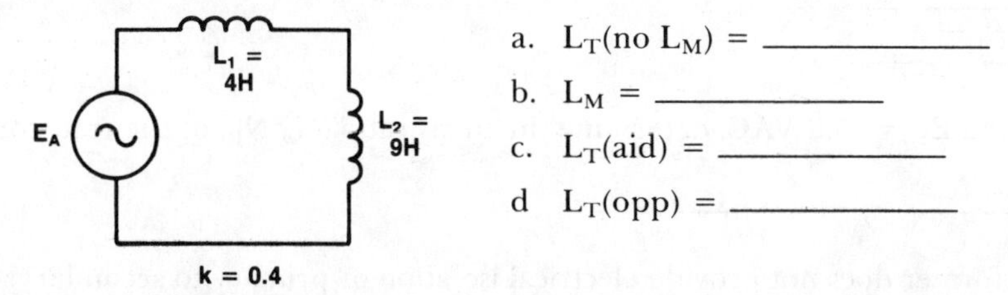

8. In the circuit shown, determine the total inductance of the two parallel-connected inductors if there is no mutual inductance. Then determine their mutual inductance and total inductance (aiding and opposing) if they have a coefficient of coupling of 0.2.

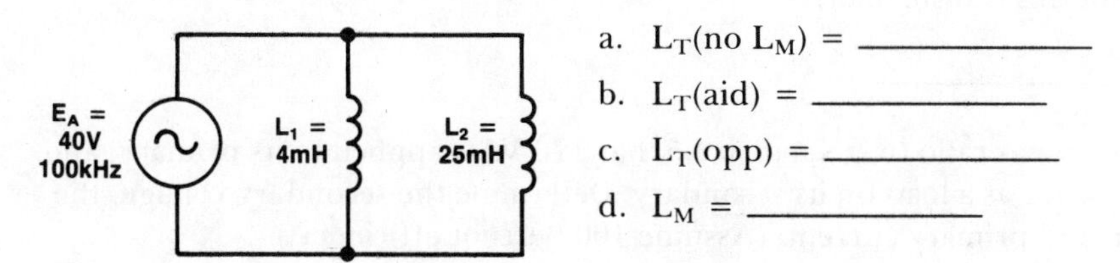

9. a. Sketch the magnetic field about the coil in the drawing. Indicate north and south poles.

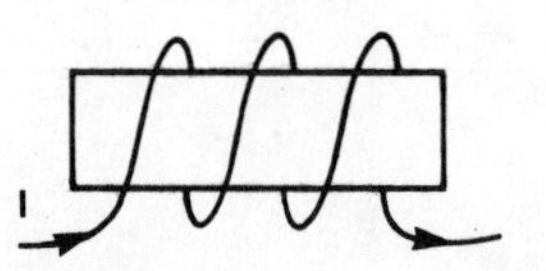

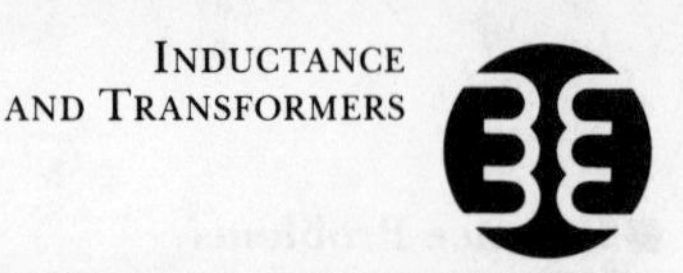

■ Practice Problems

b. Sketch the magnetic field about the conductor. Show its direction.

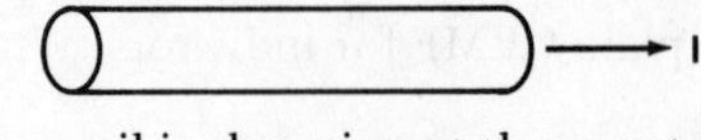

10. If the current through a coil is changing at the constant rate of 40 milliamperes every 10 seconds, determine the rate of change of the current in amperes per second. If the coil is rated at 5 millihenrys, determine the voltage across the coil.

a. ROC of i = ____________ A/sec

b. E_L = ____________

11. What coefficient of coupling is desired for transformers?

k = ____________

12. If the primary voltage is greater than the secondary voltage of a transformer, is it known as a step-up or step-down transformer?

13. What two types of core losses in a transformer are associated directly with the core?

a. ____________

b. ____________

14. If E_P = 120 VAC and E_S = 25.2 VAC, determine the turns ratio (N_S:N_P) of the transformer.

Turns ratio = ____________ : ____________

15. What type of transformer does not provide electrical isolation of primary to secondary?

16. If the primary voltage is 240 VAC with a primary current of 8 milliamperes and the secondary voltage is 50 VAC with a secondary current of 33 milliamperes, determine the percent efficiency of this transformer:

% eff = ____________

17. A transformer has a turns ratio (N_S:N_P) of 1:4.5, has 120 VAC applied to its primary, and has a 6.8 kilohm resistor as a load on its secondary. Determine the secondary voltage, the secondary current, and primary current. (Assume 100 percent efficiency.)

a. E_{sec} = ____________

b. I_{sec} = ____________

c. I_{pri} = ____________

■ Practice Problems

18. When 40 VAC is applied to the primary of a transformer, a secondary current of 8 milliamperes flows through a one kilohm resistor connected across the secondary. 2 milliamperes of primary current is present. Determine the percent efficiency of the transformer and its turns ratio.

a. % eff = ____________

b. N_S:N_P = ____________

19. Calculate X_L for a 2 millihenry coil operated at frequencies of 100 hertz, 5 kilohertz, and 1.2 megahertz.

a. X_L(f = 100 hertz) = ____________

b. X_L(f = 5 kilohertz) = ____________

c. X_L(f = 1.2 megahertz) = ____________

20. From Problem 19, you see that as the frequency applied to an inductor increases, the inductive reactance of it ____________ (increases, decreases).

21. What is the value of an inductor needed to produce a reactance of 482 kilohms at a frequency of 5 kilohertz?

L = ____________

22. What is the frequency at which an inductor of 8.5 henrys will have an inductive reactance of 1 kilohms?

f = ____________

23. Solve for the indicated values using the circuit shown. (Assume L_M = 0.)

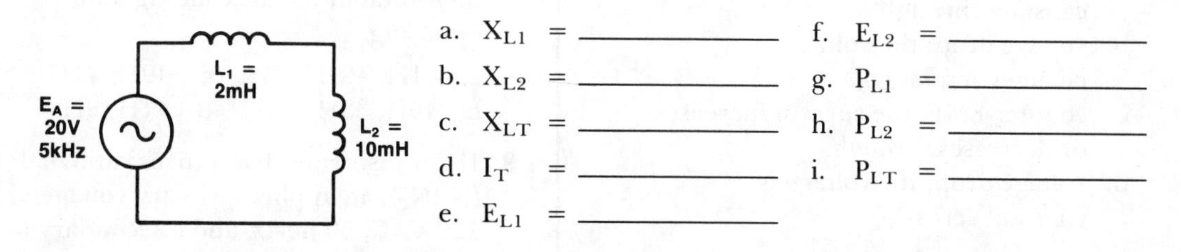

24. Solve for the values using the circuit shown. (Assume L_M = 0.)

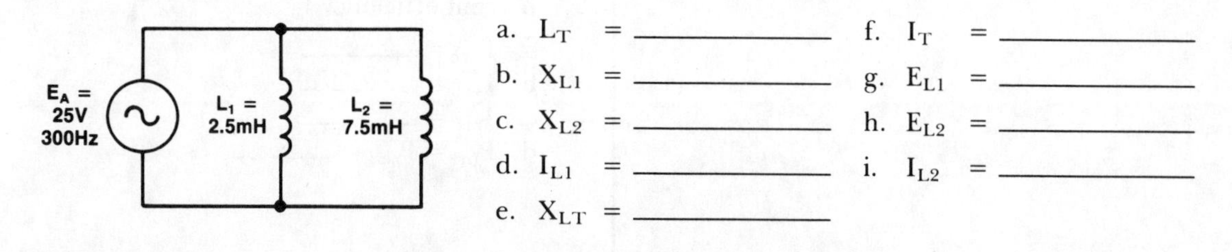

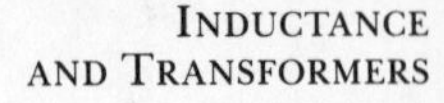

■ Quiz

1. Inductance is the property of a circuit that

a. opposes any change in voltage.
b. opposes current.
c. opposes any change in current.
d. opposes any change in frequency.

2. Which of the factors listed below does not govern the value of a coil?

a. Number of turns
b. The type of core material used
c. The size (cross-sectional area) of the coil
d. The length of the coil
e. The size of the wire used in the coil

3. The rise or fall of current through an inductor in a circuit is said to be:

a. exponential
b. logarithmic
c. linear
d. none of the above

4. The voltage that appears across an inductor in a circuit is called __________ and appears only when __________ the inductor.

a. counter EMF; the current is constant through
b. voltage drop; the voltage changes across
c. counter EMF; the current increases or decreases through
d. voltage drop; the voltage is constant across

5. The phase relationship of the voltage across an inductor and the current passing through it in an ac (sinusoidal) circuit is such that

a. the voltage lags the current by 90 degrees.
b. the current leads the voltage by 90 degrees.
c. the voltage leads the current by 90 degrees.
d. the voltage and current are in phase.

6. Determine the mutual inductance of two inductors having a coefficient of coupling of 0.8 if their values are 16 millihenrys and 5 millihenrys.

a. 64mH
b. 7.2mH
c. 20mH
d. 3.6mH

7. If two inductors are series-connected and their values are 16 henrys and 25 henrys, determine their total inductance if they have no mutual inductance.

a. 9.76H
b. 6.4H
c. 20H
d. 41H

8. If the two inductors of Question 7 have a coefficient of coupling of 0.2, determine their total inductance aiding and opposing.

a. 41H, 49H
b. 49H, 33H
c. 49H, 41H
d. 41H, 8H

9. If a transformer has a turns ratio of 1:19 (N_s:N_P), an applied primary voltage of 120 VAC, 60 hertz, and a secondary load resistance of 3.3 kilohms, determine the quantities specified below. (Assume 100 percent efficiency.)

a. E_{sec} = __________
b. I_{sec} = __________
c. I_{pri} = __________
d. P_{pri} = P_{sec} = __________

■ Quiz

10. A transformer has a greater primary current than secondary current under load conditions. Is it a step-up or step-down transformer?

11. Using the inductive reactance equation and given the data specified below, solve for the unknown quantity.

a. L = 15mH, f = 5kHz: X_L = ____

b. X_L = 20kΩ, f = 3.5MHz: L = ____

c. X_L = 600kΩ, L = 10mH: f = ____

12. Determine the requested voltages, currents and power for these two circuits. (Assume L_M = 0.)

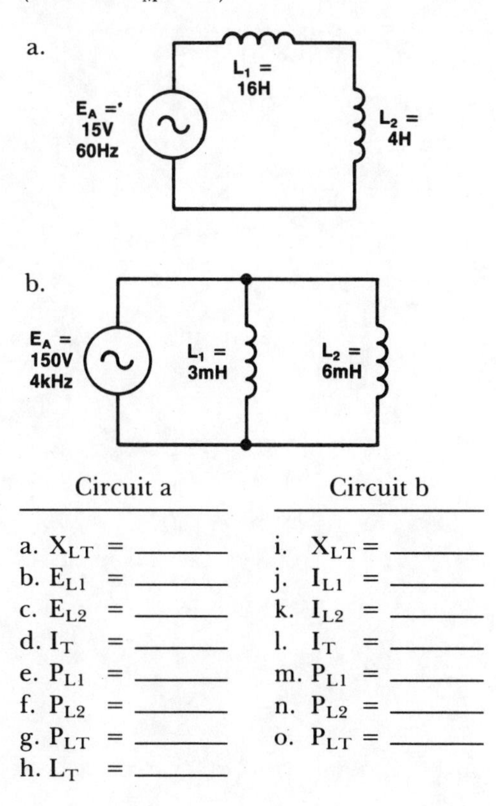

Circuit a	Circuit b
a. X_{LT} = ____	i. X_{LT} = ____
b. E_{L1} = ____	j. I_{L1} = ____
c. E_{L2} = ____	k. I_{L2} = ____
d. I_T = ____	l. I_T = ____
e. P_{L1} = ____	m. P_{L1} = ____
f. P_{L2} = ____	n. P_{L2} = ____
g. P_{LT} = ____	o. P_{LT} = ____
h. L_T = ____	

LESSON 9

RL Circuit Analysis

In Lesson 8, you were introduced to a new circuit element, the inductor. In this lesson, the methods of analysis of series and parallel circuits containing resistors and inductors will be discussed.

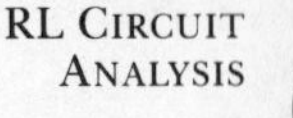

■ Objectives

At the end of this lesson you should be able to:

1. Determine the value of the inductive reactance, total impedance, total current, the component currents and voltages; real, reactive, and apparent power; and phase angle for series and parallel RL circuits with various values of R, L, applied voltage and frequency specified.

2. Draw the impedance, voltage, current, and power phasor diagrams to show the phase relationships in series and parallel RL circuits with certain circuit values specified.

3. Determine the circuit values of series and parallel RL circuits by using Pythagorean theorem relationships for vectorially adding circuit quantities.

4. Determine phase angles of RL series and parallel circuits by using the tangent trigonometric function.

5. Determine the Q of a coil when given the inductance and internal resistance of the coil.

■ Phase Relationships in a Series RL Circuit

INTRODUCTION

In the previous lesson, the inductor and its properties were introduced. However, like capacitive circuits, inductors are not usually the only component in circuits. A more common circuit is one in which inductors are combined in series, or in parallel with resistors. In this lesson you will learn techniques to solve problems for circuits containing a resistor and an inductor. You will learn how the phase relationship between voltage and current affect the analysis of such circuits. And you will learn how to compute the values for current, voltage, impedance, phase angle, and power for series and parallel circuits containing a resistor and inductor.

SERIES RL CIRCUIT

Phase Relationships in a Series RL Circuit

A typical series RL circuit is shown in *Figure 9.1*. In it one resistor is connected in series with one inductor. In a series RL circuit, as in any series circuit, the current through all the components is the same. However, the sum of the voltage drop across the resistor and the voltage drop across the inductor, do not simply add algebraically to equal the applied voltage as they would in either a purely resistive or a purely reactive circuit. This is because of the combination of resistance and reactance and the different phase relationship between the voltage and current for each component.

Recall, as shown in *Figure 9.2*, that the voltage across a resistor is in phase with the current flowing through it; but for an inductor, however, the voltage leads the current by 90 degrees. Since the components are in series with one another, the current through each component is the same. Therefore, mathematically,

$$I_R = I_L = I_T$$

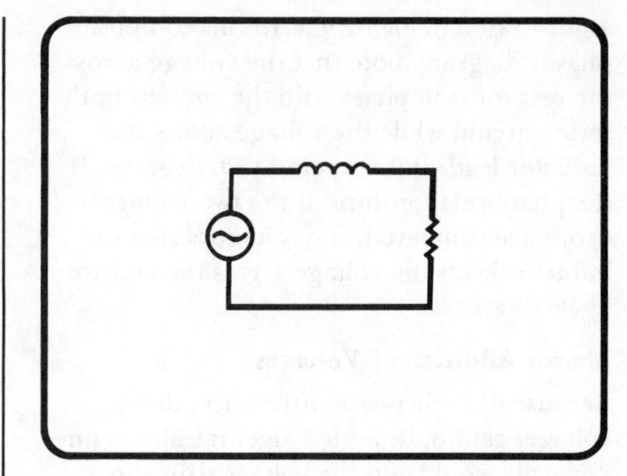

Figure 9.1 RL Series Circuit

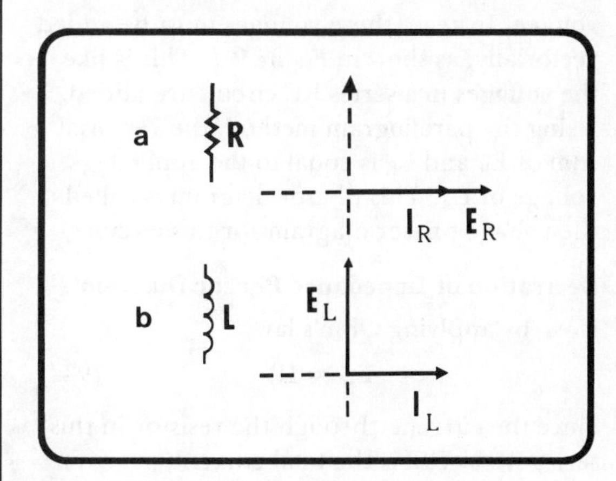

Figure 9.2 EI Phase Relationship for a Resistor, a, and an Inductor, b

Using the current then as a common basis for comparison, the two individual diagrams shown in *Figure 9.2* can be combined into one as illustrated in *Figure 9.3*. In this composite phasor diagram, note that the voltage across the resistor is in phase with the current in the series circuit, while the voltage across the inductor leads this current by 90 degrees. If the phase relationships of the two voltage drops are compared, the voltage across the inductor leads the voltage across the resistor by 90 degrees.

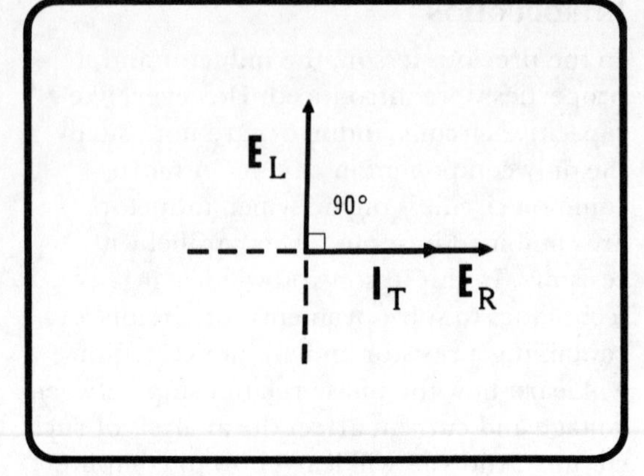

***Figure 9.3** Composite RL Circuit Phase Relationships of a Series RL Circuit*

Phasor Addition of Voltages

Because of their phase difference, these voltages cannot be added algebraically as one normally would add the voltage drops in a series circuit to obtain the total applied voltage. Instead, these voltages must be added vectorially, as show in *Figure 9.4*. This is like the voltages in a series RC circuit are added. Using the parallogram method, the vectorial sum of E_R and E_L is equal to the applied voltage of E_A. This phasor diagram is called the voltage phasor diagram for this circuit.

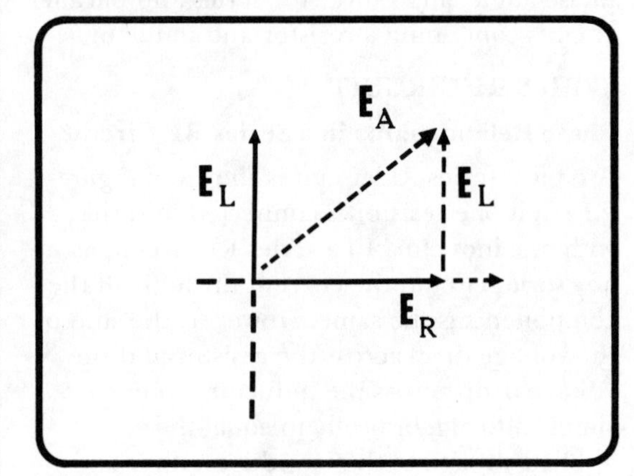

***Figure 9.4** Voltage Phasor Diagram of a Series RL Circuit*

Derivation of Impedance Phasor Diagram

Now, by applying Ohm's law,

$$\mathbf{E_R = IR} \qquad (9\text{–}1)$$

Since the current, through the resistor in this series RL circuit is the total current,

$$\mathbf{E_R = I_TR} \qquad (9\text{–}2)$$

Remember also that

$$\mathbf{E_L = I_LX_L} \qquad (9\text{–}3)$$

and that the current through the inductor is the same as the total current. Therefore, equation *9–3* could also be written as

$$\mathbf{E_L = I_TX_L} \qquad (9\text{–}4)$$

These IR and IX_L quantities can be substituted for the voltages they equal on the voltage phasor diagram as shown in *Figure 9.5a*. Since the total current is the common factor here, it may be factored out so that the phase relationship between R and X_L remains as shown in *Figure 9.5b*.

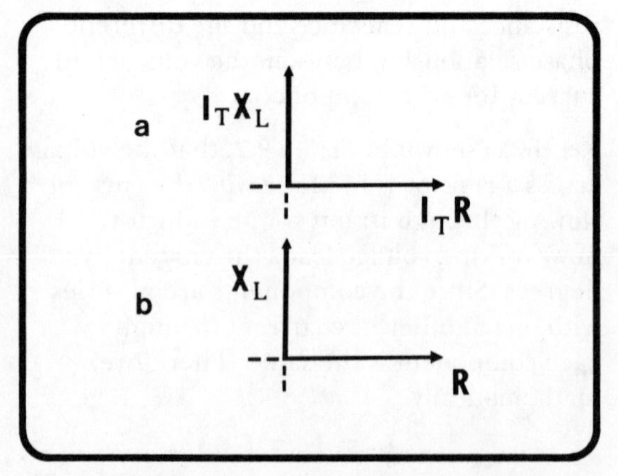

***Figure 9.5** Relationship of R and X_L*

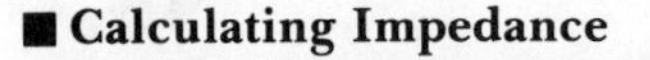

■ Calculating Impedance

Recall that in a purely resistive series circuit the sum of all ohms of resistance equals the total ohms of resistance. And in a purely reactive series circuit the sum of all ohms of reactance equals the total ohms of reactance. In an RL circuit, however, there exists a combination of resistance and reactance like in a series RC circuit, and it is called impedance, Z, measured in ohms, as illustrated in *Figure 9.6*.

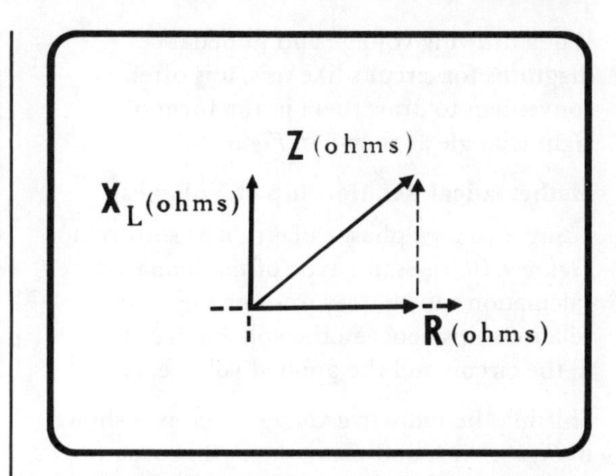

Figure 9.6 *Resistive-Inductive Circuit Phasor Diagram*

Calculating Impedance

Impedance of an RC circuit can be calculated by adding the resistance and reactance vectorially on the phasor diagram. This is accomplished by simply applying the Pythagorean theorem. As illustrated in *Figure 9.6*, the length of the reactance vector, X_L, is the same as the length between the tip of the resistance vector and the tip of the impedance vector Z. In fact, the phasor diagram can be drawn with the reactance vector placed as shown in *Figure 9.7*. Now, the right triangle of the Pythagorean theorem becomes clearly evident. Applying the Pythagorean theorem, the total impedance is equal to the square root of the resistance squared plus the inductive reactance squared.

$$\mathbf{Z = \sqrt{R^2 + X_L^2}} \qquad (9\text{–}5)$$

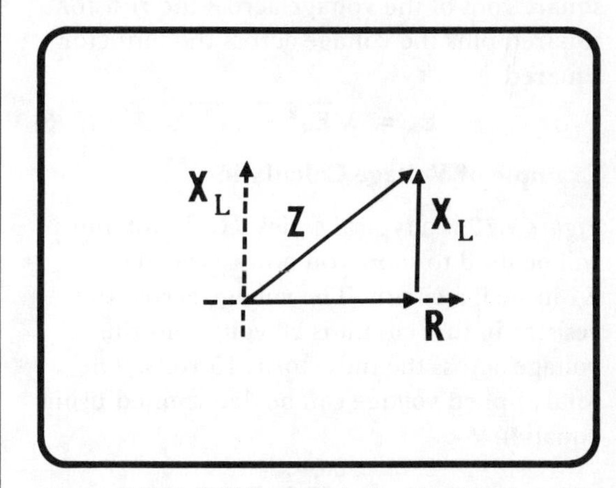

Figure 9.7 *Right Triangle Relationship of Z, R, and X_L*

For example, a typical series RL circuit is shown in *Figure 9.8*. If the resistance in that circuit is 60 ohms and the inductive reactance is 80 ohms, then the total impedance of the circuit can be determined as follows:

$$\begin{aligned} \mathbf{Z} &= \mathbf{\sqrt{R^2 + X_L^2}} \\ &= \mathbf{\sqrt{60^2 + 80^2}} \\ &= \mathbf{\sqrt{3600 + 6400}} \\ &= \mathbf{\sqrt{10{,}000}} \\ &= \mathbf{100\Omega} \end{aligned}$$

The total impedance of that circuit is 100 ohms.

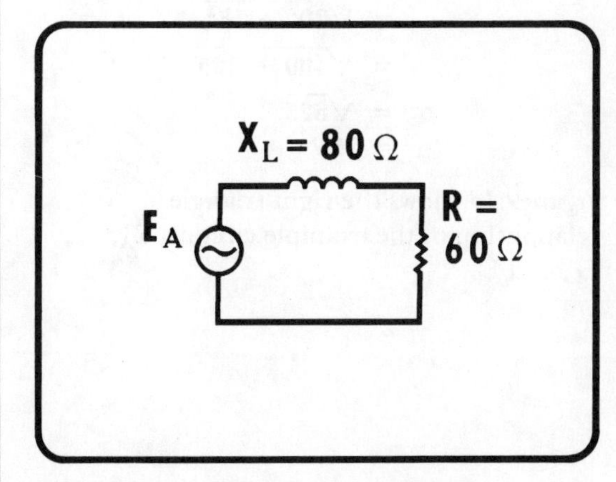

Figure 9.8 *Impedance Calculation Example*

When drawing voltage and impedance diagrams for circuits like this, it is often convenient to draw them in the form of a right triangle as shown in *Figure 9.9*.

Mathematical Relationship of Voltages

Using a voltage phasor diagram as shown in *Figure 9.10,* the same type of mathematical calculation can be used to show the relationships between the voltage drops in the circuit and the applied voltage.

Shifting the inductive voltage vector as shown in *Figure 9.11*, and using the Pythagorean theorem, the applied voltage is equal to the square root of the voltage across the resistor squared plus the voltage across the inductor squared:

$$E_A = \sqrt{E_R^2 + E_L^2} \qquad (9\text{–}6)$$

Example of Voltage Calculation

Figure 9.12 is a typical series RL circuit and will be used to show you how to calculate E_A in such a circuit. The voltage across the resistor in that circuit is 20 volts, and the voltage across the inductor is 15 volts. The total applied voltage can be determined using equation *9–6*.

$$\begin{aligned} E_A &= \sqrt{E_R^2 + E_L^2} \\ &= \sqrt{20^2 + 15^2} \\ &= \sqrt{400 + 225} \\ &= \sqrt{625} \\ &= 25V \end{aligned}$$

Figure 9.13 shows the right triangle relationship of the example circuit's E_A, E_L, and E_R.

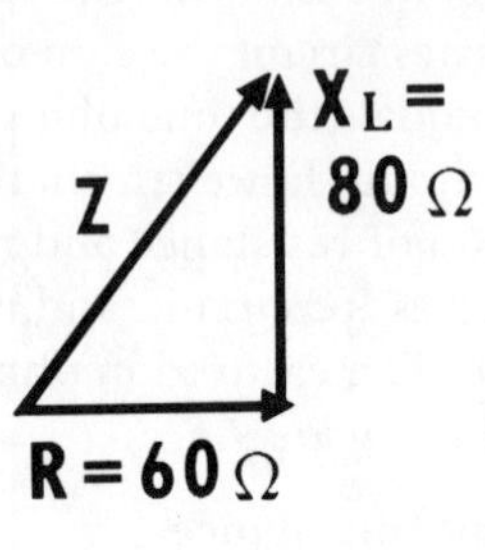

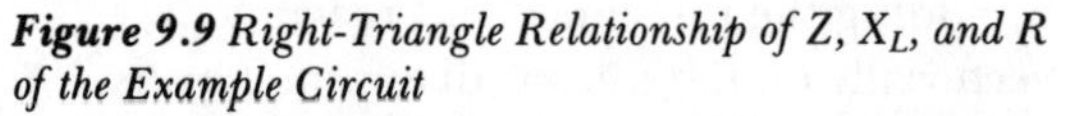
Figure 9.9 *Right-Triangle Relationship of Z, X_L, and R of the Example Circuit*

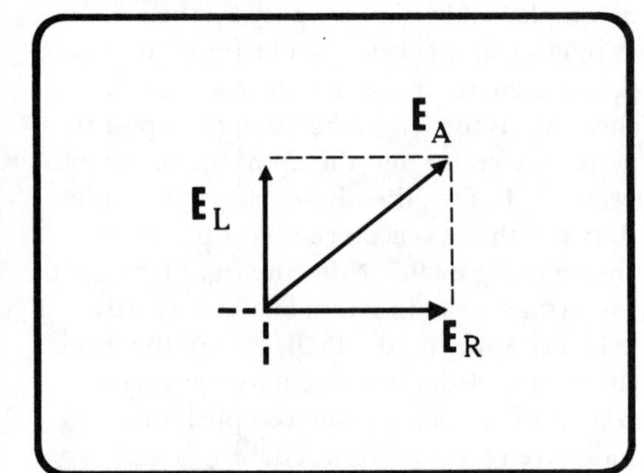

Figure 9.10 *Voltage Phasor Diagram of a Series RL Circuit*

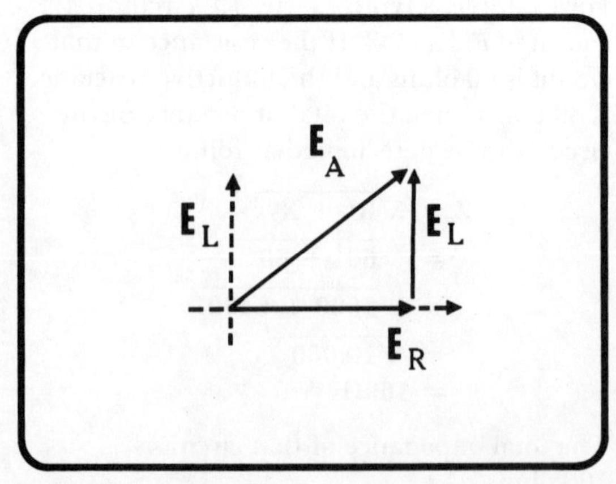

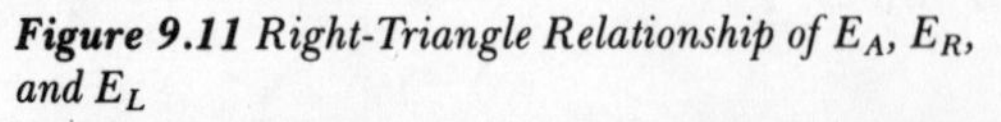
Figure 9.11 *Right-Triangle Relationship of E_A, E_R, and E_L*

Solution of a Series RL Circuit

Now, if you are given an applied voltage and values for R and L for a series RL circuit, you should be able to determine the total impedance, the total current, and the individual voltage drops across the resistor and inductor. For example, in the series RL circuit of *Figure 9.14*, the value of the resistor is 75 ohms, and the value of the inductor is 4 millihenrys. The applied voltage is 250 volts with a frequency of 4 kilohertz. To determine the total impedance of this circuit, the value of the inductive reactance must first be calculated.

$$\begin{aligned} X_L &= 2\pi fL \\ &= (6.28)\ (4kHz)\ (4mH) \\ &= 100\Omega \end{aligned}$$

The impedance phasor diagram shown in *Figure 9.15* can now be drawn to show the relationships between the values of resistance, reactance, and impedance in the circuit. The value of impedance is calculated using the Pythagorean theorem. The impedance is equal to the square root of the resistance squared plus the inductive reactance squared:

$$\begin{aligned} Z &= \sqrt{R^2 + X_L{}^2} \\ &= \sqrt{75^2 + 100^2} \\ &= \sqrt{5{,}625 + 10{,}000} \\ &= \sqrt{15{,}625} \\ &= 125\Omega \end{aligned}$$

Remember that impedance is the total opposition of the resistive-reactive circuit to the flow of alternating current. Recall that in Lesson 7, this fact was used to determine a special form of Ohm's law which is called Ohm's law for ac circuits:

$$E_A = I_T Z_T \qquad (9\text{–}7)$$

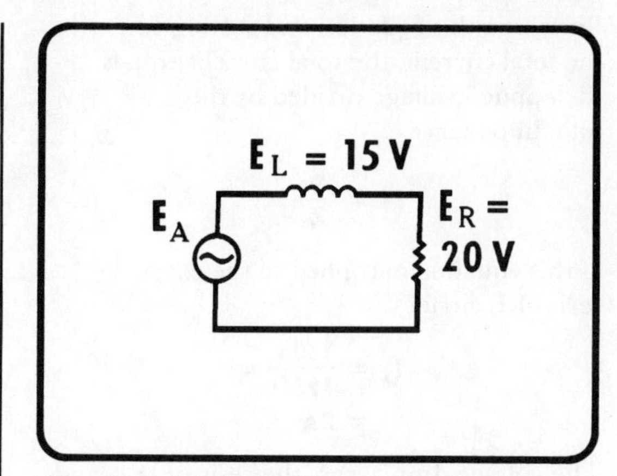

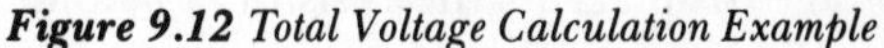
Figure 9.12 Total Voltage Calculation Example

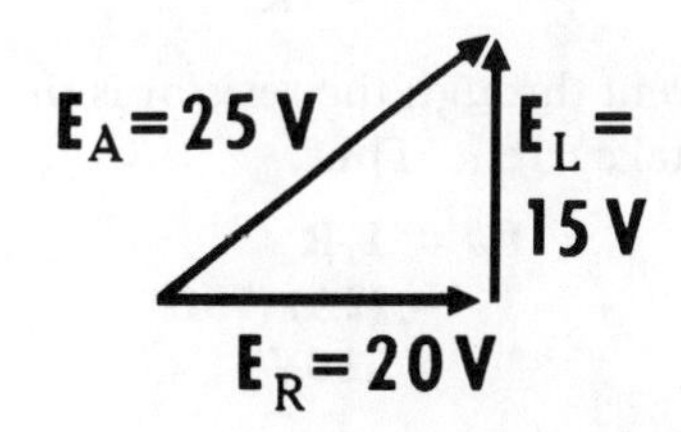

Figure 9.13 Right-Triangle Relationship of E_A, E_L and E_R of the Example Circuit

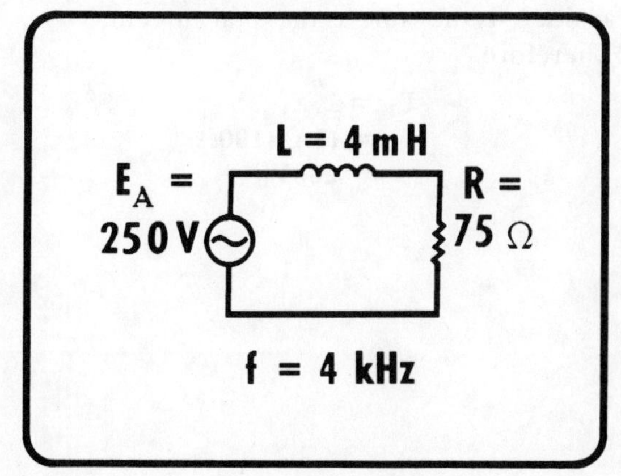

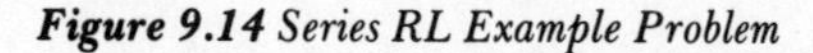
Figure 9.14 Series RL Example Problem

■ Solution of a Series RL Circuit

Rewriting this equation and solving for the total current, the total current equals the applied voltage divided by the total impedance.

$$I_T = \frac{E_A}{Z_T} \qquad (9\text{–}8)$$

If this equation is applied to the example series RL circuit:

$$\begin{aligned} I_T &= \frac{250V}{125\Omega} \\ &= 2A \end{aligned}$$

The voltage drop across the resistor, E_R, and inductor, E_L, can now be calculated as they would in any series circuit. The voltage drop across the resistor is equal to the current through it times the value of the resistance.

$$E_R = IR \qquad (9\text{–}9)$$

The current through the resistor is the same as the total current. Thus,

$$\begin{aligned} E_R &= I_T R \\ &= (2A)\,(75\Omega) \\ &= 150V \end{aligned}$$

The voltage drop across the inductor is equal to the current through it times the value of the inductive reactance.

$$E_L = I_L X_L \qquad (9\text{–}10)$$

again, I_L is the same as the total current. Therefore,

$$\begin{aligned} E_L &= I_T X_L \\ &= (2A)\,(100\Omega) \\ &= 200V \end{aligned}$$

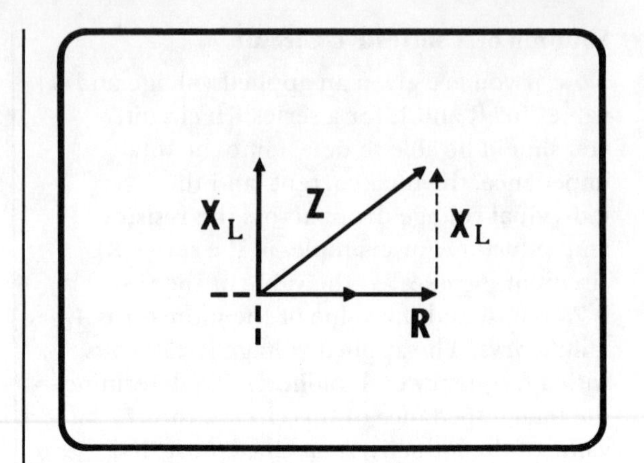

Figure 9.15 Impedance Phasor Diagram

If the voltage across the resistor, 150 volts, and the voltage across the inductor, 200 volts, are added, the result is 350 volts. That is more than the applied voltage. But remember that the voltage across the resistor and the voltage across the inductor are out of phase by 90 degrees. Therefore they must be added vectorially. Recall that vectorially, the applied voltage equals the square root of the voltage across the resistor squared, plus the voltage across the inductor squared. Therefore, for the example circuit

$$\begin{aligned} E_A &= \sqrt{E_R^2 + E_L^2} \\ &= \sqrt{150^2 + 200^2} \\ &= \sqrt{22{,}500 + 40{,}000} \\ &= \sqrt{62{,}500} \\ &= 250V \end{aligned}$$

Thus, you can see that the vector sum of the circuit voltage drops does equal the applied voltage, 250 volts.

This technique is a valuable tool for circuit analysis because the calculation can be used to check the accuracy of calculations performed to determine the various individual voltage drops in the circuit. Simply be sure that the vector sum of the circuit voltage drops equals the applied voltage.

Phase Angle in a Series RL Circuit

Recall that when originally forming the voltage phasor diagram shown in *Figure 9.16*, the current in the circuit was used as a reference quantity since it is the same throughout the circuit. This total current is in phase with the voltage across the resistor. Notice, however, that the applied voltage and the total current are out of phase. More specifically, the applied voltage leads the total current by a number of degrees. Recall that this angle by which the total applied voltage and the total current are out of phase is called the *phase angle* of the circuit. The phase angle is the number of degrees by which the current being drawn from the ac voltage source and the voltage of the ac voltage source are out of phase. The phase angle in a series RL circuit can also be recognized as the angle between the voltage across the resistor and the applied voltage, as shown in *Figure 9.16*.

Since the impedance phasor diagram is proportional to the voltage phasor diagram by the common factor of total current which cancels, X_L can be substituted for E_L, Z can be substituted for E_A, and R can be substituted for E_R as shown in *Figure 9.17*. Therefore, the phase angle in a series RL circuit is also the angle between the resistance phasor and the impedance phasor. This is similar to a series RC circuit. Recall that when determining the value of this phase angle, a trigonometric function was used called the tangent function. Recall that the tangent of an angle in a right triangle is equal to the ratio of the length of the opposite side divided by the length of the adjacent side as shown in *Figure 9.18*.

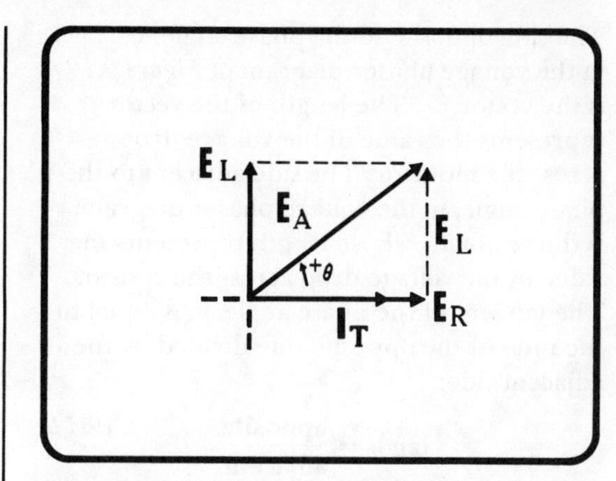

Figure 9.16 *Location of Phase Angle*

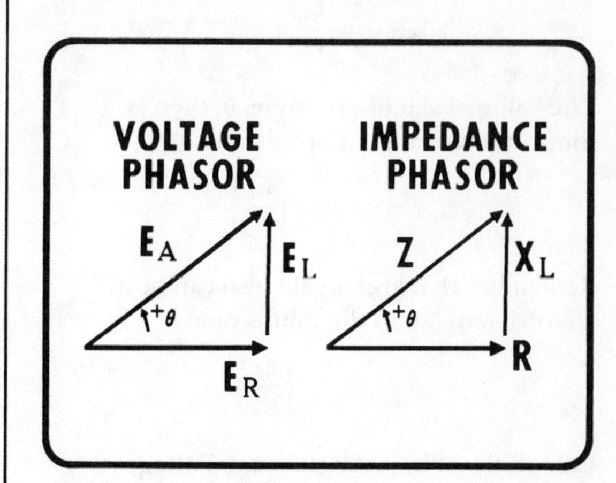

Figure 9.17 *Voltage and Impedance Phasor Diagram*

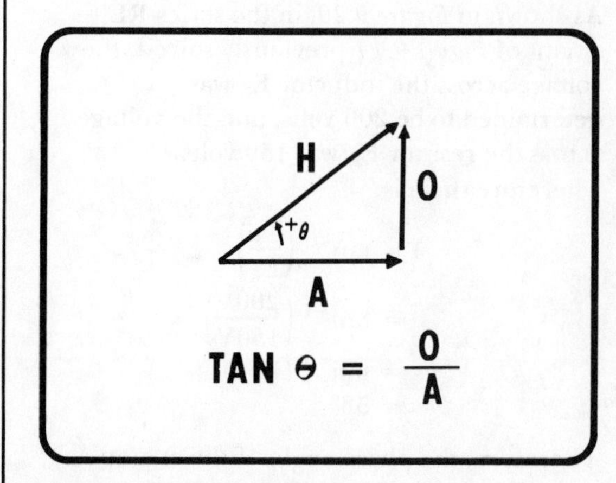

Figure 9.18 *Tangent Function*

Calculating Phase Angle from Voltage Phasor Diagram

The side opposite to the phase angle, θ, in the voltage phasor diagram of *Figure 9.19* is the vector E_L. The length of the vector represents the value of the voltage drop across the inductor. The side adjacent to the phase angle, in the voltage phasor diagram, is the vector E_R whose length represents the value of the voltage drop across the resistor. The tangent of the phase angle, θ, is equal to the ratio of the opposite side divided by the adjacent side:

$$\tan \theta = \frac{\text{opposite}}{\text{adjacent}} \quad (9\text{-}11)$$

$$\tan \theta = \frac{E_L}{E_R} \quad (9\text{-}12)$$

The value of the phase angle, θ, then is simply the arctangent of this ratio.

$$\theta = \arctan\left(\frac{E_L}{E_R}\right) \quad (9\text{-}13)$$

Remember that arctangent also can be abbreviated, tan to the minus one:

$$\theta = \tan^{-1}\left(\frac{E_L}{E_R}\right) \quad (9\text{-}14)$$

Calculating Phase Angle from Voltage Phasor Diagram

As shown in *Figure 9.20*, in the series RL circuit of *Figure 9.14* previously solved, the voltage across the inductor E_L was determined to be 200 volts, and the voltage across the resistor E_R was 150 volts. Therefore theta is:

$$\begin{aligned} \theta &= \tan^{-1}\left(\frac{E_L}{E_R}\right) \\ &= \tan^{-1}\left(\frac{200V}{150V}\right) \\ &= \tan^{-1}(1.33) \\ &= 53° \end{aligned}$$

Therefore, the phase angle of the example series RL circuit is 53 degrees, a *positive* 53 degrees.

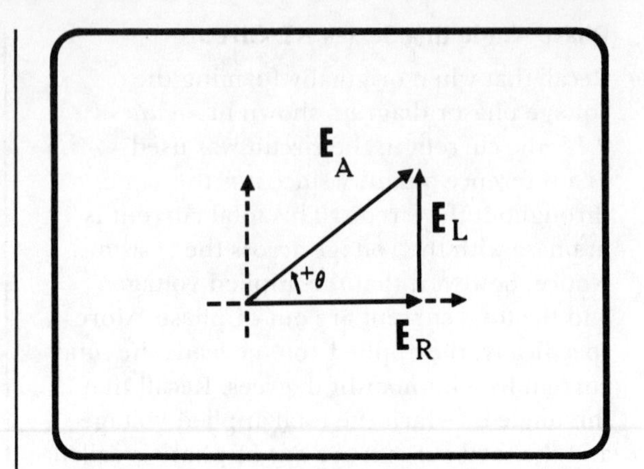

Figure 9.19 *Opposite and Adjacent Side in a Voltage Phasor Diagram*

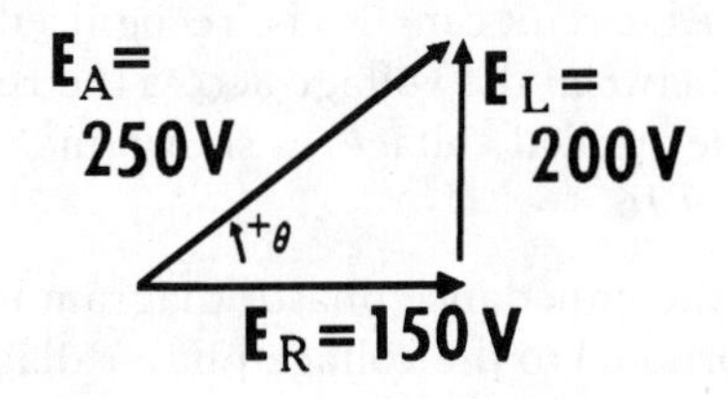

Figure 9.20 *Sample Calculation of Phase Angle Using Calculated Voltage Phasor Diagram*

Series RL Phase Angle is Positive

In the lesson about RC circuits an arbitrary standard was established for the angle of rotation of vectors for series circuits. It was established that the angle of rotation of a phasor in a counter-clockwise direction from the zero degree reference forms a positive phase angle as shown in *Figure 9.21*. Therefore, the positive sign of the phase angle, as shown in *Figure 9.22*, indicates that the applied voltage is rotated 53 degrees counter-clockwise (or up) from the current vector direction. Phase angles in series RL circuits *will always be positive phase angles.*

Calculating Phase Angle from Impedance Phasor Diagram

Recall that earlier it was described how the impedance phasor diagram is proportional to the voltage phasor diagram by a factor of the total current for a series circuit. Since these two phasor diagrams are proportional, it is possible to determine the phase angle from the impedance phasor diagram as illustrated in *Figure 9.23*. The side opposite to the phase angle, θ, in the impedance phasor diagram, is the reactance vector, X_L. The length of this vector represents the value of the inductive reactance. The side adjacent to the phase angle in the impedance phasor diagram is the resistance vector, R. Its length represents the value of the resistor. The tangent of the phase angle, then, is equal to the ratio of the opposite side, X_L, divided by the adjacent side, R.

$$\tan\theta = \frac{\text{opposite}}{\text{adjacent}}$$

$$= \frac{X_L}{R}$$

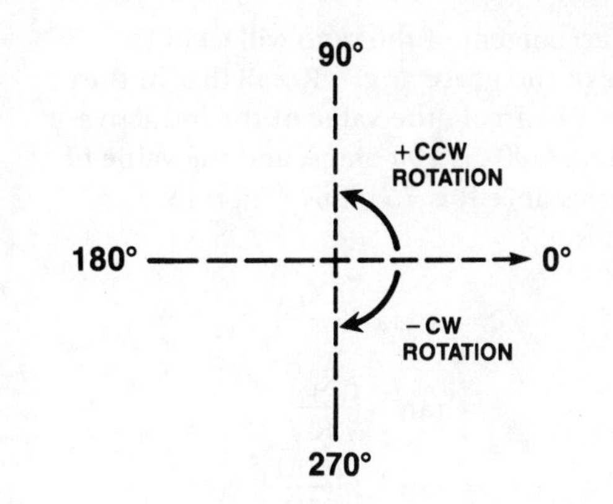

Figure 9.21 *Positive-Negative Phase Angle Convention*

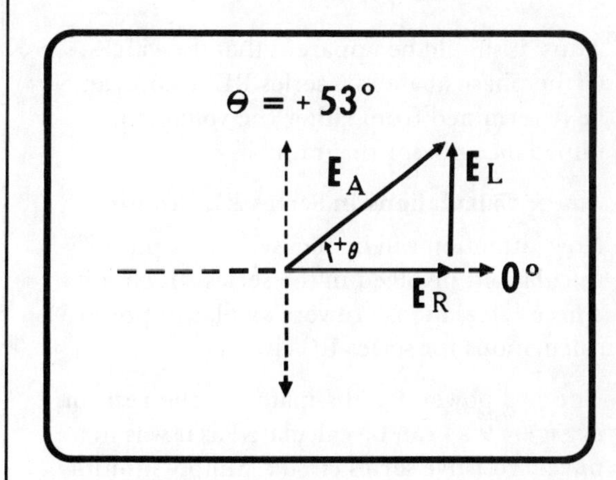

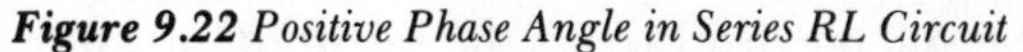

Figure 9.22 *Positive Phase Angle in Series RL Circuit*

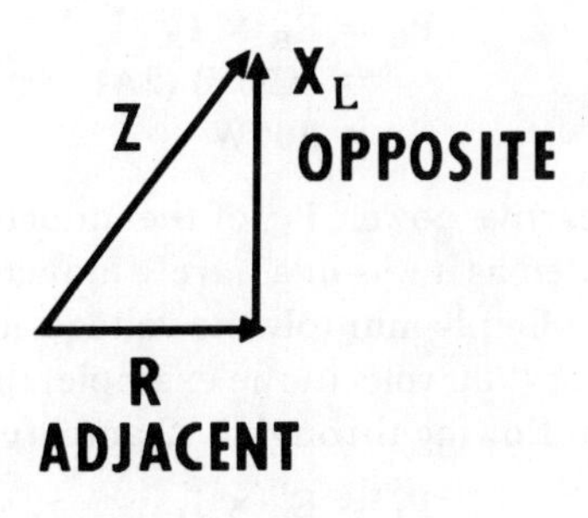

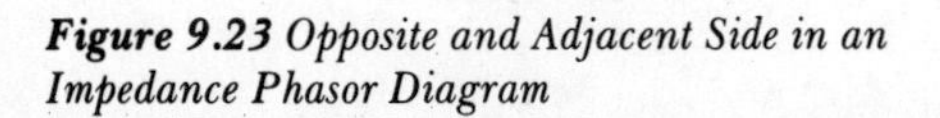

Figure 9.23 *Opposite and Adjacent Side in an Impedance Phasor Diagram*

Power Calculations in Series RL Circuit

The arctangent of the ratio will yield the value of the phase angle. Recall that in the series RL circuit, the value of the inductive reactance, X_L, is 100 ohms, and the value of the resistance R is 75 ohms. Therefore theta is:

$$\theta = \arctan\left(\frac{X_L}{R}\right)$$
$$= \tan^{-1}\left(\frac{X_L}{R}\right)$$
$$= \tan^{-1}\left(\frac{100\Omega}{75\Omega}\right)$$
$$= \tan^{-1}(1.33)$$
$$= 53°$$

Thus, it should be apparent that the value of the phase angle in a series RL circuit can be determined from either the voltage or impedance phasor diagram.

Power Calculations in Series RL Circuit

Now, attention will be focused upon power calculations involved in the series RL circuit. These calculations are very similar to power calculations for series RC circuits.

The real power, P_R, dissipated by the resistor in *Figure 9.24* can be calculated as it was in a purely resistive series circuit. Simply multiply the voltage drop across the resistor (150 volts in the example) times the current's value flowing through it (2 amperes).

$$\mathbf{P_R = E_R \times I_R}$$
$$\mathbf{= (150V)\ (2A)}$$
$$\mathbf{= 300W}$$

The reactive power, P_L, of the inductor is calculated as it was in a purely inductive circuit. Simply multiply the voltage across the inductor (200 volts in the example) times the current flowing through it (2 amperes).

$$\mathbf{P_L = E_L \times I_L}$$
$$\mathbf{= (200V)\ (2A)}$$
$$\mathbf{= 400VAR}$$

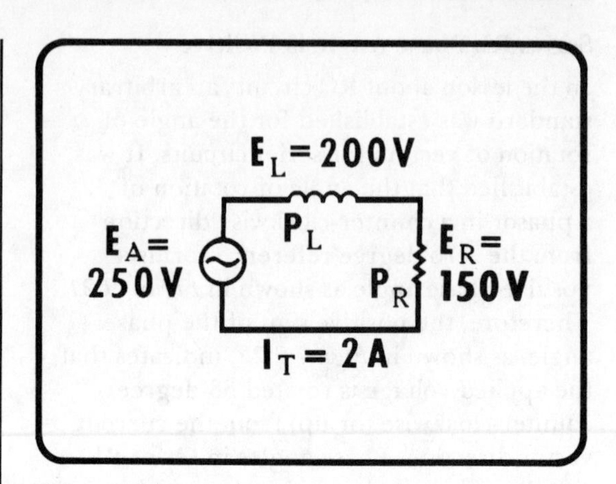

Figure 9.24 *Power in a Series RL Circuit*

Recall that the total power in a resistive-reactive circuit is called the apparent power, measured in volts-amperes. The total or apparent power can be found by multiplying the applied voltage (250 volts in the example) times the total current (2 amperes).

$$\mathbf{P_A = E_A \times I_T}$$
$$= (250V)\ (2A)$$
$$= 500VA$$

Remember, that the simple sum of the real power and the reactive power does not equal the apparent power. In the example, 500 volt-amperes does not equal 300 watts plus 400 VAR when they are added together.

$$P_A \neq P_R + P_L$$

or

$$500VA \neq 300W + 400VAR$$

■ Power Calculations in Series RL Circuit

This occurs because of the different phase relationships between the voltage and current for each component in the circuit. The phase relationships of the three power determinations are similar to the voltage phase relationships. *Figure 9.25* shows the voltage phasor diagram with its values E_A, E_R, and E_L. If each voltage is multiplied by the total current, I_T, in the circuit, it will be found that the power phasor diagram is very similar to the voltage phasor diagram because as shown in *Figure 9.26*

$$P_A = E_A I_T \qquad (9\text{–}15)$$

$$P_R = E_R I_T \qquad (9\text{–}16)$$

and

$$P_L = E_L I_T \qquad (9\text{–}17)$$

In fact, the values are proportional to the voltage diagrams by the common factor of total current, I_T.

Using the Pythagorean theorem, the apparent power, P_A, is equal to the square root of the real power, P_R, squared plus the reactive power, P_L, squared:

$$P_A = \sqrt{P_R^2 + P_L^2} \qquad (9\text{–}18)$$

Thus, the total apparent power in this circuit, 500 volt-amperes, should be equal to the square root of 300 watts squared plus 400 VAR squared. By completing the calculation, it is found that the apparent power is equal to 500 volt amperes:

$$\begin{aligned} P_A &= \sqrt{P_R^2 + P_L^2} \\ &= \sqrt{300^2 + 400^2} \\ &= \sqrt{90{,}000 + 16{,}000} \\ &= \sqrt{250{,}000} \\ &= 500\text{VA} \end{aligned}$$

This is the same value obtained by multiplying the applied voltage by total current.

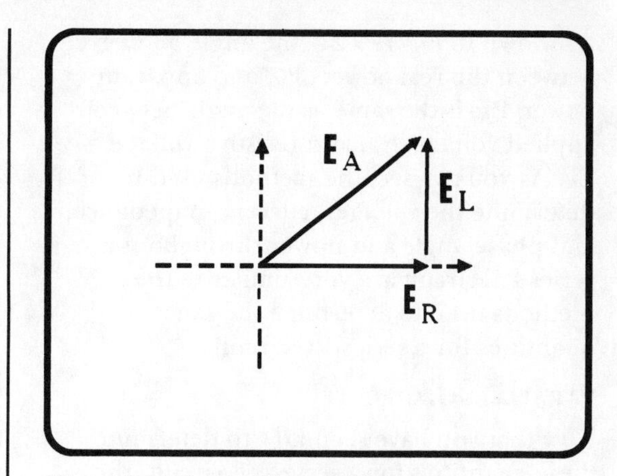

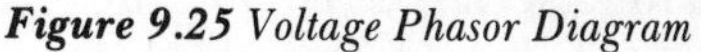

Figure 9.25 *Voltage Phasor Diagram*

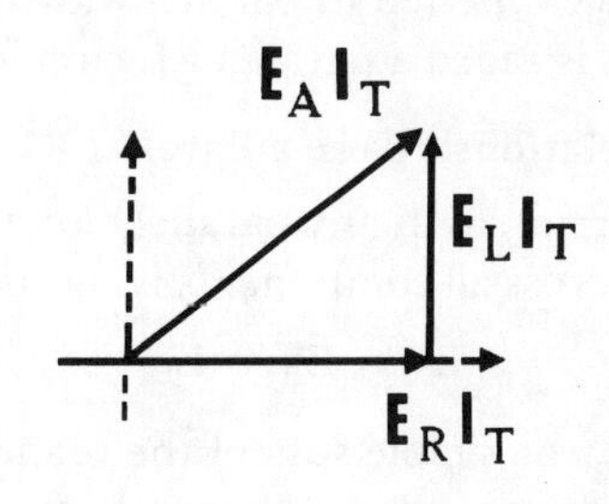

Figure 9.26 *Derivation of Power Phasor Diagram from Voltage Phasor Diagram*

As shown in *Figure 9.27*, the angle 53 degrees between the real power, P_R, and apparent power, P_A, is the same as the angle between applied voltage, E_A, and resistive voltage, E_R. As you can see, the methods used to determine the voltages, current, impedance, and phase angle and power throughout a series RL circuit are very similar to the methods used to determine the same quantities for a series RC circuit.

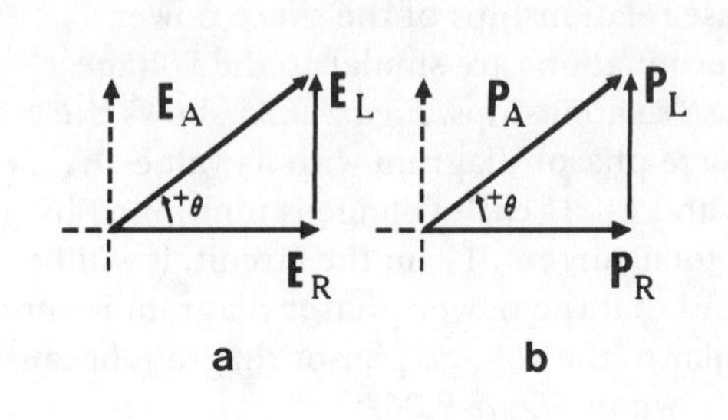

Figure 9.27 Voltage and Power Phasor Diagrams

PARALLEL RL CIRCUIT

Now that you have seen how to determine these quantities for a series RL circuit, the next step is to learn how to determine the same quantities in a parallel RL circuit. The circuit, shown in *Figure 9.28*, in which a resistor is connected in parallel with an inductor, is called a parallel RL circuit.

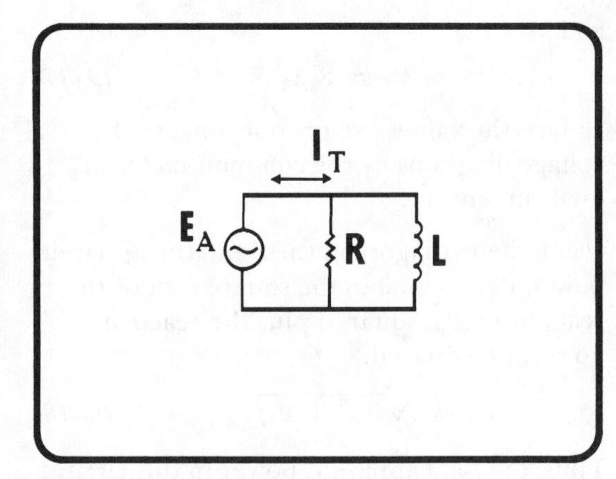

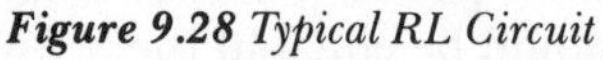

Figure 9.28 Typical RL Circuit

Phase Relationships in a Parallel RL Circuit

In this circuit, as in any parallel circuit, the voltage across all components is the same:

$$E_A = E_R = E_L$$

However, the simple sum of the branch currents does not equal the total current in the circuit:

$$I_T \neq I_R + I_L$$

This occurs because of the different phase relationships between the voltage and current of each component.

Recall that the voltage across the resistor is in phase with the current through it; in the inductor, the voltage leads the current by 90 degrees as shown in *Figure 9.29*. Since the components in this circuit are in parallel with one another, the common factor in both phasor diagrams is the voltage across the components:

$$E_A = E_R = E_L$$

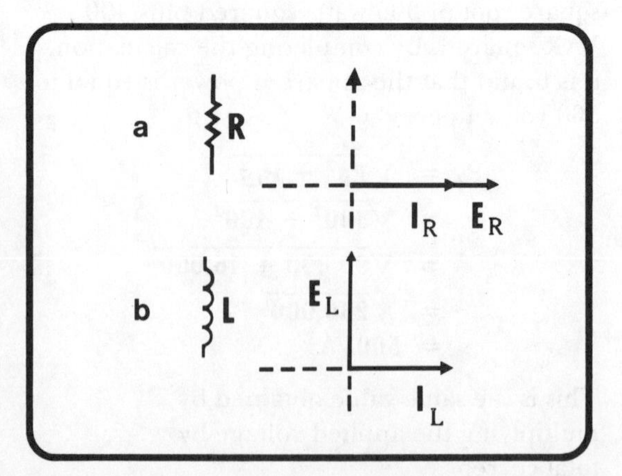

Figure 9.29 Phase Relationship for a Resistor, a, and an Inductor, b

However, to illustrate the phase relationships in a parallel circuit, E_L and I_L must be rotated 90 degrees as shown in *Figure 9.30a*. Then the two individual phasor diagrams can be combined, as shown in *Figure 9.30b*. Not that the current through the resistor is shown in phase with the applied voltage across it while the voltage across the inductor leads the current through it by 90 degrees.

Phasor Addition of Currents

Comparing the phase relationships of the two branch currents, the current through the resistor leads the current through the inductor by 90 degrees. These individual branch currents can be calculated as they were in either a purely resistive or purely inductive circuit. Simply divide the voltage across the branch by the opposition to current in that branch.

In the resistive branch of the example circuit, the opposition to flow of current is measured in ohms of resistance; thus, in the example circuit, the resistive current is determined by dividing the applied voltage across the resistor by the value of the resistor:

$$I_R = \frac{E_R}{R} = \frac{E_A}{R} \qquad (9\text{–}19)$$

In the inductive branch, the opposition to the flow of current is measured in ohms of reactance; thus, in the example circuit, the inductive branch current is determined by dividing the applied voltage across the inductor by the inductive reactance:

$$I_L = \frac{E_L}{X_L} = \frac{E_A}{X_L} \qquad (9\text{–}20)$$

These currents cannot be added algebraically to obtain the total current as the branch currents in a parallel resistive circuit can be added. The currents must be added vectorially because of the different phase

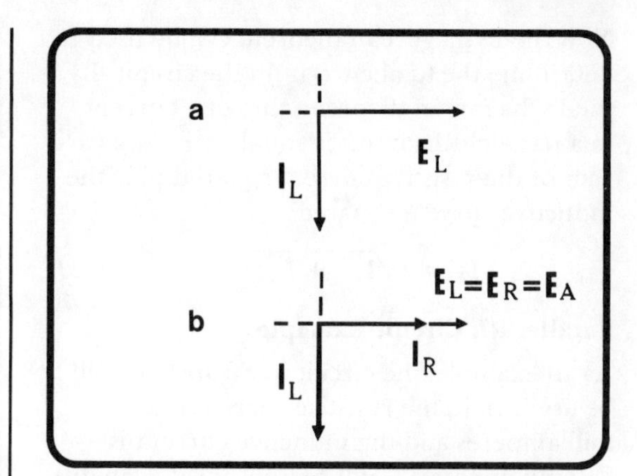

Figure 9.30 Combining Phase Relationships for Parallel RL Circuit

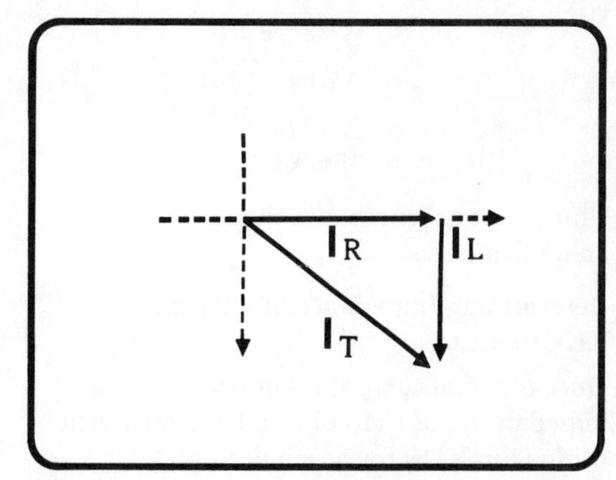

Figure 9.31 Right-Triangle Relationships of I_R, I_L, and I_T

relationship in a parallel RL circuit. The same procedure is used that was used in other diagrams to form a right triangle between the tip of the inductive current vector and the tip of the resistive current vector, with the total current vector, I_T, completing the hypotenuse of the triangle. This is illustrated in *Figure 9.31*.

■ **Parallel RL Circuit Example**
■ **Determining Impedance of a Parallel RL Circuit**

Now the Pythagorean theorem can be used to determine the total current in the circuit. By the Pythagorean theorem, the total current in a parallel RL circuit is equal to the square root of the resistive current squared plus the inductive current squared:

$$I_T = \sqrt{I_R^2 + I_L^2} \qquad (9\text{–}21)$$

Parallel RL Circuit Example

As an example, the circuit in *Figure 9.32* will be used. In it, the resistive current is 28 milliamperes and the inductive current is 21 milliamperes. Total current is calculated:

$$\begin{aligned} I_T &= \sqrt{I_R^2 + I_L^2} \\ &= \sqrt{28^2 + 21^2} \\ &= \sqrt{784 + 441} \\ &= \sqrt{1225} \\ &= 35\text{mA} \end{aligned}$$

Thus, the total current in the circuit is 35 milliamperes.

Determining Impedance of a Parallel RL Circuit

Once the total current is known, the total impedance of the circuit can be determined using Ohm's law for ac circuits:

$$E_A = I_T Z \qquad (9\text{–}22)$$

However, this time the equation must be rewritten to solve for impedance. The total impedance of the circuit is equal to the applied voltage divided by the total current:

$$Z = \frac{E_A}{I_T} \qquad (9\text{–}23)$$

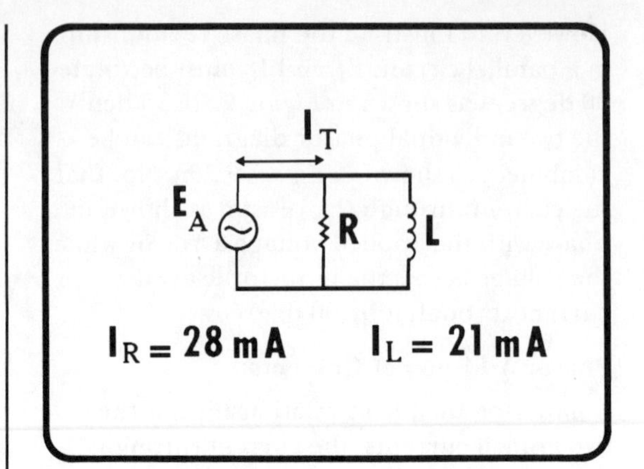

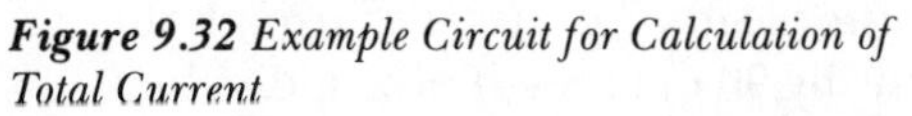

Figure 9.32 Example Circuit for Calculation of Total Current

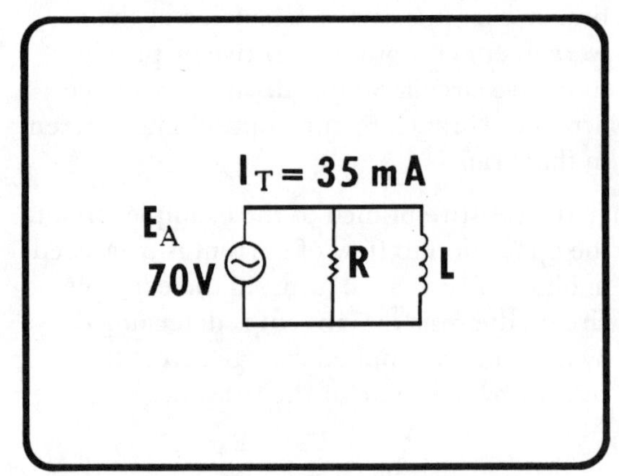

Figure 9.33 Example Circuit for Impedance Calculation

For example, if in the previous example circuit the total applied voltage was 70 volts as shown in *Figure 9.33*, then the circuit impedance is equal to the applied voltage, 70 volts, divided by the total current, 35 millamperes:

$$\begin{aligned} Z &= \frac{E_A}{I_T} \\ &= \frac{70\text{V}}{35\text{mA}} \\ &= 2\text{k}\Omega \end{aligned}$$

Phase Angle in a Parallel RL Circuit

Now the phase angle of the example RL circuit will be determined by utilizing the original current phasor diagram shown in *Figure 9.34*. Recall that the phase angle is the number of degrees of phase difference between the applied voltage and the total current. Also recall that the applied voltage, E_A, is also the voltage across the resistor, E_R, and that the current through the resistor is in phase with the voltage across it. Therefore, the applied voltage is in phase with the resistive current, I_R. The voltage across the inductor, E_L, is the same as the applied voltage and it leads the current in the inductor by 90 degrees. Therefore, I_L forms the right angle with I_R, and I_T is the hypotenuse of the right triangle.

Since the applied voltage, E_A, is in phase with the current through resistor I_R, the phase angle is the angle, θ, between I_R and I_T. And, as you can see, the phase angle is a parallel RL circuit is simply the angle between the resistive current and total current.

The value of the phase angle of a parallel RL circuit can be calculated by determining the arctangent of the ratio of the opposite and adjacent sides. The length of the opposite side represents the value of the inductive current. The length of the adjacent side represents the value of the resistive current. Mathematically, therefore:

$$\theta = \arctan\left(\frac{\text{opposite}}{\text{adjacent}}\right) \qquad (9\text{–}24)$$

or

$$\theta = \arctan\left(\frac{I_L}{I_R}\right) \qquad (9\text{–}25)$$

or

$$\theta = \tan^{-1}\left(\frac{I_L}{I_R}\right) \qquad (9\text{–}26)$$

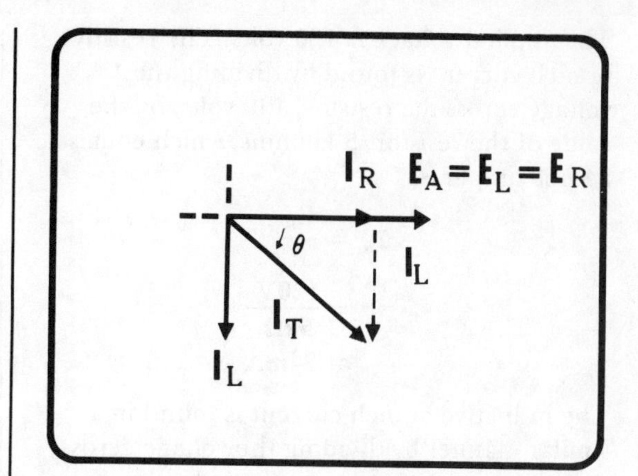

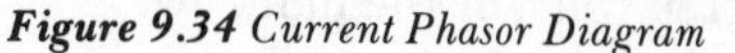
Figure 9.34 Current Phasor Diagram

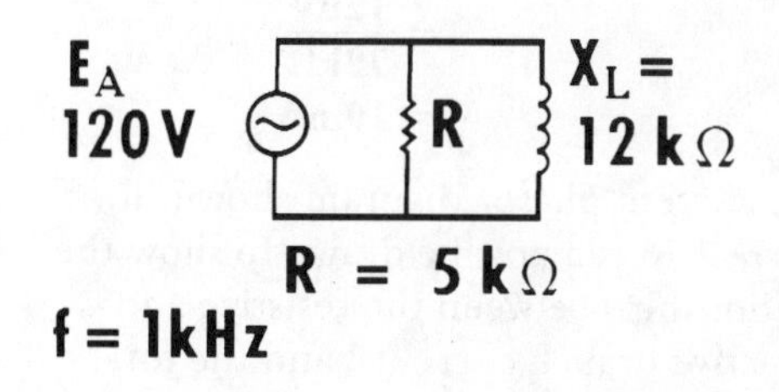

Figure 9.35 Parallel RL Circuit Example

Solution of a Parallel RL Circuit

In this section, a parallel RL circuit will be analyzed and the individual branch currents, the total current, the total impedance, and the phase angle determined.

Branch Currents

In the parallel RL circuit shown in *Figure 9.35*, the value of the resistor is 5 kilohms and the value of the inductor is 1.9 henrys. Therefore, at the applied frequency of 1,000 hertz, the inductor has an inductive reactance of 12 kilohms. This is determined by the inductive reactance equation:

$$X_L = 2\pi fL \qquad (9\text{–}27)$$

The applied voltage is 120 volts. The resistive branch current is found by dividing the voltage across the resistor, 120 volts, by the value of the resistor, 5 kilohms, which equals 24 milliamperes:

$$I_R = \frac{E_R}{R} = \frac{120V}{5k\Omega} = 24mA$$

The inductive branch current is found in a similar manner by dividing the voltage across the capacitor, 120 volts, by the inductive reactance, 12 kilohms, which equals 10 milliamperes:

$$I_L = \frac{E_L}{X_L} = \frac{120V}{12k\Omega} = 10mA$$

The current phasor diagram shown in *Figure 9.36* can now be drawn to show the relationships between the resistive and inductive branch currents, and the total circuit current.

Total Current

Because of the different phases, total current is a vector sum. Therefore, using the Pythagorean theorem, the total current is equal to the square root of the resistive branch current squared plus the inductive branch current squared. Since I_R is 24 milliamperes and I_L is 10 milliamperes, total current is calculated:

$$I_T = \sqrt{I_R^2 + I_L^2} = \sqrt{24^2 + 10^2} = \sqrt{576 + 100} = \sqrt{676} = 26mA$$

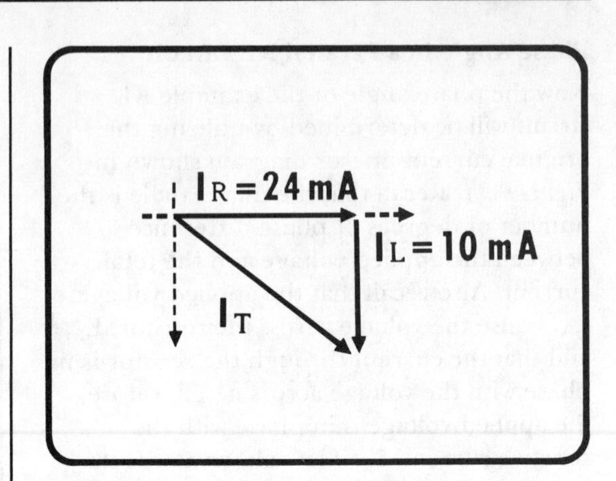

Figure 9.36 Current Phasor Diagram for Example Parallel RL Circuit

Thus, the value of the total current is 26 milliamperes. Since all of the current values were in milliamperes, the total current is measured in milliamperes also.

Impedance

The total impedance of the example parallel RL circuit can now be determined by dividing the applied voltage by the total current.

$$Z = \frac{E_A}{I_T} = \frac{120V}{26 \times 10^{-3}} = 4.6k\Omega$$

The total impedance is 4.6 kilohms.

■ **Phase Angle**
■ **Parallel RL Circuit Phase Angle is Negative**

Phase Angle

Using the current phasor diagram shown in *Figure 9.37*, it can be seen that the phase angle is the angle between the resistive current and the total current. The value of the phase angle is equal to the arctangent of the ratio of the inductive current divided by the resistive current.

$$\begin{aligned}\theta &= \arctan\left(\frac{I_L}{I_R}\right)\\ &= \arctan\left(\frac{10mA}{24mA}\right)\\ &= \arctan(0.4166)\\ &= -23°\end{aligned}$$

The phase angle is a minus 23 degrees.

Parallel RL Circuit Phase Angle is Negative

The negative sign is used with the phase angle of this particular example and the phase angle is said to be a negative 23 degrees. This is because the applied voltage in the parallel RL circuit is used as a reference at zero degrees. The negative sign is used to indicate that the total current phasor is rotated 23 degrees clockwise from the applied voltage phase, as shown in *Figure 9.38*. E_A leads I_T by 23 degrees. Or it could be said that I_T lags E_A by 23 degrees.

You should realize, therefore, that in a *series* RL circuit, the phase angle is positive as shown in *Figure 9.39*. However, in a parallel RL circuit, the phase angle is *negative*, as shown in *Figure 9.40*. The sign of the phase angle is used simply to indicate the direction of rotation from the reference at zero degrees. As a result, if you know that a circuit is an RL circuit and you know its phase angle, you can readily determine whether the resistor and inductor are connected in series or parallel by knowing the sign of the phase angle.

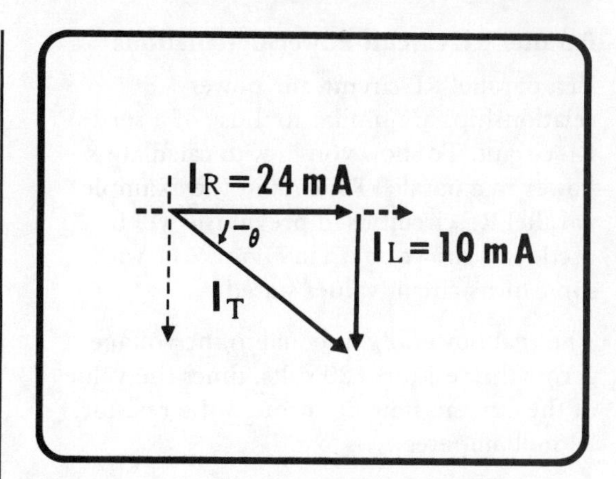

Figure 9.37 *Phase Angle in Current Phasor Diagram*

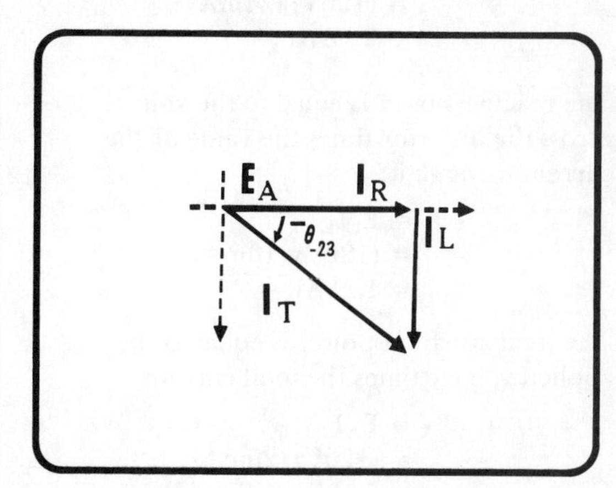

Figure 9.38 *Phase Angle of Example Circuit is −23 Degrees*

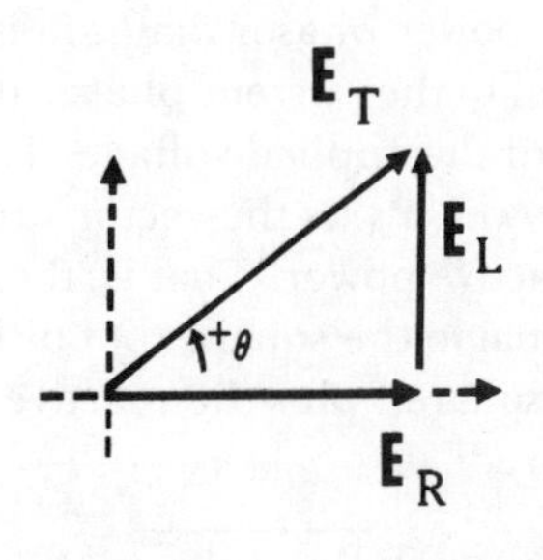

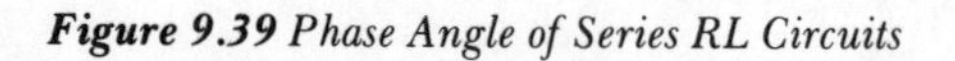

Figure 9.39 *Phase Angle of Series RL Circuits*

Parallel RL Circuit Power Calculations

In a parallel RL circuit, the power relationships are similar to those of a series RL circuit. To show you how to calculate power in a parallel RL circuit, the example parallel RL circuit used previously will be used. It is shown again in *Figure 9.41* with some of its circuit values stated.

The real power, P_R, is equal to the voltage across the resistor, 120 volts, times the value of the current flowing through the resistor, 24 milliamperes.

$$\begin{aligned} P_R &= E_R I_R \\ &= (120V)\ (24mA) \\ &= 2.88W \end{aligned}$$

The reactive power is equal to the voltage across the inductor times the value of the current through it:

$$\begin{aligned} P_L &= E_L I_L \\ &= (120V)\ (10mA) \\ &= 1.2VAR \end{aligned}$$

The total apparent power is equal to the applied voltage times the total current.

$$\begin{aligned} P_A &= E_A I_T \\ &= (120V)\ (26mA) \\ &= 3.12VA \end{aligned}$$

Since each power determination is made by multiplying the current shown in the current phasor diagram of *Figure 9.42* by the applied voltage, the power phasor diagram is proportional to the current phasor diagram by a factor of the applied voltage. The total apparent power, P_A, is the vector sum of the real and reactive power. That is, the apparent power is equal to the square root of the real power squared plus the reactive power squared:

$$P_A = \sqrt{P_R^2 + P_L^2} \qquad (9\text{–}28)$$

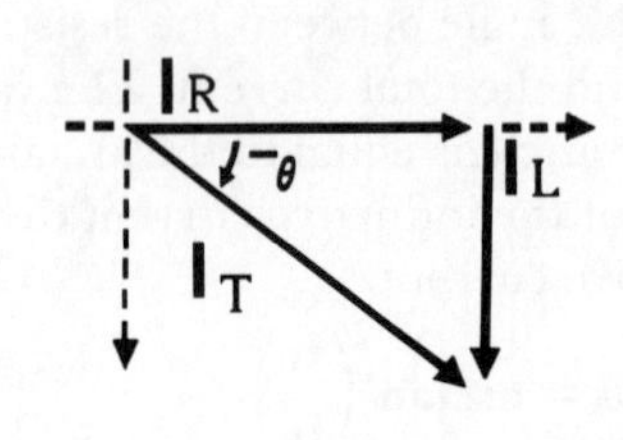

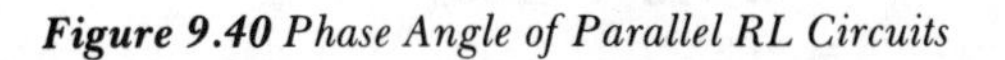

Figure 9.40 *Phase Angle of Parallel RL Circuits*

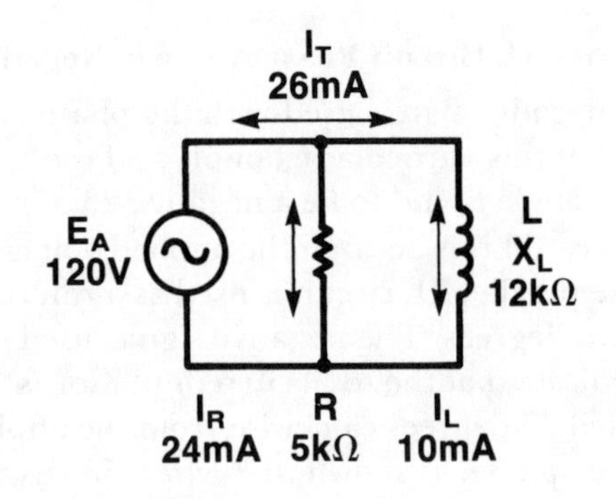

Figure 9.41 *Circuit Values for Power Calculations*

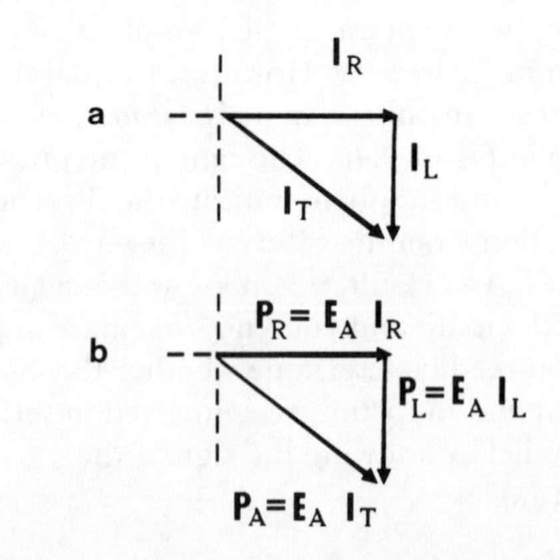

Figure 9.42 *Comparison of Current and Power Phasor Diagrams*

Therefore, P_A for the example parallel RL circuit is calculated:

$$\begin{aligned} P_A &= \sqrt{P_R^2 + P_L^2} \\ &= \sqrt{2.88^2 + 1.2^2} \\ &= \sqrt{8.29 + 1.44} \\ &= \sqrt{9.73} \\ &= 3.12\text{VA} \end{aligned}$$

The total apparent power calculated this way is 3.12 volt-amperes, the same as it was calculated using the equation $P_A = E_A I_T$.

Q OF A COIL

Now that solution methods of series and parallel RL circuits have been discussed, one other important topic concerning coils needs to be discussed—that is, the *Q or quality* factor of a coil.

Concept of Q

The inductive reactance, X_L, of a coil is an indication of the ability of a coil to produce self-induced voltage. Recall that inductive reactance is expressed mathematically as

$$X_L = 2\pi fL.$$

The expression takes into consideration the rate of change of current in the circuit in terms of frequency (f) and the size of the inductor in terms of its value (L). To manufacture a coil, however, many turns of wire are used. This results in a resistance in the coil due entirely to the length of the wire. This internal resistance was mentioned in Lesson 8 and is represented by r_i. *Figure 9.43* shows that the internal resistance of a coil is represented schematically by showing a small resistor in series with the coil. The coil represents the inductor's reactive qualities while the resistor represents its resistive qualities.

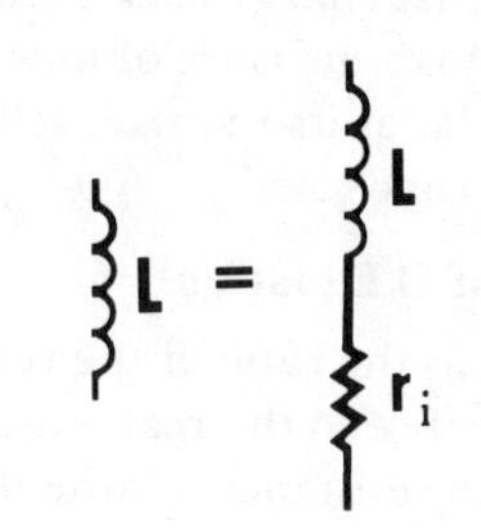

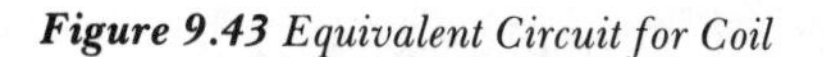

Figure 9.43 *Equivalent Circuit for Coil*

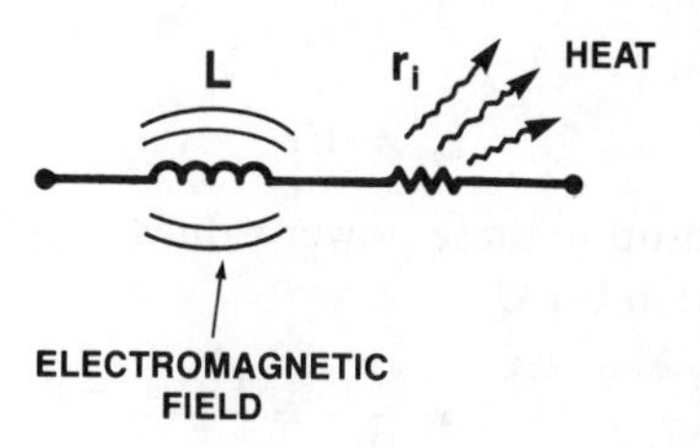

Figure 9.44 *Energy in a Coil*

When current is passed through a coil, energy is stored in the magnetic field as it expands around the coil, and then it is returned to the circuit when the magnetic field collapses *(Figure 9.44)*. No heat energy is lost in a coil due to the coil's inductive quality. But the resistance of the wire of the coil dissipates energy in the form of heat just like any other resistor would. This energy is lost to the circuit and is non-recoverable.

Basically, the quality of a coil is based on its ability to store energy in its magnetic field, then return that energy back to the circuit. Essentially, it is a measure of how efficient the coil is as far as its reactive, self-inductive qualities are concerned.

Derivation of Q Equation

Q is defined as the ratio of the reactive power in the inductance to the real power dissipated by its internal resistance. Mathematically,

$$Q = \frac{P_{Reactive}}{P_{Real}} = \frac{P_L}{P_{r_i}} \quad (9\text{–}29)$$

Figure 9.45 is the equivalent circuit of an inductor to which an ac voltage is applied. The power for the reactive and resistive properties of the coil can be written as:

$$P_L = I^2X_L \quad (9\text{–}30)$$

and

$$P_{r_i} = I^2r_i \quad (9\text{–}31)$$

Substituting these power values into the equation for Q,

$$Q = \frac{P_L}{P_{r_i}} = \frac{I^2X_L}{I^2_{r_i}} \quad (9\text{–}32)$$

Factoring out I^2,

$$Q = \frac{X_L}{r_i} \quad (9\text{–}33)$$

And writing X_L as $2\pi fL$ yields,

$$Q = \frac{2\pi fL}{r_i} \quad (9\text{–}34)$$

In the equations, r_i is the resistance measured by an ohmmeter placed across the coil. L is the inductance of the coil being considered. And f is a standard frequency chosen to compare the Q of different coils. This frequency is usually one kilohertz for large values of L.

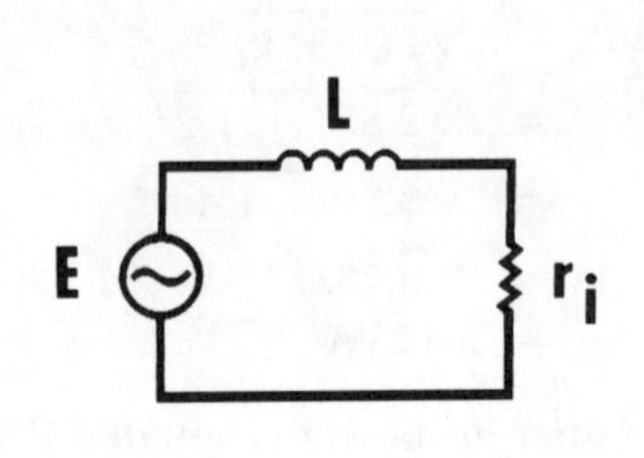

Figure 9.45 AC Voltage Applied to Equivalent Circuit of Coil

Sample Calculation of Q

Thus, to determine the Q of a coil of 8.5 henrys, r_i would be determined by measuring across it with an ohmmeter. In this example, r_i is about 400 ohms. Then using equation *9–34* and with f = one kilohertz, the coil's Q is calculated:

$$\begin{aligned} Q &= \frac{2\pi fL}{r_i} \\ &= \frac{(6.28)\,(1000\text{Hz})\,(8.5\text{H})}{400\Omega} \\ &= \frac{53{,}380\Omega}{400\Omega} \\ &= 133 \end{aligned}$$

Factors Affecting Q of a Coil

At low frequencies, r_i is simply the dc resistance of the wire used in the coil, as measured by an ohmmeter. However, as frequency increases, additional losses occur within the coil. In air-core coils, (the type used in high-frequency radio and radar circuits) this additional loss is only one, called *skin effect.* This is the tendency of high-frequency current to flow near the surface of a conductor. This results from current near the center of the conductor encountering slightly more inductive reactance due to the concentrated magnetic flux in the center compared to the surface, where part of the flux is in the air. This effect increases the effective resistance of the conductor since current flow is limited to a small cross-sectional area near the surface. Because of this effect, conductors for very high-frequency applications are often made of hollow tubing called waveguides.

In an iron core coil, greater losses occur because of eddy currents and hysteresis. These losses are due to the magnetic properties of the iron. These losses effectively increase r_i. Any increase in r_i tends to decrease the Q of the coil at high frequency even though X_L is increasing as frequency increases. At high frequency, therefore, coils with air cores, which do not have added magnetic losses, are normally used to maximize Q.

SUMMARY

In this lesson, several techniques were used to solve for values in series and parallel RL circuits. You learned how the different phase relationship between voltage and current for each component affected the method of analysis of each type of circuit.

You were shown how to determine the currents, voltages, impedance, phase angle and power for series and parallel RL circuits.

You were introduced to the concept of the Q of a coil and saw how you could determine the Q for a coil by using a simple measurement and calculation.

The concepts introduced in this lesson for solving series and parallel RL circuits may be adapted to solve any series or parallel RL circuit.

■ Worked-Out Examples

1. Calculate the voltage, current, power, and phase angle values shown for this series RL circuit.

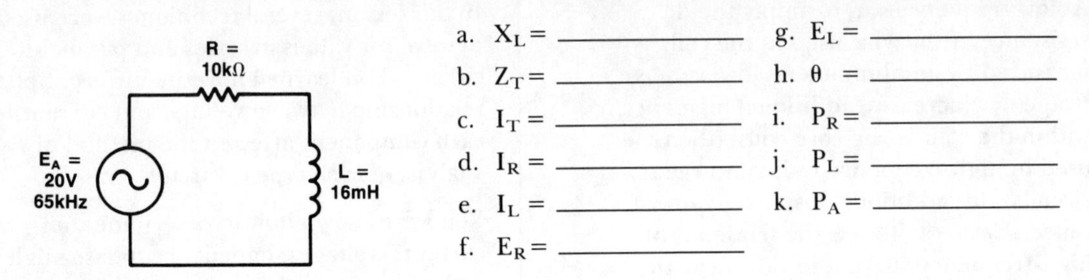

Solution:

a. $X_L = 2\pi fL = (6.28)\,(65\text{kHz})\,(16\text{mH}) = (6.28)\,(65 \times 10^3\text{Hz})\,(16 \times 10^{-3}\text{H})$

$= 6531.2\Omega = \mathbf{6.53k\Omega}$

b. $Z_T = \sqrt{R^2 + X_L{}^2} = \sqrt{10^2 + 6.53^2}\ \text{k}\Omega = \sqrt{100 + 42.6}\ \text{k}\Omega = \sqrt{142.6}\ \text{k}\Omega = \mathbf{11.9k\Omega}$

c. $I_T = \dfrac{E_A}{Z} = \dfrac{20\text{V}}{11.9\text{k}\Omega} = \mathbf{1.67mA}$

d.,e. $I_R = I_L = I_T = \mathbf{1.67mA}$ (series circuit)

f. $E_R = I_R R = (1.67\text{mA})\,(10\text{k}\Omega) = \mathbf{16.7V}$

g. $E_L = I_L X_L = (1.67\text{mA})\,(6.53\text{k}\Omega) = \mathbf{10.9V}$

h. $\theta = \arctan\left(\dfrac{X_L}{R}\right) = \arctan\left(\dfrac{6.53\text{k}\Omega}{10\text{k}\Omega}\right) = \arctan(0.653) = \mathbf{+33°}$

i. $P_R = E_R I_R = (16.7\text{V})\,(1.67\text{mA}) = \mathbf{27.9mW}$

j. $P_L = E_L I_L = (10.9\text{V})\,(1.67\text{mA}) = \mathbf{18.2mVAR}$

k. $P_A = E_T I_T = (20\text{V})\,(1.67\text{mA}) = \mathbf{33.5mVA}$

■ Worked-Out Examples

2. For the circuit of Example 1, sketch the voltage, impedance, and power phasor diagrams. Label all phasor lengths and locate the phase angle in each diagram.

Solution:

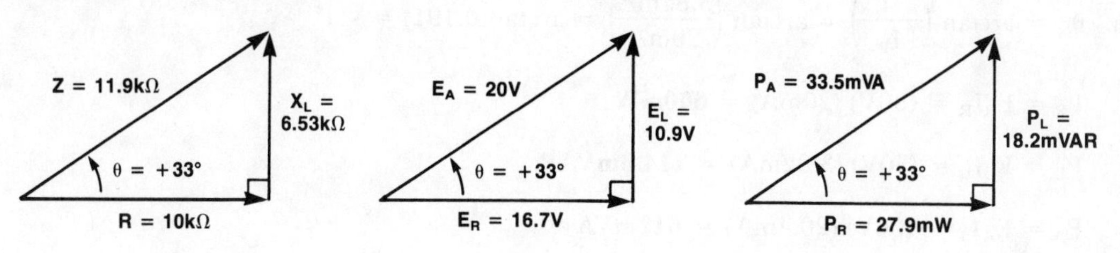

a. Impedance Phasor Diagram b. Voltage Phasor Diagram c. Power Phasor Diagram

3. Calculate the voltage, current, power, and phase angle values shown for this parallel RL circuit.

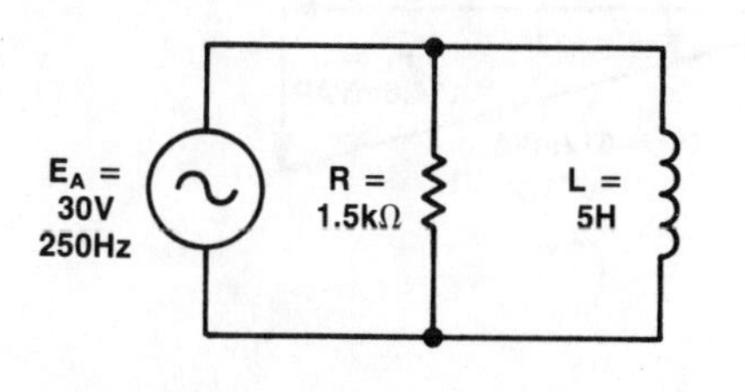

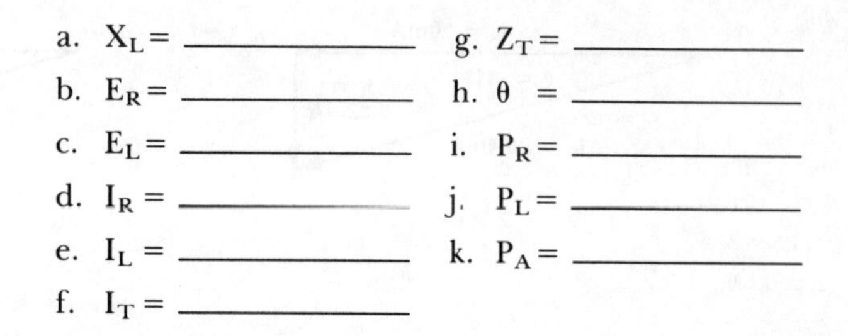

Solution:

a. $X_L = 2\pi fL = (6.28)(250\text{Hz})(5\text{H}) = 7850\Omega = \mathbf{7.85k\Omega}$

b. $E_R = E_A = \mathbf{30V}$ (parallel circuit)

c. $E_L = E_A = \mathbf{30V}$(parallel circuit)

d. $I_R = \dfrac{E_R}{R} = \dfrac{30\text{V}}{1.5\text{k}\Omega} = \mathbf{20mA}$

e. $I_L = \dfrac{E_L}{X_L} = \dfrac{30\text{V}}{7.85\text{k}\Omega} = \mathbf{3.82mA}$

f. $I_T = \sqrt{I_R{}^2 + I_L{}^2} = \sqrt{20^2 + 3.82^2}\,\text{mA} = \sqrt{400 + 14.6}\,\text{mA} = \sqrt{414.6}\,\text{mA} = \mathbf{20.4mA}$

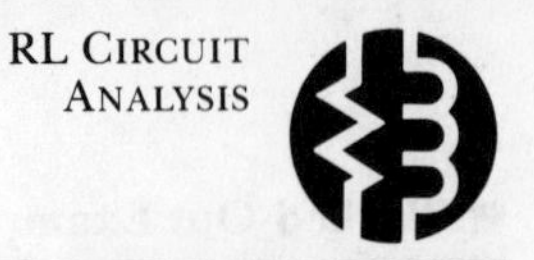

Worked-Out Examples

g. $Z_T = \frac{E_A}{I_T} = \frac{30V}{20.4mA} = \mathbf{1.47k\Omega}$

h. $\theta = \arctan\left(\frac{I_L}{I_R}\right) = \arctan\left(\frac{3.82mA}{20mA}\right) = \arctan(0.191) = -11°$

i. $P_R = E_R I_R = (30V)\,(20mA) = \mathbf{600mW}$

j. $P_L = E_L I_L = (30V)\,(3.82mA) = \mathbf{114.6mVAR}$

k. $P_A = E_A I_T = (30V)\,(20.4mA) = \mathbf{612mVA}$

4. For the circuit of Example 3, sketch the current and power phasor diagrams. Label all phasor lengths and locate the phase angle in each diagram.

Solution:

a. b.

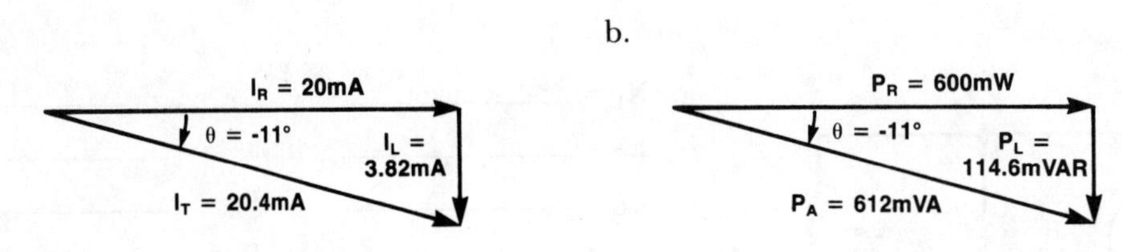

5. Determine the Q of a 10 mH coil that has an internal resistance of 7 ohms at a frequency of 1 kHz.

Solution:

$$Q = \frac{X_L}{r_i} = \frac{2\pi fL}{r_i} = \frac{(6.28)\,(1kHz)\,(10mH)}{7\Omega} = \frac{(6.28)\,(1 \times 10^3 Hz)\,(10 \times 10^{-3} H)}{7\Omega}$$

$$= \frac{62.8 \times 10^0\Omega}{7\Omega} = \frac{62.8\Omega}{7\Omega} = \mathbf{8.97}$$

■ Practice Problems

1. For the circuit shown, determine the values specified for the circuit values given.

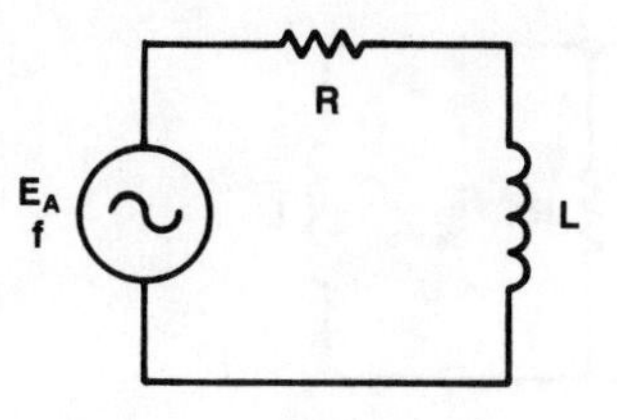

a.	b.	c.
E_A = 45V	E_A = 120V	E_A = 26V
f = 15kHz	f = 60Hz	f = 600kHz
R = 2.7kΩ	R = 1kΩ	R = 15kΩ
L = 15mH	L = 6.5H	L = 3.5mH
X_L = ______	X_L = ______	X_L = ______
E_R = ______	E_R = ______	E_R = ______
E_L = ______	E_L = ______	E_L = ______
I_R = ______	I_R = ______	I_R = ______
I_L = ______	I_L = ______	I_L = ______
I_T = ______	I_T = ______	I_T = ______
Z = ______	Z = ______	Z = ______
θ = ______	θ = ______	θ = ______
P_R = ______	P_R = ______	P_R = ______
P_L = ______	P_L = ______	P_L = ______
P_A = ______	P_A = ______	P_A = ______

2. Sketch the voltage, impedance and power phasor diagrams for Problem 1c. Label all phasor lengths, locate and identify the phase angle for each diagram.

a. Voltage Phasor Diagram   b. Impedance Phasor Diagram   c. Power Phasor Diagram

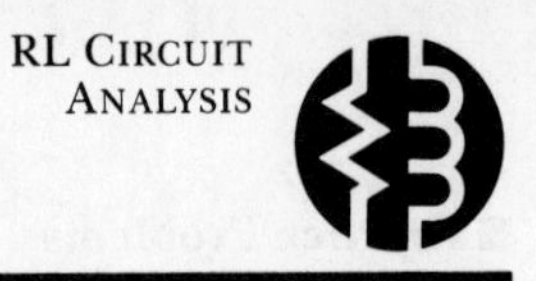

Practice Problems

3. For the circuit shown, determine the values specified for the circuit values given.

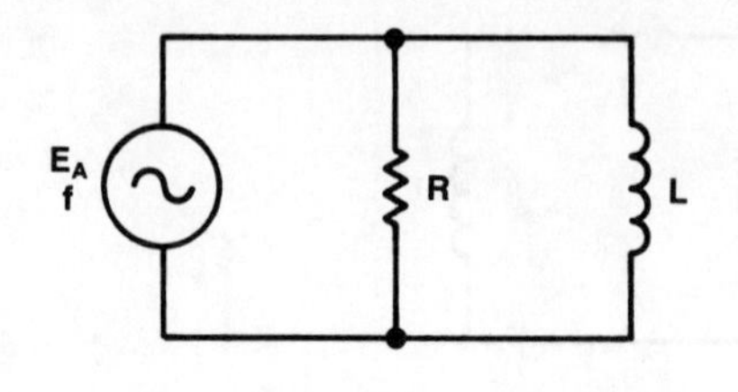

a. $E_A = 60V$
$R = 12k\Omega$
$X_L = 5k\Omega$

E_R = ______
E_L = ______
I_R = ______
I_L = ______
I_T = ______
Z = ______
θ = ______
P_R = ______
P_L = ______
P_A = ______

b. $E_A = 36V$
$R = 18M\Omega$
$X_L = 4M\Omega$

E_R = ______
E_L = ______
I_R = ______
I_L = ______
I_T = ______
Z = ______
θ = ______
P_R = ______
P_L = ______
P_A = ______

c. $E_A = 24V$
$R = 3k\Omega$
$X_L = 6k\Omega$

E_R = ______
E_L = ______
I_R = ______
I_L = ______
I_T = ______
Z = ______
θ = ______
P_R = ______
P_L = ______
P_A = ______

4. Sketch the current and power phasor diagram for Problem 3c. Label all phasor lengths, locate and identify the phase angle for each diagram.

a. Current Phasor Diagram

b. Power Phasor Diagram

5. Determine the Q of a 65 mH coil at a frequency of 1 kHz if its resistance as measured with an ohmmeter is 25Ω.

Q = ______

■ Quiz

1. For the circuit shown, determine the values specified, and sketch and label the impedance, voltage and power phasor diagrams. Label all phasor lengths. Locate and identify the phase angle in each diagram.

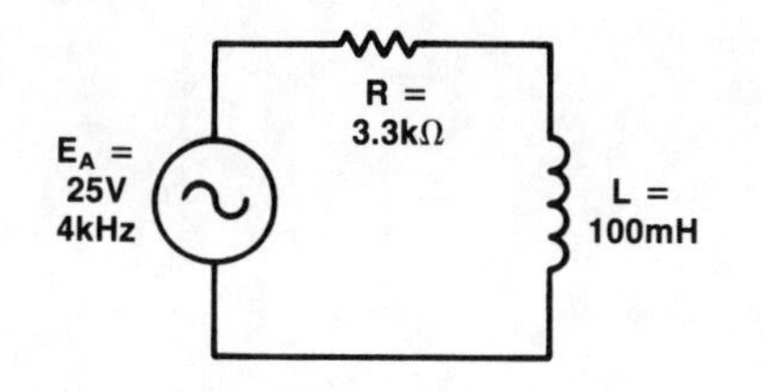

a. X_L = ________

b. Z = ________

c. E_R = ________

d. E_L = ________

e. I_R = ________

f. I_L = ________

g. I_T = ________

h. θ = ________

i. P_R = ________

j. P_L = ________

k. P_A = ________

l. Voltage Phasor Diagram

m. Impedance Phasor Diagram

n. Power Phasor Diagram

2. For the circuit shown, determine the values specified and sketch the current and power phasor diagrams. Label all phasor lengths. Locate and identify the phase angle in each diagram.

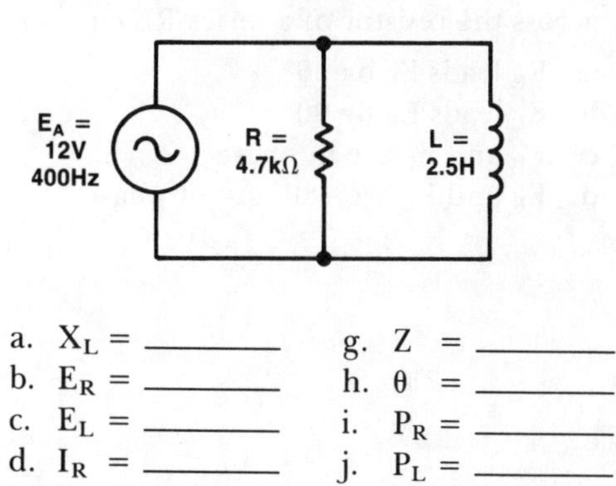

a. X_L = ________

b. E_R = ________

c. E_L = ________

d. I_R = ________

e. I_L = ________

f. I_T = ________

g. Z = ________

h. θ = ________

i. P_R = ________

j. P_L = ________

k. P_A = ________

l. Current Phasor Diagram

m. Power Phasor Diagram

3. Determine the Q of 2.5 mH coil at a frequency of 1 kHz if its internal resistance is 2Ω.

Q = ________

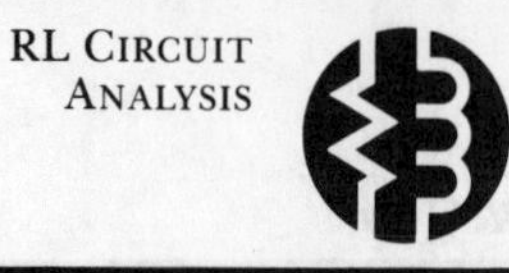

■ Quiz

4. Determine the internal resistance of a coil if its value is 8.5 H with a Q of 15 at a frequency of 1 kHz.

r_i = ________

5. What is the phase relationship of the voltage across the coil versus the voltage across the resistor of a series RL circuit?

a. E_R leads E_L by 90°
b. E_L leads E_R by 90°
c. E_R and E_L are in phase
d. E_R and E_L are 180° out of phase

LESSON 10

RC and RL Time Constants

This lesson discusses the time relationships of current and voltage in dc resistive and capacitive, and resistive and inductive circuits. A special quantity, a time constant, is introduced and RC and RL time constants are defined. Voltages and currents are calculated after various time intervals of circuit operation.

■ Objectives

At the end of this lesson you should be able to:

1. Define time constant for both RC and RL dc circuits.
2. Draw the current and voltage waveforms during the charge and discharge time of a capacitor in a dc resistive and capacitive circuit.
3. Draw the current and voltage waveforms during the time that current increases and decreases in a dc resistive and inductive circuit.
4. Given a known time interval after a capacitor begins to charge or discharge in an RC circuit, calculate the instantaneous voltage and current values in the circuit.
5. Given a known time interval after current begins to increase or decrease in an RL circuit, calculate the instantaneous voltage and current values in the circuit.
6. Convert time intervals to time constants and time constants to time intervals in RC and RL circuits.

■ Circuit Current Related to a Capacitor's Charge

INTRODUCTION

In the past several lessons, discussion was about RC and RL circuits with an ac voltage applied. Phase differences between the current, voltage and power were observed which caused phase differences between the applied voltages and total currents in the circuits. Several questions arise. What would be the response of these same circuits if a dc voltage were applied? How long does it take to charge a capacitor to the applied voltage? And how long does it take for the current in an inductive circuit to rise to its maximum value? To answer these questions, in this lesson, unit values called RC and RL time constants will be discussed, and RC and RL circuits with dc voltages applied will be analyzed.

RC DC CIRCUIT ANALYSIS: CHARGE CYCLE

Before actual dc RC and RL circuits are analyzed, however, it is necessary to review and expand on basic capacitive and inductive circuit action that was presented in previous lessons. First, the capacitor and its response to a dc voltage will be reviewed.

Circuit Current Related to a Capacitor's Charge

In the circuit of *Figure 10.1,* there is a 10 kilohm resistor in series with a 1 microfarad capacitor connected through a switch, S1, to a 10-volt dc power supply. The switch is open and the capacitor, therefore, has no charge on its plates.

Recall the basic action of a capacitor in this type of circuit was discussed in Lesson 6. When the switch first contacts position 1, at time t_1, as shown in *Figure 10.2,* the voltage is applied across the plates of the capacitor. Electrons then move from the upper plate of the capacitor through the circuit to the lower plate. This flow of electrons constitutes a current which will be called the *charge current.*

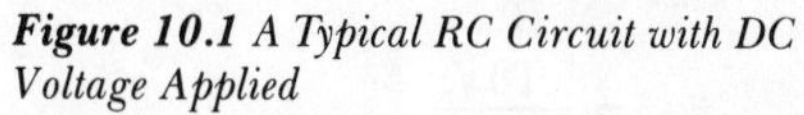

Figure 10.1 *A Typical RC Circuit with DC Voltage Applied*

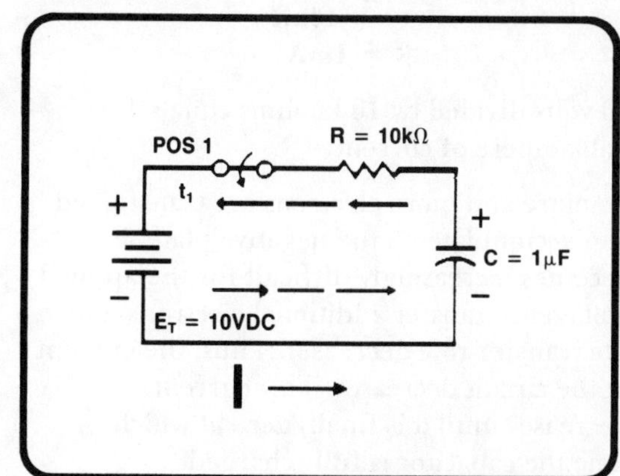

Figure 10.2 *Capacitor Charges When S1 Closes*

Calculation of I_{max}

The initial current is large, but its maximum value is limited by the value of the resistance in the circuit. The maximum value of the charge current is equal to the applied voltage, E_T, divided by the value of the resistance.

$$I_{max} = \frac{E_T}{R} \qquad (10\text{–}1)$$

In the circuit of *Figure 10.1,* E_T = 10 volts and R = 10 kilohms. Therefore,

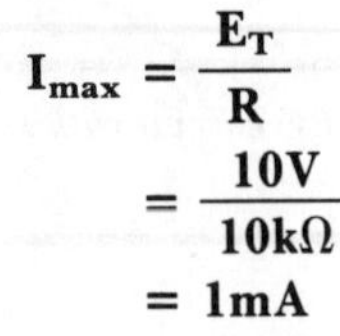

$$I_{max} = \frac{E_T}{R} = \frac{10V}{10k\Omega} = 1mA$$

10 volts divided by 10 kilohms equals 1 milliampere of current.

As more and more electrons are transferred and accumulate on the negative plate, it becomes increasingly difficult for the applied voltage to transfer additional electrons and the transfer rate decreases. Thus, the current in the circuit decreases. This current decreases until it is finally zero at which time the capacitor is fully charged.

A graphic representation of this action is shown in *Figure 10.3,* where the current is plotted versus time. At first, at time t_1 the current is at its largest value, 1 milliampere, then decreases to zero exponentially as the capacitor charges.

Variation of E_C

As this transfer of electrons occurs, as shown in *Figure 10.4,* current flows in the circuit, the charge on the capacitor builds, and a difference of potential is present across its plates. When the capacitor is fully charged, the voltage across it, E_C, is equal to the applied voltage of 10 volts.

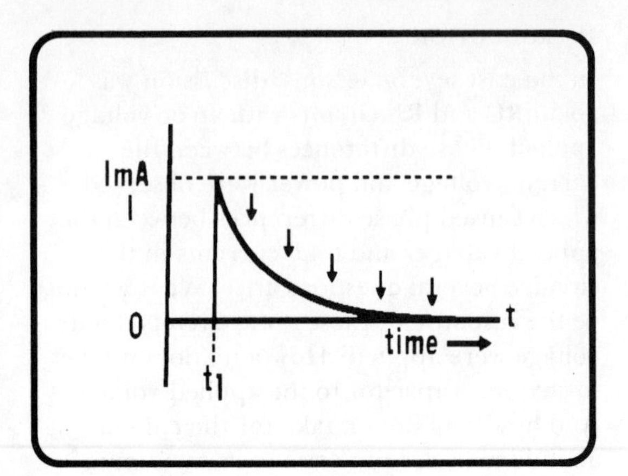

Figure 10.3 *Graph of Current During Capacitor Charge*

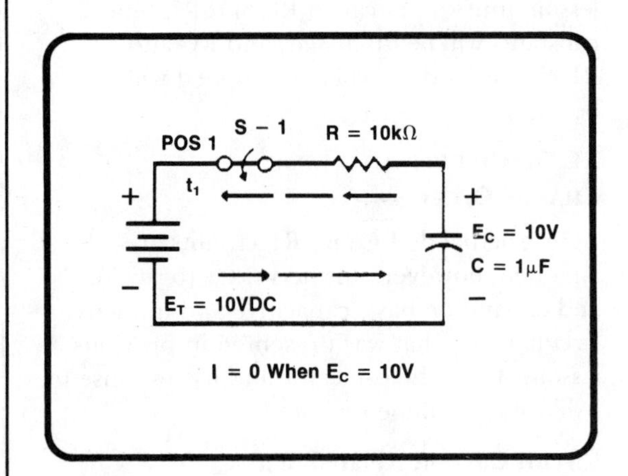

Figure 10.4 *Current Flows Until Capacitor is Fully Charged*

Figure 10.5 is a plot of that voltage, E_C. Initially, the voltage across the capacitor is zero, but as the electrons are transferred, beginning at time t_1, the capacitor charges exponentially to the applied voltage of 10 volts during a period of time. Compare *Figures 10.3* and *10.5* and note that the voltage across the capacitor increases as the current flow decreases and the capacitor charges.

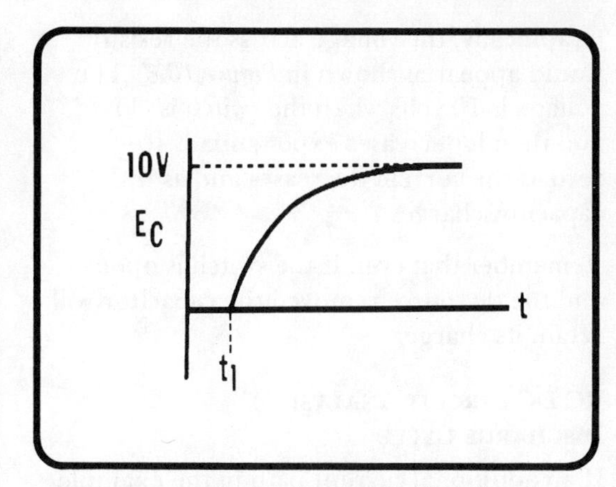

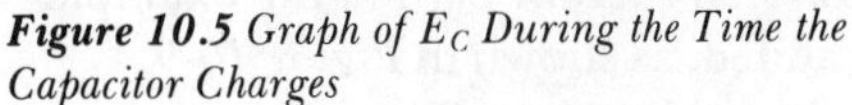
Figure 10.5 *Graph of E_C During the Time the Capacitor Charges*

Calculation of E_R

The voltage across the resistor depends upon two factors: 1) the amount of current flowing through it, and 2) the value of the resistor. More specifically, by Ohm's law,

$$E_R = IR \qquad (10\text{–}2)$$

Since the value of the resistor is constant (10 kilohms), the voltage across it depends upon the amount of current flowing in the circuit at any particular time. When the switch is first closed, as shown in *Figure 10.6,* the voltage across the resistor is equal to:

$$\begin{aligned} E_R &= IR \\ &= (1mA)\ (10k\Omega) \\ &= 10V \end{aligned}$$

So at t_1, the voltage across the resistor is 10 volts.

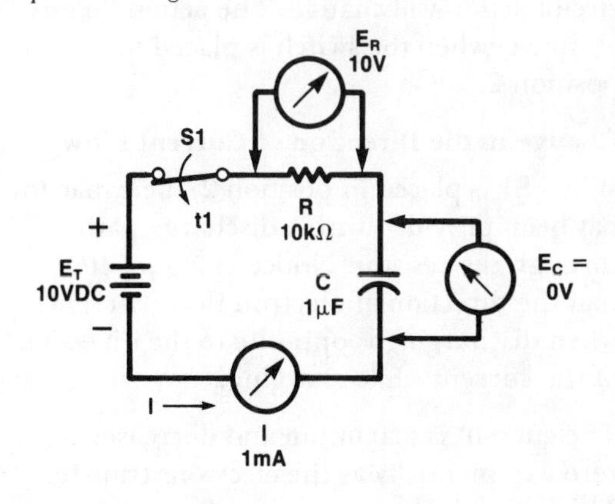

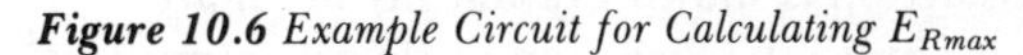
Figure 10.6 *Example Circuit for Calculating E_{Rmax}*

When the current is zero, the voltage across the resistor is equal to zero because:

$$\begin{aligned} E_R &= IR \\ &= (0mA)\ (10k\Omega) \\ &= 0V \end{aligned}$$

This is illustrated in *Figure 10.7*.

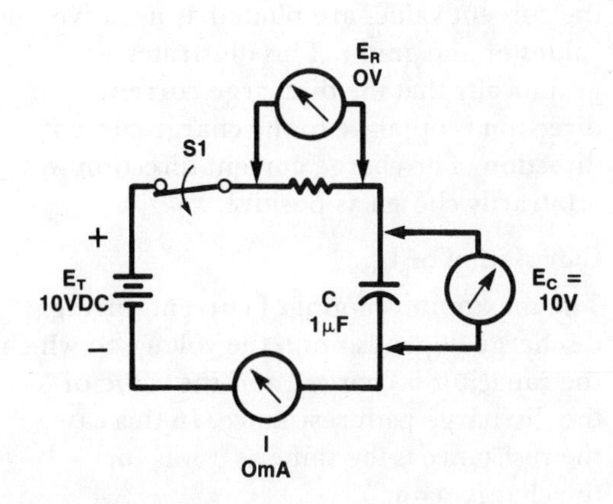

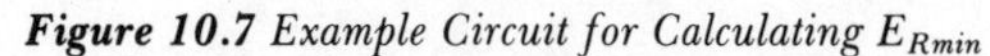
Figure 10.7 *Example Circuit for Calculating E_{Rmin}*

Graphically, the voltage across the resistor would appear as shown in *Figure 10.8*. The voltage is 10 volts when the switch is closed, and then it decreases exponentially to zero as the current decreases and as the capacitor charges.

Remember that even if the switch is open and the dc source removed the capacitor will retain its charge.

RC DC CIRCUIT ANALYSIS: DISCHARGE CYCLE

If an additional current path in the example circuit is added, as shown in *Figure 10.9*, circuit action will change. The action begins at time t_2 when the switch is placed in position 2.

Change in the Direction of Current Flow

When S1 is placed in position 2, the capacitor has been provided with a discharge path through the resistor. Notice in *Figure 10.9* that the direction of electron flow (current) when discharging is opposite to the direction of the current while charging.

The current is maximum and decreases to zero exponentially as the electrons transfer. This action is shown graphically in *Figure 10.10*. Since the direction of flow is opposite, the current values are plotted as negative values on the graph. This illustrates graphically that the discharge current direction is opposite to the charge current direction. The charge current direction was arbitrarily chosen as positive.

Calculation of I_{max}

The maximum amount of current during discharge depends upon the voltage to which the capacitor is charged and the value of the discharge path resistance. In this case, the resistance is the same as it was for the charge path.

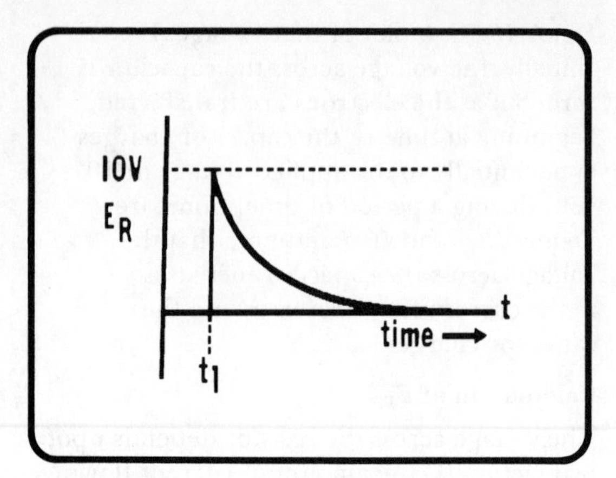

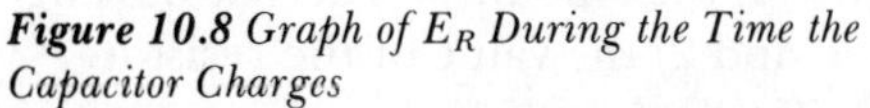
Figure 10.8 *Graph of E_R During the Time the Capacitor Charges*

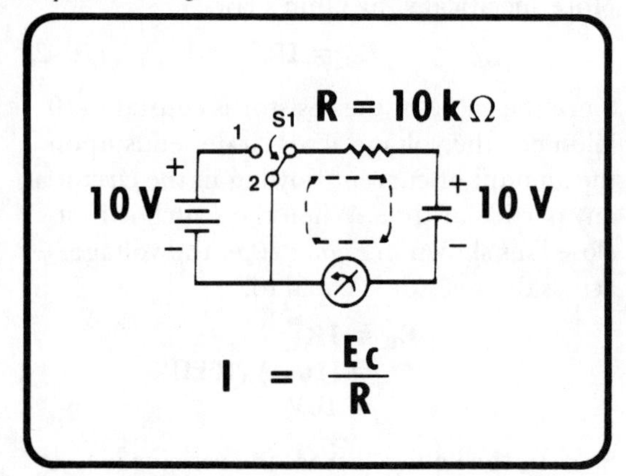

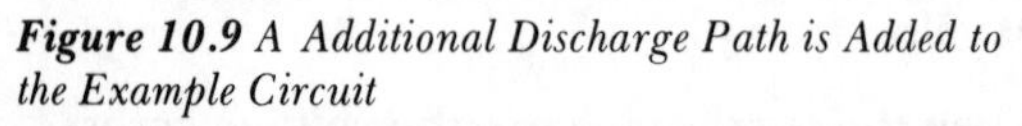
Figure 10.9 *A Additional Discharge Path is Added to the Example Circuit*

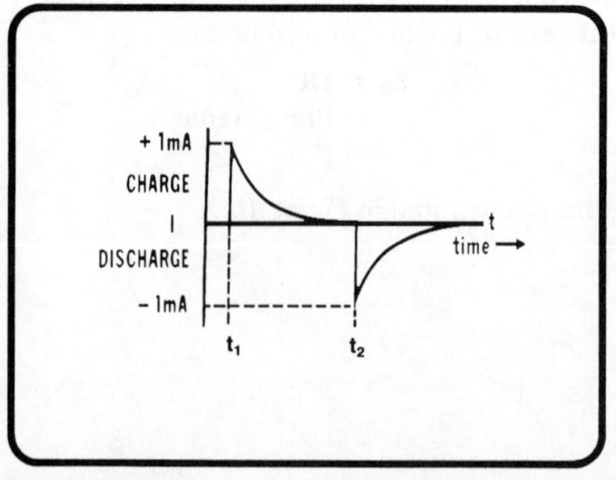

Figure 10.10 *Negative Current Values During Capacitor Discharge Time (Time from t_2)*

- Variation of E_C
- Calculation of E_R

The maximum discharge current is calculated:

$$I_{max} = \frac{E_C}{R}$$

In the example circuit, $E_C = 10$ volts and R = 10 kilohms. Therefore,

$$I_{max} = \frac{E_C}{R} = \frac{10V}{10k\Omega} = 1mA$$

1 milliamp is the discharge current. The discharge current decreases from the I_{max} value exponentially as the capacitor discharges. When the charge on the plates is balanced, no more electrons are transferred, the voltage across the capacitor is zero, and the circuit current is zero.

Variation of E_C

As the capacitor discharges, the voltage across the capacitor is 10 volts initially at t_2, and then decreases to zero exponentially, as shown graphically in *Figure 10.11*.

Calculation of E_R

The voltage across the resistor during capacitor discharge depends upon the amount of current passing through it at any particular time, as it did during capacitor charge. However, since the current during discharge flows in a direction opposite to its direction during the charging process, the polarity of the voltage across the resistor will be opposite to the polarity it had during *charging* as shown in *Figure 10.12*.

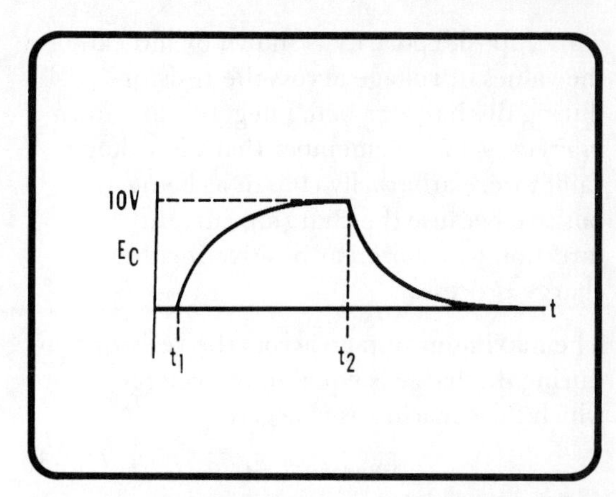

Figure 10.11 *Graph of E_C During Capacitor Discharge Time (Time from t_2)*

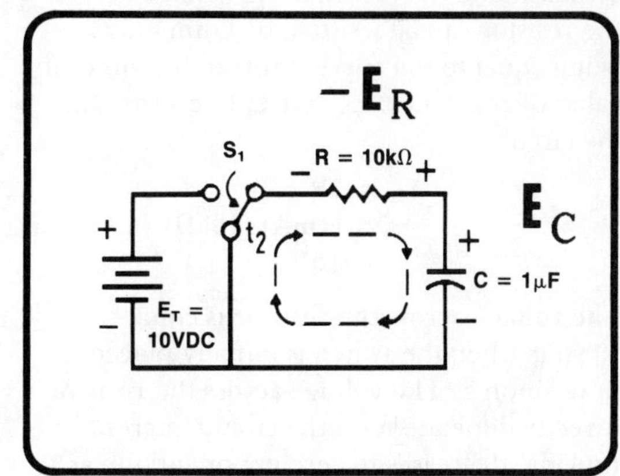

Figure 10.12 *Polarity of E_R During Capacitor Discharge Time*

This opposite polarity is shown by indicating the values of voltage across the resistor during discharge as being negative as shown in *Figure 10.13*. Remember that the voltage values were arbitrarily chosen as being positive because the charging current direction was chosen as positive for the charge process.

The maximum voltage across the resistor during discharge is equal to the voltage to which the capacitor is charged.

$$E_{R(max)} = E_C \qquad (10\text{–}3)$$

In this case, that voltage is 10 volts. This is evident from the fact that the voltage across the resistor can be written, by Ohm's law, as being equal to the current through it times its value of resistance: E_R = IR. Therefore, in the circuit,

$$E_R = IR$$
$$= (1mA)\,(10k\Omega)$$
$$= 10V$$

The voltage across the resistor is equal to 10 volts when the switch is initially placed in position 2. The voltage across the resistor directly dependent on the circuit current flowing, decreases to zero exponentially as the current decreases to zero. This shown in *Figure 10.13*. Note the shape is the same as the current waveform of *Figure 10.10*.

RC TIME CONSTANTS

A summary of the current and voltage waveforms for dc RC circuits during the charge and discharge time of the capacitor is shown in *Figure 10.14*. However, a question immediately comes to mind. How long does the charging and discharging take? The following discussion should answer that question.

RC Time Constant Defined

The time required for a capacitor to fully charge is measured in terms of a quantity called the *time constant* of the circuit.

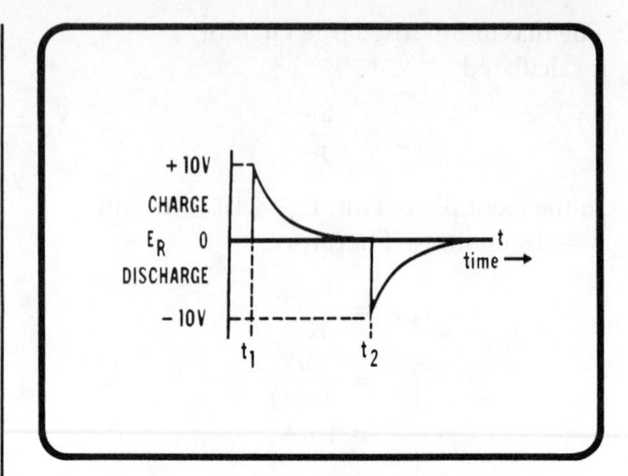

Figure 10.13 *Negative E_R Values During Capacitor Discharge Time (Time from t_2)*

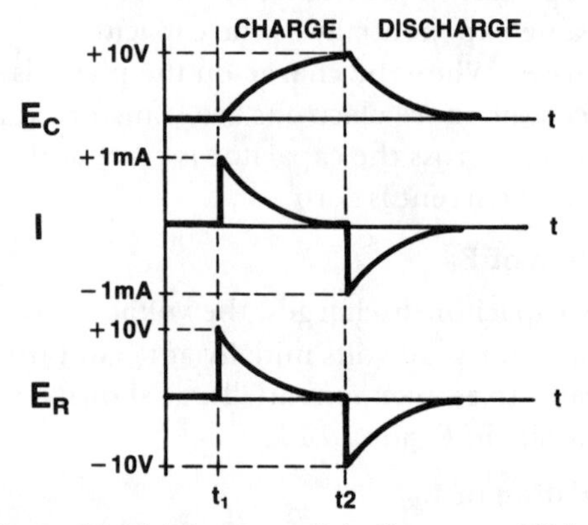

Figure 10.14 *A Summary of the Current and Voltage Waveforms for DC RC Circuits During Capacitor Charge and Discharge Times*

The *time constant* of an RC circuit is defined as the amount of time necessary for the capacitor to charge to 63.2 percent of its final voltage.

The time constant is symbolized by the Greek letter, tau, τ.

For an RC circuit, tau is equal to the value of the resistance times the value of capacitance in the circuit. Expressed mathematically,

$$\tau = RC \qquad (10\text{–}4)$$

Where tau is measured in seconds, R in ohms, and C in farads.

■ Calculation of an RC Time Constant

Calculation of an RC Time Constant

The resistance of the example RC circuit being discussed *(Figure 10.1)* is 10 kilohms. Its capacitance is 1 microfarad. The time constant for the circuit is calculated:

$$\begin{aligned} \tau &= RC \\ &= (10k\Omega)\ (1\mu F) \\ &= (10 \times 10^{3})\ (1 \times 10^{-6}) \\ &= 10 \times 10^{-3} \text{ seconds} \\ &= 10ms \end{aligned}$$

The time constant is 10 milliseconds.

In the circuit, the capacitor is charging to 10 volts. 63.2 percent of 10 volts is 6.32 volts. Thus, as shown in *Figure 10.15,* it takes 10 milliseconds or one time constant of time for the capacitor to charge to 6.32 volts.

In the next time duration of one time constant the capacitor will charge to 63.2 percent of the remaining value of voltage between the maximum, 10 volts, and its present charge, 6.32 volts. Therefore,

$$10V - 6.32V = 3.68V$$

and

$$63.2\% \text{ of } 3.68V = 2.33V$$

Thus, at the end of two time constants or after 20 milliseconds of charging, as shown in *Figure 10.16*, the capacitor will have charged to

$$6.32 + 2.33 = 8.65V$$

This process continues with the capacitor charging 63.2 percent of the remaining difference during each time constant of time until after about five time constants, the capacitor, for all practical purposes, is fully charged with a charge of 9.93 volts as shown in *Figure 10.17*. Thus, it is assumed for practical cases that it takes five time constants for the capacitor to fully charge.

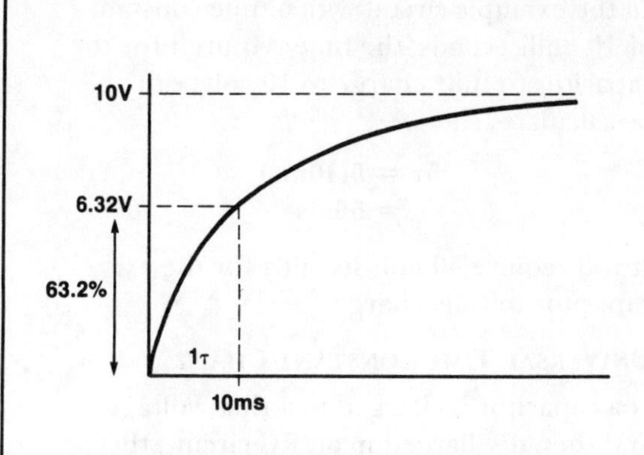

Figure 10.15 *E_C of Example Circuit After 1τ*

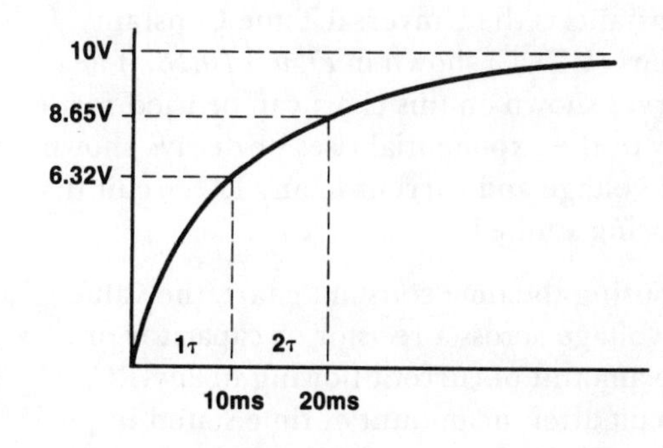

Figure 10.16 *E_C of Example Circuit After 2τ*

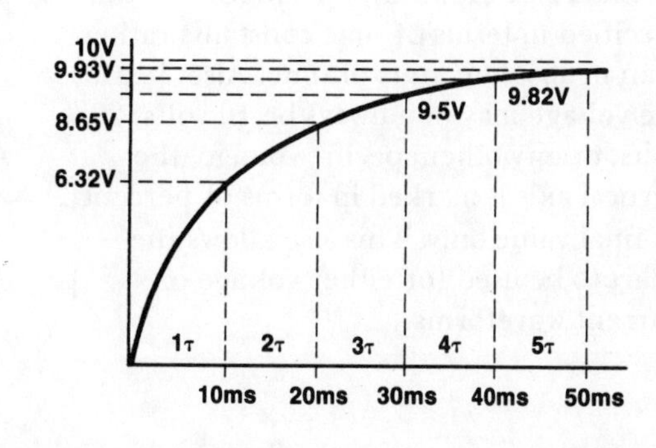

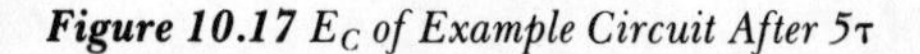

Figure 10.17 *E_C of Example Circuit After 5τ*

■ UTCC Related to E_C

In the example circuit with a time constant of 10 milliseconds, the time required for the capacitor to fully charge to 10 volts can be calculated:

$$5\tau = 5(10ms)$$
$$= 50ms$$

It will require 50 milliseconds for the capacitor to fully charge.

UNIVERSAL TIME CONSTANT CHART

If a capacitor is charged to a final voltage and then discharged in an RC circuit, the discharge curve—just like the charge curve—will be exponential and governed by time constants. A chart that relates the percent of the final value to time in terms of time constants is the Universal Time Constant Chart (UTCC) shown in *Figure 10.18*. The curves shown on this chart can be used for any of the exponential rises or decays shown for voltage and current in any RC circuit that is being studied.

By using the time constant chart, the value of voltage across a resistor or capacitor or the amount of current flowing in any RC circuit after an amount of time stated in terms of time constants for the circuit can be determined.

Because the time constant changes if either resistance or capacitance changes, the time is specified in terms of time constants rather than in absolute units of time. Also, because the voltage may not always be 10 volts, 20 volts, or any other specific voltage, the vertical axis is marked in terms of percent of final value only. This also allows the chart to be used for either voltage or current waveforms.

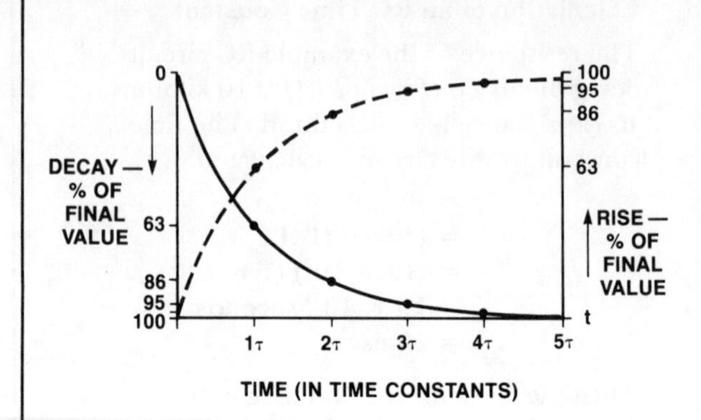

Figure 10.18 Universal Time Constant Chart

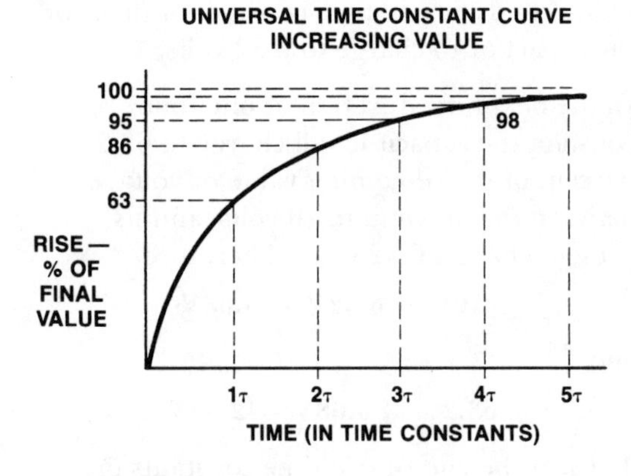

Figure 10.19 Percent E_C Versus Time (τ)

UTCC Related to E_C

Note that the increasing curve of *Figure 10.19* has a shape exactly like the shape of the rise of voltage across the capacitor. This can easily be proved true by plotting the voltages across the capacitor in the circuit of *Figure 10.1* against time measured in time constants for exact multiples of one time constant.

Initially, the voltage across the capacitor is zero when the switch is closed. The origin represents time equals zero and voltage equals zero. This is the time when the switch closes with no initial voltage across the capacitor. As calculated previously and shown in *Figure 10.17,* after one time constant, (10 milliseconds), the voltage across the capacitor equals 6.32 volts (63.2 percent of maximum).

After two time constants, 20 milliseconds, the voltage across the capacitor equals 8.65 volts or 86.5 percent of maximum. After three time constants, 30 milliseconds, E_C equals 9.5 volts or 95 percent of maximum. After four time constants, 40 milliseconds, E_C equals 9.82 volts or 98 percent of maximum. And after five time constants, the voltage across the capacitor is equal to 9.93 volts or 99 percent of maximum.

If all of these points are connected with a smooth curve, the result is the rising curve shown on the UTCC of *Figure 10.18* and *Figure 10.19*.

Exponential Characteristic of the UTCC

The shape of these curves follows a precise mathematical relationship which involves an exponent containing the time constant term. Because of the exponent containing the time constant term, they are called exponential curves. This characteristic results in the terms *exponential increase* or *exponential decrease*. Because these exponential equations are complex mathematically, the chart is used to provide a direct-read capability.

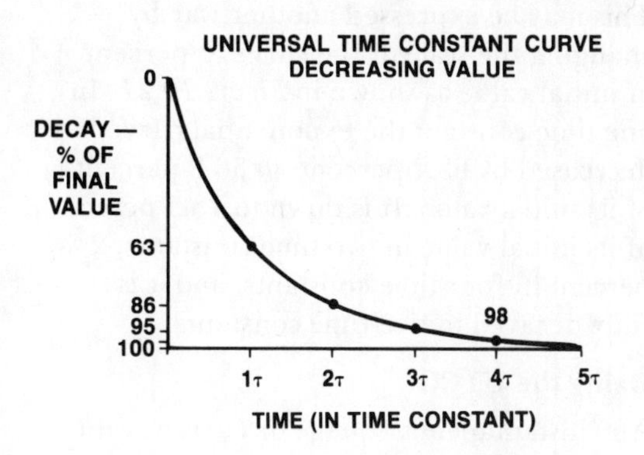

Figure 10.20 *Exponential Characteristics of Curves*

UTCC Related to Circuit Current and E_R

Recall that it was stated earlier that the exponential curve applies to decreasing waveforms as well as to increasing waveforms. An example is the circuit current waveform which is maximum initially and decreases to zero in five time constants, as shown by the curve in *Figure 10.20.* The same waveform applies to the voltage across the resistor. This decaying waveform is simply a mirror image of the rising exponential waveform shown in *Figure 10.19*.

At one time constant the decaying waveform has reduced 63.2 percent from its initial value. At two time constants the waveform has reduced 86.5 percent, 95 percent at three time constants, 98 percent at four time constants, and at five time constants an assumed reduction of 100 percent.

■ Using the UTCC

This may be expressed another way by changing the scale of the curve to percent of initial value as shown in *Figure 10.21*. In one time constant the exponential curve has decreased by 63.2 percent, to 36.8 percent of its initial value. It is down to 13.5 percent of its initial value in two time constants, 2 percent in four time constants, and it is fully decayed in five time constants.

Using the UTCC

Any instantaneous voltage or current value across or through the components of a dc resistive-capacitive circuit during the charge and discharge of the capacitor can be determined by using the Universal Time Constant Chart. (Later in this lesson, it will also be used to predict values of current and voltage while current is increasing and decreasing in dc resistive-inductive circuits.) For instance, using the UTCC the percent of full charge on the capacitor after any charging time or time constant period can be determined.

Figure 10.22 shows that after one and one-half time constants the percent of full charge, from the UTCC chart, is approximately 78 percent. If the applied voltage is 10 volts, as shown in *Figure 10.23a*, the voltage across the capacitor, E_C at this time is 78 percent of 10 volts which is 7.8 volts. This is shown graphically in *Figure 10.23b*. If 50 volts were applied, as shown in the circuit of *Figure 10.24a*, E_C is 78 percent of 50 volts which is 39 volts. This is shown graphically in *Figure 10.24b*.

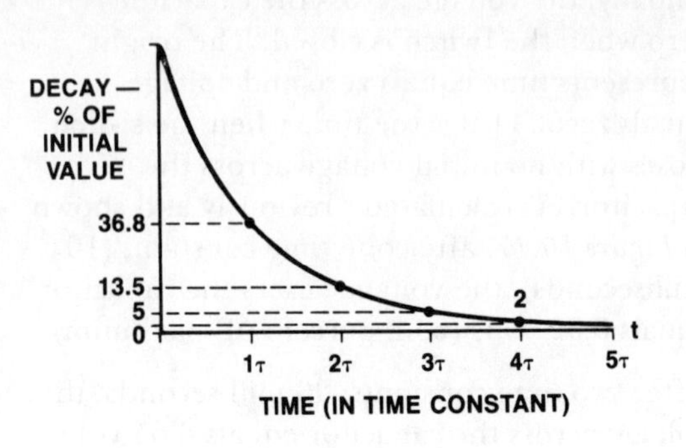

Figure 10.21 *Exponential Decrease as a Percent of Initial Value*

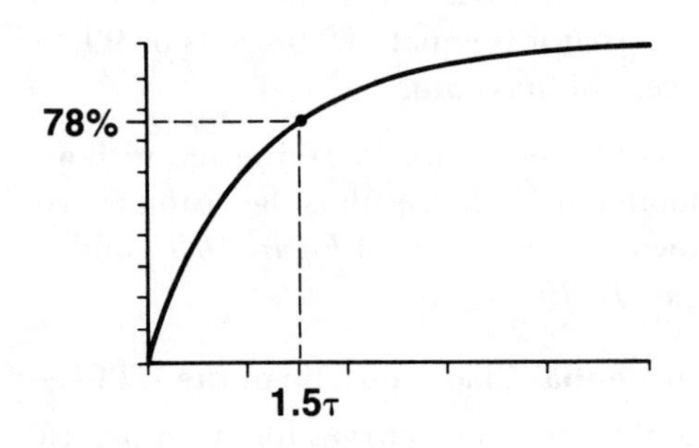

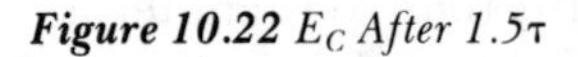

Figure 10.22 *E_C After 1.5τ*

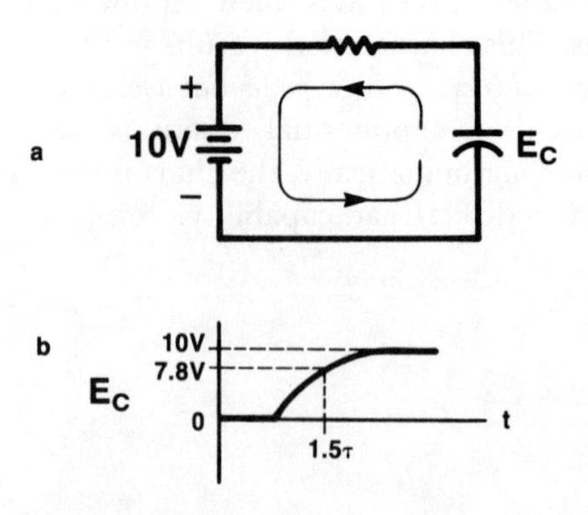

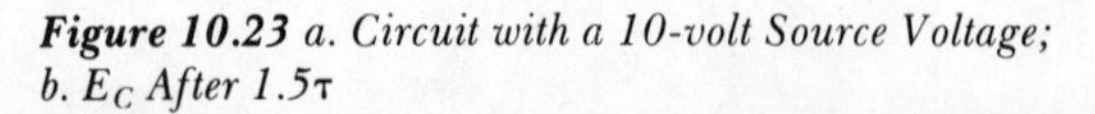

Figure 10.23 *a. Circuit with a 10-volt Source Voltage; b. E_C After 1.5τ*

If, however, one wants to know the current in the circuit or the voltage across the resistor after a specified number of time constants the percent of maximum must be read from the decaying exponential waveform. After one and one-half time constants, for example, the percent of maximum read from the chart for current or voltage across the resistor is 22 percent as shown in *Figure 10.25*. If the maximum current in the circuit is 10 milliamperes then after one and one-half time constants, the current has decreased to 22 percent of 10 milliamperes which is 2.2 milliamperes (0.22 × 10mA = 2.2mA). This shown graphically in *Figure 10.26*.

If the maximum voltage across the resistor was 30 volts, then after one and one-half time constants, the voltage across the resistor has decreased to 22 percent of 30 volts which is 6.6 volts (0.22 × 30V = 6.6V). This is shown graphically in *Figure 10.27*.

During the discharge of the capacitor, the chart again is used. However, the appropriate curve must be chosen.

Converting Time to Time Constants

Sometimes it is necessary to convert from time to time constants. The technique is as follows. To find the number of time constants that a specified time represents, simply divide the time, in seconds, by the number of seconds in one time constant.

$$\text{Number of Time Constants} = \frac{\text{t (seconds)}}{\tau\text{ (seconds)}} \quad (10\text{–}5)$$

For example, if the value of R in an RC circuit is 100 ohms and the value of C is 200 microfarads, then the time of one time constant, τ, is equal to R times C, and

$$\begin{aligned}\tau &= RC \\ &= (100\Omega)\,(200\mu F) \\ &= (1 \times 10^2)\,(20 \times 10^{-5}) \\ &= 20ms\end{aligned}$$

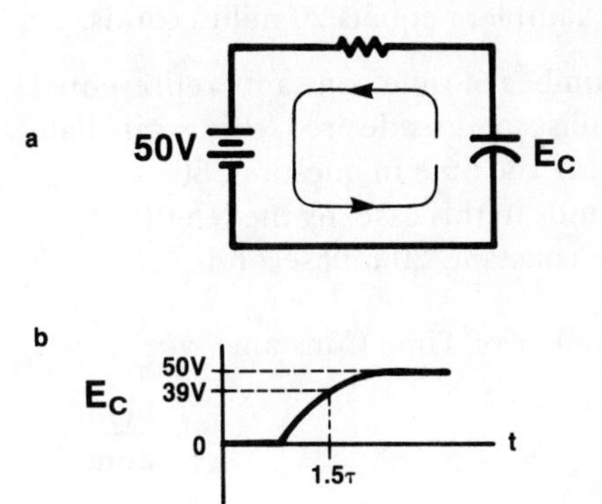

Figure 10.24 a. Circuit with a 50-volt Source Voltage; b. E_C After 1.5τ

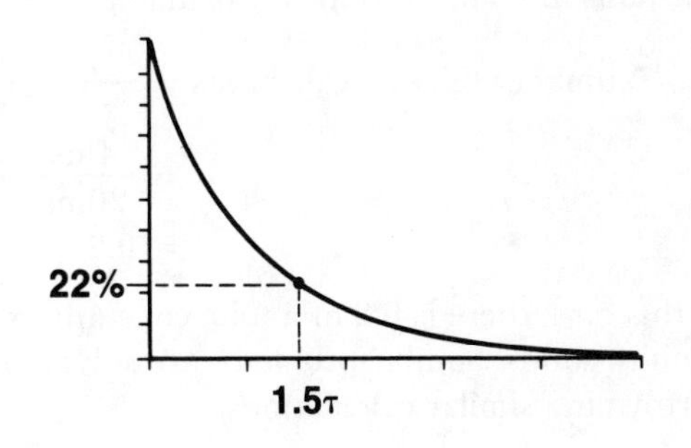

Figure 10.25 I_C After 1.5τ

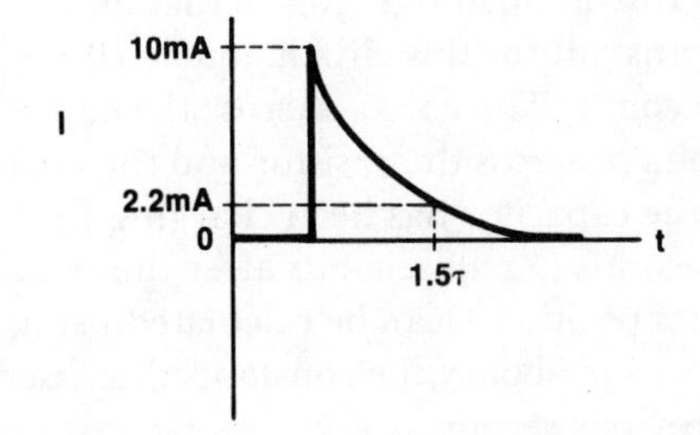

Figure 10.26 I After 1.5 τ with I_T of 10 Milliamperes

Solving for Instantaneous Voltage and Current Values: Charging

In this example, τ equals 20 milliseconds.

If the number of time constants represented by 60 milliseconds is desired, this is calculated by dividing the time in question, 60 milliseconds in this case, by the length of one time constant, 20 milliseconds.

$$\text{Number of Time Constants} = \frac{t}{\tau} = \frac{60\text{ms}}{20\text{ms}} = 3$$

In this example, there are three time constants in 60 milliseconds.

For the same R and C, the number of time constants in 4 milliseconds is calculated:

$$\text{Number Of Time Constants} = \frac{t}{\tau} = \frac{4\text{ms}}{20\text{ms}} = 0.2$$

In this case, there is 0.2 of a time constant in 4 milliseconds. Similar problems are solved by performing similar calculations.

Solving for Instantaneous Voltage and Current Values: Charging

The original circuit, shown in *Figure 10.1* and repeated in *Figure 10.28,* consists of a 10 kilohm resistor, a 1 microfarad capacitor, and a 10-volt dc source. Recall that one time constant for this circuit equals 10 milliseconds. The voltage across the capacitor, the voltage across the resistor and the current after the capacitor has been charging for 5 milliseconds (5 milliseconds after the switch contacts position 1)can be calculated using the information about time constants discussed in the previous section.

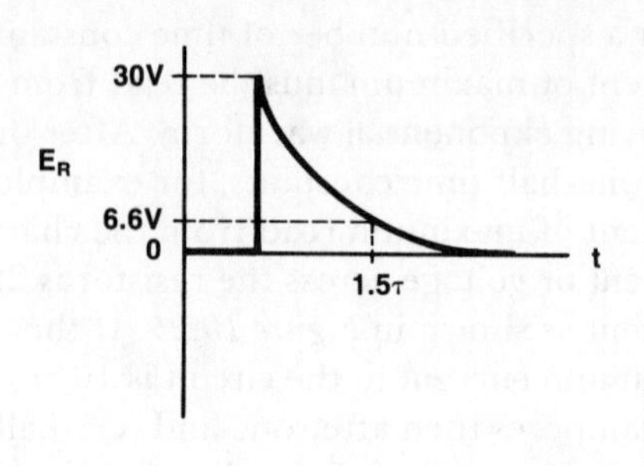

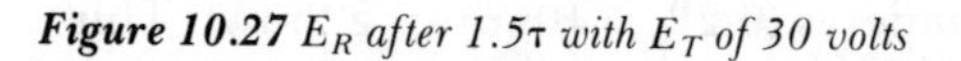
***Figure 10.27** E_R after 1.5τ with E_T of 30 volts*

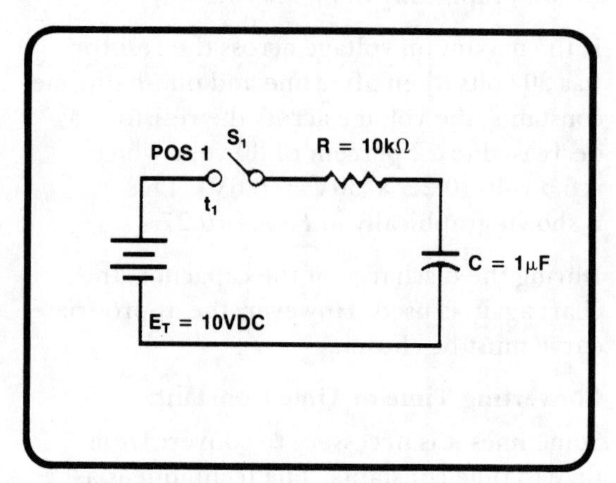

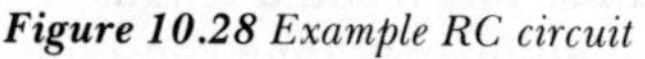
***Figure 10.28** Example RC circuit*

■ Solving for Instantaneous Voltage and Current Values: Charging

5 milliseconds in terms of time constants is calculated:

$$\textbf{Number Of Time Constants } (\tau) = \frac{t}{\tau}$$

$$= \frac{5ms}{10ms}$$

$$= 0.5$$

In the example circuit of *Figure 10.28,* 5 milliseconds equals 0.5 time constants. The waveforms for the circuit are as shown in *Figure 10.29*. The maximum capacitor charge, E_C, is 10 volts, or the applied voltage. The maximum current initially, is 10 volts ÷ 10 kilohms = 1 milliampere. The maximum voltage across the resistor is 10 volts, which is the same as the capacitor's maximum charge.

The Universal Time Constant Chart of *Figure 10.20* can be used to determine the percent of maximum voltage across the capacitor. The rising curve, repeated in *Figure 10.30*, shows that the percent of maximum voltage across the capacitor after one-half of a time constant is about 40 percent.

$$E_c = 40\%\ E_{c(max)}$$

$$= (0.40)(10V)$$

$$= 4V$$

Thus, the voltage across the capacitor after one-half of a time constant (5 milliseconds) is 4 volts.

Using the decaying curve as shown in *Figure 10.31,* the percent of maximum current in the circuit after one-half of one time constant is determined to be 60 percent. Said another way, the current has decayed 40 percent from its initial value to a 60 percent value.

$$I = 60\%\ I_{(max)}$$

$$= (0.6)(1mA)$$

$$= 0.6\ mA$$

Thus, the current in the circuit after 5 milliseconds is 0.6 milliamperes.

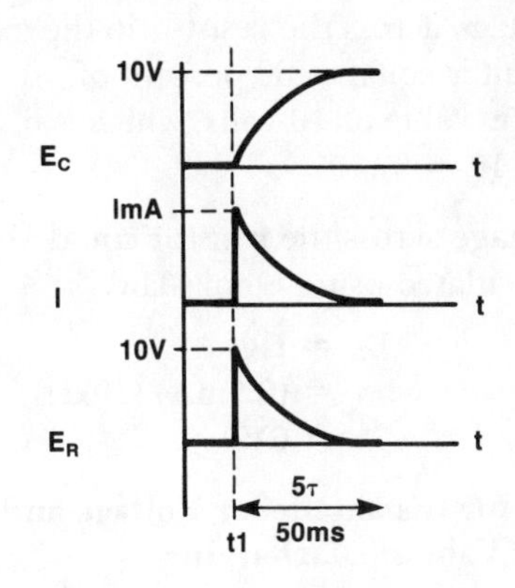

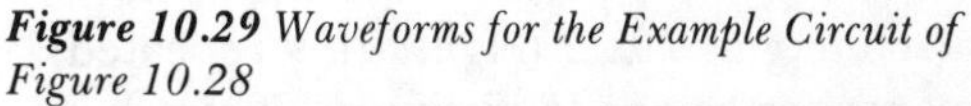

Figure 10.29 *Waveforms for the Example Circuit of Figure 10.28*

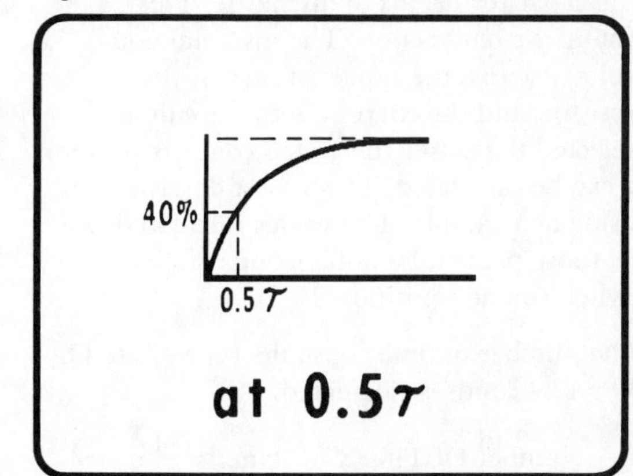

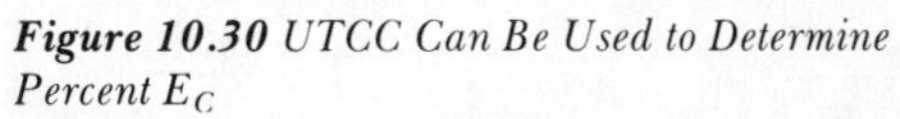

Figure 10.30 *UTCC Can Be Used to Determine Percent* E_C

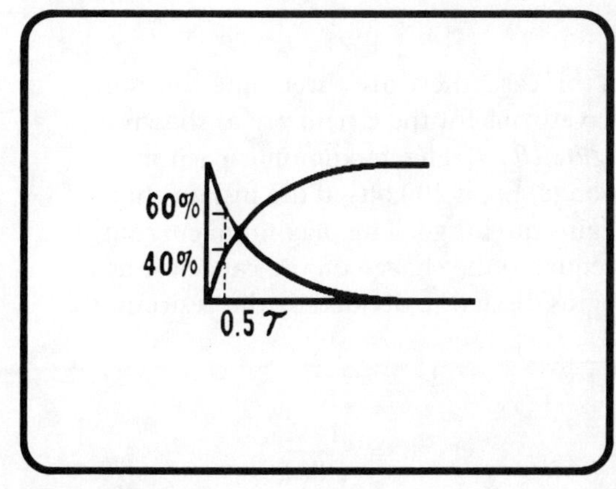

Figure 10.31 *UTCC Can Be Used to Determine Percent* I_{max}

■ Solving for Instantaneous Voltage and Current Values: Discharging

The voltage across the resistor in the example RC circuit is equal to 60 percent of its maximum value of 10 volts, which is 6 volts ($0.60 \times 10 = 6$).

The voltage across the resistor could also have been calculated using Ohm's law:

$$\begin{aligned} \mathbf{E_R} &= \mathbf{IR} \\ &= \mathbf{(0.6mA)\,(10k\Omega)} \\ &= \mathbf{6V} \end{aligned}$$

Solving for Instantaneous Voltage and Current Values: Discharging

When the switch is placed in position 2 as shown in *Figure 10.32* (*Figure 10.9* repeated) the capacitor begins to discharge. This changes circuit action. The instantaneous voltage across the capacitor, across the resistor, and the current in the circuit at a selected time after the switch contacts position 2 can be calculated. This is best described using an example. The values calculated will be those present 30 milliseconds after the switch contacts position 2.

The number of time constants represented by 30 milliseconds is calculated:

$$\begin{aligned} \text{Number Of Time Constants} &= \frac{t}{\tau} \\ &= \frac{30\text{ms}}{10\text{ms}} \\ &= 3 \end{aligned}$$

In this case, there are three time constants. Waveforms for the circuit are as shown in *Figure 10.33*. The maximum capacitor voltage, E_C, is 10 volts at the instant that it begins discharge. The maximum current, I_{max} is equal to the charge on the capacitor as it begins discharge divided by the resistance.

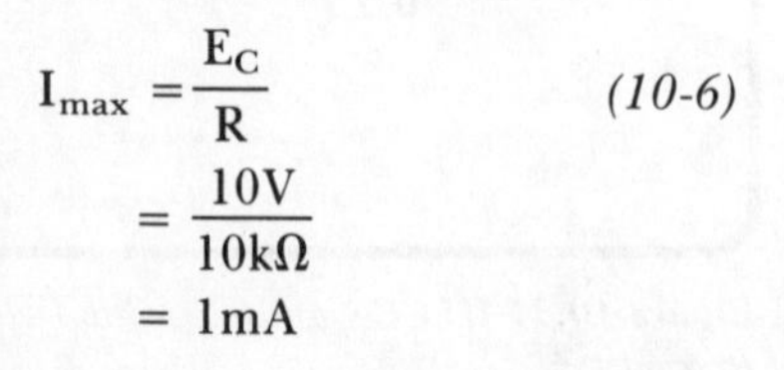

$$\begin{aligned} I_{max} &= \frac{E_C}{R} \qquad (10\text{-}6) \\ &= \frac{10\text{V}}{10\text{k}\Omega} \\ &= 1\text{mA} \end{aligned}$$

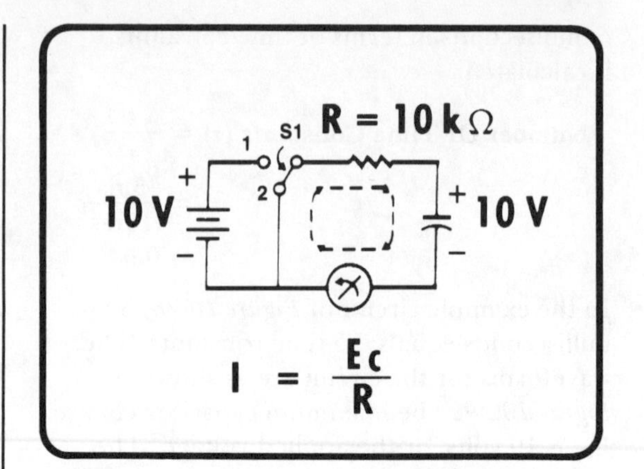

Figure 10.32 Example RC Circuit with Switch in Position 2

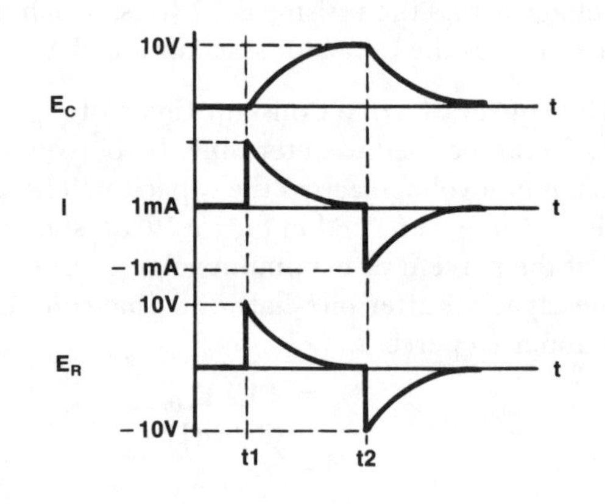

Figure 10.33 Example Circuit Waveforms

In this example, $I_{max} = 1$ milliampere. Remember, however, that the current is flowing opposite to the direction of current flow during charge. Thus, the maximum discharge current is designated as a *negative* 1 milliampere.

The voltage across the resistor during discharge is equal to the voltage across the capacitor. The voltage is specified as *negative* during discharge since it is *opposite* in polarity to the voltage which appeared across the resistor during the charge of the capacitor.

■ RC Time Constants Related to Kirchoff's Voltage Law

From the Universal Time Constant Chart in *Figure 10.34,* using the decaying curve, the percent of maximum voltage or current after three time constants is determined to be four percent. Thus, after three time constants:

$$
\begin{aligned}
E_C &= 4\%E_{Cmax} \\
&= (0.04)\ (10V) \\
&= 0.4V
\end{aligned}
$$

The charge on the capacitor has decreased from 10 volts to 0.4 volts after three time constants.

After three constants, the current will decrease from I_{max} to 4 percent of I_{max}:

$$
\begin{aligned}
I &= 4\%I_{max} \\
&= (0.04)\ (-1mA) \\
&= -0.04mA \\
&= -40\mu A
\end{aligned}
$$

Therefore, after three time constants I = −40 microamperes.

After three time constants, the voltage across the resistor will decrease from E_{Rmax} to 4 percent of E_{Rmax}:

$$
\begin{aligned}
E_R &= 4\%E_{Rmax} \\
&= (0.04)\ (-10V) \\
&= -0.4V
\end{aligned}
$$

Thus, $E_R = -0.4V$.

RC Time Constants Related to Kirchoff's Voltage Law

It is interesting to note that Kirchoff's voltage law is valid for dc RC circuits. At any instant in time the voltage across the resistor plus the voltage across the capacitor equals the applied voltage: $E_R + E_C = E_A$. Initially, at t_1, when the switch in the circuit of *Figure 10.28* is first contacted at position 1, the voltage across the resistor equals 10 volts, or the applied voltage, and the voltage across the capacitor equals zero volts. This is shown in *Figure 10.35*. The sum of these two voltages equals 10 volts, or the applied voltage.

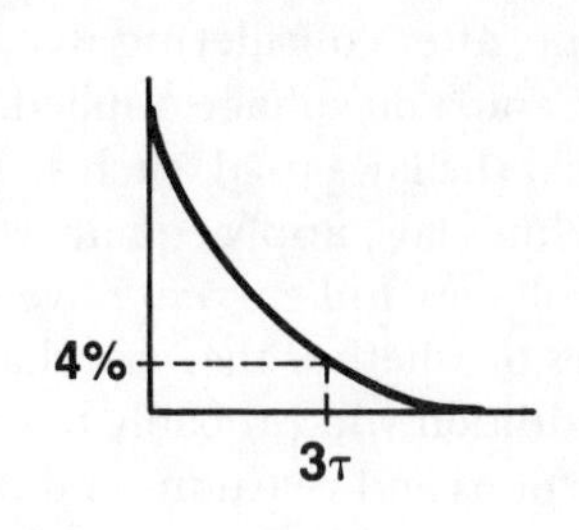

Figure 10.34 *UTCC Used to Determine Percent E_C, I, and E_R*

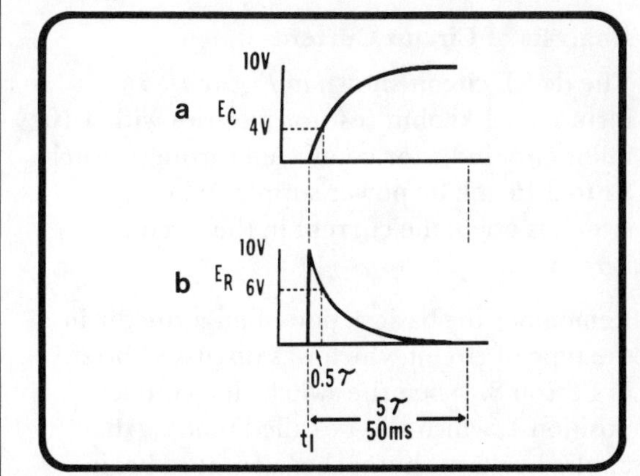

Figure 10.35 *Waveforms of E_C and E_R in Example Circuit*

After one-half time constant, as calculated previously, the voltage across the resistor is 6 volts, and the voltage across the capacitor is 4 volts. The sum of these voltages is 10 volts (6V + 4V = 10V).

After five time constants when the voltage across the resistor is 0 volts, the capacitor has charged to 10 volts. The sum of these voltage drops, too, is 10 volts (0V + 10V = 10V).

■ Analysis of Circuit Current

RL DC Circuit Analysis: Opposing a Current Increase

By this time, after considering RC circuits with an ac and a dc voltage applied, you have learned that the laws used, such as Kirchoff's law and Ohm's law, apply equally well whether voltages and current have steady-state values or whether they are changing with time. In addition you probably have realized that the actions and reactions and the circuit analysis methods of resistance-capacitive reactance circuits and resistance-inductive reactance circuits are similar. Recognizing that similarity, the discussion and analysis shifts to resistance-inductance (RL) circuits.

Analysis of Circuit Current

The dc RL circuit shown in *Figure 10.36* includes a 1 kilohm resistor in series with a 10 millihenry inductor connected through switch S1 to a 10 volt dc power supply. When the switch is open, the current in the circuit is zero.

Remember the basic action of an inductor in the type of circuit which was discussed briefly in Lesson 8. When the switch first contacts position 1, which will be called time t_1, the applied voltage causes the current to begin flowing in the direction indicated in *Figure 10.36*. Recall that if the inductance L is removed and only the resistor is present in the circuit, the current will instantaneously increase to a final value determined by Ohm's law, the applied voltage divided by the resistance in the circuit. This current flow in a resistive circuit is shown in *Figure 10.37*. The current would rise immediately to 10mA (10 volts divided by 1 kilohm). However, with an inductor in series with a resistor, the increasing current must pass through the inductor which opposes any change in current. What happens is that as the current initially attempts to increase from zero, the inductor sets up a counter-EMF. This counter EMF across the inductor inhibits the

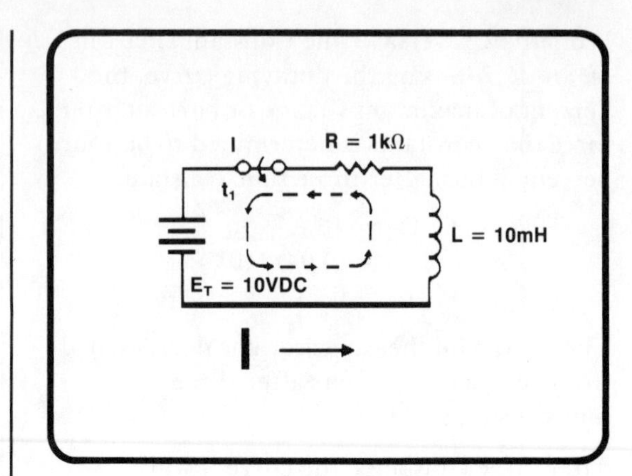

Figure 10.36 *A DC RL Circuit*

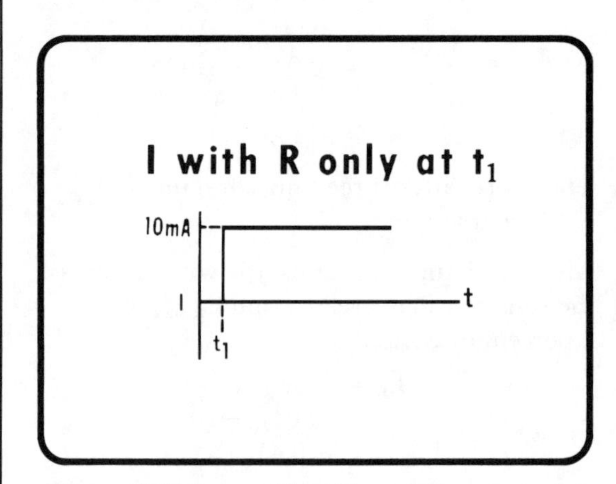

Figure 10.37 *Current Flow When a Voltage Is Applied to a Purely Resistive Circuit*

instantaneous rise of current as shown in *Figure 10.38*. The counter EMF of the inductor continues its opposition to the change of current until the current finally reaches its maximum value which is limited by the resistance in the circuit.

The rise of current in the inductive circuit follows an exponential curve as did the voltage rise across the plates of a capacitor. The maximum value of current is equal to the value of the applied voltage divided by the value of the resistance:

$$I_{max} = \frac{E_T}{R} \qquad (10\text{–}7)$$

Thus, the maximum current in the example circuit of *Figure 10.36* can be calculated:

$$I_{max} = \frac{E_T}{R} = \frac{10V}{1k\Omega} = 10mA$$

Calculation of E_R

Since the voltage across the resistor is a direct result of the current flowing in the circuit, the voltage rise will appear similar to the rise of current. The voltage across the resistor will have a maximum value which can be expressed by Ohm's law:

$$E_R = IR \qquad (10\text{–}8)$$

Thus the voltage across the resistor in the example dc RL circuit is calculated:

$$E_R = IR = (10mA)(1k\Omega) = 10V$$

For a value of maximum current equal to 10 milliamperes, E_R equals 10 volts.

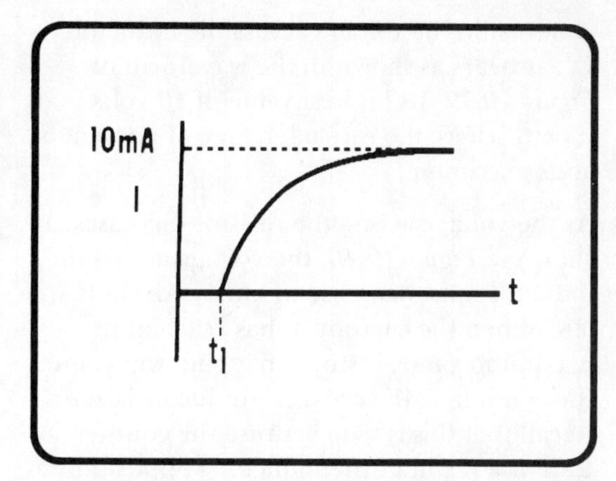

Figure 10.38 Current Flow When a Voltage Is Applied to an RL Circuit

Calculation of E_L

When the switch of the circuit in *Figure 10.36* is placed in position 1, no current flows instantly, and no voltage appears across the resistor: $I = 0$; $E_R = 0$. Remember that this is a series circuit with a dc voltage applied. Because of this, Kirchoff's voltage law applies, and the applied voltage must equal the sum of the voltage drops in the circuit which are the voltage across the resistor, E_R, plus the voltage across the inductor, E_L:

$$E_T = E_R + E_L \qquad (10\text{–}9)$$

Because the voltage across the resistor is zero, the 10 volts applied to the circuit must *all* be dropped across the inductor. That is, the counter EMF is 10 volts and $E_T = E_L$. In addition, this is physically correct because the voltage across the inductor is equal to the inductance times the rate of change of current:

$$E_L = L\left(\frac{\Delta I}{\Delta t}\right) \qquad (10\text{–}10)$$

$\frac{\Delta I}{\Delta t}$ also is expressed in many instances as $\frac{dI}{dt}$.

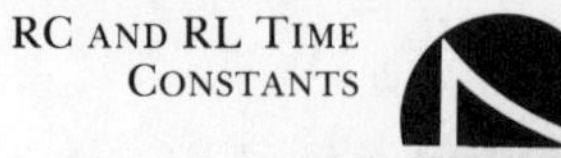

■ Calculation of E_L

As a result, the voltage across the inductor, E_L, appears as shown in the waveform of *Figure 10.39*. Its highest value of 10 volts is at time t_1 where the rate of change of current is at its maximum.

As the voltage across the resistor increases, as shown in *Figure 10.40,* the voltage across the inductor must decrease to satisfy Kirchoff's law. When the current, I, has reached its maximum value, it stops changing value and the counter EMF across the inductor is zero. Recall that this is true because the counter EMF is a result of the inductor's opposition to a *change* of current value. When the current reaches its maximum value the magnetic field surrounding the inductor is fully expanded as shown in *Figure 10.41*. The voltage across the resistor is now maximum, and it equals the applied voltage, thus obeying Kirchoff's voltage law.

DC RL Circuit Analysis: Opposing Current Decrease

What will happen to a dc RL circuit when there is no longer voltage available to continue to supply current to the inductor? The way to find out is to stop the voltage. So assume that in the example RL circuit that the switch is placed in position 2 as shown in *Figure 10.42*. (The switch is of a type that makes contact at position 2 before it breaks contact at position 1.) Effectively, the applied voltage is removed from the circuit, and the resistor and inductor are connected in parallel with each other.

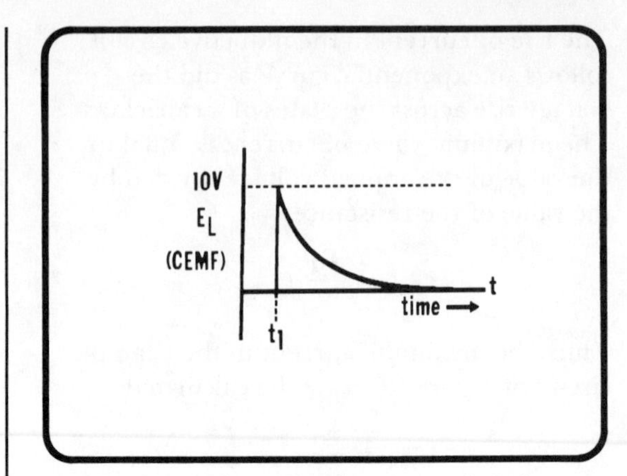

Figure 10.39 *Voltage Waveform Across the Inductor*

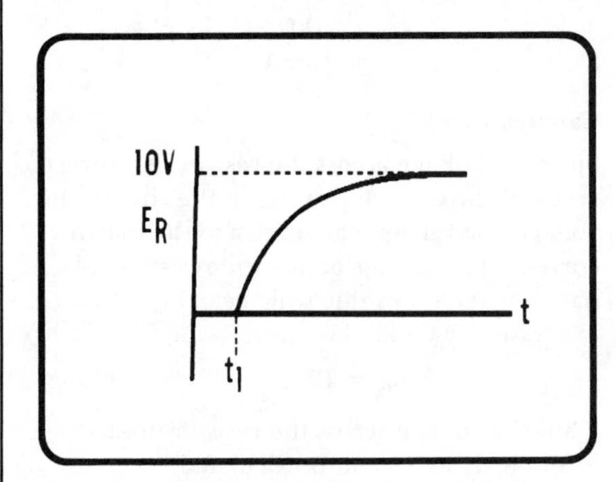

Figure 10.40 *Voltage Waveform Across the Resistor*

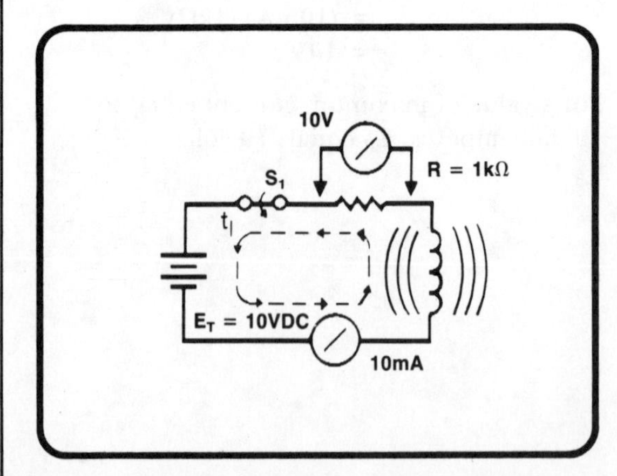

Figure 10.41 *Effect of Counter EMF in a dc RL Circuit*

Analysis of Circuit Current

Since the power supply has been disconnected, there is now no external voltage source to maintain current through the inductor and resistor. Thus, the magnetic field surrounding the inductor begins to collapse, and current begins to decrease in the circuit. As shown in *Figure 10.43*, the current direction is still the same, but it is becoming less and less in magnitude. The inductor, therefore, will generate a counter EMF to oppose the decrease in current. Thus, the collapsing magnetic field generates an EMF of the polarity shown in *Figure 10.43* to keep current flowing in the circuit— to keep current from decreasing. The decay of current is exponential, as shown in *Figure 10.44,* because of the action of the inductor.

Analysis of E_R

The voltage across the resistor, which is a result of the current, decreases as the current decreases. This is evident in the decreasing curve after time t_2 in the waveform representing E_R in *Figure 10.44*. The counter EMF across the inductor has changed polarity in order to oppose the decrease in current.

Analysis of E_L

To help clarify this fact, remember that because the magnetic field around the inductor is collapsing, the flux lines are now cutting the windings of the inductor in the opposite direction from the direction that occurred when the magnetic field was increasing. When the magnetic field was increasing because of a changing current that was increasing, the counter EMF set up opposed the increase in current. Now, however, when the current decreases, the collapsing magnetic field sets up a counter EMF that now aids the current flow— to keep it going— to keep it from decreasing.

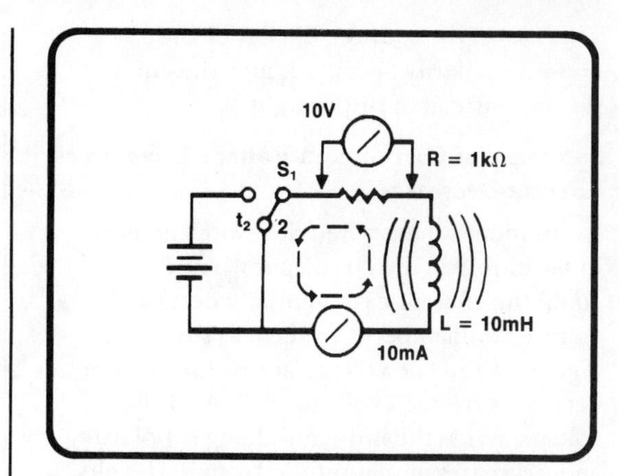

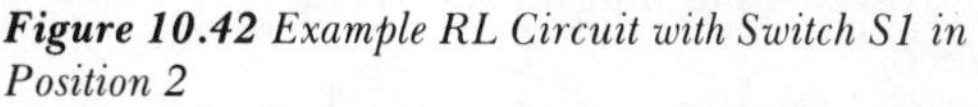
Figure 10.42 *Example RL Circuit with Switch S1 in Position 2*

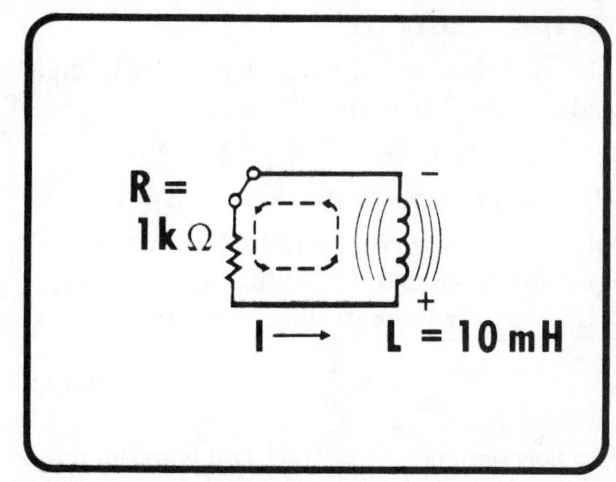

Figure 10.43 *Parallel Relationship of R and L in Example Circuit*

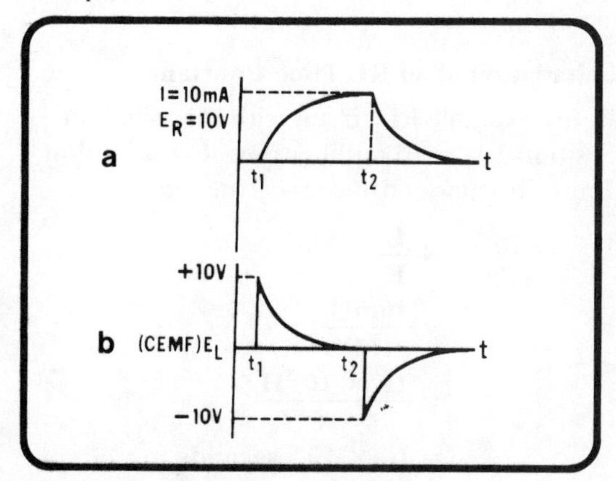

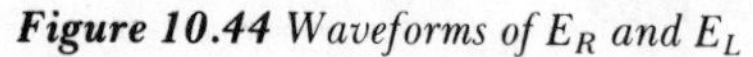
Figure 10.44 *Waveforms of E_R and E_L*

Therefore, the counter EMF is of the opposite polarity because it aids the current flowing instead of opposing it.

Summary of Current and Voltage Changes as Current Decreases

In summary, then, when the switch is placed in position 2 in the circuit shown in *Figure 10.42* the current exponentially decreases from 10 milliamperes to zero, as shown in *Figure 10.45*. The voltage across the resistor decreases from 10 volts to zero. And the voltage across the inductor changes polarity, and decreases in magnitude from − 10 volts to zero.

RL TIME CONSTANTS

The time duration of these changes of voltage and current, like in the RC circuit, are measured in terms of time constants.

RL TIME CONSTANTS DEFINED

The time constant of a series RL circuit is equal to the value of the inductor, in henrys, divided by the value of the resistor in ohms:

$$\tau = \frac{L}{R} \qquad (10\text{–}11)$$

The *time constant* of an RL circuit is defined as the amount of time necessary for the current in the circuit to reach 63.2 percent of its maximum value.

Calculation of an RL Time Constant

In the example RL circuit with the switch in position 1,L = 10 millihenries, R = 1 kilohm. Thus, the time constant is calculated:

$$\begin{aligned}\tau &= \frac{L}{R}\\ &= \frac{10\text{mH}}{1\text{k}\Omega}\\ &= \frac{10 \times 10^{-3}\text{H}}{1 \times 10^{3}\,\Omega}\\ &= 10 \times 10^{-6} \text{ seconds}\\ &= 10\mu\text{s}\end{aligned}$$

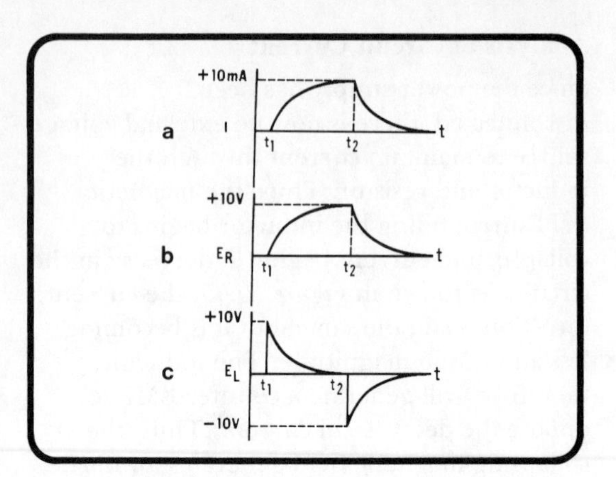

Figure 10.45 Waveforms in Example RL Circuit with Switch in Position 2

The time constant is 10 microseconds.

Five time constants are required for the current in the circuit to reach its maximum value. In the circuit the time constant is 10 microseconds. Therefore, five time constants equal 50 microseconds. Thus, it takes 50 microseconds for the current in the circuit to reach a value of 10 milliamperes. When the current decreases it also takes five time constants (50 microcseconds) for the current to decrease to zero provided that the current flows through the same resistance and inductance. *Figure 10.46* graphically illustrates the time required for the current to increase or decrease in the example RL circuit.

■ **RL Circuit Summary**

The UTCC and DC RL Circuit

Since the voltages and currents in an RL circuit increase and decrease exponentially and follow similar equations to the ones used for the RC circuit, the Universal Time Constant Chart can be used to determine voltages and/or current in the RL circuit during their rise or decay.

RL Circuit Summary

A typical RL circuit will be used to show you how to use the UTCC to determine voltages and currents in such a circuit. The RL circuit is shown in *Figure 10.47*. It is composed of an 8 henry inductor and 100 ohm resistor connected through a switch, S1, as shown, to an applied voltage of 20 volts.

The time constant is calculated:

$$\tau = \frac{L}{R} = \frac{8H}{100\Omega} = 0.08 \text{ seconds} = 80\text{ms}$$

The maximum voltage across the resistor and inductor is the applied voltage of 20 volts.

The maximum current is calculated using Ohm's law:

$$I_{max} = \frac{E_T}{R} = \frac{20V}{100\Omega} = 0.2A = 200mA$$

The current will take five time constants to reach its final value. Since it takes 80 milliseconds for one time constant, it will take 400 milliseconds to reach its maximum value of 200 milliamperes, (5 × 80ms = 400ms).

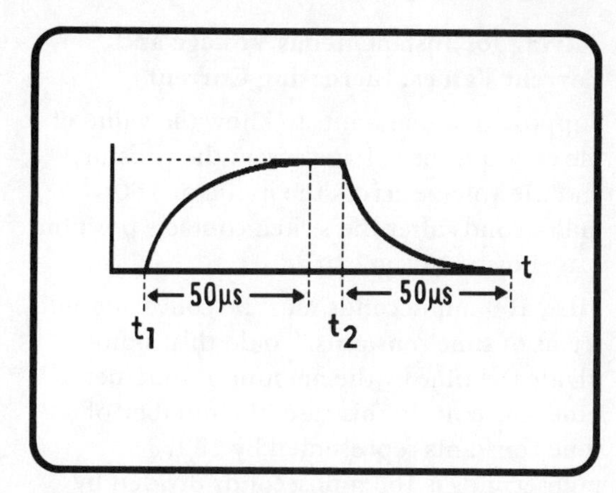

Figure 10.46 Time Required for the Current to Increase or Decrease in Example RL Circuit

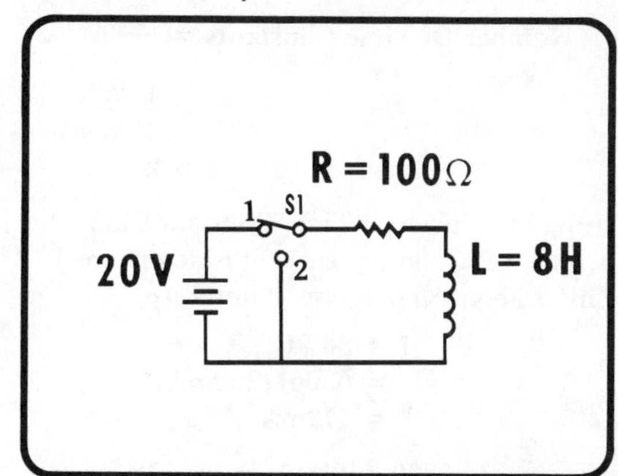

Figure 10.47 Example RL Circuit for Discussion of UTCC

Solving for Instantaneous Voltage and Current Values: Increasing Current

Suppose that one wants to know the value of the current, the voltage across the resistor, and the voltage across the inductor 160 milliseconds after the switch contacts position 1 as shown in *Figure 10.48*.

First, 160 milliseconds must be converted into terms of time constants. To do this, simply divide the time by the amount of time per time constant. In this case, the number of time constants represented by 160 milliseconds is 160 milliseconds divided by 80 milliseconds or two time constants.

$$\text{Number Of Time Constants} = \frac{t}{\tau} = \frac{160\text{ms}}{80\text{ms}} = 2$$

Using the Universal Time Constant Chart the percent of maximum current is determined. This is about 86 percent. Therefore,

$$I = 86\%I_{max} = (0.86)(200\text{mA}) = 172\text{mA}$$

86 percent of 200 milliamperes is 172 milliamperes. Thus, the current in the circuit after 160 milliseconds is 172 milliamps.

The instantaneous voltage across the resistor can be calculated in the same way. However, since the voltage across the resistor is a result of the current in the circuit, the voltage across the resistor can be determined by using Ohm's law and the value of the current.

$$E_R = IR = (172\text{mA})(100\Omega) = 17.2\text{V}$$

In this example, E_R is 17.2 volts after 160 milliseconds.

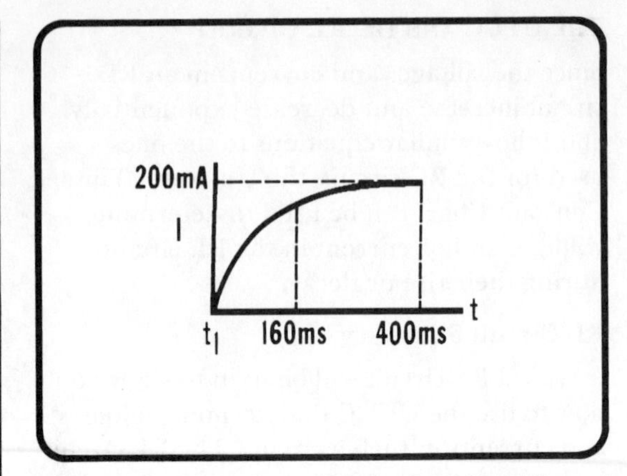

Figure 10.48 Waveform Analysis after 160 Milliseconds

The voltage across the inductor can be determined by using the chart. However, it can also be calculated. The calculation is based on the fact that the voltage across the resistor, E_R, plus the voltage across the inductor, E_L, must equal the applied voltage, E_T. Mathematically,

$$E_T = E_R + E_L$$

This expression can be rewritten to calculate E_L:

$$E_L = E_T - E_R$$

Therefore, if the voltage across the resistor is 17.2 volts:

$$E_L = E_T - E_R = 20\text{V} - 17.2\text{V} = 2.8\text{V}$$

The voltage across the inductor 160 milliseconds after t_1 is 2.8V.

Solving for Instantaneous Voltage and Current Values: Decreasing Current

Once the current has reached maximum, the switch in the circuit, as shown in *Figure 10.49*, is changed to position 2. For an example, the voltages and currents in that circuit will be calculated for a time 80 milliseconds after the switch has contacted position 2. The waveforms for E_R, E_L, and I are shown in *Figure 10.50*.

First, the number of time constants after t_2 is calculated.

$$\text{Number Of Time Constants} = \frac{t}{\tau} = \frac{80\text{ms}}{80\text{ms}} = 1$$

Using the Universal Time Constant Chart, the percent to which the maximum current has decayed after one time current is determined as 37%. As shown in *Figure 10.51*, the current value after 80 milliseconds (one time constant) can be calculated:

$$I = 37\% I_{max} = (0.37)(200\text{mA}) = 74\text{mA}$$

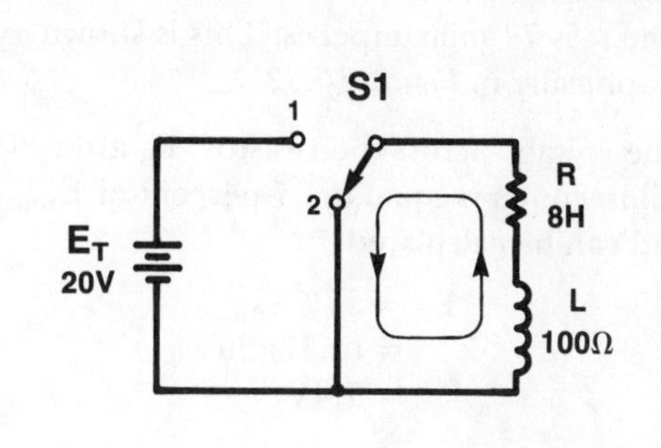

Figure 10.49 Example RL Circuit with Switch S1 in Position 2

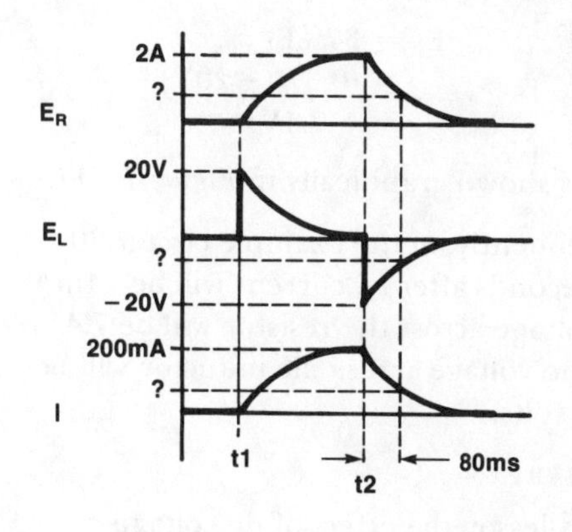

Figure 10.50 Waveforms for E_R, E_L, and I for Example RL Circuit

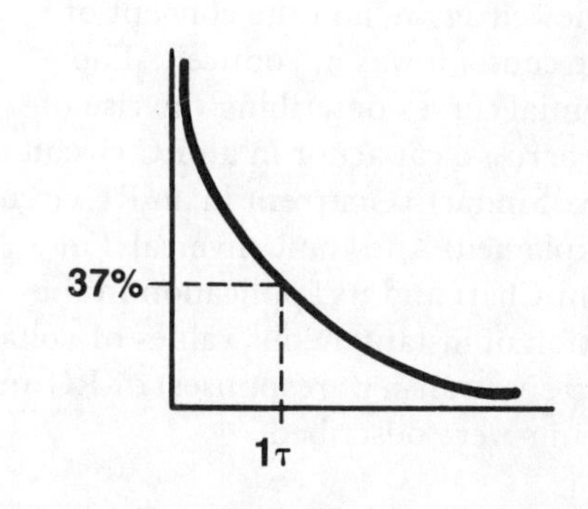

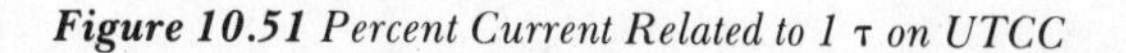

Figure 10.51 Percent Current Related to 1 τ on UTCC

Thus, the current 80 milliseconds after time t_2 is 74 milliamperes. This is shown graphically in *Figure 10.52.*

The voltage across the resistor, E_R after 80 milliseconds is equal to 37 percent of E_{Rmax} and can be calculated:

$$\begin{aligned} E_R &= 37\%E_{Rmax} \\ &= (0.37)(20V) \\ &= 7.4V \end{aligned}$$

This is shown graphically in *Figure 10.53.* Remember that Kirchoff's voltage law must be satisfied; therefore, the voltage across the inductor, E_L, after 80 milliseconds is 37 percent of E_{Lmax}, but negative.

$$\begin{aligned} E_L &= 37\%E_{Lmax} \\ &= (0.37)(-20V) \\ &= -7.4V \end{aligned}$$

This is shown graphically in *Figure 10.54.*

Consequently, in the example circuit 80 milliseconds after t_2, current will be 74mA, the voltage across the resistor will be 7.4 volts and the voltage across the inductor will be −7.4 volts.

SUMMARY

In this lesson the effect of dc voltage transitions on resistive-capacitive, and resistive-inductive circuits was discussed. The basic action of a capacitor and an inductor was reviewed again, and the concept of the time constant was introduced. The exponential curves describing the rise of voltage across a capacitor in an RC circuit or the rise of inductive current in an RL circuit were explained. And the Universal Time Constant Chart and its application in the calculation of instantaneous values of voltage and current (transient responses) of RC and RL circuits were described.

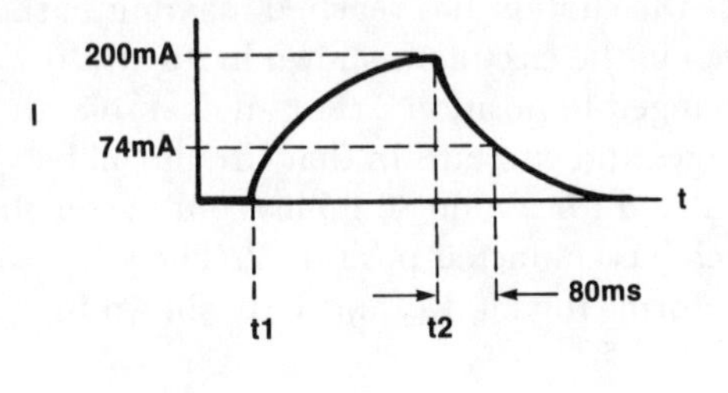

Figure 10.52 *Example RL Circuit I Waveform*

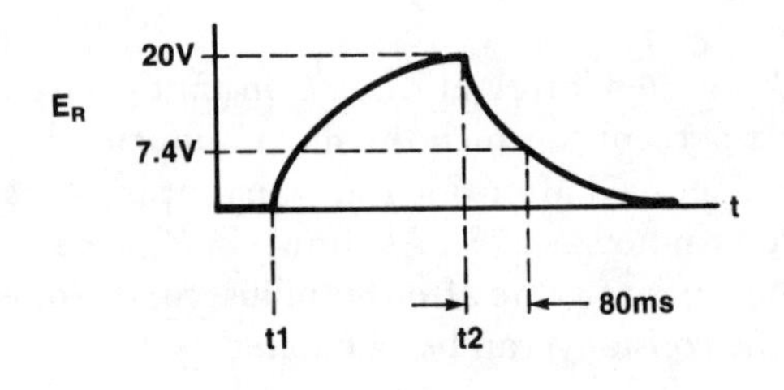

Figure 10.53 *Example RL Circuit E_R Waveform*

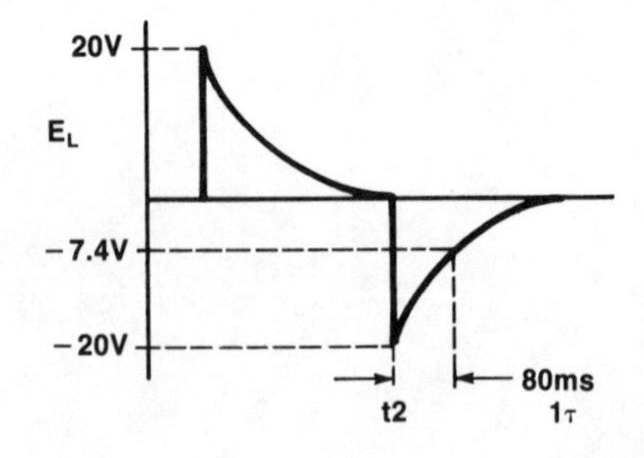

Figure 10.54 *Example RL Circuit E_L Waveform*

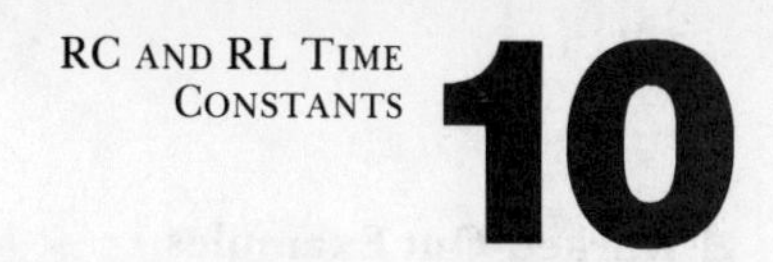

■ Worked-Out Examples

1. Given the values for R and C below, calculate the time of one time constant.

a. $R = 60k\Omega$, $C = 0.02\mu F$, $1\tau =$ ____________

b. $R = 5k\Omega$, $C = 30\mu F$, $1\tau =$ ____________

Solution:

a. $1\tau_{(s)} = R_{(\Omega)} \times C_{(F)} = 60k\Omega \times 0.02\mu F = 60 \times 10^3 \times 0.02 \times 10^{-6}$
$= 1.2 \times 10^{-3}s =$ **1.2ms**

b. $1\tau = R \times C = 5 \times 10^3 \times 30 \times 10^{-6} =$ **0.15s**

2. Given the values for 1τ and R below, calculate the value of C.

a. $1\tau = 5ms$, $R = 30k\Omega$, $C =$ ____________

b. $1\tau = 30\mu s$, $R = 47\Omega$, $C =$ ____________

Solution:

a. $C = \dfrac{1\tau}{R} = \dfrac{5ms}{30k\Omega} =$ **0.167μF**

b. $C = \dfrac{1\tau}{R} = \dfrac{30\mu s}{47\Omega} =$ **0.638μF**

3. Given the values for 1τ and C below, calculate the value of R.

a. $1\tau = 5s$, $C = 0.1\mu F$, $R =$ ____________

b. $1\tau = 5ms$, $C = 0.1\mu F$, $R =$ ____________

Solution:

a. $R = \dfrac{1\tau}{C} = \dfrac{5s}{0.1\mu F} =$ **50MΩ**

b. $R = \dfrac{1\tau}{C} = \dfrac{5ms}{0.1\mu F} =$ **50kΩ**

4. Calculate one time constant for this circuit.

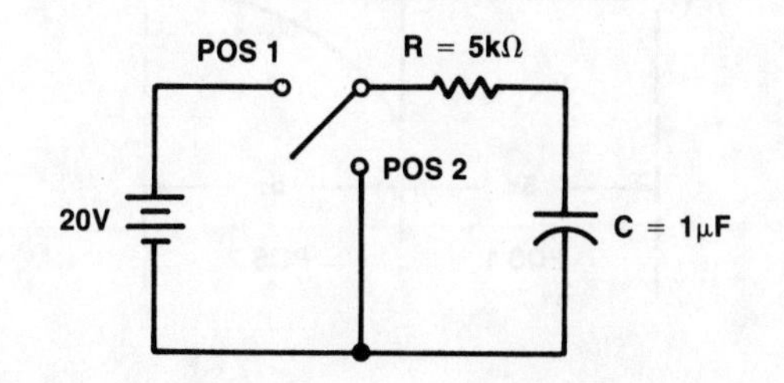

■ Worked-Out Examples

Solution:

$1\tau_{(s)} = R \times C = 5k\Omega \times 1\mu F = (5 \times 10^3)(1 \times 10^{-6}) = 5 \times 10^{-3}$ seconds or **5ms**

5. Calculate one time constant for this circuit.

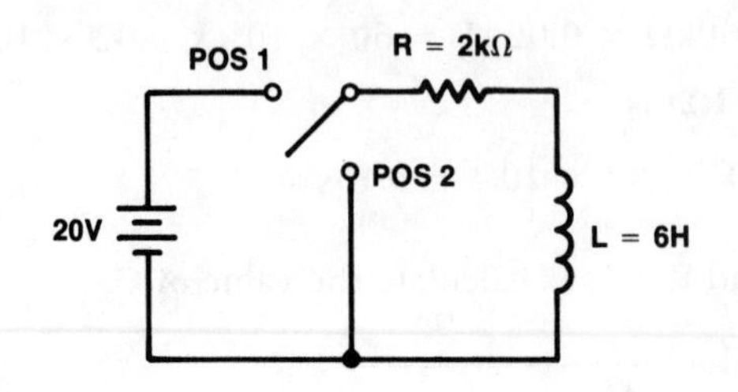

Solution:

$$1\tau_{(s)} = \frac{L}{R} = \frac{6H}{2k\Omega} = \frac{6}{2 \times 10^3} = 3 \times 10^{-3} \text{ seconds or } \mathbf{3ms}$$

6. For the circuit in Problem 4, draw the waveforms showing E_C, E_R, and I while the switch is in position 1 and position 2 for 5 time constants.

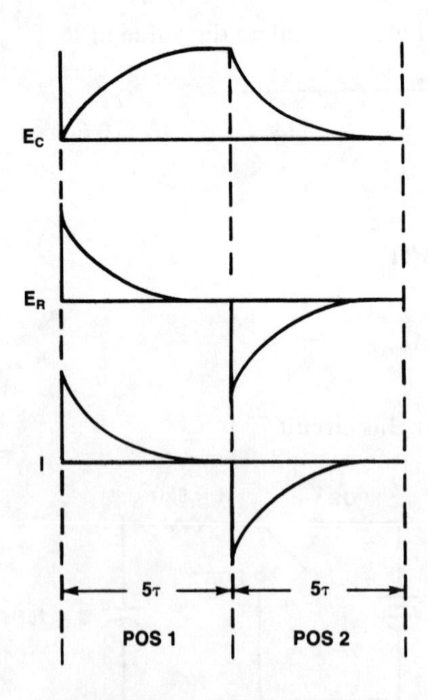

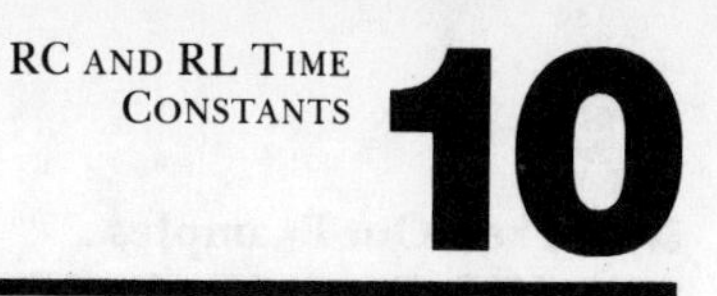

■ Worked-Out Examples

7. For the circuit in Problem 4, calculate the following:

a. I_{max} = ______________

b. E_{Cmax} = ______________

c. E_{Rmax} = ______________

d. 5τ = ______________

Solution:

a. $I_{max} = \frac{E_A}{R} = \frac{20V}{5k\Omega} = \mathbf{4mA}$

b. $E_{Cmax} = E_A = \mathbf{20V}$

c. $E_{Rmax} = E_A = \mathbf{20V}$

d. $5\tau = 5 \times 1\tau = 5 \times 5ms = \mathbf{25ms}$

8. If 1 time constant equals 20ms, calculate the number of time constants if:

a. time = 6ms
b. time = 30ms
c. time = 50ms
d. time = 80ms

Solution:

a. $\text{number of } \tau = \frac{\text{time}}{\text{time of } 1\tau} = \frac{6ms}{20ms} = \mathbf{0.3}$

b. $\text{number of } \tau = \frac{30ms}{20ms} = \mathbf{1.5}$

c. $\text{number of } \tau = \frac{50ms}{20ms} = \mathbf{2.5ms}$

d. $\text{number of } \tau = \frac{80ms}{20ms} = \mathbf{4}$

Worked-Out Examples

9. If 1 time constant equals 50μs, calculate the time interval if:

a. number of time constants equals 0.7τ
b. number of time constants equals 1.3τ
c. number of time constants equals 2.5τ
d. number of time constants equals 3.8τ

Solution:

a. time = 0.7 × 50μs = **35μs**

b. time = 1.3 × 50μs = **65μs**

c. time = 2.5 × 50μs = **125μs**

d. time = 3.8 × 50μs = **190μs**

10. For the circuit in Problem 4, calculate the following instantaneous circuit values 12ms after the switch is placed in position 1.

a. E_C = ____________
b. E_R = ____________
c. I = ____________

Solution:

(1) Calculate the number of time constant represented by 12ms:

$$\text{number of time constants} = \frac{\text{time}}{\text{time of one time constant}} = \frac{12\text{ms}}{5\text{ms}} = 2.4\tau$$

(2) Use UTCC to determine percent of maximum circuit value after 2.4τ:

2.4τ = 90% of E_C

2.4τ = 10% of E_R and I

a. E_C = 0.9 × 20V = **18V**

b. E_R = 0.1 × 20V = **2V**

c. I = 0.1 × 4ma = **0.4mA**

■ Practice Problems

1. Calculate one time constant for the circuit below.

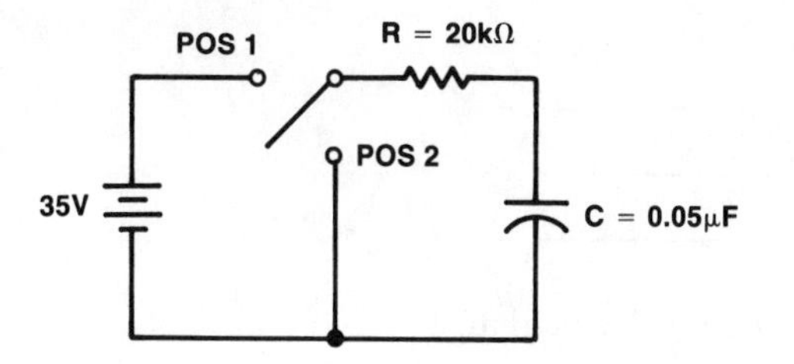

2. Calculate one time constant for the circuit below.

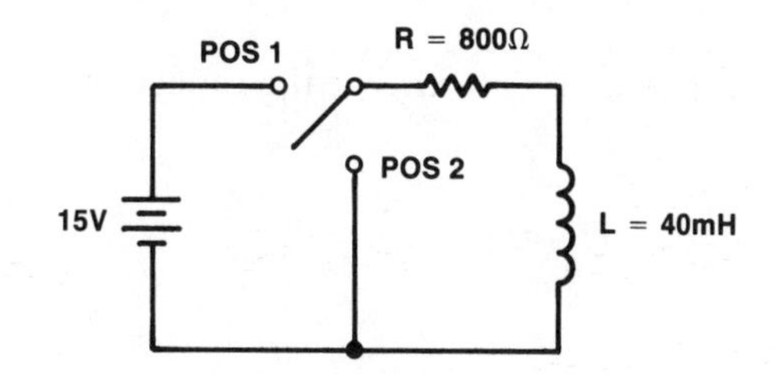

3. For the circuit in Problem 2, calculate the following:

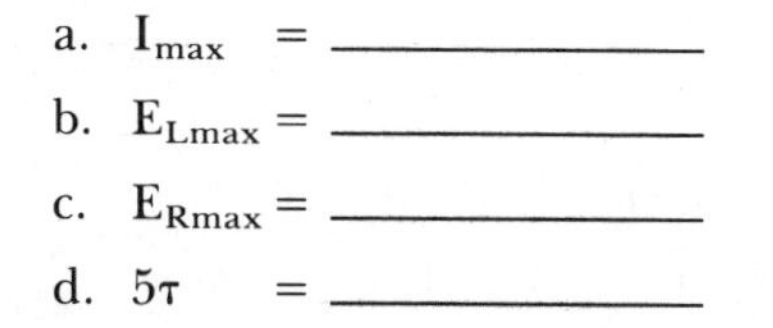

a. I_{max} = ______________

b. E_{Lmax} = ______________

c. E_{Rmax} = ______________

d. 5τ = ______________

4. For the circuit in Problem 2, draw the waveforms showing E_L, E_R, and I while the switch is in position 1 for 5 time constants.

5. If 1 time constant equals 8ms, calculate the number of time constants if:

a. time = 3ms
b. time = 10ms
c. time = 20ms
d. time = 38ms

Practice Problems

6. For the circuit in Problem 2, assume that the switch was in position 1 for 5 time constants and then moved to position 2. Calculate the following instantaneous values 100µs after the switch is moved to position 2.

   a. E_L = ______________

   b. E_R = ______________

   c. I = ______________

7. If R = 10kΩ and C = 0.004µF, calculate 1τ

8. If 1τ = 73ms and R = 360Ω, calculate C.

9. If 1τ = 471µs and C = 0.015µF, calculate R.

10. If 1τ = 37ms, calculate the time it takes to charge the capacitor in an RC time constant circuit.

■ Quiz

1. Calculate one time constant for the circuit below.

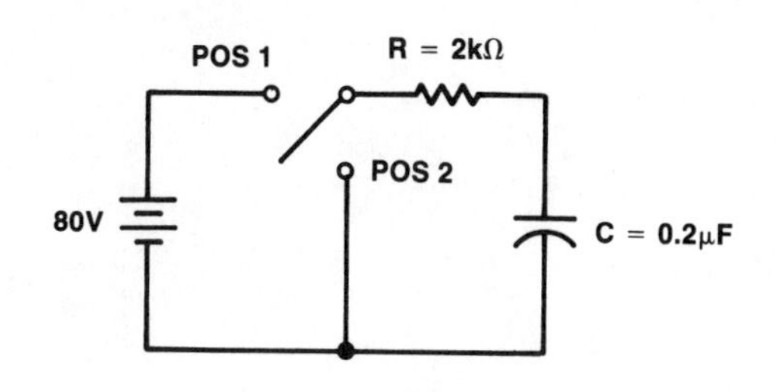

2. Calculate one time constant for the circuit below.

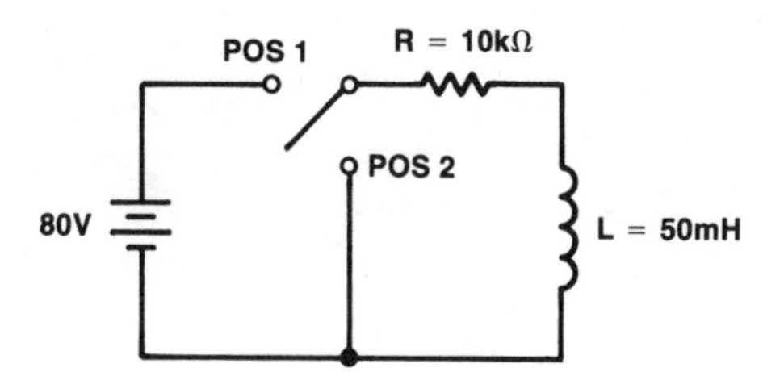

3. For the circuit in Question 1, calculate the following:

   a. I_{max} = ________
   b. E_{Cmax} = ________
   c. E_{Rmax} = ________
   d. 5τ = ________

4. For the circuit in Question 2, assume switch has been in position 1 for 5 time constants and then moved to position 2. Draw the waveforms showing E_L, E_R, and I for 5 time constants after the switch is moved to position 2.

5. If 1 time constant equals 25ms, calculate the time interval for the following time constants.

   a. 1.5τ
   b. 3τ
   c. 4.0τ
   d. 4.5τ

6. If 1 time constant equals 1.8ms, calculate the number of time constants for the following time intervals.

   a. 0.8ms
   b. 1.2ms
   c. 3.2ms
   d. 5.0ms

7. For the circuit in Question 1, calculate the following instantaneous circuit values 600μs after the switch is moved to position 1.

   a. E_C = ________
   b. E_R = ________
   c. I = ________

8. If R = 630Ω and C = 15μF, calculate 1τ.

9. If 1τ = 43μs and R = 65kΩ, calculate C.

10. If 1τ = 33ms and C = 0.005μF, calculate R.

LESSON 11

RLC Circuit Analysis

In this lesson, series and parallel circuits with resistors, capacitors, and inductors are analyzed. Many of the same techniques used in the solution of resistive-capacitive and resistive-inductive circuits are used in this analysis. Phasor diagrams provide descriptions of the circuits that lead to Pythagorean theorem solutions of certain circuit values.

■ Objectives

At the end of this lesson, you should be able to:

1. Draw phasor diagrams showing the phase relationships of various circuit values in series and parallel RLC circuits.

2. Identify the various circuit values in series and parallel RLC circuits that can be determined by Pythagorean theorem analysis.

3. Identify the positive and negative phase angles in series and parallel RLC circuits.

4. Calculate total reactance, total reactive current, total reactive voltage, and total reactive power in RLC circuits.

5. Given schematic diagrams and typical circuit values for the circuits below, calculate current, voltage, impedance and power values.

- Phase Differences
- Current as a Reference
- Development of a Circuit Phasor Diagram

INTRODUCTION

In previous lessons, series and parallel RL and RC circuits with ac voltage sources have been discussed, and you learned several useful techniques for calculating values in RL and RC circuits. In this lesson, discussion will concern an analysis of more complicated circuits consisting of series and parallel combinations of resistance, inductance, and capacitance. These circuits are called RLC circuits. In this lesson, you will simply apply the techniques you have already learned to determine RLC circuit values.

SERIES RLC CIRCUITS SUMMARIZED

The first circuit combination of resistance, capacitance and inductance to be analyzed is one that is connected in series. It is shown in *Figure 11.1*, and is called a series RLC circuit.

Phase Differences

As in any series resistive-reactive circuit, the simple sum of the voltage drops in the circuit does not equal the applied voltage.

$$E_A \neq E_R + E_L + E_C \qquad (11\text{-}1)$$

Recall that this occurs because of the different phase relationships between the voltage and current for each component.

Current as a Reference

Recall that in a series circuit as shown in *Figure 11.1*, the current is the same throughout the circuit. Therefore, as shown in *Figure 11.2*, it will be used as the reference quantity when discussing the phase relationships of the voltages in the circuit.

Development of a Circuit Phasor Diagram

Also recall that as shown in *Figure 11.3* the voltage across the resistor, E_R, is in phase with the current passing through it. This is the circuit current so the voltage across the resistor is in phase with the circuit current.

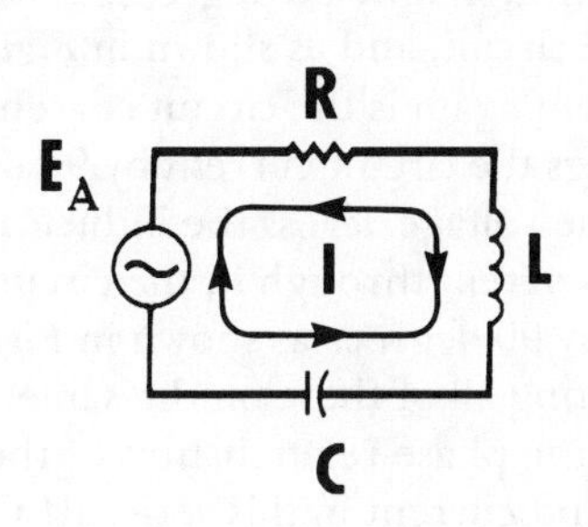

Figure 11.1 *Typical Series RLC Circuit*

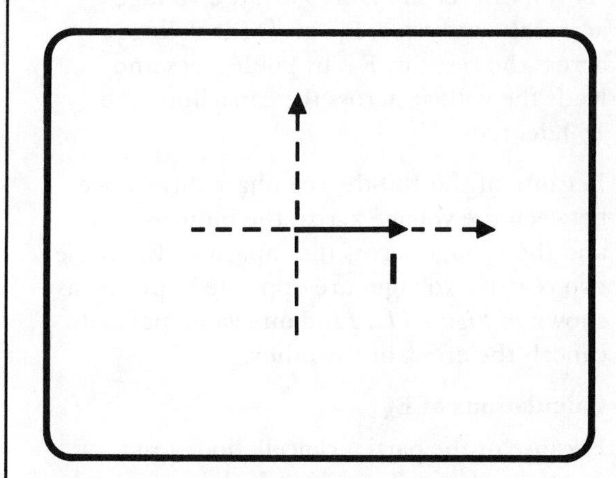

Figure 11.2 *Current is the Same Throughout a Series Circuit*

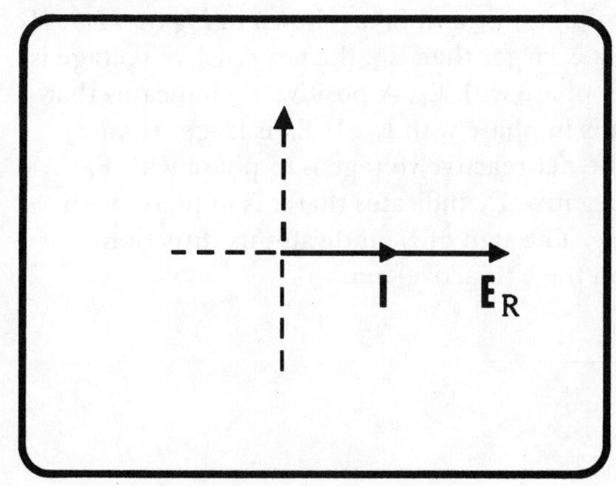

Figure 11.3 *Voltage Across the Resistor,* E_R

The voltage across a capacitor, E_C, lags the current through it by 90 degrees, as in any capacitive circuit, and as shown in *Figure 11.4.* The current again is the circuit current so that E_C lags the circuit current by 90 degrees. Finally, the voltage across the inductor, E_L, leads the current through it, the circuit current, by 90 degrees as shown in *Figure 11.5*. Plotting all of these on the same diagram, the phase relationships of the voltages and current in this series RLC circuit can be compared as shown in *Figure 11.6*.

Voltage Phasor Comparisons

As you can see in *Figure 11.6*, the voltage across the inductor, E_L, leads the voltage across the resistor, E_R, by 90 degrees and leads the voltage across the capacitor, E_c by 180 degrees.

Because of the 180-degree phase difference between the voltage across the inductor, E_L, and the voltage across the capacitor, E_C, these two reactive voltages are opposite in phase, as shown in *Figure 11.7*, and one value partially cancels the effect of the other.

Calculations of E_X

Because of the partial cancellation, a net reactive voltage, E_X, exists which is equal to the difference between the two reactive voltages, E_L and E_C as shown in *Figure 11.8*. If E_L is larger than E_C, the net reactive voltage is in phase with E_L. A positive E_X indicates that it is in phase with E_L. If E_C is larger than E_L, the net reactive voltage is in phase with E_C. A negative E_X indicates that it is in phase with E_C. The sign of E_X indicates its direction on the Y(reactive) axis.

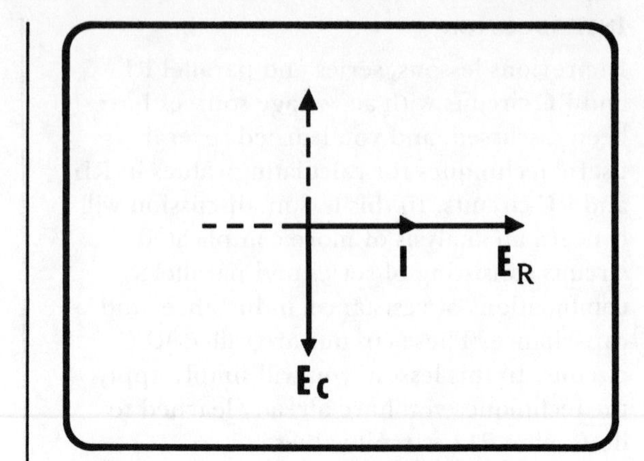

Figure 11.4 *Voltage and Current Relationship of Capacitance*

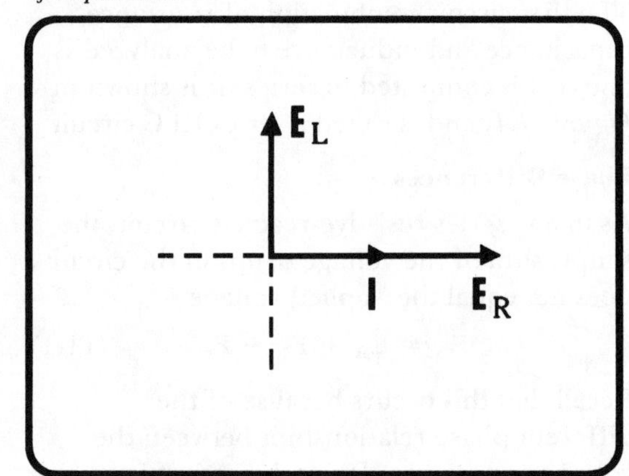

Figure 11.5 *Voltage and Current Relationship of Inductance*

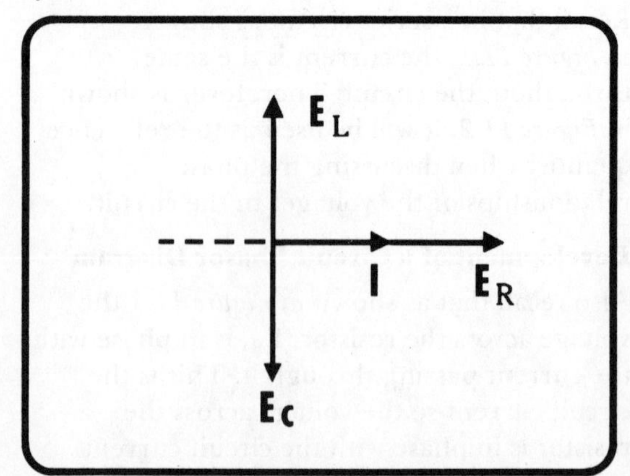

Figure 11.6 *Phase Relationships in Example RLC Series Circuits*

- Calculations of E_A
- Calculation of Circuit Voltage Values

Calculation of E_A

If E_L and E_C are known, it is possible to calculate the applied voltage, E_A. For example, assume that E_L is larger than E_C. The net reactive voltage, E_X, will, therefore, be in phase with E_L as shown in *Figure 11.8*. E_A is then calculated by extending the phasor E_A from the origin as shown in *Figure 11.9*.

Note that if the phasor E_X is shifted to the right so that it extends between the tip of E_R and the tip of E_A, the right triangle for vector addition becomes apparent. By the Pythagorean theorem, E_A equals the square root of E_R squared plus E_X squared:

$$E_A = \sqrt{E_R^2 + E_X^2} \qquad (11\text{–}2)$$

Since E_X is the net difference between E_L and E_C,

$$E_X = E_L - E_C \qquad (11\text{–}3)$$

Substituting this expression for E_X into equation *11-2,* the applied voltage is equal to the square root of E_R squared plus E_L minus E_C quantity squared as shown in equation *11–4*.

$$E_A = \sqrt{E_R^2 + (E_L - E_C)^2} \qquad (11\text{–}4)$$

This equation described the relationship between the voltages present in the series RLC circuit.

Calculation of Circuit Voltage Values

Since, by Ohm's law,

$$E = IR \qquad (11\text{–}5)$$

the voltage across the resistor is the current through it times the resistance.

$$E_R = IR \qquad (11\text{–}6)$$

Because

$$I_R = I_L = I_C = I_T$$

in a series RLC circuit, E_R can be expressed:

$$E_R = I_TR \qquad (11\text{–}7)$$

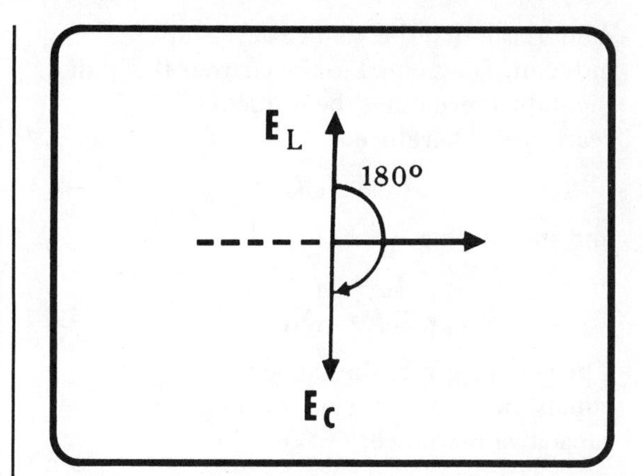

Figure 11.7 *E_L and E_C Partially Cancel the Effects of the Other*

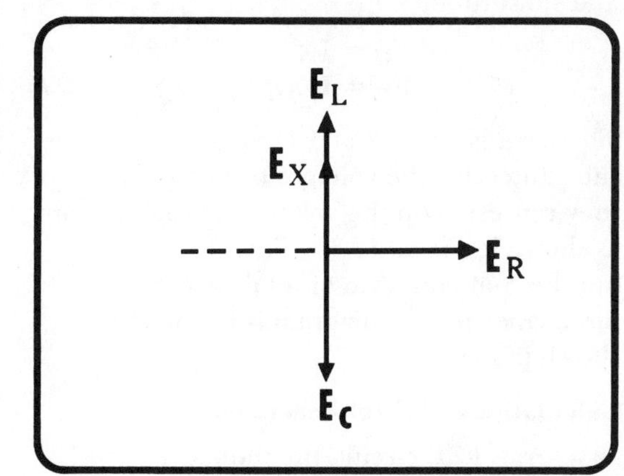

Figure 11.8 *Phase of E_X when E_L is Greater than E_C*

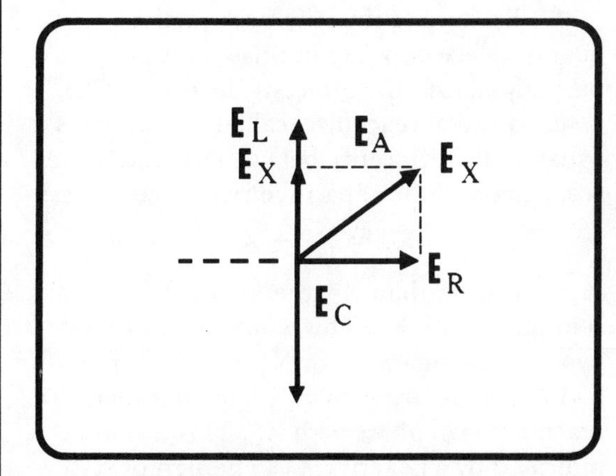

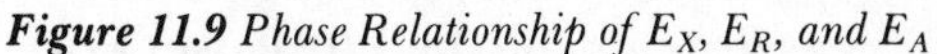

Figure 11.9 *Phase Relationship of E_X, E_R, and E_A*

■ Calculations of Total Reactance

Also, recall that the voltage across an inductor, E_L, is equal to the current through the inductance times the inductive reactance. Therefore

$$E_L = I_L X_L \quad (11\text{–}8)$$

and substituting

$$I_L = I_T,$$
$$E_L = I_T X_L \quad (11\text{–}9)$$

The voltage across the capacitor equals the capacitive current times its capacitive reactance:

$$E_C = I_C X_C \quad (11\text{–}10)$$

and substituting

$$I_C = I_T,$$
$$E_C = I_T X_C \quad (11\text{–}11)$$

These equivalent IR and IX quantities can be substituted for the voltages of *Figure 11.6* they represent on the voltage phasor diagram as shown in *Figure 11.10*. Then by factoring out the common term of total current, the impedance phasor diagram is formed as shown in *Figure 11.11*.

Calculations of Total Reactance

In a series RLC circuit, the inductive reactance and capacitive reactance are 180 degrees out of phase as shown in *Figure 11.11*. Because of this, the two reactive quantities are opposite in phase and one partially cancels the effect of the other. The result is a total reactance, called X_T, which is equal to the difference between the inductive reactance and the capacitive reactance values:

$$X_T = X_L - X_C \quad (11\text{–}12)$$

If X_L is larger than X_C, the net total reactance is in phase with X_L. This would be indicated by a positive quantity for X_T. On the other hand, if X_C is larger than X_L, the net total reactance is in phase with X_C. This would be indicated by a negative X_T. The sign of X_T indicates its direction on the Y(reactive) axis.

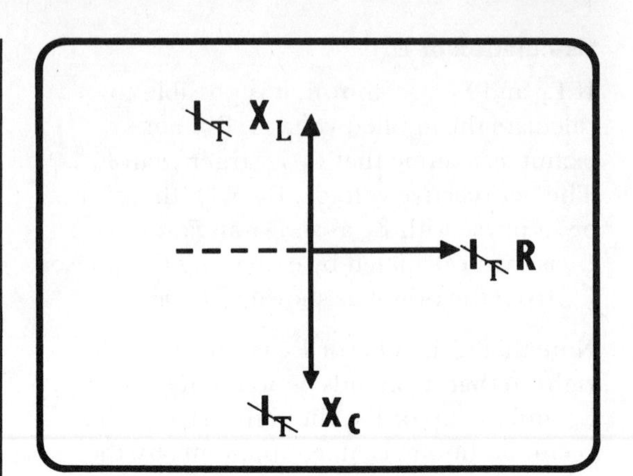

Figure 11.10 *Substituting I_R and I_X Values*

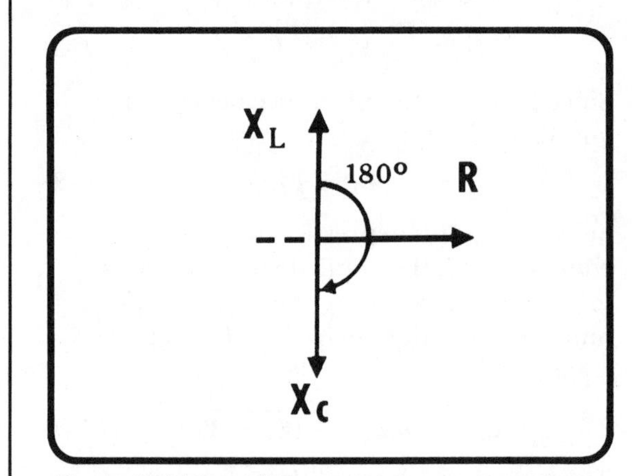

Figure 11.11 *X_L and X_C are 180 Degrees Out of Phase*

If in the voltage phasor diagram, E_L is larger than E_C to make E_X positive, then to correspond, X_L would be larger than X_C to make X_T positive on the impedance phasor diagram.

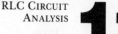

Calculation of Z_T

Recall that the total opposition to the flow of ac current in a resistive-reactive circuit is called impedance, Z, as shown in *Figure 11.12*. The total impedance of a series RLC circuit is equal to the vector sum of the total resistance and net total reactance.

Z_T is vector sum of R_T plus X_T *(11–13)*

Note in *Figure 11.12* that if phasor X_T is shifted to the right so that it extends between the tip of Z and the tip of R, the right triangle for vectorial addition appears. By the Pythagorean theorem, the total impedance of the circuit is:

$$Z_T = \sqrt{R^2 + X_T^2} \quad (11\text{–}14)$$

Since X_T is the net difference between X_L and X_C, per equation *11–12*, substituting this expression for X_T into equation *11–14* gives equation *11–15* for the total impedance.

$$Z_T = \sqrt{R^2 + (X_L - X_C)^2} \quad (11\text{–}15)$$

Calculation of I_T

Recall that the total current in a series RLC circuit is simply equal to the applied voltage divided by the total impedance of the circuit (Ohm's law for ac circuits).

$$I_T = \frac{E_A}{Z_T} \quad (11\text{–}17)$$

Once the total current is known, the voltage drops across each component may be determined by Ohm's law as stated in equations *11–7, 11–9,* and *11–11*.

Calculation of the Phase Angle

Recall that the phase angle is defined as the phase difference between the total applied voltage and the total current being drawn from that voltage supply. Returning to the voltage phasor diagram for the series RLC circuit, shown in *Figure 11.13*, remember that the total current in the circuit is used as a reference for determining phase relationships in the circuit. This total current is in phase with the voltage across the resistor. The circuit phase angle between the total applied voltage and the total current drawn from the voltage supply is represented by the angle theta. The tangent of any angle, theta, of a right triangle is:

$$\tan\theta = \frac{\text{opposite side}}{\text{adjacent side}} \quad (11\text{–}17)$$

Figure 11.12 *Impedance Phasor Diagram*

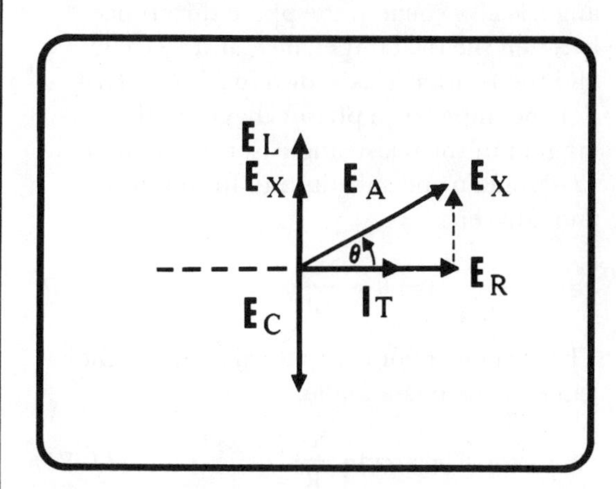

Figure 11.13 *Voltage Phasor Diagram*

■ Alternative Methods of Calculating the Phase Angle
■ Calculation of Series RLC Circuit Power Values

For the specific angle of *Figure 11.13*,

$$\tan\theta = \frac{E_X}{E_R} \quad (11\text{–}18)$$

The arctangent of this ratio equals the value of the phase angle.

$$\theta = \arctan\frac{E_X}{E_R} \quad (11\text{–}19)$$

Alternative Methods of Calculating the Phase Angle

Since the impedance phasor diagram is proportional to the voltage phasor diagram by a factor of the total current, the phase angle is also equal to the phase difference between the total impedance of the circuit and the resistance as shown in *Figure 11.14*. For the impedance phasor diagram, the tangent of the phase angle is, using equation *11–17* again and substituting for opposite and adjacent,

$$\tan\theta = \frac{X_T}{R} \quad (11\text{–}20)$$

The arctangent of this ratio also equals the value of the phase angle:

$$\theta = \arctan\frac{X_T}{R} \quad (11\text{–}21)$$

Calculation of Series RLC Circuit Power Values

Power calculations in series and parallel RLC circuits are performed in a similar manner to power calculations in series and parallel RL and RC circuits. The primary difference is that both inductive and capacitive reactive power are involved.

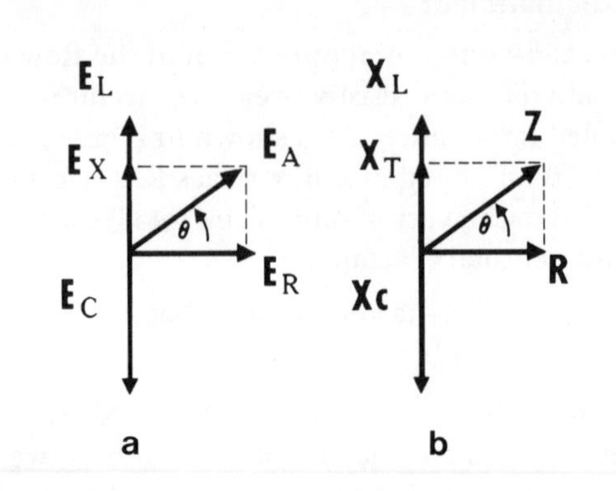

Figure 11.14 *Phase Angle Related to Voltage and Impedance Phasor Diagrams*

As you know, power is voltage times current or P = EI. The real power, P_R, in watts dissipated by the resistor, is equal to E_R times I_R.I_R is equal to I_T, the total circuit current of the series circuit. Similarly, the reactive power of the inductor, P_L in VAR, equals E_L times I_T and the reactive power of the capacitor, P_C in VAR, equals E_C times I_T.

In series RLC circuits, the power phasor diagram is proportional to the voltage phasor diagram by a factor of the total current as shown in *Figure 11.15*. The inductive power phasor and the capacitive power phasor are out of phase by 180 degrees. Thus, the resultant reactive power, which is designated P_X, is equal to the difference between the inductive reactive power and the capacitive reactive power as shown in *Figure 11.16*.

If P_L is larger than P_C, the net reactive power is in phase with P_L. If P_C is larger than P_L, the net reactive power is in phase with P_C. The sign of P_X will determine its ultimate vectorial direction.

■ Calculation of Series RLC Circuit Power Values

Recall that the total power in a resistive-reactive circuit is the apparent power measured in volts-amperes, and is the vector sum of the resistive or real power and the net reactive power.

Note in *Figure 11.17* that if the phasor P_X is shifted to the right so that it extends between the tip of the total apparent power and the tip of the real power, the right triangle for vector addition appears. By the Pythagorean theorem:

$$P_A = \sqrt{P_R^2 + P_X^2} \qquad (11\text{–}22)$$

Since P_X is equal to the difference between P_L and P_C,

$$P_X = P_L - P_C \qquad (11\text{–}23)$$

Substituting this expression for P_X into equation *11–22*, the total apparent power is:

$$P_A = \sqrt{P_R^2 + (P_L - P_C)^2} \qquad (11\text{–}24)$$

which is the relationship between the different types of power in a series RLC circuit.

The total apparent power can also be expressed in P = EI form as

$$P_A = E_A I_T \qquad (11\text{–}25)$$

where E_A is the applied voltage and I_T the circuit current. The apparent power calculated by using one of these two methods should be identical to the apparent power calculated by using the other method.

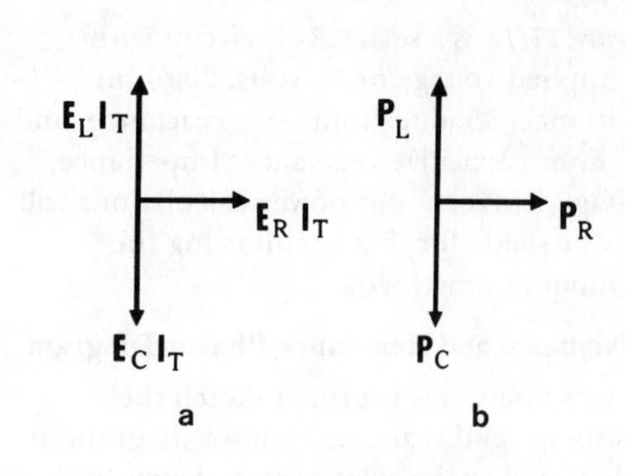

Figure 11.15 *Voltage and Power Phasor Diagrams*

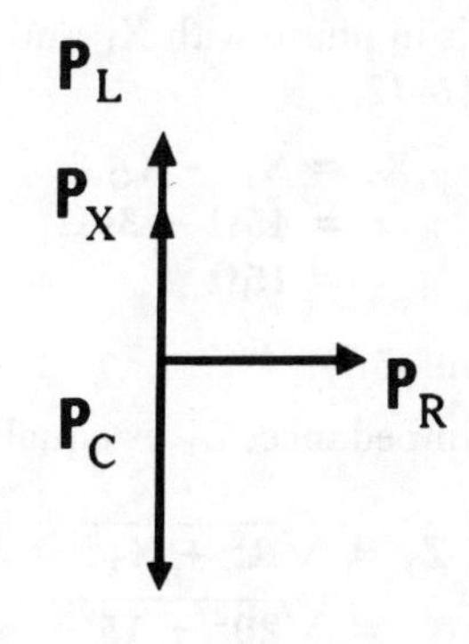

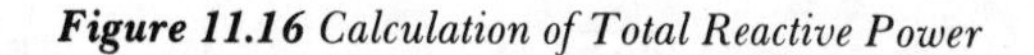

Figure 11.16 *Calculation of Total Reactive Power*

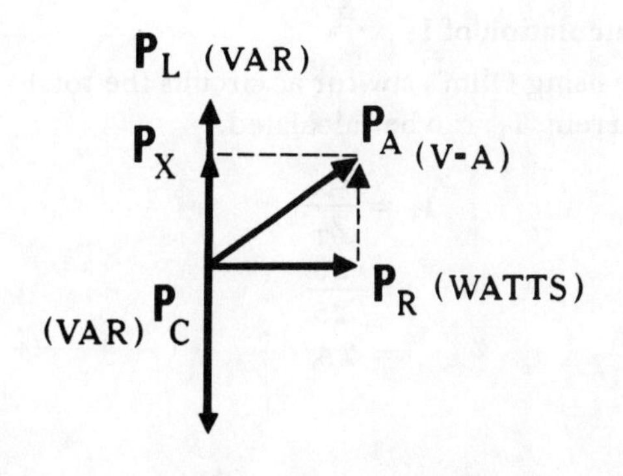

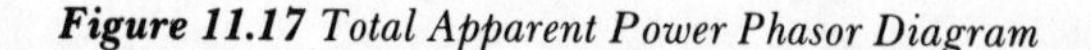

Figure 11.17 *Total Apparent Power Phasor Diagram*

Analysis of a Series RLC Circuit

Figure 11.18 is a series RLC circuit with an applied voltage of 50 volts, 20 ohms resistance, 45 ohms inductive reactance, and 30 ohms capacitive reactance. Impedance, voltage, current, and power calculations will now be made for this circuit using the techniques described.

Resistance and Reactance Phasor Diagram

At this point, it is useful to sketch the resistance and reactance phasor diagram to help visualize the relationships between the resistive and reactive quantities of the circuit. *Figure 11.19* shows such a diagram. Note that X_L is larger than X_C; thus, the net total reactance is in phase with X_L and is, using equation *11–12,*

$$X_T = X_L - X_C$$
$$= 45\Omega - 30\Omega$$
$$= 15\Omega$$

Calculation of Z_T

The total impedance, Z_T, is equal to

$$Z_T = \sqrt{R^2 + X_T{}^2}$$
$$= \sqrt{20^2 + 15^2}$$
$$= \sqrt{400 + 225}$$
$$= \sqrt{625}$$
$$= 25\Omega$$

Calculation of I_T

By using Ohm's law for ac circuits the total current, I_T, can be calculated.

$$I_T = \frac{E_A}{Z_T}$$
$$= \frac{50}{25}$$
$$= 2A$$

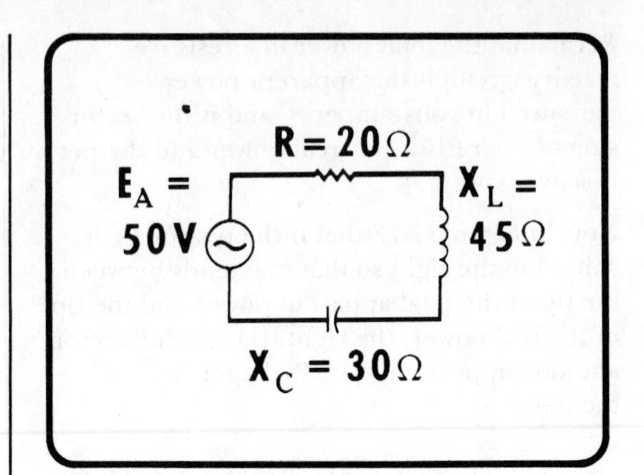

Figure 11.18 Series RLC Circuit Example

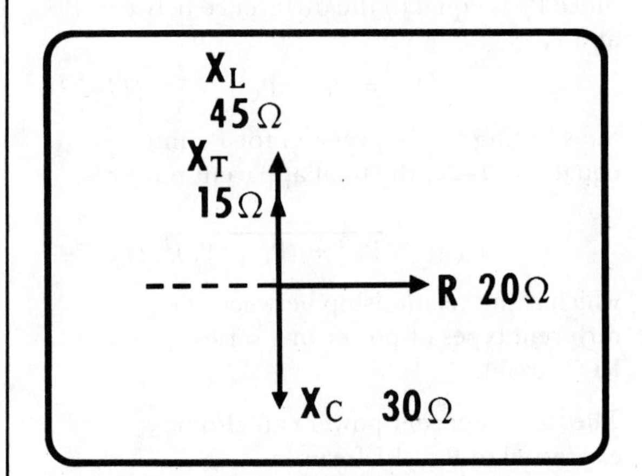

Figure 11.19 Resistance and Rectance Phasor Diagram

Calculation of Circuit Voltage Values

Using Ohm's law again, the voltage drops across each of the components in the circuit can be calculated.

The voltage across the resistor, E_R, is calculated:

$$E_R = I_T R$$
$$= (2A)\,(20\Omega)$$
$$= 40V$$

The voltage across the inductor, E_L, is calculated:

$$\begin{aligned} E_L &= I_T X_L \\ &= (2A)(45\Omega) \\ &= 90V \end{aligned}$$

The voltage across the capacitor, E_C, is calculated:

$$\begin{aligned} E_C &= I_T X_C \\ &= (2A)(30\Omega) \\ &= 60V \end{aligned}$$

Calculation of E_T

At first, as shown in *Figure 11.20,* this result appears impossible since there seems to be more voltage in the circuit than the applied voltage, E_A. However, you must keep in mind that the voltage across the inductor and the voltage across the capacitor are 180 degrees out of phase and one partially cancels the other.

Therefore, the applied voltage, E_A, is calculated using equation *11–4.*

$$\begin{aligned} E_A &= \sqrt{E_R^2 + (E_L - E_C)^2} \\ &= \sqrt{40^2 + (90 - 60)^2} \\ &= \sqrt{40^2 + 30^2} \\ &= \sqrt{1600 + 900} \\ &= \sqrt{2500} \\ &= 50V \end{aligned}$$

This calculation confirms the original applied voltage value given in the example. And this method gives you a valuable check of the accuracy of the calculation of the individual voltage drops that were calculated for each component in the circuit.

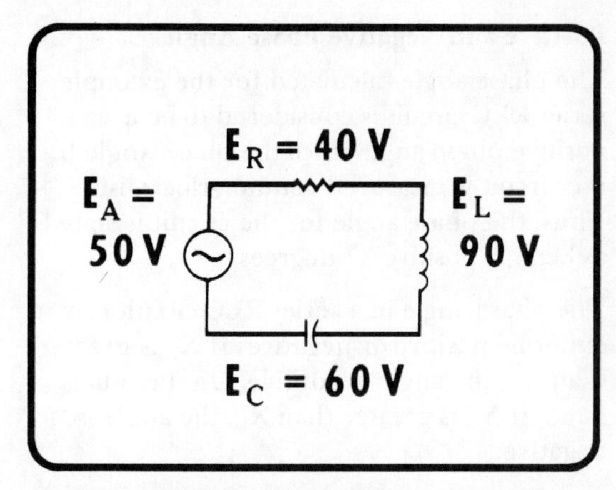

Figure 11.20 *Calculated Voltage Drops in Example Series RLC Circuit*

Calculation of Phase Angle

Using the voltage phasor diagram of *Figure 11.13,* the phase angle of the example series RLC circuit can be calculated:

$$\begin{aligned} \theta &= \arctan\left(\frac{E_X}{E_R}\right) \\ &= \arctan\left(\frac{30V}{40V}\right) \\ &= \arctan(0.75) \\ &= 37^\circ \end{aligned}$$

The angle whose tangent is 0.75 is about 37 degrees.

Alternate Method of Calculating Phase Angle

The same result may be obtained using the impedance phasor diagram. In this case the arctangent is X_T divided by R.

$$\begin{aligned} \theta &= \arctan\left(\frac{X_T}{R}\right) \\ &= \arctan\left(\frac{15\Omega}{20\Omega}\right) \\ &= \arctan(0.75) \\ &= 37^\circ \end{aligned}$$

Positive and Negative Phase Angles

The phase angle calculated for the example series RLC circuit is considered to be a positive phase angle since the phase angle for the circuit is measured counter-clockwise. Thus, the phase angle for the circuit is stated as being a positive 37 degrees.

The phase angle in a series RLC circuit can either be positive or negative. If X_L is greater than X_C, the angle is positive. On the other hand, if X_C is greater than X_L, the angle is negative.

Calculation of Component Power Values

In the example series RLC circuit, the voltage drops in the circuit are: 40 volts across the resistor, 90 volts across the inductor and 60 volts across the capacitor. The applied voltage is 50 volts and a total current of two amperes is flowing.

Multiplying the voltage times the current for each component yields these individual power values:

$$P_R = E_R I_T = (40V)(2A) = 80W$$
$$P_L = E_L I_T = (90V)(2A) = 180VAR$$
$$P_C = E_C I_T = (60V)(2A) = 120VAR$$

Calculation of Total Reactive Power

The net reactive power, P_X, is equal to the difference between P_L and P_C, or 180 VAR minus 120 VAR which equals 60 VAR. Since P_L is larger than P_C, P_X is positive and is in phase with P_L.

Calculation of Total Apparent Power

By equation *11–22*, the apparent power, P_A, is equal to the square root of P_R squared plus P_X squared:

$$\begin{aligned} P_A &= \sqrt{P_R{}^2 + P_X{}^2} \\ &= \sqrt{80^2 + 60^2} \\ &= \sqrt{6400 + 3600} \\ &= \sqrt{10{,}000} \\ &= 100VA \end{aligned}$$

The total apparent power of the example series RLC circuit is 100 volt-amperes.

This should be the same as the value obtained by multiplying the applied voltage, E_A, by the total current, I_T.

$$\begin{aligned} P_A &= E_A I_T \\ &= (50V)(2A) \\ &= (100VA) \end{aligned}$$

It is, and proves the accuracy of the first calculation.

PARALLEL RLC CIRCUITS SUMMARIZED

Now that all calculations for a series RLC circuit have been made, the method of solution of a parallel RLC circuit will be discussed. The analysis of this circuit will be similar to the analysis of either a parallel RL or a parallel RC circuit.

In a parallel RLC circuit, such as in *Figure 11.21*, the sum of the branch currents is not equal to the total current as it would be in either a purely resistive, a purely inductive, or purely capacitive circuit. That is,

$$\mathbf{I_T \neq I_R + I_L + I_C} \qquad (11\text{–}26)$$

This occurs because of the different phase relationships between the voltage and current for each component. Recall that this was also true for RL and RC parallel circuits.

- Voltage as a Reference
- Development of a Circuit Phasor Diagram
- Calculations of Branch Currents

Voltage as a Reference

Since the voltage across all components of a parallel circuit is the same as the applied voltage, E_A, it will be used as the reference quantity in discussing the phase relationships of the currents in the circuit. Therefore, $E_A = E_R = E_L = E_C$.

Development of a Circuit Phasor Diagram

Recall that the current through a resistor, I_R, is in phase with the voltage across it, E_R. The capacitive current, I_C, leads the voltage across the capacitor, E_A, by 90 degrees; and the current through an inductor, I_L, lags the voltage across it, E_A, by 90 degrees. Thus, the phase relationship of the applied voltage and currents in a parallel RLC circuit are as shown in *Figure 11.22*.

Calculations of Branch Currents

The individual branch currents in the example RLC circuit can be calculated as they are in either a purely resistive, purely capacitive, or purely inductive circuit. Simply divide the voltage across the branch by the opposition to current in the branch.

In the resistive branch the opposition to the flow of current is measured in ohms of resistance. The resistive current is determined by dividing the applied voltage by the value of the resistor:

$$I_R = \frac{E_A}{R} \qquad (11\text{–}27)$$

In the inductive branch, the opposition to the flow of current is measured in ohms of inductive reactance. The inductive current is determined by dividing the applied voltage by the reactance of the inductor:

$$I_L = \frac{E_A}{X_L} \qquad (11\text{–}28)$$

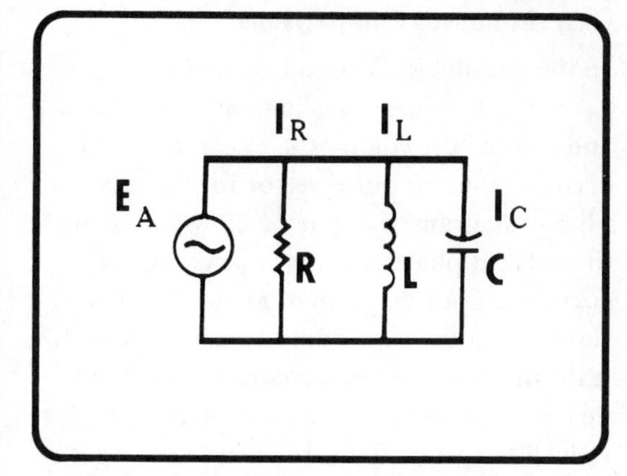

Figure 11.21 *Parallel RLC Circuit Example*

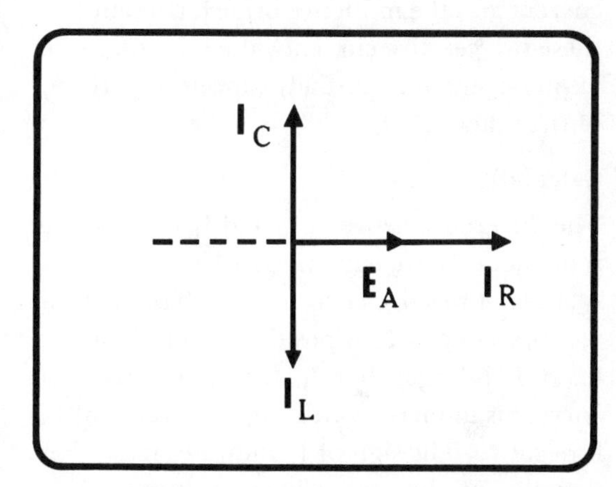

Figure 11.22 *Phase Relationships of E_A and I in a Parallel RLC Circuit*

In the capacitive branch, the opposition to the flow of current is measured in ohms of capacitive reactance. The capacitive current is determined by dividing the applied voltage by the reactance of the capacitor:

$$I_C = \frac{E_A}{X_C} \qquad (11\text{–}29)$$

- Current Phasor Comparisons
- Calculation of I_X

Current Phasor Comparisons

In the parallel RLC circuit, since $E_A = E_R = E_L = E_C$, and since I_R is in phase with E_R and thus all the voltages E_L, E_C and E_A, I_R becomes the reference vector for the current phasor diagram of *Figure 11.23*. Comparing the current phase relationships using the current phasor diagram of *Figure 11.23*, you can see that the capacitive branch current, I_C, leads the resistive branch current, I_R, by 90 degrees; the inductive branch current, I_L, lags I_R by 90 degrees. Because of the 180-degree difference between the capacitive branch current and the inductive branch current, these two reactive current values are opposite in phase, and one partially cancels the effect of the other.

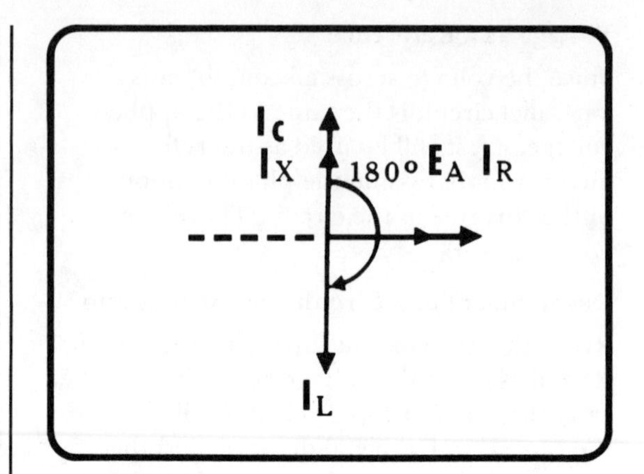

Figure 11.23 *Current Phasor Diagram for a Parallel RLC Circuit*

Calculation of I_X

The difference between I_C and I_L is I_X, which is the net reactive current. If I_C is larger than I_L, the net reactive current is in phase with I_C and the value of I_X is positive. If, on the other hand, I_L is larger than I_C, the net reactive current is in phase with I_L and the value of I_X is negative. The sign of I_X indicates its direction on the Y(reactive) axis.

By adding the total current vector, I_T, to the current phasor diagram, and shifting vector I_X to the right so that it extends between the tip of I_R and the tip of I_T, a right triangle is produced. This is shown in *Figure 11.24*. The right triangle can then be used to add the phasors I_R and I_X vectorially.

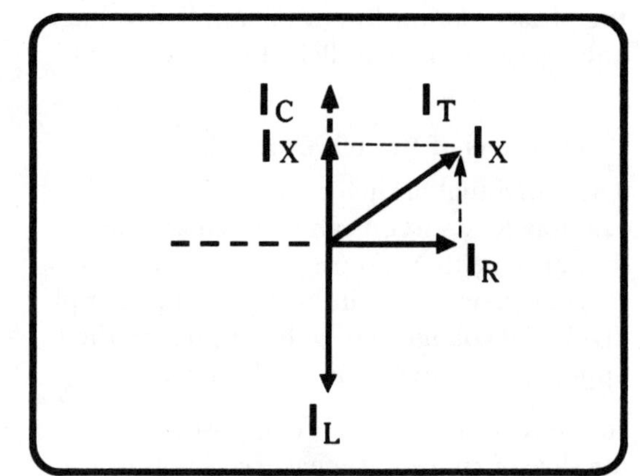

Figure 11.24 *Current Phasor Diagram with I_T Phasor*

Applying the Pythagorean theorem,

$$I_T = \sqrt{I_R^2 + I_X^2} \quad (11\text{–}31)$$

Since I_X is the net difference between I_C and I_L, as shown in equation *11–31*,

$$I_X = I_C - I_L \quad (11\text{–}31)$$

I_X is equal to I_C minus I_L. Substituting this expression for I_X into equation *11–30*, the total current is equal to

$$I_T = \sqrt{I_R^2 + (I_C - I_L)^2} \quad (11\text{–}32)$$

which is the relationship between the branch currents in a parallel RLC circuit.

Calculation of Z_T

Once the total current is known, the total impedance, Z_T, of the circuit is easily determined using Ohm's law for ac circuits:

$$Z_T = \frac{E_A}{I_T} \qquad (11\text{–}33)$$

Calculation of Phase Angle

Recall that the phase angle is the number of degrees of phase difference between the applied voltage and the total current. Also recall that the applied voltage is in phase with the resistive current. Therefore, the phase angle is measured from the horizontal vector I_R (which is also E_A) to the vector I_T; it is identified by the angle, theta, on the current phasor diagram of *Figure 11.25*.

The tangent of the phase angle, theta, is equal to the ratio of the net reactive current divided by the resistive current as shown in equation *11–34:*

$$\tan\theta = \frac{I_X}{I_R} \qquad (11\text{–}34)$$

Therefore,

$$\theta = \arctan\left(\frac{I_X}{I_R}\right) \qquad (11\text{–}35)$$

the arctangent of the net reactive current divided by the resistive current equals the value of the phase angle.

Calculation of Power

Power calculations in parallel RLC circuits are the same as power calculations in series RLC circuits. The individual power values are calculated by multiplying current through a component times voltage across a component.

Therefore,

$$P_R = E_R I_R = E_A I_R \qquad (11\text{–}36)$$

$$P_L = E_L I_L = E_A I_L \qquad (11\text{–}37)$$

$$P_C = E_C I_C = E_A I_C \qquad (11\text{–}38)$$

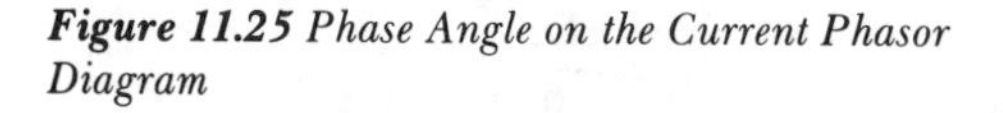

Figure 11.25 *Phase Angle on the Current Phasor Diagram*

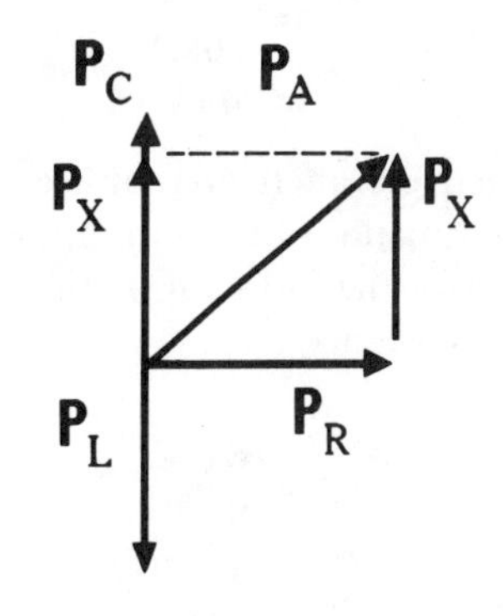

Figure 11.26 *Total Apparent Power Phasor Diagram*

The total apparent power is found by using the Pythagorean theorem solution of resistive and net reactive power shown in the power phasor diagram of *Figure 11.26* defined by these equations:

$$P_A = \sqrt{P_R^2 + P_X^2} \qquad (11\text{–}39)$$

$$P_X = P_C - P_L \qquad (11\text{–}40)$$

$$P_A = \sqrt{P_R^2 + (P_C - P_L)^2} \qquad (11\text{–}41)$$

ANALYSIS OF A PARALLEL RLC CIRCUIT

Figure 11.27 shows a typical parallel RLC circuit with 180 VAC, 60 kilohms resistance, 30 kilohms inductive reactance, and 18 kilohms capacitive reactance. Impedance, voltage, current, and power measurements will be performed for this circuit using the techniques just described.

Calculation of Branch Currents

The resistive branch current, I_R, is calculated by dividing the voltage across the resistor, 180 volts, by the value of the resistor, 60 kilohms.

$$I_R = \frac{E_A}{R} = \frac{180V}{60k\Omega} = 3mA$$

The inductive branch current is determined in a similar manner by dividing the voltage across the inductor by the inductive reactance.

$$I_L = \frac{E_A}{X_L} = \frac{180V}{30k\Omega} = 6mA$$

The capacitive branch current is calculated by dividing the voltage across the capacitor by the capacitive reactance.

$$I_C = \frac{E_A}{X_C} = \frac{180V}{18k\Omega} = 10mA$$

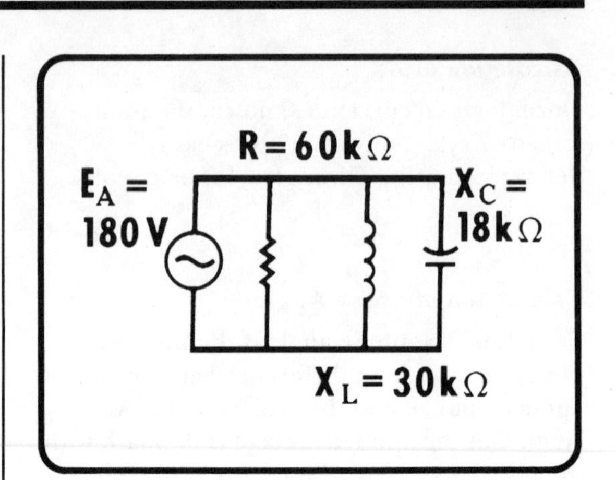

Figure 11.27 Typical Parallel RLC Circuit

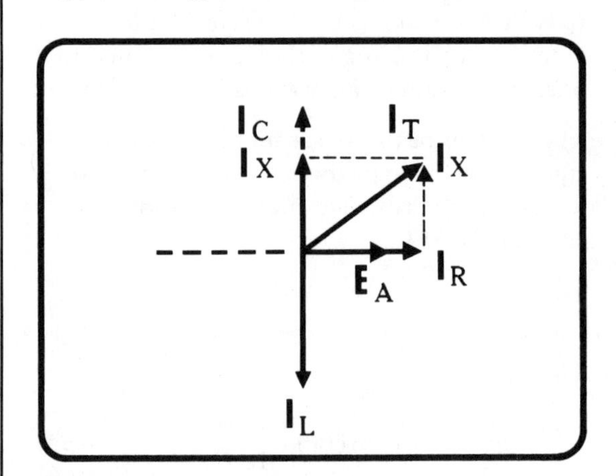

Figure 11.28 Current Phasor Diagram for Parallel RLC Circuit Example

Current Phasor Diagrams

The current phasor diagram can now be drawn to show the relationships between the resistive, inductive and capacitive branch currents as shown in *Figure 11.28*. Note that I_C is larger than I_L; therefore, the net reactive current is in phase with I_C. The net reactive current, I_X, equals

$$I_X = I_C - I_L = 10mA - 6mA = 4mA$$

■ Calculation of I_T
■ Calculation of Z
■ Calculations of the Phase Angle

Calculation of I_T

As you can see in *Figure 11.28*, by shifting the phasor I_X to the right so that it extends from the tip of the vector representing I_R to the tip of the vector for I_T, a right triangle is formed. The right triangle can then be used to determine the value of I_T. Using the Pythagorean theorem, the total current, I_T, is now equal to the vector sum of the values of I_X and I_R.

$$
\begin{aligned}
I_T &= \sqrt{I_R^2 + I_X^2} \\
&= \sqrt{3^2 + 4^2} \\
&= \sqrt{9 + 16} \\
&= \sqrt{25} \\
&= 5\text{mA}
\end{aligned}
$$

The total current in the circuit is 5 milliamperes.

Calculation of Z

The total impedance of this circuit can now be calculated by dividing the applied voltage by the total current.

$$
\begin{aligned}
Z &= \frac{E_A}{I_T} \\
&= \frac{180\text{V}}{5\text{mA}} \\
&= 36\text{k}\Omega
\end{aligned}
$$

Calculations of the Phase Angle

The phase angle, theta is the angle between the resistive current and the total current as shown on the current phasor diagram in *Figure 11.29*. The value of the phase angle is equal to the arctangent of the ratio of net reactive current to the resistive current. Therefore,

$$
\begin{aligned}
\theta &= \arctan\left(\frac{I_X}{I_R}\right) \\
&= \arctan\left(\frac{4\text{mA}}{3\text{mA}}\right) \\
&= \arctan(1.33) \\
&= 53^\circ
\end{aligned}
$$

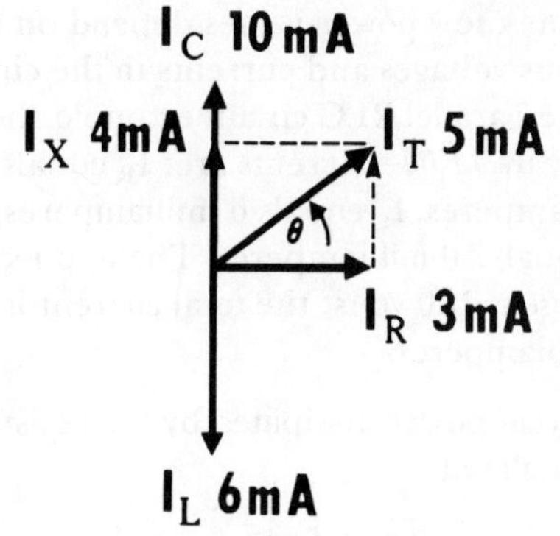

Figure 11.29 *Current Phasor Diagram for Calculating Theta*

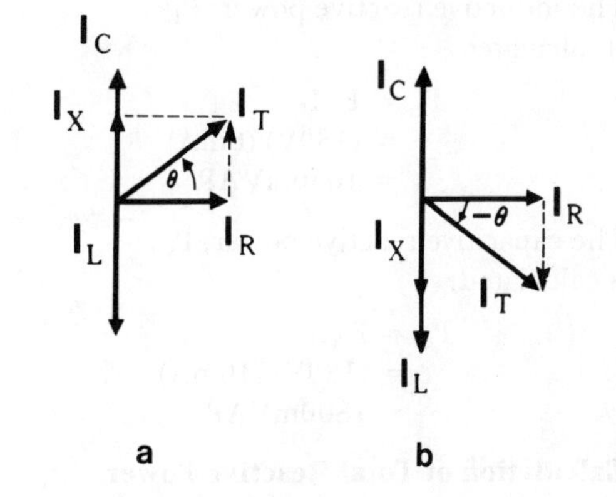

Figure 11.30 *Current Phasor Diagrams for Positive and Negative Thetas*

Thus, the phase angle of this parallel RLC circuit is approximately 53 degrees.

The phase angle is described as a positive phase angle because the total current phasor is rotated counter-clockwise from the reference. However, the phase angle of parallel RLC circuit may be either positive or negative depending upon the relationships of I_L and I_C. If I_C is greater than I_L, the phase angle theta will be positive as shown in *Figure 11.30a*. If I_L is greater than I_C, Theta will be negative as shown in *Figure 11.30b*.

Calculation of Component Power Values

As you know power values depend on the various voltages and currents in the circuit. In the parallel RLC circuit example shown in *Figure 11.31*, currents are: I_R equals 3 milliamperes, I_L equals 6 milliamperes, and I_C equals 10 milliamperes. The applied voltage is 180 volts; the total current is 5 milliamperes.

The real power dissipated by the resistor, P_R, is calculated:

$$\begin{aligned} P_R &= E_A I_R \\ &= (180V)\,(3mA) \\ &= 540mW \end{aligned}$$

The inductive reactive power, P_L is calculated:

$$\begin{aligned} P_L &= E_A I_L \\ &= (180V)\,(6mA) \\ &= 1080mVAR \end{aligned}$$

The capacitive reactive power, P_C, is calculated:

$$\begin{aligned} P_C &= E_A I_C \\ &= (180V)\,(10mA) \\ &= 1800mVAR \end{aligned}$$

Calculation of Total Reactive Power

The net reactive power, P_X is calculated:

$$\begin{aligned} P_X &= P_C - P_L \\ &= 1800mVAR - 1080mVAR \\ &= 720mVAR \end{aligned}$$

The power phasor diagram will look like the one in *Figure 11.26*.

Since P_C is larger than P_L, P_X is positive and is in phase with P_C.

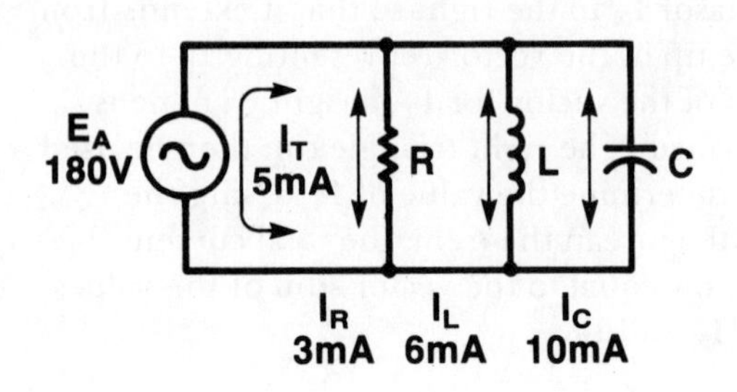

***Figure 11.31** Branch Currents in Parallel RLC Circuit Example*

Calculation of Total Power

The apparent total power, P_A, is calculated using equation *11–39*:

$$\begin{aligned} P_A &= \sqrt{P_R^2 + P_X^2} \\ &= \sqrt{540mW^2 + 720mVAR^2} \\ &= \sqrt{291{,}600 + 518{,}400} \\ &= \sqrt{810{,}000} \\ &= 900mVA \end{aligned}$$

This value should be the same as the apparent power value obtained by multiplying the applied voltage by the total current:

$$\begin{aligned} P_A &= E_A I_T \\ &= (180V)\,(5mA) \\ &= 900mVA \end{aligned}$$

SUMMARY

In this lesson, the concepts and techniques you learned in analyzing RL and RC circuits were applied to determine circuit values in RLC circuits. You were shown how to determine current, voltage, reactance, and the phase angle for any RLC circuit. Also, a new concept was introduced — that when values are in opposite phase, one value partially cancels the effect of the other. This difference is called the net effect. The net voltage, E_X, net total reactance, X_T, and net reactive current I_X, were calculated. The methods you learned in this lesson should enable you to determine equivalent circuit values in any series or parallel RLC circuit.

■ Worked-Out Examples

1. Draw a phasor diagram for this circuit showing current and impedance phasors.

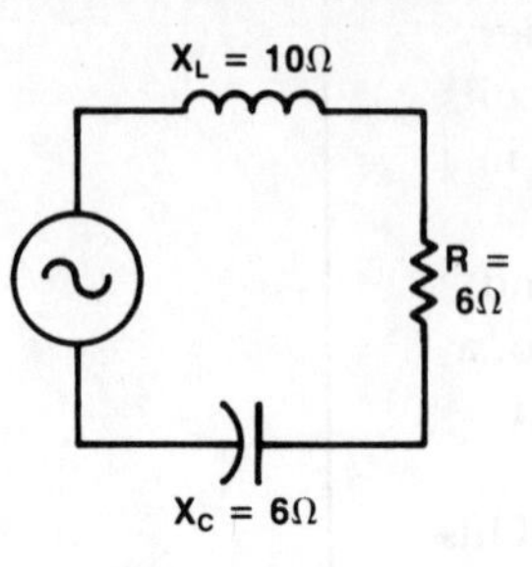

Solution:

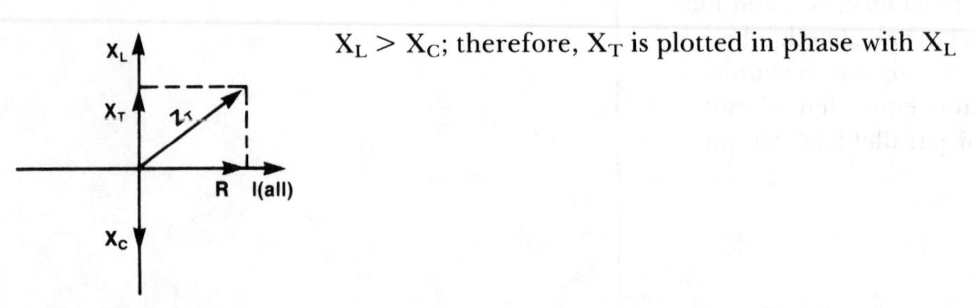

$X_L > X_C$; therefore, X_T is plotted in phase with X_L

2. Draw voltage phasor diagrams for the circuit in Example 1. Calculate the value and sign of the phase angle.

Solution:

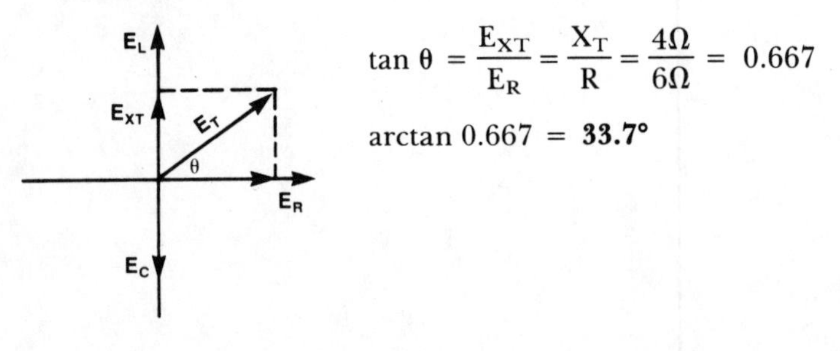

$$\tan\theta = \frac{E_{XT}}{E_R} = \frac{X_T}{R} = \frac{4\Omega}{6\Omega} = 0.667$$

$$\arctan 0.667 = \mathbf{33.7^\circ}$$

The angle is rotated counterclockwise; therefore, the sign of the angle is positive.

■ Worked-Out Examples

3. Draw phasor diagrams showing Pythagorean theorem relationships in parallel RLC circuits. Write equations for each solution.

Solution:

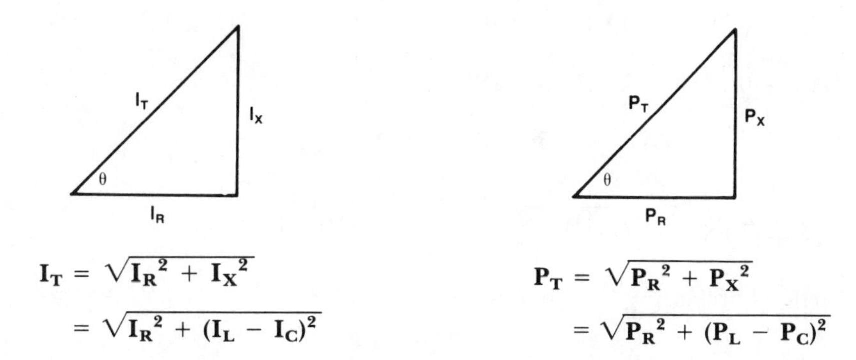

$$I_T = \sqrt{I_R^2 + I_X^2} = \sqrt{I_R^2 + (I_L - I_C)^2}$$

$$P_T = \sqrt{P_R^2 + P_X^2} = \sqrt{P_R^2 + (P_L - P_C)^2}$$

4. Given the circuit and typical circuit values shown, calculate the circuit values specified.

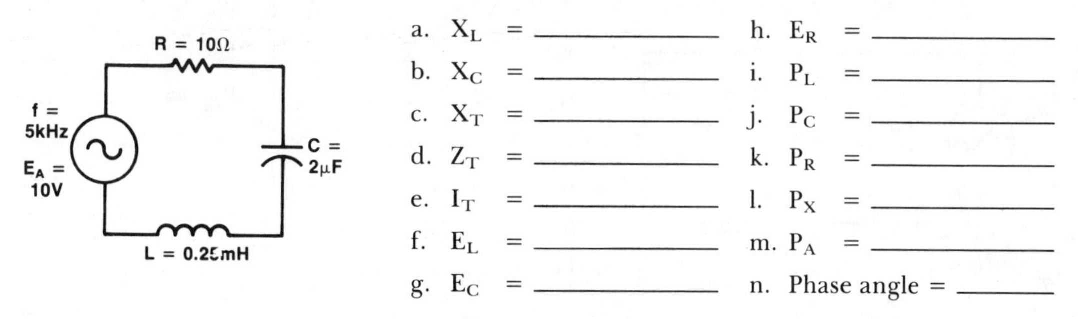

a. X_L = ______
b. X_C = ______
c. X_T = ______
d. Z_T = ______
e. I_T = ______
f. E_L = ______
g. E_C = ______
h. E_R = ______
i. P_L = ______
j. P_C = ______
k. P_R = ______
l. P_X = ______
m. P_A = ______
n. Phase angle = ______

Solution:

a. $X_L = 2\pi fL = 6.28 \times 5 \times 10^3 \times 0.25 \times 10^{-3} = \mathbf{7.85\Omega}$

b. $X_C = \dfrac{1}{2\pi fC} = \dfrac{1}{6.28 \times 5 \times 10^3 \times 2 \times 10^{-6}} = \dfrac{1}{6.28 \times 10^{-2}} = \mathbf{15.9\Omega}$

c. $X_T = X_L - X_C = 7.85\Omega - 15.9\Omega = \mathbf{8.05\Omega}$

d. $Z_T = \sqrt{R^2 + X_T^2} = \sqrt{10^2 + 8.05^2} = \sqrt{1.65 \times 10^2} = \mathbf{12.8\Omega}$

e. $I_T = \dfrac{E_A}{Z_T} = \dfrac{10\text{VAC}}{12.8\Omega} = \mathbf{0.781A_{rms}}$

f. $E_L = I_T X_L = 0.781\text{A} \times 7.85\Omega = \mathbf{6.13VAC}$

g. $E_C = I_T X_C = 0.781\text{A} \times 15.9\Omega = \mathbf{12.4VAC}$

■ Worked-Out Examples

h. $E_R = I_T R = 0.781A \times 10\Omega = \mathbf{7.81VAC}$

i. $P_L = I_T E_L = 0.781A \times 6.13V = \mathbf{4.79VAR_{rms}}$

j. $P_C = I_T E_C = 0.781A \times 12.4V = \mathbf{9.68VAR_{rms}}$

k. $P_R = I_T E_R = 0.781A \times 7.81V = \mathbf{6.1W_{rms}}$

l. $P_X = P_L - P_C = 4.79VAR - 9.68VAR = \mathbf{4.89VAR_{rms}}$

m. $P_A = \sqrt{P_R^2 + P_X^2} = \sqrt{6.1W^2 + 4.89VAR^2} = \sqrt{6.11} = \mathbf{7.82VA_{rms}}$

n. $\text{phase angle} = \arctan\left(\frac{X_T}{R}\right) = \arctan 0.805 = \mathbf{-38.8^\circ}$

5. Given this circuit and typical circuit values shown, calculate the circuit values specified.

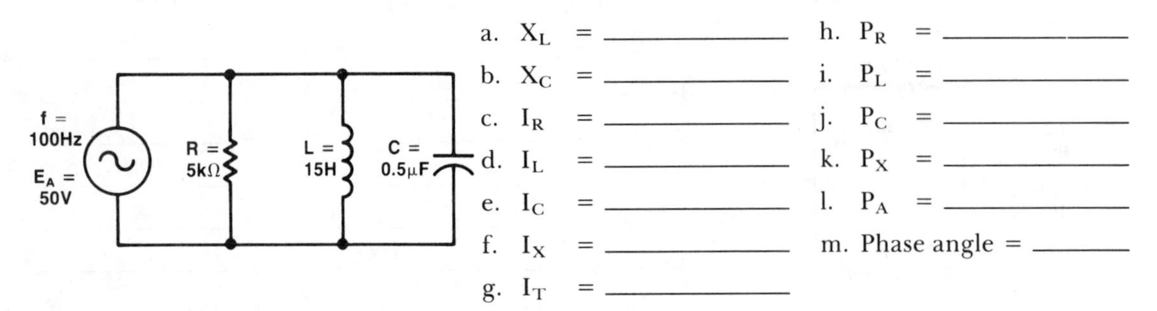

a. X_L = ________
b. X_C = ________
c. I_R = ________
d. I_L = ________
e. I_C = ________
f. I_X = ________
g. I_T = ________
h. P_R = ________
i. P_L = ________
j. P_C = ________
k. P_X = ________
l. P_A = ________
m. Phase angle = ________

Solution:

a. $X_L = 2\pi fL = 6.28 \times 100 \times 15 = \mathbf{9.42k\Omega}$

b. $X_C = \frac{1}{2\pi fC} = \frac{1}{6.28 \times 100 \times 0.5 \times 10^{-6}} = \frac{1}{3.14 \times 10^{-4}} = \mathbf{3.18k\Omega}$

c. $I_R = \frac{E_A}{R} = \frac{50VAC}{5k\Omega} = \mathbf{10mA_{rms}}$

d. $I_L = \frac{E_A}{X_L} = \frac{50VAC}{9.42k\Omega} = \mathbf{5.31mA_{rms}}$

e. $I_C = \frac{E_A}{X_C} = \frac{50VAC}{3.18k\Omega} = \mathbf{15.7mA_{rms}}$

f. $I_X = I_L - I_C = 5.31mA - 15.7mA = \mathbf{10.4mA}$

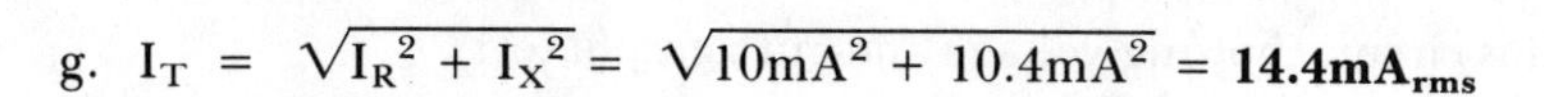

g. $I_T = \sqrt{I_R^2 + I_X^2} = \sqrt{10mA^2 + 10.4mA^2} = \mathbf{14.4mA_{rms}}$

h. $P_R = I_R E_A = 10mA \times 50V = \mathbf{0.5W_{rms}}$

i. $P_L = I_L E_A = 5.31mA \times 50V = \mathbf{0.27VAR_{rms}}$

j. $P_C = I_C E_A = 15.7mA \times 50V = \mathbf{0.79VAR_{rms}}$

k. $P_X = P_L - P_C = 0.27VAR - 0.79VAR = \mathbf{0.52VAR_{rms}}$

l. $P_A = \sqrt{P_R^2 + P_X^2} = \sqrt{0.5W^2 + 0.52VAR^2} = \sqrt{0.52} = \mathbf{0.72VA_{rms}}$ or

$P_A = I_T E_A = 14.4mA \times 50V = \mathbf{0.72VA_{rms}}$

m. $\tan\theta = \frac{I_X}{I_R} = \frac{10.4mA}{10mA} = \mathbf{1.04}$

$\arctan 1.04 = \mathbf{46.1°}$

■ Practice Problems

1. Draw a phasor diagram for this circuit showing voltage and current phasors.

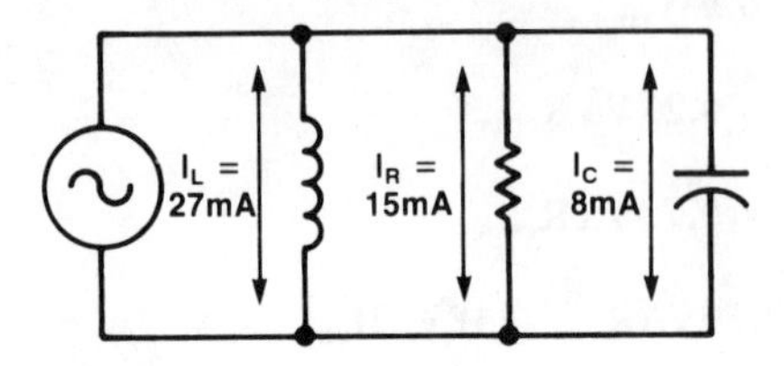

2. Draw power phasor diagrams for the circuit in problem 1. Calculate the value and sign of the phase angle.

3. Draw phasor diagrams showing Pythagoren theorem relationships in series RLC circuits. Write equations for each solution.

4. Given the circuit and typical circuit values below, calculate the circuit values specified.

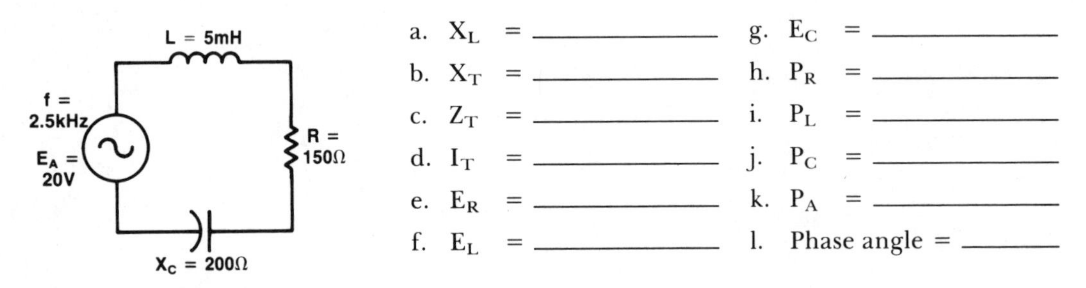

5. Given this circuit and typical circuit values shown, calculate the circuit values specified.

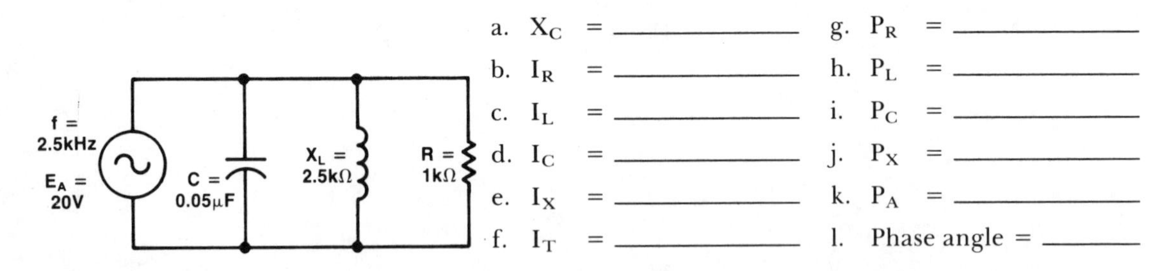

Quiz

1. Draw a phasor diagram for this circuit showing power and current phasors.

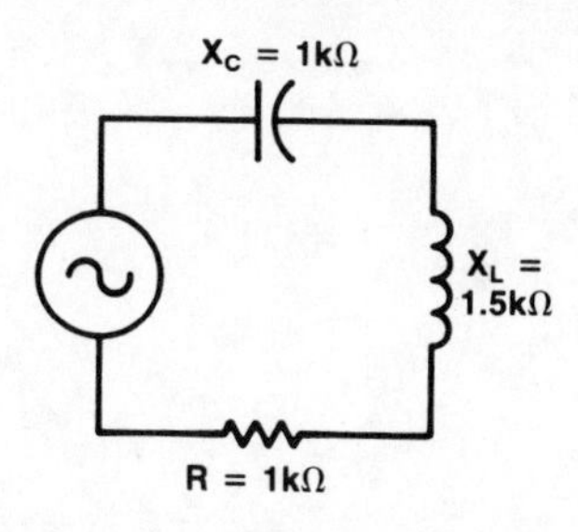

2. Draw a phasor diagram for this circuit showing power and voltage phasors.

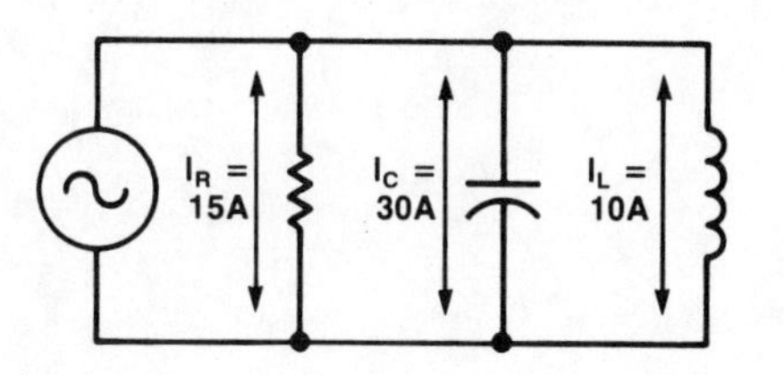

3. Calculate the value and sign of the phase angle in the circuit of Problem 1.
4. Calculate the value and sign of the phase angle in the circuit of Problem 2.
5. Calculate the value of Z_T in Problem 1.
6. Calculate the value of I_T in Problem 2.
7. Given this circuit and typical circuit values shown, calculate the circuit values specified.

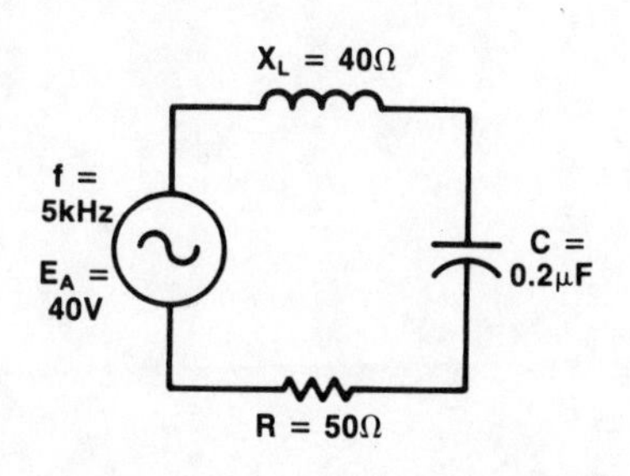

a. X_C = ________
b. X_T = ________
c. Z_T = ________
d. I_T = ________
e. E_R = ________
f. E_L = ________
g. E_C = ________
h. P_R = ________
i. P_L = ________
j. P_C = ________
k. P_A = ________
l. Phase angle = ________

8. Given this circuit and typical values shown, calculate the circuit values specified.

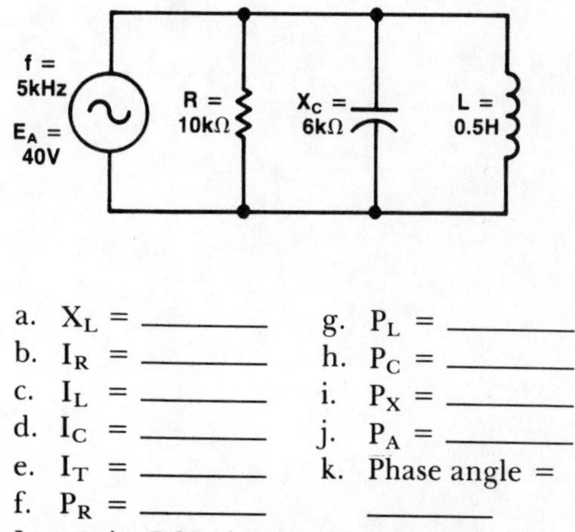

a. X_L = ________
b. I_R = ________
c. I_L = ________
d. I_C = ________
e. I_T = ________
f. P_R = ________
g. P_L = ________
h. P_C = ________
i. P_X = ________
j. P_A = ________
k. Phase angle = ________

9. In a series RCL circuit, which component will have the most voltage across it?
10. In a parallel RCL circuit, which component will have the most current through it?

LESSON 12

Phasor Algebra

This lesson provides an introduction to phasor algebra by introducing you to the use of the j-operator and its application to the solution of simple RL and RC circuits. Both rectangular and polar coordinate representations of circuit impedance are presented as well as the methods of conversion from one form to the other. These techniques are used to solve the more complex circuits of Lesson 13.

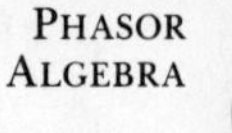

■ Objectives

At the end of this lesson you should be able to:

1. Specify both the rectangular and polar coordinates of a point on a complex plane.
2. Simplify the expression of a multiplication or division problem involving the j-operator.
3. Understand what numerical quantity j actually represents.
4. Specify what effect multiplication by j or a power of j (i.e. j^2, j^3, j^4, or $-j$) will have on a vector (phasor) located somewhere on the complex plane.
5. Write the complex impedance of an RL or RC circuit in either rectangular ($R + jX$) or polar ($Z\angle\theta$) form.
6. Convert the complex impedance of a circuit in rectangular form to polar form.
7. Convert the complex impedance of a circuit in polar form to rectangular form.
8. Specify whether a circuit that has a positive or negative j term will have a complex impedance in the rectangular form positive or negative phase angle.

■ X-Y Coordinates

INTRODUCTION

In previous lessons, when discussing RL, RC and RLC circuits, you have seen that it is necessary to work with phasor quantities in order to determine values of impedance, current, voltage and phase angle. The vectorial relationships in the circuits were simple enough to allow the use of right triangle trigonometric relationships and the Pythagorean theorem. However, if more complex ac circuits are encountered, the phase relationships become more complex and the solution of problems involving such a circuit with its various components and phase relationships could be a long process. Because of this, a different method of solution called *phasor algebra* must be used. The phasor algebra method simplifies solving complex ac circuit problems.

INTRODUCTION TO PHASOR ALGEBRA

Phasor algebra employs a simple method which can express the value or magnitude of voltage, current, reactance, resistance or impedance and their phase angles, and the phase relationships of these quantities with each other.

Before this method of circuit analysis can be used, however, it is necessary to first review some basic facts concerning numbers and graphs.

X-Y Coordinates

Recall that in an earlier lesson discussion concerned how to graph a curve on a set of X-Y axes shown in *Figure 12.1*. The point where the two axes cross is called the *origin*. This is the point from which all other points on either axis are measured. Distances to the right on the X-axis, and up on the Y-axis are designated with positive numbers. Distances to the left and down are negative numbers.

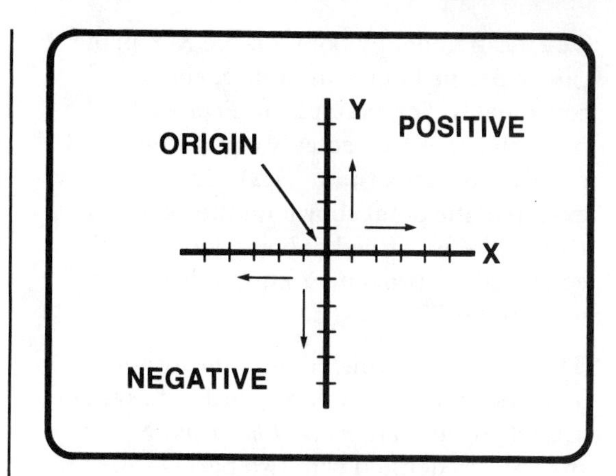

Figure 12.1 *An X-Y Coordinates Graph*

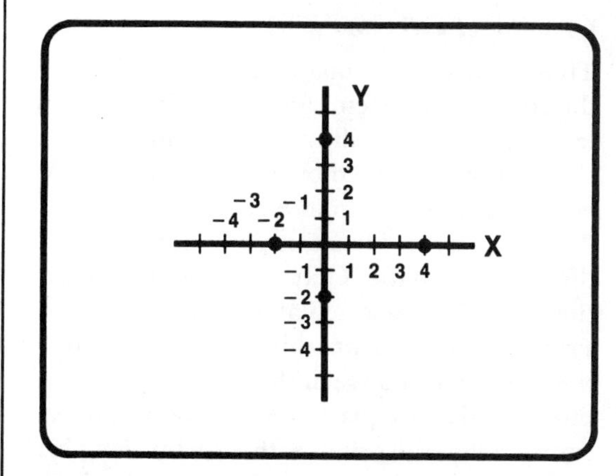

Figure 12.2 *Positive and Negative Numbers on Coordinates*

Each number represents a unique point along the number line. Thus, as shown in *Figure 12.2*, +4 represents the point that is four units to the right of the origin on the X-axis or four units above the origin on the Y-axis.

Similarly, −2 represents a point that is two units to the left of the origin on the X-axis, or two units down from the origin on the Y-axis.

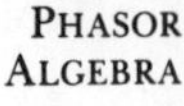

■ Location of a Vector

The location of any point on the X-Y plane can be defined in terms of its X and Y coordinates. For example, in *Figure 12.3* the point shown is located at X equals four, and Y equals two. It is written: (4,2). Or, as another example, the point shown on the graph in *Figure 12.4* is defined as being located at X equals minus two, and Y equals three. It is written: (−2,3).

These numbers which define the location of a *specific* point on the X-Y coordinate system are called the *coordinates of the point.* Note that the point is defined with two parts — an X-part and Y-part.

Location of a Vector

These indicators of unique points on the coordinate system can be used in an interesting way. To illustrate, a point is chosen that is defined by X equals four and Y equals three as shown in *Figure 12.5*. Projecting down from this point parallel to the Y axis onto the X axis it can be seen that the X axis is intersected at X equals 4. Projecting across from this point parallel to the X axis it can be seen that the Y axis is intersected at Y equals to 3. The point can be connected to the origin with a vector. For this example it is called vector H. Vector H is in a position on the X-Y coordinate system located by point (4,3). It originates at the origin and its length is a measure of the magnitude of the quantity it represents.

The reason for calling it H is because, as you will note, it is the hypotenuse of a right triangle that has been used in previous lessons to determine phase angles and for vector additions. The right triangle is formed by a vector of Y units as the opposite side and a vector of X units as the adjacent side.

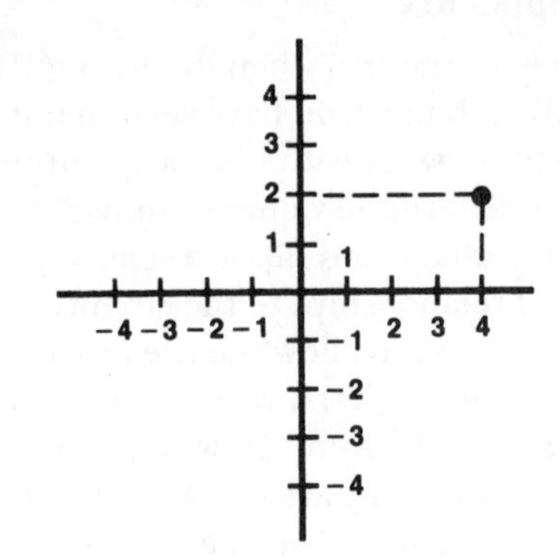

Figure 12.3 *Point (4,2) on the X-Y Coordinates*

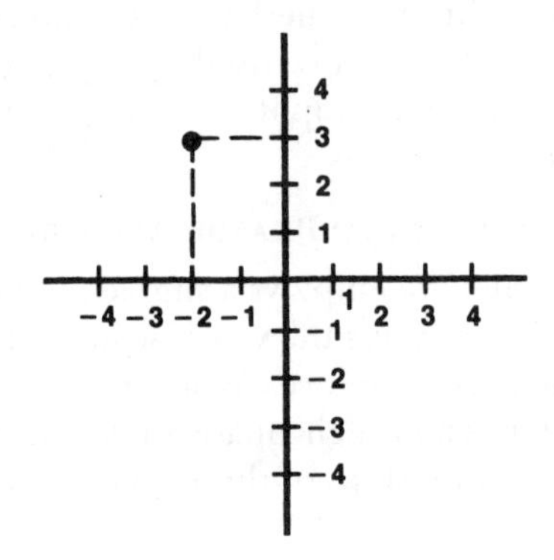

Figure 12.4 *Point (−2,3) on the X-Y Coordinates*

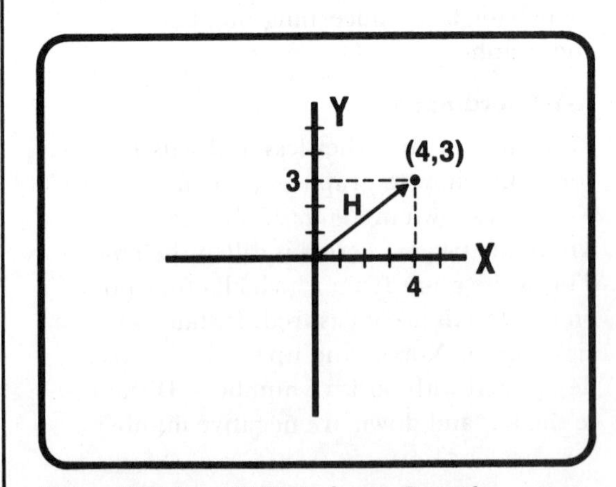

Figure 12.5 *Point (4,3) Used as an Example*

Recall that the length of a hypotenuse is equal to the square root of the sum of the squared sides. Therefore,

$$\begin{aligned} H &= \sqrt{X^2 + Y^2} \\ &= \sqrt{4^2 + 3^2} \\ &= \sqrt{16 + 9} \\ &= \sqrt{25} \\ &= 5 \end{aligned} \qquad (12\text{–}1)$$

The vector in this example has a length of 5 units. As has been shown in past lesson, the 5 units could be 5 volts, 5 amperes, 5 watts, and so forth.

Rotating a Vector

When the vector H is rotated to the X axis as shown in *Figure 12.6*, you can see that it lies on the X axis and becomes a vector of 5 units long located by the point (5,0). There is no Y part. The Y part is equal to zero.

Now the vector H is rotated back again to the original position as shown in *Figure 12.5*, note that vector H does not change in magnitude. It is still 5 units long. The parts X and Y always indicate the point in the X-Y coordinate system to which the vector is rotated. The original point from which it started is (4,3). To find the point to which vector H was rotated, you locate the +4 on the X axis, and project up parallel to the Y axis three units.

Thus, it can be seen that the X part of the number defines a vector along the X axis; the Y part of the number is a point that indicates how far the vector H has been rotated away from the X axis.

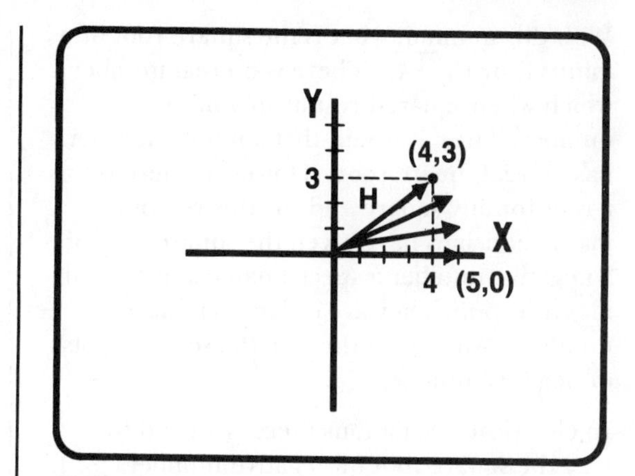

***Figure 12.6** Vector H Rotated to the X Axis*

Note that the distance from the axis to the final point is the Y part of the coordinates for the point. It is suggested that you think of the Y part of the coordinates as an *"operator"* because it is said to operate on the vector H to tell you how far the vector H has been moved from the X axis to arrive at the final point. Later in this lesson you will be told how to use this simple procedure to easily describe any vector in the coordinate system.

Square Roots of Negative Numbers

Remember that when you square any number —whether it is positive or negative— the result is always a positive quantity because a plus times a plus or a minus times a minus is always a plus. This is why plus or minus is always written when designating the square root of a positive number. There are two roots, a plus root and a minus root. For example, the square root of four could be either a plus two or a minus two:

$$\sqrt{4} = \pm 2 \qquad (12\text{–}2)$$

Real and Imaginary Axis

With this in mind, what is the square root of a minus four ($\sqrt{-4}$)? There is no real number which when squared results in a minus number. Thus, it is said that a minus number has no real square roots. However, there is a root for any square and for this reason, mathematicians have given the square root of a negative number a special name and special identification. They are called imaginary numbers—numbers that are the square roots of negative numbers.

In electrical calculations there is a need to take the square root of negative numbers. The method used when calculating the square root of a negative number is to take the square root of the number as if it were positive and then write a j before it to indicate that it is an imaginary number. For example,

$$\sqrt{9} = 3 \qquad (12\text{–}3)$$

and

$$\sqrt{-9} = j3 \qquad (12\text{–}4)$$

Similarly

$$\sqrt{-4} = j2 \qquad (12\text{–}5)$$

the

$$\sqrt{-1} = j1 = j \qquad (12\text{–}6)$$

The square root of minus one can be written as just j. It is the basic imaginary quantity—the square root of a minus one—and the special identification is j. Therefore, the square root of minus 9 could be written as the square root of minus 1 times the square root of 9:

$$(\sqrt{-1})\ (\sqrt{9}) = j3 \qquad (12\text{–}7)$$

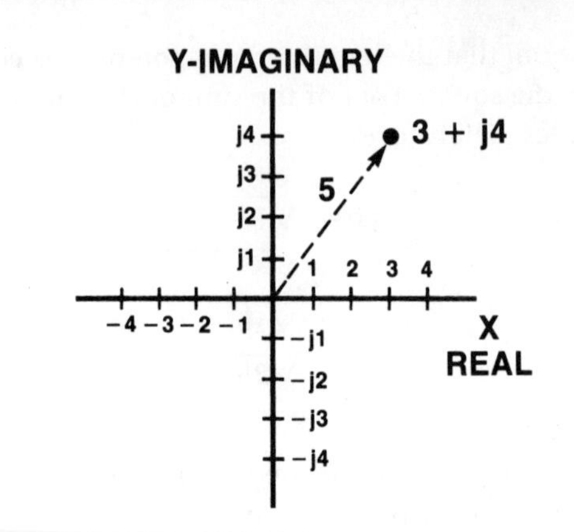

Figure 12.7 Numbers Represented by X and Y Axes

The square root of minus 4 could be written as the square root of minus 1 times the square root of 4:

$$(\sqrt{-1})\ (\sqrt{4}) = j2 \qquad (12\text{–}8)$$

Correspondingly, the square root of minus 1 could be written as the square root of minus 1 times the square root of 1:

$$(\sqrt{-1})\ (\sqrt{1}) = j1 \qquad (12\text{–}9)$$

In all cases then, the square root of a minus number can be considered as the square root of a positive number times the square root of minus one, or j.

Real and Imaginary Axes

Customarily, as shown in *Figure 12.7*, the Y-axis is used for representing imaginary numbers and is referred to as the imaginary axis, while the X-axis is referred to as the real axis. This coordinate system is used to define the location of points in terms of the distances marked off on the real and imaginary axes. For instance, the point shown in *Figure 12.7* is located at a real axis value of three and an imaginary axis value of j4. It is written simply as 3 plus j4, 3 + j4.

■ Complex Numbers
■ Polar Coordinates

Complex Numbers

If a vector is drawn from the origin (*Figure 12.7*) to this point, its magnitude is again 5 units as in the previous example ($5 = \sqrt{3^2 + 4^2}$). Thus once again the final point for the vector with a magnitude of 5 has been identified with coordinates, but in a different way. This time they are identified with a real part (X-axis) and a so-called imaginary part (Y-axis). The designations of these points on the real-imaginary coordinates represent actual numbers. However, since they are composed of a real part and an imaginary part, they are called *complex numbers.*

A complex number is defined as a number represented by the algebraic sum of a real number and an imaginary number. Consequently, the graph or plane defined by the real number and imaginary axes, in which point locations are defined by complex numbers, is known as a *complex plane*. Since the points are located by coordinates measured and plotted on lines that are perpendicular to each other, these coordinates are called rectangular coordinates.

Polar Coordinates

There is an alternate method of specifying the location of any unique point, and this is in terms of its *polar coordinates.*

The point's distance from the origin is the length (magnitude) of the vector H as shown in *Figure 12.8* and as discussed earlier. Recall it was calculated to be five units in magnitude. Two particular positions to which the vector was located are the points identified by the complex numbers 3 + j4 and 4 + j3.

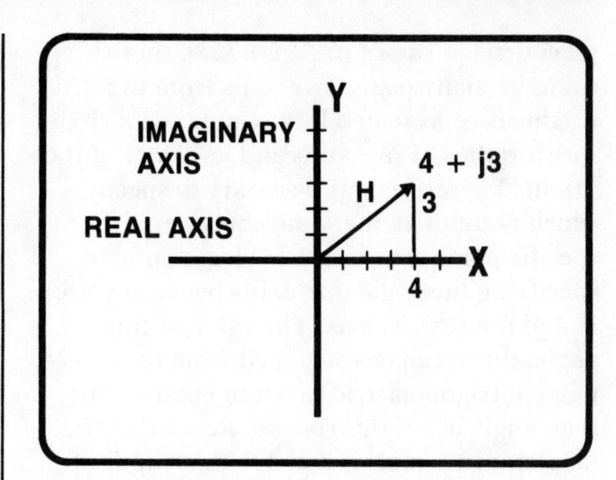

***Figure 12.8** Position of Point Location 4 + j3*

■ Polar Coordinates

Of course, as shown in *Figure 12.9*, there could be many points five units from the origin, as represented by any point on a circle with a radius of five units and its center at the origin. Therefore, it is necessary to specify which point of all these possible ones is the specific point to be located. This is done by specifying the angle that exists between vector H and the real, X, axis. The value of this angle, theta, can be calculated from the tangent trigonometric function because this is an angle in a right triangle. Recall that the tangent of an angle is equal to the length of the opposite side divided by the length of the adjacent side:

$$\tan \theta = \frac{\text{opposite side}}{\text{adjacent side}} \qquad (12\text{–}10)$$

Therefore, for the example shown in *Figures 12.8* and *12.9*, the tangent of theta is equal to three units divided by four units:

$$\tan \theta = \frac{3}{4} \qquad (12\text{–}11)$$

$$= 0.75$$

$$\theta = \arctan(0.75) \qquad (12\text{–}12)$$

$$= 37°$$

Using a calculator as in Lesson 7, the arctangent of the ratio 0.75 is determined to be an angle of approximately 37 degrees.

The specific point is plotted in *Figure 12.10* and is at a distance of five units from the origin and at an angle from the positive X-axis of 37 degrees. It is written as $5 \angle 37°$ and stated as "five at 37 degrees". Note that the angle specified is positive. That is because, as shown in *Figure 12.11*, angles measured in a counter-clockwise direction from the reference axis are positive. Note that the positive X-axis is the reference. Angles measured clockwise from the reference axis are considered to be negative.

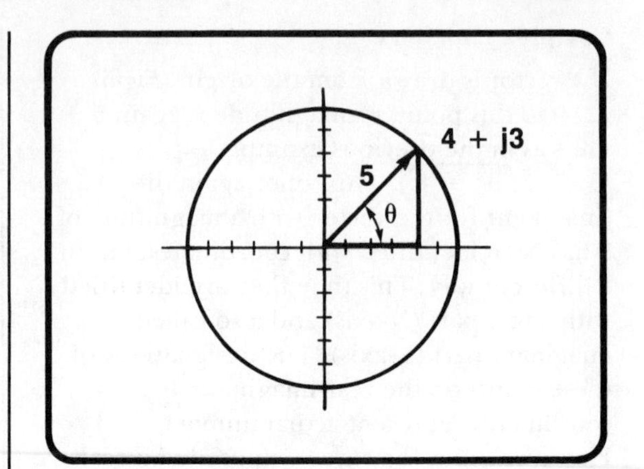

***Figure 12.9** Polar Coordinates*

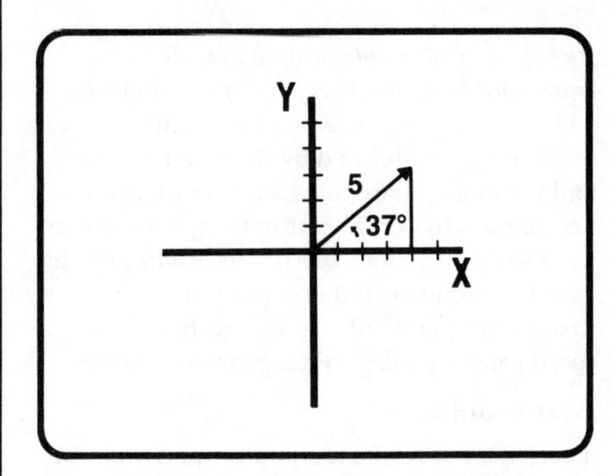

***Figure 12.10** Theta of Example Vector H is 37 Degrees*

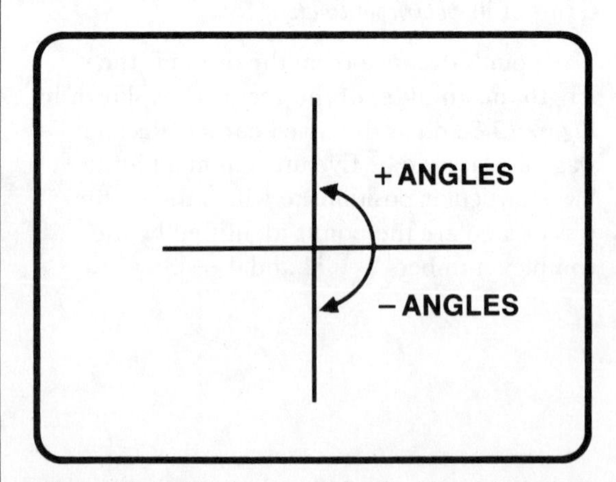

***Figure 12.11** Numbers are Either Positive or Negative*

Polar Coordinate Definition

The polar coordinates of a point in a plane specify the distance of the point from the origin at a specified angle from the X axis. Said another way, as shown in *Figure 12.12*, the distance from the origin is a vector of specified magnitude rotated from the X axis through a specific angle. In the example, 4 + j3 represents a point in *rectangular coordinates.* Five at 37 degrees represents the same point in *polar coordinates.* Four plus j3 locates a point in a plane to which a vector has been rotated; five at 37 degrees describes the same vector.

$$4 + j3 \text{ SAME AS } 5 \angle 37^\circ \qquad (12\text{-}13)$$

Relationships of Rectangular/Polar Coordinate Systems to Circuit Analysis

Note in *Figure 12.13a* and *b* that the relationship between the quantities X, Y, H and theta is identical to the relationships between resistance, reactance, impedance, and the phase angle.

Because of this identical relationship, complex number methods can be used to solve ac circuit problems.

The j Operator

Before beginning a discussion of how the actual solutions are accomplished, there are some things you should keep in mind about the term j, and the powers of j. Remember, from equation *12–6*, that j is equal to the square root of minus one.

J-squared is equal to j times j or the square root of a minus one times the square root of a minus one, which is simply minus one. Expressed in equation form, using equation *12–6*, the result is:

$$\begin{aligned} j &= \sqrt{-1} \\ j^2 &= j \times j \qquad (12\text{–}14) \\ &= \sqrt{-1} \times \sqrt{-1} \\ &= -1 \end{aligned}$$

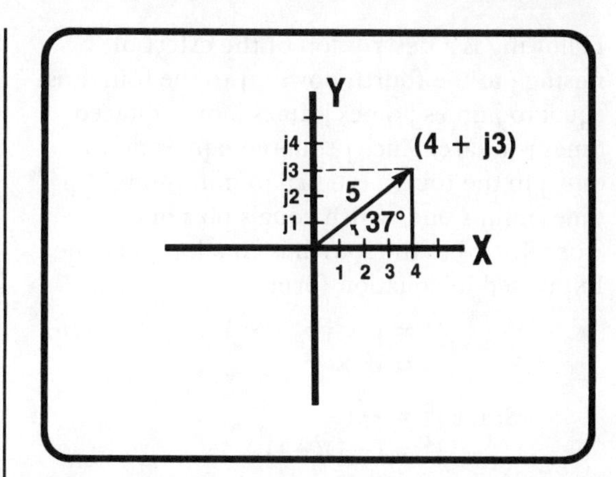

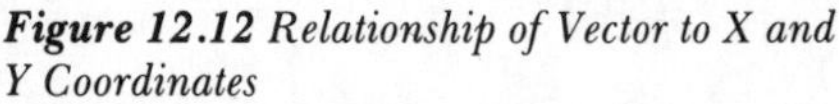
Figure 12.12 Relationship of Vector to X and Y Coordinates

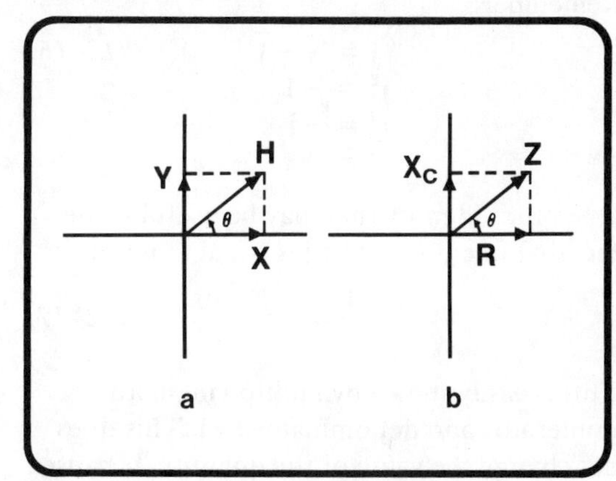

Figure 12.13 Relationships of X, Y, H, θ, Resistance, Reactance, and Impedance

Following is a description of the effect of multiplying j by itself three times. J-cubed is equal to j times j times j, or j-squared time j. Since j-squared equals minus one, j-cubed is equal to minus one times j, or simply minus j. Thus, j-cubed equals minus j. Expressed in equation form:

$$\begin{aligned} j^3 &= j \times j \times j \qquad (12\text{–}15) \\ &= j^2 \times j \end{aligned}$$

$$\begin{aligned} \text{Since } j^2 &= -1 \\ j^3 &= -1 \times j \\ &= -j \end{aligned}$$

■ Applying j-Operator Notation to Vectors
■ First Rotation

Following is a description of the effect of raising j to the fourth power. J to the fourth is equal to j times j times j times j, or j-squared times j-square. Since j-squared equals minus one, j to the fourth is equal to minus one times minus one, which equals plus one. Therefore, j to the fourth is equal to plus one. Expressed in equation form:

$$\begin{aligned} j^4 &= j \times j \times j \times j \qquad (12\text{–}16) \\ &= j^2 \times j^2 \end{aligned}$$

$$\begin{aligned} \text{Since } j^2 &= -1 \\ j^4 &= (-1)(-1) \\ &= +1 \end{aligned}$$

Remember:

$$\begin{aligned} j &= \sqrt{-1} \qquad (12\text{–}16) \\ j^2 &= -1 \\ j^3 &= -j \\ j^4 &= +1 \end{aligned}$$

One other identity that may be useful is the fact that one divided by j is equal to minus j:

$$\frac{1}{j} = -j \qquad (12\text{–}17)$$

This is easily shown by multiplying both numerator and denominator by j. This does not change the value of the quantity, because effectively one over j has been multiplied by one. Multiplying through in the numerator and the denominator, j times one equals j and j times j equals j-squared, which is a minus one. J divided by minus one is minus j. So, one divided by j equals minus j. In equation form the steps are as follows:

$$\begin{aligned} \frac{1}{j} \times \frac{j}{j} &= \frac{j}{j^2} \qquad (12\text{–}18) \\ &= \frac{j}{-1} \\ &= -j \end{aligned}$$

Therefore,

$$\frac{1}{j} = -j$$

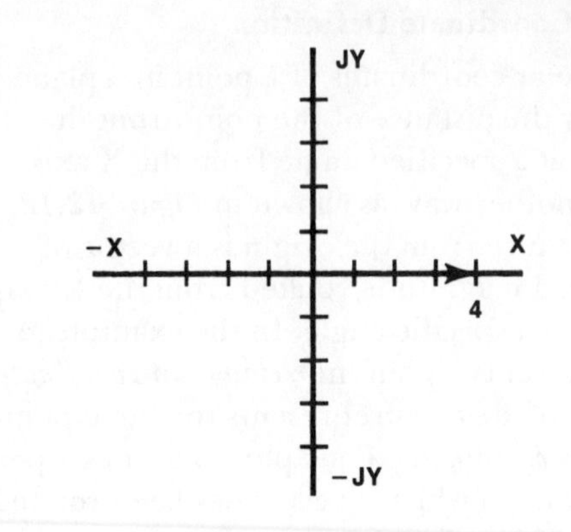

Figure 12.14 A Vector 4 Units in Length on the X Axis

Applying j-Operator Notation to Vectors

With this background in the j-operator, you will now see how the real and imaginary coordinate axes can be used to describe the length of a vector and what effect the j-term has on that vector.

If a point is designated as four units from the origin along the positive X-axis, as shown in *Figure 12.14*, a vector four units long can be drawn from the origin along the X-axis to describe the point.

First Rotation

A similar vector four units long can be drawn on the positive Y-axis as shown in *Figure 12.15*. Its length is designated by the imaginary number j4. Notice that this vector is the same length as the first vector, but it has been rotated 90 degrees. Thus, it can be said the factor j, when it multiplies a vector length, rotates that vector 90 degrees in a counter-clockwise direction. That is why the name *operator* is used. It is said that j *operates* on the vector to rotate it 90 degrees counter-clockwise. Thus, it is called the j-operator.

In reality, the j written before the length of a vector indicates only that the vector is measured off on the positive imaginary axis.

Second Rotation

Now, keeping in mind the idea that the j-operator rotates a vector 90 degrees counter-clockwise when it multiplies the vector's length, this vector, j four, is multiplied times j once more.

$$j \times j4 = j^2 4 \qquad (12\text{–}19)$$
$$= -4$$

Therefore, j × j4 will rotate the vector four units in length an additional 90 degrees further counter-clockwise as shown in *Figure 12.16*. The vector is now located on the X axis, four units in length and pointing in the negative X direction. Thus, as you can see, the vector has rotated a total of 180 degrees from the original positive X-axis direction. The vector now indicates negative four units on the X axis.

Note that, as shown in equation *12–19*, j times j four equals j-squared times four. Since j-squared equals minus one, this becomes simply minus four, which describes exactly the position and length of the vector four units on the negative X axis.

Third Rotation

Now −4 is multiplied by j. Multiplying by j rotates the vector another 90 degrees to the negative imaginary axis, as shown in *Figure 12.17*.

$$j(-4) = -j4 \qquad (12\text{–}20)$$

This vector position is now −j4 (minus j four), 270 degrees from the original positive X axis position.

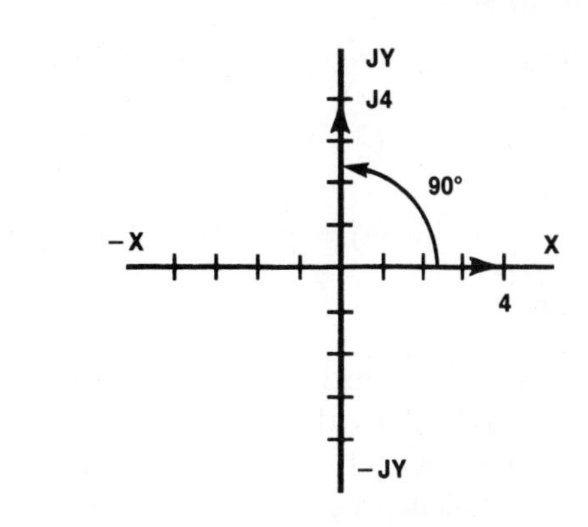

Figure 12.15 *Vector 4 Units in Length Rotated 90 Degrees*

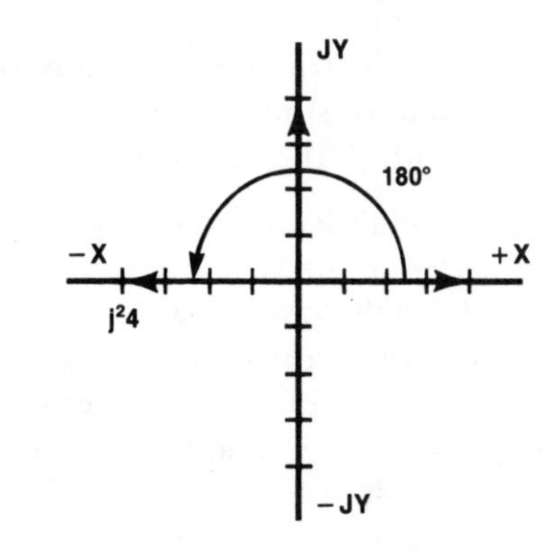

Figure 12.16 *Vector 4 Units in Length Rotated 180 Degrees*

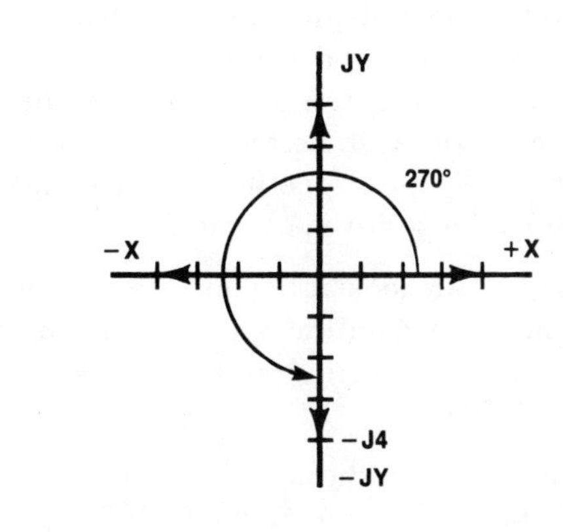

Figure 12.17 *Vector Rotated 270 Degrees from Original Position*

Fourth Rotation

Now, multiplying the $-j4$ once again by j rotates the vector back to the positive X-axis as shown in *Figure 12.18*. In equation form this is:

$$\begin{aligned} j(-j4) &= j^2(-4) \\ &= -1(-4) \\ &= +4 \end{aligned} \qquad (12\text{–}21)$$

The vector, four units long, is now positioned from the origin along the positive X axis. Four multiplications by j or four 90-degree rotations—360 degrees—places the vector back in the original positive X axis position. This is confirmed by equation *12-21* which shows that j times minus j four equals j-squared times minus four. Since j-squared is equal to minus one, this is equal to minus one times minus four or simply, four. This is the position from which the vector began.

In summary, multiplying j times a vector length operates on the vector to rotate it 90 degrees from the positive X-axis in a counter-clockwise direction. *Figure 12.19* shows how a vector of length ten is operated on by the j-operator. At point A the vector is $+10$. At point B it is $+J10$. At point C it is -10. At point D it is $-j10$. Note that $-j10$ is $-j$ times 10 and that $-j10$ is along the negative imaginary axis. For this reason it can be said that multiplying by $-j$ effectively rotates the vector 90 degrees in a *clockwise direction*—in a negative direction from the positive X axis. As indicated by the arrows in *Figure 12.19*, this is opposite to the rotation caused by the positive j-operator.

Another way to look at the operator is that a multiplier of one indicates a rotation from the positive X-axis of zero degrees as shown in *Figure 12.20a*.

A multiplier of j indicates a 90-degree angle of rotation as shown in *Figure 12.20b*.

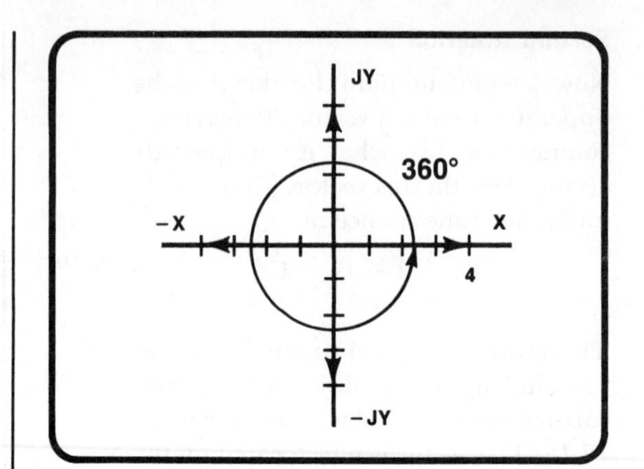

Figure 12.18 *Vector Rotated Back to Original Position*

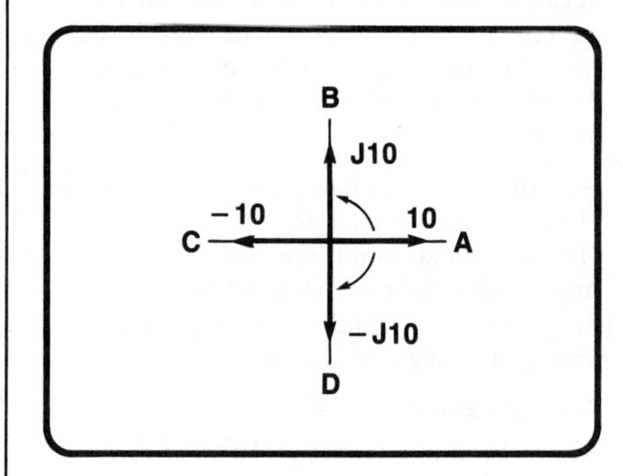

Figure 12.19 *Effect of j-Operator on a Ten-Unit Vector*

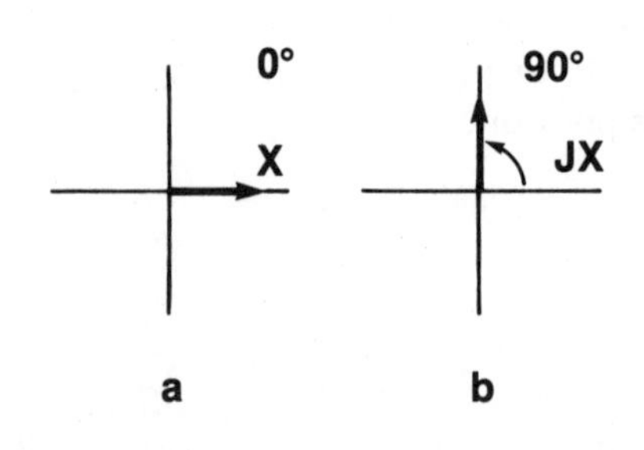

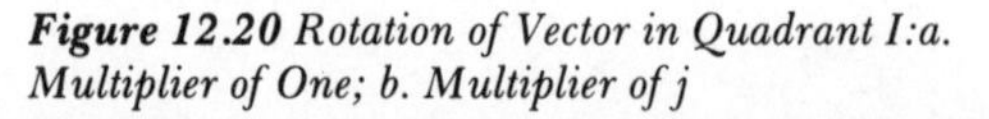

Figure 12.20 *Rotation of Vector in Quadrant I:a. Multiplier of One; b. Multiplier of j*

A multiplier of j^2 indicates a 180-degree rotation as shown in *Figure 12.21a*.

A multiplier of j^3 indicates a 270-degree rotation as shown in *Figure 12.21b*.

Note that a multiplier of j^3 is the same as a multiplier of minus j which indicates a minus 90-degree rotation as shown in *Figure 12.22*.

This type of j-operator notation is used to denote the phase relationships of quantities in ac circuits. For ac circuit analysis, the two most frequently used j-designations are j and −j, since j indicates a 90-degree phase shift in the counter-clockwise direction, *(Figure 12.19)* and minus j indicates a 90-degree phase shift in the clockwise direction. You have seen these 90° phase displacements repeatedly in the previous lessons on RC, RL and RLC circuit analysis.

Application of J-Operator to AC Circuit Analysis

It is now time to apply principles using the j operator and complex numbers to do circuit analysis. Several examples should demonstrate the technique.

RL Impedance Written in Rectangular Form

For example, *Figure 12.23* shows an RL circuit with a 20-ohm resistor and an inductive reactance of 15 ohms. The impedance phasor diagram for the circuit is as shown in *Figure 12.24*. The resistance vector can be written as 20, while the reactance vector can be written as j15, which indicates a 90-degree phase difference between R and X_L, with X_L positioned on the +j axis. The resultant vector of the two quantities added vectorially is the impedance Z of the circuit. It can be written in complex-number form:

$$Z = 20 + j15 \qquad (12\text{–}22)$$

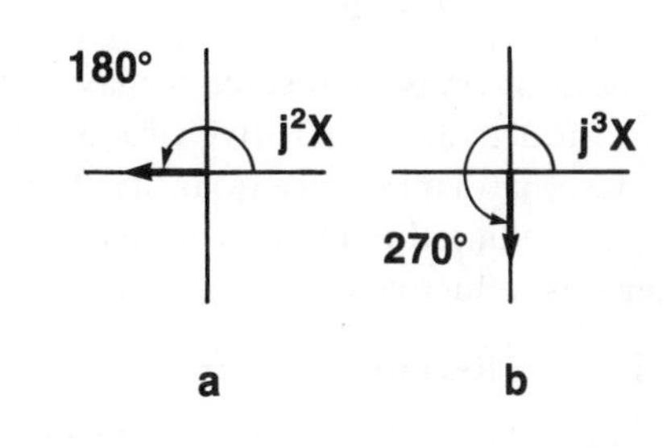

Figure 12.21 *Rotation of Vector in Quadrant II and III: a. Multiplier of j^2; b. Multiplier of j^3*

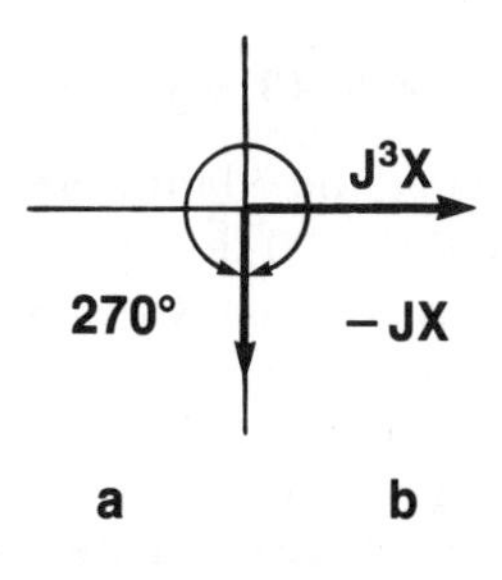

Figure 12.22 *Rotation of Vector in Quadrant IV: a. Multilplier of j^3; b. Multiplier of −j*

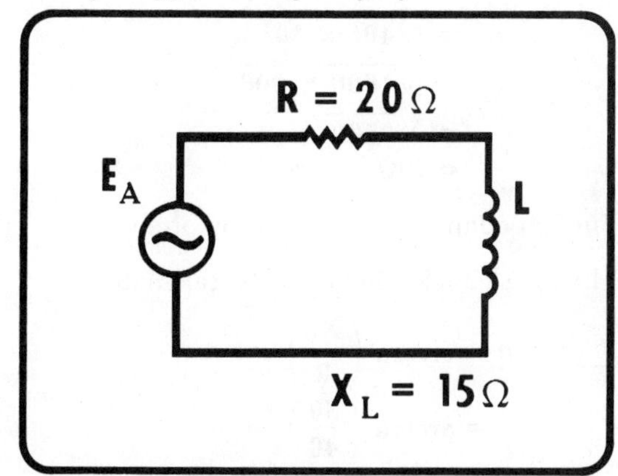

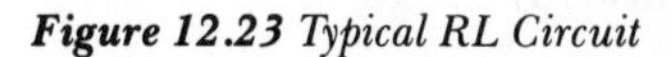
Figure 12.23 *Typical RL Circuit*

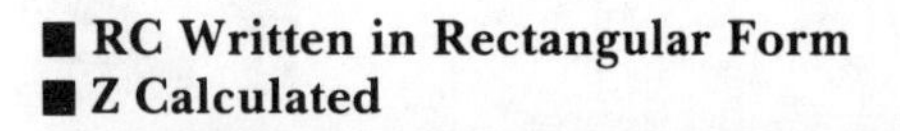

The impedance of a circuit in a more general form is written:

$$Z = R \pm jX \qquad (12\text{–}23)$$

This general form is used since X may pertain to either inductive or capacitive reactance. X simply designates a reactive quantity. Plus or minus j preceding the quantity indicates whether it is inductive or capacitive.

RC Written in Rectangular Form

For example, *Figure 12.25* shows a series resistive-capacitive circuit composed of 40 ohms of resistance and 30 ohms of capacitive reactance. There is a 90-degree phase difference between resistance and capacitive reactance as shown in the impedance phasor diagram of *Figure 12.26*. The impedance is written:

$$Z = 40 - j30 \qquad (12\text{–}24)$$

The −j preceding the 30 indicates capacitive reactance and that a 90-degree phase difference exists between R and X_C, with X_C positioned on the −j axis.

Z Calculated

The impedance can be calculated, as usual, by using the Pythagorean theorem.

$$\begin{aligned} Z &= \sqrt{R_2 \times X_C^2} \\ &= \sqrt{40^2 \times 30^2} \\ &= \sqrt{1600 \times 900} \\ &= \sqrt{2500} \\ &= 50\Omega \end{aligned}$$

The impedance magnitude is 50 ohms

The phase angle, theta, can be calculated:

$$\begin{aligned} \theta &= \arctan\left(\frac{X_C}{R}\right) \\ &= \arctan\left(\frac{30}{40}\right) \\ &= \arctan(0.75) \\ &= 37° \end{aligned}$$

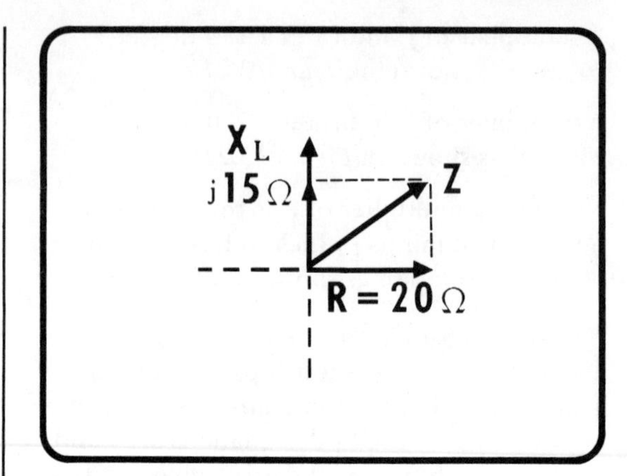

Figure 12.24 *Impedance Phasor Diagram for Circuit of Figure 12.23*

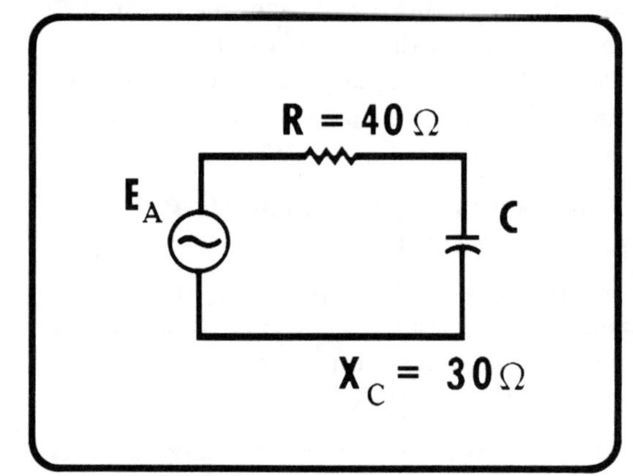

Figure 12.25 *A Typical Series RC Circuit*

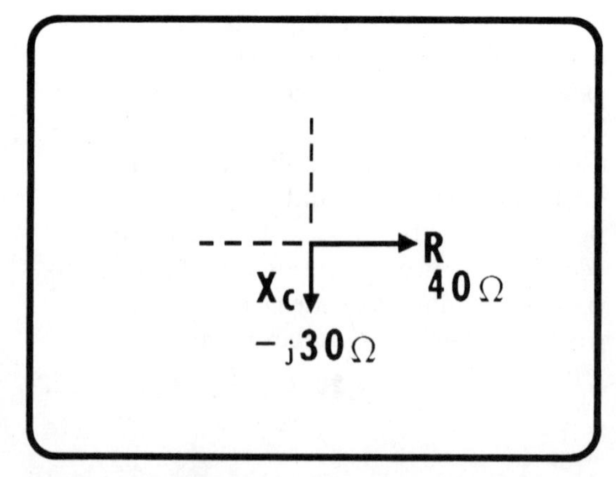

Figure 12.26 *Phase Relationship Between Resistance and Capacitive Reactance*

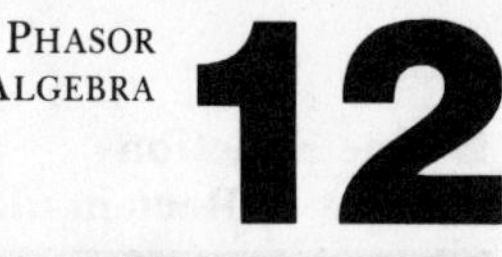

The phase angle is approximately 37 degrees.

Therefore, as shown in *Figure 12.27*, the resultant impedance in this example is a vector quantity that is neither along the X-axis as a resistance, nor along the Y-axis as a reactance. It is at an angle which you already know to be the phase angle of the circuit.

Z in Polar and Rectangular Form

To completely describe the impedance in its polar form, its magnitude must be accompanied by its direction, which is the phase angle. In this example, the impedance in its polar form is 50 ohms at an angle of minus 37 degrees, or

$$\mathbf{Z = 50\Omega \angle -37°} \qquad (12\text{–}25)$$

In its rectangular form, the impedance is written (equation *12–24*):

$$\mathbf{Z = 40 - j30.} \qquad (12\text{–}24)$$

Both represent the same impedance. 40 minus j30 describes the same resultant impedance vector as 50 at minus 37 degrees.

Using the diagram shown in *Figure 12.28*, both polar and rectangular forms of impedance can be indicated. The rectangular form of the impedance can be written R + jX, and the polar form can be written $Z \angle \theta$.

Converting Impedance From Rectangular to Polar Form

The conversion from rectangular to polar form can be written as a mathematical expression as shown in equation *12–26*.

$$\mathbf{Z \angle \theta = \sqrt{R^2 + X^2} \angle \arctan} \left(\frac{X}{R}\right) \qquad (12\text{–}26)$$

This equation says: the magnitude of Z at the angle theta equals the square root of R-squared plus X-squared, the angle theta is an angle with a tangent which is X divided by R.

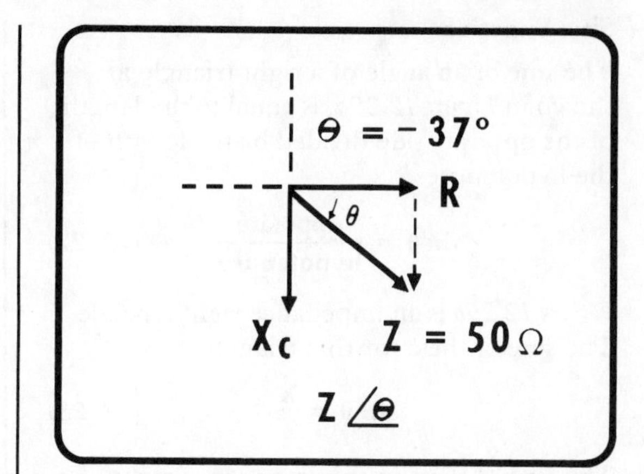

***Figure 12.27** Impedance of Series RC Circuit Example*

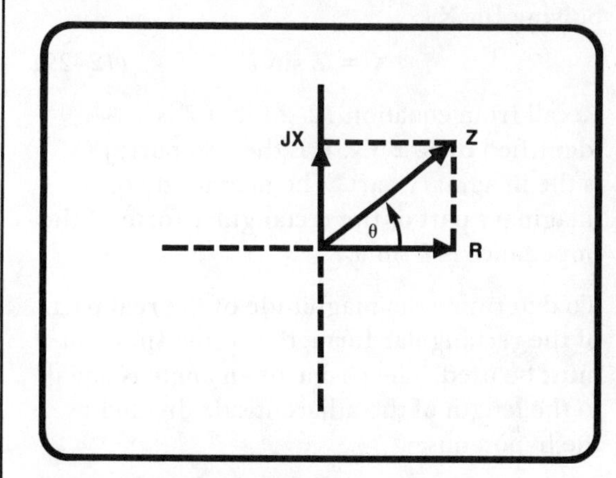

***Figure 12.28** Diagram for Indicating Polar or Rectangular Form of Impedance*

Converting Impedance From Polar to Rectangular Form

A reverse conversion is possible if the polar form of the impedance is known. This can be converted to the rectangular form using two trigonometric functions: 1) the sine function and 2) the cosine function. You have used the tangent function a great deal in these last lessons, but the sine and cosine functions haven't been used very much even though they are related to the tangent function and were discussed in previous lessons.

Sine Function

The sine of an angle of a right triangle as shown in *Figure 12.29a*, is equal to the length of the opposite side divided by the length of the hypotenuse:

$$\sin \theta = \frac{\text{opposite}}{\text{hypotenuse}} \qquad (12\text{–}27)$$

Figure 12.29b is an impedance right triangle. The sine of theta for that triangle is

$$\sin \theta = \frac{X}{Z} \qquad (12\text{–}28)$$

The sine of theta in this impedance right triangle, is equal to the ratio, X divided by Z. Solving for X,

$$X = Z \sin \theta \qquad (12\text{–}29)$$

Recall from equation *12–23* that Z is identified by R ± jX. R is the real part; jX is the imaginary part. The magnitude of the imaginary part of the rectangular form of the impedance is Z sin θ.

To determine the magnitude of the real part of the rectangular form, the cosine function must be used. The cosine of an angle is equal to the length of the adjacent side divided by the hypotenuse:

$$\cos \theta = \frac{\text{adjacent}}{\text{hypotenuse}} \qquad (12\text{–}30)$$

Therefore, in the impedance right triangle formed by R, X, and Z,

$$\cos \theta = \frac{R}{Z} \qquad (12\text{–}31)$$

Solving for the magnitude of R,

$$R = Z \cos \theta \qquad (12\text{–}32)$$

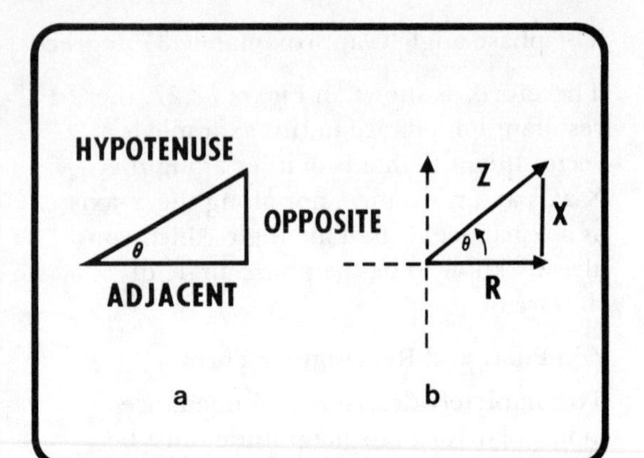

***Figure 12.29** Right Triangle Relationships: a. Right Triangle; b. Z, X. R, and θ*

The magnitude of the real part of the rectangular form of the impedance is Z cos θ.

Therefore, if the values of Z and theta are known, the magnitude of R and X can be derived using these two relationships:

$$R = Z \cos \theta \qquad (12\text{–}32)$$

$$X = Z \sin \theta \qquad (12\text{–}29)$$

The rectangular form can be expressed in the form of R plus jX or R minus jX depending on the value of θ.

Polar to Rectangular Example

For example, suppose that the polar form of the impedance of a circuit is $5 \angle 37^\circ$ as shown in *Figure 12.30*. The general expression for the impedance is given in equation *12–33*.

$$Z \angle \theta = Z \cos \theta \pm j(Z \sin \theta) \qquad (12\text{–}33)$$

Solving for R using equation *12–32*,

$$\begin{aligned} R &= Z \cos \theta \\ &= 5 \cos 37^\circ \\ &= 5(0.799) \\ &= 5(0.8) \\ &= 4\Omega \end{aligned}$$

R equals Z times the cosine of theta, or five times the cosine of 37 degrees. From trigonometric tables or by using a calculator the cosine of 37 degrees is determined to be approximately 0.8. Therefore, R equals five times 0.8, which is four ohms.

Solving for X using equation *12–29*,

$$\begin{aligned} \mathbf{X} &= \mathbf{Z \sin \theta} \\ &= \mathbf{5 \sin 37°} \\ &= \mathbf{5(0.602)} \\ &= \mathbf{5(0.6)} \\ &= \mathbf{3\Omega} \end{aligned}$$

X equals 3 ohms.

Thus, the rectangular form of the impedance of this circuit is R, the real part, equals four ohms and X, the reactive imaginary part, equals three ohms. Writing this in complex form, the impedance is four plus j3 ohms.

$$\begin{aligned} \mathbf{Z} &= \mathbf{R + jX} \\ &= \mathbf{(4 + j3)\Omega} \end{aligned}$$

If the angle had been a negative 37 degrees the impedance would be four minus j3.

The conversion formula can be written, as it was in equation *12–33*, as a general mathematical expression:

$$\begin{aligned} \mathbf{Z} &= \mathbf{Z\angle\theta} \\ &= \mathbf{R \pm jX} \\ &= \mathbf{Z \cos \theta \pm j(Z \sin \theta)} \qquad (12\text{–}33) \end{aligned}$$

Rectangular to Polar Example

To help you understand the technique of converting rectangular to polar form, another example will be analyzed. This example concerns a series RC circuit shown in *Figure 12.31* which consists of a 10-ohm resistor and a capacitive reactance of 24 ohms.

The relationship of R, X_C, and Z for this circuit is shown in *Figure 12.32*.

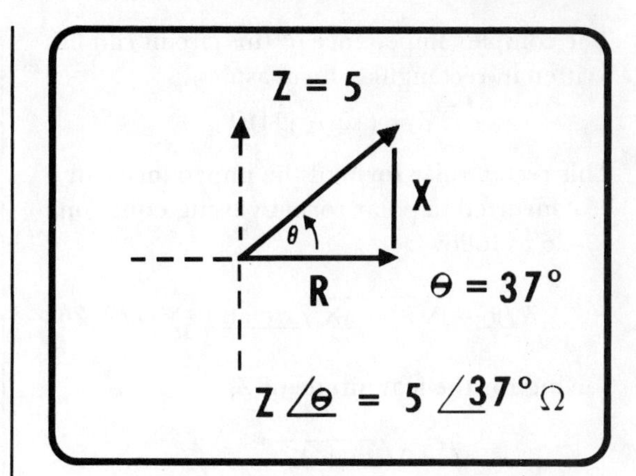

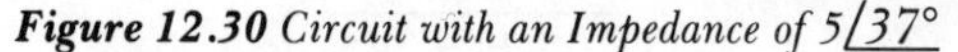
Figure 12.30 *Circuit with an Impedance of* $5\angle 37°$

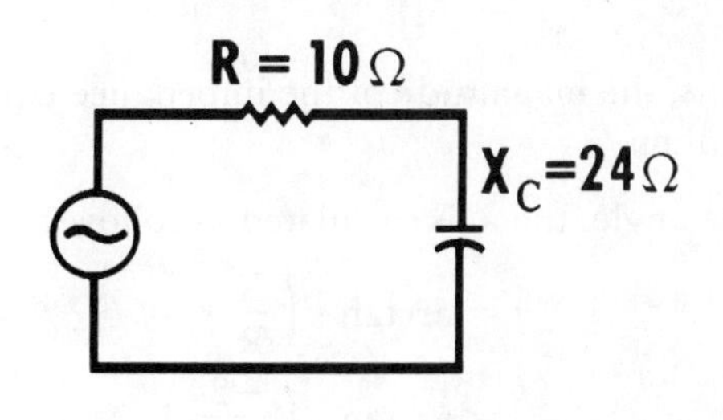

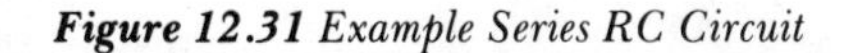
Figure 12.31 *Example Series RC Circuit*

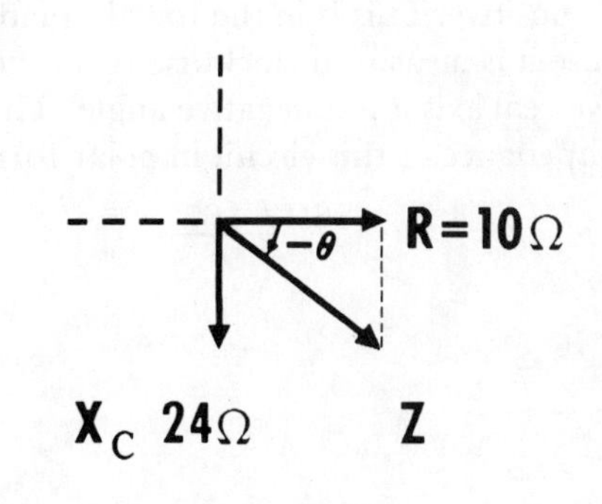

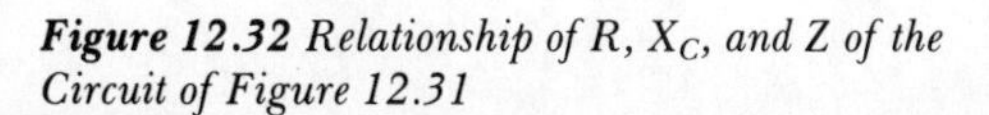
Figure 12.32 *Relationship of R,* X_C*, and Z of the Circuit of Figure 12.31*

■ Polar to Rectangular Form Example

The complex impedance of this circuit can be written in rectangular form as:

$$Z = (10 - j24)\Omega$$

This rectangular form of the impedance can be converted to polar form by using equation *12–26* as follows:

$$Z \angle\theta = \sqrt{R^2 + X^2} \angle \arctan\left(\frac{X}{R}\right) \quad (12\text{–}26)$$

Solving for the magnitude of Z,

$$\begin{aligned} Z &= \sqrt{R^2 + X^2} \\ &= \sqrt{10^2 + -24^2} \\ &= \sqrt{100 + 576} \\ &= \sqrt{676} \\ &= 26\Omega \end{aligned}$$

Thus, the magnitude of the impedance equals 26 ohms.

The angle, theta, is calculated as follows:

$$\begin{aligned} \theta &= \arctan\left(\frac{X}{R}\right) \\ &= \arctan\left(\frac{-24}{10}\right) \\ &= \arctan(-2.4) \\ &= -67^\circ \end{aligned}$$

Note that this is a negative angle. It has its opposite side on the $-j$ axis and its adjacent side is positive. This is in the fourth quadrant. Because it is measured clockwise from the postive real axis it is a negative angle. Thus, the impedance of this circuit in polar form is

$$Z = 26\Omega \angle -67^\circ$$

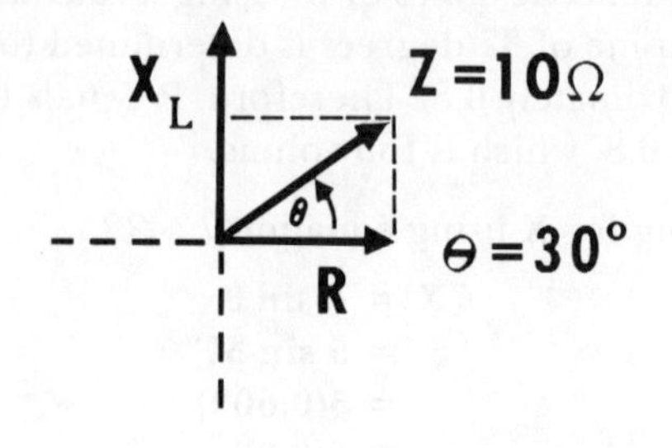

Figure 12.33 Impedance of an Example RL Circuit

Polar to Rectangular Form Example

Suppose that the impedance of a series RL circuit is as shown in *Figure 12.33*. It has been specified in polar form as $Z = 10\Omega \angle 30^\circ$. As in the first example, you have been asked to convert this specification to its rectangular form. Here is how it is done. Again, equation *12–33* gives the general form for the impedance:

$$\begin{aligned} Z \angle\theta &= Z \cos\theta + jZ \sin\theta & (12\text{–}33) \\ &= R + jX & (12\text{–}33) \end{aligned}$$

Solving for the real part you use:

$$R = Z \cos\theta \quad (12\text{–}32)$$

Solving for the imaginary part you use:

$$jX = jZ \sin\theta \quad (12\text{-}34)$$

Therefore, using the values in the example, R can be calculated:

$$\begin{aligned} R &= Z \cos\theta \\ &= 10 \cos 30^\circ \\ &= 10(0.866) \\ &= 8.66\Omega \end{aligned}$$

The resistive portion of the impedance is 8.66 ohms.

The reactive portion of the impedance is calculated similarly:

$$\begin{aligned} jX &= jZ \sin \theta \\ &= j10 \sin 30° \\ &= j10(0.5) \\ &= j5\Omega \end{aligned}$$

There are 5 ohms of inductive reactance. Thus, $10\angle 30°$ in polar form is the same impedance as 8.66 + j5 in rectangular form.

Some Observations

Note that when performing these conversions from polar to rectangular form that if the angle is positive the j-term is positive, indicating an inductive circuit as shown in *Figure 12.34a.* If the angle is negative, indicating a capacitive circuit as shown in *Figure 12.34b,* a negative j-term results.

Conversely, when converting from rectangular to polar form, if the j-term is positive, a positive angle results; if the j-term is negative, a negative angle results.

The examples emphasized another important fact. When applying complex number analysis to ac resistive-reactive circuits, the resistance is the real term of the impedance, the reactance of an inductor is an imaginary term and is written as a positive j-term, and the reactance of a capacitor, another imaginary term, is written as a negative j-term. This is shown in *Figure 12.35.* This is also shown in the equations for impedance: for a series RL circuit,

$$Z = R + jX_L \qquad (12\text{–}35)$$

and for a series RC circuit,

$$Z = R - jX_C \qquad (12\text{-}36)$$

Remember that the plus or minus j simply indicates the 90-degree phase difference between R and X.

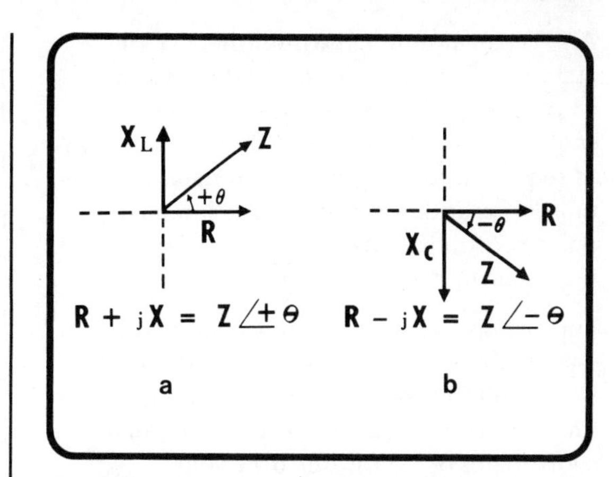

Figure 12.34 *Converting to Polar from Rectangular: a. An Inductive Circuit Causes a Positive Angle; b. A Capacitive Circuit Causes a Negative Angle*

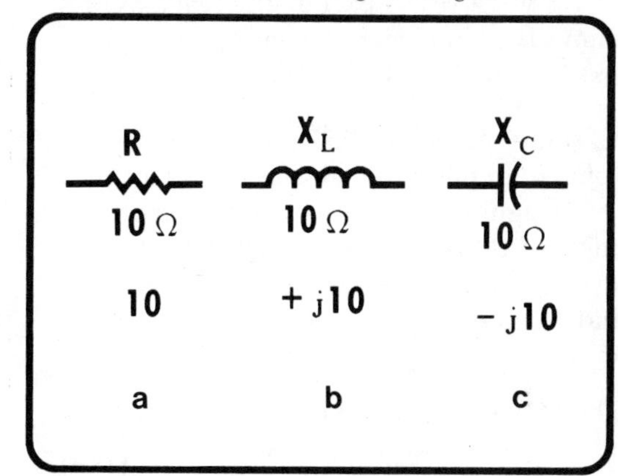

Figure 12.35 *Terms Used with Different AC Circuit Quantities*

Complex Problem Analysis

You will find that when analyzing more complex ac circuits in following lessons that designating impedance in rectangular coordinates as complex numbers with real and imaginary parts allows the problem solutions to be accomplished essentially in two parts. One part concerns all the real values, and another part concerns all the imaginary values with their j operators. Adding, subtracting, multiplying, and dividing these complex numbers becomes almost as simple as solving purely resistive circuits.

■ Complex Problem Analysis

Any vector positioned within the X-Y plane at any angle from zero to 360 degrees can be represented with complex number coordinates and manipulated mathematically as required. Rectangular coordinates are used mostly for addition and subtraction, and polar coordinates are used mostly for multiplication and division.

For example, a series ac circuit has two voltage drops which must be added together to obtain total voltage. V_1 shown in *Figure 12.36* is one voltage. It is a vector in polar coordinates of magnitude 6.71 volts positioned at an angle of 116.6 degrees, $6.71\angle 116.60$. In rectangular coordinates V_1 consists of a vector part on the real axis of minus three and an imaginary part of plus j six, $V_1 = -3 + j6$.

V_2 shown in *Figure 12.37* is the second voltage. In polar coordinates it is a voltage with magnitude 6.71 volts positioned at an angle of 243.4 degrees, $6.71\angle 234.40$. It has a vector part on the real axis of minus three and an imaginary part of minus j six, $V_2 = -3 - j6$.

The total voltage drop is:

$$V_T = V_1 + V_2 \qquad (12\text{–}37)$$

In order to find the rectangular coordinates of V_T the real part and imaginary parts of V_1 and V_2 are simply added separately.

$$\begin{aligned} V_1 &= -3 + j6 \qquad (12\text{–}38)\\ V_2 &= -3 - j6\\ V_T &= -6 + j0 \end{aligned}$$

Minus three of V_1 plus minus three of V_2 gives a sum of minus 6 for the real part of V_T.

Plus j six and minus j six results in a sum of j zero for the imaginary part of V_T. The result is a vector with a rectangular coordinate of minus six plus j zero.

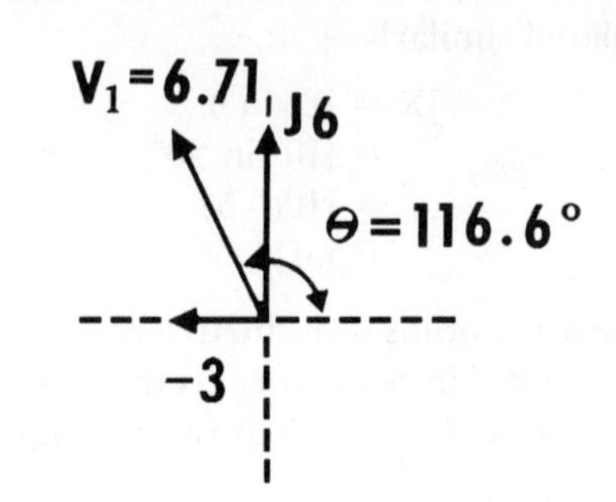

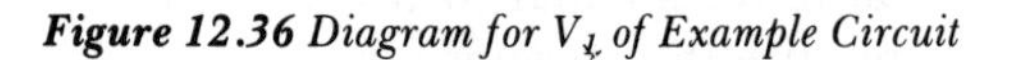
Figure 12.36 *Diagram for V_1 of Example Circuit*

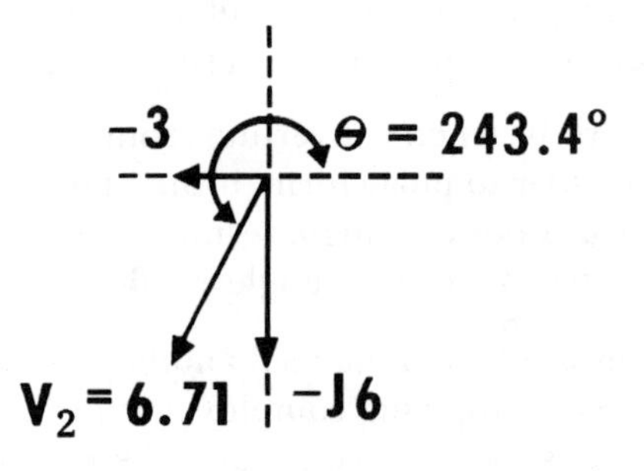

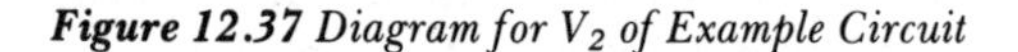
Figure 12.37 *Diagram for V_2 of Example Circuit*

■ Summary

As shown in *Figure 12.38,* the total voltage, V_T, in polar form is a voltage with magnitude of six volts positioned at an angle of 180 degrees, $6\angle 180°$. Any resultant vector can be obtained from the addition of any number of vectors in a similar fashion.

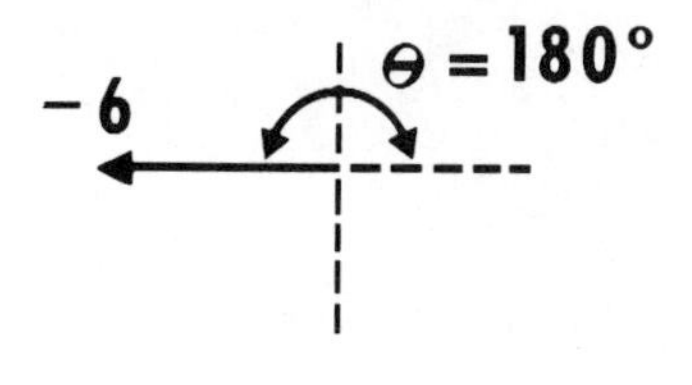

Figure 12.38 *Diagram for V_T of Example Circuit*

You probably have noted in this lesson that when performing calculations with complex numbers only addition has been done. Subtraction of complex numbers can also be done using the rectangular form. However, multiplication and division of complex numbers are performed more easily if the numbers are in polar form. The subjects of subtraction, multiplication, and division of complex numbers are discussed in detail in lesson 13.

SUMMARY

In this lesson, the concept of imaginary numbers, complex numbers, and the j-operator were introduced. You learned how to express a complex number in rectangular and polar form, and how to convert from one to the other. In the next lesson, these ideas and techniques will be applied to solve for circuit values of complex RLC circuits.

■ Worked-Out Examples

1. For the points shown on the complex plane shown, specify both rectangular and polar coordinates of the points.

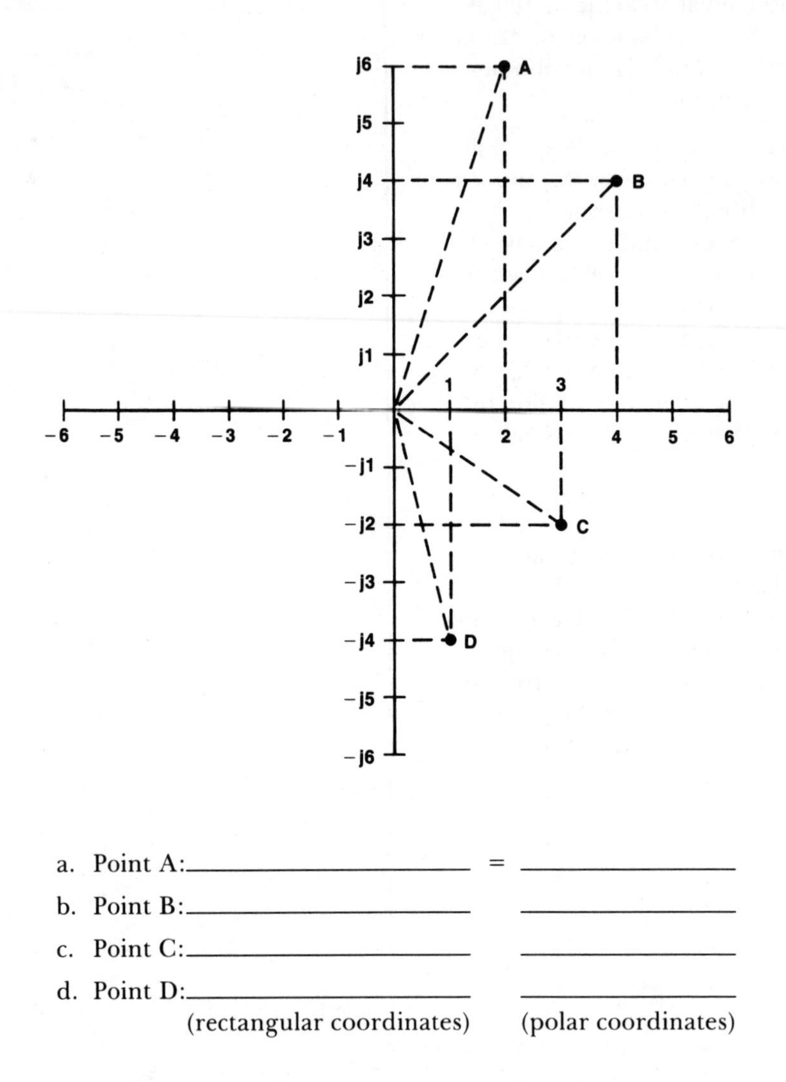

a. Point A:______________________ = __________________

b. Point B:______________________ __________________

c. Point C:______________________ __________________

d. Point D:______________________ __________________

(rectangular coordinates) (polar coordinates)

■ Worked-Out Examples

Solutions:

a. **Point A:** 2 + j6 (rectangular coordinates)

To determine polar coordinates, use conversion formulas: R = 2, X = 6

$$Z = \sqrt{R^2 + X^2} = \sqrt{2^2 + 6^2} = \sqrt{4 + 36} = \sqrt{40} = 6.32$$

$$\theta = \arctan\left(\frac{X}{R}\right) = \arctan\left(\frac{6}{2}\right) = \arctan(3) = 71.6°$$

Therefore:

Point A: 2 + j6 = 6.32 $\angle$71.6°

b. **Point B:** 4 + j4 (rectangular coordinates)

To determine polar coordinates: R = 4, X = 4

$$Z = \sqrt{R^2 + X^2} = \sqrt{4^2 + 4^2} = \sqrt{16 + 16} = \sqrt{32} = 5.66$$

$$\theta = \arctan\left(\frac{X}{R}\right) = \arctan\left(\frac{4}{4}\right) = \arctan(1) = 45°$$

Therefore:

Point B: 4 + j4 = 5.66$\angle$45°

c. **Point C:** 3 − j2 (rectangular coordinates)

To determine polar coordinates: R = 3, X = −2

$$Z = \sqrt{R^2 + X^2} = \sqrt{3^2 + (-2)^2} = \sqrt{9 + 4} = \sqrt{13} = 3.6$$

$$\theta = \arctan\left(\frac{X}{R}\right) = \arctan\left(\frac{-2}{3}\right) = -\arctan(0.67) = -33.7°$$

Therefore:

Point C: 3 − j2 = 3.6$\angle$−33.7°

d. **Point D:** 1 − j4 (rectangular coordinates)

To determine polar coordinates: R = 1, X = −4

$$Z = \sqrt{R^2 + X^2} = \sqrt{1^2 + (-4)^2} = \sqrt{1 + 16} = \sqrt{17} = 4.12$$

$$\theta = \arctan\left(\frac{X}{R}\right) = \arctan\left(\frac{-4}{1}\right) = -\arctan(4) = -76°$$

Therefore:

Point D: 1 − j4 = 4.12$\angle$−76°

■ Worked-Out Examples

2. Simplify to the lowest-order expressions the following multiplication and division problems below involving the j-operator.

a. $(j)(j)$

b. j^3

c. $\frac{1}{j^2}$

d. $\sqrt{\frac{-1(j)}{j^2}}$

c. $(j^3)(j^2)$

f. $(j^3)(\frac{1}{j^2})(\frac{1}{j})$

g. $\left(\frac{1}{j}\right)(j^2)\left(\frac{j^2}{j}\right)$

h. j^4

i. $\frac{1}{j^3}$

j. $j^3\left(\frac{1}{j}\right)$

k. $j\ (\frac{1}{j^2})$

l. $(j^2)(j^3)(j)(j^4)j$

■ Worked-Out Examples

Solutions:

a. $(j)(j) = (\sqrt{-1})(\sqrt{-1}) = \mathbf{-1}$

b. $(j^3) = (j)(j^2) = j(-1) = \mathbf{-j}$

c. $\left(\frac{1}{j^2}\right) = \frac{1}{-1} = \mathbf{-1}$

d. $\left[\frac{\sqrt{-1}(j)}{j^2}\right] = \frac{(\sqrt{-1})\ (\sqrt{-1})}{-1} = \frac{-1}{-1} = \mathbf{+1}$

e. $(j^3)(j^2) = (-j)(-1) = \mathbf{j}$

f. $(j^3)\left(\frac{1}{j^2}\right)(j) = (-j)\left(\frac{1}{-1}\right)(j) = (-j^2)(-1) = (+1)(-1) = \mathbf{-1}$

g. $\left(\frac{1}{j}\right)(j^2)\left(\frac{j^2}{j}\right) = (-j)(j^2)(j) = (-j^2)(j^2) = -j^4 = \mathbf{-1}$

h. $j^4 = (j^2)(j^2) = (-1)(-1) = \mathbf{+1}$

i. $\frac{1}{j^3} = \left(\frac{1}{j}\right)\left(\frac{1}{j^2}\right) = (-j)(-1) = \mathbf{+j}$

j. $j^3\left(\frac{1}{j}\right) = \frac{j^3}{j} = j^2 = \mathbf{-1}$

k. $j\left(\frac{1}{j^2}\right) = j\left(\frac{1}{-1}\right) = j(-1) = \mathbf{-j}$

l. $(j^2)(j^3)(j)(j^4)j = (-1)(-j)(j)(+1)(j) = +j^3 = \mathbf{-j}$

■ Worked-Out Examples

3. Multiplying by j terms rotates a vector. What j term would have to be used as a multiplier to move the vector shown from its initial position to its final (result) position.

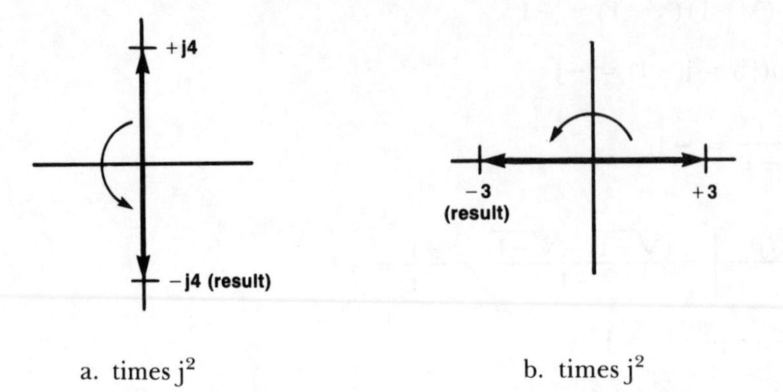

a. times j^2

b. times j^2

4. Write the complex impedance form of the following circuits in rectangular form, then convert it to polar form.

a.

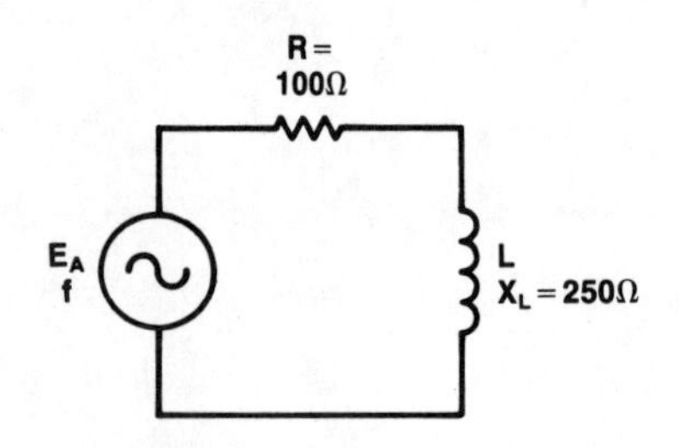

$Z = R \pm jX$ (rectangular) $=$ **100 + j250**

Converting to polar:

$$Z = \sqrt{R^2 + X^2} = \sqrt{100^2 + 250^2} = \sqrt{72{,}500} = 269.3\Omega$$

$$\theta = \arctan\left(\frac{X}{R}\right) = \arctan\left(\frac{250}{100}\right) = \arctan(2.5) = 68.2°$$

$\therefore Z = \mathbf{268.3\angle 68.2°}$

■ **Worked-Out Examples**

b.

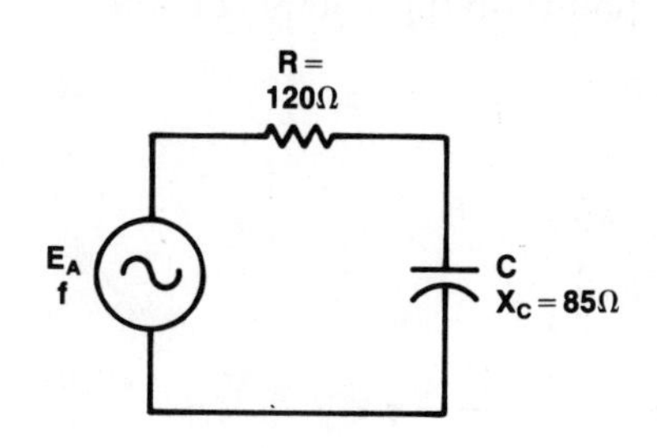

$Z = R \pm jX$ (rectangular)

$Z = \mathbf{120 - j85}$

Converting to polar:

$$Z = \sqrt{R^2 + X^2} = \sqrt{120^2 + 85^2} = \sqrt{21{,}625} = 147\Omega$$

$$\theta \arctan \left(\frac{X}{R}\right) = \arctan\left(\frac{-85}{120}\right) = \arctan(-0.7083) = -35.3°$$

$$\therefore Z = \mathbf{147\angle -35.3°}$$

5. Convert the following complex impedances in polar form to rectangular form.

a. $Z = 150\angle 25°$

$$R \pm jX = Z\cos\theta \pm jZ\sin\theta = 150(\cos 25°) + j150(\sin 25°)$$
$$= 150(0.9063) + j150(0.4226) = \mathbf{135.9 + j63.4\Omega}$$

b. $Z = 480\angle -80°$

$$R \pm jX = Z\cos\theta \pm jZ\sin\theta = 480\cos(-80°) + j480\sin(-80°)$$
$$= 480(0.1736) + j480(-0.9848) = \mathbf{83.3 - j472.7\Omega}$$

■ Practice Problems

1. Specify the rectangular and polar coordinates of the points on this complex plane.

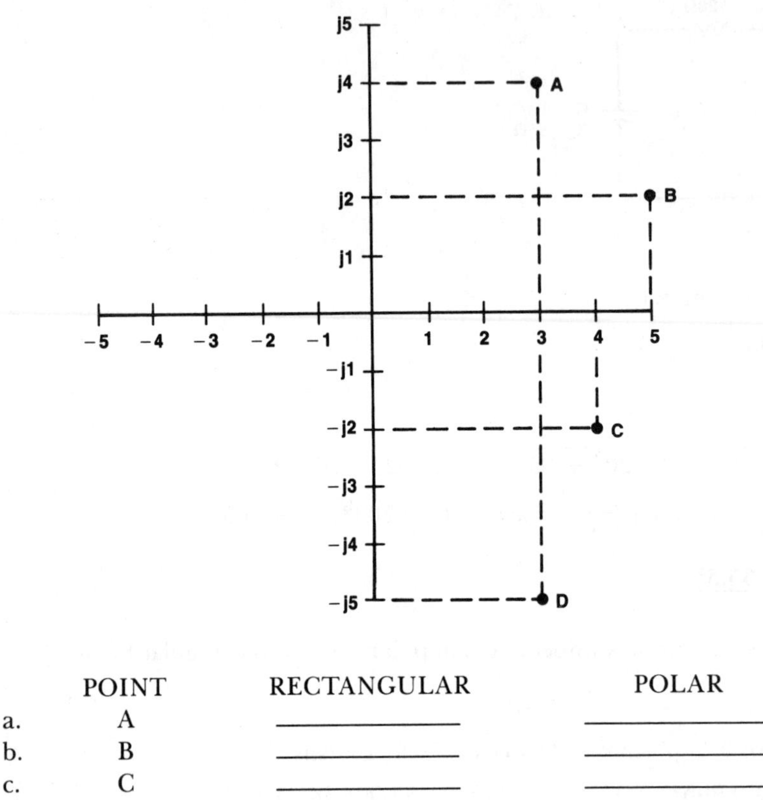

	POINT	RECTANGULAR	POLAR
a.	A	____________	____________
b.	B	____________	____________
c.	C	____________	____________
d.	D	____________	____________

2. Simplify the following j-operator expressions:

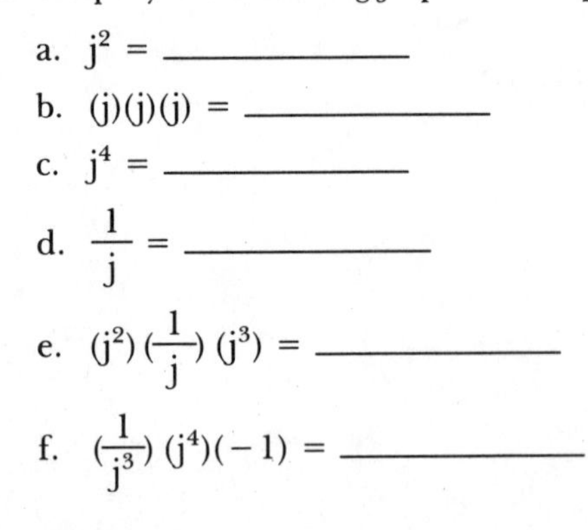

a. $j^2 =$ ____________

b. $(j)(j)(j) =$ ____________

c. $j^4 =$ ____________

d. $\frac{1}{j} =$ ____________

e. $(j^2)(\frac{1}{j})(j^3) =$ ____________

f. $(\frac{1}{j^3})(j^4)(-1) =$ ____________

■ Practice Problems

3. Show where the resulting vector will be located on the following complex plane if the vector shown is multiplied by the indicated j term.

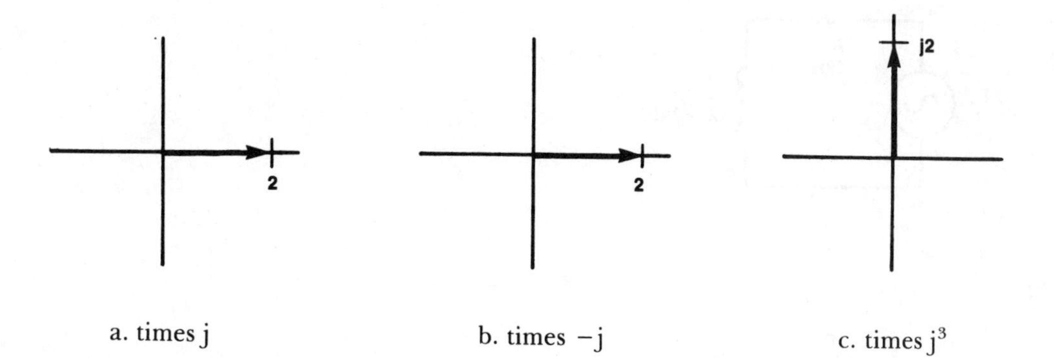

a. times j b. times $-j$ c. times j^3

4. Write the complex impedance form of the following circuits in rectangular form:

a. Z = ______________

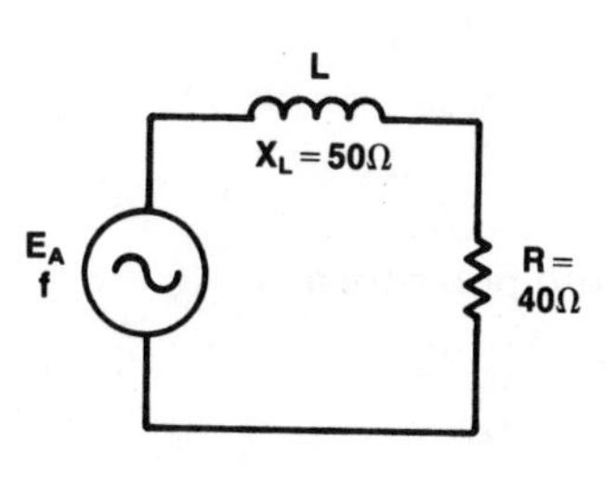

b. Z = ______________

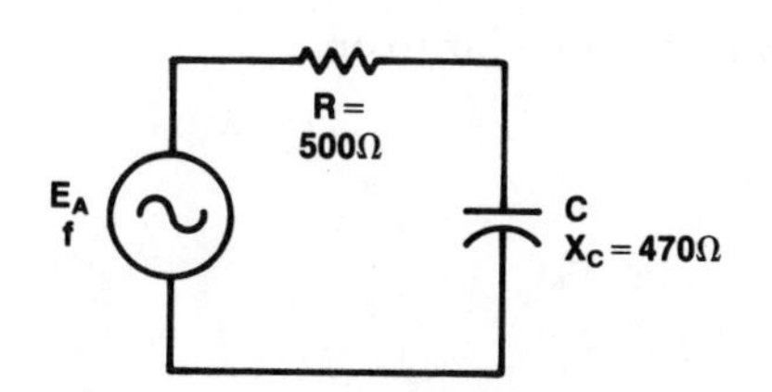

■ Practice Problems

c. Z = ____________

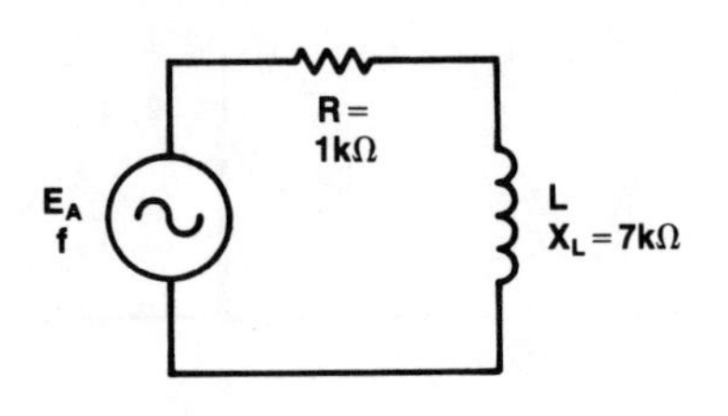

d. Z = ____________

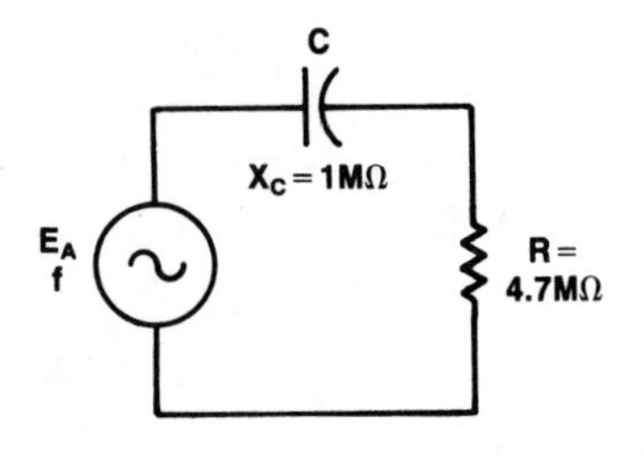

5. Convert the following rectangular impedance expressions into polar form.

a. 50 + j30 = ____________

b. 3.2k − j4.7k = ____________

c. 370 − j150 = ____________

d. 3M − j500k = ____________

6. Convert the following polar impedance expressions into rectangular form.

a. $550\angle 28°$ = ____________

b. $100k\angle -60°$ = ____________

c. $68M\angle 34°$ = ____________

d. $18k\angle -25°$ = ____________

■ Quiz

1. Specify the rectangular and polar coordinates of the points on the following complex plane.

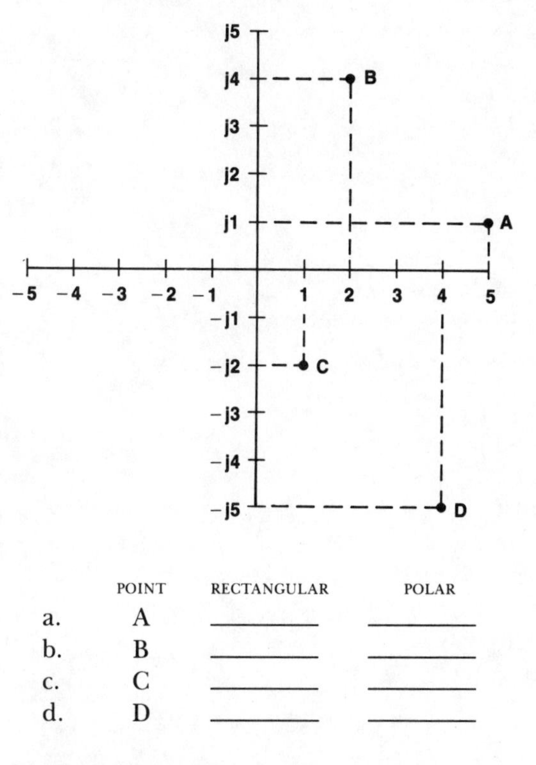

	POINT	RECTANGULAR	POLAR
a.	A	________	________
b.	B	________	________
c.	C	________	________
d.	D	________	________

2. Simplify the following j-operator expressions:

a. j = ________
b. j^2 = ________
c. j^3 = ________
d. j^4 = ________

3. Show where the resulting vector will be located on the following complex plane if the vector shown is multiplied by the indicated j term.

a.

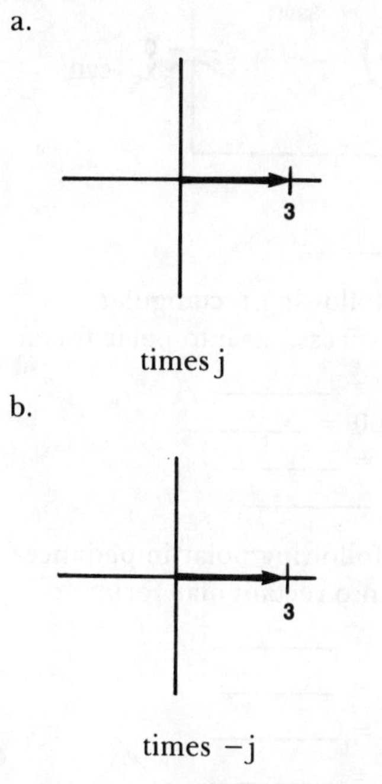

4. Write the complex impedance form of the following circuits in rectangular form:

a.

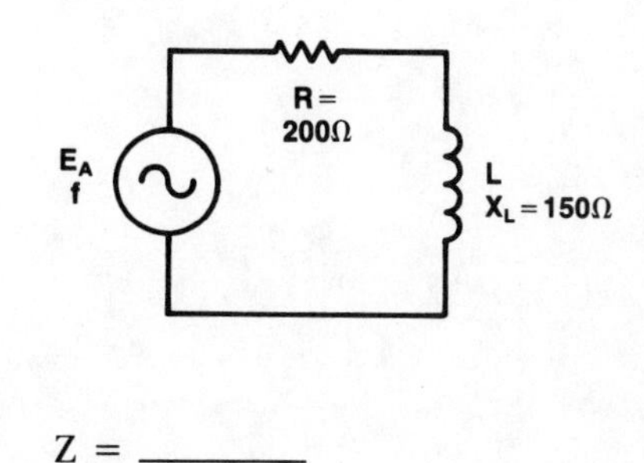

Z = ________

■ Quiz

b.

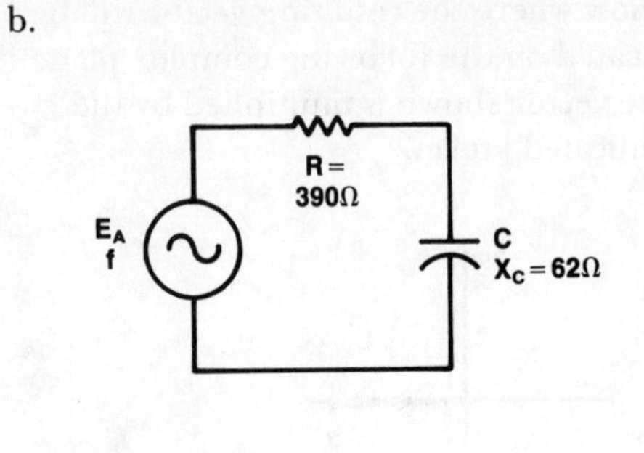

Z = ________

5. Convert the following rectangular impedance expressions into polar form:

a. 10 + j50 = ________
b. 370 − j600 = ________
c. 29 + j65 = ________
d. 7 − j9 = ________

6. Convert the following polar impedance expressions into rectangular form:

a. $68\angle+15°$ = ________
b. $34\angle-24°$ = ________
c. $16\angle+72°$ = ________
d. $27\angle-32°$ = ________

LESSON 13

Complex RLC Circuit Analysis

Lesson 12 introduced phasor algebra techniques that could be used to solve ac circuit problems. Imaginary numbers, complex numbers, and the j-operator were introduced. This lesson will show how to apply those concepts to solve RLC circuits which would be very difficult to solve using other methods.

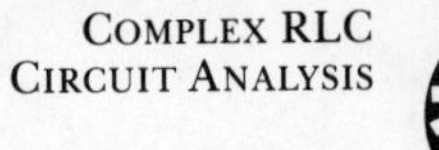

■ Objectives

At the end of this lesson you should be able to:

1. Calculate the sum, product, difference, or quotient of two or more j-operator terms.

2. Determine the sum, product, difference, or quotient of two or more complex numbers in rectangular form (R + jX).

3. Determine the product or quotient of two or more complex numbers in polar form ($Z \angle \theta$).

4. Express the impedance of a series, parallel, or series-parallel circuit in rectangular or polar form.

5. Write the impedance of a passive element (resistor, inductor, or capacitor) using j-operator notation or polar notation when given the reactance or resistance of that passive element.

6. Determine the circuit impedance, currents, and voltages of an RLC circuit and express them in either rectangular or polar form.

Introduction

In the previous lesson, you were introduced to the concept of imaginary numbers, complex numbers and the j-operator, and told how to express a complex number in rectangular or polar form. You were told how to convert from rectangular form to polar form and from polar form to rectangular form. The purpose of learning how to use complex numbers is to be able to simplify the solution of complicated ac circuits.

Writing Impedances with J-Operators

In the last lesson, it was stated that complex numbers, in either rectangular or polar form, are simply a code which can be used to express the value or magnitude of quantitites and their phase relationships in an ac circuit. It was also stated that the primary advantage of using complex numbers in the solution of ac circuits is that they could be solved using resistive circuit methods.

A Series Circuit

A good way to begin this lesson is by applying what you have already learned. Therefore, first a series RLC circuit shown in *Figure 13.1* will be discussed. It consists of two resistors of 10 ohms and 4 ohms connected to an inductor with a reactance of 8 ohms and a capacitor with a reactance of 20 ohms. To illustrate the technique that will be used throughout this lesson, the circuit first will be divided into two complex impedances as shown in *Figure 13.2*.

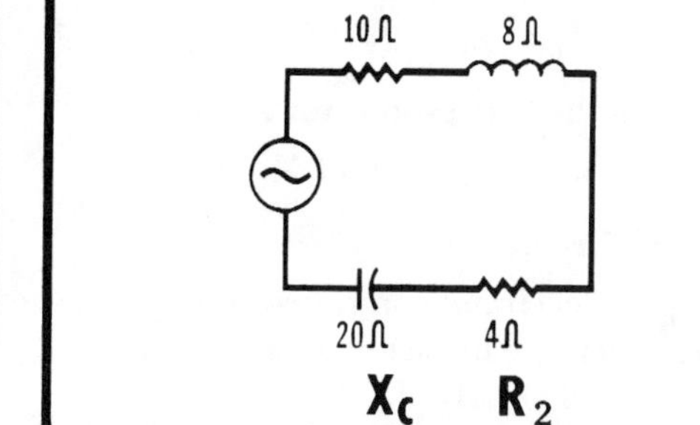

Figure 13.1 *A Series RLC Circuit*

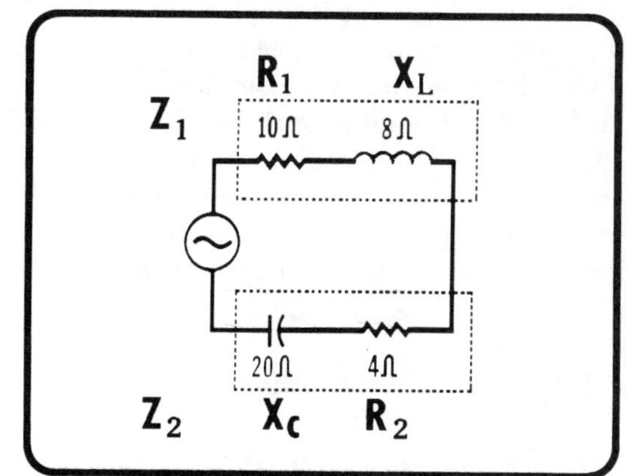

Figure 13.2 *The Circuit of Figure 13.1 Divided into Two Complex Impedances*

The impedance of the series combination of R_1 and L is called Z_1. It can be written:

$$Z_1 = R_1 + jX_L \qquad (13\text{–}1)$$

or

$$Z_1 = 10 + j8$$

The impedance of the series combination of R_2 and C, is called Z_2. It can be written:

$$Z_2 = R_2 - jX_c \qquad (13\text{–}2)$$

or

$$Z_2 = 4 - j20$$

jX_L is a positive quantity because the vector X_L leads the vector R by 90 degrees on the impedance phasor diagram shown in *Figure 13.3*. jX_C is a negative quantity because the vector X_C lags the vector R on the same diagram.

The two series impedances, Z_1 and Z_2, can be treated as if they were series resistances. The total impedance of the circuit is:

$$Z_T = Z_1 + Z_2 \qquad (13\text{–}3)$$

Substituting the expressions for Z_1 and Z_2,

$$Z_T = (10 + j8) + (4 - j20)$$

A Parallel Circuit

Next, a parallel circuit composed of two resistive-reactance branches as shown in *Figure 13.4* will be solved. The RL branch has a resistance of 3 ohms and an inductive reactance of 4 ohms. The RC branch has a resistance of 6 ohms and capacitive reactance of 10 ohms. The same technique outlined in the previous example will be used. First divide the circuit into two complex impedances, Z_1 and Z_2, as shown in *Figure 13.5*. Then write the complex number representation of the impedance for each branch. Using equation *13–1* the resistive-inductive branch, Z_1, can be written as:

$$Z_1 = 3 + j4$$

The impedance of the resistive-capacitive branch, Z_2, can be written:

$$Z_2 = 6 - j10$$

The total impedance of the circuit is the parallel combination of Z_1 and Z_2. Recall from earlier lessons that when two resistors are connected in parallel, the total resistance can be calculated using the product-over-sum equation:

$$R_T = \frac{R_1 \times R_2}{R_1 + R_2} \qquad (13\text{–}4)$$

The total impedance for a circuit can be calculated using a similar equation:

$$Z_T = \frac{Z_1 \times Z_2}{Z_1 + Z_2} \qquad (13\text{–}5)$$

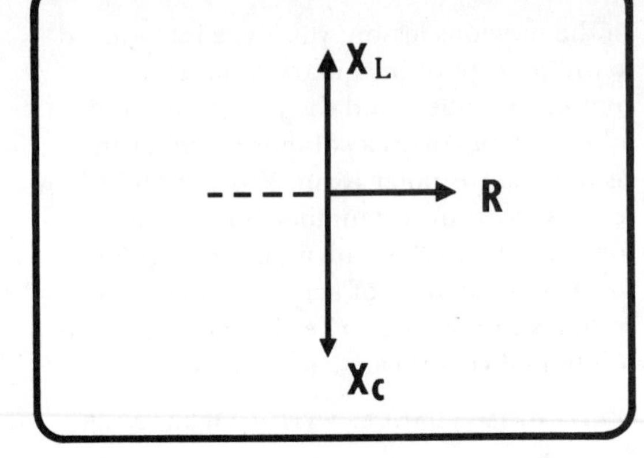

Figure 13.3 *Impedance Phasor Diagram for the Circuit of Figure 13.1*

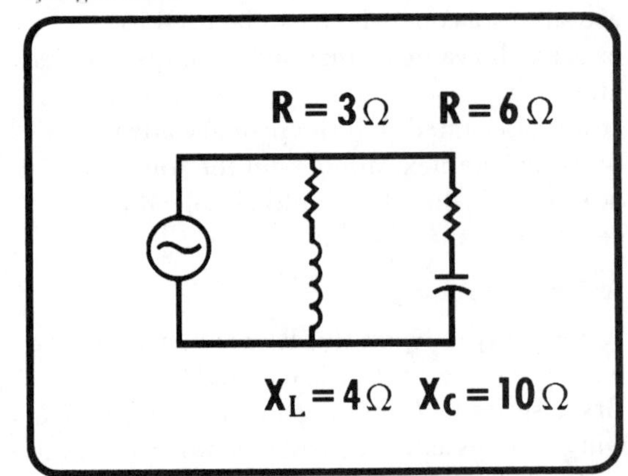

Figure 13.4 *A Parallel RLC Circuit*

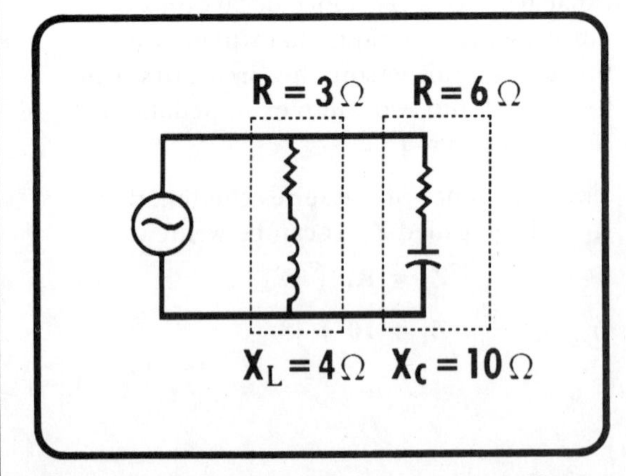

Figure 13.5 *The Circuit of Figure 13.4 Divided into Two Complex Impedances*

Substituting the rectangular impedance for Z_1 and Z_2, this equation becomes:

$$Z_T = \frac{(3 + j4)(6 - j10)}{(3 + j4) + (6 - j10)}$$

Once the impedance of the circuits has been converted to this form, addition, subtraction, multiplication, and division can be used to reduce the circuits to a single complex number representing the value of the total impedance. Because the j operator is included, a discussion of these arithmetic operators will be useful before you continue with the complete problem solution.

Math Operations Involving J-Operators

You already know how to perform math operations using real numbers. That is basic arithmetic. These same operations can be performed using imaginary numbers; addition, subtraction, multiplication, and division are performed with j-terms as you would add, subtract, multiply and divide using algebraic techniques. Simply treat the term j as you would the variable X in an algebraic expression and make sure that you recognize the positive and negative quantities. For example, addition and subtraction using imaginary numbers are performed like this:

$$j2 + j6 = j8 \qquad j3 - j5 = -j2$$
$$j4 - j3 = j1 \qquad -j4 + j7 = j3$$

Here are some additional examples:

$$-j1 - j9 = -j10 \qquad -j12 + j8 = -j4$$

Similarly, multiplication and division are performed like this:

$$j2 \times j4 = j^2 8 \qquad \frac{j^2 6}{j2} = j3$$
$$\frac{j8}{2} = j4$$

Here are some additional examples:

$$-j3 \times j4 = -j^2 12 \qquad \frac{-j8}{j2} = -4$$
$$-j8 \times -j4 = j^2 32$$
$$(j5)(-j2) = -j^2 10 \qquad \frac{j^4 20}{-j^2 5} = -j^2 4$$

And since one over j equals minus j,

$$\frac{16}{j2} = \frac{1}{j} \times \frac{16}{2} = -j8$$

or

$$\frac{-j^2 9}{j^3 3} = -\frac{3}{j} = -\frac{1}{j} \times 3 = j3$$

Once division or multiplication has been performed, terms containing a power of j can be simplified using the following basic identities which were introduced in Lesson 12:

$$j^2 = -1$$
$$j^3 = -j$$
$$j^4 = +1$$

For example, $j^2 8$ may be simplified:

$$j^2 8 = (-1)8 = -8$$

MATH OPERATIONS WITH COMPLEX NUMBERS — RECTANGULAR FORM

Now that basic mathematical equations using imaginary numbers have been illustrated, let's see how these operations may be performed using complex numbers.

Addition

To add two rectangular form complex numbers, simply add the two real parts and add the two imaginary parts. The result is expressed as another complex number called the *complex sum* which has both real and imaginary parts.

$$(R_1 + jX_1) + (R_2 + jX_2) = (R_1 + R_2) + j(X_1 + X_2) \qquad (13\text{–}6)$$

For example, suppose the two complex numbers 4 + j2 and 6 + j1 must be added:

$$\begin{aligned} &(4 + j2) + (6 + j1) \\ &= (4 + 6) + (j2 + j1) \\ &= 10 + j3 \end{aligned}$$

Thus, the sum of 4 + j2 and 6 + j1 equals 10 + j3.

As another example,

$$\begin{aligned} &(-8 + j2) + (10 - j5) \\ &= (-8 + 10) + (j2 - j5) \\ &= 2 - j3 \end{aligned}$$

Subtraction

Subtraction using rectangular form complex numbers is very similar to addition. To subtract two complex numbers, first find the difference between the real parts, then find the difference between the imaginary parts. Remember, this must be the algebraic difference. The result is a complex number with real and imaginary parts.

$$\begin{aligned} &(R_1 + jX_1) - (R_2 + jX_2) \\ &= (R_1 - R_2) + j(X_1 - X_2) \end{aligned} \qquad (13\text{–}7)$$

For example, 4 + j2 is subtracted from 8 + j8.

$$\begin{aligned} &(8 + j8) - (4 + j2) \\ &= (8 - 4) + (j8 - j2) \\ &= 4 + j6 \end{aligned}$$

Effectively, the subtraction process is the same as any algebraic subtraction, you change the sign of the number that is being subtracted and then perform an addition.

As a second example,

$$\begin{aligned} &(10 - j7) - (2 - j3) \\ &= (10 - 2) + j(-7 + 3) \\ &= 8 - j4 \end{aligned}$$

Written in a little different form may make it easier to understand.

$$\begin{array}{r} 10 - j7 \\ -(2 - j3) \\ \hline \end{array}$$

2 − j3 is subtracted from 10 − j7 by changing the sign of 2 − j3 and adding

$$\begin{array}{r} 10 - j7 \\ +(-2 + j3) \\ \hline 8 - j4 \end{array}$$

Multiplication

The multiplication of two rectangular form complex numbers may be performed by using basic binomial algebraic methods. That is, the complex numbers are in the form $(R_1 + jX_1)$ and $(R_2 + jX_2)$. When you multiply one by the other the result is obtained by multiplying each term in the first number times each term in the second number and adding the resulting products:

$$\begin{aligned} &(R_1 + jX_1)(R_2 + jX_2) \\ &= R_1R_2 + jR_1X_2 + jR_2X_1 + j^2X_1X_2 \end{aligned} \qquad (13\text{–}8)$$

You can then group like terms. At this time, terms 2 and 3 both have j-prefixes and can be grouped:

$$(R_1R_2) + j(R_1X_2 + R_2X_1) + j^2(X_1X_2)$$

Since $j^2 = -1$ in term 4, the expression simplifies to:

$$(R_1R_2 - X_1X_2) + j(R_1X_2 + R_2X_1) \qquad (13\text{–}9)$$

As an example the product of 9 + j3 and 4 − j2 is:

$$\begin{aligned} &(9 + j3)(4 - j2) \\ &= 36 - j18 + j12 - j^26 \end{aligned}$$

Combining like terms results in:

$$= 36 - j6 - j^26$$

Since the $j^2 = -1$, the result simplifies to:

$$\begin{aligned} &= 36 - j6 - (-1)6 \\ &= 36 - j6 + 6 \end{aligned}$$

Finally, combining real terms:

$$= 42 - j6$$

Notice that the product reduces to the real plus imaginary complex number form. It is important to remember to carry along the sign of the reactance terms and apply the rules for the product of signed terms.

$$+ \times + = +$$
$$+ \times - = -$$
$$- \times - = +$$
$$- \times + = -$$

Division

The division of two rectangular form complex numbers may also be performed by using algebraic methods. The form is

$$\frac{R_1 + jX_1}{R_2 + jX_2} \qquad (13\text{–}10)$$

This process, however, becomes quite involved since the division of a real number by an imaginary number is not possible. Because of *this*, the denominator must be converted first to a real number by a process called *rationalization* of the fraction. To do this, both the numerator and denominator must be multiplied by what is called the *conjugate* of the denominator. The conjugate of a complex number is simply the number with the j-term having an opposite sign. For example, the conjugate of 2 + j2 is 2 – j2. Similarly, the conjugate of 6 – j4 is 6 + j4. Following this rule to divide 3 + j2 by 2 + j2, you must multiply both the numerator and denominator by the conjugate of the denominator 2 – j2:

$$\frac{(3 + j2)}{(2 + j2)} \times \frac{(2 - j2)}{(2 - j2)}$$

In doing so, you do not change the value of the fraction since (2 – j2) divided by (2 – j2) times the fraction is like multiplying the fraction times one.

$$\frac{(3 + j2)}{(2 + j2)} \times 1$$

Multiplying numerators and denominators:

$$\frac{(3 + j2)}{(2 + j2)} \times \frac{(2 - j2)}{(2 - j2)} = \frac{6 - j6 + j4 - j^2 4}{4 - j4 + j4 - j^2 4}$$

Combining j-terms in the numerator and denominator, the fraction becomes:

$$= \frac{6 - j2 - j^2 4}{4 - j^2 4}$$

Since $j^2 = -1$, the expression can be reduced as follows:

$$\frac{6 - j2 - (-1)4}{4 - (-1)4} = \frac{6 - j2 + 4}{4 + 4} = \frac{10 - j2}{8}$$

Notice that all imaginary terms have been eliminated from the denominator. Recall that this is the purpose of rationalization. Now, the fraction can be split up and the division performed:

$$= \left(\frac{10}{8}\right) - \left(\frac{j2}{8}\right) = 1.25 - j0.25$$

The final result is again a complex number with real and imaginary parts. Again, it must be remembered to follow all of the rules that pertain to multiplication and division of signed terms.

MULTIPLICATION AND DIVISION IN POLAR FORM

Complex numbers, expressed in rectangular form, can be multiplied and divided by using the basic algebraic methods discussed. However, it is much easier to multiply and divide complex numbers by converting them from rectangular form to polar form and then performing the operations in polar form.

Multiplication

To multiply in polar form, one need only multiply the magnitudes and add the angles:

$$Z_1 \angle \theta_1 \times Z_2 \angle \theta_2 \qquad (13\text{–}8)$$
$$= Z_1 \times Z_2 \angle(\theta_1 + \theta_2)$$

For example,

$$2\,\underline{/30^\circ} \times 8\,\underline{/20^\circ}$$
$$= 2 \times 8\,\underline{/30^\circ + 20^\circ} = 16\underline{/50^\circ}$$

additional examples:

$$5\,\underline{/150^\circ} \times 3\,\underline{/-75^\circ} = 15\underline{/75^\circ}$$
$$15\underline{/-200^\circ} \times 6\underline{/180^\circ} = 90\underline{/-20^\circ}$$

The algebraic rules for signs apply to the addition of the angles. The polar magnitude is always positive. If in some solution it were to come out negative it means that the angle should be changed by 180°.

Division

Division of polar complex numbers is similar. Simply divide the numerator's magnitude by the denominator's magnitude and subtract the denominator's angle from the numerator's angle.

$$\frac{Z_1\underline{/\theta_1}}{Z_2\underline{/\theta_2}} = \left(\frac{Z_1}{Z_2}\right)\underline{/(\theta_1 - \theta_2)} \qquad (13\text{–}9)$$

For example,

$$\frac{14\,\underline{/60^\circ}}{7\,\underline{/20^\circ}} = \frac{14}{7}\underline{/60^\circ - 20^\circ} = 2\,\underline{/40^\circ}$$

Additional examples show how the algebraic convention of signs apply to the subtraction of the angles. The polar magnitude is always positive with the same provision as in multiplication.

$$\frac{15\,\underline{/-45^\circ}}{5\,\underline{/-45^\circ}} = \frac{15}{5}\,\underline{/-45^\circ - (-45^\circ)} = 3\,\underline{/0^\circ}$$

$$\frac{40\,\underline{/-120^\circ}}{20\,\underline{/60^\circ}} = \frac{40}{20}\,\underline{/-120^\circ - 60^\circ} = 2\,\underline{/-180^\circ}$$

$$\frac{16\underline{/60^\circ}}{4\underline{/-60^\circ}} = \frac{16}{4}\,\underline{/60^\circ - (-60^\circ)} = 4\,\underline{/120^\circ}$$

USE OF RECTANGULAR FORM VERSUS POLAR FORM

It should be noted that addition and subtraction of complex numbers in polar form is usually not convenient or possible.

Therefore, simply remember that addition and subtraction are performed most easily using the rectangular form of the complex number while multiplication and division are performed most easily using the polar form of the complex number.

CIRCUIT ANALYSIS USING J-OPERATOR NOTATION

Now, with these basic mathematical operations for complex numbers in mind, the impedances of several RLC circuits will be simplified.

Simplifying Impedance for A Series Circuit

Suppose the impedance of a series RLC circuit can be written as:

$$\mathbf{Z = 80 + j60 - j20}$$

Combining the j-terms gives:

$$\mathbf{Z = 80 + j40}$$

The resultant impedance of this circuit can be written simply as 80 + j40 in rectangular form. Notice that the j40 term indicates that the total reactance of the circuit is an *inductive* 40 ohms. The components of this series RLC circuit are shown on the phasor diagram shown in *Figure 13.6*. The various phasors are drawn with magnitudes to scale and positioned in phase.

The polar form of this impedance can be found using the conversion equation *13–13*.

$$Z\underline{/\theta} = \sqrt{R^2 + X^2}\,\underline{/\arctan\left(\frac{X}{R}\right)} \qquad (13\text{–}13)$$

Substituting equation *13–13* for the values X and R gives:

$$Z\underline{/\theta} = \sqrt{80^2 + 40^2}\,\underline{/\text{arctan}\left(\frac{40}{80}\right)}$$

$$= \sqrt{6400 + 1600}\,\underline{/\text{arctan}\ (0.5)}$$

$$= \sqrt{8000}\,\underline{/\text{arctan}\ (0.5)}$$

$$= 89.4\ \underline{/27^\circ}$$

Thus, the impedance of 80 + j40 in rectangular form equals 89.4 at 27 degrees in polar form. This polar form corresponds to a vector whose magnitude is the value of the impedance and whose angle is the phase angle of the circuit. This is shown in *Figure 13.7.* It is the same answer that would be obtained by drawing the vector to scale and using the Pythagorean theorem. Both methods yield the same result.

Since this circuit was simple enough to solve by right triangle phasor analysis, it should also be possible to use complex numbers to arrive at valid solutions of even more complex circuits.

Series Circuit with Two Complex Impedances

In the circuit of *Figure 13.1* and *13.2*, the total impedance is the series combination of two complex impedances. This was defined previously as,

$$Z_T = (10 + j8) + (4 - j20)$$

Combining real terms and imaginary terms, the total impedance of this circuit can be expressed as:

$$Z_T = (10 + 4) + j(8 - 20)$$
$$= 14 - j12$$

Thus, the total impedance of the circuit can be expressed in rectangular form as 14 − j12.

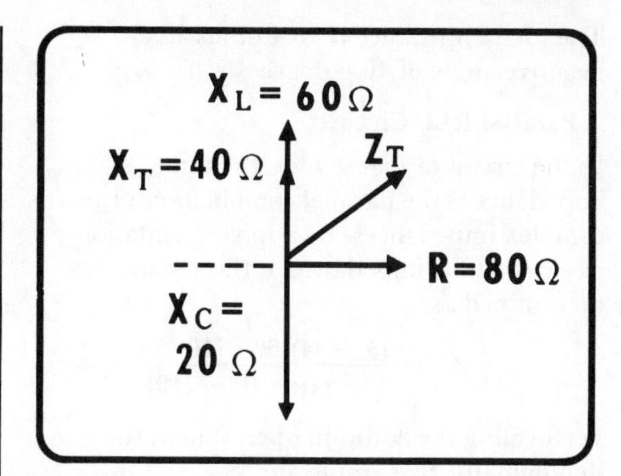

Figure 13.6 *Phasor Diagram for a Series RLC Circuit*

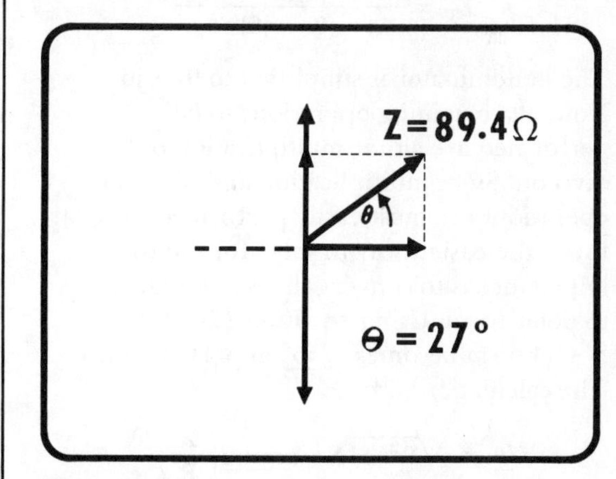

Figure 13.7 *Impedance and Phase Angle for the RLC Circuit*

This can be converted to polar form which indicates the magnitude of the impedance and the phase angle of the circuit:

$$Z\underline{/\theta} = \sqrt{R^2 + X^2}\,\underline{/\text{arctan}\left(\frac{X}{R}\right)}$$

$$= \sqrt{14^2 + 12^2}\,\underline{/\text{arctan}\left(\frac{-12}{14}\right)}$$

$$= \sqrt{196 + 144}\ \ \underline{/\text{arctan}\ -(0.86)}$$

$$= \sqrt{340}\,\underline{/\text{arctan}\ -0.86}$$

$$= 18.4\ \underline{/-40.6^\circ}$$

The total impedance is 18.4 ohms at a negative angle of 40.6 degrees.

A Parallel RLC Circuit

In the circuit of *Figure 13.4*, the total impedance is the parallel combination of two complex impedances. By applying equation *13–5*, the total impedance of this circuit was determined as,

$$Z_T = \frac{(3 + j4)(6 - j10)}{(3 + j4) + (6 - j10)}$$

Performing the addition operation in the denominator first results in

$$Z_T = \frac{(3 + j4)(6 - j10)}{(9 - j6)}$$

The denominator is simplified to 9 − j6. Now, all remaining operations to be performed are either multiplication or division. Since multiplication and division operations are most easily performed in polar form, the easiest way to solve for the total impedance is to convert all rectangular terms to polar form. Using equation *13–13*, the 3 + j4 term becomes $5\angle 53°$ in polar form. The calculations are:

$$Z\angle\theta = \sqrt{R^2 + X^2}\angle \arctan\left(\frac{X}{R}\right)$$

$$3 + j4 = \sqrt{3^2 + 4^2}\angle \arctan\left(\frac{4}{3}\right)$$

$$= \sqrt{9 + 16}\angle \arctan (1.33)$$

$$= \sqrt{25}\angle \arctan (1.33)$$

$$= 5\angle 53°$$

Similar calculations,

$$6 - j10 = \sqrt{6^2 + 10^2}\angle \arctan\left(\frac{-10}{6}\right)$$

$$= \sqrt{36 + 100}\angle \arctan (-1.67)$$

$$= \sqrt{136}\angle \arctan (-1.67)$$

$$= 11.7\angle -59°$$

result in a polar form of $11.7\angle -59°$ for the 6 − j10 term. The angle is negative because the j-term is negative.

The 9 − j6 term in polar form is $10.8\angle -34°$ arrived at as follows:

$$9 - j6 = \sqrt{9^2 + 6^2}\angle \arctan\left(\frac{-6}{9}\right)$$

$$= \sqrt{81 + 36}\angle \arctan (-0.67)$$

$$= \sqrt{117}\angle \arctan (-0.67)$$

$$= 10.8\angle -34°$$

The angle, again, is negative since the j-term is negative. Thus, 9 − j6 is equal to 10.8 $\angle -34°$ in polar form.

Substituting these equivalent polar-form impedances for their corresponding rectangular-form impedances in the original equation results in the following equation for Z_T.

$$Z_T = \frac{(5\angle 53°)(11.7\angle -59°)}{(10.8\angle -34°)}$$

Now the total impedance can be determined simply by the multiplication and division of polar-form terms. Performing the multiplication operation in the numerator first:

$$Z_T = \frac{58.5\angle -6°}{10.8\angle -34°}$$

Now, performing the division:

$$Z_T = \left(\frac{58.5}{10.8}\right)\angle(-6°) - (-34°) = 5.4\angle 28°$$

Therefore, the total impedance of the circuit is equal to 5.4 ohms at a phase angle of 28 degrees.

Solving for Impedance of RLC Circuit

Now these phasor algebra techniques will be applied to solve for the total impedance of the series-parallel ac circuit shown in *Figure 13.8*. First divide the circuit into several complex impedances as you have seen done previously to simplify its solution as illustrated in *Figure 13.9*. Let R_1 and L_1 be impedance Z_1, R_2 and C be impedance Z_2 and R_3 and L_2 be impedance Z_3.

Expressed in rectangular form these impedances are:

$$Z_1 = 2 + j6$$
$$Z_2 = 3 - j4$$
$$Z_3 = 1 + j2$$

Looking at the circuit simply in terms of these three impedances, the total impedance of the circuit can be written as Z_1 in series with the parallel combination of Z_2 and Z_3. That is,

$$Z_T = Z_1 + \frac{Z_2 \times Z_3}{Z_2 + Z_3} \qquad (13\text{–}14)$$

If the complex impedances are substituted Z_T is:

$$Z_T = (2 + j6) + \frac{(3 - j4)(1 + j2)}{(3 - j4) + (1 + j2)}$$

Rectangular form addition can be performed in the denominator. This makes Z_T:

$$Z_T = (2 + j6) + \frac{(3 - j4)(1 + j2)}{(4 - j2)}$$

Before the first and second term can be added in rectangular form, the second term must have its rectangular components converted to polar form so the multiplication and division can be performed.

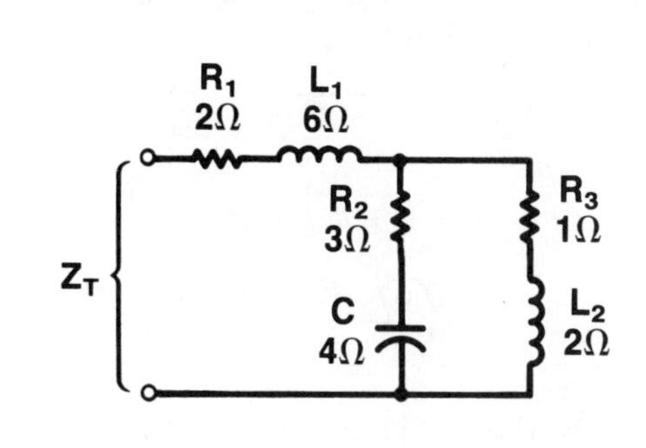

Figure 13.8 Series-Parallel RLC Circuit

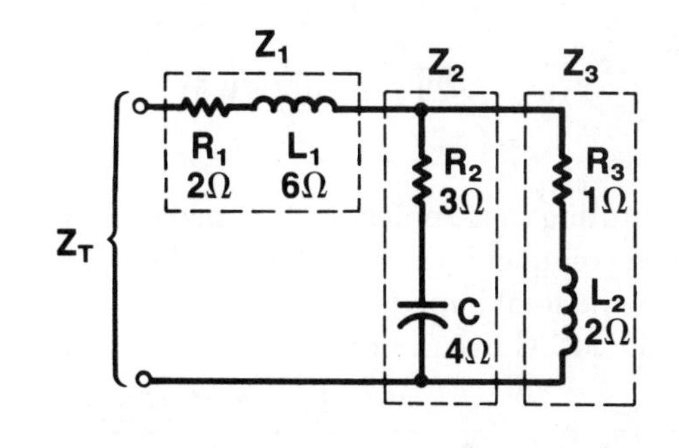

Figure 13.9 The Circuit of Figure 13.8 Divided Into Three Complex Impedances

The $3 - j4$ term is $5 \angle -53°$ in polar form. The calculations are as follows:

$$Z\angle\theta = \sqrt{3^2 + 4^2} \angle \arctan\left(\frac{-4}{3}\right)$$
$$= \sqrt{9 + 16} \angle \arctan -1.33$$
$$= \sqrt{25} \angle -\arctan 1.33)$$
$$= 5 \angle -53°$$

The angles are rounded to two significant figures. The angle is negative because the j term is negative.

Solving for Impedance of RLC Circuit

The 1 + j2 term is 2.24 $\underline{/63^\circ}$ in polar form calculated as follows:

$$Z\underline{/\theta} = \sqrt{1^2 + 2^2}\,\underline{/\arctan\left(\frac{2}{1}\right)}$$

$$= \sqrt{1 + 4}\,\underline{/\arctan\ (2)}$$

$$= \sqrt{5}\,\underline{/\arctan\ (2)}$$

$$= 2.24\,\underline{/63^\circ}$$

Similarly, the 4 − j2 term is 4.47$\underline{/27^\circ}$ in polar form.

$$Z\underline{/\theta} = \sqrt{4^2 + (-2)^2}\,\underline{/\arctan\left(\frac{-2}{4}\right)}$$

$$= \sqrt{16 + 4}\,\underline{/\arctan\ (-0.5)}$$

$$= \sqrt{20}\,\underline{/\arctan\ (-0.5)}$$

$$= 4.47\,\underline{/-27^\circ}$$

Substituting these polar-form impedances for the rectangular-form impedances which they equal, the total impedance equation for the complex RLC circuit now is:

$$Z_T = (2 + j6) + \frac{(5\underline{/-53^\circ})(2.24\,\underline{/63^\circ})}{(4.47\underline{/-27^\circ})}$$

Performing the multiplication in the numerator results in,

$$Z_T = (2 + j6) + \frac{(11.2\underline{/10^\circ})}{(4.47\underline{/-27^\circ})}$$

Performing the division results in the total impedance of

$$Z_T = (2 + j6) + 2.5\underline{/37^\circ}$$

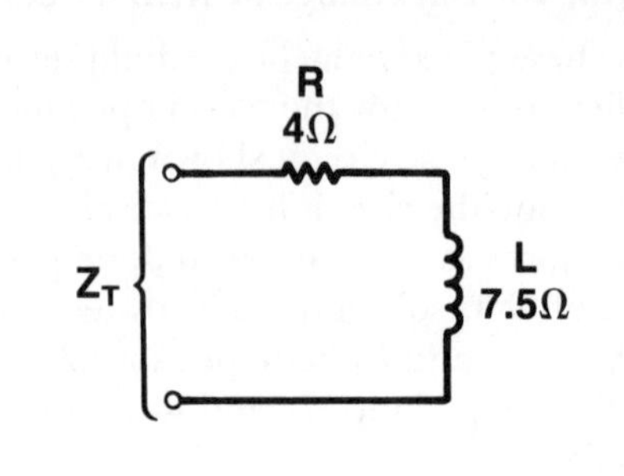

Figure 13.10 *Simple RL Circuit*

To perform the indicated addition, first the second term of the impedance in polar form should be converted to its equivalent rectangular form. Equation *13–15*, discussed in previous lessons, is used for this purpose.

$$R + jX = Z\cos\theta + jZ\sin\theta \qquad (13\text{–}15)$$

$$2.5\underline{/37^\circ} = (2.5\cos 37^\circ) + j(2.5\sin 37^\circ)$$

$$= 2.5\,(0.7986) + j2.5(0.6018)$$

$$= 2 + j1.5$$

Thus, the rectangular form of 2.5 $\underline{/37^\circ}$ equals 2 + j1.5.

Substituting this rectangular form into the total impedance equation, and adding Z_T becomes,

$$Z_T = (2 + j6) + (2 + j1.5)$$

$$Z_T = 4 + j7.5$$

This is the total impedance of this series-parallel circuit expressed in rectangular form. Note as shown in *Figure 13.10* that this impedance indicates that the circuit would appear to a power supply as a 4-ohm resistor in series with an inductor possessing an inductive reactance of 7.5 ohms at the applied frequency. Obviously, since reactances will change with changes in frequency, at a different frequency the circuit may appear to be either an RL or RC circuit.

The total impedance of the circuit can also be converted to its polar form to yield the magnitude of the impedance and the phase angle. It is $8.5\underline{/62°}$ shown as follows:

$$Z_T = 4 + j7.5$$

$$Z\underline{/\theta} = \sqrt{4^2 + 7.5^2}\underline{/\arctan\left(\frac{7.5}{4}\right)}$$

$$= \sqrt{16 + 56.25}\underline{/\arctan\ (1.875)}$$

$$= \sqrt{72.25}\underline{/\arctan\ (1.875)}$$

$$= 8.5\underline{/62°}$$

This form is usually the most-desirable form since it not only gives the magnitude of the circuit's impedance, but also its phase angle.

Solution of RLC Circuit

These phasor algebra techniques can now be used to solve for the impedance, total current and voltage drops in the series RLC circuit illustrated in *Figure 13.11*. The circuit contains 15 ohms resistance, 30 ohms capacitive reactance, 50 ohms of inductive reactance, and a second capacitive reactance of 40 ohms. The applied voltage, E_A, is 50 volts. If the reactances were not given, you would have to use the frequency of the applied voltage to calculate the capacitive and inductive reactances. Writing in j-operator notation, the reactance of C_1 is $-j30$ ohms, the reactance of the inductor is j50 ohms, and the reactance of C_2 is $-j40$ ohms.

Impedance

The total impedance of the circuit can be written:

$$Z_T = 15 - j30 + j50 - j40$$

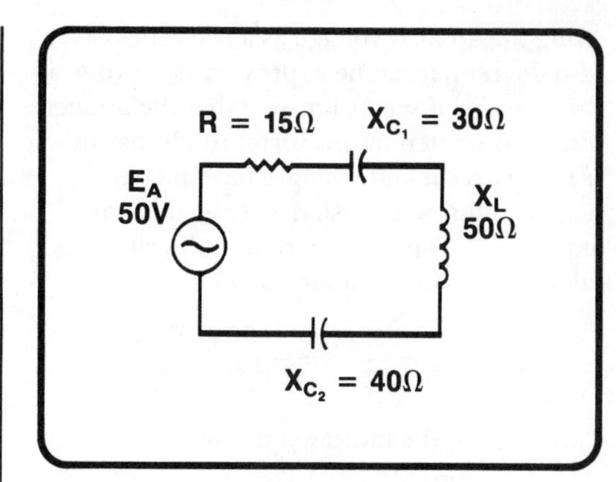

Figure 13.11 *Series RLC Circuit Example*

The ohms are understood since only impedance terms are being considered. By addition of j terms the total impedance of the circuit reduces to

$$Z_T = 15 + j20 - j40$$
$$= 15 - j20$$

Converting this to polar form results in the total impedance equal to $25\underline{/-53°}$. The calculations are as follows:

$$Z_T = 15 - j20$$

$$Z\underline{/\theta} = \sqrt{15^2 + 20^2}\underline{/\arctan\left(\frac{-20}{15}\right)}$$

$$= \sqrt{225 + 400}\underline{/\arctan\ (-1.33)}$$

$$= \sqrt{625}\underline{/\arctan\ (-1.33)}$$

$$= 25\underline{/-53°}$$

Since the j-term is negative, the angle, θ, is negative.

Total Current

The total current of any ac circuit is equal to the applied voltage divided by the total impedance of the circuit:

$$I_T = \frac{E_A}{Z_T} \qquad (13\text{–}16)$$

Using the applied voltage as a reference at zero degrees, it can be expressed as 50 volts at 0 degrees. If it seems unusual that the applied voltage is written in this form, recall that in reality, current and voltages have phase relationships, not resistance, reactance, or impedance. The total current in the circuit is calculated using equation *13–16*.

$$I_T = \frac{E_A}{Z_T} = \frac{50V\angle 0°}{25\Omega\angle -53°}$$

Carrying out the indicated division,

$$I_T = \frac{50V}{25\Omega}\angle(0°) - (-53°) = 2A\angle 53°$$

Thus, the total current equals $2A\angle 53°$.

Voltage Drops

Recall that according to Ohm's law the voltage drop across any resistor is equal to the current through it times the value of the resistor:

$$E_R = IR \qquad (13\text{–}17)$$

Also recall that the voltage drop across any reactance is equal to the current through it times the value of the reactance:

$$E_X = I_X X \qquad (13\text{–}18)$$

Since the circuit is a series circuit, the current through all components is the same as the total current, I_T. Thus, to determine the voltage drop across any component in the circuit, simply multiply the total current, $2A\angle 53°$, times the resistance, R, or reactance, X, of the component in the circuit.

In order to perform multiplication of two complex numbers, both must be expressed in the same form, either rectangular or polar. The current is expressed in polar form, therefore, to multiply conveniently by the polar form of the current, you should convert all resistances and reactances in the circuit to polar form.

Pure Resistance Polar Form

This is easily accomplished in terms of individual resistances or reactances. For example, R_1, the resistance term of 15 ohms in the circuit of *Figure 13.11*, can be written in polar form as simply $15\Omega\angle 0°$ which means the impedance is all resistive with no reactance. This might be more clearly understood when written in rectangular form. In the form $R + jX$, $R = 15$ ohms and $X = 0$ because it is all resistive. The polar form can be obtained by use of the conversion equation:

$$Z\angle\theta = \sqrt{R^2 + X^2}\angle\arctan\left(\frac{X}{R}\right)$$

$$Z\angle\theta = \sqrt{15^2 + 0^2}\angle\arctan\left(\frac{0}{15}\right)$$

$$= \sqrt{225 + 0}\angle\arctan(0)$$

$$= \sqrt{225}\angle\arctan(0)$$

$$= 15\angle 0°$$

Pure Reactance Polar Form

The polar form of the reactance of C_1, $-j30$, is $30\Omega\angle -90°$. Its rectangular form is really $0 - j30$ which is a vector of a magnitude of 30 plotted on the $-j$ axis. The polar form $30\angle -90°$ can also be verified using the conversion equation. Similarly, the reactance of the inductor, $j50$, can be written in polar form as $50\Omega\angle 90°$, and the reactance of C_2, $-j40$, can be written in polar form as $40\Omega\angle -90°$. In summary, the polar forms are:

$$X_{C1} = 30\Omega\angle -90°$$
$$X_L = 50\Omega\angle +90°$$
$$X_{C2} = 40\Omega\angle -90°$$

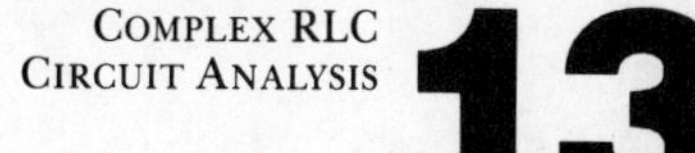

■ **Phasor Diagram**

Notice in *Figure 13.12* that these angles of zero degrees, 90 degrees and −90 degrees denote the phase relationships of the resistances and reactances just like the j-operator did for the rectangular forms.

Now that the resistances and reactances of the circuit are written in polar form, they can be multiplied easily by the polar-form current to obtain the voltage drops in the circuit. The voltage across the resistor, E_R, equals the current times the resistance:

$$\begin{aligned} E_R &= I_T R \\ &= (2A\angle 53^\circ)(15\Omega\angle 0^\circ) \\ &= (2A \times 15\Omega)\angle 53^\circ + (0^\circ) \\ &= 30V\angle 53^\circ \end{aligned}$$

The voltage across C_1, E_{C1}, equals the current times the capacitive reactance of C_1:

$$\begin{aligned} E_{C1} &= I_T X_{C1} \\ &= (2A\angle 53^\circ)(30\Omega\angle -90^\circ) \\ &= 2A \times 30\Omega\ \angle 53^\circ + (-90^\circ) \\ &= 60V\angle -37^\circ \end{aligned}$$

The voltage across the inductor, E_L, equals the current times the inductive reactance:

$$\begin{aligned} E_L &= I_T X_L \\ &= (2A\angle 53^\circ)(50\Omega\angle 90^\circ) \\ &= (2A \times 50\Omega)\angle 53^\circ + 90^\circ \\ &= 100V\angle 143^\circ \end{aligned}$$

The voltage across C2, E_{C2}, equals the current times the capacitive reactance of C2:

$$\begin{aligned} E_{C2} &= I_T X_{C2} \\ &= (2A\angle 53^\circ)(40\Omega\angle -90^\circ) \\ &= (2A \times 40\Omega)\angle 53^\circ + (-90^\circ) \\ &= 80V\angle -37^\circ \end{aligned}$$

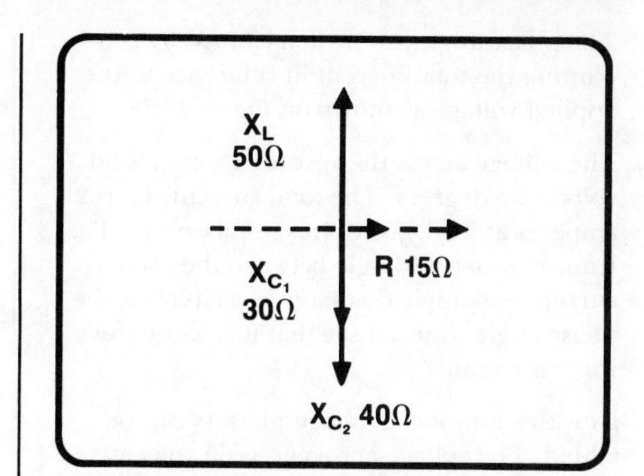

Figure 13.12 Phase Relationship of the Resistances and Reactances of Figure 13.11

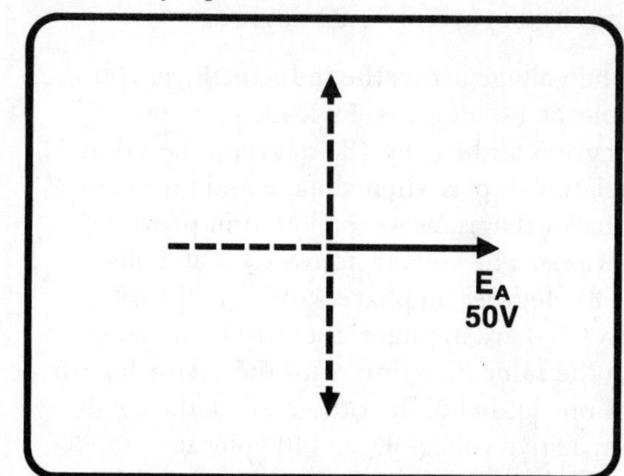

Figure 13.13 E_A as a Reference at 0°

Phasor Diagram

Since the applied voltage was used as a reference at 0°, and the total current and voltage drops in the circuit were derived using the applied voltage as reference, all of these angles expressed in polar form in these calculations are measured with reference to the applied voltage as shown in *Figure 13.13*.

The phase angle can be determined by plotting the total current in reference to the applied voltage as shown on *Figure 13.14*.

The voltage across the resistor E_R equals 30 volts at 53 degrees. The total current, I_T, is 2 amperes at 53 degrees. It is in phase with E_R. Knowing that the angle between the total current and applied voltage in a circuit is the phase angle, you can see that it is 53 degrees for this circuit.

Now the additional voltage phasors can be added. The voltage across C_1 is 60 volts at -37 degrees as shown in *Figure 13.15*. Note that this lags E_R by 90 degrees because it lags the circuit current by 90 degrees.

The voltage across the inductor E_L is 100 volts at 143 degrees. E_L leads I_T by 90 degrees and E_{C1} by 180 degrees, the same relationship as when voltage and impedance phasor diagrams were plotted in previous lessons. The voltage across C_2 is 80 volts at -37 degrees, in phase with E_{C1}. Since E_{C1} and E_{C2} are in phase, they are both shown in the same direction, with the vector length of one added to the other. Thus, the total capacitive voltage E_C is 140 volts at -37 degrees.

Using this method provides the phase of every voltage or current in the circuit with respect to the reference quantity used, in this example, the applied voltage E_A.

With the same procedures the values of voltage, current and impedance can be solved for even the most complex RLC circuit.

SUMMARY

In this lesson you were introduced to applications for using complex numbers to solve for the total impedance of an ac circuit. You were told how to perform the mathematical operations of addition, subtraction, multiplication and division

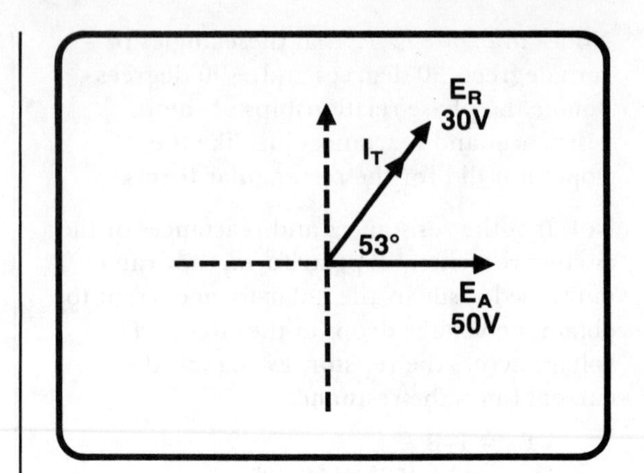

Figure 13.14 *Voltage Phasor Diagram with E_A at 0°*

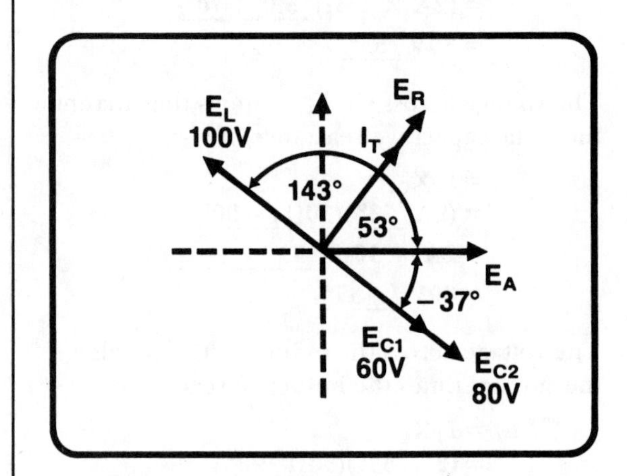

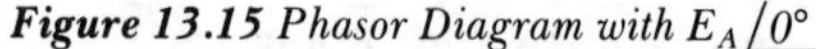

Figure 13.15 *Phasor Diagram with $E_A \angle 0°$*

involving complex numbers in rectangular or polar form.

You used these basic operations to simplify complex impedances of ac circuits, and then applied phasor algebra techniques to solve for voltage current and impedance values in a series RLC circuit. The results of the calculations were used to draw the voltage phasor diagram. Analyzing and understanding the phasor diagram should make you realize that the circuit solutions were similar to results that were obtained with much simpler circuits and that, using the demonstrated techniques, any complex RLC circuit can be analyzed.

■ Worked-Out Examples

1. Perform the following operations involving j-operator terms:

a. $j2 + j8 = \mathbf{j10}$

b. $j6 - j3 = \mathbf{j3}$

c. $j8 \times j5 = \mathbf{j^2 40}$

d. $j10 \div j2 = \mathbf{5}$

e. $(j2) + (-j3) = j2 - j3 = -j1 = \mathbf{-j}$

f. $(j4) - (-j3) = j4 + j3 = \mathbf{j7}$

g. $j^2 6 \div j2 = \dfrac{j^2 6}{j2} = \left(\dfrac{j^2}{j}\right)\left(\dfrac{6}{2}\right) = \mathbf{j3}$

h. $j26 \times j^2 4 = j^3(26 \times 4) = \mathbf{j^3 104}$

2. Perform the following operations involving rectangular form complex numbers; show your work.

a. $(6 + j2) + (3 - j4) = (6 + 3) + j(2 + (-4))$
$= 9 + j(2 - 4) = 9 + j(-2) = \mathbf{9 - j2}$

b. $(4 + j3)(2 - j3) = (4 \times 2) + (4 \times -j3) + (2 \times j3) + (j3 \times -j3)$
$= (8) + (-j12) + (j6) + (-j^2 9)$
$= 8 - j12 + j6 - j^2 9 = 8 - j6 - j^2 9$
since $j^2 = -1$ the expression becomes:
$= 8 - j6 - (-1)9 = 8 - j6 + 9 = \mathbf{17 - j6}$

c. $(6 + j4) - (10 - j8) = (6 - 10) + j(4 - (-8)) = (-4) + j(4 + 8) = \mathbf{-4 + j12}$

d. $\dfrac{(8 + j3)}{(9 + j4)} = \dfrac{(8 + j3)}{(9 + j4)} \times \dfrac{(9 - j4)}{(9 - j4)} = \dfrac{72 - j32 + j27 - j^2 12}{81 - j36 + j36 - j^2 16}$

$= \dfrac{72 - j5 - j^2 12}{81 - j^2 16}$

since $j^2 = -1$

$= \dfrac{72 - j5 - (-1)12}{81 - (-1)16} = \dfrac{72 - j5 + 12}{81 + 16}$

$= \dfrac{84 - j5}{97} = \left(\dfrac{84}{97}\right) - j\left(\dfrac{5}{97}\right) = \mathbf{(0.87) - j(0.05)}$

■ Worked-Out Examples

3. Perform the indicated operations shown below involving polar form complex numbers.

a. $10\angle 16° \times 8\angle -40° = (10 \times 8)\angle (16°) + (-40°) = \mathbf{80\angle -24°}$

b. $\dfrac{20\angle 15°}{5\angle 10°} = \left(\dfrac{20}{5}\right)\angle 15° - 10° = \mathbf{4\angle 5°}$

c. $3\angle 15° \times 6\angle 20° = (3 \times 6)\angle 15° + 20° = \mathbf{18\angle 35°}$

d. $\dfrac{16\angle 40°}{8\angle -10°} = \left(\dfrac{16}{8}\right)\angle 40° - (-10°) = 2\angle 40° + 10° = \mathbf{2\angle 50°}$

4. Given the following RLC circuits, express the impedance of the circuit in both polar and rectangular form.

a.

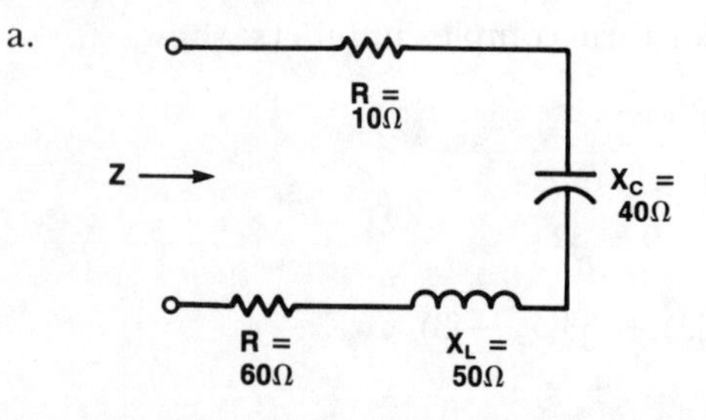

$$Z = 10\Omega - j40\Omega + j50\Omega + 60\Omega = 10 - j40 + j50 + 60$$

$$= 70 - j40 + j50 = \mathbf{70 + j10 (Rectangular)}$$

$$Z = 70 + j10 = \sqrt{70^2 + 10^2}\angle \arctan\left(\frac{10}{70}\right)$$

$$= \sqrt{5000}\angle \arctan(0.1429) = \mathbf{70.7\angle 8.1° \ (Polar)}$$

b.

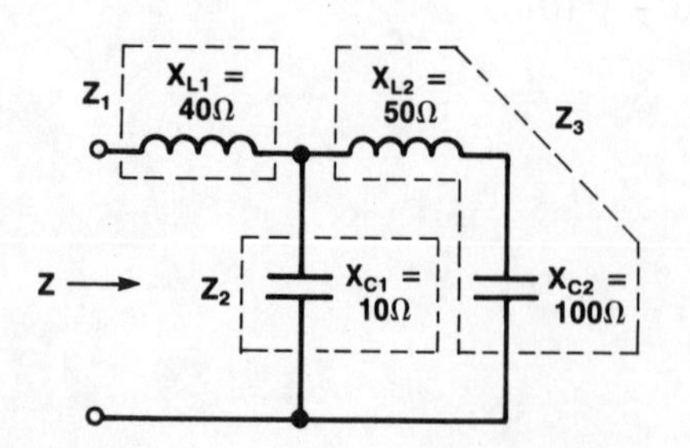

Dividing sections of circuit b into Z_1, Z_2 and Z_3

$Z_1 = 0+j40$

$Z_2 = 0-j10$

$Z_3 = (0 + j50) + (0 - j100) = j50 - j100 = -j50$

Now Z_2 and Z_3 are in parallel and their combination is in series with Z_1, Thus

$$Z_T = Z_1 + Z_2 \| Z_3 = Z_1 + \frac{Z_2 \times Z_3}{Z_2 + Z_3} = j40 + \frac{(-j10)(-j50)}{(-j10) + (-j50)} = j40 + \frac{-j^2 500}{-j60}$$

Since minus divided by a minus yields a positive result, j^2 divided by $j = j$ and 500 divided by 60 equals 8.33, the expression simplifies to

$Z_T = j40 + j8.33 =$ **(0+j48.33) (Rectangular)** or

$Z_T =$ **$48.33 \angle 90°$ (Polar)**

5. Express the resistance or reactance of the following passive elements in rectangular (j-operator) form and then in polar form.

a.

10Ω
R

Rectangular	**(10 + j0)Ω**
Polar	**$10\Omega \angle 0°$**

b.

$X_L = 50\Omega$
L

Rectangular	**(0 + j50)Ω**
Polar	**$50\Omega \angle +90°$**

c.

$X_C = 40\Omega$
C

Rectangular	**(0 − j40)Ω**
Polar	**$40\Omega \angle -90°$**

■ Worked-Out Examples

6. Determine the values indicated below for the RLC circuit shown. Express all answers in polar form. Use $E_A = 36V\angle 0°$ for reference.

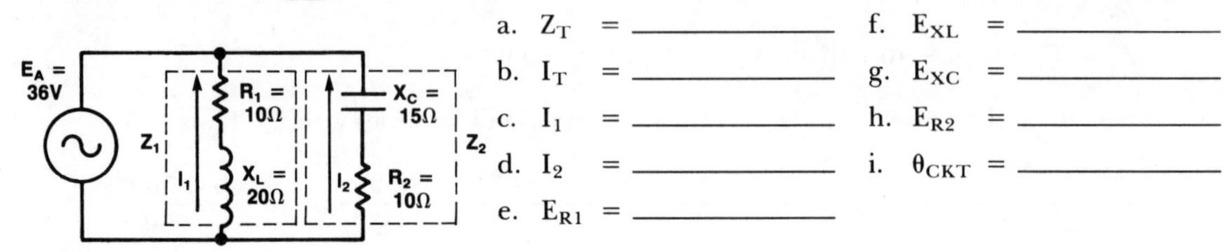

a. Z_T = ____________
b. I_T = ____________
c. I_1 = ____________
d. I_2 = ____________
e. E_{R1} = ____________
f. E_{XL} = ____________
g. E_{XC} = ____________
h. E_{R2} = ____________
i. θ_{CKT} = ____________

Solution:

a. $$Z_1 = R_1 + jX_L = \overset{\text{Rect.}}{10 + j20} = \sqrt{10^2 + 20^2}\angle \arctan\left(\frac{20}{10}\right) = \overset{\text{Polar}}{22.4\angle +63.4°}$$

$$Z_2 = R_2 - jX_C = 10 - j15 = \sqrt{10^2 + (-15)^2}\angle \arctan\left(\frac{-15}{10}\right) = 18\angle -56.3°$$

$$Z_T = Z_1 \parallel Z_2 = \frac{Z_1 \times Z_2}{Z_1 + Z_2} = \frac{(10 + j20)(10 - j15)}{(10 + j20) + (10 - j15)}$$

$$= \frac{(10 + j20)(10 - j15)}{(20 + j5)^*} = \frac{(22.4\angle +63.4°)(18\angle -56.3°)}{(20.6\angle 14°)}$$

$$^*\textit{Note: } 20 + j5 = \sqrt{20^2 + 5^2}\angle \arctan\left(\frac{5}{20}\right) = 20.6\angle 14°)$$

$$Z_T = \frac{403.2\angle 7.1°}{20.6\angle 14°} = \mathbf{19.6\Omega\angle -6.9°}$$

b. $$I_T = \frac{E_A}{Z_T} = \frac{36V\angle 0°}{19.6\Omega\angle -6.9°} = \mathbf{1.84A\angle +6.9°}$$

c. $$I_1 = \frac{E_A}{Z_1} = \frac{36V\angle 0°}{22.4\Omega\angle 63.4°} = \mathbf{1.6A\angle -63.4°}$$

d. $$I_2 = \frac{E_A}{Z_2} = \frac{36V\angle 0°}{18\Omega\angle -56.3°} = \mathbf{2A\angle 56.3°}$$

■ **Worked-Out Examples**

e. $E_{R1} = I_1R_1 = (1.6A\angle -63.4°)(10\Omega\angle 0°) = \mathbf{16V\angle -63.4°}$

f. $E_{XL} = I_1X_L = (1.6A\angle -63.4°)(20\Omega\angle +90°) = \mathbf{32V\angle 26.6°}$

g. $E_{XC} = I_2X_C = (2A\angle 56.3°)(15\Omega\angle -90°) = \mathbf{30V\angle -33.7}$

h. $E_{R2} = I_2R_2 = (2A\angle 56.3°)(10\Omega\angle 0°) = \mathbf{20V\angle 56.3°}$

i. $^*\theta_{CKT} = \mathbf{6.9°}$ (Polar Coordinate Angle of Current)

*Recall that the phase angle of a circuit is the angle of phase difference between the total current and the applied voltage. Since E_A is referenced at 0° and the current, I_T, is at 6.9°, the angular difference is the phase angle: 6.9°.

Practice Problems

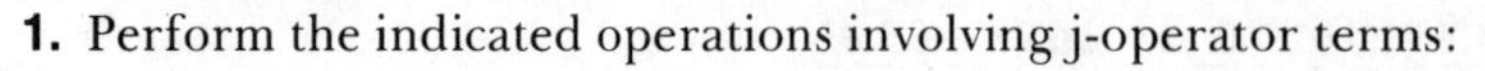

1. Perform the indicated operations involving j-operator terms:

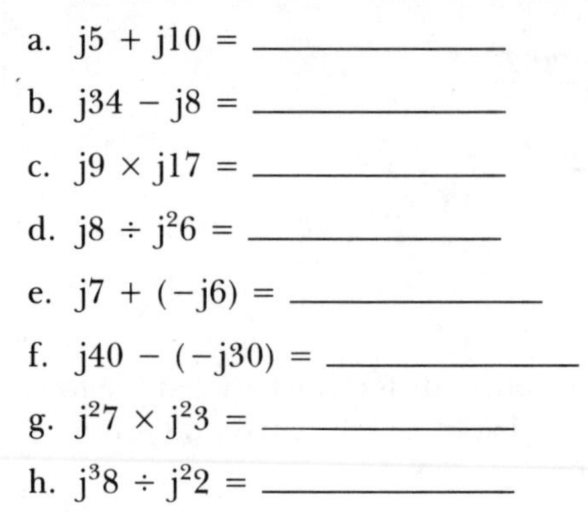

   a. $j5 + j10 =$ ____________
   b. $j34 - j8 =$ ____________
   c. $j9 \times j17 =$ ____________
   d. $j8 \div j^2 6 =$ ____________
   e. $j7 + (-j6) =$ ____________
   f. $j40 - (-j30) =$ ____________
   g. $j^2 7 \times j^2 3 =$ ____________
   h. $j^3 8 \div j^2 2 =$ ____________

2. Perform the indicated operations involving rectangular form complex numbers.
   a. $(6 + j4)(3 + j2) =$ ____________
   b. $(6 - j1)(4 - j3) =$ ____________
   c. $(10 - j2) + (8 + j3) =$ ____________
   d. $(4 + j6) - (3 + j2) =$ ____________
   e. $\dfrac{9 + j2}{8 - j3} =$ ____________
   f. $(16 + j2) - (3 - j6) =$ ____________

3. Perform the indicated operations shown below involving polar form complex numbers.
   a. $4\angle 10^\circ \times 6\angle 5^\circ =$ ____________
   b. $\dfrac{10\angle 16^\circ}{2.5\angle -4^\circ} =$ ____________
   c. $6\angle 10^\circ \times 5\angle -8^\circ =$ ____________
   d. $\dfrac{40\angle 60^\circ}{8\angle 20^\circ} =$ ____________

■ **Practice Problems**

4. Given the following RLC circuit, express the impedance of the circuit in both polar and rectangular form.

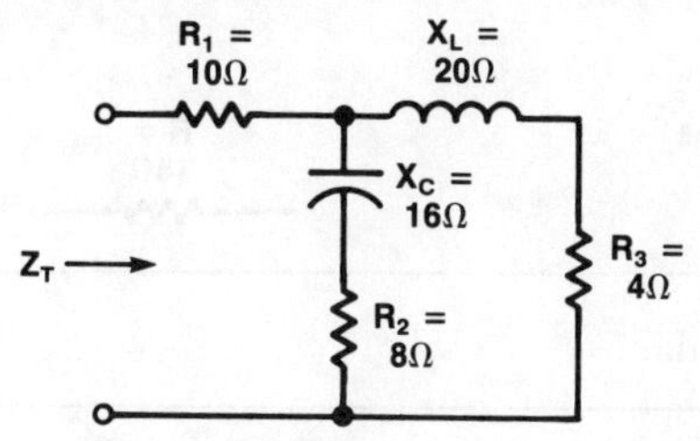

5. Given the following RLC circuit, express the impedance of the circuit in both polar and rectangular form.

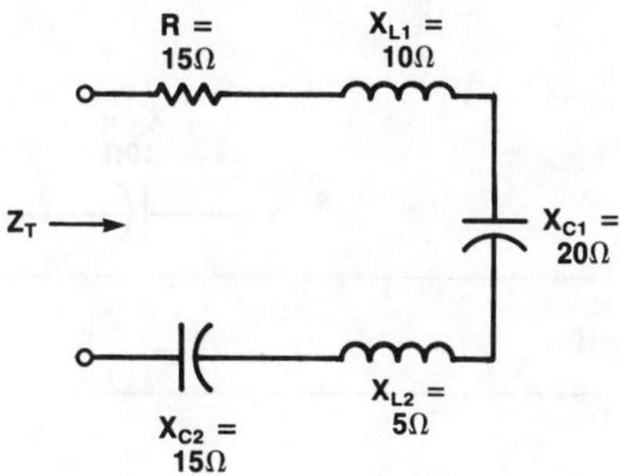

6. Given the following RLC circuit, express the impedance of the circuit in both polar and rectangular form.

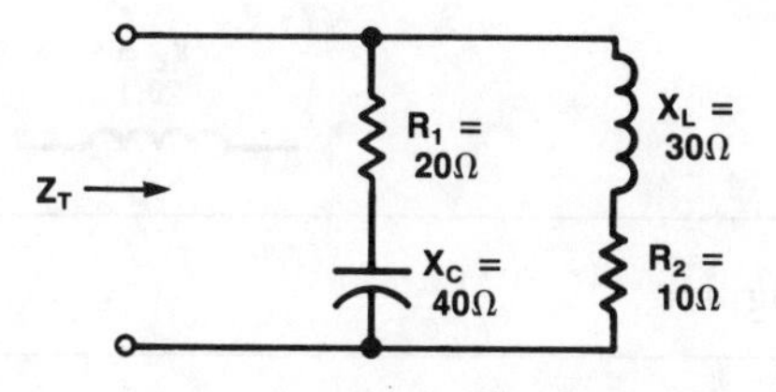

7. Given the following RLC circuit, express the impedance of the circuit in both polar and rectangular form.

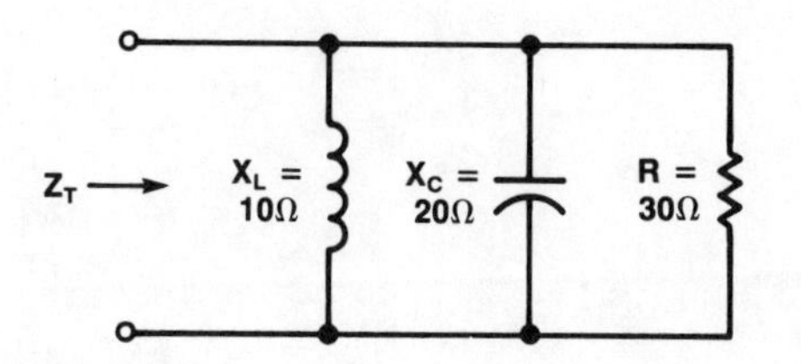

■ Practice Problems

8. Express the resistance or reactance of the following passive elements in rectangular (j-operator) form and then in polar form.

a.

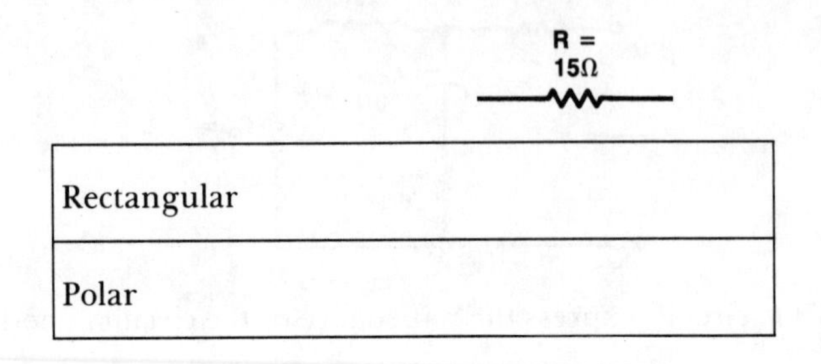

Rectangular	
Polar	

b.

X_C =
10Ω

Rectangular	
Polar	

c.

X_L =
20Ω

Rectangular	
Polar	

■ Practice Problems

9. Determine the following values for the RLC circuit shown. Express all answers in polar form. Use E_A as reference at 0°.

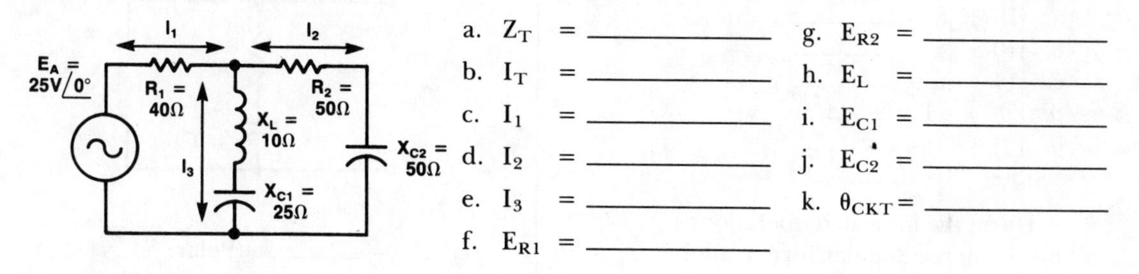

a. Z_T = ______
b. I_T = ______
c. I_1 = ______
d. I_2 = ______
e. I_3 = ______
f. E_{R1} = ______
g. E_{R2} = ______
h. E_L = ______
i. E_{C1} = ______
j. E_{C2} = ______
k. θ_{CKT} = ______

10. Determine the following values for the RC circuit shown. Express all answers in polar form. Use E_A reference at 0°.

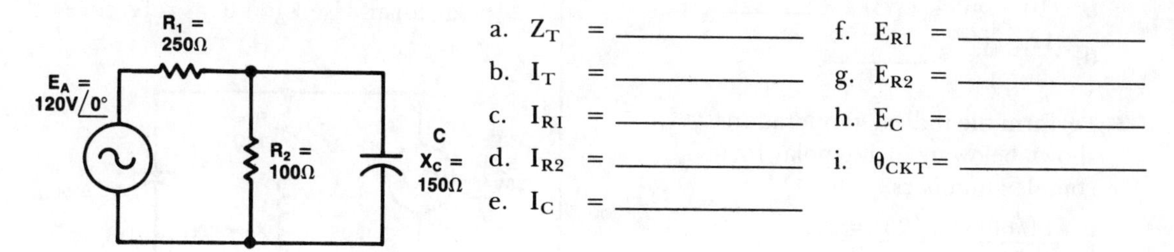

a. Z_T = ______
b. I_T = ______
c. I_{R1} = ______
d. I_{R2} = ______
e. I_C = ______
f. E_{R1} = ______
g. E_{R2} = ______
h. E_C = ______
i. θ_{CKT} = ______

■ Quiz

1. Perform the indicated operations involving j-operator terms; simplify.
   a. $j10 + j8 =$ ______
   b. $j10 - j8 =$ ______
   c. $j14 \times j6 =$ ______
   d. $j28 \div j^2 4 =$ ______
   e. $j^2 6 \times j8 =$ ______
   f. $3 \div j2 =$ ______

2. Perform the indicated operations involving rectangular form complex numbers; simplify.
   a. $(5 + j6) + (2 - j4) =$ ______
   b. $(8 - j2) - (10 - j6) =$ ______
   c. $(10 + j6)(4 - j3) =$ ______
   d. $\dfrac{12 - j4}{8 - j3} =$ ______

3. Perform the indicated operations shown below involving polar form complex numbers.
   a. $14\angle 60° \times 8\angle 25° =$ ______
   b. $\dfrac{36\angle 15°}{12\angle 18°} =$ ______
   c. $5\angle 12° \times 6\angle -8° =$ ______
   d. $\dfrac{15\angle -40°}{3\angle -10°} =$ ______

4. Given the following RLC circuits, express the impedance of the circuit in both polar and rectangular form.

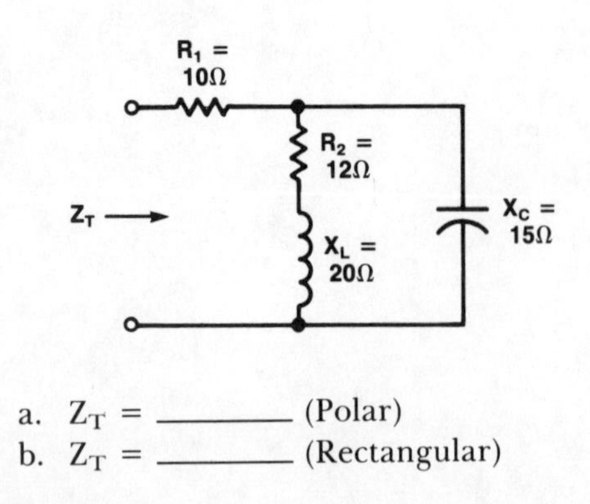

   a. $Z_T =$ ______ (Polar)
   b. $Z_T =$ ______ (Rectangular)

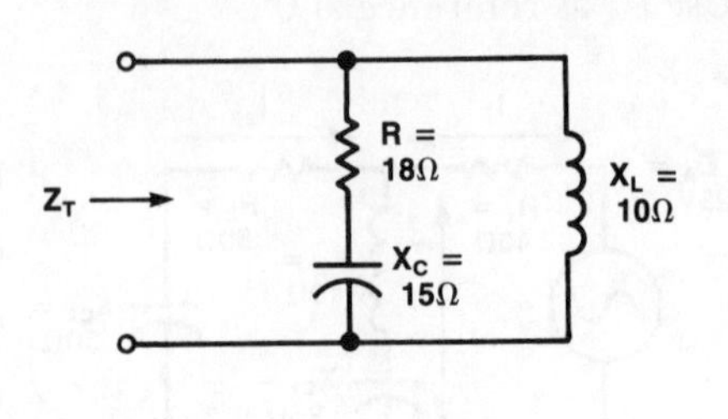

   c. $Z_T =$ ______ (Polar)
   d. $Z_T =$ ______ (Rectangular)

5. Determine the values indicated for the following circuits. Express all answers in polar form. Use E_A at 0° as reference.

I.

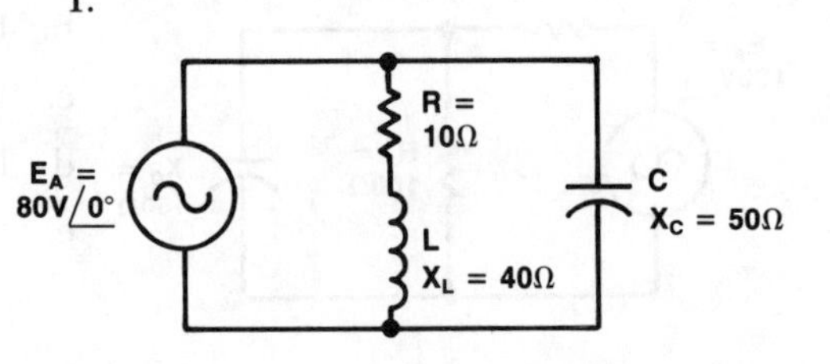

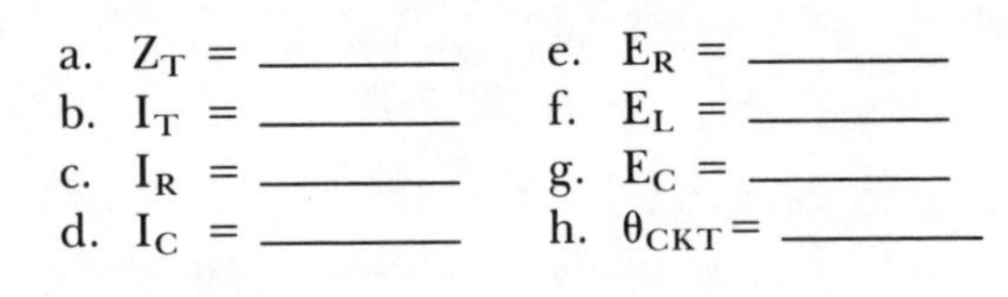

   a. $Z_T =$ ______
   b. $I_T =$ ______
   c. $I_R =$ ______
   d. $I_C =$ ______
   e. $E_R =$ ______
   f. $E_L =$ ______
   g. $E_C =$ ______
   h. $\theta_{CKT} =$ ______

II.

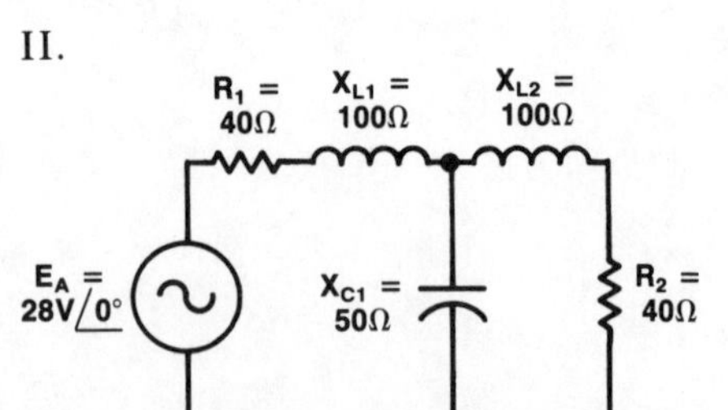

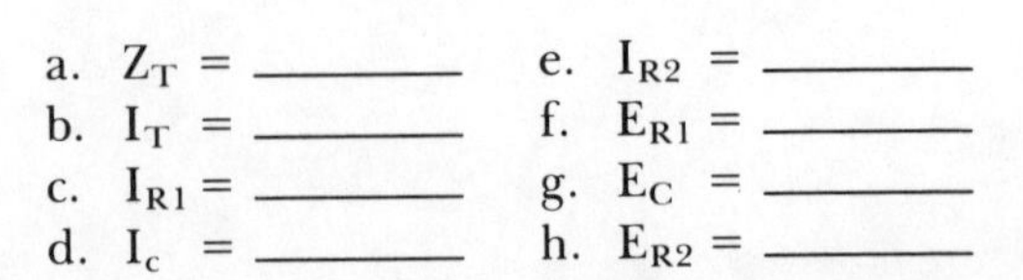

   a. $Z_T =$ ______
   b. $I_T =$ ______
   c. $I_{R1} =$ ______
   d. $I_c =$ ______
   e. $I_{R2} =$ ______
   f. $E_{R1} =$ ______
   g. $E_C =$ ______
   h. $E_{R2} =$ ______

LESSON 14

Resonance

In this lesson the property of resonance will be introduced. Its effects on the impedance, currents, and voltages of a circuit will be examined. The response of RLC circuits to frequencies near resonance and the concept of the frequency response of a circuit will be investigated. The Q, or quality factor of a circuit, and its interaction with frequency response will be discussed.

Objectives

At the end of this lesson you should be able to:

1. State the basic definition of resonance.

2. Determine the resonant frequency of a circuit with a capacitor and inductor in parallel or series with one another.

3. Determine the value of the inductance needed to achieve resonance when given the value of a typical capacitor and the desired resonant frequency.

4. Determine the value of a capacitor needed to achieve resonance when given the value of a typical inductance and the desired resonant frequency.

5. Determine the resonant frequency, the impedance at resonance, the total circuit current at resonance, the Q of the circuit, the expected phase angle, bandwidth, and the voltage across the inductor or capacitor at resonance when given the schematic of a series RLC circuit.

6. Determine the effect on a circuit's frequency response when the resistance, inductance or capacitance is changed.

7. Determine the resonant frequency, the impedance at resonance, the total circuit current at resonance, the Q of the circuit, the expected phase angle, current through the capacitor or inductor at resonance, and the bandwidth when given the schematic drawing of a parallel RLC circuit.

8. Determine the effect on the frequency response of changing the circuit resistance, inductance or the capacitance of a parallel resonant circuit.

9. Determine the upper and lower cutoff frequency at the half-power points of the frequency response when given the bandwidth and resonant frequency of a resonant circuit.

■ **Inductive Versus Capacitive Reactance**

INTRODUCTION

In Lessons 11 and 13 RLC circuits containing series and parallel combinations of resistors, inductors, and capacitors were analyzed. The circuits had a particular value of inductive reactance and capacitive reactance which depended upon the frequency of the applied voltage.

In this lesson, the concept of *resonance* will be introduced. A circuit is said to be resonant at the frequency where its *inductive reactance* and *capacitive reactance* are the same value. This frequency is called the *resonant frequency* of the circuit. Resonance plays a very important role in the operation of many circuits used to transmit and receive radio and television signals. In fact, radio and television could not operate without resonant circuits.

CONCEPT OF RESONANCE

Inductive Versus Capacitive Reactance

First, let's review some basic considerations regarding inductive and capacitive reactance. Recall that both inductive and capacitive reactance are not only dependent upon the value of the inductor or capacitor, but they are also dependent upon the applied frequency. This is evident in the equations shown, *14–1* for inductive reactance and *14–2* for capacitive reactance.

$$X_L = 2\pi fL \quad (14\text{–}1)$$

and

$$X_C = \frac{1}{2\pi fC} \quad (14\text{–}2)$$

As was shown previously for an inductor, as the applied frequency increases, the reactance of the inductor (equation *14–1*) increases if the value of the inductance remains constant.

$$X_L\uparrow = 2\pi f\uparrow L$$

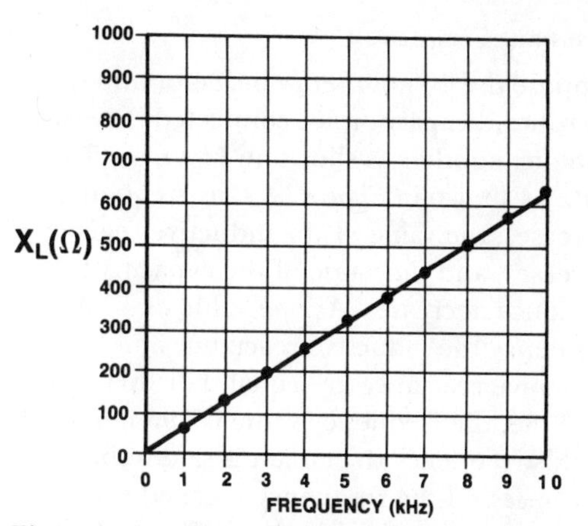

Figure 14.1 *X_L Versus f for a 10 Millihenry Inductor*

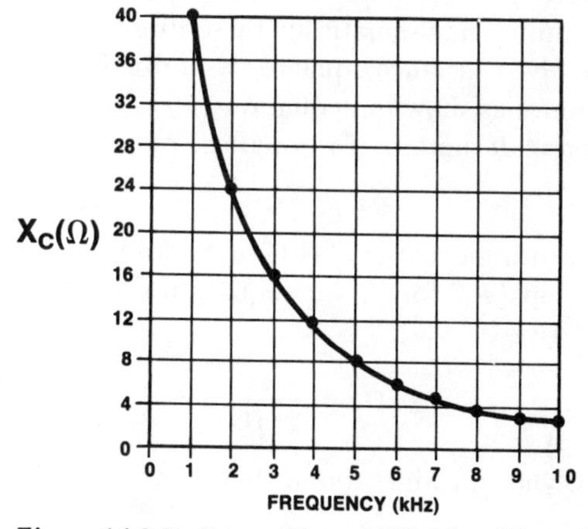

Figure 14.2 *X_C Versus f for a 4 Microfarad Capacitor*

This is shown again in *Figure14.1*, a graph for a 10 millihenry inductor. For a capacitor, as the applied frequency increases, the reactance of the capacitor decreases if the value of the capacitor remains constant.

$$X_C\downarrow = \frac{1}{2\pi f\uparrow C}$$

Figure 14.2 shows how capacitive reactance of a 4 microfarad capacitor decreases as the applied frequency increases.

■ Resonant Frequency
■ Equivalent Resonant Frequency Equations

Resonant Frequency

Suppose the 10 millihenry inductor and 4 microfarad capacitor are connected in series with one another as shown in *Figure 14.3*. Now, as shown in *Figure 14.4*, as frequency increases, the value of the inductive reactance increases, and the value of the capacitive reactance decreases. At one value of frequency the inductive reactance and capacitive reactance are equal. For this example, the frequency value is 796 hertz. At this frequency the circuit is said to be *at resonance* and the frequency is called the *resonant frequency* of this circuit. The value of the resonant frequency of this circuit can be determined mathematically by setting the capacitive reactance equal to the inductive reactance as shown in equation *14–3*. This is the basic definition of resonance.

$$X_L = X_C \qquad (14\text{–}3)$$

Substituting equation *14–1* for X_L and equation *14–2* for X_C into equation *14–3* gives,

$$2\pi fL = \frac{1}{2\pi fC}$$

By solving for f it is found that

$$f = \frac{1}{2\pi\sqrt{LC}}$$

where L is the value of the inductor in henrys, C is the value of the capacitor in farads, 2π is the constant 6.28, and f is the resonant frequency in hertz. Since this is the resonant frequency, a subscript "r" is used to distinguish it from all other frequencies:

$$f_r = \frac{1}{2\pi\sqrt{LC}} \qquad (14\text{–}4)$$

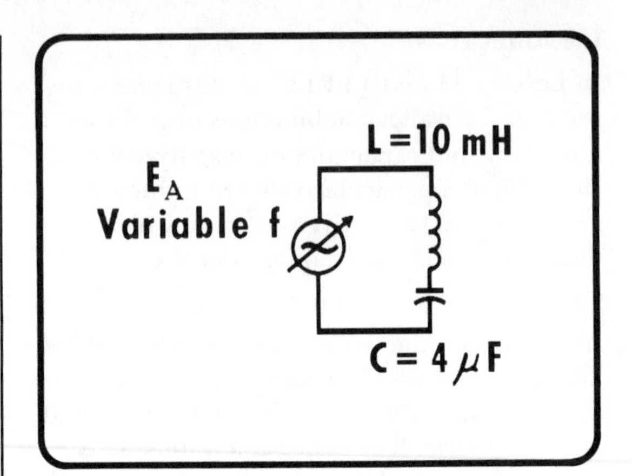

Figure 14.3 *Series LC Circuit*

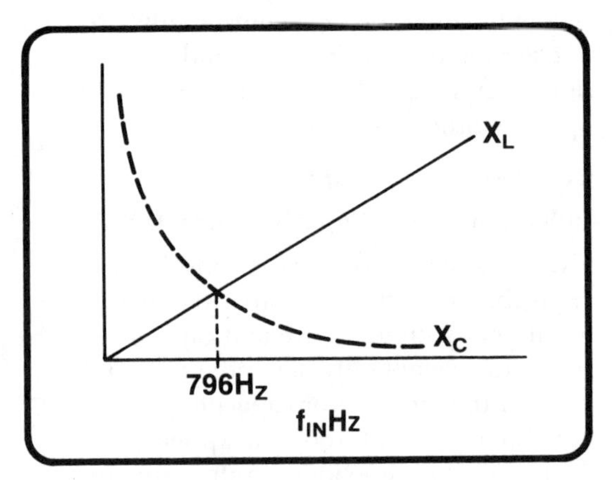

Figure 14.4 *X_L and X_C for Series LC Circuit with Resonance at 796 Hertz*

Equivalent Resonant Frequency Equations

For a specific inductance and capacitance, equation *14–4* is used to determine the frequency at which their reactances are equal in value—the resonant frequency.

Suppose, however, that you want to know what value of inductor will provide a certain resonant frequency when connected with a given capacitor or vice-versa. These values can be determined by rewriting equation *14–4* into equivalent forms;

$$L = \frac{1}{(2\pi)^2 f_r^{\,2} C} \qquad (14\text{–}5)$$

or similarly,

$$C = \frac{1}{(2\pi)^2 f_r^2 L} \qquad (14\text{–}6)$$

SERIES RESONANCE

Impedance

Now, consider the effect of resonance on a series RLC circuit such as the one illustrated in *Figure 14.5*. You know that the general expression for the impedance of any series RLC circuit may be expressed as:

$$Z = \sqrt{R^2 + (X_L - X_C)^2} \qquad (14\text{–}7)$$

At resonance, from equation *14–3*,

$$X_L = X_C$$

and the net reactance of the circuit is zero. The impedance, then, at resonance, is

$$Z = \sqrt{R^2 + 0^2} \qquad (14\text{–}8)$$

$$= \sqrt{R^2}$$

$$= R$$

Thus, at resonance, the impedance of the series RLC circuit appears entirely resistive and is equal to the resistance in the circuit.

To further clarify this fact, note that the impedance phasor diagram for a series RLC circuit is as shown in *Figure 14.6*. If X_L is equal in value to X_C, as in the case at resonance, the net reactance of the circuit is zero, and the impedance phasor falls directly upon the resistance phasor. This indicates that the impedance is equal to the value of the resistance at resonance.

Minimum Impedance At Resonance

Since X_L equals X_C, the net reactance is zero and thus there is no reactive term in the impedance equation, equation *14–7*. As shown in *Figure 14.4* if the applied frequency to the circuit is varied either side of the resonant frequency, then a difference will exist between X_L and X_C. This difference would result in a net reactance and would add vectorially to the resistance in the circuit to create a larger impedance. Such circuits were the type discussed in previous lessons. Thus, it should be quite clear that at the resonant frequency of a series circuit the impedance expressed by equation *14–7* is at its lowest level, as expressed by equation *14–8*.

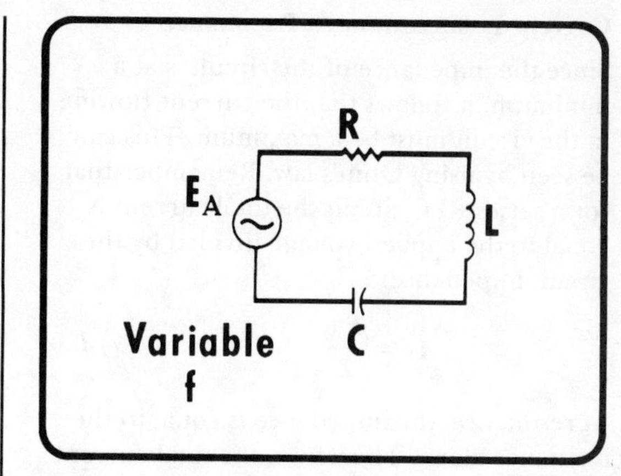

Figure 14.5 *Series RLC Circuit*

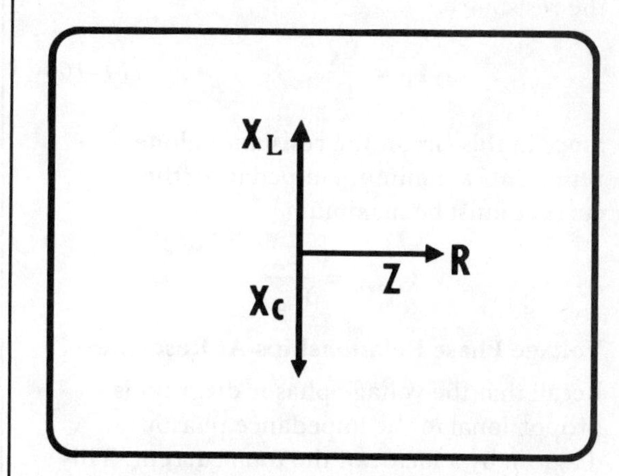

Figure 14.6 *Impedance Phasor Diagram for a Series RLC Circuit at Resonance*

Current Is Maximum At Resonance

Since the impedance of the circuit is at a minimum, it follows that the current flowing in the circuit must be a maximum. This can be seen by using Ohm's law. Remember that for a series RLC circuit the total current is equal to the applied voltage divided by the circuit impedance:

$$I_T = \frac{E_A}{Z} \qquad (14\text{–}9)$$

At resonance, the impedance is equal to the resistance alone. Therefore, the total current is equal to the applied voltage divided by the resistance.

$$I_T = \frac{E_A}{R} \qquad (14\text{–}10)$$

Since in this circuit the resistance alone represents a minimum impedance, the current must be maximum.

$$I_{Tmax} = \frac{E_A}{R_{min}}$$

Voltage Phase Relationships At Resonance

Recall that the voltage phasor diagram is proportional to the impedance phasor diagram by a factor of the total current. This is shown in *Figure 14.7*. At resonance, when $X_L = X_C$, then the voltage across the inductor, E_L, must equal the voltage across the capacitor, E_C. If $E_L = E_C$, these voltages cancel because they are 180 degrees out of phase. As a result the applied voltage, E_A, is in phase with and equal to the voltage across the resistor, as shown in *Figure 14.8*.

Phase Angle Equals Zero

Assume the total current is taken as reference at zero degrees. Since the current is in phase with the voltage across the resistor, and at resonance the circuit becomes entirely resistive, then the applied voltage is in phase with the total current and the phase angle is zero. This also is shown in *Figure 14.8*.

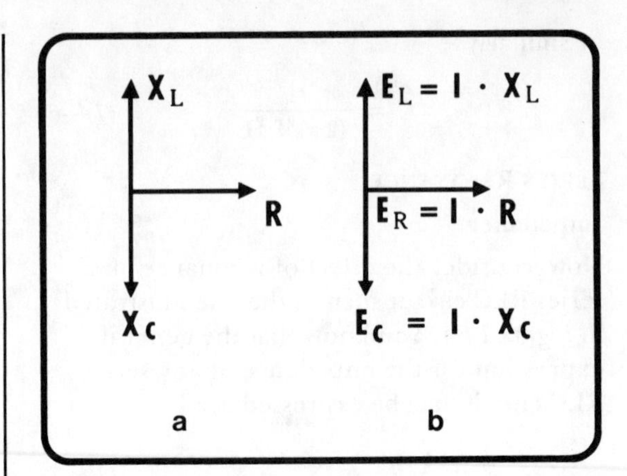

Figure 14.7 *Impedance and Voltage Phasor Diagrams for Series RLC Circuit at Resonance*

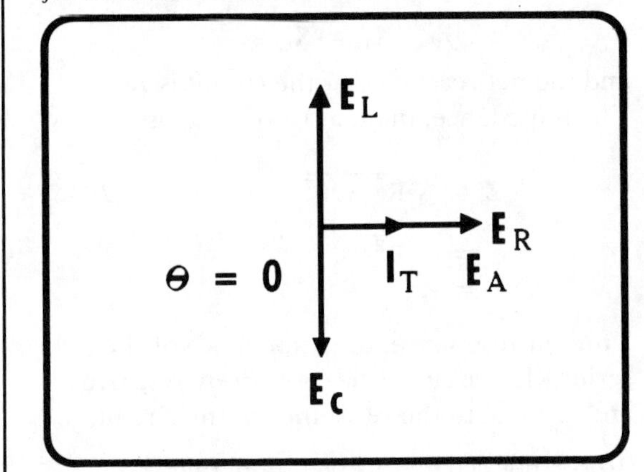

Figure 14.8 *Phase Angle Equals Zero at Resonance*

■ Q of a Circuit
■ Voltage Magnification

Q of a Circuit

A circuit is sensitive to changes when used at the resonant frequency. This is determined by the *Q* or *quality factor* of the circuit. Recall that in the discussion of inductance in Lesson 8 that the Q of a circuit was defined as the ratio of the reactive power of a reactive component to the real power dissipated in the resistance in the circuit. This is expressed again in equation *14–11*.

$$Q = \frac{P_{X\ (\text{reactive power})}}{P_{R\ (\text{real power})}} \qquad (14\text{–}11)$$

Let's consider the reactive power of the inductor in a series RLC circuit, such as the one shown in *Figure 14.5*. The reactive power of the inductor is equal to the voltage across the inductor times the current through it:

$$P_{XL} = E_L I_L \qquad (14\text{–}12)$$

Since $E_L = I_L X_L$, equation *14–12* can be rewritten as the inductive current squared times the value of inductive reactance. The result is equation *14–13*.

$$P_{XL} = I_L^2 X_L \qquad (14\text{–}13)$$

The value of the real power dissipated by the resistance in the circuit is equal to the voltage across the resistance times the current through it (equation *14–14*).

Using the same reasoning, this can be rewritten:

$$P_R = E_R I_R$$
$$P_R = I_R^2 R \qquad (14\text{–}14)$$

Substituting these two equations into equation *14.11* for Q,

$$Q = \frac{I_L^2 X_L}{I_R^2 R} \qquad (14\text{–}15)$$

In the series circuit, the current through the inductor equals the current through the resistor, thus,

$$I_L = I_R$$

and

$$I_L^2 = I_R^2 \qquad (14\text{–}16)$$

As a result equation *14–15* simplifies to equation *14–17*,

$$Q = \frac{X_L}{R} \qquad (14\text{–}17)$$

If the capacitive reactance had been used, the equation would be:

$$Q = \frac{X_C}{R} \qquad (14\text{–}18)$$

Thus, the Q of a resonant series RLC circuit is equal to the value of the inductive or capacitive reactance at the resonant frequency divided by the value of the series resistance in the circuit.

Voltage Magnification

Using the original equation *14–11* for Q,

$$Q = \frac{P_X}{P_R}$$

and substituting the E-I equivalents for the power values, equation *14–19* results.

$$Q = \frac{E_L I_L}{E_R I_R} \qquad (14\text{–}19)$$

Since the current through the inductor, the current through the resistor, and the total current in the series circuit are the same, $I_T = I_L = I_R$, both I_L and I_R in equation *14–19* can be replaced by I_T.

$$Q = \frac{E_L I_T}{E_R I_T} \qquad (14\text{–}20)$$

Cancelling I_T, equation *14–20* simplifies to:

$$Q = \frac{E_L}{E_R} \qquad (14\text{–}21)$$

Q is equal to E_L divided by E_R.

■ Measuring Q

Earlier you saw in the voltage phasor diagram of *Figure 14.8* that the voltage across the resistor and the applied voltage in a series circuit were equal at resonance. Substituting $E_R = E_A$ into equation *14–21* results in:

$$Q = \frac{E_L}{E_A} \qquad (14\text{–}22)$$

If both sides of equation *14–22* are multiplied by EA and rearranged, the result is equation *14–23:*

$$E_AQ = \frac{E_LE_A}{E_A}$$

$$E_L = QE_A \qquad (14\text{–}23)$$

Equation *14–23* relates the fact that at resonance, the voltage across the inductor will be Q times larger than the applied voltage. This is referred to as the *voltage magnification* in a series resonant circuit, and Q is sometimes referred to as the *magnification factor*.

Measuring Q

One of the easiest methods that can be used to measure the actual Q of a resonant circuit is based on the fact stated in equation *14–23*. The first step in the method is to connect a frequency generator to the circuit as shown in *Figure 14.9*. Next, the circuit is brought to resonance by adjusting the frequency of the generator. With the circuit at resonance, a voltmeter is used to measure the applied voltage and the voltage across the inductor or capacitor as shown in *Figure 14.9*. (Either the inductor or capacitor can be used because the voltage across either is the same at resonance.) The value of Q is determined by equation *14–22* by simply dividing the voltage across the reactive component by the applied voltage.

The rise in voltage across the inductor or capacitor at resonance also provides a means of determining when a circuit is resonant. For example, if the voltage across the inductor is

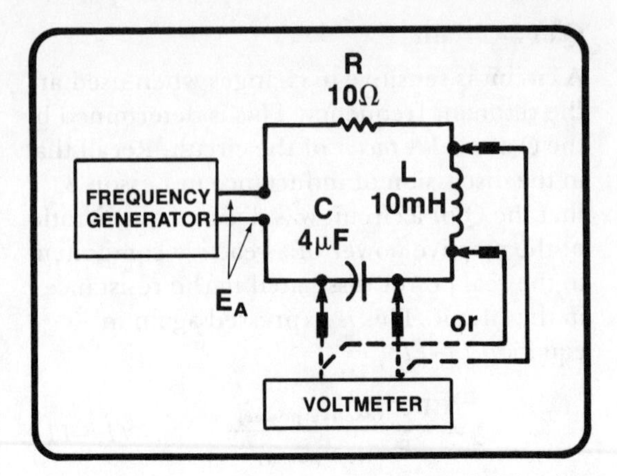

Figure 14.9 *Setup for Determining the Q of a Circuit*

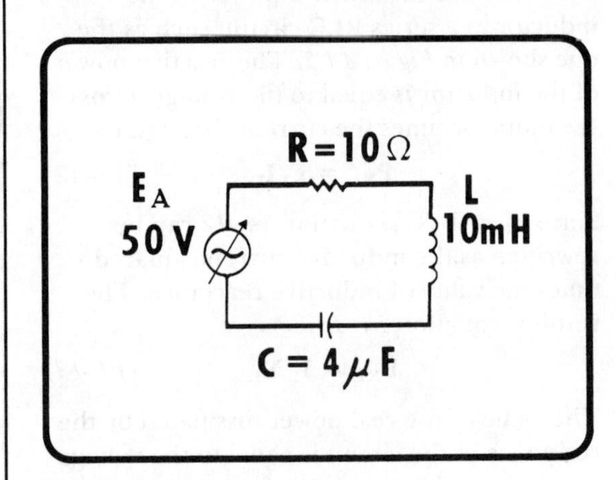

Figure 14.10 *Series Resonant Circuit*

monitored as the frequency is varied, when the circuit becomes resonant the current reaches a peak value and the voltage across the inductor rises to a maximum of Q times the applied voltage.

Series Resonant Circuit Example

With these facts about a series resonant circuit in mind, let's solve the series circuit shown in *Figure 14.10*. It consists of a 10-ohm resistor, a 10-millihenry inductor and a 4-microfarad capacitor connected to a 50-volt variable-frequency ac source. The frequency of the ac voltage is adjusted to the resonant frequency for the circuit. The values of impedance, current, circuit voltages, and Q of the circuit will be determined at the resonant frequency.

Resonant Frequency

Since L is 10 millihenrys and C is 4 microfarads the resonant frequency is determined by using equation *14–4*.

$$f_r = \frac{1}{2\pi\sqrt{LC}} \qquad (14\text{–}4)$$

Substituting values,

$$\begin{aligned} f_r &= \frac{1}{(6.28)\sqrt{(10mH)(4\mu F)}} \\ &= \frac{1}{(6.28)\sqrt{(10 \times 10^{-3}H)(4 \times 10^{-6}F)}} \\ &= \frac{1}{(6.28)\sqrt{40 \times 10^{-9}}} \\ &= \frac{1}{(6.28)\sqrt{4 \times 10^{-8}}} \\ &= \frac{1}{6.28(2 \times 10^{-4})} \\ &= \frac{1}{12.56 \times 10^{-4}} \\ &= 796Hz \end{aligned}$$

The resonant frequency of the circuit is 796 hertz.

Checking $X_L = X_C$ at Resonance

Recall that it was stated that at the resonant frequency, the reactance of the inductor and the capacitor are equal. To prove this for the 10-millihenry inductor and the 4-microfarad capacitor, their individual reactances at the resonant frequency of 796 hertz are calculated.

The reactance of the inductor at 796 hertz is calculated using equation *14-1*.

$$\begin{aligned} X_L &= 2\pi fL \\ &= (6.28)(796Hz)(10mH) \\ &= (6.28)(796)(10 \times 10^{-3}) \\ &= 50\Omega \end{aligned}$$

Next, the capacitive reactance at 796 hertz is calculated using equation *14–2*.

$$\begin{aligned} X_C &= \frac{1}{2\pi fC} \\ &= \frac{1}{(6.28)(796Hz)(4\mu F)} \\ &= \frac{1}{(6.28)(796Hz)(4 \times 10^{-6})} \\ &= 50\Omega \end{aligned}$$

Thus, at the resonant frequency, $X_L = X_C =$ 50 ohms.

Impedance

The magnitude of the impedance of this circuit in its most general form is calculated using equation *14–7:*

$$Z = \sqrt{R^2 + (X_L - X_C)^2} \qquad (14\text{–}7)$$

Substituting R = 10 ohms and X_L = 50 ohms and X_C = 50 ohms:

$$\begin{aligned} Z &= \sqrt{10^2 + (50 - 50)^2} \\ &= \sqrt{10^2 + 0^2} \\ &= \sqrt{100} \\ &= 10\Omega \end{aligned}$$

The impedance of this series circuit, when resonant, is 10 ohms, which is the value of the circuit resistance.

Current

The value of the series circuit current at resonance is determined by dividing the applied voltage by the impedance of the circuit:

$$I_T = \frac{E_A}{Z} = \frac{E_A}{R} = \frac{50V}{10\Omega} = 5A$$

Circuit Q

The Q of the circuit is found equal to 5 by using equation *14–17:*

$$Q = \frac{X_L}{R} = \frac{50\Omega}{10\Omega} = 5$$

E_L And E_C

The voltage across the inductor equals the voltage across the capacitor at resonance and both are equal to Q times the applied voltage (equation *14–23):*

$$E_L = E_C = QE_A = 5(50V) = 250V$$

To verify this, the value of this voltage in the traditional manner can be calculated. Recall in that calculation the voltage across the inductor equals the current through the inductor times the value of the inductive reactance:

$$E_L = I_L X_L = (5A)(50\Omega) = 250V$$

Since the capacitor has the same reactance at resonance, the voltage drop across it will be identical to the voltage across the inductor.

$$E_C = I_C X_C = (5A)(50\Omega) = 250V$$

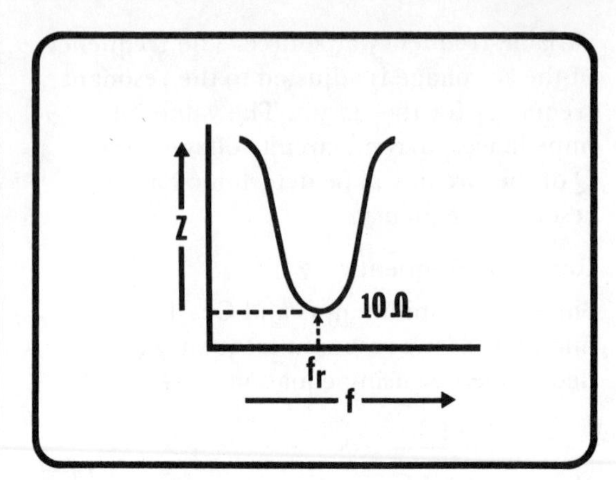

Figure 14.11 *Frequency Response of Impedance*

Frequency Responses

In *Figure 14.4,* inductive reactance and capacitive reactance versus frequency were graphed. With resonant circuits, it is usually helpful to graph the response of the circuit *current* or *impedance* against frequency as shown in *Figures 14.11* and *14.12*. These graphs are called the frequency responses of the circuit.

As shown in *Figure 14.11* the impedance, Z, of a series resonant circuit has a minimum value at the resonant frequency, f_r, when $X_L = X_C$, and the total reactance of the circuit is zero. The impedance increases on either side of the resonant frequency because X_L and X_C are not equal and do not result in a net reactance of zero. In the series resonant circuit just solved, the impedance had a value of 10 ohms at resonance and would be plotted on the impedance response graph as shown in *Figure 14.11*.

The current, on the other hand, has a maximum value at resonance and varies inversely with the impedance as shown in *Figure 14.12*. That is, as the impedance increases, the current decreases. For the series resonant circuit just solved the current frequency response appears as shown in *Figure 14.12* with a maximum current of 5 amperes plotted at the resonant frequency, f_r.

■ Frequency Responses

BANDWIDTH

As indicated by the frequency response of *Figure 14.12*, the effect of resonance is most predominant at the resonant frequency. However, if you examine the responses of the circuit in greater detail by observing the effect of varying the frequency above and below the resonant frequency, you will find that for a band of frequencies the circuit exhibits very nearly the same effects as at resonance. This is shown in *Figure 14.13*.

The band of frequencies over which the effect exists is called the *bandwidth*, and the end points of the band have been defined. The lower point is called f-lower, and the upper point is called f-upper. f-lower and f-upper are the frequencies at which the current has a value of 70.7 percent of its maximum value at resonance. The bandwidth can be determined mathematically by subtracting f-lower from f-upper:

$$\mathbf{BW = f_{upper} - f_{lower}} \qquad (14\text{–}24)$$

For the example circuit, of *Figure 14.14*, f_r is 1000 hertz, f-upper is 1200 hertz, and f-lower is 800 hertz. Therefore, using equation *14–24*,

$$\mathbf{BW = 1200Hz - 800Hz}$$
$$\mathbf{= 400Hz}$$

The bandwidth is 400 hertz.

The points f-lower and f-upper are also called the cutoff or edge frequencies because they define the points at which the resonance of a circuit begins to cut off—the points where resonance begins to lose its effect.

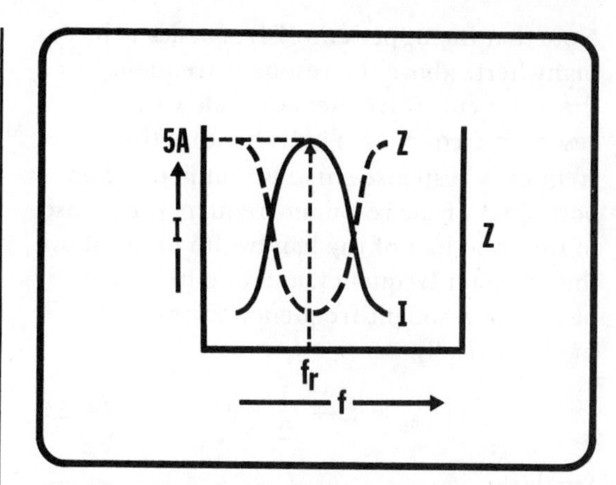

Figure 14.12 *Frequency Response of Current of Series Resonant Circuit*

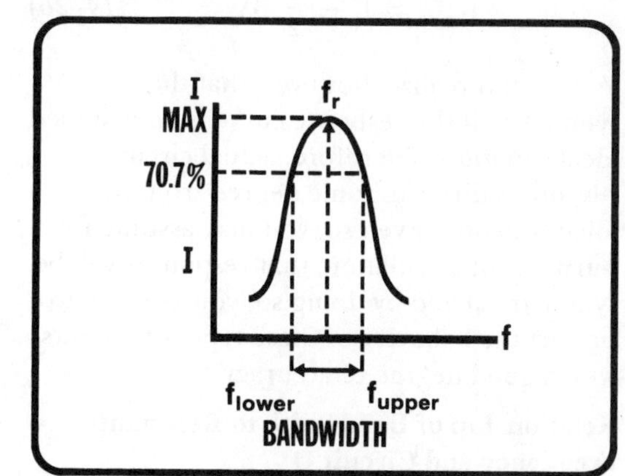

Figure 14.13 *Bandwidth of Frequency Response*

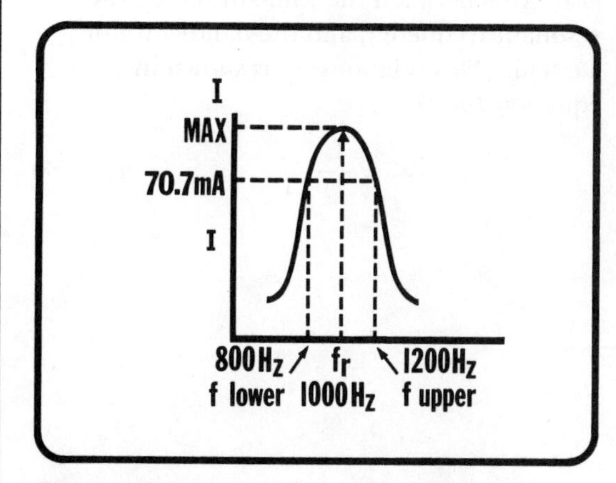

Figure 14.14 *Frequency Response Bandwidth of 400 Hertz*

Relationship of Bandwidth to Resonant Frequency and Circuit Q

Note that the upper cutoff frequency is just as many hertz above the resonant frequency as the lower cutoff frequency is below the resonant frequency. This is because the frequency response curve is symmetrical on both sides of the resonant frequency. Because of this, one-half of the bandwidth exists above the resonant frequency, and one-half exists below the resonant frequency. Expressed mathematically,

$$f_{upper} = f_r + \frac{1}{2}\,BW \qquad (14\text{–}25)$$

Similarly,

$$f_{lower} = f_r - \frac{1}{2}\,BW \qquad (14\text{–}26)$$

You should realize, however, that the symmetry of the response curve is an assumed ideal situation. Therefore, actual circuit responses differ to some degree from the ideal situation. Even so, you may assume for purposes of calculation, that responses will be symmetrical and by doing so, you are able to predict the behavior of these types of circuits with a good degree of accuracy.

Relationship of Bandwidth to Resonant Frequency and Circuit Q

An interesting fact is that a relationship also exists between the value of the Q, the resonant frequency, and the bandwidth of a circuit. This relationship is shown in equation *14–27:*

$$BW = \frac{f_r}{Q} \qquad (14\text{–}27)$$

Equation *14–27* means that if the resonant frequency and the Q of a circuit are known, the bandwidth can be calculated.

In the series circuit of *Figure 14.10* solved *previously,* the resonant frequency was 796 hertz and the Q had a value of 5. The bandwidth of that circuit, then, is calculated using equation *14–27:*

$$\begin{aligned} BW &= \frac{f_r}{Q} \\ &= \frac{796Hz}{5} \\ &= 159.2Hz \end{aligned}$$

Thc upper and lower cutoff frequencies can be calculated using equations *14–25* and *14–26* as:

$$\begin{aligned} f_{upper} &= f_r + \frac{1}{2}BW \\ &= 796Hz + \frac{1}{2}(159.2Hz) \\ &= 796Hz + 79.6Hz \\ &= 875.6Hz \end{aligned}$$

$$\begin{aligned} f_{lower} &= f_r - \frac{1}{2}BW \\ &= 796Hz - \frac{1}{2}(159.2Hz) \\ &= 796Hz - 79.6Hz \\ &= 716.4Hz \end{aligned}$$

■ **Half-Power Points**

These values are graphed in *Figure 14.15*. Recall that these frequencies define the points on the response curve at which the current is 70.7 percent of its maximum value. Therefore, in the series resonant circuit example of *Figure 14.10* in which the current has a maximum value of 5 amperes, the value of the current at these cutoff frequencies is 3.535 amperes, calculated as follows:

$$\begin{aligned} I(\text{cutoff frequency}) &= 70.7\% I_{Max} \quad (14\text{–}28) \\ &= 70.7\%(5A) \\ &= 0.707(5A) \\ &= 3.535A \end{aligned}$$

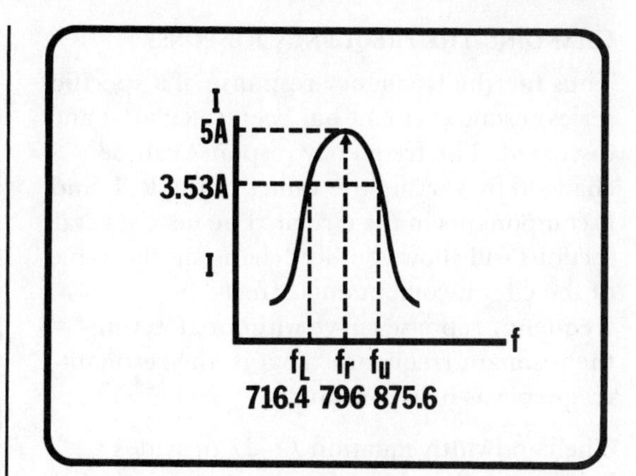

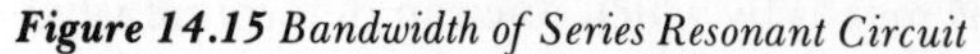
Figure 14.15 *Bandwidth of Series Resonant Circuit*

Half-Power Points

The upper and lower frequency cutoff points on the response curve are sometimes referred to as the *half-power points* of the frequency response. This is because at these points on the response curve, the real power dissipation in the circuit is exactly one-half of what it is at the resonant frequency. To illustrate this, the series circuit of *Figure 14.10*, which is repeated in *Figure 14.16*, will be used. Recall that the real power is equal to the current squared times the value of the resistance:

$$P_R = I^2R \quad (14\text{–}29)$$

In the example circuit the series resistance is 10 ohms, and at resonance the maximum current is 5 amperes. Therefore,

$$\begin{aligned} P_R &= I^2R \\ &= (5A)^2(10\Omega) \\ &= (25)(10) \\ &= 250W \end{aligned}$$

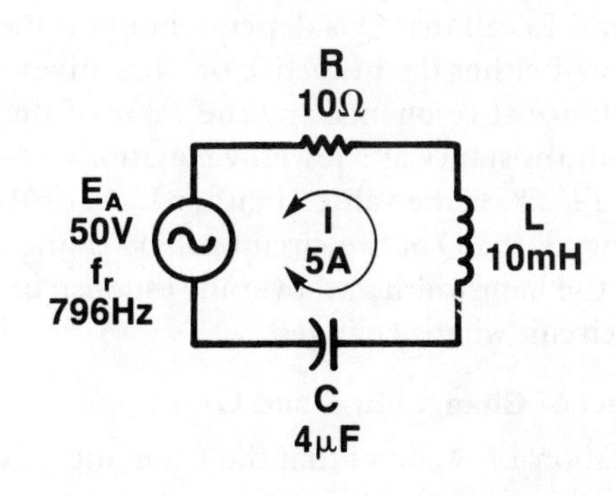

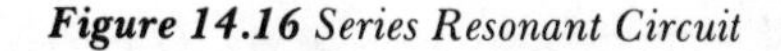
Figure 14.16 *Series Resonant Circuit*

At resonance, the real power dissipated is 250 watts. At the cut-off frequency points the current from equation *14–28*, is 3.535 amperes. Using this value the power is calculated as:

$$\begin{aligned} P_R &= I^2R \\ &= (3.535A)^2(10\Omega) \\ &= (12.5)(10) \\ &= 125W \end{aligned}$$

Therefore, at the cutoff frequencies, the real power dissipated in the circuit is 125 watts, one-half the real power dissipated at the resonant frequencies.

CHANGING THE FREQUENCY RESPONSE

Thus far, the frequency response of a specific series resonant circuit has been calculated and discussed. The frequency response can be changed by varying the values of the R, L and C components in the circuit. The next several sections will show you how changing the value of the circuit components affects the frequency reponse curve without affecting the resonant frequency. That is, the resonant frequency is held constant.

The bandwidth equation *14–27* provides a basic means by which one can determine what effect changing the value of a component will have upon the frequency response of the circuit. Recall that Q is dependent upon the value of either the inductive or capacitive reactance at resonance and the value of the circuit resistance as shown by equations *14–17* and *14–18*. If the value of either L, C, or R is changed, the Q of the circuit will be changed, and the bandwidth and overall response of the circuit will be changed.

Effect of Changes in L and C

Equation *14–4* shows that the resonant frequency is dependent upon the

$$f_r = \frac{1}{2\pi\sqrt{LC}} \qquad (14\text{–}4)$$

values of L and C. Changing either L or C changes the frequency response, but the value of either cannot be changed independently without changing the resonant frequency. However, since the resonant frequency is dependent upon the product of L and C, the resonant frequency can be held constant by changing both L and C without changing their product. For example, if a 6-henry inductor and a 2-farad capacitor are used initially, their LC product is 6 times 2 or 12 ((6H)(2F) = 12).

If the values are changed to a 4 henry inductor and a 3 farad capacitor, the LC product remains the same at 12 ((4H)(3F) = 12), and the resulting resonant frequency will be the same. Note that to keep the product the same and therefore, the resonant frequency, the value of the inductor was *decreased* while the value of the capacitor was *increased*. Obviously, many other possible combinations of L and C will yield the same product, and thus, the same resonant frequency.

L/C Ratio

Although changing the values of L and C in this manner keeps the resonant frequency constant, it does have a definite effect upon the frequency response of the circuit. The curvature (steepness) of the sides of the frequency reponse curve is changed. This change in curvature represents a change in bandwidth. In many cases, this change is referred to as "changing skirts" of the response curve.

For a series resonant circuit, the degree of steepness of the response curve (bandwidth) is determined by the L/C ratio. This is the ratio of the inductance divided by the capacitance:

$$\text{L/C ratio} = \frac{L}{C} \qquad (14\text{–}30)$$

When the value of L is increased, the ratio is increased; when the value of L is decreased, the ratio is decreased. Increasing L increases the L/C ratio and increases X_L, the inductive reactance (equation *14–1*).

$$X_L\uparrow\ = 2\pi f_r L\uparrow$$

From equation *14–17*, an increase in X_L will increase Q if the resistance of the circuit is held constant. Correspondingly, Q will decrease if X_L is decreased.

$$\uparrow Q = \frac{X_L \uparrow}{R} \qquad (14\text{–}17)$$

From equation *14–27*, an increase or decrease in Q will affect the bandwidth. The resonant frequency, f_r, is being held constant, therefore, an increase in the value of Q will cause a decrease in bandwidth:

$$\downarrow BW = \frac{f_r}{Q \uparrow} \qquad (14\text{–}27)$$

Therefore, when L is increased, X_L increases, the L/C ratio is increased, Q increases, the bandwidth is decreased and the sides of the response curve shown in *Figure 14.17* become steeper. Conversely, when L is decreased, X_L decreases, the L/C ratio is decreased, Q decreases, the bandwidth is increased and the sides of the response curve become less steep as shown in *Figure 14.18*.

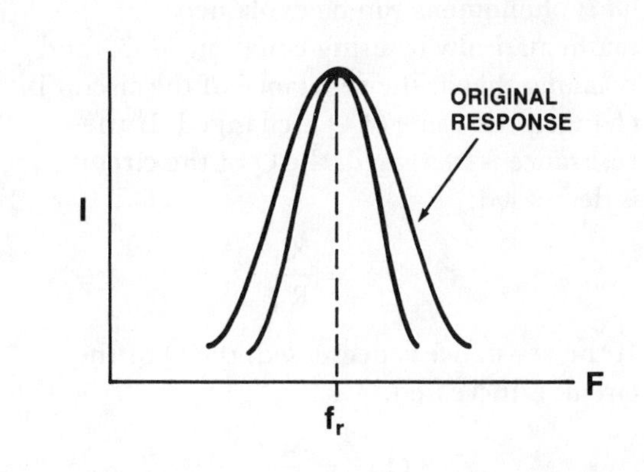

Figure 14.17 *Series Resonant Circuit Frequency Response Curve with Increased L*

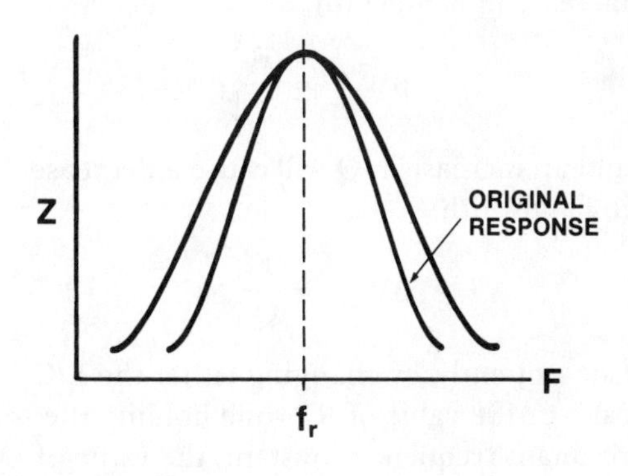

Figure 14.18 *Series Resonant Circuit Frequency Response Curves with Decreased L*

Effect of R

Another way to change the circuit's frequency response is to change the circuit's resistance. Recall from equation *14–4* that the resonant frequency of a series RLC circuit is not dependent on the value of the resistance, therefore, changing R will not change the resonant frequency. Changing R, affects the amount of maximum current at resonance. If L and C remain constant, a decrease in resistance causes an increase in the maximum current at resonance. As shown in *Figure 14.19*, this has the effect of increasing the slope of the sides of the response curve and decreasing the bandwidth. Conversely, an increase in resistance will result in a decrease of the maximum current value at resonance and a corresponding increase in bandwidth. This also is shown in *Figure 14.19*.

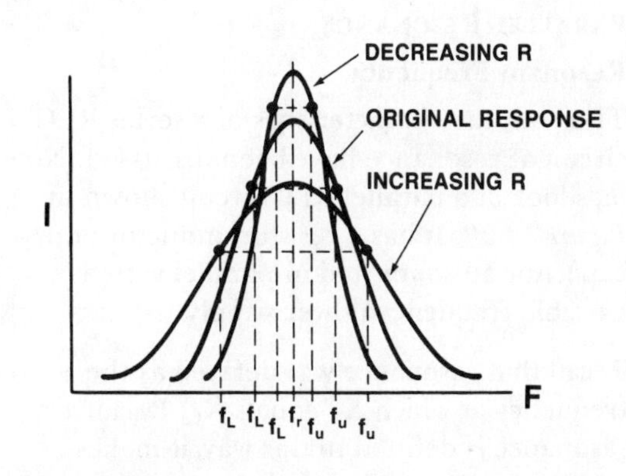

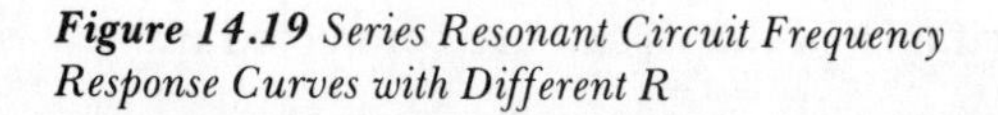

Figure 14.19 *Series Resonant Circuit Frequency Response Curves with Different R*

■ Resonant Frequency
■ Total Current

This phenomena can be explained mathematically by using equation *14–17* and recalling that as the resistance of the circuit is changed the value of Q is changed. If the resistance is increased, the Q of the circuit is decreased:

$$Q\downarrow = \frac{X_L}{R\uparrow}$$

If the resistance is decreased, the Q of the circuit is increased:

$$Q\uparrow = \frac{X_L}{R\downarrow}$$

As before, if the resonant frequency remains constant, a decrease in Q will cause an increase in bandwidth,

$$BW\uparrow = \frac{F_r}{Q\downarrow}$$

and an increase in Q will cause a decrease in bandwidth.

$$BW\downarrow = \frac{F_r}{Q\uparrow}$$

Consequently, by changing either the L/C ratio or the value of R, while holding the resonant frequency constant, the value of Q, and therefore the bandwidth and frequency response, can be changed.

PARALLEL RESONANCE

Resonant Frequency

Thus far, the characteristics of a series RLC circuit at resonance have been discussed. Now let's look at a parallel RLC circuit shown in *Figure 14.20*. It has a resistor, inductor and capacitor all connected in parallel with a variable frequency power supply.

Recall that resonance was defined as the frequency at which X_L equals X_C. Because resonance is defined in this way, it makes no difference whether the inductance and capacitance of the circuit are connected in series or in parallel. Therefore, equation *14–14* for determining the resonant frequency is the same for a parallel circuit as it was for a series circuit:

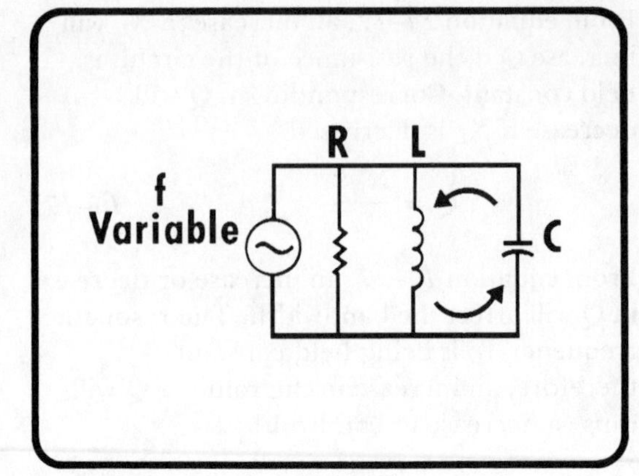

***Figure 14.20** Parallel RLC Circuit*

$$f_r = \frac{1}{2\pi\sqrt{LC}} \qquad (14\text{–}4)$$

At the resonant frequency for this parallel circuit, when X_L and X_C are equal, the inductive and capacitive branch currents will be equal, $I_L = I_C$, and 180 degrees out of phase.

Total Current

Recall that the total current, I_T, in a parallel RLC circuit is equal to the vector sum of the branch currents I_C, I_L and I_R and it is calculated using equation *14–31*.

$$I_T = \sqrt{I_R^2 + (I_C - I_L)^2} \qquad (14\text{–}31)$$

The phasor diagram is shown in *Figure 14.21*. I_R is the current in the resistive branch, I_L is the current in the inductive branch and I_C is the current in the capacitive branch. At resonance, when I_L and I_C are equal, the net reactive current is equal to zero and equation *14–31* becomes,

$$I_T = \sqrt{I_R^2 + 0^2} \qquad (14\text{–}32)$$

and can be rewritten as

$$I_T = \sqrt{I_R^2} \qquad (14\text{–}33)$$

Solving for I_T results in equation *14–34*.

$$I_T = I_R \text{ (minimum)} \qquad (14\text{–}34)$$

At resonance the total current in a parallel resonant RLC circuit is equal to the resistive branch current because as shown in *Figure 14.22a*, $I_L = I_C$ and is 180 degrees out of phase with I_C. Since at resonance the total current seen by the power source is resistive current, the current must be at its lowest level, at its minimum. If the applied frequency is varied either side of the resonant frequency, as shown by the phasor diagrams of *Figure 14.22*, then a difference will exist between the value of I_L and I_C. This difference will result in a net reactive current which would add vectorially to the resistive current of the circuit and create a larger value of total current.

The phasor diagram of the parallel circuit at resonance is shown in *Figure 14.23*. Since the applied voltage is taken as reference at zero degrees, and since at resonance the resistive current is in phase with the applied voltage, the total current is in phase with the applied voltage and the phase angle is zero.

Impedance

The total impedance of the parallel circuit is equal to the applied voltage divided by the total current.

$$Z = \frac{E_A}{I_T} \qquad (14\text{–}35)$$

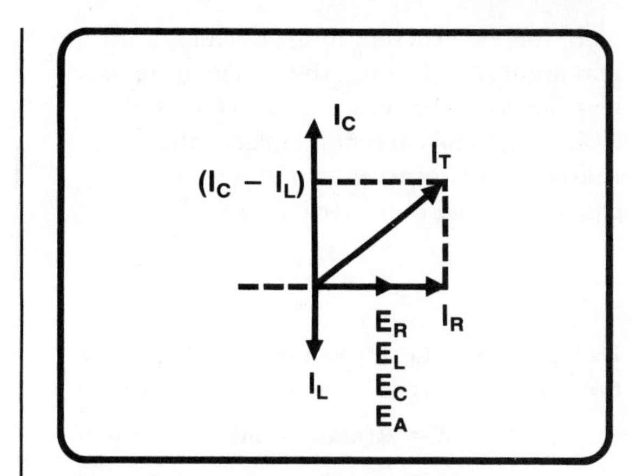

Figure 14.21 *Total Current Phasor Diagram*

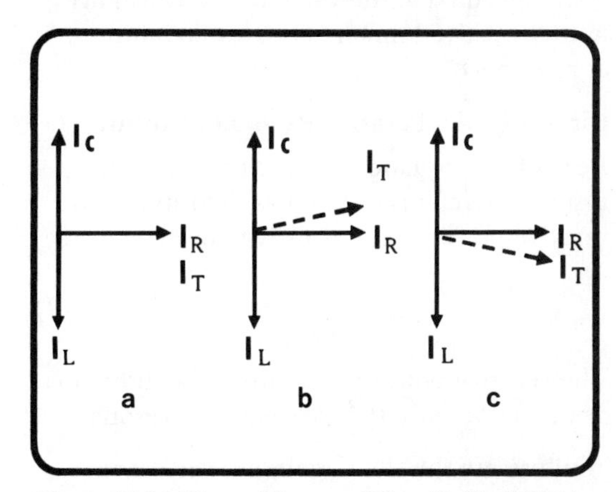

Figure 14.22 *Phasor Diagrams When f is Varied Either Side of Resonance*

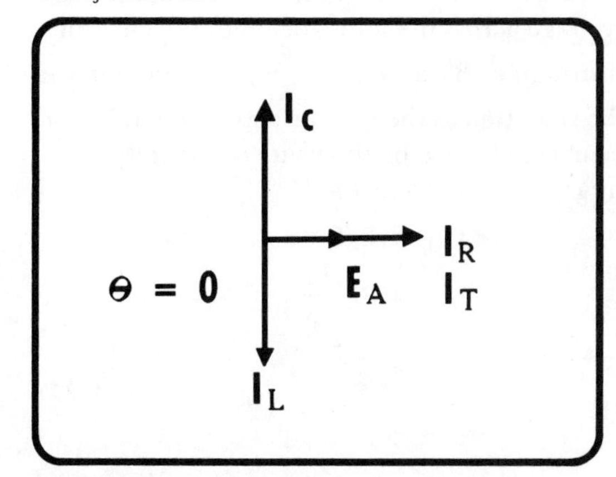

Figure 14.23 *Phase Angle Equals Zero Degrees at Resonance*

■ Circuit Q of a Parallel-Resonant Circuit
■ Current Magnification

Since the total current in the circuit is a minimum at resonance, the impedance must be at its maximum value. From equation *14–34*, the total current is equal to the resistive branch current, and therefore, equation *14–35* can be rewritten:

$$Z = \frac{E_A}{I_R}$$

By Ohm's law, the applied voltage divided by the resistive current is equal to resistance:

$$Z = R(\text{maximum}) \qquad (14\text{–}36)$$

In other words, the impedance at resonance is simply equal to the value of the resistance in the resistive branch and is a maximum at resonance.

Circuit Q of a Parallel-Resonant Circuit

Recall from equation *14–11* that the Q of a resonant circuit is defined as the ratio of the reactive power to the resistive power.

$$Q = \frac{P_X}{P_R} \qquad (14\text{–}11)$$

The reactive power is for either the inductor or capacitor since they are equal at resonance.

To illustrate this, let's again use the reactive power of the inductor as an example. The reactive power of the inductor is equal to the voltage across the inductor times the current through it. Because $I_L = \frac{E_L}{X_L}$, the equation can be rewritten as the voltage across the inductor squared divided by the value of the inductive reactance (equation *14–37*).

$$\begin{aligned} P_{XL} &= E_L I_L \\ &= E_L \frac{E_L}{X_L} \\ &= \frac{E_L^2}{X_L} \end{aligned} \qquad (14\text{–}37)$$

The real power dissipated by the resistor is equal to the voltage across the resistor times the current through it. Because $I_R = \frac{E_R}{R}$, the equation can be rewritten as the voltage across the resistor squared divided by the value of the resistor (equation *14–38*).

$$\begin{aligned} P_R &= E_R I_R \\ &= E_R\left(\frac{E_R}{R}\right) \\ &= \frac{E_R^2}{R} \end{aligned} \qquad (14\text{–}38)$$

Substituting equations *14–37* and *14–38* into equation *14–11*,

$$\begin{aligned} Q &= \frac{P_X}{P_R} \\ &= \frac{(E_L^2/X_L)}{(E_R^2/R)} \end{aligned} \qquad (14\text{–}39)$$

In the parallel circuit, E_L, the voltage across the inductor equals E_R, the voltage across the resistor. As a result equation *14–39* simplifies to equation *14–40*.

$$Q = \frac{(1/X_L)}{(1/R)} = \frac{R}{X_L} \qquad (14\text{–}40)$$

The Q of a parallel resonant circuit of the type shown in *Figure 14.20* is equal to the value of the parallel circuit resistance divided by the value of the parallel circuit inductive reactance calculated at the resonant frequency. Since the capacitive reactance equals the inductive reactance at resonance, the same value of Q may be obtained by dividing the resistance by the capacitive reactance:

$$Q = \frac{R}{X_C} \qquad (14\text{–}41)$$

Current Magnification

Now, let's take the original equation for Q, equation *14–11*, and use it in a slightly different way. In equation *14–42*, the E and I

■ Current Magnification

equivalents are substituted for the power values, P_X and P_R.

$$Q = \frac{P_X}{P_R} = \frac{E_L I_L}{E_R I_R} \qquad (14\text{–}42)$$

Since the voltage across the inductor, E_L, and voltage across the resistor, E_R, are the same in the parallel circuit under consideration, and they both equal E_A, they cancel each other and the equation simplifies to equation *14–43*.

$$Q = \frac{E_A I_L}{E_A I_R} = \frac{I_L}{I_R} \qquad (14\text{–}43)$$

Q equals the inductive branch current divided by the resistive branch current.

Earlier you saw in the current phasor diagrams of *Figure 14.22* and proven by equation *14–34* that the current through the resistor and the total current were equal at resonance. As a result of the substitution $I_R = I_T$, equation *14–43* for Q can be rewritten:

$$Q = \frac{I_L}{I_R}$$

$$Q = \frac{I_L}{I_T} \qquad (14\text{–}44)$$

If both sides of equation *14–44* are multiplied by I_T and rearranged, the equation can be rewritten as:

$$I_T Q = \left(\frac{I_L}{I_T}\right) I_T$$

I_T cancels on the right side, therefore,

$$I_L = QI_T \qquad (14\text{–}45)$$

The equation relates the fact that at resonance, the current through the inductor will be Q times larger than the total current.

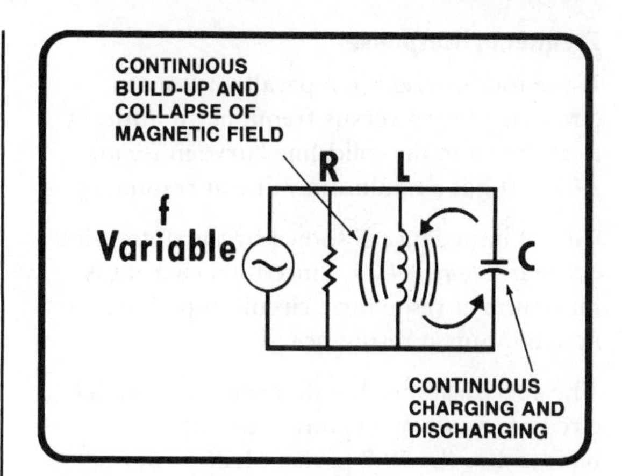

Figure 14.24 *Circulatng Current in L and C Branches at the Resonant Frequency*

Since the capacitive current is equal to the inductive current, the capacitive current is also Q times the total circuit current at resonance (equation *14–46*).

$$I_L = I_C = QI_T \qquad (14\text{–}46)$$

This is referred to as the *current magnification* in a parallel resonant circuit. Q, again, is known as the *magnification factor.*

You may find this unusual but, the current as expressed by equations *14–45* and *14–46* is much larger in the inductive and capacitive branches of a parallel circuit at resonance than the total current. This is shown in *Figure 14.24*. The larger current in the inductive-capacitive branches is a result of the continual charging and discharging of the capacitor and the continual building-up and collapsing of the magnetic field of the inductor at the resonant frequency. It is a characteristic of resonance and is known as the *circulating current.* The parallel combination of the inductor and capacitor is often called a *tank circuit,* often shortened simply to *tank.* Although the circulating current is large, the *total current* drawn by the circuit has a *minimum value* at resonance.

- **Frequency Response**
- **Resonant Frequency**
- **$X_L = X_C$**

Frequency Response

If the total current in a parallel resonant circuit is plotted versus frequency it appears as is shown in the solid-line curve in *Figure 14.25*. It has a minimum value at resonance.

Circuit impedance is shown by the dotted-line curve in *Figure 14.25*. Since total current is minimum at resonance, circuit impedance is at maximum at resonance.

The specifications that describe the parallel circuit frequency response, such as bandwidth, the 70.7 percent half-power points, and upper and lower cutoff frequencies are determined exactly as they were for a series resonant circuit.

PARALLEL RESONANT CIRCUIT EXAMPLE

With these facts in mind concerning a parallel resonant circuit, let's solve the parallel circuit shown in *Figure 14.26*. It consists of a 1-kilohm resistor, a 10-millihenry inductor and a 4-microfarad capacitor connected in parallel to a 100 VAC variable-frequency ac source adjustable to the circuit's resonant frequency. At resonance the values of currents, impedance, voltages, and Q of the circuit will be determined, and the circuit's frequency response curve will be drawn.

Resonant Frequency

In this circuit where L is 10 millihenrys and C is 4 microfarads, the resonant frequency is:

$$f_r = \frac{1}{2\pi\sqrt{LC}}$$
$$= \frac{1}{(6.28)\sqrt{(10mH)(4\mu F)}}$$
$$= 796Hz$$

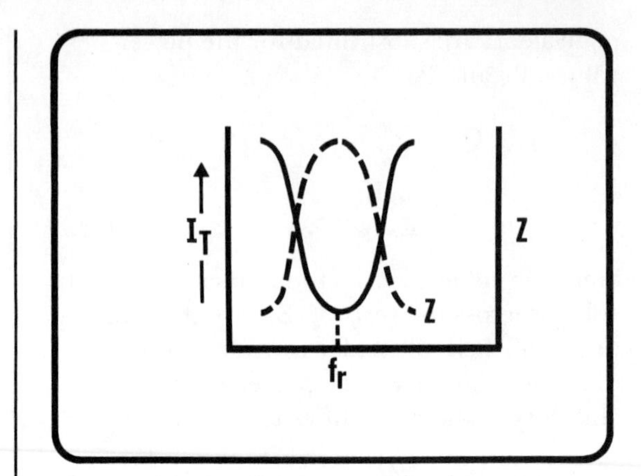

Figure 14.25 *Parallel Resonance Current and Impedance Frequency Responses*

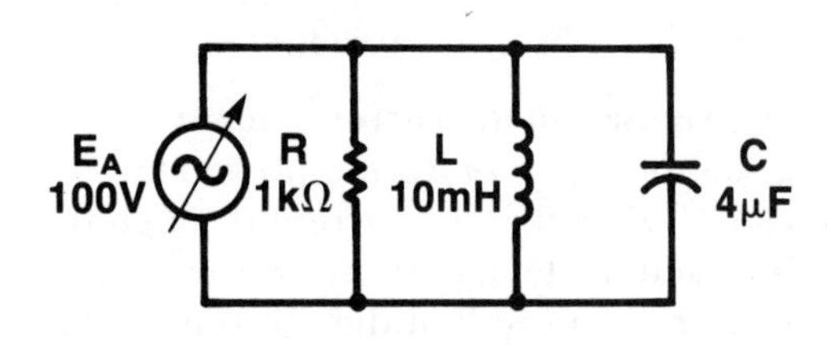

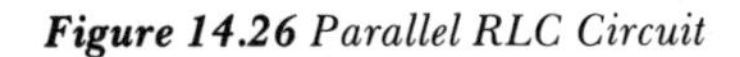

Figure 14.26 *Parallel RLC Circuit*

$X_L = X_C$

At the resonant frequency of 796 hertz, the value of the inductive and capacitive reactance are equal and, as before, both are equal to 50 ohms.

$$X_L = 2\pi fL$$
$$= (6.28)(796Hz)(10mH)$$
$$= 50\Omega$$

$$X_C = \frac{1}{2\pi fC}$$
$$= \frac{1}{(6.28)(796Hz)(4\mu F)}$$
$$= 50\Omega$$

Branch Currents

The branch current can be calculated using Ohm's law for each branch. The voltage across each branch is E_A which is equal to 100 volts. I_L is calculated first.

$$\begin{aligned} I_L &= \frac{E_A}{X_L} \\ &= \frac{100V}{50\Omega} \\ &= 2A \end{aligned}$$

In a similar manner, the capacitive branch current I_C, is determined:

$$\begin{aligned} I_C &= \frac{E_A}{X_C} \\ &= \frac{100V}{50\Omega} \\ &= 2A \end{aligned}$$

The resistive branch current is the last to be calculated.

$$\begin{aligned} I_R &= \frac{E_A}{R} \\ &= \frac{100V}{1K\Omega} \\ &= 100mA \end{aligned}$$

Total Current

The total current for the example circuit in its most general form is:

$$I_T = \sqrt{I_R{}^2 + (I_C - I_L)^2}$$

Substituting values, $I_R = 100mA$, $I_L = 2A$, $I_C = 2A$,

$$\begin{aligned} I_T &= \sqrt{(100mA)^2 + (2A - 2A)^2} \\ &= \sqrt{(100mA)^2 + 0^2} \\ &= \sqrt{(100mA)^2} \\ &= 100mA \end{aligned}$$

As was determined previously for parallel RLC circuits the total current at resonance is simply equal to the resistive branch current. In this case it has a minimum value of 100 milliamperes.

Impedance

The value of the impedance is equal to the applied voltage divided by the total current or,

$$\begin{aligned} Z &= \frac{E_A}{I_T} \\ &= \frac{100V}{100mA} \\ &= 1k\Omega \end{aligned}$$

The circuit impedance at resonance is a maximum value and equal to 1 kilohm, the value of the parallel branch resistance.

Circuit Q

The Q of the parallel circuit can now be determined by substituting $R = 1$ kilohm and $X_L = 50$ ohms.

$$\begin{aligned} Q &= \frac{R}{X_L} \\ &= \frac{1000\Omega}{50\Omega} \\ &= 20 \end{aligned}$$

Circulating Current

It was established previously that the capacitive or inductive circulating current was Q times the total current at resonance. It is calculated as follows:

$$\begin{aligned} I_C = I_L &= QI_T \\ &= (20)(100mA) \\ &= 2A \end{aligned}$$

Frequency Response and Bandwidth

This 2 amperes is identical to the value calculated for the inductive and capacitive branch currents determined earlier and verifies that calculation.

Frequency Response and Bandwidth

The frequency response of the circuit is shown in *Figure 14.27*. It shows that impedance is at a maximum of 1 kilohm at resonance: 796 hertz. The bandwidth from equation *14-27* is equal to the resonant frequency divided by the Q of the circuit. It is calculated as follows:

$$\begin{aligned} BW &= \frac{f_r}{Q} \\ &= \frac{796\text{Hz}}{20} \\ &= 39.8\text{Hz} \end{aligned}$$

The upper and lower cut-off frequencies are calculated using equation *14-25* and *14-26*. From equation *14-25*,

$$\begin{aligned} f_{upper} &= f_r + \frac{1}{2}BW \\ &= 796\text{Hz} + \frac{1}{2}(39.8\text{Hz}) \\ &= 796\text{Hz} + 19.9\text{Hz} \\ &= 815.9\text{Hz} \end{aligned}$$

From equation *14-26*,

$$\begin{aligned} f_{lower} &= f_r - \frac{1}{2}BW \\ &= 796\text{Hz} - \frac{1}{2}(39.8\text{Hz}) \\ &= 796\text{Hz} - 19.9\text{Hz} \\ &= 776.1\text{Hz} \end{aligned}$$

Remember that these upper and lower cut-off frequencies are the frequencies at which the frequency response is down to 70.7 percent of its maximum value. In this example, the value of the impedance at these cutoff frequencies is equal to 70.7 percent of 1 kilohm.

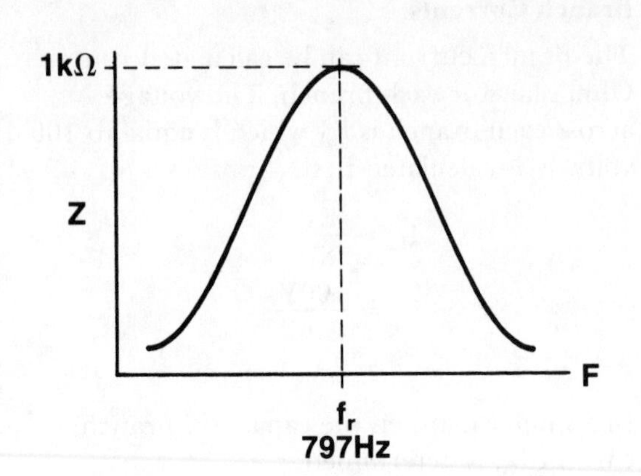

Figure 14.27 Impedance Frequency Response of Parallel RLC Circuit

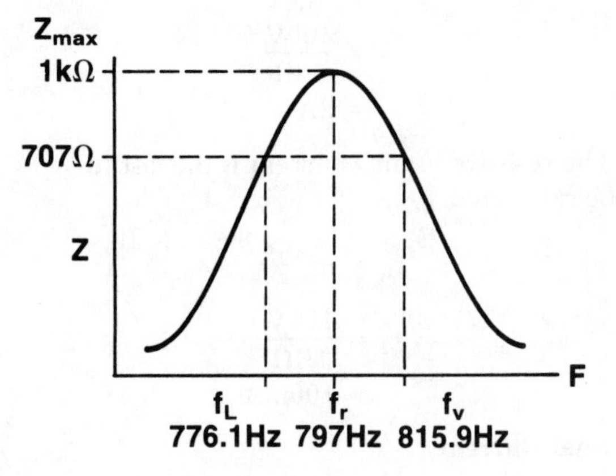

Figure 14.28 Parallel RLC Frequency Response

$$\begin{aligned} Z(\text{cutoff frequency}) &= 70.7\%\ Z_{max} \\ &= 70.7\%\ (1k\Omega) \\ &= 0.707\ (1000\Omega) \\ &= 707\Omega \end{aligned}$$

All of these values are presented as they apply to the frequency response of the impedance in *Figure 14.28*.

■ Effect of Changing L/C Ratio
■ Effect of Changing R Value

ALTERING THE FREQUENCY RESPONSE

As in a series resonant circuit, the frequency response of a parallel resonant circuit can be altered by changing the value of R, L and C.

In all of the following discussion, the resonant frequency is held constant. Component values are changed accordingly, as in series resonant circuits.

Effect of Changing L/C Ratio

Impedance plotted against frequency is shown in *Figure 14.29*. It also demonstrates the effect of changing the L over C ratio on the frequency response of a parallel resonant circuit. The effect of increasing the L/C ratio for a parallel resonant circuit has the opposite effect as it did for a series resonant circuit. An increase in the L/C ratio causes a decrease in Q, a decrease in the steepness of the response curve, and an increase in bandwidth.

This effect can be substantiated mathematically by reasoning that in increasing the L/C ratio, L is increased causing an increase in the inductive reactance at the resonant frequency. Assuming the applied voltage is held constant, an increase in X_L in this parallel circuit causes a decrease in the inductive branch current, I_L, indicating a decrease in the circuit Q..

From equation *14–40* an increase in X_L causes a decrease in the Q of the circuit, and from equation *14-27* a decrease in Q causes an increase in the bandwidth. Conversely, a decrease in the L/C ratio causes an increase in Q, an increase in the steepness of the response curve, and a decrease in bandwidth as shown in *Figure 14.30*.

Effect of Changing R Value

The effect of changing the resistance of the circuit on the frequency response of a parallel resonant circuit is shown in *Figures 14.31* and *14.32*.

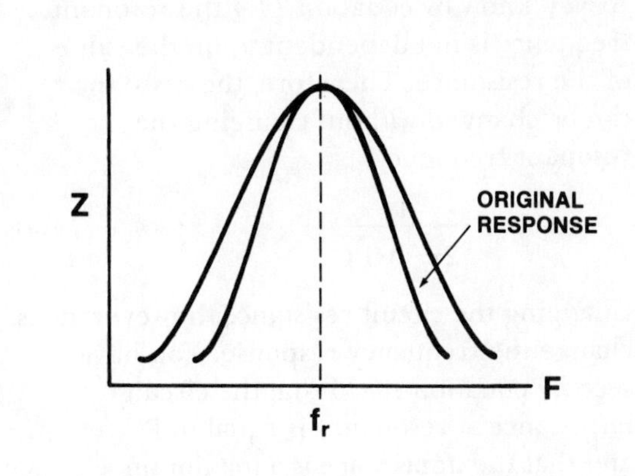

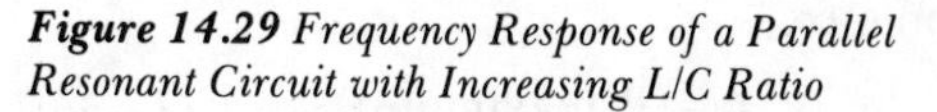
Figure 14.29 *Frequency Response of a Parallel Resonant Circuit with Increasing L/C Ratio*

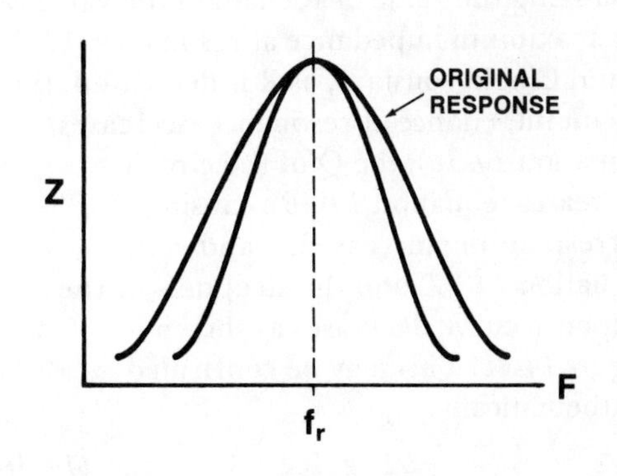

Figure 14.30 *Frequency Response of a Parallel Resonant Circuit with Decreasing L/C Ratio*

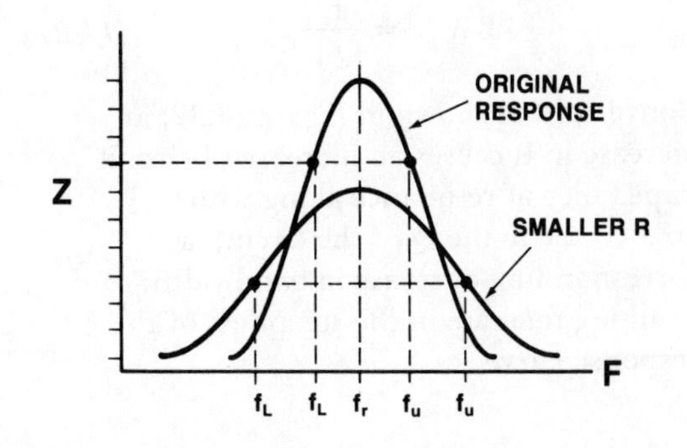

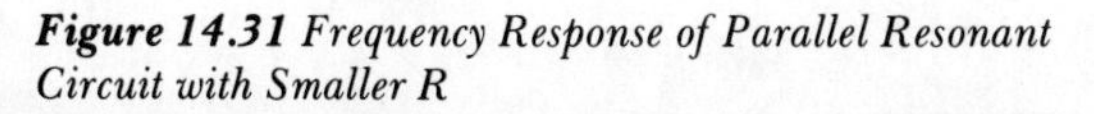
Figure 14.31 *Frequency Response of Parallel Resonant Circuit with Smaller R*

■ **Summary**
■ **Course Summary**

As you know by equation *14-4* the resonant frequency is not dependent upon the value of the resistance. Therefore, the resistance can be changed without changing the resonant frequency.

$$f_r = \frac{1}{2\pi\sqrt{LC}} \qquad (14\text{-}4)$$

Changing the circuit resistance, however, does change the frequency response. You have seen by equation *14-36* that the circuit impedance at resonance is equal to R, and that the impedance is a maximum at resonance.

$$Z = R(\text{maximum}) \qquad (14\text{-}36)$$

Changing the value of R changes the value of the maximum impedance at resonance. With L and C held constant, as R is decreased, the circuit impedance at resonance decreases (equation *14-36*), the Q of the circuit decreases (equation *14-40*), causing a corresponding increase in bandwidth (equation *14-27*) and the steepness of the response curve decreases as shown in *Figure 14.31*. This may be confirmed mathematically:

$$Z\downarrow = R\downarrow \qquad (14\text{-}36)$$

$$Q\downarrow = \frac{R\downarrow}{X_L} \qquad (14\text{-}40)$$

$$BW\uparrow = \frac{f_r}{Q\downarrow} \qquad (14\text{-}27)$$

Conversely, as shown in *Figure 14.32*, an increase in R causes an increase of circuit impedance at resonance along with an increase in the Q of the circuit, a corresponding decrease in bandwidth, a resulting increase in the steepness of the response curve.

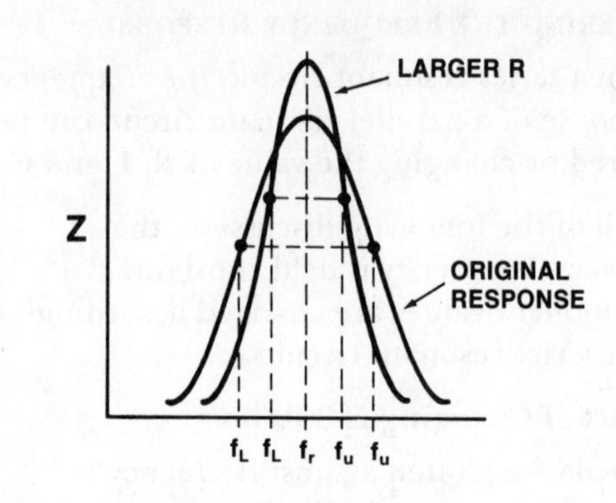

Figure 14.32 *Effect on Frequency Response of Parallel Resonant Circuit with Larger R*

SUMMARY

In this lesson, the concept of resonance was introduced and its effect on the impedance, currents, and voltages of a series RLC circuit was discussed. The concept of Q, the frequency response and the bandwidth of a resonant circuit were discussed. You were shown how the Q of a circuit could be changed by changing the components of the circuit and what effect changing its value had upon the frequency response of the circuit. Corresponding discussions showed you the same effects for a parallel RLC circuit.

COURSE SUMMARY

Throughout this text, the basic concepts of alternating current circuits have been developed from the simplest circuit to very complex RLC circuits. Basic tools and techniques have been explained and applied with worked-out examples such that a student completing this material should be able to analyze and solve the most complex ac circuit application.

■ Worked-Out Examples

1. Calculate the resonant frequency of the following L-C combinations:

a. L = 250μH, C = 10pF f_r = ____________

b. L = 16mH, C = 0.22μF f_r = ____________

Solutions:

a. Using the resonant frequency equation:

$$f_r = \frac{1}{2\pi\sqrt{LC}} = \frac{1}{(6.28)\sqrt{(250\mu H)(10pF)}} = \frac{1}{(6.28)\sqrt{(250 \times 10^{-6}H)(10 \times 10^{-12}F)}}$$

$$= \frac{1}{(6.28)\sqrt{2.5 \times 10^{-15}}} = \frac{1}{6.28(5 \times 10^{-8})} = \frac{1}{3.14 \times 10^{-7}}$$

$$= 3.18 \times 10^{6}Hz = \mathbf{3.18MHz}$$

b. Using the resonant frequency equation again:

$$f_r = \frac{1}{2\pi\sqrt{LC}} = \frac{1}{(6.28)\sqrt{(16mH)(0.22\mu F)}} = \frac{1}{(6.28)\sqrt{(16 \times 10^{-3}H)(0.22 \times 10^{-6}F)}}$$

$$= \frac{1}{(6.28)\sqrt{3.52 \times 10^{-9}}} = \frac{1}{(6.28)(5.93 \times 10^{-5})} = \frac{1}{3.73 \times 10^{-4}} = 2.68 \times 10^{3}Hz = \mathbf{2.68kHz}$$

2. Determine the value of the inductor needed to provide the resonant frequency specified if used in conjunction with the capacitor given:

a. f_r = 55kHz, C = 1.5μF L = ____________

b. f_r = 27MHz, C = 10pF L = ____________

Solutions:

$$a.\ L = \frac{1}{4\pi^2 f_r^2 C} = \frac{1}{4(\pi)^2(55kHz)^2(1.5\mu F)} = \frac{1}{4(3.14)^2(55 \times 10^3)^2(1.5 \times 10^{-6})}$$

$$= \frac{1}{4(9.86)(3.03 \times 10^9)(1.5 \times 10^{-6})} = \frac{1}{1.79 \times 10^5} = 5.57 \times 10^{-6}H = \mathbf{5.57\mu H}$$

$$b.\ L = \frac{1}{4\pi^2 f_r^2 C} = \frac{1}{4(\pi)^2(27MHz)^2(10pF)} = \frac{1}{4(3.14)^2(27 \times 10^6)^2(10 \times 10^{-12})}$$

$$= \frac{1}{4(9.86)(7.29 \times 10^{14})(10 \times 10^{-12})} = \frac{1}{2.88 \times 10^5} = 3.48 \times 10^{-6}H = \mathbf{3.48\mu H}$$

Worked-Out Examples

3. Determine the value of capacitor needed to provide the resonant frequency specified if used in conjunction with the inductor given:

a. $f_r = 40\text{kHz}, L = 18\text{mH}$ $\qquad C =$ ________

b. $f_r = 1\text{MHz}, L = 0.05\text{mH}$ $\qquad C =$ ________

Solutions:

a. $$C = \frac{1}{4\pi^2 f_r^2 L} = \frac{1}{4(\pi)^2(40\text{kHz})^2(18\text{mH})} = \frac{1}{4(3.14)^2(40 \times 10^3)^2(18 \times 10^{-3})}$$
$$= \frac{1}{4(9.86)(1.6 \times 10^9)(18 \times 10^{-3})} = \frac{1}{1.14 \times 10^9} = 8.80 \times 10^{-10}\text{F} = \mathbf{880pF}$$

b. $$C = \frac{1}{4\pi^2 f_r^2 L} = \frac{1}{4(\pi)^2(1\text{MHz})^2(0.05\text{mH})} = \frac{1}{4(3.14)^2(1 \times 10^6)^2(0.05 \times 10^{-3})}$$
$$= \frac{1}{4(9.86)(1 \times 10^{12})(0.05 \times 10^{-3})} = \frac{1}{1.97 \times 10^9} = 5.07 \times 10^{-10}\text{F} = \mathbf{507pF}$$

4. For the following circuit, determine the resonant frequency of the circuit; then determine the values specified at the resonant frequency.

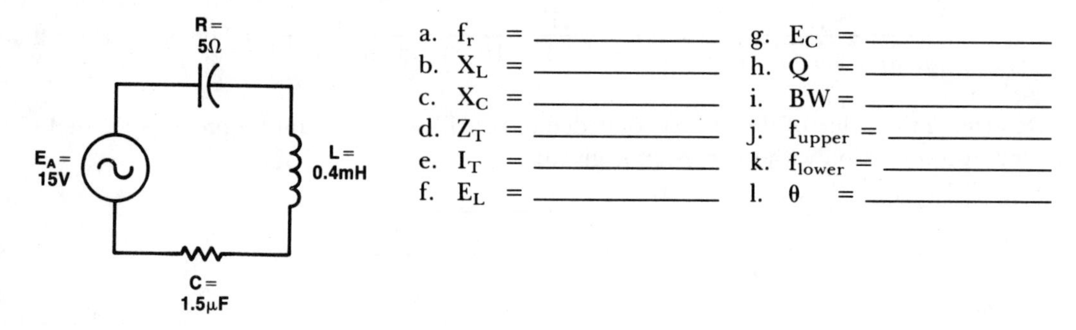

a. f_r = ________
b. X_L = ________
c. X_C = ________
d. Z_T = ________
e. I_T = ________
f. E_L = ________
g. E_C = ________
h. Q = ________
i. BW = ________
j. f_{upper} = ________
k. f_{lower} = ________
l. θ = ________

Solutions:

a. $$f_r = \frac{1}{2\pi\sqrt{LC}} = \frac{1}{(6.28)\sqrt{(0.4\text{mH})(1.5\mu\text{F})}} = \mathbf{6.50kHz}$$

b. $X_L = 2\pi f L = (6.28)(6.5\text{kHz})(0.4\text{mH}) = (6.28)(6.5 \times 10^3)(0.4 \times 10^{-3}) = \mathbf{16.3\Omega}$

c. $X_C = X_L = \mathbf{16.3\Omega}$

d. $Z_T = R = 5\Omega$

e. $$I_T = \frac{E_A}{Z_T} = \frac{E_A}{R} = \frac{15\text{V}}{5\Omega} = \mathbf{3A}$$

f. $E_L = I_T X_L = (3\text{A})(16.3\Omega) = \mathbf{49V}$

■ **Worked-Out Examples**

g. $E_C = E_L = \mathbf{49V}$

h. $Q = \frac{X_L}{R} = \frac{16.3\Omega}{5\Omega} = \mathbf{3.26}$

i. $BW = \frac{f_r}{Q} = \frac{6.5kHz}{3.26} = \mathbf{2kHz}$

j. $f_{upper} = f_r + \frac{1}{2}BW = 6.5kHz + \frac{1}{2}(2kHz) = 6.5kHz + 1kHz = \mathbf{7.5kHz}$

k. $f_{lower} = f_r - \frac{1}{2}BW = 6.5kHz - \frac{1}{2}(2kHz) = 6.5kHz - 1kHz = \mathbf{5.5kHz}$

l. $\theta = \arctan\left(\frac{X_T}{R}\right)$ where $X_T = \left| X_L - X_C \right|$

$= \arctan\left(\frac{0}{5\Omega}\right) = \arctan(0) = \mathbf{0°}$

5. Sketch the frequency response of I versus frequency for the circuit of Problem 4. Show the location of the resonant frequency, upper and lower edge (cutoff) frequencies, bandwidth, and values of current at resonance and at the edge (cutoff) frequencies.

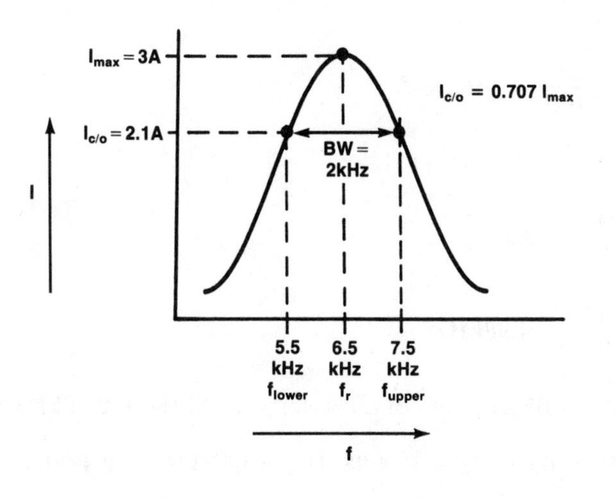

■ Worked-Out Examples

6. For the following circuit, determine the resonant frequency of the circuit; then determine the values specified at the resonant frequency.

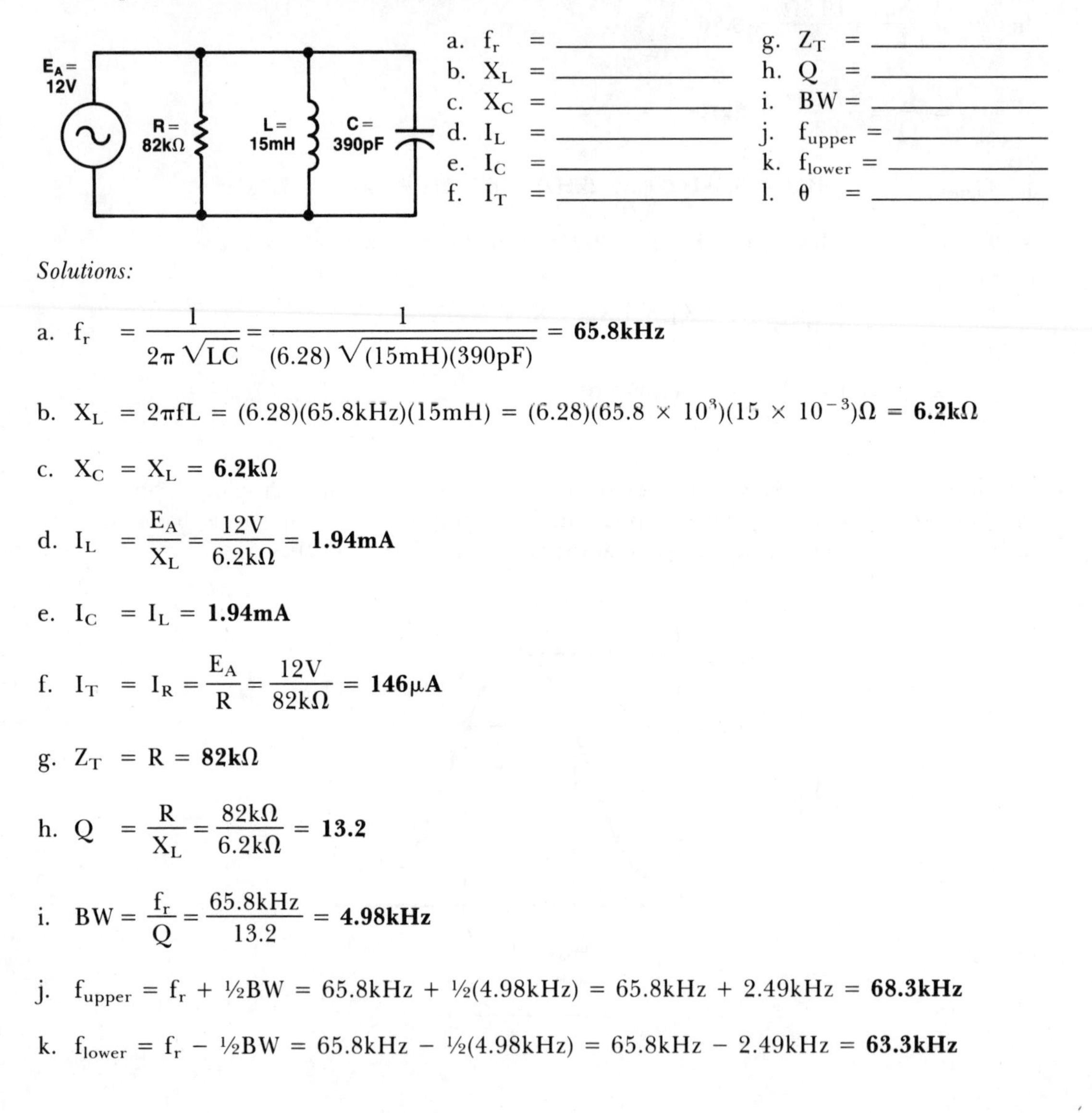

a. f_r = ______ g. Z_T = ______
b. X_L = ______ h. Q = ______
c. X_C = ______ i. BW = ______
d. I_L = ______ j. f_{upper} = ______
e. I_C = ______ k. f_{lower} = ______
f. I_T = ______ l. θ = ______

Solutions:

a. $f_r = \frac{1}{2\pi\sqrt{LC}} = \frac{1}{(6.28)\sqrt{(15mH)(390pF)}} = \mathbf{65.8kHz}$

b. $X_L = 2\pi fL = (6.28)(65.8kHz)(15mH) = (6.28)(65.8 \times 10^3)(15 \times 10^{-3})\Omega = \mathbf{6.2k\Omega}$

c. $X_C = X_L = \mathbf{6.2k\Omega}$

d. $I_L = \frac{E_A}{X_L} = \frac{12V}{6.2k\Omega} = \mathbf{1.94mA}$

e. $I_C = I_L = \mathbf{1.94mA}$

f. $I_T = I_R = \frac{E_A}{R} = \frac{12V}{82k\Omega} = \mathbf{146\mu A}$

g. $Z_T = R = \mathbf{82k\Omega}$

h. $Q = \frac{R}{X_L} = \frac{82k\Omega}{6.2k\Omega} = \mathbf{13.2}$

i. $BW = \frac{f_r}{Q} = \frac{65.8kHz}{13.2} = \mathbf{4.98kHz}$

j. $f_{upper} = f_r + \frac{1}{2}BW = 65.8kHz + \frac{1}{2}(4.98kHz) = 65.8kHz + 2.49kHz = \mathbf{68.3kHz}$

k. $f_{lower} = f_r - \frac{1}{2}BW = 65.8kHz - \frac{1}{2}(4.98kHz) = 65.8kHz - 2.49kHz = \mathbf{63.3kHz}$

■ **Worked-Out Examples**

7. Sketch the frequency response of Z versus frequency for the circuit of Problem 6. Show the location of the resonant frequency, upper and lower edge (cutoff) frequencies, bandwidth, and values of impedance at resonance and at the edge (cutoff) frequencies.

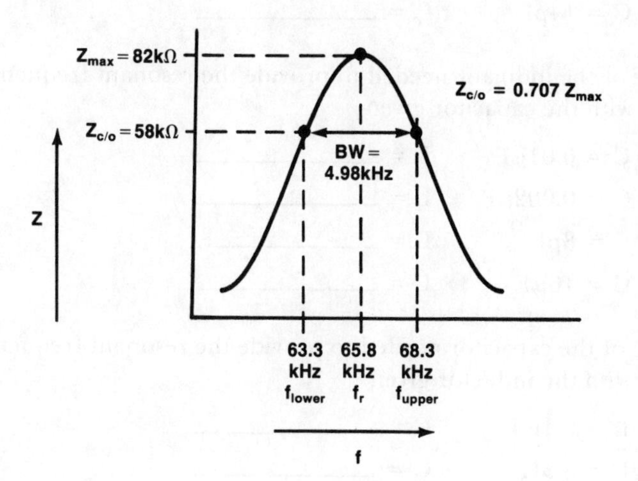

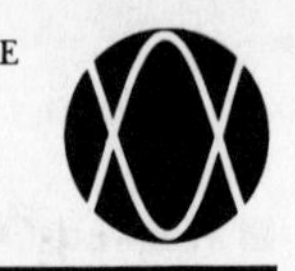

Practice Problems

1. Calculate the resonant frequency of the following L-C combinations:

a. L = 10mH C = 14μF f_r = ________

b. L = 150μH C = 8μF f_r = ________

c. L = 2.5mH C = 0.22μF f_r = ________

d. L = 18μH C = 14pF f_r = ________

2. Determine the value of the inductor needed to provide the resonant frequency specified if used in conjunction with the capacitor given:

a. f_r = 100kHz C = 0.01μF L = ________

b. f_r = 1.5MHz C = 0.002μF L = ________

c. f_r = 8kHz C = 8pF L = ________

d. f_r = 450Hz C = 16μF L = ________

3. Determine the value of the capacitor needed to provide the resonant frequency specified if used in conjunction with the inductor given:

a. f_r = 75kHz L = 0.1mH C = ________

b. f_r = 100Hz L = 8.5H C = ________

c. f_r = 11MHz L = 50μH C = ________

d. f_r = 320kHz L = 0.5mH C = ________

4. For the following circuit, determine the resonant frequency; then determine the values specified at the resonant frequency:

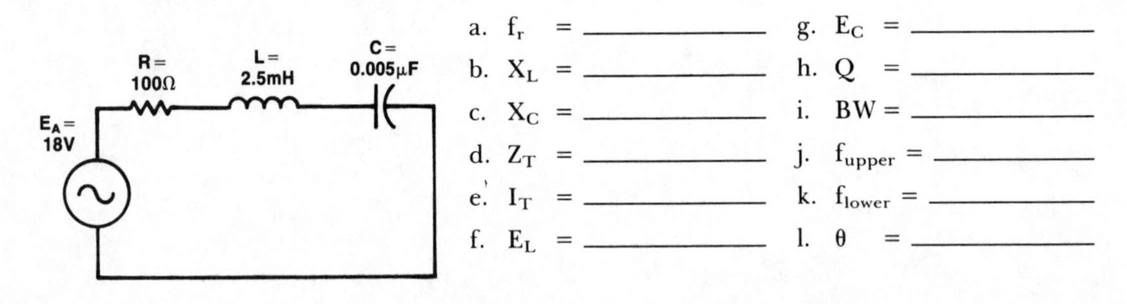

a. f_r = ________

b. X_L = ________

c. X_C = ________

d. Z_T = ________

e. I_T = ________

f. E_L = ________

g. E_C = ________

h. Q = ________

i. BW = ________

j. f_{upper} = ________

k. f_{lower} = ________

l. θ = ________

■ Practice Problems

5. For the following circuit, determine the resonant frequency; then determine the values specified at the resonant frequency.

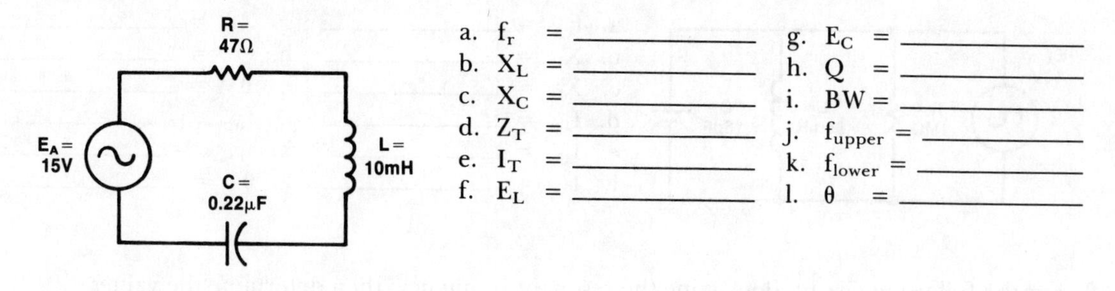

a. f_r = ____________
b. X_L = ____________
c. X_C = ____________
d. Z_T = ____________
e. I_T = ____________
f. E_L = ____________
g. E_C = ____________
h. Q = ____________
i. BW = ____________
j. f_{upper} = ____________
k. f_{lower} = ____________
l. θ = ____________

6. Sketch the frequency responses of I versus frequency for the circuits of Problems 4 and 5. Show the location of the resonant frequency, upper and lower edge (cutoff) frequencies, bandwidth, and values of current at resonance and at the edge (cutoff) frequencies.

a. Current frequency response of the circuit of Problem 4:

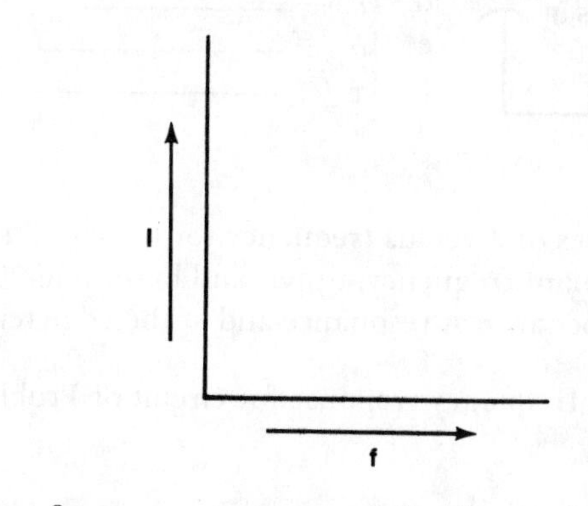

b. Current frequency response of the circuit of Problem 5:

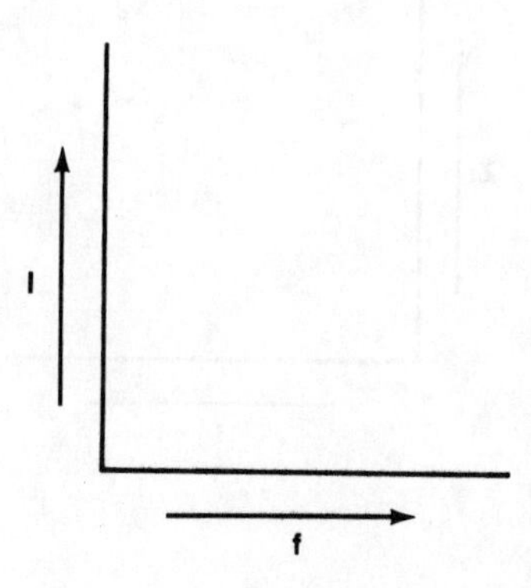

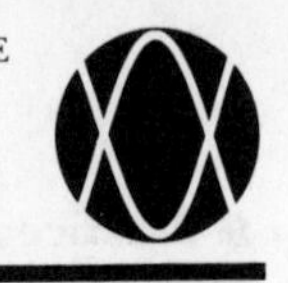

Practice Problems

7. For the following circuit, determine the resonant frequency; then determine the values specified at the resonant frequency.

$E_A =$ 10V, R = 1MΩ, L = 85mH, C = 15pF

a. f_r = ______
b. X_L = ______
c. X_C = ______
d. I_L = ______
e. I_C = ______
f. I_T = ______
g. Z_T = ______
h. Q = ______
i. BW = ______
j. f_{upper} = ______
k. f_{lower} = ______
l. θ = ______

8. For the following circuit, determine the resonant frequency; then determine the values specified at the resonant frequency.

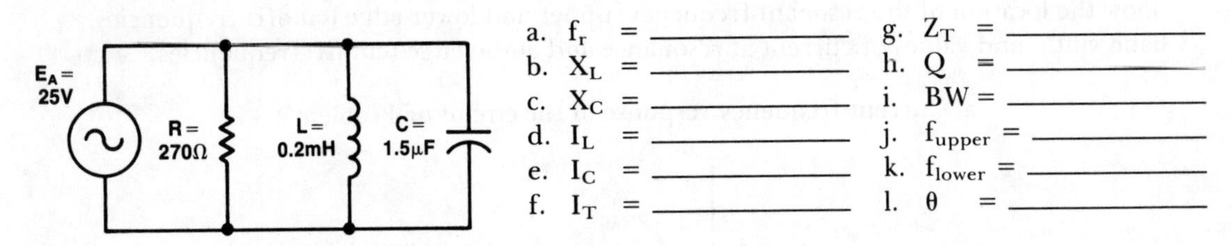

9. Sketch the frequency responses of Z versus frequency for the circuits of Problems 7 and 8. Show the location of the resonant frequency, upper and lower edge (cutoff) frequencies, bandwidth, and values of impedance at resonance and at the edge (cutoff) frequencies.

a. Impedance frequency response for circuit of Problem 7:

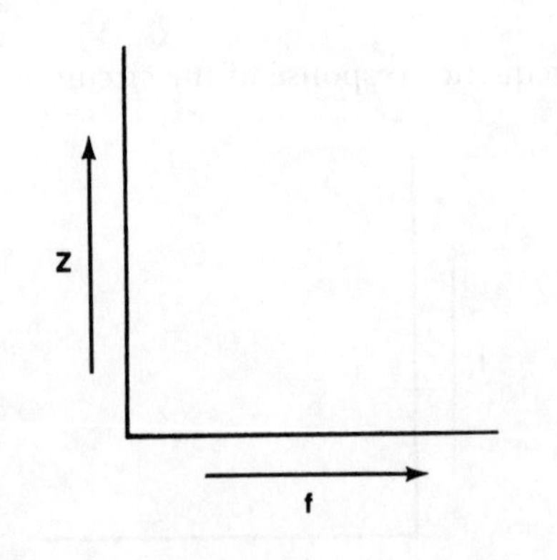

b. Impedance frequency response of circuit of Problem 8:

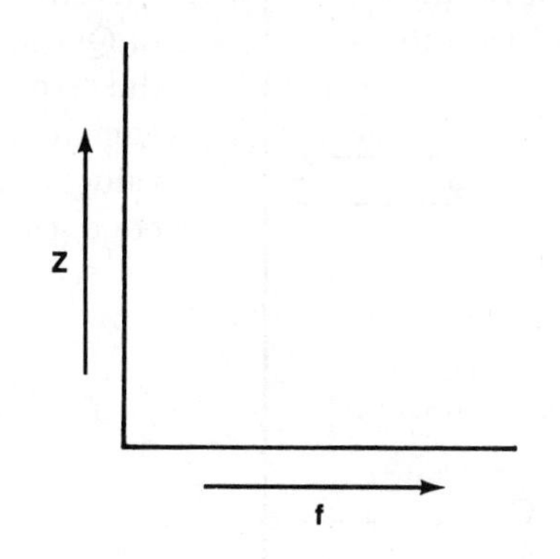

10. In the circuit shown below, it is desired to be able to change its resonant frequency from 540kHz to 1600kHz by adjusting the value of the capacitor. Determine the range of values over which the capacitor must be variable to achieve this.

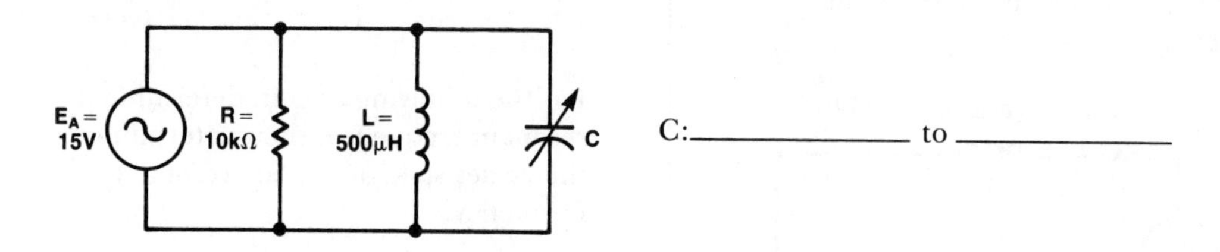

C:____________ to ____________

Quiz

1. Determine the value of the inductor needed to provide the resonant frequency specified if used in conjunction with the capacitor given:

   a. f_r = 100kHz C = 25pF L = ______

   b. f_r = 15kHz C = 0.04μF L = ______

2. Determine the value of the capacitor needed to provide the resonant frequency specified if used in conjunction with the inductor given:

   a. f_r = 6.5MHz L = 0.4mH C = ______

   b. f_r = 850kHz L = 300μH C = ______

3. For the following circuit, determine the resonant frequency; then determine the values specified at the resonant frequency.

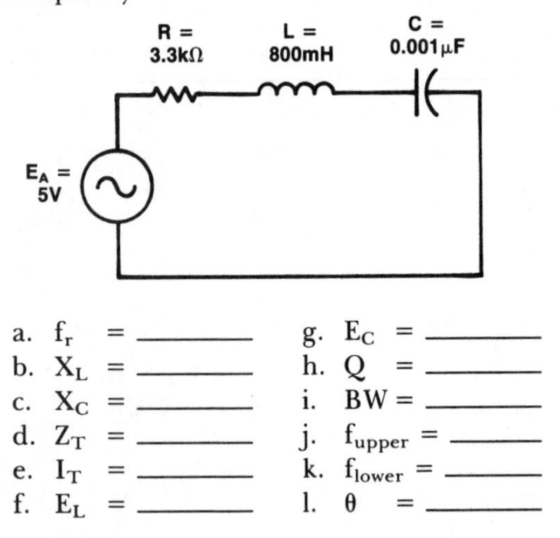

a. f_r = ______		g. E_C = ______
b. X_L = ______		h. Q = ______
c. X_C = ______		i. BW = ______
d. Z_T = ______		j. f_{upper} = ______
e. I_T = ______		k. f_{lower} = ______
f. E_L = ______		l. θ = ______

4. Sketch the frequency response of current versus frequency for the circuit of Question 3. Show the location of the resonant frequency, upper and lower edge (cutoff) frequencies, bandwidth, and values of current at resonance and at the edge (cutoff) frequencies.

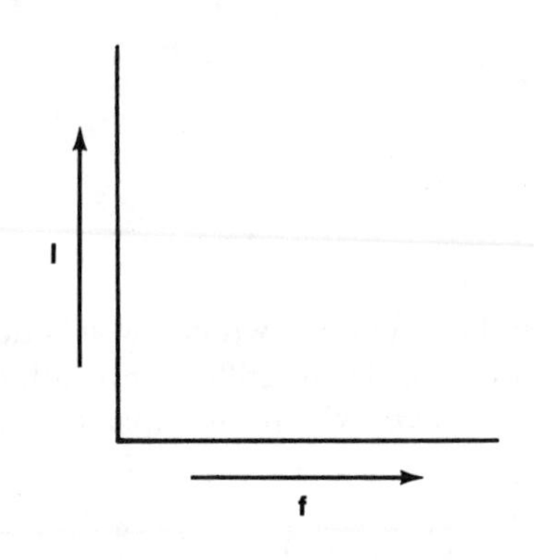

5. For the following circuit, determine the resonant frequency; then determine the values specified at the resonant frequency.

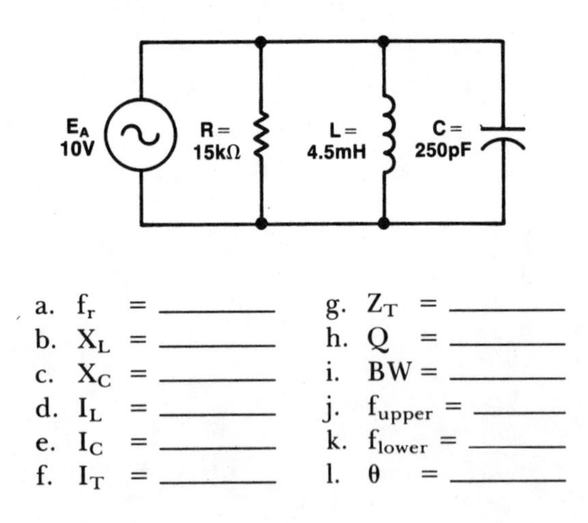

a. f_r = ______	g. Z_T = ______
b. X_L = ______	h. Q = ______
c. X_C = ______	i. BW = ______
d. I_L = ______	j. f_{upper} = ______
e. I_C = ______	k. f_{lower} = ______
f. I_T = ______	l. θ = ______

■ **Quiz**

6. Sketch the frequency response of impedance versus frequency for the circuit of Question 5. Show the location of the resonant frequency, upper and lower edge (cutoff) frequencies, bandwidth, and values of impedance at resonance and at the edge (cutoff) frequencies.

Lesson 1. Practice Problem Answers

1. a. dc
b. dc
c. dc
d. ac
e. ac
f. ac

2. a. 5¾ cycles
b. ¾ cycle
c. 3 cycles
d. 1½ cycles
e. 2 cycles
f. 2½ cycles

3. a. 1000Hz
b. 222Hz
c. 12.2kHz
d. 33.3Hz
e. 200Hz
f. 6.67kHz

4. a. 0.025ms
b. 667μs
c. 15.9μs
d. 8.47μs
e. 1.25μs
f. 0.667μs

5.

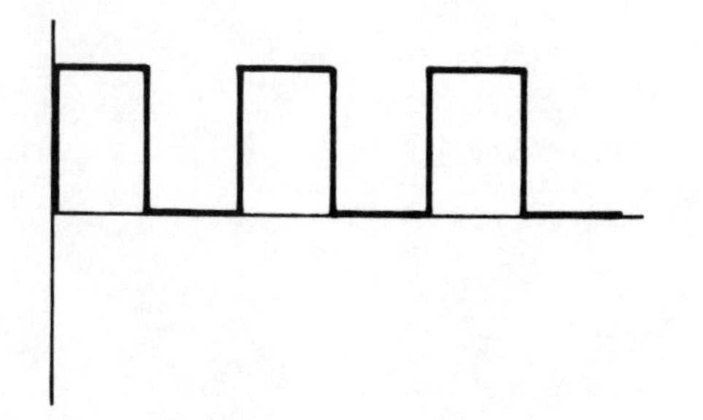

Lesson 2. Practice Problem Answers

1. a. 159μs
b. 3.62ms
c. 26.5μs
d. 132ns

2. a. 20Hz
b. 175Hz
c. 29.6kHz
d. 213kHz

3. a. (1.) $185V_{pk}$
(2.) 130.8VAC
b. (1.) $8.9V_{pk}$
(2.) 6.29VAC
c. (1.) $3.75mV_{pk}$
(2.) 2.65VAC
d. (1.) $16mA_{pk}$
(2.) $11.3mA_{rms}$

4. a. (1.) $14.1V_{pk}$
(2.) $28.3V_{pp}$
b. (1.) $170V_{pk}$
(2.) $340V_{pp}$
c. (1.) $11mV_{pk}$
(2.) $22.1mV_{pp}$
d. (1.) $3.54mA_{pk}$
(2.) $7.07mA_{pp}$

5. a. (1.) 0.593
(2.) 0.805
(3.) 0.737
b. (1.) 0.447
(2.) 0.894
(3.) 0.500

6. a. 71.8°
b. 60°
c. 45°
d. 30°
e. 83.1°
f. 4°

7. a. 20
b. 20
c. 14.1
d. 103.5

8.

	0	A
a.	34.2	94
b.	64.3	76.6
c.	86.6	50.0
d.	98.5	17.4

9. a. II
b. I
c. III
d. II
e. IV

10. a. 46°
b. 80°
c. 25°
d. 60°

■ Lesson 3. Practice Problem Answers

1. a. V
b. H
c. H
d. V
e. M
f. H
g. V
h. V
i. M
j. M
k. H
l. H
m. H
n. H
o. H
p. M
q. M
r. V
s. M
t. V
u. H

2. 4

3. 12.5

4. 3.2

5. 2

6. a. 0.8V = 800mV
b. 0.4V = 400mV
c. 0.283V = 283mV
d. 16ms
e. 62.5Hz

7. a. 30V
b. 15V
c. 10.6V
d. 0.04μs = 40ns
e. 25MHz

8. a. 0.4V = 400mV
b. 0.2V = 200mV
c. 0.14V = 141mV
d. 200μS
e. 5kHz

9. a. 0.028V = 28mV
b. 14mV
c. 9.89mV
d. 28μs
e. 35.7kHz

10. a. 14V
b. 7V
c. 4.95V
d. 14ms
e. 71.4Hz

■ Lesson 4. Practice Problem Answers

1. a. 24.6V
b. 11.4V
c. 32V
d. −82V
e. −53.6V
f. 50.8V

2. a. −20.6V
b. 19.2V
c. 281.5V
d. 44.7V
e. −27.5V
f. 169.6V

3.

θ	sin θ	E_{pk}
0	0	0
18	0.3090	7.4V
36	0.5878	14.1V
54	0.8090	19.4V
72	0.9511	22.8V
90	1	24V
108	0.9511	22.8V
126	0.8090	19.4V
144	0.5878	14.1V
162	0.3090	7.4V
180	0	0
198	−0.3090	−7.4V
216	−0.5878	−14.1V
234	−0.8090	−19.4V
252	−0.9511	−22.8V
270	−1	−24V
288	−0.9511	−22.8V
306	−0.8090	−19.4V
324	−0.5878	−14.1V
342	0.3090	−7.4V
360	0	0

These values must now be plotted as E_{pk} versus θ.

4. a. 172°
b. 298°
c. 131.8°
d. 235°
e. 103°
f. 34.4°

5. a. 0.663 rad
b. 1.75 rad
c. 4.03 rad
d. 4.97 rad
e. 6.00 rad
f. 1.47 rad

6. a. leads, 90°
b. lags, 90°
c. leads, 135°
d. lags, 45°

7. a.

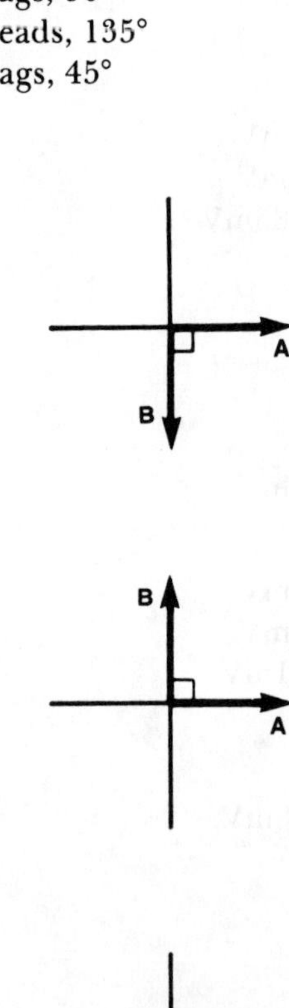

b.

c.

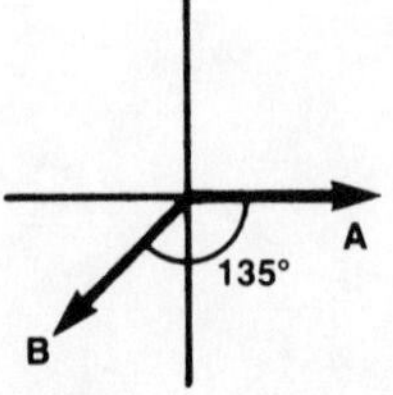

■ Lesson 4. Practice Problem Answers

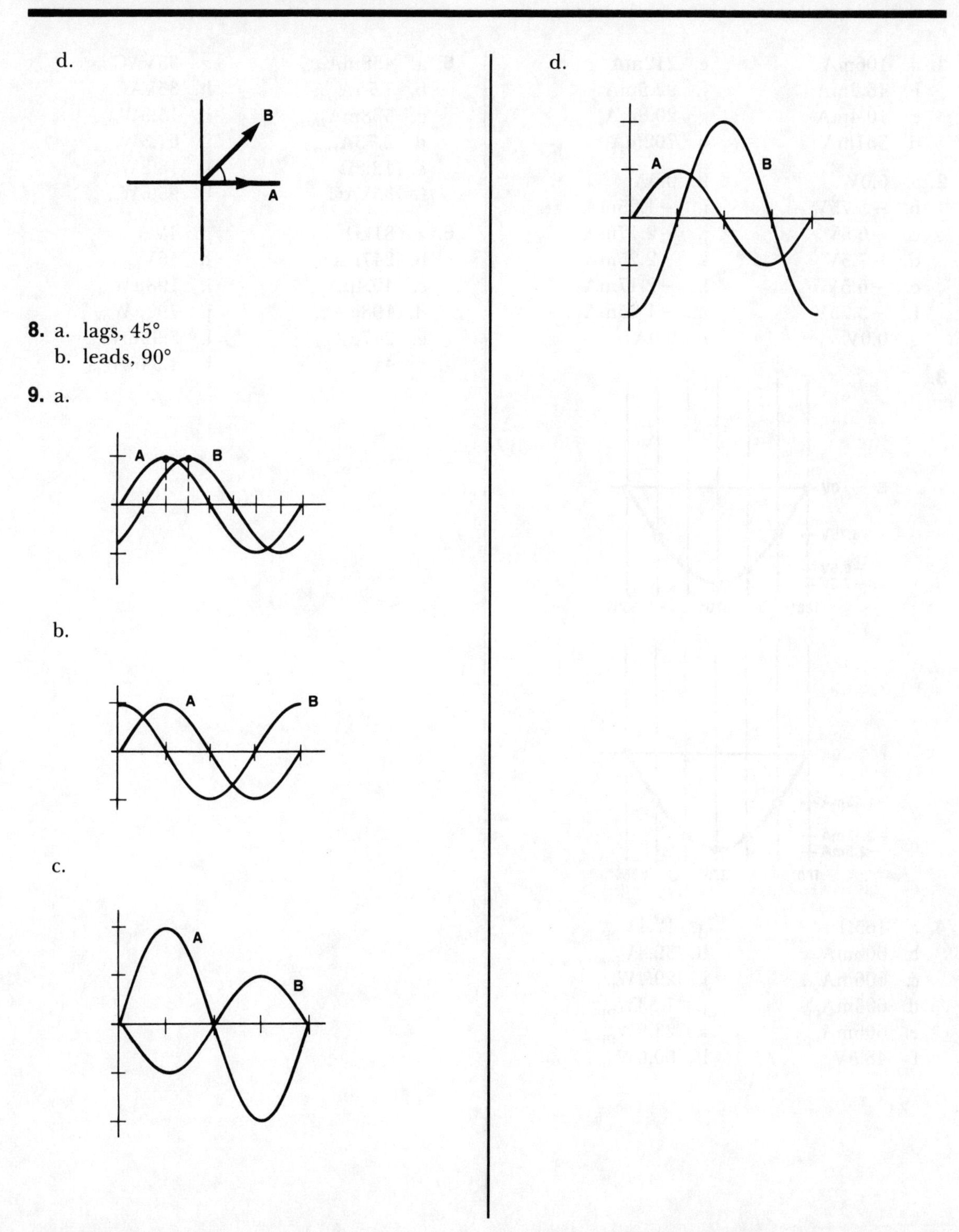

d.

8. a. lags, 45°
b. leads, 90°

9. a.

b.

c.

d.

■ Lesson 5. Practice Problem Answers

1. a. 106mA
b. 46.2mA
c. 10.4mA
d. 351mA
e. 212mA
f. 92.5mA
g. 20.8mA
h. 702mA

2. a. 0.0V
b. −3.75V
c. −6.5V
d. −7.5V
e. −6.5V
f. −3.75V
g. 0.0V
h. 0.0A
i. −1.25mA
j. −2.17mA
k. −2.25mA
l. −2.17mA
m. −1.25mA
n. 0.0A

3.

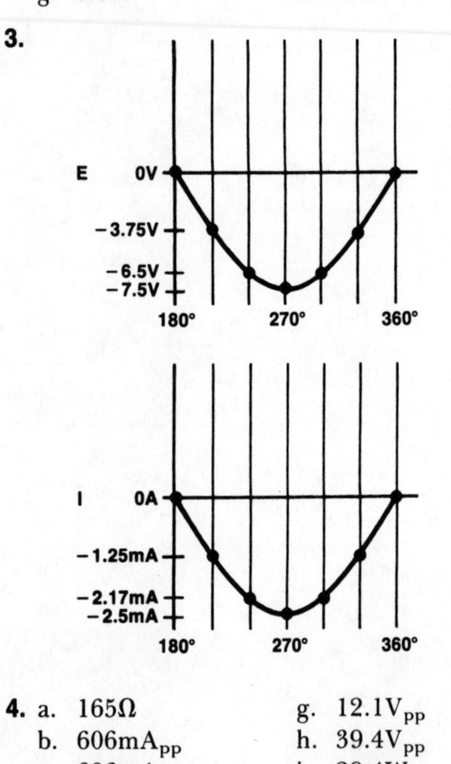

4. a. 165Ω
b. $606mA_{pp}$
c. $606mA_{pp}$
d. $606mA_{pp}$
e. $606mA_{pp}$
f. $48.5V_{pp}$
g. $12.1V_{pp}$
h. $39.4V_{pp}$
i. $29.4W_{pp}$
j. $7.33W_{pp}$
k. $23.9W_{pp}$
l. $60.6W_{pp}$

5. a. $438mA_{rms}$
b. $1.75A_{rms}$
c. $538mA_{rms}$
d. $2.73A_{rms}$
e. 12.8Ω
f. 35VAC
g. 35VAC
h. 35VAC
i. $15.3W_{rms}$
j. $61.3W_{rms}$
k. $18.8W_{rms}$
l. $95.6W_{rms}$

6. a. 81kΩ
b. $247\mu A_{pk}$
c. $49.4\mu A_{pk}$
d. $198\mu A_{pk}$
e. $247\mu A_{pk}$
f. $4V_{pk}$
g. $4V_{pk}$
h. $16V_{pk}$
i. $198\mu W_{pk}$
j. $792\mu W_{pk}$
k. $3.95mW_{pk}$
l. $4.94mW_{pk}$

Lesson 6. Practice Problem Answers

1. Available in medium values from about 0.001 microfarads to 2 microfarads; has a WVDC of about 50 to 2000 volts; and has a typical dielectric constant of 2 to 6 with a dielectric strength of 400 to 1250 volts per mil.

2. a. C charges up to the applied 20 volts over a period of time.
   b. C retains the 20-volt charge.
   c. C discharges to a neutral (zero volts) charge over a period of time.

3. A capacitor in an ac circuit acts as a variable resistance with the resistance being indirectly related to the source frequency. Specifically,

$$X_C = \frac{1}{2\pi fXC}$$

4. Connecting capacitors in parallel is like increasing the area of the plates of the capacitor; both increase capacitance.

5. 2.11kΩ

6. 4.5kpF

7. 0.0127μF

8. X_{CT} = 5.88 kΩ
   C_T = 270 pF

9. a. 1.25 μF
   b. 3.18Ω
   c. 1.59Ω
   d. 7.96Ω
   e. 12.7Ω
   f. 0.394A
   g. 0.394A
   h. 0.394A
   i. 1.25VAC
   j. 0.626VAC
   k. 3.14VAC
   l. $E_{CT} = E_A$ = 5VAC
   m. 0.493VAR
   n. 0.247VAR
   o. 1.24VAR
   p. 1.97VAR

10. a. 2.86Ω
    b. 3.18μF
    c. 1.59μF
    d. 6.37μF
    e. 11.1μF
    f. 10VAC
    g. 10VAC
    h. 10VAC
    i. $1A_{rms}$
    j. $0.5A_{rms}$
    k. $2A_{rms}$
    l. $3.5A_{rms}$
    m. 10VAR
    n. 5VAR
    o. 20VAR
    p. 35VAR

11. a. 0.796μF
    b. 0.398μF
    c. 3.98Ω
    d. 1.59Ω
    e. 14.9Ω
    f. 0.534μF
    g. 13.4VAC
    h. 4.45VAC
    i. 4.45VAC
    j. 2.13VAC
    k. $1.34A_{rms}$
    l. $1.12A_{rms}$
    m. $0.223A_{rms}$
    n. $1.34A_{rms}$
    o. $1.34A_{rms}$
    p. 18.0VAR
    q. 4.98VAR
    r. 0.992VAR
    s. 2.85VAR
    t. 26.8VAR

■ Lesson 7. Practice Problem Answers

1. a. 0.268
b. 0.577
c. 1.0
d. 1.73
e. 3.73

2. a. 5.71°
b. 41.98°
c. 63.43°
d. 75.96°
e. 82.87°

3. The arctangent is an angle in degrees whose tangent is the ratio given.

4. a. 8.06 units
b. 6.63mA
c. 50 feet
d. 12.25V

5. a. 60.3°
b. 33.9°
c. 36.9°
d. 56.4°

6.

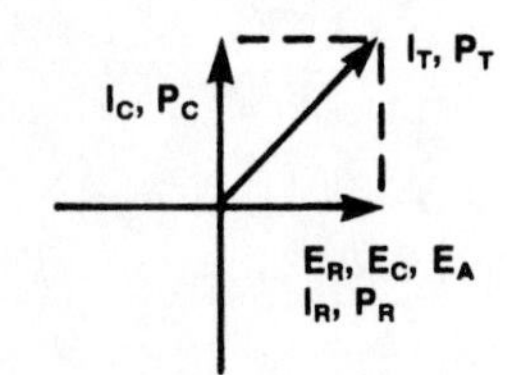

7. a. Parallel RC Circuit
b. 19.6mA
c. 23.4°

8. a. 31.8Ω
b. 67.9Ω
c. $0.295A_{pk}$
d. $0.295A_{pk}$
e. $0.295A_{pk}$
f. $9.38V_{pk}$
g. $17.7V_{pk}$
h. $2.77VAR_{pk}$
i. $5.22W_{pk}$
j. $5.9VA_{pk}$
k. −27.9°

9. a. 1.25mA
b. $0.5mA_{pk}$
c. $1.35mA_{pk}$
d. 3.7kΩ
e. $5V_{pk}$
f. $5V_{pk}$
g. $6.25mVAR_{pk}$
h. $2.5mW_{pk}$
i. $6.75mVA_{pk}$
j. 68.2°

10. a. 796Ω
b. 1990Ω
c. 569Ω
d. 750Ω
e. 941Ω
f. $31.9mA_{pp}$
g. $31.9mA_{pp}$
h. $31.9mA_{pp}$
i. $22.9mA_{pp}$
j. $9.15mA_{pp}$
k. 4.79V
l. $19.1V_{pp}$
m. $18.2V_{pp}$
n. $18.2V_{pp}$
o. $0.153W_{pp}$
p. $0.609W_{pp}$
q. $0.417VAR_{pp}$
r. $0.167VAR_{pp}$
s. $0.957VA_{pp}$
t. −37.2°

Lesson 8. Practice Problem Answers

1. Inductance is the property of a circuit that opposes any change in current.

2. Oersted and Faraday

3. Decrease. This happens because the permeability of iron is more than that of air and as the iron core is extracted, the permeability of the core is reduced; Thus, the value of the inductance of the coil is decreased.

4. Increase

5. $k = \frac{\phi \text{ Common}}{\phi \text{ Total}} = \frac{3500}{4000} = 0.875$

6. 0 to 1 (k = 0, no mutual inductance to k = 1, unity coupling)

7. a. 13H
   b. 2.4H
   c. 17.8H
   d. 8.2H

8. a. 3.45mH
   b. 4.9mH
   c. 1.84mH
   d. 2mH

9. a.

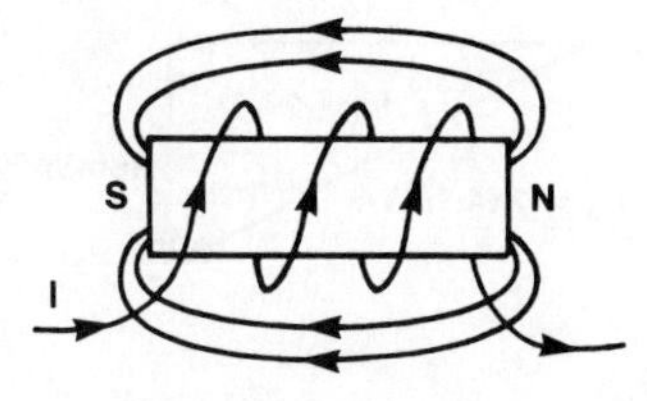

b.

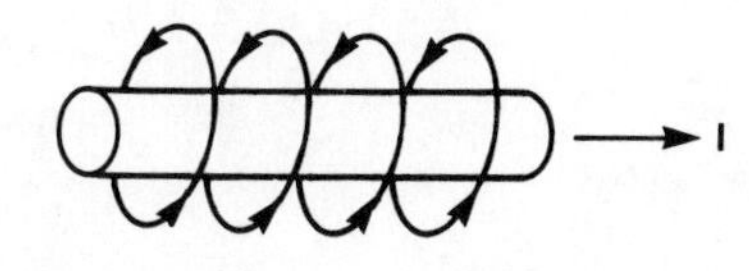

10. a. ROC of i = 4×10^{-3}A/sec
    b. $E_L = 20\mu V$

11. k = 1

12. Step-down

13. a. hysteresis
    b. eddy currents

14. $\frac{N_S}{N_P} = \frac{E_S}{E_P} = \frac{25.2V}{120V} = \frac{1}{4.76}$

    $N_S:N_P = 1:4.76$

15. Autotransformer

16. $P_{pri} = 1920mW$; $P_{sec} = 1650mW$

    $$\% \text{ eff} = \frac{P_{sec}}{P_{pri}} \times 100$$

    $$= \frac{1650mW}{1920mW} \times 100 = 85.9\%$$

17. a. 26.67V
    b. 3.92mA
    c. 0.87mA

18. a. 80%
    b. 1:5 ($N_S:N_P$)

19. a. 1.26Ω
    b. 62.8Ω
    c. 15kΩ

20. Increases

21. 15.4H

22. 18.7Hz

23. a. 62.8Ω
    b. 314Ω
    c. 376.8Ω
    d. 53mA
    e. 3.3V
    f. 16.7V
    g. 175mVAR
    h. 885mVAR
    i. 1060mVAR

24. a. 1.875mH
    b. 4.71Ω
    c. 14.13Ω
    d. 5.3A
    e. 3.53Ω
    f. 7.07A
    g. 25V
    h. 25V
    i. 1.77A

■ Lesson 9. Practice Problem Answers

1. a. X_L = 1.413kΩ, E_R = 39.8V, E_L = 20.8V, I_R = 14.75mA, I_L = 14.75mA, I_T = 14.75mA
Z = 3.05kΩ, θ = +27.6°, P_R = 587mW, P_L = 306.8mVAR, P_A = 663.75mVA

b. X_L = 2.45kΩ, E_R = 45.3V, E_L = 111V, I_R = 45.3mA, I_L = 45.3mA, I_T = 45.3mA
Z = 2.65kΩ, θ = +67.8°, P_R = 2.05W, P_L = 5.03VAR, P_A = 5.43VA

c. X_L = 13.2kΩ, E_R = 19.5V, E_L = 17.2V, I_R = 1.3mA, I_L = 1.3mA, I_T = 1.3mA
Z = 20kΩ, θ = +41.3°, P_R = 25.4mW, P_L = 22.4mVAR, P_A = 33.8mVA

2. a. Voltage Phasor Diagram

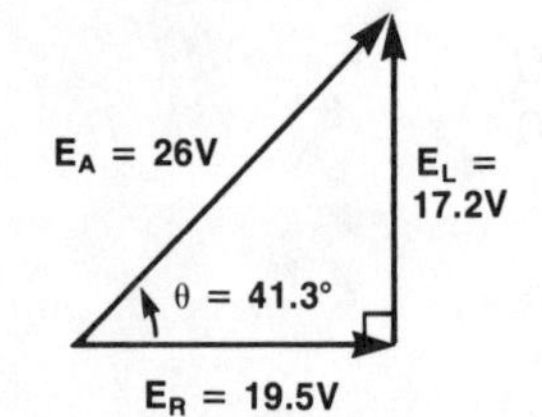

b. Impedance Phasor Diagram

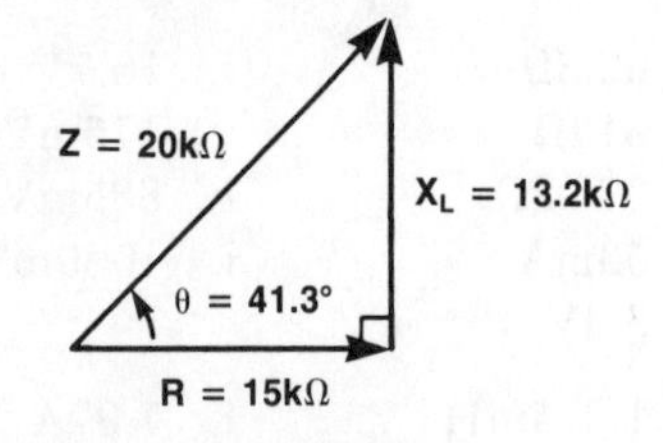

c. Power Phasor Diagram

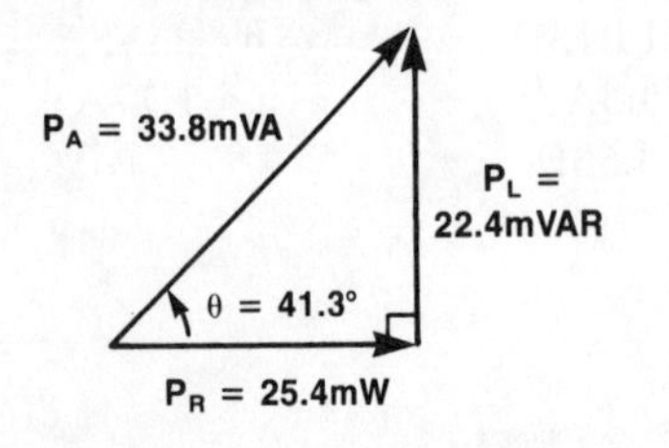

3. a. E_R = 60V, E_L = 60V, I_R = 5mA, I_L = 12mA, I_T = 13mA
Z = 4.6kΩ, θ = -67.4°, P_R = 300mW, P_L = 720mVAR, P_A = 780mVA

b. E_R = 36V, E_L = 36V, I_R = 2μA, I_L = 9μA, I_T = 9.2μA
Z = 3.9MΩ, θ = −77.5°, P_R = 72μW, P_L = 324μVAR, P_A = 331μVA

c. E_R = 24V, E_L = 24V, I_R = 8mA, I_L = 4mA, I_T = 8.94mA
Z = 2.68kΩ, θ = -26.6°, P_R = 192mW, P_L = 96mVAR, P_A = 214.7mVA

4. a. Current Phasor Diagram

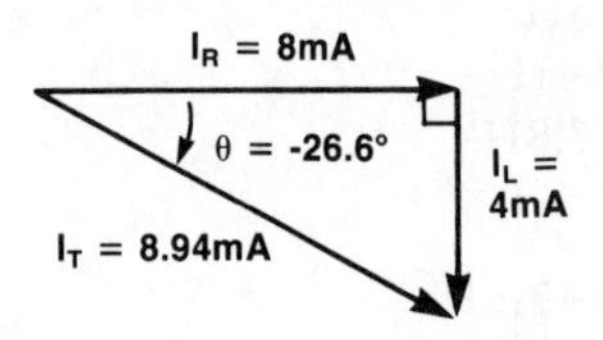

b. Power Phasor Diagram

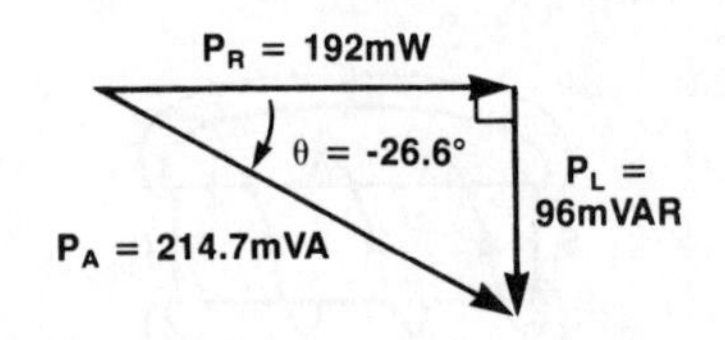

5. Q = 16.3

■ Lesson 10. Practice Problem Answers

1. $1\tau = 1ms$

2. $1\tau = 50\mu s$

3. a. 18.8mA
b. 15V
c. 15V
d. 250μs

4.

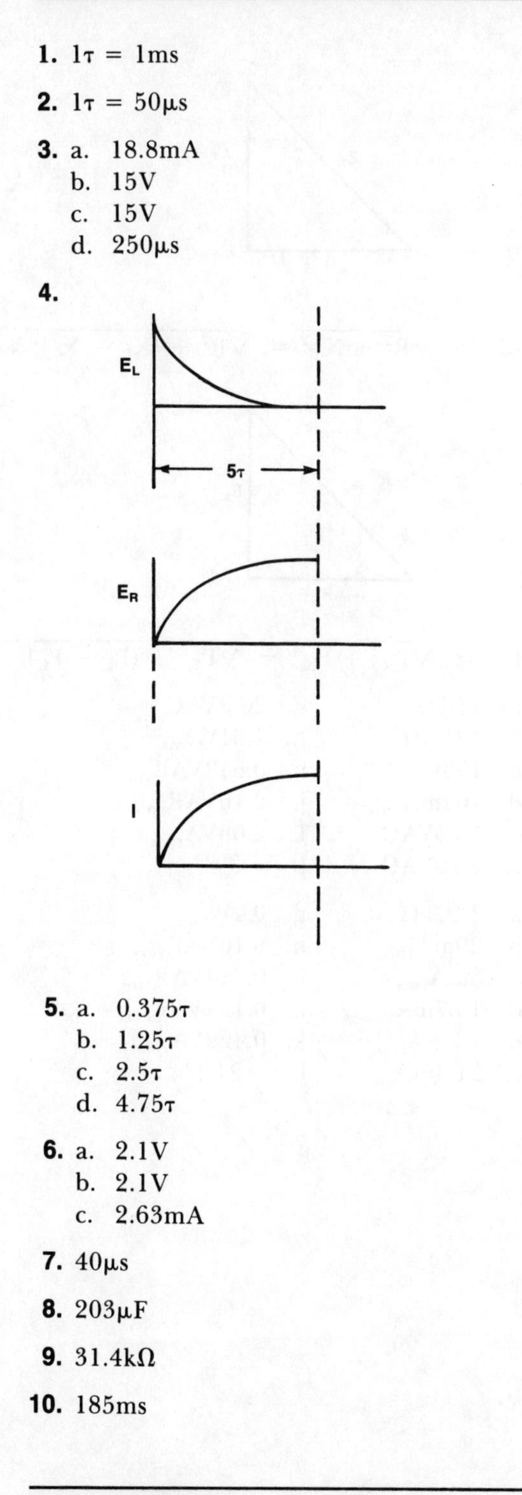

5. a. 0.375τ
b. 1.25τ
c. 2.5τ
d. 4.75τ

6. a. 2.1V
b. 2.1V
c. 2.63mA

7. 40μs

8. 203μF

9. 31.4kΩ

10. 185ms

■ Lesson 11. Practice Problem Answers

1. $I_L > I_C$; therefore, I_X is plotted in phase with I_L.

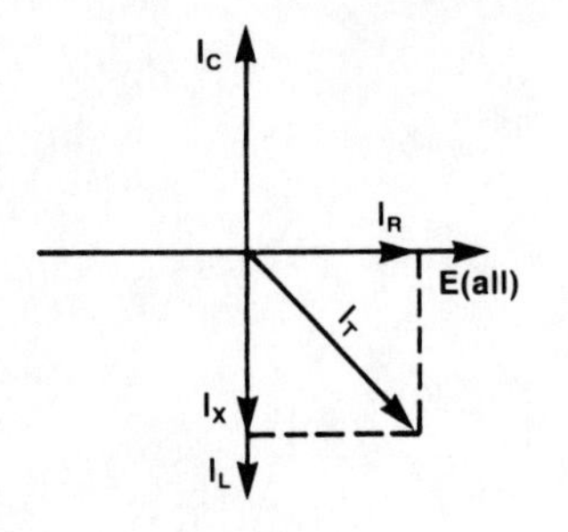

2.

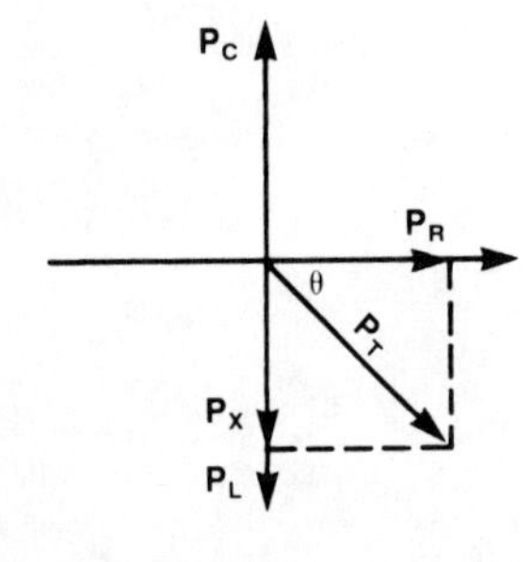

$$\tan\theta = \frac{I_X}{I_R} = \frac{19\text{mA}}{15\text{mA}} = 1.27$$

arctan 1.27 = **−51.7°**

The angle is rotated clockwise; therefore, the sign of the angle is negative.

3.

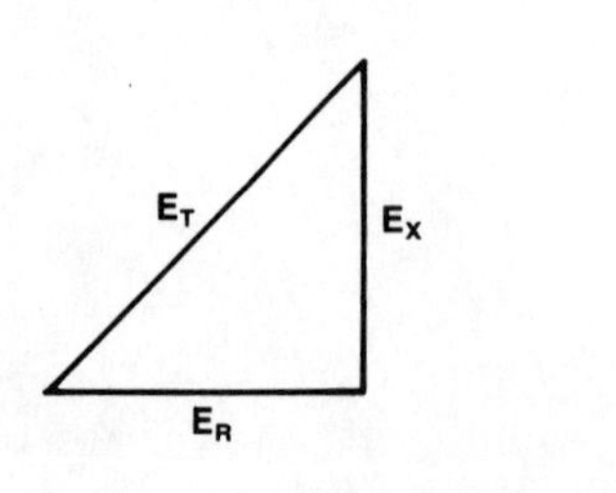

$$E_T = \sqrt{E_R^2 + E_X^2} = \sqrt{E_R^2 + (E_L - E_C)^2}$$

Z_T
X_T
R

$$Z_T = \sqrt{R^2 + X_T^2} = \sqrt{R^2 + (X_L - X_C)^2}$$

P_A
P_X
P_R

$$P_A = \sqrt{P_R^2 + P_X^2} = \sqrt{P_R^2 + (P_L - P_C)^2}$$

4. a. 78.5Ω
b. 121.5Ω
c. 193Ω
d. 104mA_{rms}
e. 15.5VAC
f. 8.16VAC
g. 20.8VAC
h. 1.61W_{rms}
i. 0.849VAR_{rms}
j. 2.16VAR_{rms}
k. 2.08VA_{rms}
l. −39°

5. a. 1.27kΩ
b. 20mA_{rms}
c. 8mA_{rms}
d. 15.7mA_{rms}
e. 7.7mA_{rms}
f. 21.4mA_{rms}
g. 0.4W_{rms}
h. 0.16VAR_{rms}
i. 0.314VAR_{rms}
j. 0.154VAR_{rms}
k. 0.429VA_{rms}
l. +21.1°

■ Lesson 12. Practice Problem Answers

1. a. $3 + j4 = 5\angle 53°$
b. $5 + j2 = 5.38\angle 21.8°$
c. $4 - j2 = 4.47\angle -26.6°$
d. $3 - j5 = 5.83\angle -59°$

2. a. -1
b. $-j$
c. $+1$
d. $-j$
e. $+1$
f. $-j$

3. a.

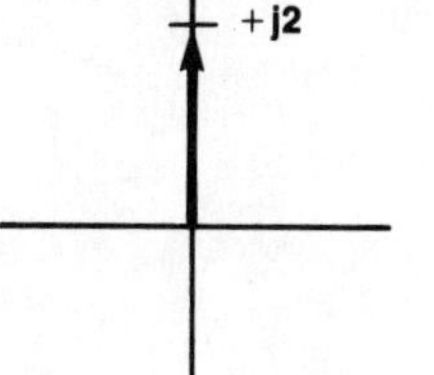

b.

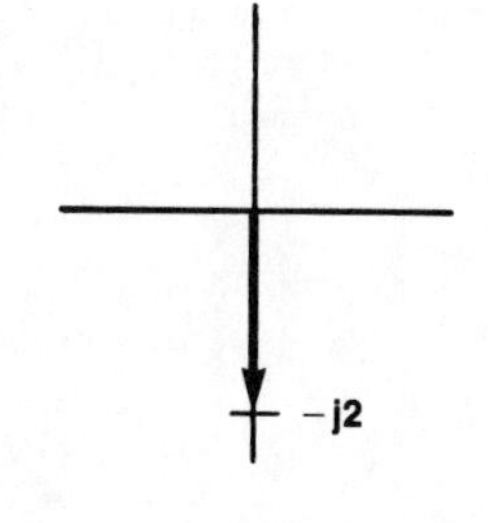

c.

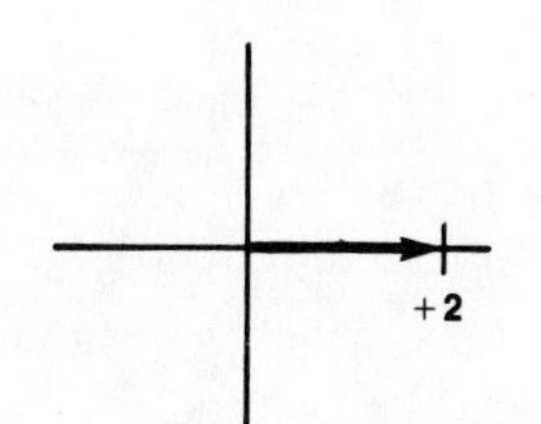

4. a. $40 + j50$
b. $500 - j470$
c. $1000 + j7000$
d. $4.7M - j1M$

5. a. $58.3\angle 31°$
b. $5.69k\angle 55.8°$
c. $399\angle -22.1°$
d. $3.04M\angle -9.46°$

6. a. $485.6 + j258.2$
b. $50k - j86.6k$
c. $56.4M + j38M$
d. $16.3k - j7.6k$

Lesson 13. Practice Problem Answers

1. a. j15
b. j26
c. $j^2 153 = -153$
d. −j1.33
e. j1 = j
f. j70
g. $j^4 21 = 21$
h. j4

2. a. 10 + j24
b. 21 − j22
c. 18 + j1
d. 1 + j4
e. 0.9 + j0.6
f. 13 + j8

3. a. $24\angle 15°$
b. $4\angle 20°$
c. $30\angle 2°$
d. $5\angle 40°$

4. $(38.9 - j1.57)\Omega = 38.9\Omega\angle -2.3°$

5. $(15 - j20)\Omega = 25\Omega\angle -53°$

6. $(40 + j20)\Omega = 44.7\Omega\angle 26.6°$

7. $(9.27 + j13.9)\Omega = 16.7\angle 56.3°$

8. a. $15\Omega = 15\Omega\angle 0°$
b. $-j10\Omega = 10\Omega\angle -90°$
c. $+j20\Omega = 20\Omega\angle +90°$

9. a. $43.6\Omega\angle -17°$
b. $573mA\angle 17°$
c. $573mA\angle 17°$
d. $105mA\angle -20.6°$
e. $493mA\angle 24.4°$
f. $22.9V\angle 17°$
g. $5.23V\angle -20.6°$
h. $4.93V\angle 114.4°$
i. $12.3V\angle -65.6°$
j. $5.23V\angle 110.6°$
k. 17°

10. a. $322.5\Omega\angle -8.2°$
b. $372mA\angle 8.2°$
c. $372mA\angle 8.2°$
d. $310mA\angle -25.5°$
e. $207mA\angle 64.5°$
f. $93V\angle 8.2°$
g. $31V\angle -25.5°$
h. $31V\angle -25.5°$
i. 8.2°

■ Lesson 14. Practice Problem Answers

1. a. 425.4kHz
b. 4.59kHz
c. 6.79kHz
d. 10MHz

2. a. 253μH
b. 5.63μH
c. 49.5H
d. 7.82mH

3. a. 0.045μF
b. 0.298μF
c. 4.19pF
d. 475pF

4. a. 45kHz
b. 707Ω
c. 707Ω
d. 100Ω
e. 180mA
f. 127V
g. 127V
h. 7.07
i. 6.36kHz
j. 48.2kHz
k. 41.8kHz
l. 0°

5. a. 3.39kHz
b. 213Ω
c. 213Ω
d. 47Ω
e. 319mA
f. 68V
g. 68V
h. 4.54
i. 747Hz
j. 3.76kHz
k. 3.02kHz
l. 0°

6. a.

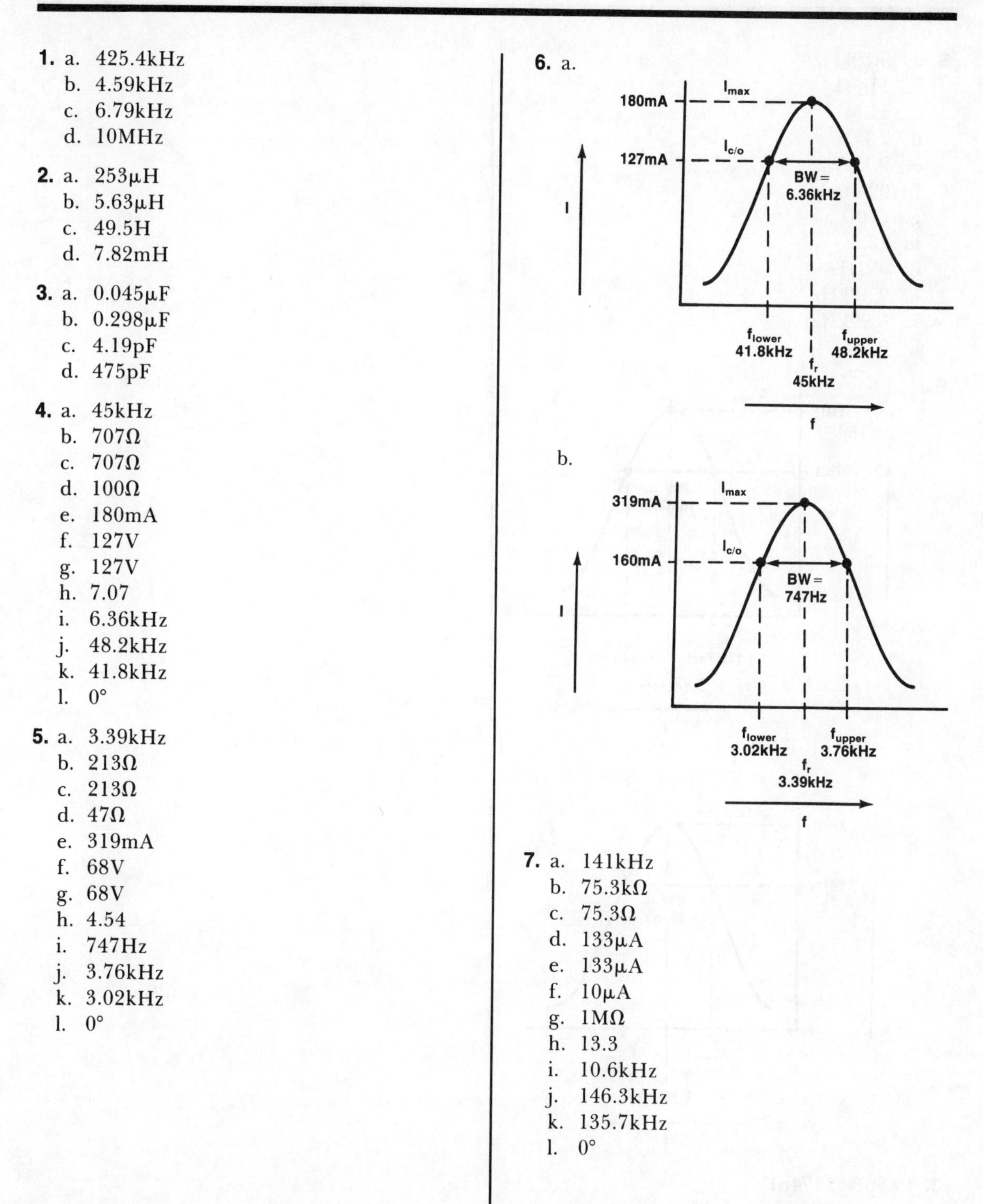

7. a. 141kHz
b. 75.3kΩ
c. 75.3Ω
d. 133μA
e. 133μA
f. 10μA
g. 1MΩ
h. 13.3
i. 10.6kHz
j. 146.3kHz
k. 135.7kHz
l. 0°

Lesson 14. Practice Problem Answers

8. a. 9.19kHz
b. 11.5Ω
c. 11.5Ω
d. 2.17A
e. 2.17A
f. 92.6mA
g. 270Ω
h. 23.5
i. 391Hz
j. 9.39kHz
k. 8.99kHz
l. 0°

9. a.

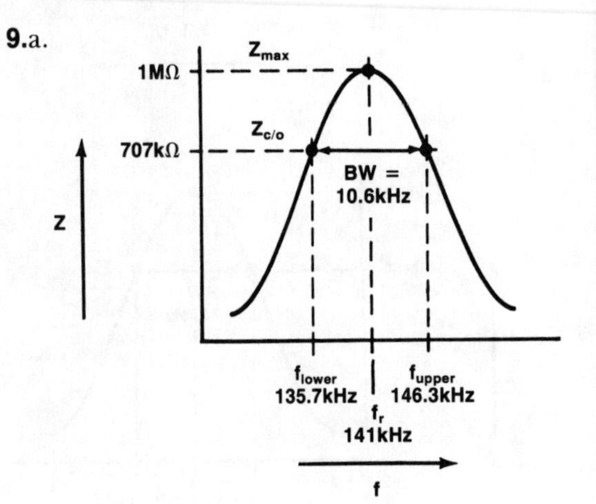

b.

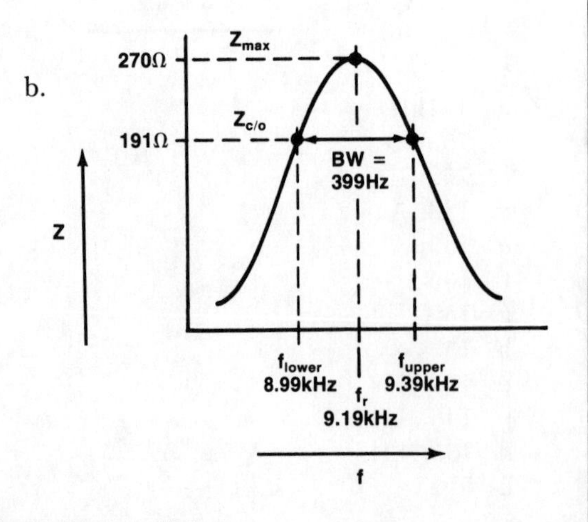

10. 19.8pF to 174pF

■ Lesson 1. Quiz Answers

1. a. dc
b. dc
c. ac
d. ac
e. dc
f. ac

2. a. 4 cycles
b. 1¼ cycles
c. 2½ cycles
d. 3½ cycles
e. 2½ cycles
f. 2½ cycles

3. a. 66.7Hz
b. 26.9Hz
c. 26.9kHz
d. 125kHz
e. 20.8Hz
f. 136kHz

4. a. 33.3ms
b. 3.33ms
c. 15.7μs
d. 50μs
e. 588μs
f. 45.5ns

5.

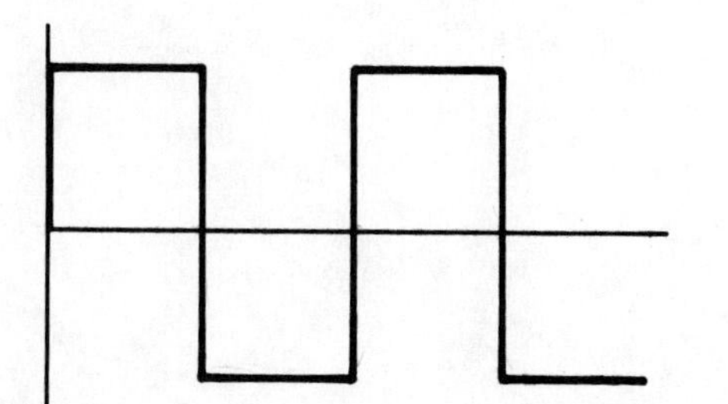

Lesson 2. Quiz Answers

1. a. 189ns
b. 6.13μs
c. 27Hz
d. 357kHz

2. a. (1.) 74V
(2.) 26.2V
b. (1.) 150mV
(2.) 53mV
c. (1.) 1.08mA
(2.) 382μA

3. a. (1.) 84.8V
(2.) 169.7V
b. (1.) 9.5mA
(2.) 18.9mA
c. (1.) 4.5A
(2.) 8.9A

4. a. (1.) 0.880
(2.) 0.475
(3.) 1.85
b. (1.) 0.287
(2.) 0.957
(3.) 0.300

5. a. less than
b. equal to
c. greater than

6. c

7. b

8. c

9. a

10. a. 43°
b. 52°
c. 53°

■ Lesson 3. Quiz Answers

1. a. P
b. A
c. G
d. N
e. H
f. O
g. Q
h. M
i. K
j. J
k. C
l. E
m. S
n. R
o. X
p. F
q. D
r. T
s. B
t. U
u. V
v. W
w. I
x. L

2. c

3. c

4. b

5. c

6. b

7. a

8. c

9. b

10. b

11. c

12. a

13. d

14. a

15. c

16. b

17. d

18. b

19. d

20. c

21. c

22. 6 vertical divisions
5 horizontal divisions

23. a. 1.4V
b. 0.7V
c. 0.5V
d. 40ms
e. 25Hz

24. a. 800mV
b. 400mV
c. 282mV
d. 24μs
e. 41.7kHz

25. a. 5.6V
b. 2.8V
c. 1.98V
d. 560μs
e. 1.79kHz

26. a. 10.0V
b. 5.0V
c. 3.5V
d. 8.4ms
e. 119Hz

27. 25V

■ Lesson 4. Quiz Answers

1. a. 36.4V
b. − 16.4V

2. a. 19.4V
b. 11.4V

3. a. 223.5°
b. 315.2°

4. a. 0.77 rad
b. 4.33 rad

5. a. lags, 90°
b. leads, 45°
c.
d.

6. a.
b.
c.

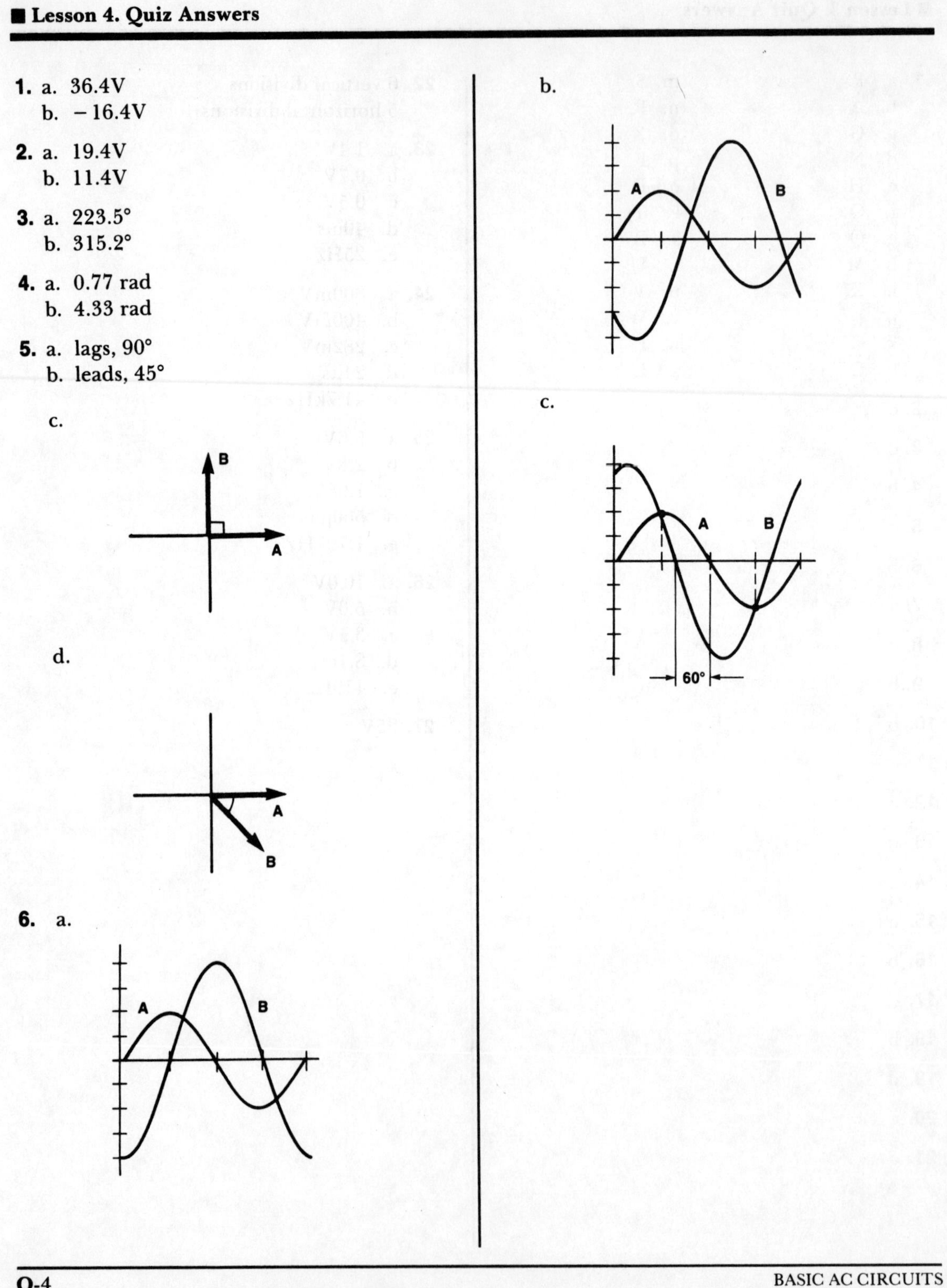

■ Lesson 5. Quiz Answers

1. a. 88mA
b. 1.45mA
c. 32.6V
d. 219.4V
e. 31.2mA
f. 407μA
g. 11.6V
h. 77.9V

2. a. 0.0V
b. 35.4V
c. 50V
d. 35.4V
e. 0.0V
f. −35.4V
g. −50V
h. −35.4V
i. 0.0V
j. 0.0A
k. 1.77mA
l. 2.5mA
m. 1.77mA
n. 0.0mA
o. −1.77mA
p. −2.5mA
q. −1.77mA
r. 0.0A

3.

+50V
+35.4V
E 0V
−35.4V
−50V
0° 45° 90° 135° 180° 225° 270° 315° 360°

+2.5mA
+1.77mA
I 0A
−1.77mA
−2.5mA
0° 45° 90° 135° 180° 225° 270° 315° 360°

4. a. 2.56kΩ
b. $11.7mA_{pk}$
c. $11.7mA_{pk}$
d. $11.7mA_{pk}$
e. $11.7mA_{pk}$
f. $9.59V_{pk}$
g. $14.0V_{pk}$
h. $6.32V_{pk}$
i. $112mW_{pk}$
j. $164mW_{pk}$
k. $73.9mW_{pk}$
l. $351mW_{pk}$

5. a. $97.6mA_{pp}$
b. $66.7mA_{pp}$
c. $148mA_{pp}$
d. $312mA_{pp}$
e. 256Ω
f. $80V_{pp}$
g. $80V_{pp}$
h. $80V_{pp}$
i. $7.81W_{pp}$
j. $5.34W_{pp}$
k. $11.8W_{pp}$
l. $25W_{pp}$

6. a. 1.19kΩ
b. $8.39mA_{rms}$
c. $8.39mA_{rms}$
d. $2.6mA_{rms}$
e. 5.78mA
f. 6.88VAC
g. 3.12VAC
h. 3.12VAC
i. $57.7mW_{rms}$
j. $8.11mW_{rms}$
k. $18.0mW_{rms}$
l. $83.9mW_{rms}$

Lesson 6. Quiz Answers

1. a. teflon
b. ceramic
c. electrolytic
d. paper

2. a. Charge to the 50-volt source voltage over a period of time.
b. Retain the 50-volt charge.
d. Discharge to zero volts over a period of time.

3.

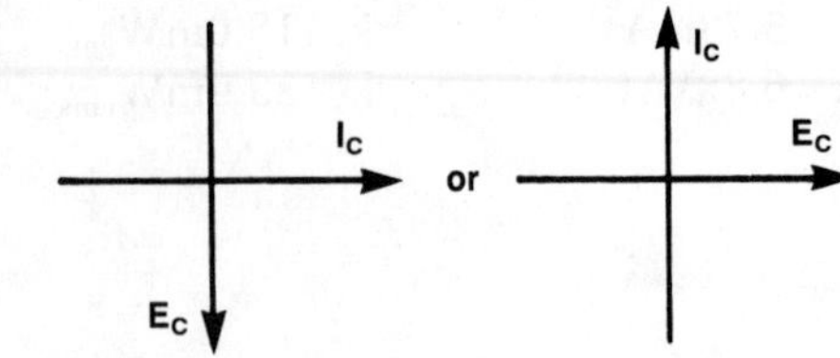

4.

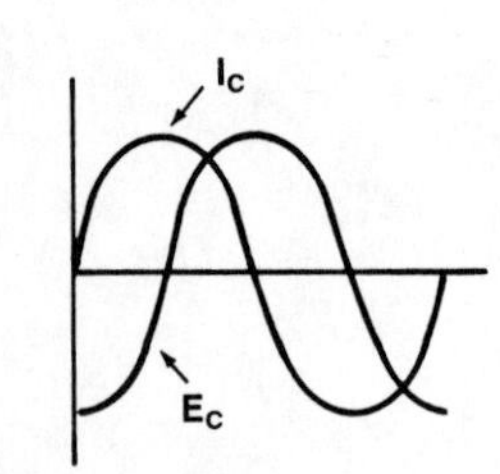

5. d.

6. a. 159Hz
b. 395Ω
c. 2.53 μF

7. a. 0.008μF
b. 39.8kΩ
c. 1.99kΩ
d. 7.96kΩ
e. 49.8kΩ
f. 0.201mA (rms)
g. 0.201mA (rms)
h. 0.201mA (rms)
i. 0.201mA (rms)
j. 8VAC
k. 0.4VAC
l. 1.6VAC
m. 1.61mVAR (rms)
n. 80.4μVAR (rms)
o. 0.322mVAR (rms)
p. 2.01mVAR (rms)

8. a. 10.7Ω
b. 2.65μF
c. 3.18μF
d. 1.59μF
e. 7.42μF
f. 50VAC
g. 50VAC
h. 50VAC
i. 1.67A (rms)
j. 2A (rms)
k. 1A (rms)
l. 4.67A (rms)
m. 83.5VAR (rms)
n. 100VAR (rms)
o. 50VAR (rms)
p. 233VAR (rms)

9. a. 1.59μF
b. 2.65μF
c. 7.96Ω
d. 4.0Ω
e. 12.2Ω
f. 1.31μF
g. 2.05A (rms)
h. 0.616A (rms)
i. 0.616A (rms)
j. 1.44A (rms)
k. 2.05A (rms)
l. 16.3VAC
m. 6.16VAC
n. 2.45VAC
o. 8.61VAC
p. 33.4VAR (rms)
q. 3.79VAR (rms)
r. 1.51VAR (rms)
s. 12.4VAR (rms)
t. 51.1VAR (rms)

10. rms

■ Lesson 7. Quiz Answers

1. a. 5 feet
b. 21.63Ω
c. 25.08V
d. 70.7mA

2. a. 53.1°
b. 33.7°
c. 66.5°
d. 45°

3.

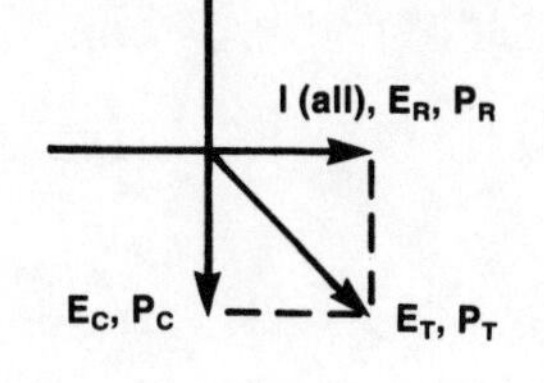

4. a. Series RC circuit
b. 25.3V
c. −50.8°

5. a. 79.6Ω
b. 144Ω
c. $69.4mA_{rms}$
d. $69.4mA_{rms}$
e. $69.4mA_{rms}$
f. 5.53VAC
g. $8.33mA_{rms}$
h. $384mVAR_{rms}$
i. $578mVAR_{rms}$
j. $694VA_{rms}$
k. −33.6°

6. a. $57.1mA_{pk}$
b. $57.1mA_{pk}$
c. $80.8mA_{pk}$
d. 495Ω
e. $2.29VAR_{pk}$
f. $2.29W_{pk}$
g. $3.23VA_{pk}$
h. 45°

7. a. 16Ω
b. 20Ω
c. 25.6Ω
d. $0.273A_{p\text{-}p}$
e. $0.273A_{p\text{-}p}$
f. $0.273A_{p\text{-}p}$
g. $0.182A_{p\text{-}p}$
h. $0.091A_{p\text{-}p}$
i. $2.73V_{p\text{-}p}$
j. $1.64V_{p\text{-}p}$
k. $5.46V_{p\text{-}p}$
l. $5.46V_{p\text{-}p}$
m. $0.745W_{p\text{-}p}$
n. $0.448W_{p\text{-}p}$
o. $0.994VAR_{p\text{-}p}$
p. $0.497VAR_{p\text{-}p}$
q. $1.91VA_{p\text{-}p}$
r. 51.3°

8. rms

9. peak

10. peak-to-peak

■ Lesson 8. Quiz Answers

1. c.

2. e.

3. a.

4. c.

5. c.

6. b.

7. d.

8. b.

9. a. 6.3VAC
b. 1.9mA
c. 0.1mA
d. 12mW

10. Step-up

11. a. 471Ω
b. 0.9mH
c. 9.55MHz

12. a. 7.5kΩ
b. 12V
c. 3V
d. 2mA
e. 24mVAR
f. 6mVAR
g. 30mVAR
h. 20H
i. 50Ω
j. 2A
k. 1A
l. 3A
m. 300VAR
n. 150VAR
o. 450VAR

■ Lesson 9. Quiz Answers

1. a. $X_L = 2512\Omega = 2.51k\Omega$
b. $Z = 4147\Omega = 4.15k\Omega$
c. $E_R = 19.9V$
d. $E_L = 15.1V$
e. $I_R = 6.03mA$
f. $I_L = 6.03mA$
g. $I_T = 6.03mA$
h. $\theta = +37.3°$
i. $P_R = 120mW$
j. $P_L = 91.3mVAR$
k. $P_A = 150.75mVA$
l. Voltage Phasor Diagram

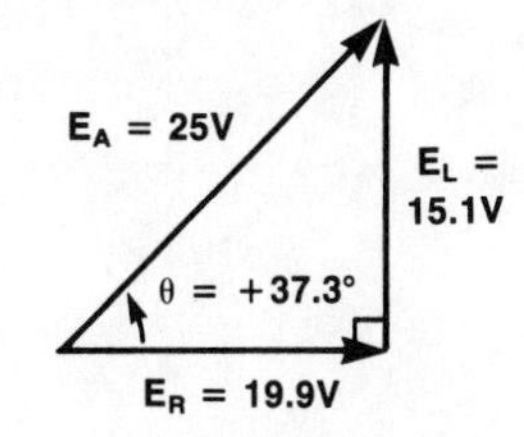

m. Impedance Phasor Diagram

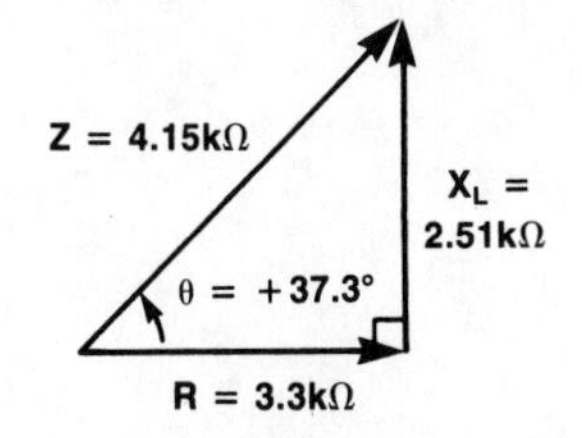

n. Power Phasor Diagram

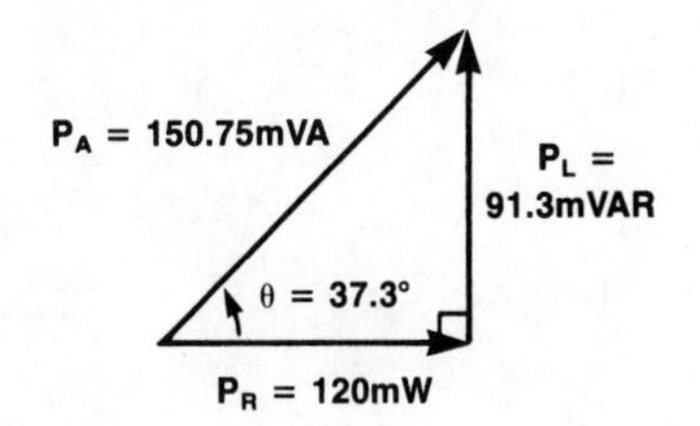

2. a. $X_L = 6.28k\Omega$
b. $E_R = 12V$
c. $E_L = 12V$
d. $I_R = 2.55mA$
e. $I_L = 1.91mA$
f. $I_T = 3.19mA$
g. $Z = 3.76k\Omega$
h. $\theta = -36.8°$
i. $P_R = 30.6mW$
j. $P_L = 22.9mVAR$
k. $P_A = 38.3mVA$
l. Current Phasor Diagram

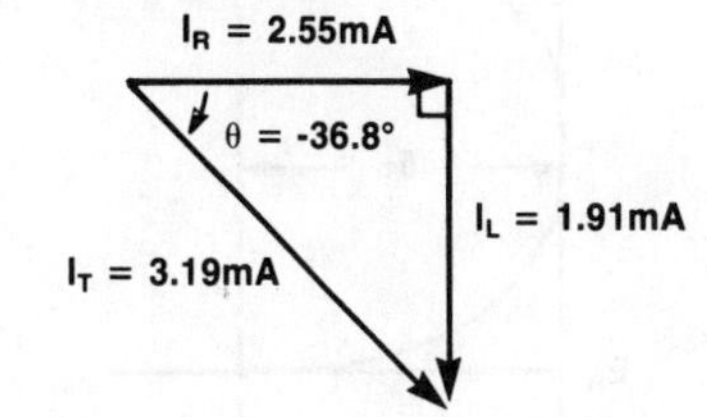

m. Power Phasor Diagram

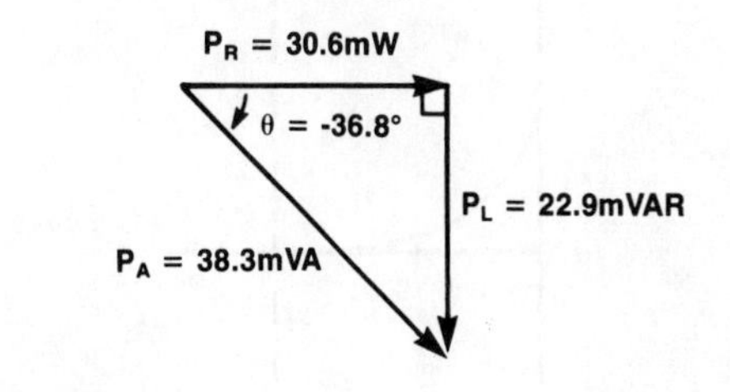

3. $Q = 7.85$

4. $3.56k\Omega$

5. b.

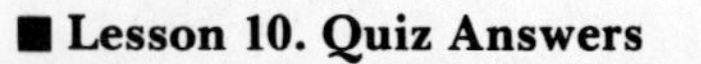

Lesson 10. Quiz Answers

1. 400μs

2. 5μs

3. a. 40mA
b. 80V
c. 80V
d. 2ms

4.

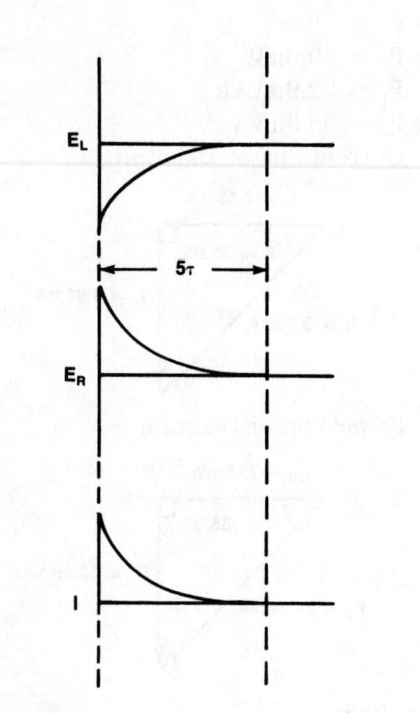

5. a. 37.5ms
b. 75ms
c. 100ms
d. 113ms

6. a. 0.444τ
b. 0.667τ
c. 1.78τ
d. 2.78τ

7. a. 60V
b. 20V
c. 10mA

8. 9.45ms

9. 662pF

10. 6.6MΩ

Lesson 11. Quiz Answers

1.

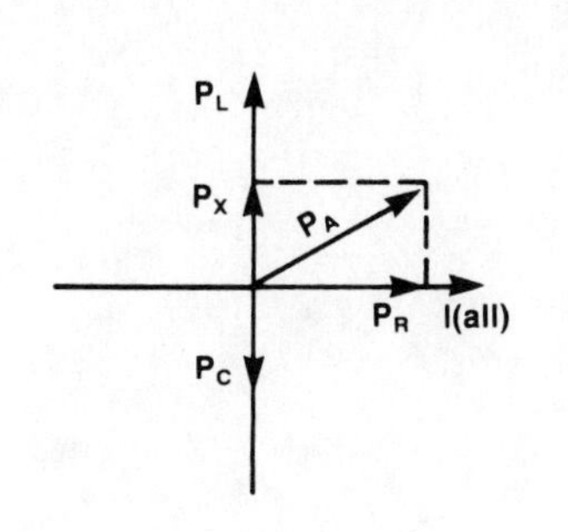

2.

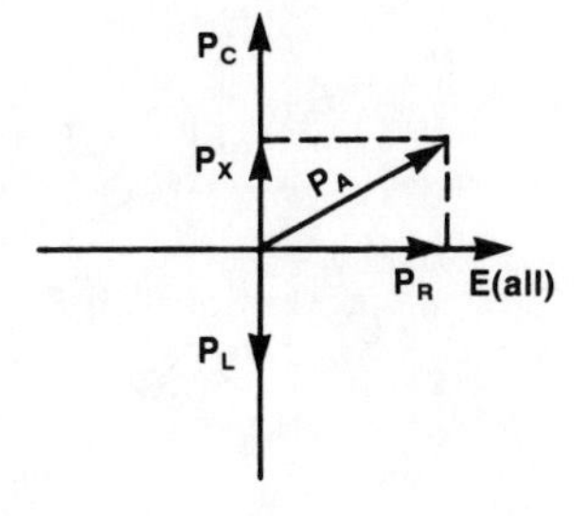

3. +26.6°

4. +53.1°

5. 1.12kΩ

6. 25A

7. a. 159Ω
   b. 119Ω
   c. 129Ω
   d. $0.31A_{rms}$
   e. 15.5VAC
   f. 12.4VAC
   g. 49.3VAC
   h. $4.81W_{rms}$
   i. $3.84VAR_{rms}$
   j. $15.3VAR_{rms}$
   k. $12.4VA_{rms}$
   l. −67.2°

8. a. 15.7kΩ
   b. $4mA_{rms}$
   c. $2.55mA_{rms}$
   d. $6.67mA_{rms}$
   e. $5.74mA_{rms}$
   f. $0.16W_{rms}$
   g. $0.102VAR_{rms}$
   h. $0.267VAR_{rms}$
   i. $0.165VAR_{rms}$
   j. $0.23VA_{rms}$
   k. +45.8°

9. The component with the largest value of resistance or reactance.

10. The component with the smallest value of resistance or reactance.

■ Lesson 12. Quiz Answers

1. a. $5 + j1 = 5.1\underline{/11.3°}$
b. $2 + j4 = 4.47\underline{/63.4°}$
c. $1 - j2 = 2.23\underline{/-63.4°}$
d. $4 + j5 = 6.4\underline{/-51.3°}$

2. a. $\sqrt{-1}$
b. -1
c. $-j = -\sqrt{-1}$
d. $+1$

3. a.

b.

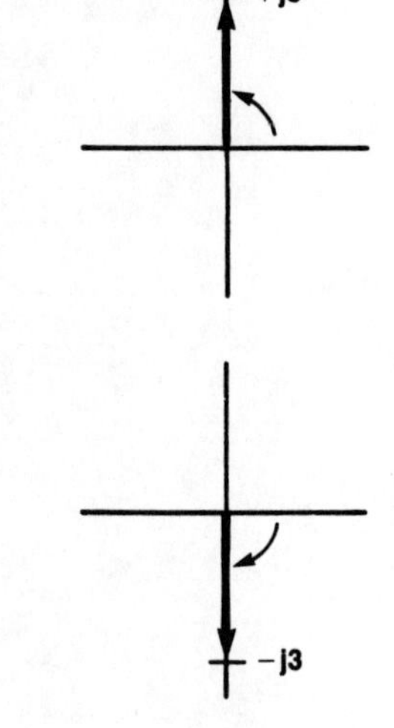

4. a. $200 + j150$
b. $390 - j62$

5. a. $51\underline{/78.7°}$
b. $705\underline{/-58.3°}$
c. $71.2\underline{/66°}$
d. $11.4\underline{/-52°}$

6. a. $65.7 + j17.6$
b. $31.1 - j13.8$
c. $4.94 + j15.2$
d. $22.9 - j14.3$

■ Lesson 13. Quiz Answers

1. a. $j18$
b. $j2$
c. $j^2 84 = -84$
d. $-j7$
e. $j^3 48 = -j48$
f. $-j1.5$

2. a. $7 + j2$
b. $-2 + j4$
c. $58 - j6$
d. $1.48 + j0.055$

3. a. $112\angle 85°$
b. $3\angle -3°$
c. $30\angle 4°$
d. $5\angle -30°$

4. a. $33.82\angle -39.8°$
b. $25.98 - j21.66$
c. $12.5\angle 65.6°$
d. $5.16 + j11.4$

5. I. a. $145.8\Omega\angle 31°$
b. $548.7\text{mA}\angle -31°$
c. $1.94\text{A}\angle -76°$
d. $1.6\text{A}\angle 90°$
e. $19.4\text{V}\angle -76°$
f. $77.6\text{V}\angle 14°$
g. $80\text{V}\angle 0°$
h. $-31°$

II. a. $67.4\Omega\angle 16.8°$
b. $416\text{mA}\angle -16.8°$
c. $416\text{mA}\angle -16.8°$
d. $0.7\text{A}\angle +0.1°$
e. $323\text{mA}\angle -158.1°$
f. $16.6\text{V}\angle -16.8°$
g. $34.8\text{V}\angle -89.9°$
h. $12.9\text{V}\angle -158.1°$

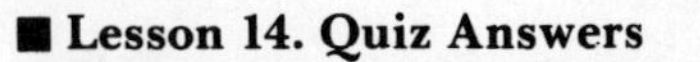

■ Lesson 14. Quiz Answers

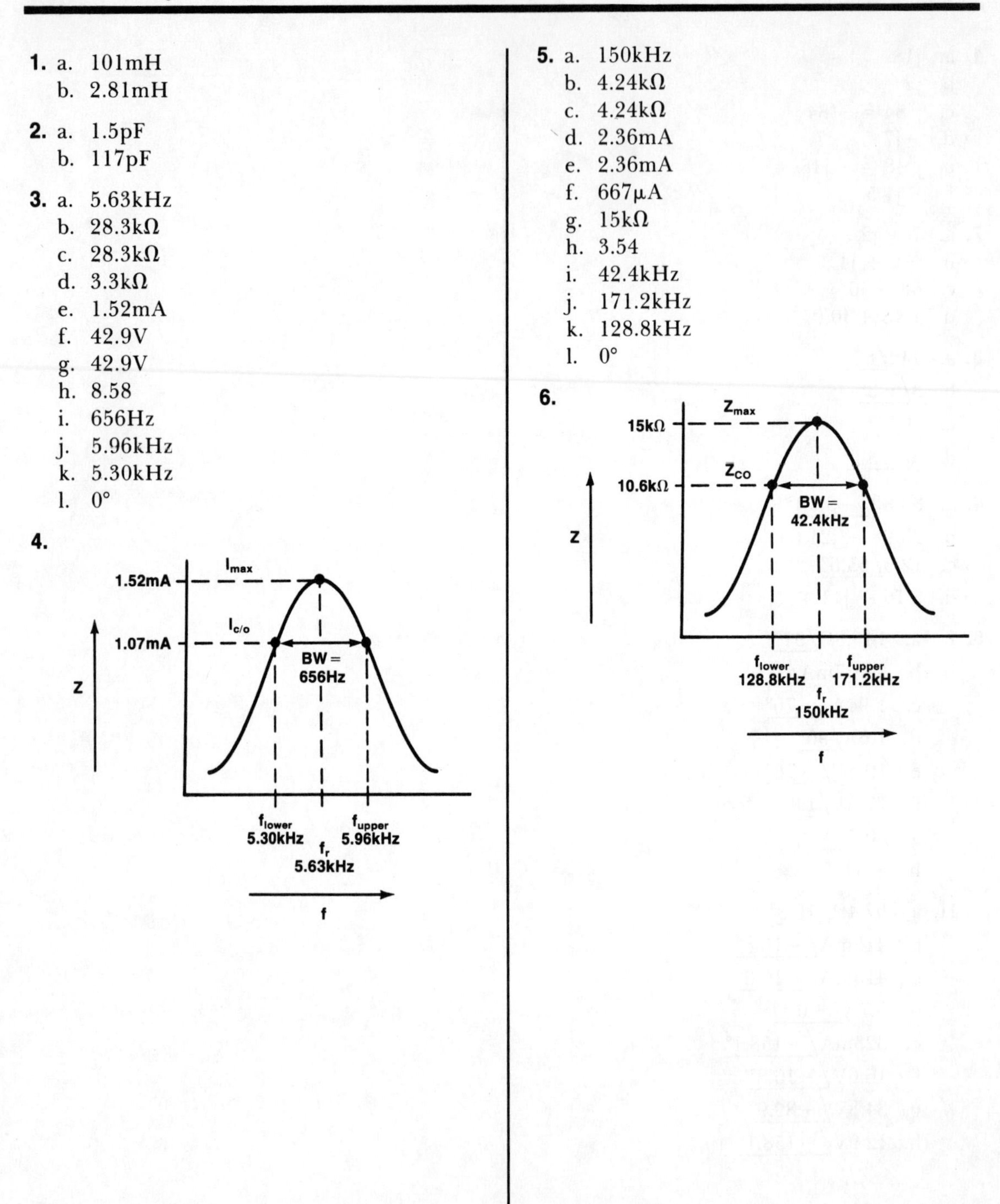

1. a. 101mH
b. 2.81mH

2. a. 1.5pF
b. 117pF

3. a. 5.63kHz
b. 28.3kΩ
c. 28.3kΩ
d. 3.3kΩ
e. 1.52mA
f. 42.9V
g. 42.9V
h. 8.58
i. 656Hz
j. 5.96kHz
k. 5.30kHz
l. 0°

4.

5. a. 150kHz
b. 4.24kΩ
c. 4.24kΩ
d. 2.36mA
e. 2.36mA
f. 667μA
g. 15kΩ
h. 3.54
i. 42.4kHz
j. 171.2kHz
k. 128.8kHz
l. 0°

6.

Appendices

The following charts are listed to give a convenient method for comparing various common English and metric units to allow easy conversion from one unit to another. These comparisons are for common values of lengths, areas, volume, speed, and electric resistivity. Included also is a listing of several other miscellaneous unit comparisons.

Length Comparisons

To use this chart to compare (and thus convert) one unit to another, find the existing measurement in the From column and then find the desired unit in the vertical headings (TO). Where these two intersect will give you the conversion of *one* existing unit (From) into *one* new unit (To). For example, if you have *one inch* and you need this *in centimeters*; find "1 inch" in the From column (4th line down) and go over to the vertical column labeled cm; and you find that 1 inch = 2.54 cm. Then, if you wanted to convert 25 inches (or any value of inches) into centimeters you would simply multiply 25 (or any given number of inches) by 2.54 for 63.5 centimeters.

Length Comparisons

From \ To	cm	meter	km	in	ft	mile	naut. mile
1 Centimeter	1	1×10^{-2}	1×10^{-5}	0.3937	3.281×10^{-2}	6.214×10^{-6}	5.40×10^{-6}
1 Meter	100	1	1×10^{-3}	39.37	3.281	6.214×10^{-4}	5.40×10^{-4}
1 Kilometer	1×10^{5}	1×10^{3}	1	3.937×10^{4}	3281	0.6214	0.540
1 Inch	2.54	2.54×10^{-2}	2.54×10^{-5}	1	8.333×10^{-2}	1.578×10^{-5}	1.371×10^{-5}
1 Foot	30.48	0.3048	3.048×10^{-4}	12	1	1.894×10^{-4}	1.646×10^{-4}
1 Statute Mile	1.609×10^{5}	1609	1.609	6.336×10^{4}	5280	1	0.8670
1 Nautical Mile	1.852×10^{5}	1852	1.852	7.293×10^{4}	6076.1	1.1508	1

The charts that follow are used in the same manner as the length comparison chart with the "From" in the left column and the "To" conversions listed in the following vertical columns.

Area Comparison

From \ To	meter^2	cm^2	ft^2	in^2	circ mil
1 square meter	1	1 X 10^4	10.76	1550	1.974 X 10^9
1 square centimeter	1 X 10^{-4}	1	1.076 X 10^{-3}	0.1550	1.974 X 10^5
1 square foot	9.290 X 10^{-2}	929.0	1	144	1.833 X 10^8
1 square inch	6.452 X 10^{-4}	6.452	6.944 X 10^{-3}	1	1.273 X 10^6
1 circular mil	5.067 X 10^{-10}	5.067 X 10^{-6}	5.454 X 10^{-9}	7.854 X 10^{-7}	1

Volume Comparison

From \ To	meter^3	cm^3	1	ft^3	in^3
1 cubic meter	1	1 X 10^6	1000	35.31	6.102 X 10^4
1 cubic centimeter	1 X 10^{-6}	1	1. X 10^{-3}	3.531 X 10^{-5}	6.102 X 10^{-2}
1 liter	1.000 X 10^{-3}	1000	1	3.531 X 10^{-2}	61.02
1 cubic foot	2.832 X 10^{-2}	2.832 X 10^4	28.32	1	1728
1 cubic inch	1.639 X 10^{-5}	16.39	1.639 X 10^{-2}	5.787 X 10^{-4}	1

Speed Comparison

From \ To	ft/sec	km/hr	meter/sec	miles/hr	cm/sec	knot
1 foot per second	1	1.097	0.3048	0.6818	30.48	0.5925
1 kilometer per hour	0.9113	1	0.2778	0.6214	27.78	0.540
1 meter per second	3.281	3.6	1	2.237	100	1.944
1 mile per hour	1.467	1.609	0.4470	1	44.70	0.8689
1 centimeter per second	3.281 X 10^{-2}	3.6 X 10^{-2}	0.01	2.237 X 10^{-2}	1	1.944 X 10^{-2}
1 knot	1.688	1.852	0.5144	1.151	51.44	1

Electric Resistivity Comparison

From \ To	μohm-cm	ohm-cm	ohm-m	ohm-circ mil/ft
1 micro-ohm-centimeter	1	1×10^{-6}	1×10^{-8}	6.015
1 ohm-centimeter	1×10^{6}	1	0.01	6.015×10^{6}
1 ohm-meter	1×10^{8}	100	1	6.015×10^{8}
1 ohm-circular mil per foot	0.1662	1.662×10^{-7}	1.662×10^{-9}	1

Miscellaneous Unit Comparisons

1 fathom = 6 ft
1 yard = 3 ft
1 rod = 16.5 ft
1 U.S. gallon = 4 U.S. fluid quarts
1 U.S. quart = 2 U.S. pints
1 U.S. pint = 16 U.S. fluid ounces
1 U.S. gallon = 0.8327 British imperial gallon
1 British imperial gallon = 1.2 U.S. gallons

1 liter = 1000 cm^3
1 knot = 1 nautical mile/hr
1 mile/min = 88 ft/sec = 60 miles/hr
1 meter = 39.4 in = 3.28 ft
1 inch = 2.54 cm
1 mile = 5280 ft = 1.61 km
1 angstrom unit = 10^{-10} meters
1 horsepower = 550 ft-lb/sec = 746 watts

The Greek Alphabet

(Including common use of symbols in basic electricity)

Letter	Capital	Common Use of Symbol	Lower	Common Use of Symbol
Alpha	A		α	
Beta	B		β	
Gamma	Γ		γ	
Delta	Δ	change in	δ	change in
Epsilon	E		ϵ	base of natural logs
Zeta	Z		ζ	
Eta	H		η	
Theta	Θ		θ, ϑ	angle (phase angle)
Iota	I		ι	
Kappa	K		κ	dielectric constant
Lambda	Λ		λ	wavelength
Mu	M		μ	micro
Nu	N		ν	frequency
Xi	Ξ		ξ	
Omicron	O		o	
Pi	Π		π	3.14159
Rho	P		ρ	specific resistance, resistivity
Sigma	Σ	sum of terms	σ, ς	
Tau	T		τ	time constant
Upsilon	Υ		υ	
Phi	Φ		ϕ, φ	magnetic flux
Chi	X		χ	
Psi	Ψ		ψ	
Omega	Ω	ohms	ω	angular frequency
(Reversed Omega)	(℧)	mho		

Symbol	Device	Symbol	Device
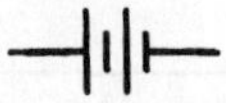	Battery or DC Power Supply		Push Button Normally Open (PBNO)
	Resistor		Push Button Normally Closed (PBNC)
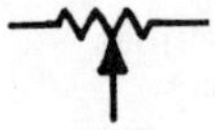	Potentiometer		Earth Ground
	Rheostat		Chassis Ground
	Tapped Resistor	 +	Capacitor
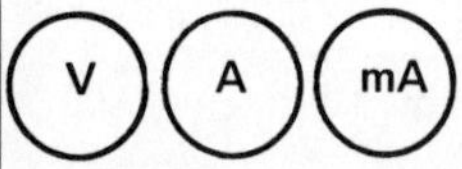 V A mA	Meters – Symbol to Indicate Function	 +	Capacitor, Polarized (Electrolytic)
	Lamp		Coil, Air Core
	Switch SPST		Coil, Iron Core
	Switch SPDT		Fuse
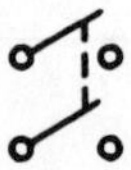	Switch DPST		Conductor, General
		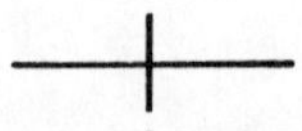	No Connection
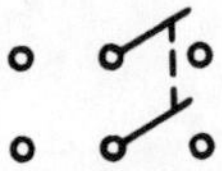	Switch DPDT	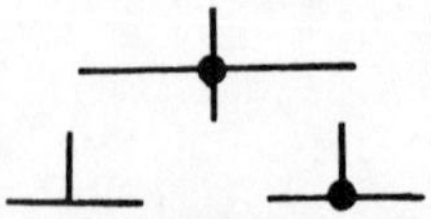	Connection
	AC Power Supply		Transformer

TERM	UNIT	SYMBOL	FORMULA: SERIES		FORMULA: PARALLEL
Charge	Coulomb	Q	1 coulomb = 6.28×10^{18} electrons		
Voltage (Potential difference, EMF)	Volt (V)	E	$E_T = E_1 + E_2 + E_3 + \ldots$	E / I R; $E = IR$	$E_T = E_1 = E_2 = E_3 \ldots$
Current (Flow of charge)	Ampere (Amp) (A)	I	$I_T = I_1 = I_2 = I_3 \ldots$	E / I R; $I = E/R$	$I_T = I_1 + I_2 + I_3 \ldots$
Resistance	Ohm (Ω)	R	$R_T = R_1 + R_2 + R_3 + \ldots$	E / I R; $R = E/I$; $R = 1/G$	$R_T = \frac{1}{1/R_1 + 1/R_2 + 1/R_3 + \ldots}$; $R_T = \frac{R_1 R_2}{R_1 + R_2}$; $R_T = \frac{R_s}{N}$
Conductance	Mho ($\mho$)	G	$G_T = 1/R_T$	$G = 1/R$	$G_T = G_1 + G_2 + G_3 \ldots$
Power	Watt (W)	P	$P = IE$; $P = E^2/R$; $P = I^2R$	P / I E; E^2 / R P; P / I^2 R	$P = IE$; $P = E^2/R$; $P = I^2R$
Capacitance	Farad (F)	C	$C_T = \frac{1}{1/C_1 + 1/C_2 + 1/C_3 + \ldots}$	$C = Q/E$; $\tau = RC$	$C_T = C_1 + C_2 + C_3 + \ldots$
Inductance	Henry (H)	L	$L_T = L_1 + L_2 + L_3 + \ldots$	$\tau = L/R$	$L_T = \frac{1}{1/L_1 + 1/L_2 + 1/L_3 + \ldots}$

TERM	UNIT	SYMBOL	RC CIRCUITS	
			SERIES	PARALLEL
frequency	hertz	f	$f = \frac{1}{T}$	
Voltage	Volt (V)	E	$E_R = I_TR$ $E_C = I_TX_C$ $E_A = I_TZ_T$ or $\sqrt{E_R^2 + E_C^2}$ $I_TZ_T = \sqrt{I_T^2R^2 + I_T^2X_C^2}$	$E_A = E_R = E_C$
Current	Ampere (Amp) (A)	I	$I_T = \frac{E_A}{Z}$ or $I_T = I_C = I_R$	$I_R = \frac{E_R}{R} = \frac{E_A}{R}$ $I_C = \frac{E_C}{X_C} = \frac{E_A}{X_C}$ $I_T = \sqrt{I_R^2 + I_C^2}$
Impedance	Ohm (Ω)	Z	$Z = \sqrt{R^2 + X_C^2}$	$Z = \frac{E_A}{I_T}$ or $Z = \frac{1}{\sqrt{\left(\frac{1}{R}\right)^2 + \left(\frac{1}{X_C}\right)^2}}$
Capacitive Reactance	Ohm (Ω)	X_C	$X_C = \frac{1}{2\pi fC}$	
Inductive Reactance	Ohm (Ω)	X_L	$X_L = 2\pi fL$	
Real Power	Watt (W)	P_R	$P_R = E_RI_R$	$P_R = E_RI_R = E_AI_R$
Apparent Power	Volt-Ampere (VA)	P_T or P_A	$P_T = E_AI_T$ or $P_T = \sqrt{P_R^2 + P_C^2}$	$P_T = E_AI_T$ or $P_T = \sqrt{P_R^2 + P_C^2}$
Reactive Power	Volt-Ampere-Reactive (VAR)	P_C or P_L	$P_C = E_CI_C$	$P_C = E_CI_C = E_AI_C$
Phase Angle	Degree (°)	θ	$\theta = \arctan\left(\frac{E_C}{E_R}\right)$ or $\theta = \tan^{-1}\left(\frac{E_C}{E_R}\right)$ or $\theta = \arctan\left(\frac{X_C}{R}\right)$ or $\theta = \tan^{-1}\left(\frac{X_C}{R}\right)$	$\theta = \arctan\left(\frac{I_C}{I_R}\right)$ or $\theta = \tan^{-1}\left(\frac{I_C}{I_R}\right)$

RL CIRCUITS		RLC CIRCUITS	
SERIES	PARALLEL	SERIES	PARALLEL
$f = \frac{1}{T}$			
$E_R = I_TR$ $E_L = I_TX_L$ $E_A = I_TZ$ or $\sqrt{E_R^2 + E_L^2}$ $I_TZ_T = \sqrt{I_T^2R^2 + I_T^2X_L^2}$	$E_A = E_R = E_L$	$E_R = I_TR$ $E_L = I_TX_L$ $E_C = I_TX_C$ $E_X = E_L - E_C$ $E_A = \sqrt{E_R^2 + E_X^2}$ or $E_A = \sqrt{E_R^2 + (E_L - E_C)^2}$	$E_A = E_R = E_L = E_C$
$I_T = \frac{E_A}{Z}$ $I_T = I_L = I_R$	$I_R = \frac{E_R}{R} = \frac{E_A}{R}$ $I_L = \frac{E_L}{X_L} = \frac{E_A}{X_L}$ $I_T = \sqrt{I_R^2 + I_L^2}$	$I_T = \frac{E_A}{Z}$ $I_T = I_C = I_L = I_R$	$I_R = \frac{E_R}{R} = \frac{E_A}{R}$ $I_L = \frac{E_L}{X_L} = \frac{E_A}{X_L}$ $I_C = \frac{E_C}{X_C} = \frac{E_A}{X_C}$ $I_X = I_C - I_L$ $I_T = \sqrt{I_R^2 + I_X^2}$ or $I_T = \sqrt{I_R^2 + (I_C - I_L)^2}$
$Z = \sqrt{R^2 + X_L^2}$	$Z = \frac{E_A}{I_T}$	$Z = \sqrt{R^2 + X_T^2}$ or $Z = \sqrt{R^2 + (X_L - X_C)^2}$	$Z = \frac{E_A}{I_T}$
$X_C = \frac{1}{2\pi fC}$			
$X_L = 2\pi fL$			
$P_R = E_RI_R$	$P_R = E_RI_R = E_AI_R$	$P_R = E_RI_R$	$P_R = E_RI_R = E_AI_R$
$P_A = E_AI_T$ or $P_A = \sqrt{P_R^2 + P_L^2}$	$P_A = E_AI_T$ or $P_A = \sqrt{P_R^2 + P_L^2}$	$P_A = E_AI_T$ or $P_A = \sqrt{P_R^2 + P_X^2}$ or $P_A = \sqrt{P_R^2 + (P_L - P_C)^2}$	$P_A = E_A \times I_T$ or $P_A = \sqrt{P_R^2 + P_X^2}$ or $P_A = \sqrt{P_R^2 + (P_C - P_L)^2}$
$P_L = E_LI_L$	$P_L = E_LI_L = E_AI_L$	$P_L = E_LI_L$	$P_C = E_CI_C = E_AI_C$ $P_L = E_LI_L = E_AI_L$
$\theta = \arctan\left(\frac{E_L}{E_R}\right)$ or $\theta = \tan^{-1}\left(\frac{E_L}{E_R}\right)$ or $\theta = \arctan\left(\frac{X_L}{R}\right)$ or $\theta = \tan^{-1}\left(\frac{X_L}{R}\right)$	$\theta = \arctan\left(\frac{I_L}{I_R}\right)$ or $\theta = \tan^{-1}\left(\frac{I_L}{I_R}\right)$	$\theta = \arctan\left(\frac{E_X}{E_R}\right)$ or $\theta = \tan^{-1}\left(\frac{E_X}{E_R}\right)$ or $\theta = \arctan\left(\frac{X_T}{R}\right)$ or $\theta = \tan^{-1}\left(\frac{X_T}{R}\right)$	$\theta = \arctan\left(\frac{I_X}{I_R}\right)$ or $\theta = \tan^{-1}\left(\frac{I_X}{I_R}\right)$

APPENDIX 6. RIGHT TRIANGLE INFORMATION AND EQUIVALENT FIRST–QUADRANT ANGLES

Right Triangle Functions

$$\text{sine}\ \theta = \frac{\text{opposite}}{\text{hypotenuse}}$$

$$\text{cosine}\ \theta = \frac{\text{adjacent}}{\text{hypotenuse}}$$

$$\text{tangent}\ \theta = \frac{\text{opposite}}{\text{adjacent}}$$

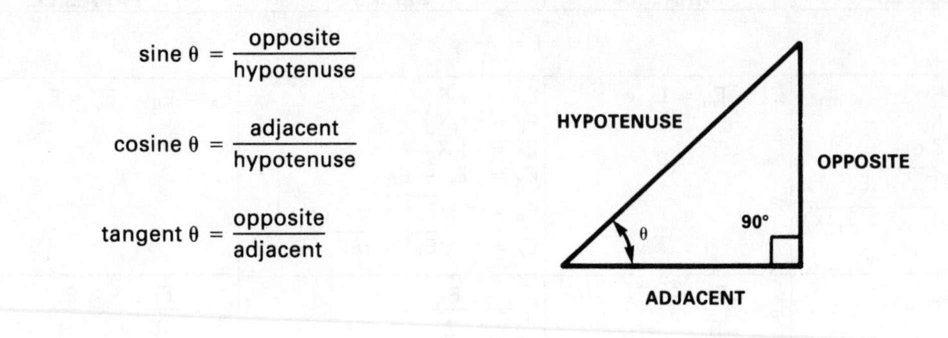

Equivalent First-Quadrant Angles

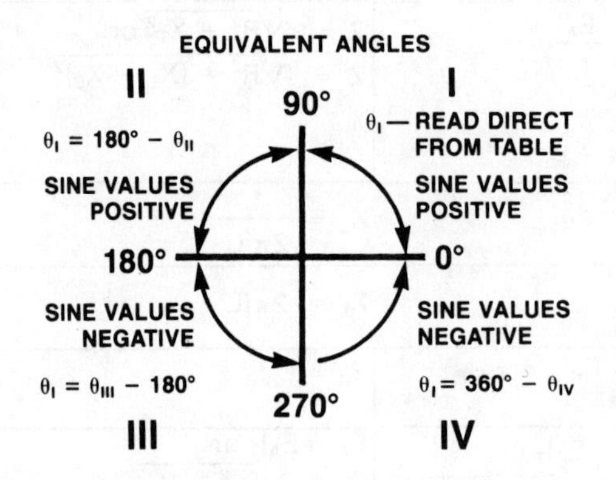

APPENDIX 7. 0-90° TRIGONOMETRIC FUNCTIONS

ANGLE	SIN	COS	TAN	ANGLE	SIN	COS	TAN
0 °	0.0000	1.000	0.0000	45 °	0.7071	0.7071	1.0000
1	.0175	.9998	.0175	46	.7193	.6947	1.0355
2	.0349	.9994	.0349	47	.7314	.6820	1.0724
3	.0523	.9986	.0524	48	.7431	.6691	1.1106
4	.0698	.9976	.0699	49	.7547	.6561	1.1504
5	.0872	.9962	.0875	50	.7660	.6428	1.1918
6	.1045	.9945	.1051	51	.7771	.6293	1.2349
7	.1219	.9925	.1228	52	.7880	.6157	1.2799
8	.1392	.9903	.1405	53	.7986	.6018	1.3270
9	.1564	.9877	.1584	54	.8090	.5878	1.3764
10	.1736	.9848	.1763	55	.8192	.5736	1.4281
11	.1908	.9816	.1944	56	.8290	.5592	1.4826
12	.2079	.9781	.2126	57	.8387	.5446	1.5399
13	.2250	.9744	.2309	58	.8480	.5299	1.6003
14	.2419	.9703	.2493	59	.8572	.5150	1.6643
15	.2588	.9659	.2679	60	.8660	.5000	1.7321
16	.2756	.9613	.2867	61	.8746	.4848	1.8040
17	.2924	.9563	.3057	62	.8829	.4695	1.8807
18	.3090	.9511	.3249	63	.8910	.4540	1.9626
19	.3256	.9455	.3443	64	.8988	.4384	2.0503
20	.3420	.9397	.3640	65	.9063	.4226	2.1445
21	.3584	.9336	.3839	66	.9135	.4067	2.2460
22	.3746	.9272	.4040	67	.9205	.3907	2.3559
23	.3907	.9205	.4245	68	.9272	.3746	2.4751
24	.4067	.9135	.4452	69	.9336	.3584	2.6051
25	.4226	.9063	.4663	70	.9397	.3420	2.7475
26	.4384	.8988	.4877	71	.9455	.3256	2.9042
27	.4540	.8910	.5095	72	.9511	.3090	3.0777
28	.4695	.8829	.5317	73	.9563	.2924	3.2709
29	.4848	.8746	.5543	74	.9613	.2756	3.4874
30	.5000	.8660	.5774	75	.9659	.2588	3.7321
31	.5150	.8572	.6009	76	.9703	.2419	4.0108
32	.5299	.8480	.6249	77	.9744	.2250	4.3315
33	.5446	.8387	.6494	78	.9781	.2079	4.7046
34	.5592	.8290	.6745	79	.9816	.1908	5.1446
35	.5736	.8192	.7002	80	.9848	.1736	5.6713
36	.5878	.8090	.7265	81	.9877	.1564	6.3138
37	.6018	.7986	.7536	82	.9903	.1392	7.1154
38	.6157	.7880	.7813	83	.9925	.1219	8.1443
39	.6293	.7771	.8098	84	.9945	.1045	9.5144
40	.6428	.7660	.8391	85	.9962	.0872	11.43
41	.6561	.7547	.8693	86	.9976	.0698	14.30
42	.6691	.7431	.9004	87	.9986	.0523	19.08
43	.6820	.7314	.9325	88	.9994	.0349	28.64
44	.6947	.7193	.9657	89	.9998	.0175	57.29
				90	1.0000	.0000	∞

APPENDIX 8. HOW TO USE SQUARE ROOT TABLES

The following table can be used to find the square root or square of most any number. Numbers from 1 to 120 can be read directly from the table. But what about a number such as 150? How can its square or square root be found? The secret to the use of this table is in the understanding of *factoring*. Factoring a number means to break the original number up into two smaller numbers, that, when multiplied together, give you back the original. For example, 150 is equal to 10 times 15. Ten and 15 are said to be *factors* of 150. If 10 times 15 is equal to 150, then the square root of 10 times the square root of 15 is equal to the square root of 150. Both 10 and 15 are listed on the square and square root table. The square root of 10 from the table is equal to 3.162. The square root of 15 is equal to 3.873; 3.162 times 3.873 is equal to 12.246426, which should be the square root of 150. You can test this number by multiplying it by itself. Thus, 12.246426 squared is equal to 149.97, etc., – very close to 150. (Small errors due to rounding will normally occur when using the tables.) The factoring procedure written out mathematically would then be:

$$150 = 10 \times 15$$

$$\sqrt{150} = \sqrt{10} \times \sqrt{15} \quad \text{(Look up } \sqrt{10}, \sqrt{15} \text{ in tables)}$$

$$\sqrt{150} = 3.162 \times 3.873$$

$$\sqrt{150} = 12.246\ldots\ldots$$

Try another number now, say, 350. First, factor 350:

$$350 = 35 \times 10$$

The square root of 350 must equal the square root of 35 times the square root of 10.

$$\sqrt{350} = \sqrt{35} \times \sqrt{10}$$

Go to the tables and look up the square roots of 10 and 35:

$$\sqrt{350} = 5.9161 \times 3.162$$

Multiply the square roots of 10 and 35, and you have found the square root of 350.

$$\sqrt{350} = 18.706\ldots\ldots$$

To check the accuracy of your calculations, multiply 18.706 by itself.

$$18.706^2 = 349.91$$

Again, very close to the original number.

Try one more number, this time 1150.

First, factor 1150.

$$1150 = 115 \times 10$$

The square root of 1150 must equal the square root of 115 times the square root of 10.

$$\sqrt{1150} = \sqrt{115} \times \sqrt{10}$$

Look up the square roots of 115 and 10 from the tables.

$$\sqrt{1150} = 10.7238 \times 3.162$$

Multiply the square roots of 115 and 10, and you have the square root of 1150.

$$\sqrt{1150} = 33.908$$

To check the validity of this number, square it. It should be very close to 1150.

N	$\sqrt{N}$	N^2	N	$\sqrt{N}$	N^2	N	$\sqrt{N}$	N^2
1	1.000	1	41	6.4031	1681	81	9.0000	6561
2	1.414	4	42	6.4807	1764	82	9.0554	6724
3	1.732	9	43	6.5574	1849	83	9.1104	6889
4	2.000	16	44	6.6332	1936	84	9.1652	7056
5	2.236	25	45	6.7082	2025	85	9.2195	7225
6	2.449	36	46	6.7823	2116	86	9.2736	7396
7	2.646	49	47	6.8557	2209	87	9.3274	7569
8	2.828	64	48	6.9282	2304	88	9.3808	7744
9	3.000	81	49	7.0000	2401	89	9.4340	7921
10	3.162	100	50	7.0711	2500	90	9.4868	8100
11	3.3166	121	51	7.1414	2601	91	9.5394	8281
12	3.4641	144	52	7.2111	2704	92	9.5917	8464
13	3.6056	169	53	7.2801	2809	93	9.6437	8649
14	3.7417	196	54	7.3485	2916	94	9.6954	8836
15	3.8730	225	55	7.4162	3025	95	9.7468	9025
16	4.0000	256	56	7.4833	3136	96	9.7980	9216
17	4.1231	289	57	7.5498	3249	97	9.8489	9409
18	4.2426	324	58	7.6158	3364	98	9.8995	9604
19	4.3589	361	59	7.6811	3481	99	9.9499	9801
20	4.4721	400	60	7.7460	3600	100	10.0000	10000
21	4.5826	441	61	7.8102	3721	101	10.0499	10201
22	4.6904	484	62	7.8740	3844	102	10.0995	10404
23	4.7958	529	63	7.9373	3969	103	10.1489	10609
24	4.8990	576	64	8.0000	4096	104	10.1980	10816
25	5.0000	625	65	8.0623	4225	105	10.2470	11025
26	5.0990	676	66	8.1240	4356	106	10.2956	11236
27	5.1962	729	67	8.1854	4489	107	10.3441	11449
28	5.2915	784	68	8.2462	4624	108	10.3923	11664
29	5.3852	841	69	8.3066	4761	109	10.4403	11881
30	5.4772	900	70	8.3666	4900	110	10.4881	12100
31	5.5678	961	71	8.4261	5041	111	10.5357	12321
32	5.6569	1024	72	8.4853	5184	112	10.5830	12544
33	5.7446	1089	73	8.5440	5329	113	10.6301	12769
34	5.8310	1156	74	8.6023	5476	114	10.6771	12996
35	5.9161	1225	75	8.6603	5625	115	10.7238	13225
36	6.0000	1296	76	8.7178	5776	116	10.7703	13456
37	6.0828	1369	77	8.7750	5929	117	10.8167	13689
38	6.1644	1444	78	8.8318	6084	118	10.8628	13924
39	6.2450	1521	79	8.8882	6241	119	10.9087	14161
40	6.3246	1600	80	8.9443	6400	120	10.9545	14400

THE DIVIDE-AND-AVERAGE METHOD TO FIND SQUARE ROOTS

STEP	PRESS	DISPLAY
1. CHOOSE A NUMBER. LET'S USE 89.	[MC] 89	89.
2. ESTIMATE A SQUARE ROOT; DIVIDE BY IT.	[÷] 9	9.
3. ADD YOUR ESTIMATE TO DISPLAY	[+]	9.8888888
	9	9.
4. DIVIDE BY 2.	[÷]	18.888888
	2	2.
	[=] [M+]	9.444444 (HALF WAY BETWEEN)
5. CHECK TO SEE IF IT IS A ROOT.	[MR] [X] [MR] [=]	89.197522
6. IF NOT, REPEAT THE PROCESS	89 [÷] [MR]	9.444444
	[+]	9.4235298
	[MR]	9.444444
	[÷] 2 [=] [MC] [M+]	9.4339865
	[MR] [X] [MR] [=]	89.000101

*This procedure is for use with calculators that do not have a square root key but do have a memory function. The calculator that was used to keystroke this example was a TI 1750 which does have a square root key (as do most scientific calculators) and the above result could have been obtained by entering the number 89 and pressing the square root key.

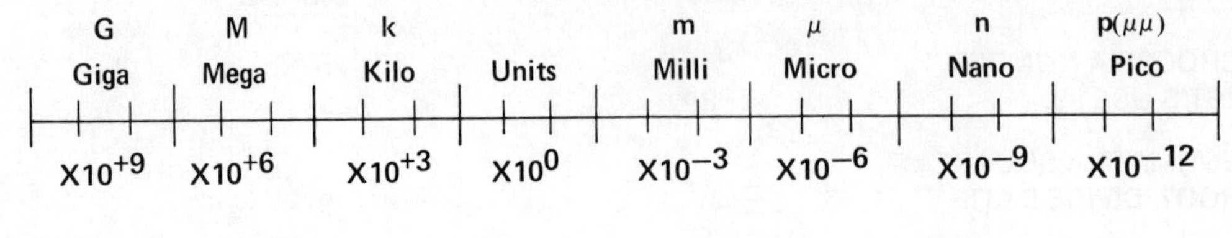

1 unit = 1 .

. 0 0 1 = 1 milli

1 kilo = 1 0 0 0 .

. 0 0 0 0 0 1 = 1 micro

1 mega = 1 0 0 0 0 0 0 .

. 0 0 0 0 0 0 0 0 1 = 1 nano

1 giga = 1 0 0 0 0 0 0 0 0 0 .

. 0 0 0 0 0 0 0 0 0 0 0 1 = 1 pico

STANDARD FORM: X.XX × $10^{+\text{exponent}}$

Symbol	Prefix	Value	Power of 10
G	giga	1,000,000,000.	$X10^{+9}$
M	mega	1,000,000.	$X\ 10^{+6}$
k	kilo	1,000.	$X10^{+3}$
—	(units)	1.	$X10^{0}$
m	milli	.001	$X10^{-3}$
μ	micro	.000 001	$X10^{-6}$
n	nano	.000 000 001	$X10^{-9}$
p (μμ)	pico	.000 000 000 001	$X10^{-12}$

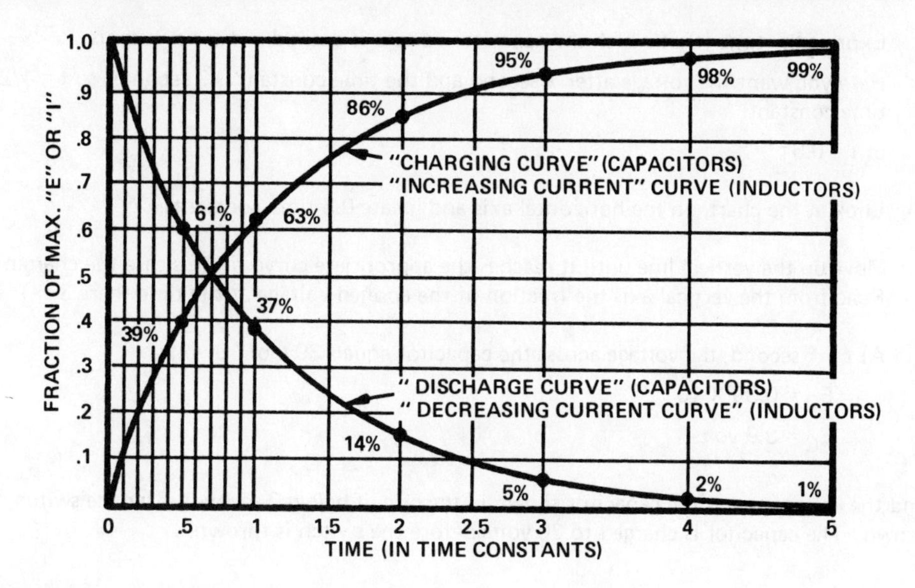

How to Use This Chart

This chart can be used to graphically determine the voltage or current at any point in time for an RC or L/R circuit, during charging (or current buildup), or discharge (or current collapse).

The examples shown below illustrate the use of the chart.

1. Find the voltage across the capacitor shown in the circuit below, 1 second after the switch is thrown.

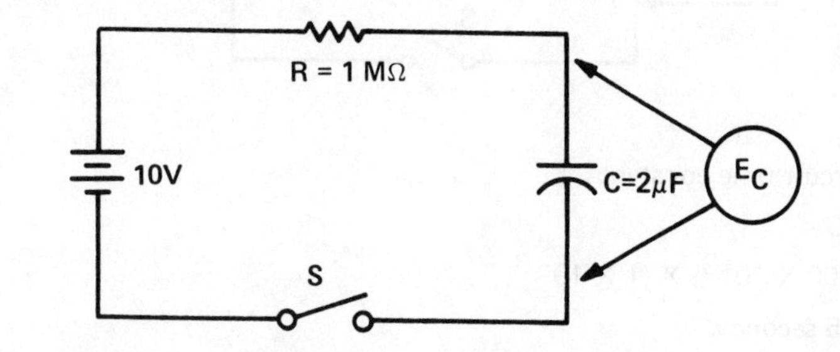

Solution

a. First find the circuit time constant

$\tau = RC$

$\tau = (1 \times 10^6) \times (2 \times 10^{-6}) = 2$ seconds

The voltage at any point along a charge or discharge curve may be calculated by using one of these two mathematical formulas:

Charge: e (at time t) = $E_{app}\,(1 - \epsilon^{-t/RC})$

Discharge: e (at time t) = $E_{app}\,(\epsilon^{-t/RC})$

b. Express the time (t) at which the capacitor voltage is desired in *time constants*.

Here you want the voltage after 1 second and the time constant is 2 seconds, so t = 1/2 (the time constant)

or $t = 0.5\tau$

c. Look at the chart, on the horizontal axis and locate 0.5 time constants.

d. Move up the vertical line until it reaches the appropriate curve (in this case the charging curve). Read from the vertical axis the fraction of the applied voltage at the time (here 39%).

e. At t = 1 second, the voltage across the capacitor equals 39% of 10 volts or

$E_C = 0.39 \times 10$

$E_C = 3.9$ volts

2. Find the voltage across the capacitor shown in the circuit below 2 seconds after the switch, S, is thrown. The capacitor is charged to 20 volts before the switch is thrown.

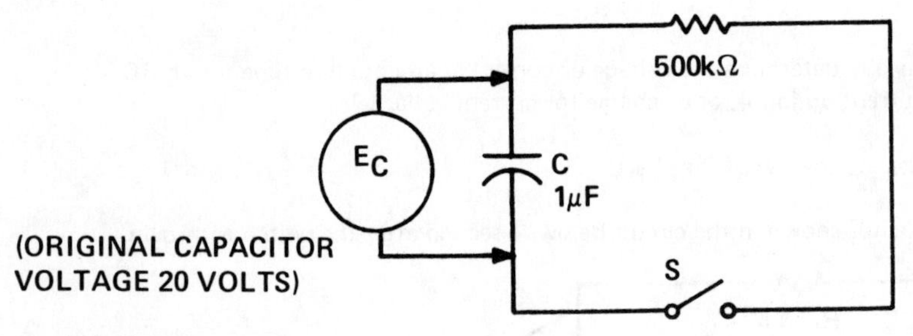

Solution

a. Find the circuit time constant

$\tau = RC$

$\tau = (500 \times 10^{+3}) \times (1 \times 10^{-6})$

$\tau = 0.5$ seconds

b. Express the time at which the capacitor's voltage is desired in time constants. Here, 2 seconds divided by 0.5 seconds is 4; 2 seconds is 4 time constants for this circuit.

$t = 4\tau$

c. Look at the chart, locate 4 time constants on the horizontal axis.

d. Move up the vertical line until it reaches the appropriate curve (the discharge curve). Read the fraction of the original voltage from the vertical axis (2%).

e. At t = 2 seconds, the voltage across the capacitor is at 2% of the original voltage or is at 2% of 20 volts.

$E_C = 0.02 \times 20$

$E_C = 0.40$ volts

Remember that 5 time constants is required for a 100% charge (full charge or discharge for RC circuits, maximum or zero current for L/R circuits).

APPENDIX 12. PEAK, PEAK-TO-PEAK, AND RMS CONVERSION CHART

This chart contains factors to easily convert one ac value of a voltage or current to the other two types of values.

IF YOU WANT \ IF YOU HAVE	RMS	PK	PP
RMS		$= 0.707_{pk}$	$= 0.3535_{pp}$
PK	$= 1.414_{rms}$		$= 0.5_{pp}$
PP	$= 2.828_{rms}$	$= 2_{pk}$	

	With f_r Constant						
	R	L	X_L	Q	BW	L/C	RESPONSE
Series Resonance	*	↑	↑	↑	↓	↑	
	*	↓	↓	↓	↑	↓	
	↑	*	*	↓	↑	*	
	↓	*	*	↑	↓	*	
Parallel Resonance	*	↑	↑	↓	↑	↑	
	*	↓	↓	↑	↓	↓	
	↑	*	*	↑	↓	*	
	↓	*	*	↓	↑	*	

Series Resonance

$Q = \frac{X_L}{R}$ $\quad f_r = \frac{1}{2\pi\sqrt{LC}}$

$BW = \frac{f_r}{Q}$

Parallel Resonance

$Q = \frac{R}{X_L}$ $\quad f_r = \frac{1}{2\pi\sqrt{LC}}$

$BW = \frac{f_r}{Q}$

Common

$BW = f_U - f_L$

$f_U = f_r + ½BW$

$f_L = f_r - ½BW$

*Factor held constant

Half-power point—Real power is exactly one-half of what it is at the resonant frequency, f_r.

Bandwidth—Those frequency values where the frequency response is equal to or greater than 70.7 percent of the value at the mid-band frequency, in this case the resonant frequency.

■ Resistor and Capacitor Color Codes

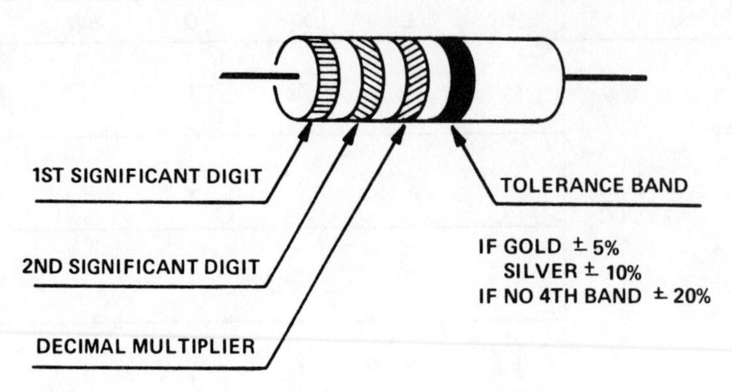

RESISTORS AND CAPACITORS				CAPACITORS ONLY	
	Significant Digit	Decimal Multiplier (Put These Zeros Behind First Two Digits)	(Power of Ten)	Tolerance In %	Voltage Rating (V)
Black	0	1	10^0	20	
Brown	1	1 0	10^1	1	100
Red	2	1 00	10^2	2	200
Orange	3	1 000	10^3	3	300
Yellow	4	1 0000	10^4	4	400
Green	5	1 00000	10^5	5	500
Blue	6	1 000000	10^6	6	600
Violet	7	1 0000000	10^7	7	700
Gray	8	1 00000000	10^8	8	800
White	9	1 000000000	10^9	9	900
Gold	—	Multiply by 0.1	10^{-1}		1000
Silver	—	Multiply by 0.01	10^{-2}	10	2000
			No Color	20	500

PREFERRED VALUES FOR RESISTORS AND CAPACITORS

The numbers listed in the chart below, and *decimal multiples* of these numbers, are the commonly available resistor values at 5%, 10%, and 20% tolerance. Capacitors generally fall into the same values, except 20, 25, and 50 are very common, and any of the values can have a wide range of tolerances available.

20% Tolerance (*No* 4th Band)	10% Tolerance (*Silver* 4th Band)	5% Tolerance (*Gold* 4th Band)
10*	10	10
		11
	12	12
		13
15	15	15
		16
	18	18
		20
22	22	22
		24
	27	27
		30
33	33	33
		36
	39	39
47	47	47
		51
	56	56
		62
68	68	68
		75
	82	82
		91
100	100	100

Mica Capacitor Color Codes

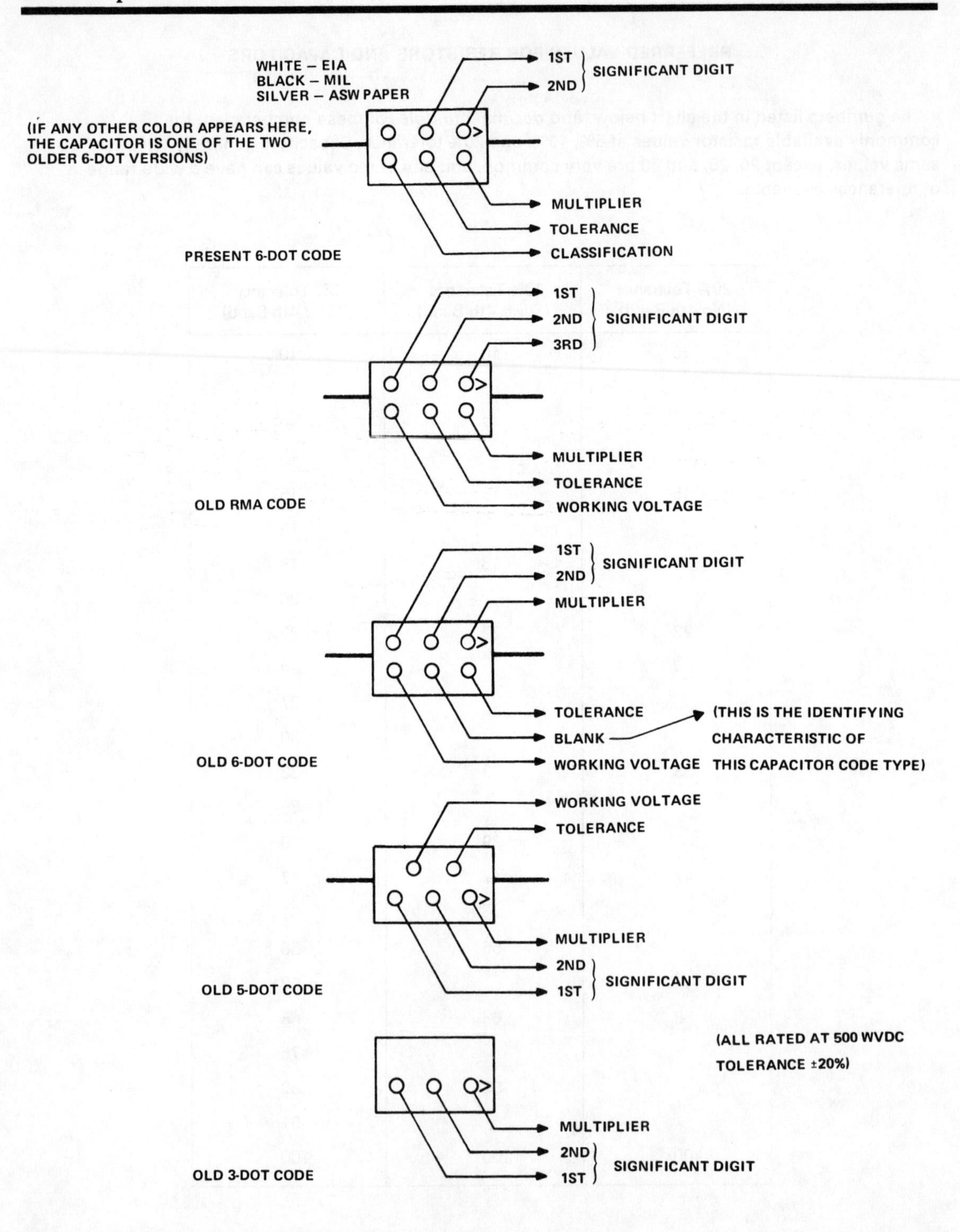

Tubular Capacitor Color Codes

All Values Are Read in Picofarads

Color	Significant Digit	Decimal Multiplier	Tolerance		Temperature Coefficient ppm/°C
			Above 10 pF (in %)	Below 10 pF (in pF)	
Black	0	1	20	2.0	0
Brown	1	10	1		−30
Red	2	100	2		−80
Orange	3	1000			−150
Yellow	4				−220
Green	5		5	0.5	−330
Blue	6				−470
Violet	7				−750
Gray	8	0.01		0.25	30
White	9	0.1	10	1.0	500

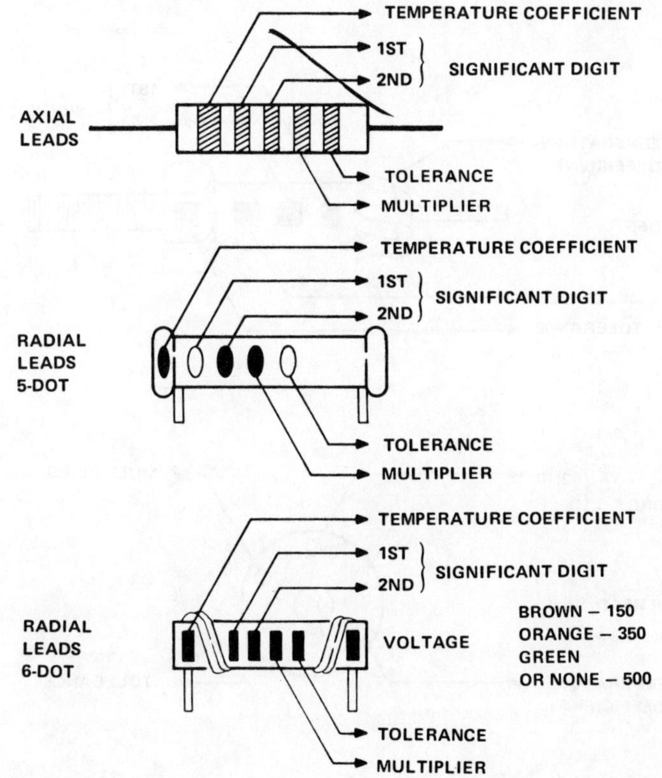

■ Ceramic Capacitor Color Codes

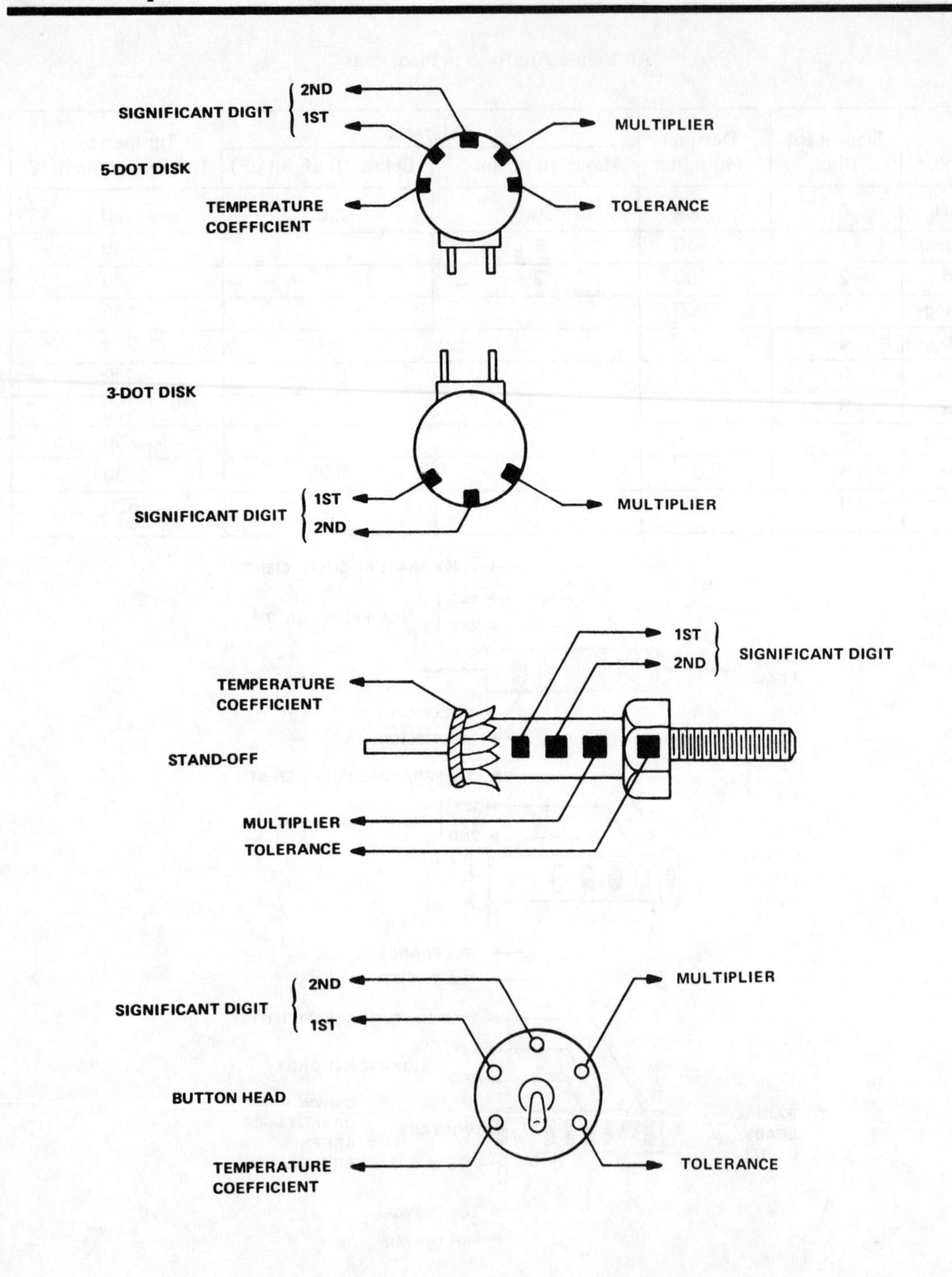

Chassis and Transformer Wiring Color Code

Most of the following color codes are standardized by the Electronic Industries Association (EIA). Although members are not required to adhere to the color codes, it is industry practice to do so where practical.

Chassis Wiring

In electronic systems wires are usually color-coded to ease assembly and speed tracing connections when troubleshooting the equipment. Usually the colors of the wires are in accordance with the following system.

COLOR	CONNECTED TO
Red	B+ voltage supply
Blue	Plate of amplifier tube or collector of transistor
Green	Control grid of amplifier tube or base of transistor (also for input to diode detector)
Yellow	Cathode of amplifier tube or emitter of transistor
Orange	Screen grid
Brown	Heaters or filaments
Black	Chassis ground return
White	Return for control grid (AVC bias)

I-F Transformers

Blue—plate
Red—B$^+$
Green—control grid or diode detector
White—control grid or diode return
Violet—second diode lead for duodiode detector

A-F Transformers

Blue—plate lead (end of primary winding)
Red—B+ (center-tap on push-pull transformer)
Brown—plate lead (start of primary winding on push-pull transformer)
Green—finish lead of secondary winding
Black—ground return of secondary winding
Yellow—start lead on center-tapped secondary

Power Transformers (Figure 1)

Primary without tap—black
Tapped primary:
    Common—black
    Tap—black and yellow stripes
    Finish—black and red stripes
High-voltage secondary for plates of rectifier—red
    Center tap—red and yellow stripes
Low-voltage secondary for rectifier filament—yellow
Low-voltage secondary for amplifier heaters—green, brown, or slate
    Center tap—same color with yellow stripe

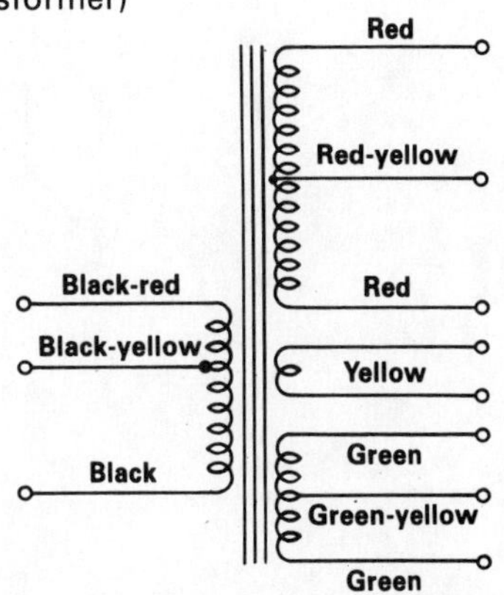

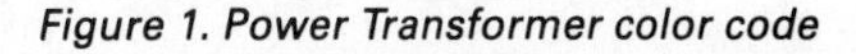
Figure 1. Power Transformer color code

Bibliography

Adams, J. E., *Electrical Principles and Practices,* New York, N.Y.: McGraw-Hill Book Co., 1963.

Angerbauer, G., *Principles of DC and AC Circuits,* North Scituate, Massachusetts: Duxbury Press, 1978.

Bell, D. A., *Fundamentals of Electric Circuits,* Reston, Virginia: Reston Publishing Co., Inc., 1978.

Blitzer, R., *Basic Electricity for Electronics,* New York, N.Y.: John Wiley and Sons., Inc., 1974.

Bureau of Naval Personnel, *Basic Electricity,* New York, N.Y.: Dover Publications, Inc., 1970.

Carper, D., *Basic Electronics,* Columbus, Ohio: Charles E. Merrill Publishing Co., 1975.

Churchman, L. W., *Introduction to Circuits,* New York, N.Y.: Holt, Rinehart, and Winston, 1976.

Cooke, N. M., *Basic Mathematics for Electronics,* Second Edition, New York, N.Y.: McGraw-Hill Book Co., 1960.

DeFrance, J. J., *Electrical Fundamentals,* Englewood Cliffs, New Jersey: Prentice-Hall, Inc., 1969.

Doyle, J. M., *An Introduction to Electrical Wiring,* Reston, Virginia: Reston Publishing Co., Inc., 1975.

Fiske, K. A., and Harter, J. H., *Direct Current Circuit Analysis Through Experimentation,* Third Edition, Seal Beach, California: The Technical Education Press, 1970.

Gillie, A. C., *Electrical Principles of Electronics,* Second Edition, New York, N.Y.: McGraw-Hill Book Co., 1969.

Gothmann, W. H., *Electronics: A Contemporary Approach,* Englewood Cliffs, N.J.: Prentice Hall, Inc., 1980.

Graf, R. F., *Modern Dictionary of Electronics,* Indianapolis, Indiana: Howard W. Sams and Co., Inc., 1968.

Graham, K. C., *Fundamentals of Electricity,* Fifth Edition: Chicago, Illinois: American Technical Society, 1968.

Grob, B., *Basic Electronics,* Second Edition, New York, N.Y.: McGraw-Hill Book Co., 1971.

Halliday, D., and Resnick, R., *Physics for Students of Science and Engineering, Part II,* New York, N.Y.: John Wiley and Sons, 1960.

Harris, N. C., and Hemmerling, E. M., *Introductory Applied Physics,* Third Edition, New York, N.Y.: McGraw-Hill Book Co., 1972.

Herrick, C. M., *Unified Concepts of Electronics,* Englewood Cliffs, N.J.: Prentice-Hall, Inc., 1970.

Jackson, H. W., *Introduction to Electric Circuits,* Englewood Cliffs, N.J.: Prentice Hall, Inc., 1976.

Mileaf, H., *Electricity One-Seven,* New York, N.Y.: Hayden Book Co., Inc. 1966.

Mileaf, H., *Electronics, One-Seven,* Rochelle Park, N.J.: Hayden Book Co., 1967.

Naval Air Technical Training Command, *Basic Electricity,* Memphis, Tennessee: NavAirTech Training Center, 1967.

■ Bibliography

Oppenheimer, S. L., Hess, F. R., Borchers, J. P., *Direct and Alternating Currents,* Second Edition, New York, N.Y.: McGraw-Hill Book Co., 1973.

Philco Education Operations, *Basic Concepts and DC Circuits,* Fort Washington, Pennsylvania: Philco Corporation, 1960.

Shrader, R. L., *Electrical Fundamentals for Technicians,* New York, N.Y.: McGraw-Hill Book Co., 1969.

Singer, B. B., *Basic Mathematics for Electricity and Electronics,* Second Edition, New York, N.Y.: McGraw-Hill Book Co., 1965.

Siskind, C. S., *Electrical Circuits, Direct and Alternating Current,* Second Edition, New York, N.Y.: McGraw-Hill Book Co., 1965.

Staff of Buck Engineering Co., Inc. (DeVito, M. J., Project Supervisor), *Introduction to Electricity and Electronics,* Farmingdale, New Jersey: Buck Engineering Co., Inc., 1971.

Staff of Electrical Technology Department, New York Institute of Technology, Schure, A., Project Director, *A Programmed Course in Basic Electricity,* New York, N.Y.: McGraw-Hill Book Co., 1970.

Suffern, M. G., *Basic Electrical and Electronic Principles,* Third Edition, New York, N.Y.: McGraw-Hill Book Co., 1962.

Thomson, C. M., *Fundamentals of Electronics,* Englewood Cliffs, N.J.: Prentice-Hall, Inc., 1979.

Timbie, W. H. *Essentials of Electricity,* Third Edition, New York, N.Y.: John Wiley and Sons, Inc. 1963.

Tocci, R. J., *Introduction to Electric Circuit Analysis,* Columbus, Ohio: Charles E. Merrill Publishing Co., 1974.

Trejo, P. E., *DC Circuits,* Palo Alto, California: Westinghouse Learning Press, 1972.

Turner, R. P., *Basic Electricity,* Second Edition, New York, N.Y.: Holt, Rinehart, and Wilson, 1963.

Weick, C. B., *Principles of Electronic Technology,* New York, N.Y.: McGraw-Hill Book Co., 1969.

Wellman, W. R., *Elementary Electricity,* Second Edition, New York, N.Y.: Van Nostrand Reinhold Company, 1971.

■ Glossary

adjacent side
The side of a right triangle beside or adjacent to an included angle theta.

algebraic sum
All positive quantities in an expression added together and each negative quantity subtracted from that result.

alternating current
An electrical current which changes in both magnitude and direction.

alternating current generator
A device which generates an alternating voltage by rotating a loop of conductor material through a magnetic field.

ammeter
A meter connected in series with a circuit, branch, or component which measures the current flowing through that circuit, branch, or component.

ampere
The unit of measure for current flow which equals 1 coulomb of electrons passing one point in a circuit in 1 second.

arctangent
Inverse of the tangent function. Arctangent of an angle theta means that theta is an angle whose tangent is the arctangent value.

bandwidth
The band of frequencies over which ac quantities remain within specified limits. In the case of resonant circuits, it is the band of frequencies over which the resonant effect exists.

branch
Path for current flow in a circuit.

bridge circuit
A special type of parallel-series circuit in which the voltages in each branch may be balanced by adjustment of one component. A special version called a Wheatstone bridge may be used to accurately measure resistance.

capacitance
The ability of a nonconductor to store a charge. Equal to the quantity of stored charge (Q) divided by the voltage (E) across the device when that charge was stored. Unit of measurement is the farad.

■ Glossary

capacitive reactance
The changing opposition of a capacitor to the flow of alternating electrical current at the applied frequency. It is inversely related to the source frequency. It is measured in ohms and has the symbol Ω.

capacitor
A device that can store a charge on conducting plates through the action of an electrostatic field between the plates.

cell
A single unit device which converts chemical energy into electrical energy.

chassis
A metal frame used to secure and house electrical components and associated circuitry.

circuit
A complete path for current flow from one terminal to the other of a source such as a battery or a power supply or an alternating current generator.

circuit analysis
A technique of examining components in circuits to determine various values of voltage, current, resistance, power, etc.

circuit reduction
A technique of circuit analysis whereby a complex combination of circuit components is replaced by a single equivalent component or several equivalent components.

circuit sense
An ability to recognize series or parallel portions of complex circuits to apply series and parallel circuit rules to those portions of the circuit for circuit analysis.

circulating current
A characteristic of resonance, it is the larger current in the inductive-capacitive branches of a resonant circuit, the result of the continual charging and discharging of the capacitor and the continual expansion and collapse of the magnetic field of an inductor at the resonant frequency.

coefficient of coupling
The fraction of the total magnetic flux lines produced by two coils which is common to both coils.

coil
A number of turns of wire wrapped around a core used to oppose changes in current flow. (Also called an inductor.)

■ Glossary

combining like terms
Algebraic addition of parts of an equation that each contain the same unknown quantity.

common point
A voltage reference point in a circuit. A point which is "common" to many components in the circuit.

complex number
A number represented by the algebraic sum of a real number and an imaginary number.

condenser
See capacitor.

conductance
The ability to conduct or carry current. Conductance is equivalent to the reciprocal of (or one over) the resistance.

conductor
A material with many free electrons that will carry current.

conjugate of a complex number
A complex number with the j-operator having an opposite sign from its mate. For example, the conjugate of $2 + j2$ is $2 - j2$ and the conjugate of $6 - j4$ is $6 + j4$.

coulomb
A large quantity of electrons that form a unit that is convenient when working with electricity and equals 6.25 billion, billion electrons (or 6.25×10^{18} electrons).

counter electromotive force (CEMF)
The voltage that appears across an inductor with a changing current flowing through it due to a property called self-inductance.

current
"Electron current" is the flow of electrons (negative charges) through a material from a negative potential to a positive potential. "Conventional current" is the flow of positive charges from a positive potential to negative potential. Current flow is a general term often used to mean either of the above. Symbol is I, unit is ampere.

current magnification
The increase in total circuit current caused by the Q factor.

cutoff frequency
Specified end frequency points that define bandwidth. In the case of resonance, the frequency at which the effects of resonance fall outside specified limits.

■ Glossary

dielectric
An insulating material with properties that enable its use between the two plates of a capacitor.

dielectric breakdown (in a capacitor)
Failure of an insulator to prevent current flow from one plate of a capacitor through the insulator to the other plate. This often causes permanent damage to the capacitor.

dielectric constant
A factor which indicates how much more effective (compared to air) a material is in helping a capacitor store a charge when the insulating material is between the capacitor's plates.

dielectric strength
A factor which indicates how well a dielectric resists breakdown under high voltages.

direct current
Current that flows in only one direction. Its magnitude may change but its direction does not.

direct relationship
One in which two quantities both increase or both decrease while other factors remain constant.

earth ground
A point that is at the potential of the earth or something that is in direct electrical connection with the earth such as water pipes.

efficiency (of a transformer)
The ratio of the power in a secondary circuit divided by the power in a primary circuit.

effective value
Also referred to as rms value. (See root-mean-square value.)

electricity
The flow of electrons through simple materials and devices.

electrolyte
A chemical (liquid or paste) which reacts with metals in a cell to produce electricity.

electromagnetic field (magnetic field)
A field of force produced around a conductor whenever there is current flowing through it. This field can be visualized with magnetic lines of force called magnetic flux.

■ Glossary

electromotive force (EMF)
A force that makes a current flow in a circuit measured by the amount of work done on a quantity of electricity passing from one point of electrical potential to a higher or lower point of electrical potential. Measured in volts (V).

electron
Negatively charged particles surrounding the nucleus of an atom which determine chemical and electrical properties of the atom.

electrostatic force
A force which exists between any two charged objects. If the two objects each have the same type of charge, the force is a repulsion. If the two objects each have different types of charge, the force is an attraction. Unlike charges attract; like charges repel.

energy
The ability to do work. Unit commonly used in measuring energy is the joule which is equal to the energy supplied by a 1-watt power source in 1 second.

equivalent resistance
The value of one single resistor that can be used to replace a more complex connection of several resistors.

exponent
A number written above and to the right of another number called the base. Example: 10^2, 10 is the base, 2 is the exponent. A number which indicates how many times the base is multiplied by itself. $10^2 = 10 \times 10 = 100$.

farad
The unit of capacitance. A capacitor has 1 farad of capacitance when it can store 1 coulomb of charge with a 1-volt potential difference placed across it.

free electrons
Electrons which are not bound to a particular atom but circulate among the atoms of the substance.

galvanometer
An ammeter with a center scale value of zero amperes.

giga
The metric prefix meaning one billion or 10^9. Abbreviated G.

graticule
The scale on the face of the cathode-ray tube of an oscilloscope.

■ Glossary

ground

A voltage reference point in a circuit which may be connected to earth ground.

half–power points

The upper and lower frequency points of a frequency response curve at which the real power dissipation in the circuit is exactly one-half of what it is at the mid-band frequencies. For resonant circuits the mid-band frequency is the resonant frequencies.

henry

Unit of measure for inductance. A 1-henry coil produces 1 volt when the current through it is changing at a rate of 1 ampere per second. Abbreviated H.

horizontal sweep

A proportional amount of time required for the spot on the cathode-ray tube face of an oscilloscope to travel from one side of the face to the other side.

hypotenuse

The longest side of a right triangle.

imaginary axis

Customarily the Y-axis used to represent an imaginary number in a complex number system.

impedance

The opposition to the flow of alternating electrical current. The impeding of the current. It is measured in ohms and has the symbol Ω.

inductance

The ability of a coil to store energy and oppose changes in current flowing through it . A function of cross sectional area, number of turns on coil, length of coil, and core material.

inductive reactance

A quantity that represents the opposition that a given inductance present to a changing ac current in a circuit. It is measured in ohms. It is a direct function of the frequency of the applied ac voltage and the value of the inductor.

inductor

A number of turns of wire wrapped around a core used to provide inductance in a circuit. (Also called a coil.)

in-phase

Two or more waveforms in which there is a zero-degree phase difference between the waveforms.

■ Glossary

instantaneous value
The value of voltage or current at a specific instant in the cycle of an ac signal (e.g. a sine wave).

insulator
A material with very few free electrons. A nonconductor.

inverse relationship
A relationship between two quantities in which an increase in one quantity causes a decrease in the other quantity while other factors are held constant.

junction
A connection common to more than two components in a circuit. Also called a node. (See also common point.)

kilo
A metric prefix meaning 1000 or 10^3. Abbreviated k.

Kirchhoff's current law
One of many tools of circuit analysis which states that the sum of the currents arriving at any point in a circuit must equal the sum of the currents leaving that point.

Kirchhoff's voltage law
Another tool of circuit analysis which states that the algebraic sum of all the voltages encountered in any loop equals zero.

leakage resistance
The normally high resistance of an insulator such as a dielectric between the plates of a capacitor.

load
A device such as a resistor which receives electrical energy from a source and that draws current and/or provides opposition to current, requires voltage, or dissipates power.

lagging waveform
A waveform whose cycle begins after another waveform cycle.

leading waveform
A waveform whose cycle begins before another waveform cycle.

leakage current
The small electron flow discharge between plates of a capacitor due to the fact an insulator is not a perfect nonconductor.

■ Glossary

loop
A closed path for current flow in a circuit.

loop equation
The algebraic sum of all the voltages in a loop set equal to zero.

lower cutoff frequency
The end frequency point defining the lower end of the bandwidth. For resonant circuits, the frequency below the resonant frequency at which the effect of resonance is outside of specified limits.

magnetic flux
Magnetic lines of force in a material.

mega
A metric prefix meaning one million or 1,000,000 or 10^6. Abbreviated M.

micro
A metric prefix meaning one millionth of 1/1,000,000 or 10^{-6}. Abbreviated with the Greek letter mu(μ).

milli
A metric prefix meaning one thousandth or 1/1000 or 10^{-3}. Abbreviated m.

mutual inductance
A measure of the voltage induced in a coil due to a changing current flowing in another coil close by. Measured in units called henrys (H).

nano
A metric prefix meaning one billionth or 1/1,000,000,000 or 10^{-9}. Abbreviated n.

negative ion
An atom which has gained one or more electrons.

node
A junction. A connection common to more than two components in a circuit.

node current equations
A mathematical expression of Kirchhoff's current law at a junction or node.

node voltage
The voltage at a node with respect to some reference point in the circuit.

Glossary

non-sinusoidal waveform
A waveform that cannot be expressed mathematically by using the sine function.

ohm
The unit of resistance. Symbol Ω.

ohmmeter
An instrument used to measure resistance.

Ohm's law
A basic tool of circuit analysis which states that, in simple materials, the amount of current through the material varies directly with the applied voltage and varies inversely with the resistance of the material. Gives rise to three common equations for use in circuit analysis: $E = IR$, $R = E/I$, $I = E/R$.

open circuit
A circuit interruption that causes an incomplete path for current flow.

opposite side
The side of a right triangle across from or opposite to the included angle theta.

oscilloscope
An electronic measuring instrument that can visually display rapidly varying electrical signals as a function of time. Often used to measure voltage or current.

out of phase
Two or more waveforms in which there is a finite number of degrees of phase difference between the waveforms.

parallel circuit
A circuit that has two or more paths (or branches) for current flow.

parallel-series circuit
A circuit with several branches wired in parallel. Each branch contains one or more components connected in series, but no single component carries the total circuit current.

peak amplitude
The maximum positive or negative deviation of an electrical signal (e.g. sinewave) from a zero reference level.

peak-to-peak amplitude
The distance between an ac signal's maximum positive and maximum negative peaks.

■ Glossary

percent
A ratio of one part to the total amount. One part of a hundred.

permeability of core material
The ability of a material to conduct magnetic lines of force or magnetic flux.

phase
The term used to describe the relative position of ac quantities in time reference to each other.

phase angle
The angular difference in electrical degrees between the total applied voltage and total current being drawn from the voltage supply in an ac circuit.

phasor
A phase vector.

phasor algebra
A mathematical method which expresses the value or magnitude of ac quantities such as current, reactance, resistance, or impedance and their respective phase angles, and indicates the phase relationships of these quantities with each other.

phasor diagram
A diagram showing the relationships of vectors used to represent phase relationships of ac circuit quantities.

phosphorescence
The ability of a material to emit light after being struck with electrons.

pico
A metric prefix meaning one million millionth or 10^{-12}. Abbreviated p.

polar coordinators
A means of identifying an ac quantity as a vector with a given magnitude (length) and a given direction (phase angle) in a quadrant coordinate system.

polarity of voltage
A means of describing a voltage with respect to some reference point, either positive or negative.

positive ion
An atom that has lost one or more electrons.

■ Glossary

potential difference
A measure of force produced between charged objects that moves free electrons. Also called voltage or electromotive force. Symbol is E, unit is the volt (abbreviated V).

power
The rate at which work is done or the rate at which heat is generated (abbreviated P). The unit of power is the watt (abbreviated W), which is equal to one joule per second.

power (real)
The power dissipated in purely resistive circuit components. It is measured in units called watts (W).

power (reactive)
The product of the voltage across and current through purely reactive circuit components. It is measured in units called volts-amperes-reactive (VAR).

power (apparent)
A combination of real and reactive power added vectorially, calculated using total current and total voltage values. It is measured in units called volt-amperes (VA).

power dissipated
Power which escapes from components in the form of heat by the convection of air moving around the component.

power rating of a resistor (or component)
How much power a resistor can dissipate (give off) safely in the form of heat in watts.

power supply
A device which is usually plugged into a wall outlet and can replace a battery in many applications by providing a known potential difference between two convenient terminals.

Pythagorean theorem
A mathematical theorem describing the relationships between the lengths of the sides of a right triangle: the square of the length of the hypotenuse of a right triangle equals the sum of the squares of the lengths of the other two sides.

quadrant
One-fourth of a circle.

quality factor (Q)
The ratio of the reactive power in an inductance to the real power dissipated by its internal resistance. It is a measure of the ability of a coil to store energy in its magnetic field and what part of that energy is returned back to the circuit containing the inductance.

■ Glossary

radian
The angle included within an arc on the circumference equal to the radius of a circle.

rationalization
The conversion of the denominator of a fractional number to a real number.

real axis
Customarily the X-axis used to represent real numbers in a complex number system.

reciprocal
Mathematical "inverse". The reciprocal of any number is simply that number divided into one.

rectangular coordinates
A means of identifying an ac quantity with two numbers which define the location of a specific point on a rectangular X-Y coordinate system.

reference point
An arbitrarily chosen point in a circuit to which all other points in the circuit are compared, usually when measuring voltages. Also called reference node.

relay
A switch (or combination of switches) activated by an electromagnetic coil.

repetitious waveform
A waveform in which each following cycle is identical to the previous cycle.

resistance
Opposition to current flow which is a lot like friction because it opposes electron motion and generates heat. Symbol is R. Unit is the ohm (Ω).

resonant frequency
A frequency at which a circuit's inductive reactance and capacitive reactance are the same value. At this frequency the circuit is said to be at resonance.

resultant
A vector which represents the sum of two vectors.

root-mean-square
The square root of the mean of squared values. Mean is an average of the sum of the squares of instantaneous values of a voltage or current waveform. Root is the square root of the mean. It is abbreviated rms and is sometimes referred to as an effective value.

Glossary

root-mean-square (rms) value

An ac voltage value equivalent to the value of a dc voltage which causes an equal amount of power dissipation due to the circuit current flowing through a resistance. The rms value of a sinusoidal waveform is 70.7 percent or 0.707 of its peak amplitude value. RMS value is also referred to as effective value. It is the effective value of an ac voltage that produces the same amount of resistor power dissipation in heat as a specific dc voltage.

rounding off

A procedure by which a number with many digits can be reduced to a number with only a selected number of significant digits. For example, if three significant digit rounding is desired, the first three significant digits are kept, and the fourth examined. If the fourth digit is 5 or greater, the third significant digit is raised by one. If the fourth digit is 4 or less, the first three digits are kept unchanged.

scientific notation

A type of shorthand used to keep track of decimal places which utilizes powers of the number 10. Standard form for scientific notation is $D.DD \times 10^{E}$, where D represents each of the first 3 significant digits, and E represents the exponent, or power of ten.

series circuit

A circuit with only one path through which current can flow.

series-parallel circuit

A group of series and parallel components in which at least one circuit element lies in series in the path of the total current.

short circuit

A path with little or no resistance connected across the terminals of a circuit element.

shunt

Another term which means parallel. Often also refers to the low value of parallel resistance used in an ammeter for determining or changing the "range" of the meter.

significant digits

Those digits within a number which have the greatest weight. In the decimal system digits to the left of any designated digit are more significant than those to the right.

sign of a voltage

A notation, either positive (+) or negative (−), in front of a voltage. (Important in solving circuit equations when analyzing circuits and helps determine the voltage that aids or opposes current flow in a circuit, especially a dc circuit.)

■ **Glossary**

sinusoidal waveform
A waveform that can be expressed mathematically by using the sine function. A waveform produced by an alternating current generator which constantly varies in magnitude and direction as determined by the sine trigonometric function.

skin effect
The tendency of high frequency current to flow near the surface of a conductor.

solenoid
A term used to mean coil or inductor, also used to mean a type of relay such as that used to switch the starter current in an automobile.

source
A device, such as a battery or power supply, which supplies the potential difference and electrical energy to the circuit.

square root of a number
A number which must be multiplied by itself to obtain the original number.

square of a number
A number multiplied by itself.

store
A calculator operation where the number in the display is transferred to the memory where it is held until it is recalled. Identified with STO, M+ or other such memory keys.

substitute
To replace one part of a formula or equation with another quantity which is its equal.

switch
A device that is used to open or close circuits, thereby stopping or allowing current flow in a circuit or through a component.

symmetrical
Parts on opposite sides of a dividing line or median plane correspond in size, shape, and relative position.

tank circuit
A parallel combination of an inductor and capacitor

terminal
A connection point on a device or component.

■ Glossary

time constant
The time it takes in seconds for a capacitor to charge up to approximately 63 percent of the applied voltage or the time it takes for a fully charged capacitor to discharge from 100 percent down to approximately 37 percent of full charge. Equal to the product of R (in ohms) times C (in farads) in a resistive-capacitive circuit. Also a measure of the current rise and fall in inductive circuits. Equal to the quotient of L/R in resistive-inductive circuits, L in henries, R in ohms. Symbol is Ω.

total current
The total current supplied by a voltage source applied to a circuit.

transposing (rearranging)
Moving a quantity from one side of an equation across the equal sign to the other side of the equation and changing its sign.

triggering
The act of starting a horizontal sweep on an oscilloscope.

trigonometry
The study of triangles and their relationships and functions.

troubleshooting
A technique used to locate a problem in a circuit.

upper cutoff frequency
The end frequency point defining the upper end of the bandwidth. For resonant circuits, the frequency above the resonant frequency at which the effect of resonance is considered to be outside of specified limits.

vector
A line whose length represents a magnitude and whose direction represents its phase with respect to some reference.

vector addition
The sum of two or more vectors.

vertical deflection
The direction the trace on a scope will travel up and down periodically from a center reference point.

volt
The unit of voltage or potential difference. Abbreviated V.

■ Glossary

voltage
A measure of the push or potential difference which makes each electron move. Symbol E. Unit is the volt.

voltage divider
A type of circuitry that provides an economical way to obtain one or several lower voltages from a single higher voltage supply.

voltage drop
Change in voltage available between points in a circuit produced by current flow through circuit components that provide opposition to current flow. Also called an IR, IX or IZ drop. Unit is the volt.

voltmeter
An instrument used to measure voltage between two points in a circuit.

watt
The unit of power. Abbreviated W. Equal to one joule per second.

waveform amplitude
The height of a measured electrical signal (e.g. sine wave) of voltage, current, or impedance on a scale representing the magnitude of the signal, the signal value.

waveform cycle
A waveform that begins at any electrical degree point and progresses through a 360-degree change.

waveform frequency
The number of waveform cycles occurring within one second of time.

waveform period (also referred to as waveform time)
The time required to complete one cycle of the waveform.

working voltage
The recommended maximum voltage at which a capacitor should be operated.

■ **Index**

Index

Index

■ Index

■ Index